Third Edition

Statistical Reasoning
for everyday life

Jeffrey O. Bennett
University of Colorado at Boulder

William L. Briggs
University of Colorado at Denver

Mario F. Triola
Dutchess Community College

PEARSON
Addison Wesley

Boston San Francisco New York
London Toronto Sydney Tokyo Singapore Madrid
Mexico City Munich Paris Cape Town Hong Kong Montreal

Editor in Chief: Deirdre Lynch
Associate Editor: Sara Oliver Gordus
Editorial Assistant: Christina Lepre
Senior Managing Editor: Karen Wernholm
Senior Production Supervisor: Tracy Patruno
Senior Designer: Barbara T. Atkinson
Photo Researcher: Beth Anderson
Digital Assets Manager: Marianne Groth
Senior Media Producer: Cecilia Fleming
Software Development: Edward Chappell and Marty Wright
Marketing Manager: Wayne Parkins
Marketing Assistants: Caroline Celano and Kathleen DeChavez
Senior Author Support/Technology Specialist: Joe Vetere
Senior Prepress Supervisor: Caroline Fell
Rights and Permissions Advisor: Dana Weightman
Manufacturing Manager: Evelyn Beaton
Text Design: Leslie Haimes
Production Coordination: Lifland et al., Bookmakers
Composition: Progressive Information Technologies
Illustrations: Scientific Illustrators

Cover photos: Shutterstock

Library of Congress Cataloging-in-Publication Data

Bennett, Jeffrey O.
Statistical reasoning for everyday life / Jeffrey O. Bennett, William L. Briggs, Mario F. Triola. – 3rd ed.
p. cm.
Includes index.
ISBN-13 978-0-321-28672-7 ISBN-10 0-321-28672-3
1. Statistics. I. Briggs, William L. II. Triola, Mario F. III. Title.

QA276.12.B45 2008
519.5–dc22

2007060107

5 6 7 8 9 10—RRD-JC—11 10

This book is dedicated to everyone
who will try to make the world
a better place. We hope that your
study of statistics will be useful
to your efforts.

And it is dedicated to those who
make our own lives brighter,
especially Lisa, Grant, Brooke, Julie,
Katie, Ginny, Marc, and Scott.

Contents

Preface

Why Study Statistics?

Science fiction writer H. G. Wells once wrote, "Statistical thinking will one day be as necessary for efficient citizenship as the ability to read and write." The future that Wells imagined is here. Statistics is now an important part of everyday life, unavoidable whether you are starting a new business, deciding how to plan for your financial future, or simply watching the news on television. Statistics comes up in everything from opinion polls to economic reports to the latest research on cancer prevention. Understanding the core ideas behind statistics is therefore crucial to your success in the modern world.

What Kind of Statistics Will You Learn in This Book?

Statistics is a rich field of study—so rich that it is possible to study it for a lifetime and still feel as if there's much left to learn. Nevertheless, you can understand the core ideas of statistics with just a quarter or semester of academic study. This book is designed to help you learn these core ideas. The ideas you'll study in this book represent the statistics that you'll *need* in your everyday life—and that you can reasonably learn in one course of study. In particular, we've designed this book with three specific purposes:

1. To provide you with the understanding of statistics you'll need for **college** courses, particularly in social sciences such as economics, psychology, sociology, and political science.
2. To help you develop the ability to reason using statistical information—an ability that is crucial to almost any **career** in the modern world.
3. To provide you with the power to evaluate the many news reports of statistical studies that you encounter in your daily **life**, thereby helping you to form opinions about their conclusions and to decide whether the conclusions should influence the way you live.

Who Should Read This Book?

We hope this book will be useful to everyone, but it is designed primarily for students who are *not* planning to pursue advanced course work in statistics. In particular, this book should provide a suitable introduction to statistics for students majoring in a broad range of fields that require statistical literacy, including most disciplines in the humanities and social sciences. The level of this text should be appropriate to anyone who has completed two years of high school mathematics.

Approach

This book takes an approach designed to help you understand important statistical ideas qualitatively, using quantitative techniques only when they clarify those ideas. Here are a few of the key pedagogical strategies that guided the creation of this book.

START WITH THE BIG PICTURE. Most people entering a statistics course have little prior knowledge of the subject, so it is important to keep sight of the overall purpose of statistics while learning individual ideas or methods. We therefore begin this book with a broad overview of statistics in Chapter 1, in which we explain the relationship between samples and populations, discuss sampling methods and the various types of statistical study, and show numerous examples designed to help you decide whether to believe a statistical study. This "big picture" overview of statistics provides a solid foundation for the more in-depth study of statistical ideas in the rest of the book.

Build Ideas Step by Step. The goal of any course in statistics is to help students understand real statistical issues. However, it is often easier to begin by investigating simple examples in order to build step-by-step understanding that can then be applied to more complex studies. We apply this strategy within every section and every chapter, gradually building toward real examples and case studies.

Use Computations to Enhance Understanding. The primary goal of this book is to help students understand statistical concepts and ideas, but we firmly believe that this goal is best achieved by doing at least some computation. We therefore include computational techniques wherever they will enhance understanding of the underlying ideas.

Connect Probability to Statistics. Many statistics courses include coverage of probability, but to students the concept of probability often seems disconnected from the rest of the subject matter. This is a shame, since probability plays such an integral role in the science of statistics. We discuss this point beginning in Chapter 1, with the basic structure of statistical studies, and then revisit it throughout the book—particularly in Chapter 6, where we present many ideas of probability. For those courses in which coverage of probability is not emphasized, Chapter 6 is designed to be optional.

Stay on Goal: Applying Statistical Reasoning to Everyday Life. Because statistics is such a rich subject, it can be difficult to decide how far to go with any particular statistical topic. In making such decisions for this book, we always turned back to the goal reflected in the title: This book is supposed to help you with the statistical reasoning needed in everyday life. If we felt that a topic was not often encountered in everyday life, we left it out. In the same spirit, we included a few topics—such as a discussion of percentages in Chapter 2 and an in-depth study of graphics in Chapter 3—that are not often covered in statistics courses but are a major part of the statistics encountered in daily life.

Modular Structure

Although we have written this book so that it can be read as a narrative from beginning to end, we recognize that many instructors might wish to teach material in a different order than we have chosen or to cover only selected portions of the text as time allows for classes of different length or with students at different levels. We have therefore organized the book with a modular structure that allows instructors to create a customized course. The 10 chapters are organized broadly by conceptual areas. Each chapter, in turn, is divided into a set of self-contained sections, each devoted to a particular topic or application. In most cases, you may cover the sections or chapters in any order or skip sections that do not fit well into your course. Please note the following specific structure within each chapter:

Learning Goals. Each chapter begins with a one-page overview of its subject matter, including a list mapping each section to learning goals.

Numbered Sections. Each chapter is subdivided into a set of numbered sections (e.g., Sections 1.1, 1.2, . . .). To facilitate use of these sections in any order, each section ends with its own set of exercises specific only to that section. Exercises are divided into subgroups with headings that should be self-explanatory, including "Statistical Literacy and Critical Thinking," "Concepts and Applications," "Projects for the Internet and Beyond," and "In the News." Answers to most odd-numbered exercises appear in the back of the book. The numbered sections also include the following pedagogical features:

- **Examples and Case Studies.** Numbered examples within each section are designed to build understanding and to offer practice with the types of questions that appear in the exercises. Case studies, which always focus on real issues, go into more depth than the numbered examples.
- **Time Out to Think.** The "Time Out to Think" features pose short conceptual questions designed to help students reflect on important new ideas. They also serve as excellent starting points for classroom discussions and, in some cases, can be used as a basis for clicker questions.

Chapter Review Exercises. The main part of each chapter ends with a brief set of exercises designed to tie many of the chapter ideas together. These exercises are designed primarily for self-study, with answers to all of them appearing in the back of the book.

Focus Topics. Each chapter concludes with two sections entitled "Focus on . . ." that go into depth on important statistical issues of our time. The topics of these sections were chosen to demonstrate the great variety of fields in which statistics plays a role, including history, environmental studies, agriculture, and economics. Each of these Focus sections includes a set of questions for assignment or discussion.

About the Third Edition

We've developed this Third Edition of *Statistical Reasoning for Everyday Life* with the help of many users and reviewers. In addition to editing and redesigning the entire book to make it even more student-friendly, we have made the following major changes for this edition:

- Because this book is intended to show the relevance of statistics to everyday life, it is critical that discussions and examples be up to date. We have therefore revised or replaced many dozens of in-text and numbered examples and case studies to be sure they reflect the latest data and topics of interest. We have also replaced four of the twenty in-depth Focus sections and updated all the others.
- We have thoroughly reworked the exercise sets, completely replacing 63% of the exercises and revising or updating data in most of the others. There are 1,412 exercises, divided into the following categories: Statistical Literacy and Critical Thinking, Concepts and Applications, Projects for the Internet and Beyond, In the News, Chapter Review, and—new to this edition—Chapter Quiz.
- In those chapters that discuss calculations, the final numbered section is followed by a feature called "Using Technology," which gives a brief overview of how the calculations can be handled with the popular software packages SPSS®, Excel®, and STATDISK®.
- The third edition contains over thirty annotated figures. The annotations highlight relevant information about visual displays of data.
- We have almost completely rewritten Chapter 9 to provide a simpler and more focused introduction to hypothesis testing.
- We have added two new sections, both in Chapter 10, covering *t* distributions and one-way ANOVA. With these additions, Chapter 10 now builds upon the ideas of hypothesis testing introduced in Chapter 9.
- We have moved and revised two sections that formerly appeared in Chapter 10. The section on Statistical Paradoxes (formerly Section 10.2) now appears at the end of Chapter 4 as Section 4.4; the section on Risk and Life Expectancy (formerly Section 10.1) now appears as Section 6.4.
- New to this edition is a companion CD-ROM that contains a printable Technology Manual and Workbook authored by Mario F. Triola, data sets, statistical software, and programs for the TI-83/84 Plus® calculator.

Acknowledgments

Writing a textbook requires the efforts of many people besides the authors. This book would not have been possible without the help of many people. We'd particularly like to thank our publisher and editor at Addison-Wesley, Greg Tobin and Deirdre Lynch, whose faith allowed us to create this book. We'd also like to thank the rest of the team at Addison-Wesley who helped produce this book, including Sara Oliver Gordus, Christina Lepre, Tracy Patruno, Ceci Fleming, Caroline Celano, Beth Anderson, Barbara Atkinson, and our production manager, Sally Lifland.

For helping to ensure the accuracy of this text, we thank Matthew Bognar, University of Iowa, and Sheila O'Leary Weaver, University of Vermont. For reviewing this text and providing invaluable advice, we thank the following individuals:

Reviewers of this edition

Jennifer Beineke
Western New England College

Matthew Bognar
University of Iowa

Pat Buchanan
Pennsylvania State University

Antonius H. N. Cillessen
University of Connecticut

Robert Dobrow
Carleton College

Beverly J. Ferrucci
Keene State College

Jack R. Fraenkel
San Francisco State University

Susan Janssen
University of Minnesota–Duluth

Becky Ladd
Arizona State University

Christopher Leary
SUNY Geneseo

Carrie M. Margolin
The Evergreen State College

Craig McCarthy
Ohio University

Thomas Petee
Auburn University

William S. Rayens
University of Kentucky

Pali Sen
University of North Florida

Donald Hugh Smith
Old Dominion University

Elizabeth Walters
Loyola College of Maryland

Sheila O'Leary Weaver
University of Vermont

Reviewers of previous editions

Dale Bowman
University of Mississippi

Patricia Buchanan
Penn State University

Robert Buck
Western Michigan University

Olga Cordero-Brana
Arizona State University

Terry Dalton
University of Denver

Jim Daly
California Polytechnic State University

Mickle Duggan
East Central University

Juan Estrada
Metropolitan State University, Minneapolis–St. Paul

Jack R. Fraenkel
San Francisco State University

Frank Grosshans
West Chester University

Silas Halperin
Syracuse University

Golde Holtzman
Virginia Polytechnic Institute and State University

Colleen Kelly
San Diego State University

Jim Koehler

Stephen Lee
University of Idaho

Kung-Jong Lui
San Diego State University

Judy Marwick
Prairie State College

Richard McGrath
Penn State University

Abdelelah Mostafa
University of South Florida

Todd Ogden
University of South Carolina

Nancy Pfenning
University of Pittsburgh

Steve Rein
California Polytechnic State University

Lawrence D. Ries
University of Missouri–Columbia

Larry Ringer
Texas A&M University

John Spurrier
University of South Carolina

Gwen Terwilliger
University of Toledo

David Wallace
Ohio University

Larry Wasserman
Carnegie Mellon University

Sheila Weaver
University of Vermont

Robert Wolf
University of San Francisco

Fancher Wolfe
Metropolitan State University, Minneapolis–St. Paul

Ke Wu
University of Mississippi

Supplements

STUDENT SUPPLEMENTS

STUDENT'S SOLUTIONS MANUAL. This manual provides detailed, worked-out solutions to all odd-numbered text exercises and chapter quiz problems. ISBN-13: 978-0-321-28706-9; ISBN-10: 0-321-28706-1.

COMPANION CD. The CD that accompanies new copies of this text contains a printable Technology Manual and Workbook authored by Mario F. Triola, which helps students learn to use SPSS®, Excel®, and STATDISK®. Also included on the CD are STATDISK Statistical Software (Version 10.4); the Excel add-in DDXL™; programs for the TI-83/84 Plus calculator; plus data sets stored as text files, Minitab® worksheets, SPSS® files, SAS® files, Excel workbooks, and a TI-83/84 Plus application.

COMPANION WEBSITE. The companion website contains additional resources for students, including data sets and web links from the text. The URL is www.aw-bc.com/bbt.

INSTRUCTOR SUPPLEMENTS

INSTRUCTOR'S EDITION. This version of the text includes the answers to all exercises and quiz problems. (The Student Edition contains answers to only the odd-numbered ones.)

INSTRUCTOR'S SOLUTIONS MANUAL. This comprehensive manual contains solutions to all text exercises and chapter quizzes.

PRINTED TEST BANK. The Test Bank contains four tests to accompany every chapter of the text.

TESTGEN®. TestGen enables instructors to build, edit, print, and administer tests using a computerized bank of questions developed to cover all the objectives of the text. TestGen is algorithmically based, allowing instructors to create multiple but equivalent versions of the same question or test with the click of a button. Instructors can also modify test bank questions or add new questions. Tests can be printed or administered online. The software and test bank are available for download from Pearson Education's online catalog.

ACTIVE LEARNING QUESTIONS. Formatted as PowerPoint® slides, these questions can be used with classroom response systems. Several multiple-choice questions are available for each section of the book, allowing instructors to quickly assess mastery of material in class. Slides are available to download from within MyStatLab and from Pearson's Instructor Resource Center (www.aw-bc.com/irc).

POWERPOINT® LECTURE SLIDES. These slides present key concepts and definitions from the text. Slides are available to download from within MyStatLab and from Pearson's Instructor Resource Center (www.aw-bc.com/irc).

TECHNOLOGY RESOURCES

MYSTATLAB™. MyStatLab (part of the MyMathLab® and MathXL® product family) is a text-specific, easily customizable online course that integrates interactive multimedia instruction with the textbook content. Powered by CourseCompass™ (Pearson Education's online teaching and learning environment) and MathXL (our online homework, tutorial, and assessment system), MyStatLab gives you the tools you need to deliver all or a portion of your course online, whether your students are in a lab setting or working from home. MyStatLab provides a rich and flexible set of course materials, featuring free-response tutorial exercises for unlimited practice and mastery. Students can also use online tools, such as animations and a multimedia textbook, to independently improve their understanding and performance. Instructors can use MyStatLab's homework and test managers to select and assign online exercises correlated directly to the textbook, and they can also create and assign their own online exercises and import TestGen tests for added flexibility. MyStatLab's online gradebook—designed specifically for mathematics and statistics—automatically tracks students' homework and test results and gives the instructor control over how to calculate final grades. Instructors can also add offline (paper-and-pencil) grades to the gradebook. MyStatLab also includes access to Pearson's Tutor Center, which provides students with tutoring via toll-free phone, fax, email, and interactive Web sessions. MyStatLab is available to qualified adopters. For more information, visit our website at www.mystatlab.com, or contact your sales representative.

MATHXL® FOR STATISTICS. MathXL for Statistics is a powerful online homework, tutorial, and assessment system that accompanies Pearson Education's textbooks in statistics. With MathXL for Statistics, instructors can create, edit, and assign online homework and tests using algorithmically generated exercises correlated at the objective level to the textbook. They can also create and assign their own online exercises and import TestGen tests for added flexibility. All student work is tracked in MathXL's online gradebook. Students can take chapter tests in MathXL and receive personalized study plans based on their test results. The study plan diagnoses weaknesses and links students directly to tutorial exercises for the objectives they need to study and retest. Students can also access supplemental animations directly from selected exercises. MathXL for Statistics is available to qualified adopters. For more information, visit our website at www.mathxl.com, or contact your sales representative.

To the Student

How to Succeed in Your Statistics Course

If you are reading this book, you probably are enrolled in a statistics course of some type. The keys to success in your course include approaching the material with an open and optimistic frame of mind, paying close attention to how useful and enjoyable statistics can be in your life, and studying effectively and efficiently. The following sections offer a few specific hints that may be of use as you study.

USING THIS BOOK

Before we get into more general strategies for studying, here are a few guidelines that will help you use *this* book most effectively.

- Before doing any assigned exercises, read assigned material *twice*.
  —On the first pass, read quickly to gain a "feel" for the material and concepts presented.
  —On the second pass, read the material in more depth and work through the examples carefully.
- During the second reading, take notes that will help you when you go back to study later. In particular:
  —*Use the margins!* The wide margins in this textbook are designed to give you plenty of room for making notes as you study.
  —Don't highlight—underline! Using a pen or pencil to underline material requires greater care than highlighting and therefore helps keep you alert as you study.
- You'll learn best by *doing*, so after you complete the reading be sure to do plenty of the end-of-section exercises and the end-of-chapter review exercises. In particular, try some of the exercises that have answers in the back of the book, in addition to any exercises assigned by your instructor.
- If you have access to MyStatLab with this book, be sure to take advantage of the many additional study resources available on this website.

BUDGETING YOUR TIME

A general rule of thumb for college classes is that you should expect to study about 2 to 3 hours per week *outside* class for each unit of credit. Based on this rule of thumb, a student taking 15 credit hours should expect to spend 30 to 45 hours each week studying outside of class. Combined with time in class, this works out to a total of 45 to 60 hours spent on academic work—not much more than the time required of a typical job, and you get to choose your own hours. Of course, if you are working while you attend school, you will need to budget your time carefully. Here are some rough guidelines for how you might divide your studying time.

If your course is	Time for reading the assigned text (per week)	Time for homework assignments (per week)	Time for review and test preparation (average per week)	Total study time (per week)
3 credits	1 to 2 hours	3 to 5 hours	2 hours	6 to 9 hours
4 credits	2 to 3 hours	3 to 6 hours	3 hours	8 to 12 hours
5 credits	2 to 4 hours	4 to 7 hours	4 hours	10 to 15 hours

If you find that you are spending fewer hours than these guidelines suggest, you could probably improve your grade by studying more. If you are spending more hours than these guidelines suggest, you may be studying inefficiently; in that case, you should talk to your instructor about how to study more effectively.

GENERAL STRATEGIES FOR STUDYING

- Don't miss class. Listening to lectures and participating in discussions is much more beneficial than reading someone else's notes. Active participation will help you retain what you are learning.
- Budget your time carefully. Putting in an hour or two each day is more effective, and far less painful, than studying all night before homework is due or before exams.
- If a concept gives you trouble, do additional reading or problem solving beyond what has been assigned. And if you still have trouble, ask for help; you surely can find friends, colleagues, or teachers who will be glad to help you learn.
- Working together with friends can be valuable in helping you to solve difficult problems. However, be sure that you learn *with* your friends and do not become dependent on them.

PREPARING FOR EXAMS

- Rework exercises and other assignments; try additional exercises to be sure you understand the concepts. Study your performance on assignments, quizzes, or exams from earlier in the semester.
- Study your notes from lectures and discussions. Pay attention to what your instructor expects you to know for an exam.
- Reread the relevant sections in the textbook, paying special attention to notes you have made in the margins.
- Study individually before joining a study group with friends. Study groups are effective only if *every* individual comes prepared to contribute.
- Don't stay up too late before an exam. Don't eat a big meal within an hour of the exam (thinking is more difficult when blood is being diverted to the digestive system).
- Try to relax before and during the exam. If you have studied effectively, you are capable of doing well. Staying relaxed will help you think clearly.

Applications Index

(CS = Case Study, E = Example, F = Focus, IE = In-Text Example, P = Problem, PR = Project)

Miscellaneous (Including Sports)

1

As a general rule, the most successful [person] in life is the [person] who has the best information.

—Benjamin Disraeli

Speaking of Statistics

IS YOUR DRINKING WATER SAFE? DO MOST PEOPLE approve of the President's tax plan? Are we getting good value for our health care dollars? Questions like these can be addressed only through statistical studies. In this first chapter, we will discuss basic principles of statistical research and lay a foundation for the more detailed study of statistics that follows in the rest of this book. Along the way, we will consider a variety of examples that show how well-designed statistical studies can provide guidance for social policy and personal decisions, as well as a few cases in which statistics can be misleading.

LEARNING GOALS

1.1 What Is/Are Statistics?
Understand the two meanings of the term *statistics* and the basic ideas behind any statistical study, including the relationships among the study's population, sample, sample statistics, and population parameters.

1.2 Sampling
Understand the importance of choosing a representative sample and become familiar with several common methods of sampling.

1.3 Types of Statistical Studies
Understand the differences between observational studies and experiments; recognize key issues in experiments, including the selection of treatment and control groups, the placebo effect, and blinding.

1.4 Should You Believe a Statistical Study?
Be able to evaluate statistical studies that you hear about in the media, so that you can decide whether the results are meaningful.

1.1 What Is/Are Statistics?

If you are like most students using this textbook, you are new to the study of statistics. You may not be sure why you are studying statistics, and you may be feeling anxious about what lies ahead. But as you begin reading, we hope you'll put your concerns aside and prepare to be pleasantly surprised.

Statistical thinking will one day be as necessary for efficient citizenship as the ability to read and write.

—H. G. Wells

The subject of statistics is often stereotyped as dry or technical, but it touches on almost everything in modern society. Statistics can tell us whether a new drug is effective in treating cancer, it can help agricultural inspectors ensure that our food is safe, and it's the key to all opinion polls. Businesses use statistics in market research and advertising. We even use statistics in sports, often as a way of ranking teams and players. Indeed, you'll be hard-pressed to think of any topic that is not linked with statistics in some important way.

The primary goal of this book is to help you learn the core ideas behind statistical methods. These basic ideas are not difficult to understand, although mastery of the details and theory behind them can require years of study. One of the great things about statistics is that even the small amount of theory covered in this book will give you the power to understand the statistics you encounter in the news, in your classes or workplace, and in your everyday life.

A good place to start is with the term *statistics* itself, which can be either singular or plural and has different meanings in the two cases. When it is singular, *statistics* is the *science* that helps us understand how to collect, organize, and interpret numbers or other information about some topic; we refer to the numbers or other pieces of information as *data*. When it is plural, *statistics* are the actual data that describe some characteristic. For example, if there are 30 students in your class and they range in age from 17 to 64, the numbers "30 students," "17 years," and "64 years" are all statistics that describe your class in some way.

Two Definitions of Statistics

- Statistics is the *science* of collecting, organizing, and interpreting data.
- Statistics are the *data* (numbers or other pieces of information) that describe or summarize something.

Note that although you'll sometimes see the word *data* used as a singular synonym for *information*, technically it is plural: One piece of information is called a *datum*, and two or more pieces are called *data*.

How Statistics Works

Did you watch the Super Bowl? Advertisers want to know, because the cost of commercial time during the big game is approaching $3 million for a 30-second spot. This advertising can be well worth the price if enough people are watching. For example, news reports stated that 93.2 million Americans watched the Indianapolis Colts win Super Bowl XLI. But you may wonder: Who counted all these people?

The answer is *no one*. The claim that 93.2 million people watched the Super Bowl came from statistical studies conducted by a company called Nielsen Media Research. This company publishes the results of its studies as the famous *Nielsen ratings*. Remarkably, Nielsen compiles these ratings by monitoring the television viewing habits of people in only 5,000 homes.

By the Way ...

Statistics originated with the collection of census and tax data, which are affairs of state. That is why the word *state* is at the root of the word *statistics*.

If you are new to the study of statistics, Nielsen's conclusion may seem like a stretch. How can anyone draw a conclusion about millions of people by studying just a few thousand? However, statistical science shows that this conclusion can be quite accurate, as long as the statistical study is conducted properly. Let's take the Nielsen ratings of the Super Bowl as an example, and ask a few key questions that will illustrate how statistics works in general.

What Is the Goal of the Research?

Nielsen's goal is to determine the total number of Americans who watched the Super Bowl. In the language of statistics, we say that Nielsen is interested in the **population** of all Americans. The number that Nielsen hopes to determine—the number of people who watched the Super Bowl—is a particular characteristic of the population. In statistics, characteristics of the population are called **population parameters**.

Although we usually think of a population as a group of people, a statistical population can be any kind of group—people, animals, or things. For example, in a study of automobile safety, the population might be *all cars on the road*. Similarly, the term *population parameter* can refer to any characteristic of a population. In the case of automobile safety, the population parameters might include the total number of cars on the road during a certain time period, the accident rate among cars on the road, or the range of weights of cars on the road.

By the Way ...

Arthur C. Nielsen founded his company and invented market research in 1923. He introduced the Nielsen Radio Index to rate radio programs in 1942 and extended his methods to television programming in the 1960s.

Definitions

The **population** in a statistical study is the *complete* set of people or things being studied.

Population parameters are specific characteristics of the population.

EXAMPLE 1 Populations and Population Parameters

For each of the following situations, describe the population being studied and identify some of the population parameters that would be of interest.

a. You work for Farmers Insurance and you've been asked to determine the average amount paid to accident victims in cars without side-impact air bags.

b. You've been hired by McDonald's to determine the weights of the potatoes delivered each week for french fries.

c. You are a business reporter covering Genentech Corporation and you are investigating whether their new treatment is effective against childhood leukemia.

Solution

a. The population consists of people who have received insurance payments for accidents in cars that lacked side-impact air bags. The relevant population parameter is the mean (average) amount paid to these people. (See Chapter 4 for a more detailed discussion of the mean and other measures of "average.")

b. The population consists of all the potatoes delivered each week for french fries. Relevant population parameters include the mean weight of the potatoes and the variation of the weights (for example, are most of them close to or far from the mean?).

c. The population consists of all children with leukemia. Important population parameters are the percentage of children who recover *without* the new treatment and the percentage of children who recover with the new treatment.

What Actually Gets Studied?

If researchers at Nielsen were all-powerful they might determine the number of people watching the Super Bowl by asking every individual American. But no one can do that, so instead they try to estimate the number of Americans watching by studying a relatively small group of people. In other words, Nielsen attempts to learn about the population of all Americans by carefully monitoring the viewing habits of a much smaller **sample** of Americans. More specifically, Nielsen has devices (called "people meters") attached to televisions in about 5,000 homes, so the roughly 13,000 people who live in these homes are the sample of Americans that Nielsen studies. (To keep up with rapidly changing viewing habits that now include cable TV and Internet viewing, Nielsen has been increasing its number of people meters, but the 5,000 households given here is still representative of the general ratings process.)

The individual measurements that Nielsen collects from the people in the 5,000 homes constitute the **raw data**. Nielsen collects much raw data—for example, when and how long each TV in the household is on, what show it is tuned to, and who in the household is watching it. Nielsen then consolidates these raw data into a set of numbers that characterize the sample, such as the percentage of viewers in the sample who watched each individual television show or the total number of people in the sample who watched the Super Bowl. These numbers are called **sample statistics**.

Definitions

A **sample** is a subset of the population from which data are actually obtained.

The actual measurements or observations collected from the sample constitute the **raw data**.

Sample statistics are characteristics of the sample found by consolidating or summarizing the raw data.

EXAMPLE 2 Unemployment Survey

The U.S. Labor Department defines the *civilian labor force* as all those people who are either employed or actively seeking employment. Each month, the Labor Department reports the unemployment rate, which is the percentage of people actively seeking employment within the entire civilian labor force. To determine the unemployment rate, the Labor Department surveys 60,000 households. For the unemployment reports, describe the

a. population **b.** sample **c.** raw data **d.** sample statistics **e.** population parameters

Solution

a. The *population* is the group that the Labor Department wants to learn about, which is all the people who make up the civilian labor force.

b. The *sample* consists of all the people among the 60,000 households surveyed.

c. The *raw data* consist of all the information collected in the survey.

d. The *sample statistics* summarize the raw data for the sample. In this case, the relevant sample statistic is the percentage of people in the sample who are actively seeking employment. (The Labor Department also calculates similar sample statistics for subgroups in the population, such as the percentages of teenagers, men, women, and veterans who are unemployed.)

e. The *population parameters* are the characteristics of the entire population that correspond to the sample statistics. In this case, the relevant population parameter is the actual unemployment rate. Note that the Labor Department does *not* actually measure this population parameter, since data are collected only for the sample and then are used to estimate the population parameter.

By the Way ...

By the Labor Department definition, someone who is not working is not necessarily unemployed. For example, stay-at-home moms and dads are not counted among the unemployed unless they are actively trying to find a job, and people who tried to find work but gave up in frustration are not counted as unemployed.

How Do Sample Statistics Relate to Population Parameters?

Suppose Nielsen finds that 31% of the people in the 5,000 homes in its sample watched the Super Bowl. This "31%" is a sample statistic, because it characterizes the sample. But what

Nielsen really wants to know is the corresponding population parameter, which is the percentage of all Americans who watched the Super Bowl.

There is no way for Nielsen researchers to know the exact value of the population parameter, because they've studied only a sample. However, Nielsen researchers hope that they've done their work so that the sample statistic is a good estimate of the population parameter. In other words, they would like to conclude that because 31% of the sample watched the Super Bowl, approximately 31% of the population also watched the Super Bowl. One of the primary purposes of statistics is to help researchers assess the validity of this type of conclusion.

TIME OUT TO THINK

Suppose Nielsen concludes that 30% of Americans watched the Super Bowl. How many people does this represent? (The population of the United States is approximately 300 million.)

Statistical science provides methods that enable researchers to determine how well a sample statistic estimates a population parameter. For example, results from surveys or opinion polls are usually quoted along with something called the **margin of error**. By adding and subtracting the margin of error from the sample statistic, we find a range of values, or **confidence interval**, that is *likely* to contain the population parameter. In most cases, the margin of error is defined so that we can have 95% confidence that this range contains the population parameter. We'll discuss the precise meaning of "likely" and "95% confidence" in Chapter 8, but for now you might find an explanation given by the *New York Times* useful (Figure 1.1). In the case of the Nielsen ratings, the margin of error is about 1 percentage point. Therefore, if 31% of the sample was watching the Super Bowl, then we can be 95% confident that the range 30% to 32% contains the actual percentage of the population watching the Super Bowl.

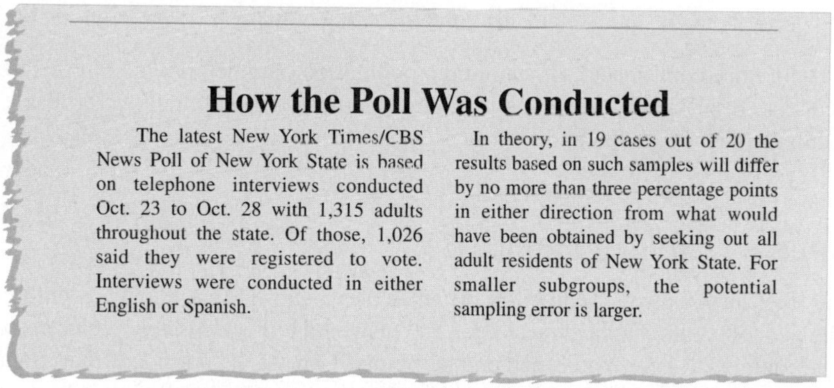

How the Poll Was Conducted

The latest New York Times/CBS News Poll of New York State is based on telephone interviews conducted Oct. 23 to Oct. 28 with 1,315 adults throughout the state. Of those, 1,026 said they were registered to vote. Interviews were conducted in either English or Spanish.

In theory, in 19 cases out of 20 the results based on such samples will differ by no more than three percentage points in either direction from what would have been obtained by seeking out all adult residents of New York State. For smaller subgroups, the potential sampling error is larger.

Figure 1.1 The margin of error in a survey or opinion poll usually describes a range that is likely (with 95% confidence, meaning in 19 out of 20 cases) to contain the population parameter. This excerpt from the *New York Times* explains a margin of error of 3 percentage points.

One of the most remarkable findings of statistical science is that it is possible to get meaningful results from surprisingly small samples. Nevertheless, larger sample sizes are better (when they are feasible), because the margin of error is generally smaller for larger samples. For example, the margin of error for a 95% confidence interval in a well-conducted poll is typically about 5 percentage points for a sample size of 400, but drops to 3 percentage points for a sample size of 1,000 and to 1 percentage point for a sample of 10,000. (See Chapter 8 to understand how margins of error are calculated.)

Definition

The **margin of error** in a statistical study is used to describe the range of values, or **confidence interval,** likely to contain the population parameter. We find this confidence interval by adding and subtracting the margin of error from the sample statistic obtained in the study. That is, the range of values likely to contain the population parameter is

from (sample statistic − margin of error)
to (sample statistic + margin of error)

The margin of error is usually defined to give a 95% confidence interval, meaning that 95% of samples of the size used in the study would contain the actual population parameter (and 5% would not).

EXAMPLE 3 People on Mars?

The Pew Research Center for the People and the Press interviewed 1,546 adult Americans about their attitudes toward the future. Asked whether humans would land on Mars within the next 50 years, 76% of these 1,546 people said either *definitely yes* or *probably yes*. The margin of error for the poll was 3 percentage points. Describe the population and the sample for this survey, and explain the meaning of the sample statistic of 76%. What can we conclude about the percentage of the population that thinks humans will land on Mars within the next 50 years?

Solution The population is all adult Americans and the sample consists of the 1,546 people who were interviewed. The sample statistic of 76% is the *actual* percentage of people in the sample who answered that humans would definitely or probably land on Mars in the next 50 years. The 76% sample statistic and the margin of error of 3 percentage points tell us that the range of values

$$\text{from } 76\% - 3\% = 73\%$$
$$\text{to } 76\% + 3\% = 79\%$$

is likely (with 95% confidence) to contain the population parameter, which in this case is the true percentage of all adult Americans who think humans will definitely or probably land on Mars within the next 50 years.

EXAMPLE 4 Cloning Humans?

The same Pew survey also asked people whether they believed that humans would be cloned within the next 50 years. On this question, 51% answered either *definitely yes* or *probably yes*. Again, the margin of error was 3 percentage points. Can we be confident that a majority of adult Americans think that humans will be cloned in the next 50 years?

Solution No. To find the range of values likely to contain the percentage of *all* adult Americans who think human cloning will definitely or probably occur, we add and subtract the margin of error of 3 percentage points from the sample statistic of 51%. This gives a range of values from 48% to 54%. Because this range includes values on both sides of 50%, we cannot be confident that the majority (that is, greater than 50%) of adult Americans think that humans will be cloned in the next 50 years.

TIME OUT TO THINK

Look for a report on an opinion poll in this week's news. Does the report give a margin of error? What does it mean in this case?

By the Way ...

A clone is an exact genetic copy of its parent. For example, a clone of *you* would be genetically identical to you, but would be born as a baby and have different life experiences than you. The first successful cloning of an adult mammal came in 1997, when Ian Wilmut and his colleagues in Scotland cloned a sheep. The clone, named Dolly, has no father, since *all* her genes came from the mother. Scientists have now cloned many other mammals.

Putting It All Together: The Process of a Statistical Study

The process used by Nielsen Media Research is similar to that used in many statistical studies. Figure 1.2 and the box below summarize the basic steps in a statistical study. Keep in mind that these steps are somewhat idealized, and the actual steps may differ from one study to another. Moreover, the details hidden in the basic steps are critically important. For example, a poorly chosen sample in Step 2 can render the entire study meaningless, and great care must be taken in inferring conclusions about a population from results found for the much smaller sample of that population.

Basic Steps in a Statistical Study

Step 1. State the goal of your study precisely; that is, determine the population you want to study and exactly what you'd like to learn about it.

Step 2. Choose a sample from the population. (Be sure to use an appropriate sampling technique, as discussed in the next section.)

Step 3. Collect raw data from the sample and summarize these data by finding sample statistics of interest.

Step 4. Use the sample statistics to make inferences about the population.

Step 5. Draw conclusions; determine what you learned and whether you achieved your goal.

By the Way ...

Statisticians often divide their subject into two major branches: **descriptive statistics**, which deals with *describing* raw data in the form of graphics and sample statistics, and **inferential statistics**, which deals with *inferring* (or estimating) population parameters from sample data. In this book, Chapters 2 through 5 primarily cover descriptive statistics, while Chapters 6 through 10 focus on inferential statistics.

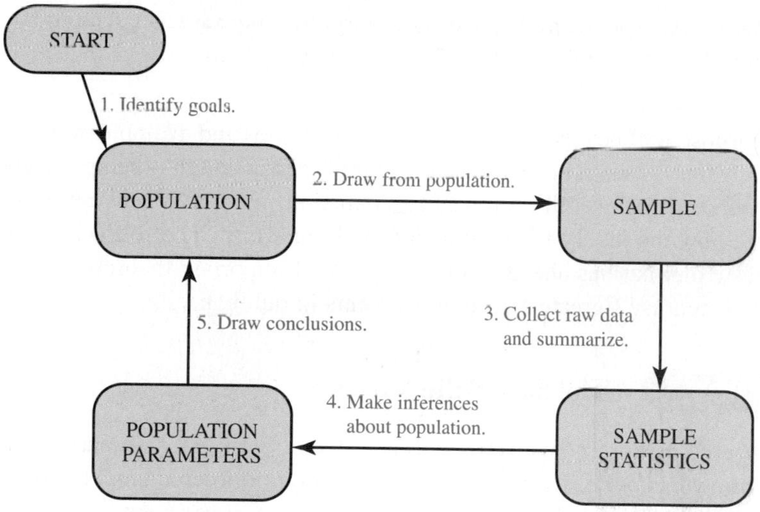

Figure 1.2 The process of a statistical study.

EXAMPLE 5 Identifying the Steps

Consider the Pew Research Center survey described in Examples 3 and 4. Identify how researchers applied the five basic steps in a statistical study.

Solution The steps apply as follows.

1. The researchers had a goal of learning about specific attitudes of Americans toward the future. They chose adult Americans as the population, deliberately leaving out children.
2. They chose 1,546 adult Americans for their sample. Although we are not told how the sample was drawn, we will assume that it was drawn so that the 1,546 adult Americans are typical of the entire adult American population.

3. They collected the raw data by asking carefully chosen questions of the people in the sample. The raw data are the individual responses to the questions. They summarized these data with sample statistics, such as the overall percentages of people in the sample who answered *yes* or *no* to each question.
4. Techniques of statistical science allowed the researchers to infer population characteristics. In this case, the inference consisted of estimating certain population parameters and calculating the margins of error.
5. By making sure that the study was conducted properly and interpreting the estimates of the population parameters, the researchers drew overall conclusions about Americans' attitudes toward the future.

Statistics: Decisions for an Uncertain World

The Nielsen ratings and most of the examples we've discussed so far involve surveys or polls. However, the subject of statistics encompasses much more, including experiments designed to test new medical treatments, analyses of the dangers of global warming, and even assessments of the value of a college education. Indeed, it is fair to say that the primary purpose of statistics is to help us make good decisions whenever we are confronted with a variety of possible options.

The Purpose of Statistics

Statistics has many uses, but perhaps its most important purpose is to help us make good decisions about issues that involve uncertainty.

This purpose will be clear in most of the case studies and examples we consider in this book, but occasionally we'll have to discuss a bit of theory that may seem somewhat abstract at first. If you keep the overall purpose of statistics in mind, you'll be rewarded in the end when you see how the theory helps us understand our world. The following case study will give you a taste of what lies ahead. It involves several important theoretical ideas that led to one of the 20th century's greatest accomplishments in public health.

By the Way ...

Polio quickly became rare in the United States after the development of the Salk vaccine, but it remained common in less-developed countries. A global effort to vaccinate children against polio began in 1998 and has achieved great success, though it has not yet reached its goal of completely eradicating the disease.

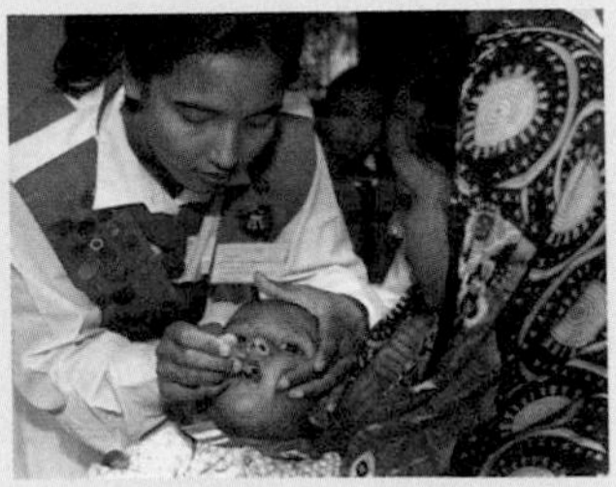

The greatest reward for doing is the opportunity to do more.

—Jonas Salk

CASE STUDY The Salk Polio Vaccine

If you had been a parent in the 1940s or 1950s, one of your greatest fears would have been the disease known as polio. Each year during this long polio epidemic, thousands of young children were paralyzed by the disease. In 1954, a large experiment was conducted to test the effectiveness of a new vaccine created by Dr. Jonas Salk (1914–1995). The experiment involved a sample of 400,000 children chosen from the population of all children in the United States. Half of these 400,000 children received an injection of the Salk vaccine. The other half received an injection that contained only salt water. (The salt water injection was a *placebo*; see Section 1.3.) Among the children receiving the Salk vaccine, only 33 contracted polio. In contrast, there were 115 cases of polio among the children who did not get the Salk vaccine. Using techniques of statistical science that we'll study later, the researchers concluded that the vaccine was effective at preventing polio. They therefore decided to launch a major effort to improve the Salk vaccine and distribute it to the population of *all* children. As a result, children in the United States and other developed countries began to get the vaccine routinely, and the horror of polio is now largely a thing of the past.

Section 1.1 Exercises

Statistical Literacy and Critical Thinking

1. **Population and Sample.** What is a population, what is a sample, and what is the difference between them?

2. **Statistic and Statistics.** Suppose that, in a discussion, one person refers to baseball statistics and another refers to the use of statistics in showing that a particular drug is an effective treatment. Do both uses of the term *statistics* have the same meaning? If not, how do they differ?

3. **Statistics and Parameters.** What is a sample statistic, what is a population parameter, and what is the difference between them?

4. **Margin of Error.** What is the margin of error in a statistical study and why is it important?

Does It Make Sense? For Exercises 5–10, decide whether the statement makes sense (or is clearly true) or does not make sense (or is clearly false). Explain clearly; not all of these statements have definitive answers, so your explanation is more important than your chosen answer.

5. **Statistics and Parameters.** My professor conducted a statistical study in which he was unable to measure any sample statistics, but he succeeded in determining the population parameters with a very small margin of error.

6. **Poor Poll.** A poll conducted two weeks before the election found that Smith would get 70% of the vote, with a margin of error of 3%, but he ended up losing the election anyway.

7. **Poll Certainty.** There is no doubt that Johnson won the election, because an exit poll showed that she received 54% of the vote and the margin of error was only 3 percentage points.

8. **Beating Nielsen.** A new startup company intends to compete with Nielsen Media Research by providing data with a larger margin of error for the same price.

9. **Depression Sample.** The goal of my study is to learn about depression among people who have suffered through a family tragedy, so I plan to choose a sample from the population of everyone who has been sick in the past month.

10. **New Product.** Our market research department surveyed 1,000 consumers on their attitudes toward our new product. Because the people in this sample were so enthusiastic about their desire to purchase the product, we decided to roll out a nationwide advertising campaign.

Concepts and Applications

Sample, Population, Statistic, and Parameter. Exercises 11–14 each describe a statistical study. In each case, identify the sample, the population, the sample statistic, and the population parameter.

11. **Stem Cell Research.** In a *Newsweek* poll conducted by Princeton Survey Research Associates International, 1,002 adults in the United States were asked whether they favor or oppose the use of "federal tax dollars to fund medical research using stem cells obtained from human embryos." Among the respondents, 48% said that they were in favor.

12. **Friday the 13th.** In a Gallup poll of 1,236 adults, 9% said that bad luck happens on a Friday that is the 13th day of a month.

13. **Galactic Distances.** Astronomers determine the distance to a far-away galaxy by measuring the distances of just a few stars within it and taking the mean of these distance measurements.

14. **Allergy Drug.** Nasonex is a drug used to treat allergy symptoms. In a test of Nasonex, 374 children aged 3 to 11 years were treated with 100 micrograms of Nasonex, and 17% of them experienced headaches.

Identifying the Range of Values. In Exercises 15–18, use the given statistic and margin of error to identify the range of values (confidence interval) likely to contain the true value of the population parameter.

15. **Stem Cell Research.** In a *Newsweek* poll conducted by Princeton Survey Research Associates International, 1,002 adults in the United States were asked whether they favor or oppose the use of "federal tax dollars to fund medical research using stem cells obtained from human embryos." Based on the poll results, it is estimated that 48% of adults are in favor, with a margin of error of 3 percentage points.

16. **ANWR Oil.** In a CBS News/*New York Times* poll, 1,241 adults in the United States were asked whether they approve or disapprove of drilling for oil and natural gas in Alaska's Arctic National Wildlife Refuge (ANWR). Based on the poll results, it is estimated that 45% of adults disapprove, with a margin of error of 3 percentage points.

17. **Body Mass Index.** Forty adult men in the United States are randomly selected and measured for their body mass index (BMI). Based on that sample, it is estimated that the

average (mean) BMI for men is 26.0, with a margin of error of 3.4.

18. **Body Temperatures.** Researchers randomly selected 106 adults and measured their body temperatures. Based on that sample, it is estimated that the average (mean) body temperature is 98.2°F, with a margin of error of 0.1°F.

19. **Interpreting Poll Results.** A poll is conducted the day before a state election for Senator. There are only two candidates running for this office. The poll results show that 58% of the voters favor the Republican candidate, with a margin of error of 3 percentage points. Should the Republican expect to win? Why or why not?

20. **Interpreting Taste Test Results.** In a nationally televised live taste test, a sample of regular drinkers of Budweiser was given blind samples of Michelob and Schlitz. Based on the results, we estimate that 52% of such subjects had a preference for Michelob, with a margin of error of 10%. Can we conclude that the majority of Budweiser drinkers prefer Michelob when given tastes of Michelob and Schlitz? Why or why not?

21. **Do People Lie About Voting?** In a survey of 1,002 people, 701 (or 70%) said that they voted in a particular presidential election (based on data from ICR Research Group). The margin of error for this survey was 3 percentage points. However, actual voting records show that only 61% of all eligible voters actually did vote. Does this imply that people lied when they responded in the survey? Explain.

22. **Why the Discrepancy?** In an Eagleton Institute poll, surveyed men were asked if they agreed with this statement: "Abortion is a private matter that should be left to the woman to decide without government intervention." Among the men who were interviewed by women, 77% agreed with the statement. Among the men who were interviewed by men, 70% agreed with the statement. Assuming that the discrepancy is significant, how might that discrepancy be explained?

Interpreting Real Studies. For each of Exercises 23–26, do the following:

a. Based on the given information, state what you think was the goal of the study. Identify a possible population and the population parameter of interest.

b. Briefly describe the sample, raw data, and sample statistic for the study.

c. Based on the sample statistic and the margin of error, identify the range of values (confidence interval) likely to contain the population parameter of interest.

23. **Changing Careers.** In a Korn/Ferry International survey of 1,733 executives, it was found that 51% of them said that if they could start their careers again, they would choose a different field. The margin of error was 3 percentage points.

24. **Superpowers.** In a Caravan survey of 1,018 adults, it was found that 11% would choose the ability to be invisible as the superpower that they would most prefer. The margin of error was 3.2 percentage points.

25. **Unemployment Rate.** Based on a recent survey of adults in 60,000 households, the U.S. Labor Department reported an unemployment rate of 4.6%. The margin of error was 0.2 percentage point.

26. **For a Rainy Day.** A Roper Organization survey of 2,000 adults in the United States showed that 64% of those surveyed kept money in regular savings accounts. The margin of error for the survey was 2.0 percentage points.

Five Steps in a Study. Describe how you would apply the five basic steps in a statistical study (as listed in the box on p. 7) to the issues in Exercises 27–30.

27. **Cell Phones and Driving.** You want to determine the percentage of licensed drivers who used a cell phone at least once while they were driving during the last week.

28. **Credit Scores.** FICO (Fair Isaac Corporation) scores are routinely used to rate the quality of consumer credit. You want to determine the average (mean) FICO score of all adults in the United States.

29. **Passenger Weight.** Recognizing that overloading commercial aircraft would lead to unsafe flights, you want to determine the average (mean) weight of airline passengers.

30. **Pacemaker Batteries.** Because the batteries used in heart pacemakers are so critically important, you want to determine the average (mean) length of time that such batteries last before failure.

Projects for the Internet and Beyond

For useful links, select "Links for Internet Projects" for Chapter 1 at www.aw.com/bbt.

31. **Current Nielsen Ratings.** Find the Nielsen ratings for the past week. What were the three most popular television shows? Explain the meaning of the "rating" and the "share" for each show.

32. **Nielsen Methods.** Nielsen Media Research frequently revises the details of its data collection methods. Visit its

Web site and read about its current strategies for rating television shows. Summarize these methods in a bulleted list format.

33. Comparing Airlines. The U.S. Department of Transportation routinely publishes on-time performance, lost baggage rates, and other statistics for different airline companies. Find a recent example of such statistics. Based on what you find, is it fair to say that any particular airline stands out as better or worse than others? Explain.

34. Labor Statistics. Use the Bureau of Labor Statistics Web site to find monthly unemployment rates over the past 12 months. If you assume that the monthly survey has a margin of error of about 0.2 percentage point, has there been a noticeable change in the unemployment rate over the past year? Explain.

35. Statistics and Safety. Identify a study that has been done (or should be done) to improve the safety of car drivers and passengers. Briefly describe the importance of statistics to the study.

36. Pew Research Center. The Pew Research Center for the People and the Press studies public attitudes toward the press, politics, and public policy issues. Go to its Web site and find the latest survey about attitudes. Select a particular recent survey, write a summary of what was surveyed, how the survey was conducted, and what was found.

IN THE NEWS

37. Statistics in the News. Identify three stories from the past week that involve statistics in some way. In each case, write a brief statement describing the role of statistics in the story.

38. Statistics in Your Major. Write a brief description of some ways in which you think that the science of statistics can be used in your major field of study. (If you have not yet selected a major, answer the question for a major that you are considering.)

39. Statistics and Entertainment. The Nielsen ratings are well known for their role in gauging television viewing. Identify another way that statistics are used in the entertainment industry. Briefly describe the role of statistics in this application.

40. Statistics in Sports. Choose a sport and describe at least three different statistics commonly tracked by participants in or spectators of the sport. In each case, briefly describe the importance of the statistic to the sport.

41. Economic Statistics. The government regularly publishes many different economic statistics, such as the unemployment rate, the inflation rate, and the surplus or deficit in the federal budget. Study recent newspapers and identify five important economic statistics. Briefly explain the purpose of each of these five statistics.

1.2 Sampling

The only way to know the true value of a population parameter is to observe *every* member of the population. For example, to learn the exact mean height of all students at your school, you'd need to measure the height of every student. A collection of data from every member of a population is called a **census**. Unfortunately, conducting a census is often impractical. In some cases, the population is so large that it would be too expensive or time-consuming to collect data from every member. In other cases, a census may be ruled out because it would interfere with a study's overall goals. For example, a study designed to test the quality of candy bars before shipping could not involve a census because that would mean testing a piece of every candy bar, leaving none intact to sell.

Not everything that can be counted counts, and not everything that counts can be counted.

—Albert Einstein

Definition

A **census** is the collection of data from *every* member of a population.

Fortunately, most statistical studies can be done without going to the trouble of conducting a census. Instead of collecting data from every member of the population, we collect data from a sample and use the sample statistics to make inferences about the population. Of course, the inferences will be reasonable only if the members of the sample represent the population fairly, at least in terms of the characteristics under study. That is, we seek a **representative sample** of the population.

Definition

A **representative sample** is a sample in which the relevant characteristics of the sample members are generally the same as the characteristics of the population.

EXAMPLE 1 A Representative Sample for Heights

Suppose you want to determine the mean height of all students at your school. Which is more likely to be a representative sample for this study: the men's basketball team or the students in your statistics class?

Solution The men's basketball team is not a representative sample for a study of height, both because it consists only of men and because basketball players tend to be taller than average. The mean height of the students in your statistics class is much more likely to be close to the mean height of all students, so the members of your class make a more representative sample than the members of the men's basketball team.

Bias

Imagine that, for the 5,000 homes in its sample, Nielsen chose only homes in which the primary wage earners worked a late-night shift. Because late-night workers aren't home to watch late-night television, Nielsen would find late-night shows to be unpopular among the homes in this sample. Clearly, this sample would *not* be representative of all American homes, and it would be wrong to conclude that late-night shows were unpopular among all Americans. We say that such a sample is *biased* because the homes in the sample differed in a specific way from "typical" American homes. (In reality, Nielsen takes great care to avoid such obvious bias in the sample selection.) More generally, the term **bias** refers to any problem in the design or conduct of a statistical study that tends to favor certain results. We cannot trust the conclusions of a biased study.

By the Way ...

Many medical studies are experiments designed to test whether a new drug is effective. In an article published in the *Journal of the American Medical Association*, the authors found that studies with positive results (the drug is effective) are more likely to be published than studies with negative results (the drug is not effective). This "publication bias" tends to make new drugs, as a group, seem more effective than they really are.

Definition

A statistical study suffers from **bias** if its design or conduct tends to favor certain results.

Bias can arise in many ways. For example:

- A sample is biased if the members of the sample differ in some specific way from the members of the general population. In that case, the results of the study will reflect the unusual characteristics of the sample rather than the actual characteristics of the population.
- A researcher is biased if he or she has a personal stake in a particular outcome. In that case, the researcher might intentionally or unintentionally distort the true meaning of the data.
- The data set itself is biased if its values were collected intentionally or unintentionally in a way that makes the data unrepresentative of the population.
- Even if a study is done well, it may be reported in a biased fashion. For example, a graph representing the data may tell only part of the story or depict the data in a misleading way (see Section 3.4).

Preventing bias is one of the greatest challenges in statistical research. Looking for bias is therefore one of the most important steps in evaluating a statistical study or media reports about a statistical study.

EXAMPLE 2 Why Use Nielsen?

Nielsen Media Research earns money by charging television stations and networks for its services. For example, NBC pays Nielsen to provide ratings for its television shows. Why doesn't NBC simply do its own ratings, instead of paying a company like Nielsen to do them?

Solution The cost of advertising on a television show depends on the show's ratings. The higher the ratings, the more the network can charge for advertising—which means NBC would have a clear bias if it conducted its own ratings. Advertisers therefore would not trust ratings that NBC produced on its own. By hiring an independent source, such as Nielsen, NBC can provide information that advertisers are more likely to believe.

TIME OUT TO THINK

The fact that NBC pays Nielsen for its services might seem to give Nielsen a financial incentive to make NBC look good in the ratings. How does the fact that Nielsen also provides ratings for other networks help prevent it from being biased toward NBC?

Sampling Methods

A good statistical study *must* have a representative sample. Otherwise the sample is biased and conclusions from the study are not trustworthy. Let's examine a few common sampling methods that, at least in principle, can provide a representative sample.

Simple Random Samples

In most cases, the best way to obtain a representative sample is by choosing *randomly* from the population. A **random sample** is one in which every member of the population has an equal chance of being selected to be part of the sample. For example, you could obtain a random sample by having everyone in a population roll a die and choosing those people who roll a 6. In contrast, the sample would not be random if you chose everyone taller than 6 feet, because not everyone would have an equal chance of being selected.

In statistics, we usually decide in advance the sample size that is needed. With **simple random sampling**, every possible sample of a particular size has an equal chance of being selected. For example, to choose a simple random sample of 100 students from all the students in your school, you could assign a number to each student in your school and choose the sample by drawing 100 of these numbers from a hat. As long as each student's number is in the hat only once, every sample of 100 students has an equal chance of being selected. As a faster alternative to using a hat, you might choose the student numbers with the aid of a computer or calculator that has a built-in *random number generator*.

TIME OUT TO THINK

Look for the random number key on a calculator. (Nearly all scientific calculators have one.) What happens when you push it? How could you use the random number key to select a sample of 100 students?

Because simple random sampling gives every sample of a particular size the same chance of being chosen, it is likely to provide a representative sample, as long as it is large enough. The larger the simple random sample, the more likely it is to be representative of the population.

EXAMPLE 3 Telephone Book Sampling

You want to conduct an opinion poll in which the population is all the residents in a town. Could you choose a simple random sample by randomly selecting names from the local telephone book?

Solution A sample drawn from a telephone book is not a simple random sample of the town population because phone books invariably are missing a lot of names, and anyone whose name is missing has no chance of being selected. For example, the phone book will be missing names when two or more people share the same phone number but have only one listing, when people choose to have an unlisted phone number or to rely exclusively on a cell phone, or when people (such as the homeless) don't have a telephone.

Systematic Sampling

Simple random sampling is effective, but in many cases we can get equally good results with a simpler technique. Suppose you are testing the quality of microchips produced by Intel. As the chips roll off the assembly line, you might decide to test every 50th chip. This ought to give a representative sample because there's no reason to believe that every 50th chip has any special characteristics compared to other chips. This type of sampling, in which we use a system such as choosing every 50th member of a population, is called **systematic sampling**.

EXAMPLE 4 Museum Assessment

When the National Air and Space Museum wanted to test possible ideas for a new solar system exhibit, a staff member interviewed a sample of visitors selected by systematic sampling. She interviewed a visitor exactly every 15 minutes, choosing whoever happened to enter the current solar system exhibit at that time. Why do you think she chose systematic sampling rather than simple random sampling? Was systematic sampling likely to produce a representative sample in this case?

Solution Simple random sampling might occasionally have selected two visitors so soon after each other that the staff member would not have had time to interview each of them. The systematic process of choosing a visitor every 15 minutes prevented this problem from arising. Because there's no reason to think that the people entering at a particular moment are any different from those who enter a few minutes earlier or later, this process is likely to give a representative sample of the population of visitors during the time of the sampling.

By the Way ...

The sampling described in Example 4 was undertaken prior to the construction of the Voyage Scale Model Solar System, a permanent scale model exhibit that stretches along the National Mall from the National Air and Space Museum to the Smithsonian Castle. The photo below shows the son of one of the exhibit creators (also an author of this textbook) touching the scale model Sun.

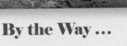

EXAMPLE 5 When Systematic Sampling Fails

You are conducting a survey of students in a co-ed dormitory in which males are assigned to odd-numbered rooms and females are assigned to even-numbered rooms. Can you obtain a representative sample when you choose every 10th room?

Solution No. If you start with an odd-numbered room, every 10th room will also be odd-numbered (such as room numbers 3, 13, 23, . . .). Similarly, if you start with an even-numbered room, every 10th room will also be even-numbered. You will therefore obtain a sample consisting of either all males or all females, neither of which is representative of the co-ed population.

TIME OUT TO THINK

Suppose you chose every 5th room, rather than every 10th room, in Example 5. Would the sample then be representative?

Convenience Samples

Systematic sampling is easier than simple random sampling, but may also be impractical in many cases. For example, suppose you want to know the proportion of left-handed students at your school. It would take great effort to select a simple random sample or a systematic sample, because both require drawing from all the students in the school. In contrast, it would be easy to use the students in your statistics class as your sample—you could just ask the left-handed students to raise their hands. This type of sample is called a **convenience sample** because it is chosen for convenience rather than by a more sophisticated procedure. For trying to find the proportion of left-handed people, the convenience sample of your statistics class is probably fine; there is no reason to think that there would be a different proportion of left-handed students in a statistics class than anywhere else. But if you were trying to determine the proportions of students with different majors, this sample would be biased because some majors require a statistics course and others do not. In general, convenience sampling tends to be more prone to bias than most other forms of sampling.

EXAMPLE 6 Salsa Taste Test

A supermarket wants to decide whether to carry a new brand of salsa, so it offers free tastes at a stand in the store and asks people what they think. What type of sampling is being used? Is the sample likely to be representative of the population of all shoppers?

Solution The sample of shoppers stopping for a taste of the salsa is a convenience sample because these people happen to be in the store and are willing to try the new product. (This type of convenience sample, in which people choose whether or not to be part of the sample, is also called a *self-selected sample*. We will study self-selected samples further in Section 1.4.) This sample is unlikely to be representative of the population of all shoppers, because different types of people may shop at different times (for example, stay-at-home parents are more likely to shop at mid-day than are working parents) and only people who like salsa are likely to participate. The data may be still be useful, however, because the opinions of people who like salsa are probably the most important ones in this case.

Cluster Samples

Cluster sampling involves the selection of *all* members in randomly selected groups, or *clusters*. Imagine that you work for the Department of Agriculture and wish to determine the percentage of farmers who use organic farming techniques. It would be difficult and costly to collect a simple random sample or a systematic sample because either would require visiting many individual farms that are located far from one another. A convenience sample of farmers in a single county would be biased because farming practices vary from region to region. You might therefore decide to select a few dozen counties at random from across the United States and survey *every* farmer in each of those counties. We say that each county contains a *cluster* of farmers, and the sample consists of *every* farmer within the randomly selected clusters.

EXAMPLE 7 Gasoline Prices

You want to know the mean price of gasoline at gas stations located within a mile of rental car locations at airports. Explain how you might use cluster sampling in this case.

Solution You could randomly select a few airports around the country. For these airports, you would check the gasoline price at *every* gas station within a mile of the rental car location.

By the Way ...

As mandated by the U.S. Constitution, voting for the President is actually done by a small group of people called *electors*. Each state may select as many electors as it has members of Congress (counting both Senators and Representatives). When you cast a ballot for President, you actually cast a vote for your state's electors, each of whom has promised to vote for a particular presidential candidate. The electors cast their votes a few weeks after the general election.

Stratified Samples

Suppose you are conducting a poll to predict the outcome of the next U.S. presidential election. The population under study is all likely voters, so you might choose a simple random sample from this population. However, because presidential elections are decided by electoral votes cast on a state-by-state basis, you'll get a better prediction if you determine voter preferences within each state. Your overall sample should therefore consist of separate random samples from each of the 50 states. In statistical terminology, the populations of the 50 states represent subgroups, or **strata**, of the total population. Because your overall sample consists of randomly selected members from each stratum, you've used **stratified sampling**.

EXAMPLE 8 Unemployment Data

The U.S. Labor Department surveys 60,000 households each month to compile its unemployment report (see Example 2 in Section 1.1). To select these households, the Department first groups cities and counties into about 2,000 geographic areas. It then randomly selects households to survey within these geographic areas. How is this an example of stratified sampling? What are the strata? Why is stratified sampling important in this case?

Solution The unemployment survey is an example of stratified sampling because it first breaks the population into subgroups. The subgroups, or strata, are the people in the 2,000 geographic regions. Stratified sampling is important in this case because unemployment rates are likely to differ in different geographic regions. For example, unemployment rates in rural Kansas may be very different from those in Silicon Valley. By using stratified sampling, the Labor Department ensures that its sample fairly represents all geographic regions.

Summary of Sampling Methods

The following box and Figure 1.3 summarize the five sampling methods we have discussed. No single method is "best," as each one has its uses. (Some studies even combine two or more types of sampling.) But regardless of how a sample is chosen, keep in mind the following three key ideas:

- A study can be successful only if the sample is representative of the population.
- A biased sample is unlikely to be a representative sample.
- Even a well-chosen sample may still turn out to be unrepresentative just because of bad luck in the actual drawing of the sample.

Common Sampling Methods

- **Simple random sampling:** We choose a sample of items in such a way that every sample of the same size has an equal chance of being selected.
- **Systematic sampling:** We use a simple system to choose the sample, such as selecting every 10th or every 50th member of the population.
- **Convenience sampling:** We use a sample that happens to be convenient to select.
- **Cluster sampling:** We first divide the population into groups, or clusters, and select some of these clusters at random. We then obtain the sample by choosing *all* the members within each of the selected clusters.
- **Stratified sampling:** We use this method when we are concerned about differences among subgroups, or *strata*, within a population. We first identify the strata and then draw a random sample within each stratum. The total sample consists of all the samples from the individual strata.

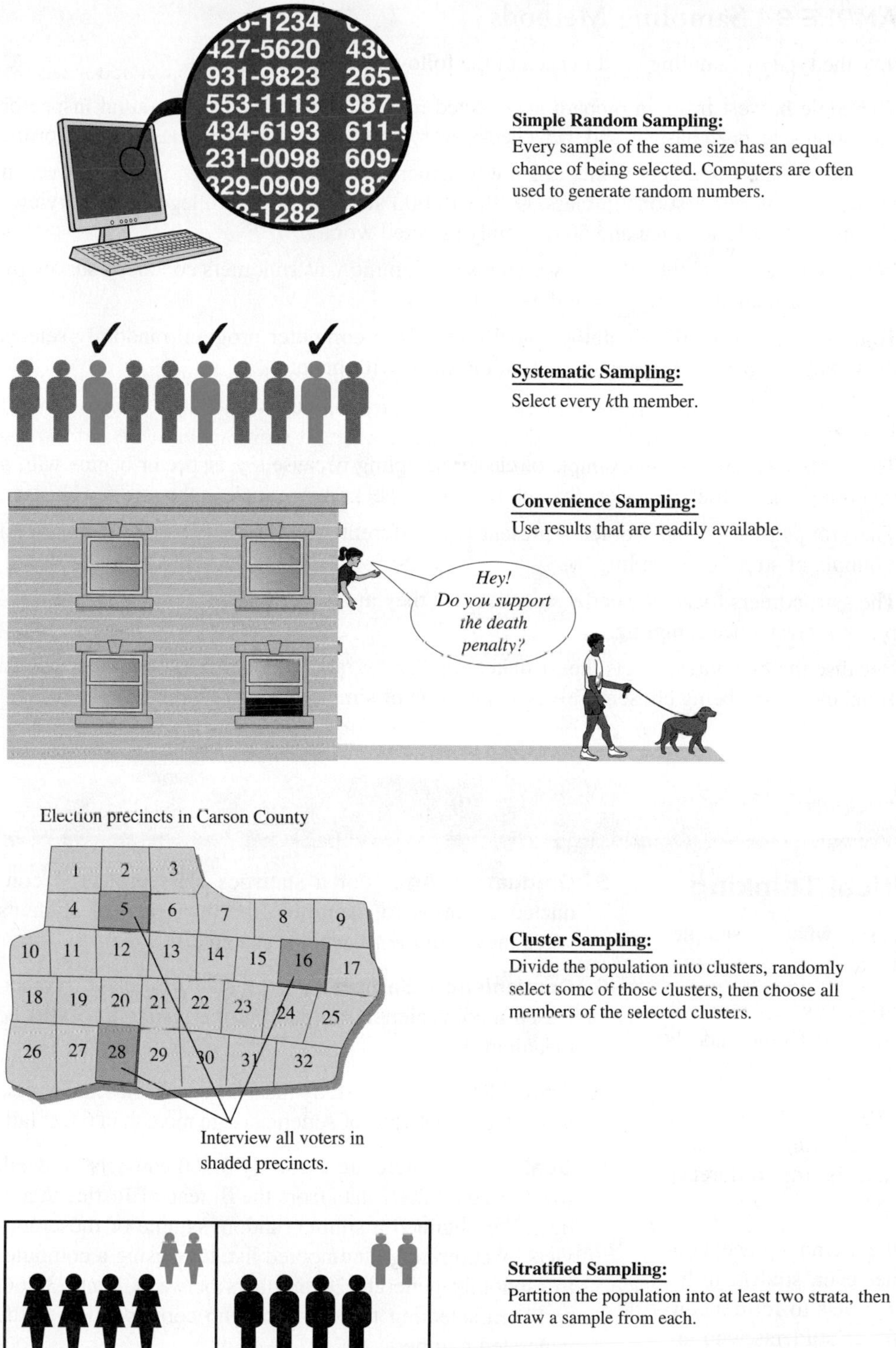

-1234
427-5620 43
931-9823 265-
553-1113 987-
434-6193 611-
231-0098 609-
329-0909 98
-1282
Simple Random Sampling:
Every sample of the same size has an equal chance of being selected. Computers are often used to generate random numbers.
Systematic Sampling:
Select every *k*th member.
Convenience Sampling:
Use results that are readily available.
Hey! Do you support the death penalty?
Election precincts in Carson County
1 2 3 4 5 6 7 8 9 10 11 12 13 14 15 16 17 18 19 20 21 22 23 24 25 26 27 28 29 30 31 32
Interview all voters in shaded precincts.
Cluster Sampling:
Divide the population into clusters, randomly select some of those clusters, then choose all members of the selected clusters.
Stratified Sampling:
Partition the population into at least two strata, then draw a sample from each.

Figure 1.3

EXAMPLE 9 Sampling Methods

Identify the type of sampling used in each of the following cases.

a. The apple harvest from an orchard is collected in 1,200 baskets. An agricultural inspector randomly selects 25 baskets and then checks every apple in each of these baskets for worms.

b. An educational researcher wants to know whether, at a particular college, men or women tend to ask more questions in class. Of the 10,000 students at the college, she interviews 50 randomly selected men and 50 randomly selected women.

c. In trying to learn whether planetary systems are common, astronomers conduct a survey by looking for planets among 100 of the nearest stars.

d. To determine who will win autographed footballs, a computer program randomly selects the ticket numbers of 11 people in a stadium filled with people.

Solution

a. The apple inspection is an example of cluster sampling because the inspector begins with a randomly selected set of clusters (baskets) and then checks every apple in the selected clusters.

b. The groups of men and women represent two different strata for this study, so this is an example of stratified sampling.

c. The astronomers focus on nearby stars because they are easier to study, so this is an example of convenience sampling.

d. Because the computer selects the 11 ticket numbers at random, every ticket number has an equal chance of being chosen. This is an example of simple random sampling.

Section 1.2 Exercises

Statistical Literacy and Critical Thinking

1. **Census and Sample.** What is a census, what is a sample, and what is the difference between them?

2. **Biased Sample.** In a pre-election survey, a sample is drawn randomly from a list of registered Democrats. Is there anything wrong with this sampling method?

3. **Cluster and Stratified Sampling.** Cluster sampling and stratified sampling both involve selecting subjects in subgroups of the population. What is the difference between these two types of sampling?

4. **Sample of Students.** One of the authors once conducted a study by recording whether each student in his class was right-handed. The objective was to form a conclusion about the proportion of college students who are right-handed. What type of sample was obtained? Is this sample likely to be biased? Why or why not?

Does It Make Sense? For Exercises 5–8, decide whether the statement makes sense (or is clearly true) or does not make sense (or is clearly false). Explain clearly; not all of these statements have definitive answers, so your explanation is more important than your chosen answer.

5. **Graduation Age.** For a statistics class project, I conducted a census to determine the mean age of students when they earn their bachelor's degree.

6. **Convenience Sample.** For a statistics class project, I used a convenience sample, but the results may still be meaningful.

7. **Biased Sample.** The study must have been biased, because it concluded that 75% of Americans are more than 6 feet tall.

8. **Death Row.** There are currently 3,366 convicts on death row (based on 2007 data from the Bureau of Justice Statistics). We obtained a simple random sample of those convicts by compiling a numbered list, then using a computer to randomly generate 20 numbers between 1 and 3,366, and then selecting the convicts who corresponded to the generated numbers.

Concepts and Applications

Census. In Exercises 9–12, determine whether a census is practical in the situation described. Explain your reasoning.

9. **Laker Heights.** You want to determine the mean height of all basketball players on the LA Lakers team.

10. High School Heights. You want to determine the mean height of all high school basketball players in the United States.

11. IQ Scores. You want to determine the mean IQ score of all statistics instructors in the United States.

12. Instructor Ages. You want to determine the mean age of all statistics instructors at the University of Colorado.

Representative Samples? In Exercises 13–16, identify the sample, population, and sampling method. Then comment on whether you think it is likely that the sample is representative of the population.

13. Senate Terms. A political scientist randomly selects 4 of the 100 Senators currently serving in Congress and then finds the lengths of time that they have served.

14. Super Bowl. During the Super Bowl game, Nielsen Media Research conducts a survey of 5,108 randomly selected households and finds that 44% of them have television sets tuned to the Super Bowl.

15. Cloning. In a Gallup poll of 1,012 randomly selected adult Americans, 89% said that cloning of humans should not be allowed.

16. Mail Survey. A graduate student at the University of Newport conducts a research project about how adult Americans communicate. She begins with a survey mailed to 500 of the adults she knows. She asks them to mail back a response to this question: "Do you prefer to use e-mail or snail mail (the U.S. Postal Service)?" She gets back 65 responses, with 42 of them indicating a preference for snail mail.

Evaluating Sample Choices. Exercises 17 and 18 each describe the goal of a study and then offer you four possible samples. In each case, decide which sample is most likely to be a representative sample, and explain why. Then explain why each of the other choices is *not* likely to be a representative sample for the study.

17. Credit Card Debt. You want to determine the mean amount of credit card debt owed by adult consumers in Florida.

a. The Florida drivers who own and have registered Land Rover vehicles

b. The first 1,000 Florida residents listed in the Fort Lauderdale phone book

c. The first 1,000 Florida residents in a complete list of all Florida telephone numbers

d. The Florida residents who mail back a survey printed in the *Miami Herald*

18. California Voters. You want to conduct a survey to determine the proportion of eligible voters in California likely to vote for the Democratic presidential candidate in the next election.

a. All eligible voters in San Diego County

b. All eligible voters in the city of Sonoma

c. All eligible voters who respond to an America OnLine (AOL) Internet survey

d. Every 1000th person on a complete list of all eligible voters in California

Bias. Are there sources of bias in the situations described in Exercises 19–22? Explain.

19. Movie Critic. A film critic for ABC News gives her opinion of the latest movie from Disney, which also happens to own ABC.

20. Car Reviews. *Consumer Reports* magazine, which does not accept free products or advertising from anyone, prints a review of new cars.

21. GMO Soybeans. Monsanto hires independent university scientists to determine whether its new, genetically engineered soybean poses any threat to the environment.

22. Drug Study Funding. The *Journal of the American Medical Association* prints an article evaluating a drug, and some of the physicians who wrote the article received funding from the pharmaceutical company that produces the drug.

Sampling Methods. In Exercises 23–38, identify each sample as a simple random sample, systematic sample, convenience sample, stratified sample, or cluster sample. In each case, state whether you think the sampling method is likely to yield a representative sample or a biased sample, and explain why.

23. Clinical Trial. In phase II testing of a new drug designed to increase the red blood cell count, a researcher obtains envelopes with the names and addresses of all treated subjects. She wants to increase the dosage in a sub–sample of 12 subjects, so she thoroughly mixes all of the envelopes in a bin and then pulls 12 of those envelopes to identify the subjects to be given the increased dosage.

24. Sobriety Checkpoint. One of the authors was an observer at a police sobriety checkpoint at which every 5th driver was stopped and interviewed.

25. Exit Polls. On the day of the presidential election, the news media organize an exit poll in which specific polling stations are randomly selected and all voters are surveyed as they leave the premises.

26. Education and Sports. A researcher for the Spalding athletic equipment company is studying the relationship between level of education and participation in any sport. He conducts a survey of 40 randomly selected golfers, 40 randomly selected tennis players, and 40 randomly selected swimmers.

27. Ergonomics. An engineering student measures the strength of fingers used to push buttons by testing family members.

28. Tax Cheating. An Internal Revenue Service researcher investigates cheating on income tax reports by surveying all waiters and waitresses at 20 randomly selected restaurants.

29. MTV Survey. A marketing expert for MTV is planning a survey in which 500 people will be randomly selected from each of the age groups 10–19, 20–29, and so on.

30. Credit Card Data. One of the authors surveyed all of his students to obtain sample data consisting of the number of credit cards students possess.

31. Fund Raising. Fundraisers for the College of Newport test a new telemarketing campaign by obtaining an alphabetical list of all alumni and calling every 100th name on that list.

32. Telephone Poll. In a Gallup poll of 1,059 adults, the interview subjects were selected by using a computer to randomly generate telephone numbers that were then called.

33. Market Research. A market researcher has partitioned all California residents into categories of unemployed, employed full-time, and employed part-time. She is surveying 50 people from each category.

34. Student Drinking. Motivated by the death of a student from binge drinking, the College of Newport conducts a study of student drinking by randomly selecting 10 different classes and interviewing all of the students in each of those classes.

35. Magazine Survey. *People* magazine chooses its "best-dressed celebrities" by compiling responses from readers who mail in a survey printed in the magazine.

36. Heart Transplants. A medical researcher at Johns Hopkins University obtains a numbered list of all patients waiting for a heart transplant, uses a computer to randomly generate 50 numbers, and then select the patients corresponding to the 50 numbers.

37. Quality Control. A sample of manufactured CDs is obtained by using a computer to randomly generate a number between 1 and 1,000 for each CD and then selecting the CD if the generated number is 1,000.

38. Seat Belts. Every 500th seat belt is tested by stressing it until it fails.

Choosing a Sampling Method. For each of Exercises 39–42, suggest a sampling method that is likely to produce a representative sample. Explain why you chose this method over other methods.

39. Student Election. You want to predict the winner of an upcoming election for student body president.

40. Blood Type. You want to determine the percentage of people in this country in each of the four major blood groups (A, B, AB, and O).

41. Heart Deaths. You want to determine the percentage of deaths due to heart disease each year.

42. Mercury in Tuna. You want to determine the average mercury content of the tuna fish consumed by U.S. residents.

Projects for the Internet and Beyond

For useful links, select "Links for Internet Projects" for Chapter 1 at www.aw.com/bbt.

43. Public Opinion Poll. Use information available on the Web site of a polling organization, such as Gallup, Harris, Pew, or Yankelovich, to answer the following questions.

a. How exactly is a sample of subjects selected?

b. Based on what you have learned, do you think the poll results are reliable? If so, why? If not, why not?

44. Unemployment Sample. Use the Bureau of Labor Statistics Web page to find details on how the Bureau chooses the sample of households in its monthly survey. Write a short summary of the procedure and why it is likely to yield a representative sample.

45. Selective Voting. The Academy Awards, the Heisman Trophy, and the *New York Times* "Bestseller List" are just three examples of cases in which selections are determined by the votes of specially selected individuals. Pick one of these selection processes and describe who votes and how those people are chosen. Discuss sources of bias in the process.

IN THE NEWS

46. **Sampling in the News.** Find a recent news report about a statistical study that you find interesting. Write a short summary of how the sample for the study was chosen, and briefly discuss whether you think the sample was representative of the population under study.

47. **Opinion Poll Sample.** Find a recent news report about an opinion poll carried out by a news organization (such as Gallup, Harris, *USA Today, New York Times*, or CNN). Briefly describe the sample and how it was chosen. Was the sample chosen in a way that was likely to introduce bias? Explain.

48. **Political Polls.** Find results from a recent poll conducted by a political organization (such as the Republican or Democratic party or an organization that seeks to influence Congress on some particular issue). Briefly describe the sample and how it was chosen. Was the sample chosen in a way that was likely to introduce bias? Should you be more concerned about bias in such a poll than you would be in a poll conducted by a news organization? Explain.

1.3 Types of Statistical Studies

Statistical studies are conducted in many different ways. In all cases, the people, animals (or other living things), or objects chosen for the sample are called the **subjects** of the study. If the subjects are people, it is common to refer to them as **participants** in the study.

Definition

The **subjects** of a study are the people, animals (or other living things), or objects chosen for the sample; if the subjects are people, they may also be called the **participants** in the study.

There are two basic types of statistical study: observational studies and experiments. In an **observational study**, we observe or measure specific characteristics while trying to be careful to avoid influencing or modifying the characteristics we are observing. The Nielsen ratings are an example of an observational study, because Nielsen uses its "people meters" to *observe* what the subjects are watching on TV, but does not try to influence what they watch.

You can observe a lot by just watching.

—Yogi Berra

Note that an observational study may involve activities that go beyond the usual definition of *observing*. Measuring people's weights requires interacting with them, as in asking them to stand on a scale. But in statistics, we consider these measurements to be observations because the interactions do not change people's weights. Similarly, an opinion poll in which researchers conduct in-depth interviews is considered observational as long as the researchers attempt only to learn people's opinions, not to change them.

In contrast, consider a medical study designed to test whether large daily doses of vitamin C help prevent colds. To conduct this study, the researchers must ask some people in the sample to take large doses of vitamin C every day. This type of statistical study is called an **experiment**. The purpose of an experiment is to study the effects of some **treatment**—in this case, large daily doses of vitamin C.

Two Basic Types of Statistical Study

There are two basic types of statistical study:

1. In an **observational study,** researchers observe or measure characteristics of the subjects, but do not attempt to influence or modify these characteristics.
2. In an **experiment,** researchers apply some **treatment** and observe its effects on the subjects of the experiment.

EXAMPLE 1 Type of Study

Identify the study as an observational study or an experiment.

a. The Salk polio vaccine study (see the Case Study on page 8)

b. A poll in which people are asked for whom they plan to vote in the next election

Solution

a. The Salk vaccine study was an *experiment* because researchers tested a treatment—in this case, the vaccine—to see whether it reduced the incidence of polio.

b. The poll is an *observational study* because it attempts to determine voting preference but does not try to sway votes.

Identifying the Variables

Statistical studies—whether observations or experiments—generally are attempts to measure what we call **variables of interest**. The term *variable* simply refers to an item or quantity that can vary or take on different values, and variables of interest are those we seek to learn about. For example, variables of interest in the Nielsen studies of television viewing habits include *show being watched* and *number of viewers*. The variable *show being watched* can take on different values such as "Super Bowl," "*60 Minutes*," or "*Lost*." The variable *number of viewers* depends on the popularity of a particular show. In essence, the raw data in any statistical study are the different values of the variables of interest.

In cases where we think cause and effect may involved, we sometimes subdivide the variables of interest into two categories. For example, each person in the study of vitamin C and colds may take a different daily dose of vitamin C and may end up with a different number of colds over some period of time. Because we are trying to learn if vitamin C causes a lower number of colds, we say that *daily dose of vitamin C* is an **explanatory variable**—it may explain or cause a change in the number of colds. Similarly, we say that *number of colds* is a **response variable**, because we expect it to respond to changes in the explanatory variable (the dose of vitamin C).

Definitions

A **variable** is any item or quantity that can vary or take on different values.

The **variables of interest** in a statistical study are the items or quantities that the study seeks to measure.

When cause and effect may be involved, an **explanatory variable** is a variable that may explain or cause the effect, while a **response variable** is a variable that responds to changes in the explanatory variable.

EXAMPLE 2 Identify the Variables

Identify the variables of interest for each study.

a. The Salk polio vaccine study

b. A poll in which people are asked for whom they plan to vote in the next election

Solution

a. The two variables of interest in the Salk vaccine study are *vaccine* and *polio*. They are variables because they can take on two different values: A child either did or did not get the vaccine and either did or did not contract polio. In this case, because the study seeks to determine whether the vaccine prevents polio, we say that *vaccine* is the explanatory variable (it may explain a change in the incidence of polio) and *polio* the response variable (it is supposed to change in response to the vaccine).

b. The variables of interest in the pre-election poll might be called *candidate's name*, since people must choose one of the candidates as the one they plan to vote for, and *proportion of supporters*, meaning the fraction of the people polled who plan to vote for a particular candidate. (There is no cause and effect involved in this study, so we do not need to decide whether the variable is explanatory or response.)

Observational Studies

The observational studies we have discussed up to this point, such as Nielsen ratings, opinion polls, and determining the mean height of students, are studies in which the data are all generally collected around the same time. Sometimes, however, observational studies look at past data or are designed to look at future data over a long period of time.

A **retrospective study** (sometimes called a *case-control* study) is an observational study that uses data from the past—such as official records or past interviews—to learn about some issue of concern. Retrospective studies are especially valuable in cases where it may be impractical or unethical to perform an experiment. For example, suppose we want to learn how alcohol consumed during pregnancy affects newborn babies. Because it is already known that consuming alcohol during pregnancy can be harmful, it would be unethical to ask pregnant mothers to test the "treatment" of consuming alcohol. However, because many mothers consumed alcohol in past pregnancies (either before the dangers were known or choosing to ignore the dangers), we can do a retrospective study in which we compare children born to those mothers to children born to mothers who did not consume alcohol.

Sometimes, the data we need to reach clear conclusions are not available in past records. In those cases, researchers may set up a **prospective study** (sometimes called a *longitudinal* study) designed to collect observations in the future from groups that share common factors. A classic example of a prospective study is the Harvard Nurses' Health Study, which was started in 1976 in order to collect data about how different lifestyles affect women's health (see the "Focus on Public Health" section at the end of this chapter). The study, still ongoing today, has followed thousands of nurses over more than three decades, collecting data about their lifestyles and health.

Variations on Observational Studies

The most familiar observational studies are those in which data are collected all at once (or as close to that as possible). Two variations on observational studies are also common:

1. A **retrospective** (or **case-control**) **study** uses data from the past, such as official records or past interviews.
2. A **prospective** (or **longitudinal**) **study** is set up to collect data in the future from groups that share common factors.

EXAMPLE 3 Observational Study

You want to know whether children born prematurely do as well in elementary school as children born at full term. What type of study should you do?

Solution An observational, retrospective study is the only real option in this case. You would collect data on past births and compare the elementary school performance of those born prematurely to that of those born at full term.

Experiments

Because experiments require active intervention, such as applying a treatment, we must take special care to ensure that they are designed in ways that will provide the information we seek. Let's examine a few of the issues that arise in the design of experiments.

The Need for Controls

Consider an experiment that gives some participants in a study vitamin C to determine its effect on colds. Suppose the people taking vitamin C daily get an average of 1.5 colds in a three-month period. How can the researchers know whether the subjects would have gotten more colds without the vitamin C? To answer this type of question, the researchers must conduct their experiment with two (or more) groups of subjects: One group takes large doses of vitamin C daily and another group does not. As we'll discuss shortly, in most cases it is important that participants be randomly assigned to the two groups.

The group of people who are randomly assigned to take vitamin C is called the **treatment group** because its members receive the treatment being tested (vitamin C). The group of people who do *not* take vitamin C is called the **control group**. The researchers can be confident that vitamin C is an effective treatment only if the people in the treatment group get significantly fewer colds than the people in the control group. The control group gets its name from the fact that it helps control the way we interpret experimental results.

Treatment and Control Groups

The **treatment group** in an experiment is the group of subjects who receive the treatment being tested.

The **control group** in an experiment is the group of subjects who do *not* receive the treatment being tested.

In most cases, it is important to choose the members of the two groups by random selection from the available pool of subjects.

EXAMPLE 4 Treatment and Control

Look again at the Salk polio vaccine Case Study (page 8). What was the treatment? Which group of children constituted the treatment group? Which constituted the control group?

Solution The treatment was the Salk vaccine. The treatment group consisted of the children who received the Salk vaccine. The control group consisted of the children who did not get the Salk vaccine and instead got an injection of salt water.

EXAMPLE 5 Mozart Treatment

A study divided college students into two groups. One group listened to Mozart or other classical music before being assigned a specific task and the other group simply was assigned the task. Researchers found that those listening to the classical music performed the task slightly better, but only if they did the task within a few minutes of listening to the music (the two groups performed equally on tasks given later). Identify the treatment and the control and treatment groups.

Solution The treatment was the classical music. The treatment group consisted of the students who listened to the music. The control group consisted of the students who did not listen to the music.

By the Way ...

The so-called Mozart effect holds that listening to Mozart can make babies smarter. The supposed effect spawned an entire industry of Mozart products for children. The state of Georgia even began passing out Mozart CDs to new mothers. However, more recent studies of the Mozart effect have been unable to substantiate the claimed effect.

Confounding Variables

Using control groups helps to ensure that we account for known variables that could affect a study's results. However, researchers may be unaware of or be unable to account for other important variables. Consider an experiment in which a statistics teacher seeks to determine whether students who study collaboratively (in study groups with other students) earn higher grades than students who study independently. The teacher chooses five students who will study collaboratively (the treatment group) and five others who will study independently (the control group). To ensure that the students all have similar abilities and will study diligently, the teacher chooses only students with high grade-point averages. At the end of the semester, the teacher finds that the students who studied collaboratively earned higher grades.

The variables of interest for this study are *collaborative study* (whether they do so or not) and *final grade*. But suppose that, unbeknownst to the teacher, the collaborative students all lived in a dormitory where a curfew ensured that they got plenty of sleep. This fact introduces a new variable—which we might call *amount of sleep*—that might partially explain the results. In other words, the experiment's conclusion may *seem* to support the benefits of collaborative study, but this conclusion is not justified because the teacher did not account for how much students slept.

In statistical terminology, this study suffers from **confounding**. The higher grades may be due either to the variable of interest (*collaborative study*) or to the differing amounts of sleep or to a combination of both. Because the teacher did not account for differences in the amount of sleep, we say that *amount of sleep* is a **confounding variable** for this study. You can probably think of other potentially confounding variables that could affect a study like this one.

Definition

A study suffers from **confounding** if the effects of different variables are mixed so we cannot determine the specific effects of the variables of interest. The variables that lead to the confusion are called **confounding variables.**

CASE STUDY Confounding Drug Results

Pfizer Corporation developed a new drug (called fluconazole) designed to prevent fungal infections in hospital patients. Several studies found the new drug to be more effective than an older drug (called amphotericin B). However, a subsequent analysis by other researchers found that the older drug had been administered orally when it was supposed to be given by injection. This introduced a source of confounding into the studies: The original researchers thought the results showed the new drug to be more effective than the old drug, but they had not taken into account the confounding effects of how the old drug was administered. Once the new researchers took this effect into account (by comparing only cases in which both drugs were administered properly), they found that the new drug was no more effective than the old drug.

Assigning Treatment and Control Groups

As the collaborative study experiment illustrates, results are almost sure to suffer from confounding if the treatment and control groups differ in important ways (other than receiving or not receiving the treatment). Researchers generally employ two strategies to prevent such differences and thereby ensure that the treatment and control groups can be compared fairly. First, they assign participants to the treatment and control groups *at random*, meaning that they use a technique designed to ensure that each participant has an equal chance of being assigned to either group. When the participants are randomly assigned, it is less likely that the people in the treatment and control groups will differ in some way that will affect the study results.

Second, researchers try to ensure that the treatment and control groups are sufficiently large. For example, in the collaborative study experiment, including 50 students in each group rather than 5 would have made it much less likely that all the students in one group would live in a special dormitory.

Strategies for Selecting Treatment and Control Groups

1. **Select groups at random.** Make sure that the subjects of the experiment are assigned to the treatment or control group at random, meaning that each subject has an equal chance of being assigned to either group.
2. **Use sufficiently large groups.** Make sure that the treatment and control groups are both sufficiently large that they are unlikely to differ in a significant way (aside from the fact that one group gets the treatment and the other does not).

EXAMPLE 6 Salk Study Groups

Briefly explain how the two strategies for selecting treatment and control groups were used in the Salk polio vaccine study.

Solution A total of about 400,000 children participated in the study, with half receiving an injection of the Salk vaccine (the treatment group) and the other half receiving an injection of salt water (the control group). The first strategy was implemented by choosing children for the two groups randomly from among all the children. The second strategy was implemented by using a large number of participants (200,000 in each group) so that the two groups were unlikely to differ by accident.

The Placebo Effect

When an experiment involves people, effects can occur simply because people know they are part of the experiment. For example, suppose you are testing the effectiveness of a new anti-depression drug. You find 500 people who suffer from depression and randomly divide them into a treatment group that receives the new drug and a control group that does not. A few weeks later, interviews with the patients show that people in the treatment group tend to be feeling much better than people in the control group. Can you conclude that the new drug works?

A strong imagination brings on the [placebo effect]. . . . Everyone feels its impact, but some are knocked over by it. . . . [Doctors] know that there are men on whom the mere sight of medicine is operative.

—French philosopher Michel de Montaigne (1533–1592)

Unfortunately, it's quite possible that the mood of people receiving the drug improved simply because they were glad to be getting some kind of treatment, which means you cannot be sure that the drug really helped. This type of effect, in which people improve because they believe that they are receiving a useful treatment, is called the **placebo effect**. (The word *placebo* comes from the Latin "to please.")

To distinguish between results caused by a placebo effect and results that are truly due to the treatment, researchers try to make sure that the participants do not know whether they are part of the treatment or control group. To accomplish this, the researchers give the people in the control group a **placebo**: something that looks or feels just like the treatment being tested, but lacks its active ingredients. For example, in a test of a drug that comes in pill form, the placebo might be a pill of the same shape and size that contains sugar instead of the real drug. In a test of an injected vaccine, the placebo might be an injection that contains only a saline solution (salt water) instead of the real vaccine. In a recent test of the effectiveness of acupuncture, the placebo consisted of treatment with needles as in real acupuncture, except the needles were not put in the special places that acupuncturists claim to be important.

As long as the participants do not know whether they received the real treatment or a placebo, the placebo effect ought to affect the treatment and control groups equally. If the results for the two groups are significantly different, the differences can be attributed to the treatment. For example, in the study of the anti-depression drug, we would conclude that the drug was effective only if the control group received a placebo and members of the treatment group improved much more than members of the control group. For even better control, some experiments use three groups: a treatment group, a placebo group, and a control group. The placebo group is given a placebo while the control group is given nothing.

By the Way ...

The placebo effect can be remarkably powerful. In some studies, up to 75% of the participants receiving the placebo actually improve. For some patients, the effect is so powerful that they plead to continue their treatment even after being told they were given a placebo rather than the real treatment. Nevertheless, different researchers disagree about the strength of the placebo effect, and some even question the reality of the effect.

Definitions

A **placebo** lacks the active ingredients of a treatment being tested in a study, but looks or feels like the treatment so that participants cannot distinguish whether they are receiving the placebo or the real treatment.

The **placebo effect** refers to the situation in which patients improve simply because they believe they are receiving a useful treatment.

Note: Although participants should not know whether they belong to the treatment or control group, for ethical reasons it is very important that participants be told that some of them will be given a placebo, rather than the real treatment.

TIME OUT TO THINK

In decades past, researchers often told all the participants in a study that they were receiving the real treatment, but actually gave a placebo to half the participants. Discuss a few reasons why this would now be considered unethical. Should researchers be allowed to use results of past studies that do not meet today's ethical criteria? Defend your opinion.

EXAMPLE 7 Vaccine Placebo

What was the placebo in the Salk polio vaccine study? Why did researchers use a placebo in this experiment?

Solution The placebo was the salt water injection given to the children in the control group. To understand why the researchers used a placebo for the control group, suppose that a placebo had *not* been used. When improvements were observed in the treatment group, it would have been impossible to know whether the improvements were due to the vaccine or to the placebo effect. In order to remove this confounding, all participants had to believe that they were being treated in the same way. This ensured that any placebo effect would occur in both groups equally, so that researchers could attribute any remaining differences to the vaccine.

By the Way ...

A related effect, known as the *Hawthorne effect*, occurs when treated subjects somehow respond differently simply because they are part of an experiment—regardless of the particular way in which they are treated. The Hawthorne effect gets it name from the fact that it was first observed in a study of factory workers at Western Electric's Hawthorne plant.

Experimenter Effects

Even if the study subjects don't know whether they received the real treatment or a placebo, the *experimenters* may still have an effect. In testing an anti-depression drug, for example, experimenters will probably interview patients to find out whether they are feeling better. But if the experimenters know who received the real drug and who received the placebo, they may inadvertently smile more at the people in the treatment group. Their smiles might improve those participants' moods, making it seem as if the treatment worked when in fact the improvement was caused by the experimenter. This type of confounding, in which the experimenter somehow influences the results, is called an **experimenter effect** (or a Rosenthal effect). The only way to avoid experimenter effects is to make sure that the experimenters don't know which subjects are in which group.

Definition

An **experimenter effect** occurs when a researcher or experimenter somehow influences subjects through such factors as facial expression, tone of voice, or attitude.

EXAMPLE 8 Child Abuse?

In a famous case, two couples from Bakersfield, California, were convicted of molesting dozens of preschool-age children at their daycare center. The evidence for the abuse came primarily from interviews with the children. However, the conviction was overturned—after one man had served 14 years in prison—when a judge re-examined the interviews and concluded that the children had given answers that they thought the interviewers wanted to hear. If we think of the interviewers as experimenters, this is an example of an experimenter effect because the interviewers influenced the children's answers through the tone and style of their questioning.

By the Way ...

Many similar cases of supposedly widespread child abuse at daycare centers and preschools are being re-examined to see if experimenter effects (by those who interviewed the children) may have led to wrongful convictions. Similar claims of experimenter effects have been made in cases involving repressed memory, in which counseling supposedly helped people retrieve lost memories of traumatic events.

Blinding

In statistical terminology, the practice of keeping people in the dark about who is in the treatment group and who is in the control group is called **blinding**. A **single-blind** experiment is one in which the participants don't know which group they belong to, but the experimenters do know. If neither the participants nor the experimenters know who belongs to each group, the study is said to be **double-blind**. Of course, *someone* has to keep track of the two groups in order to evaluate the results at the end. In a double-blind experiment, the researchers conducting the study typically hire experimenters to make any necessary contact with the participants. The researchers thereby avoid any contact with the participants, ensuring that they cannot influence them in any way. The Salk polio vaccine study was double-blind because neither the participants (the children) nor the experimenters (the doctors and nurses giving the injections and diagnosing polio) knew who got the real vaccine and who got the placebo.

Blinding in Experiments

An experiment is **single-blind** if the participants do not know whether they are members of the treatment group or members of the control group, but the experimenters do know.

An experiment is **double-blind** if neither the participants nor any experimenters know who belongs to the treatment group and who belongs to the control group.

EXAMPLE 9 What's Wrong with This Experiment?

For each of the experiments described below, identify any problems and explain how the problems could have been avoided.

a. A new drug for attention deficit disorder (ADD) is supposed to make affected children more polite. Randomly selected children suffering from ADD are divided into treatment and control groups. The experiment is single-blind. Experimenters evaluate how polite the children are during one-on-one interviews.

b. Educational researchers wonder if listening to classical music when studying improves learning. They give two groups of students an identical 2-hour lesson and then allow time to study for a short exam. One group, made up of 50 students who told the researchers that they like classical music, listens to classical music while they study. The other group, made up of 50 other students who told the researchers that they don't like classical music, studies in silence. The results show that the students who listened to classical music did better on the test.

c. Researchers wonder if the effects of a rare degenerative disease can be slowed by exercise. They identify 6 people suffering from the disease and randomly assign 3 to a treatment group that exercises every day and 3 to a control group that avoids exercise. After six months, they compare the amounts of degeneration in each group.

d. A chiropractor performs adjustments on 25 patients with back pain. Afterward, 18 of the patients say they feel better. He concludes that the adjustments are an effective treatment.

Solution

a. The experimenters assess politeness in interviews, but because they know which children received the real drug, they may inadvertently speak differently to these children during the interviews. Or, they might interpret the children's behavior differently since they know which subjects received the real drug. These are experimenter effects that can confound the study results. The experiment should have been double-blind.

b. The problem with this study is that students were not assigned to the two groups at random. By placing students who like classical music in one group and students who don't like it in the other, the researchers created a situation in which the two groups do not share the same general characteristics.

c. The results of this study will be difficult to interpret because the sample sizes are not sufficiently large. Of course, with a rare disease, finding people to participate in an experiment may be difficult.

d. The 25 patients who receive adjustments represent a treatment group, but this study lacks a control group. The patients may be feeling better because of a placebo effect rather than any real effect of the adjustments. The chiropractor might have improved his study by hiring an actor to do a fake adjustment (one that feels similar, but doesn't actually conform to chiropractic guidelines) on a control group. Then he could have compared the results in the two groups to see whether a placebo effect was involved.

EXAMPLE 10 Identifying the Study Type

For each of the following questions, what type of statistical study is most likely to lead to an answer? Be as specific as possible.

a. What is the mean income of stock brokers?

b. Do seat belts save lives?

c. Can lifting weights improve runners' times in a 10-kilometer (10 K) race?

d. Does skin contact with a particular glue cause a rash?

e. Can a new herbal remedy reduce the severity of colds?

f. Do supplements of resveratrol (an extract from red grapes) increase life span?

With proper treatment, a cold can be cured in a week. Left to itself, it may linger for seven days.

—Medical Folk Saying

Solution

a. An *observational study* can tell us the mean income of stock brokers. We need only survey the brokers, and the survey itself will not change their incomes.

b. It would be unethical to do an experiment in which some people were told to wear seat belts and others were told *not* to wear them. A study to determine whether seat belts save lives must therefore be *observational*. Because some people choose to wear seat belts and others choose not to, we can conduct a *retrospective study*. By comparing the death rates in accidents between those who do and do not wear seat belts, we can learn whether seat belts save lives. (They do.)

c. We need an *experiment* to determine whether lifting weights can improve runners' 10 K times. We select randomly from a group of runners to create a treatment group of runners who are put on a weight-lifting program and a control group that is asked to stay away from weights. We must try to ensure that all other aspects of their training are similar. Then we can see whether the runners in the lifting group improve their times more than those in the control group. We cannot use blinding in this experiment because there is no way to prevent participants from knowing whether they are lifting weights.

d. An *experiment* can help us determine whether skin contact with the glue causes a rash. In this case, it's best to use a *single-blind experiment* in which we apply the real glue to participants in one group and apply a placebo that looks the same, but lacks the active ingredient, to members of the control group. There is no need for a double-blind experiment because it seems unlikely that the experimenters could influence whether a person gets a rash. (However, if the question of whether the subject *has* a rash is subject to interpretation, the experimenter's knowledge of who got the real treatment could affect this interpretation.)

e. We should use a *double-blind experiment* to determine whether a new herbal remedy can reduce the severity of colds. Some participants get the actual remedy, while others get a placebo. We need the double-blind conditions because the severity of a cold may be affected by mood or other factors that researchers might inadvertently influence. In the double-blind experiment, the researchers do not know which participants belong to which group and thus cannot treat the two groups differently.

f. Resveratrol has been identified and made available in supplement form only recently; we will need many years of data to determine whether it has an effect on life span. We therefore should use a prospective study designed to monitor participants over many years. The participants could keep written records regarding whether and how much of the supplement they take, and eventually researchers could analyze the data to see if resveratrol has an effect on life span.

By the Way ...

Experiments with mice show that the mice live longer and have greater endurance when taking large doses of resveratrol, an extract from the skin of red grapes. However, it is not yet known whether high doses of resveratrol would have the same effects for humans, nor is it known whether resveratrol is safe for humans. Moreover, the doses given to the mice are so large that it may be impractical to give equivalent doses to humans, with our larger body sizes.

Meta-Analysis

All individual statistical studies are either observational studies or experiments. In recent years, however, statisticians have found it useful to "mine" groups of past studies to see if we can learn something that we were unable to learn from the individual studies. For example, hundreds of studies have considered the possible effects of vitamin C on colds, so researchers might decide to review the data from many of these studies as a group. This type of study, in which researchers review many past studies as a group, is called a **meta-analysis**.

Definition

In a **meta-analysis,** researchers review many past studies. The meta-analysis considers these studies as a combined group, with the aim of finding trends that were not evident in the individual studies.

CASE STUDY **Drugs to Fight Depression: A Meta-Analysis**

Government researchers estimate that 1 in 5 Americans suffers from depression at some time in their lives and that the annual cost to the economy in lost productivity is at least $40 billion. Until the 1990s, the only drugs available to combat depression belonged to a class known as tricyclic antidepressants. But in the past decade, several new drugs have come into common use. The most famous is Prozac, which has been prescribed so widely that some people claim we now live in a "Prozac nation." But do Prozac and the other new drugs work? To answer this question, researchers from the Agency for Health Care Policy and Research conducted a meta-analysis.

The researchers began by searching medical literature for studies about treatment of depression. They found more than 8,000 studies reported over a nine-year period. By looking for studies that met certain special criteria, such as observing patients for at least six weeks and comparing new and old types of anti-depression drugs, they narrowed the list to about 300 studies. They also considered about 600 studies dealing with side effects of the drugs. Their meta-analysis consisted of analyzing the results of all these studies as a group so as to look for trends that might not have been evident in individual studies.

They found that, overall, the new drugs are more effective than placebos for severely depressed people. About 50% of such people responded to the new drugs, compared to about 32% who responded to a placebo. However, the old tricyclic antidepressants were equally effective—and were often available at lower prices. The old and new drugs had different side effects, but the degrees of severity of these side effects were roughly equal. While these results can be interpreted as both good and bad news for the makers and users of the new drugs, the meta-analysis probably was more valuable for what it could *not* say. The researchers found that the data were inadequate to determine whether the new drugs were effective for mild depression or for children, despite the fact that they are commonly prescribed for both. They also found the data inadequate to evaluate herbal treatments, such as kava and St. John's wort. Thus, the meta-analysis pointed out the directions in which new research is needed.

TIME OUT TO THINK

Look again at the percentage responses to the new drugs (50%) and to the placebo (32%) in the meta-analysis of anti-depression drugs. Does the relatively large response to the placebo suggest that patients should be given a placebo before being given a real drug with potential side effects? Why or why not? If you were a psychologist, what would you suggest to someone who was part of the 50% who did *not* respond to the real drug?

Section 1.3 Exercises

Statistical Literacy and Critical Thinking

1. **Placebo.** What is a placebo, and why is it important in an experiment to test the effectiveness of a drug?

2. **Blinding.** What is blinding and why is it important in an experiment to test the effectiveness of a drug?

3. **Confounding.** In testing the effectiveness of a new vaccine, suppose that researchers used males for the treatment group and females for the placebo group. What is confounding, and how would it affect such an experiment?

4. **Withholding Treatment.** In one infamous study, 399 African American men with syphilis were *not* given a treatment that could have cured them. Others were given treatments. The intent was to learn about the effects of syphilis on African American men. Is this an experiment or an observational study? Why is this study morally and criminally wrong?

5. **Clothing Color.** A researcher plans to investigate the belief that people are more comfortable in the summer sun when they wear clothing with light colors instead of clothing with dark colors. Does it make sense to use a double-blind

experiment in this case? Is it easy to implement blinding in this case? Explain.

6. **Lawn Treatment.** A researcher plans to test the effectiveness of a new fertilizer on grass growth. Does it make sense to use a double-blind experiment in this case?

7. **Measuring Pain.** Many studies involve methods for measuring pain after some treatment. Describe how an experimenter effect might adversely affect results, and describe how the effect can be avoided.

8. **Treating Baldness.** In one actual study of a treatment for baldness, it was claimed that the placebo group had better results in growing new hair than the group treated with a drug. Are such results possible? What would such results suggest about the treatment?

Concepts and Applications

Type of Study. For Exercises 9–20, state the type of study and identify the variables of interest. Be as specific as possible.

9. **Touch Therapy.** Nine-year-old Emily Rosa became an author of an article in the *Journal of the American Medical Association* after she tested professional touch therapists (see the "Focus on Education" section in Chapter 10). Using a cardboard partition, she held her hand above one of the therapist's hands, and the therapist was asked to identify the hand that Emily chose.

10. **Quality Control.** The Federal Drug Administration randomly selects a sample of Bayer aspirin tablets. The amount of aspirin in each tablet is measured for accuracy.

11. **Magnetic Bracelets.** Some cruise ship passengers are given magnetic bracelets, which they agree to wear in an attempt to eliminate or diminish the effects of motion sickness. Others are given similar bracelets that have no magnetism.

12. **Gender Selection.** In a study of the YSORT gender selection method developed by the Genetics & IVF Institute, 152 couples had 127 baby boys and 25 baby girls.

13. **Twins.** In a study of hundreds of Swedish twins, it was determined that the level of mental skills was more similar in identical twins (twins coming from a single egg) than in fraternal twins (twins coming from two separate eggs) (*Science*).

14. **Heart Disease.** A European study of 1,500 men and women with exceptionally high levels of the amino acid homocysteine found that these individuals had double the risk of heart disease. However, the risk was substantially lower for those in the study who took vitamin B supplements (*Journal of the American Medical Association*).

15. **Prostate Cancer.** An analysis of 11 individual studies attempted to determine if there is a conclusive link between vasectomies and the incidence of prostate cancer (*Chance*).

16. **AOL Poll.** An America OnLine (AOL) poll resulted in 38,410 responses to the question "How much stock do you put in long range weather forecasts?" Among those respondents, 47% said "very little" or "none."

17. **GMO Corn.** Researchers at New York University found that the genetically modified corn known as Bt corn releases an insecticide through its roots into the soil (*Nature*).

18. **Stem Cell Research.** In a *Newsweek* poll conducted by Princeton Survey Research Associates, 48% of the 1,002 adult respondents said that they were in favor of using federal tax money to fund medical research using stem cells obtained from embryos.

19. **Magnet Treatment.** In a study of the effects of magnets on back pain, some subjects were treated with magnets while others were given nonmagnetic devices with a similar appearance. The magnets did not appear to be effective in treating back pain (*Journal of the American Medical Association*).

20. **Power Lines and Cancer.** Hundreds of scientific and statistical studies have been done to determine whether high-voltage overhead power lines increase the incidence of cancer among those living nearby. A summary study based on many previous studies concluded that there is no significant link between power lines and cancer (*Journal of the American Medical Association*).

What's Wrong with This Experiment? For each of the studies described in Exercises 21–28, identify any problems that are likely to cause confounding and explain how the problems could be avoided. Discuss any other problems that might affect the results.

21. **Poplar Tree Growth.** An experiment is designed to evaluate the effectiveness of irrigation and fertilizer on poplar tree growth. Fertilizer is used with one group of poplar trees in a moist region, and irrigation is used with poplar trees in a dry region.

22. **Internet Shopping.** Two hundred volunteers are recruited through an ad in *PC Magazine.* They are to be used for a study to compare the total amounts spent by those who use the Internet and those who do not. Each person is allowed to choose whether to be in the Internet user group or the group that does not use the Internet. After one month, the amounts of money spent by the two groups are compared.

23. **Octane Rating.** In a comparison of gasoline with different octane ratings, 24 vans are driven with 87-octane gasoline, while 28 sport utility vehicles are driven with

91-octane gasoline. After a vehicle has been driven for 250 miles, the amount of gasoline consumed is measured.

24. Aspirin Trial. In phase I of a clinical trial designed to test the effectiveness of aspirin in preventing myocardial infarctions, aspirin is given to 3 people and a placebo to 7 other people.

25. Hypertension Trial. In a clinical trial of the effectiveness of a drug to treat hypertension, subjects are told whether they are getting the actual drug or a placebo.

26. Athlete's Foot. In a clinical trial of the effectiveness of a lotion used to treat tinea pedis (athlete's foot), the physicians who evaluate the results know which subjects were given the treatment and which were given a placebo.

27. Weight Lifting. In a test of the effects of lifting heavy weights on blood pressure, one group undergoes a treatment consisting of a weight-lifting program while another group lifts tennis balls.

28. Paint Mixtures. In durability tests of two different paint mixtures, the researchers who evaluate the results know which samples are from each of the two different mixtures.

Analyzing Experiments. Exercises 29–32 present questions that might be addressed in an experiment. If you were to design the experiment, how would you choose the treatment and control groups? Should the experiment be single-blind, double-blind, or neither? Explain your reasoning.

29. Beethoven and Intelligence. Does listening to Beethoven make infants more intelligent?

30. Lipitor and Cholesterol. Does the drug Lipitor result in lower cholesterol levels?

31. Ethanol and Mileage. Does an ethanol additive in gasoline result in reduced mileage?

32. Home Siding. Does aluminum siding last longer on a home than wood siding?

Projects for the Internet and Beyond

For useful links, select "Links for Internet Projects" for Chapter 1 at www.aw.com/bbt.

33. Experimenter Effects in Repressed Memory Cases. Search the Web for articles and information about the controversy regarding recovering repressed memories. Briefly summarize one or two of the most interesting cases and, based on what you read, express your own opinion as to whether the allegedly recovered memories are being influenced by experimenter effects.

34. Ethics in Experiments. Ethical standards change from era to era. One notoriously unethical case was a study of syphilis conducted in Tuskegee, Alabama, from 1932 to 1972. (See Exercise 4.) In this study, African American males were told that they were receiving treatment for syphilis, but in fact they were not. The researchers' hidden goal was to study the long-term effects of the disease. Use the Web to learn about the history of the Tuskegee syphilis study. Hold a class discussion about the ethical issues involved in this case, or write a short essay summarizing the case and its ethical lessons.

35. Debate: Should We Use Data from Unethical Experiments? Past research often did not conform to today's ethical standards. In extreme cases, such as research conducted by doctors in Nazi Germany, the researchers sometimes killed the subjects of their experiments. While this past unethical research clearly violated the human rights of the experimental subjects, in some cases it led to insights that could help people today. Is it ethical to use the results of unethical research?

36. Study Stopped Early. It sometimes happens that a study is stopped before its completion. Use the Internet to find an example of such a study. Why was the study stopped? Should it have been stopped, or would it have been better to complete the study?

IN THE NEWS

37. Observational Studies. Look through newspapers from the past few weeks and find an example of a statistical study that was observational. Briefly describe the study and summarize its conclusions.

38. Experimental Studies. Look through newspapers for the past few weeks and find an example of a statistical study that involved an experiment. Briefly describe the study and summarize its conclusions.

39. Retrospective Studies. Look through newspapers for the past few weeks and find an example of an observational, retrospective study. Briefly describe the study and summarize its conclusions.

40. Meta-Analysis. Look through newspapers for the past few weeks and find an example of a meta-analysis. Briefly describe the study and summarize its conclusions.

1.4 Should You Believe a Statistical Study?

Much of the rest of this book is devoted to helping you build a deeper understanding of the concepts and definitions we've studied up to this point. But already you know enough to achieve one of the major goals of this book: being able to answer the question "Should you believe a statistical study?"

If I have ever made any valuable discoveries, it has been owing more to patient attention, than to any other talent.

—Isaac Newton

Most researchers conduct their statistical studies with honesty and integrity, and most statistical research is carried out with diligence and care. Nevertheless, statistical research is sufficiently complex that bias can arise in many different ways, making it very important that we always examine reports of statistical research carefully. There is no definitive way to answer the question "Should I believe a statistical study?" However, in this section we'll look at eight guidelines that can be helpful. Along the way, we'll also introduce a few more definitions and concepts that will prepare you for discussions to come later.

Eight Guidelines for Critically Evaluating a Statistical Study

1. Identify the goal of the study, the population considered, and the type of study.
2. Consider the source, particularly with regard to whether the researchers may be biased.
3. Examine the sampling method to decide whether it is likely to produce a representative sample.
4. Look for problems in defining or measuring the variables of interest, which can make it difficult to interpret any reported results.
5. Watch out for confounding variables that can invalidate the conclusions of a study.
6. Consider the setting and wording in surveys or polls, looking for anything that might tend to produce inaccurate or dishonest responses.
7. Check that results are fairly represented in graphics and concluding statements, because both researchers and media often create misleading graphics or jump to conclusions that the results do not support.
8. Stand back and consider the conclusions. Did the study achieve its goals? Do the conclusions make sense? Do the results have any practical significance?

News reports do not always provide enough information for you to apply all of the above guidelines, but you can usually find additional information on the Web. Look for clues such as "reported by NASA" or "published in the *New England Journal of Medicine*" to help you track down original sources or other relevant information.

Guideline 1: Identify the Goal, Population, and Type of Study

The first step in evaluating a statistical study is to understand the goal and approach of the study. Based on what you hear or read about a study, try to answer these basic questions:

- What was the study designed to determine?
- What was the population under study? Was the population clearly and appropriately defined?
- Was the study an observational study, an experiment, or a meta-analysis? If it was an observational study, was it retrospective? If it was an experiment, was it single- or double-blind, and were the treatment and control groups properly randomized? Given the goal, was the type of study appropriate?

EXAMPLE 1 Appropriate Type of Study?

Imagine the following (hypothetical) newspaper report: "Researchers gave 100 participants their individual astrological horoscopes and asked whether the horoscopes were accurate. 85% of the participants said their horoscopes were accurate. The researchers concluded that horoscopes are valid most of the time." Analyze this study according to Guideline 1.

Solution The goal of the study was to determine the validity of horoscopes. Based on the news report, it appears that the study was *observational*: The researchers simply asked the participants about the accuracy of the horoscopes. However, because the accuracy of a horoscope is somewhat subjective, this study should have been a controlled experiment in which some people were given their actual horoscope and others were given a fake horoscope. Then the researchers could have looked for differences between the two groups. Moreover, because researchers could easily influence the results by how they questioned the participants, the experiment should have been double-blind. In summary, the type of study was inappropriate to the goal and its results are meaningless.

By the Way ...

Surveys show that nearly half of Americans believe their horoscopes. However, in controlled experiments, the predictions of horoscopes come true no more often than would be expected by chance.

TIME OUT TO THINK

Try your own test of horoscopes. Find yesterday's horoscope for each of the 12 signs and put each one on a separate piece of paper, without anything identifying the sign. Shuffle the pieces of paper randomly, and ask a few people to guess which one was supposed to be their personal horoscope. How many people choose the right one? Discuss your results.

EXAMPLE 2 Does Aspirin Prevent Heart Attacks?

A study reported in the *New England Journal of Medicine* (Vol. 318, No. 4) sought to determine whether aspirin is effective in preventing heart attacks. It involved 22,000 male physicians considered to be at risk for heart attacks. The men were divided into a treatment group that took aspirin and a control group that did not. The results were so convincing in favor of the benefits of aspirin that the experiment was stopped for ethical reasons before it was completed, and the subjects were informed of the results. Many news reports led with the headline that taking aspirin can help prevent heart attacks. Analyze this headline according to Guideline 1.

Solution The study was an experiment, which is appropriate, and its results appear convincing. However, the fact that the sample consisted only of men means that the results should be considered to apply only to the population of men. Because results of medical tests on men do not necessarily apply to women, the headlines misstated the results when they did not qualify the population.

By the Way ...

Many recent studies show substantial differences in the ways men and women respond to the same medical treatments. For example, aspirin is more effective at thinning blood in men than in women (thinning is thought to help prevent heart attacks in some people). Morphine controls pain better in women, but ibuprofen is more effective for men. And women reject heart transplants more often than men do. These differences may stem from interactions with hormones that differ in men and women or from differences in the rates at which men and women metabolize different drugs.

Guideline 2: Consider the Source

Statistical studies are supposed to be objective, but the people who carry them out and fund them may be biased. It is therefore important to consider the source of a study and evaluate the potential for biases that might invalidate the study's conclusions.

Bias may be obvious in cases where a statistical study is carried out for marketing, promotional, or other commercial purposes. For example, a toothpaste advertisement that claims "4 out of 5 dentists prefer our brand" appears to be statistically based, but we are given no details about how the survey was conducted. Because the advertisers obviously want to say good things about their brand, it's difficult to take the statistical claim seriously without much more information about how the result was obtained.

Other cases of bias may be more subtle. For example, suppose that a carefully conducted study concludes that a new drug helps cure cancer. On the surface, the study might seem quite believable. But what if the study was funded by a drug company that stands to gain billions of dollars in sales if the drug is proven effective? The researchers may well have carried out their work with great integrity despite the source of funding, but it might be worth a bit of extra investigation to be sure.

Major statistical studies are usually evaluated by unbiased experts. For example, the process by which scientists examine each other's research is called **peer review** (because the scientists who do the evaluation are *peers* of those who conducted the research). Reputable scientific journals require all research reports to be peer reviewed before the research is accepted for publication. Peer review does not guarantee that a study is valid, but it lends credibility because it implies that other experts agree that the study was carried out properly.

Definition

Peer review is a process in which several experts in a field evaluate a research report before the report is published.

By the Way ...

After decades of arguing to the contrary, in October 1999 the Philip Morris Company—the world's largest seller of tobacco products—publicly acknowledged that smoking causes lung cancer, heart disease, emphysema, and other serious diseases. Shortly thereafter, Philip Morris changed its name to Altria.

EXAMPLE 3 Is Smoking Healthy?

By 1963, research had so clearly shown the health dangers of smoking that the Surgeon General of the United States publicly announced that smoking is bad for health. Research done since that time built further support for this claim. However, while the vast majority of studies showed that smoking is unhealthy, a few studies found no dangers from smoking and perhaps even health *benefits*. These studies generally were carried out by the Tobacco Research Institute, funded by the tobacco companies. Analyze these studies according to Guideline 2.

Solution Even in a case like this, it can be difficult to decide whom to believe. However, the studies showing smoking to be unhealthy came primarily from peer-reviewed research. In contrast, the studies carried out at the Tobacco Research Institute had a clear potential for bias. The *potential* for bias does not mean the research was biased, but the fact that it contradicts virtually all other research on the subject should be cause for concern.

EXAMPLE 4 Press Conference Science

Suppose the nightly TV news shows scientists at a press conference announcing that they've discovered evidence that a newly developed chemical can stop the aging process. The work has not yet gone through the peer review process. Analyze this study according to Guideline 2.

Solution Scientists often announce the results of their research at a press conference so that the public may hear about their work as soon as possible. However, a great deal of expertise may be required to evaluate their study for possible biases or other errors—which is the goal of the peer review process. Until the work is peer reviewed and published in a reputable journal, any findings should be considered preliminary—especially about an astonishing claim such as being able to stop the aging process.

Guideline 3: Examine the Sampling Method

A statistical study cannot be valid unless the sample is representative of the population under study. Poor sampling methods almost guarantee a biased sample that makes the study results useless.

Biased samples can arise in many ways, but two closely related problems are particularly common. The first problem, called **selection bias** (or a **selection effect**), occurs whenever researchers *select* their sample in a way that tends to make it unrepresentative of the population. For example, a pre-election poll that surveys only registered Republicans has selection bias because it is unlikely to reflect the opinions of voters of all parties (and independents).

The second problem, called **participation bias**, can arise when people *choose* to be part of a study—that is, when the participants are volunteers. The most common form of participation bias occurs in **self-selected surveys** (or **voluntary response surveys**)—surveys or polls in which people decide for themselves whether to participate. In such cases, people who feel strongly about an issue are more likely to participate, and their opinions may not represent the opinions of the larger population that has less emotional attachment to the issue.

Definitions

Selection bias (or a **selection effect**) occurs whenever researchers *select* their sample in a biased way.

Participation bias occurs any time participation in a study is voluntary.

A **self-selected survey** (or **voluntary response survey**) is one in which people decide for themselves whether to be included in the survey.

CASE STUDY The 1936 Literary Digest Poll

The *Literary Digest*, a popular magazine of the 1930s, successfully predicted the outcomes of several elections using large polls. In 1936, editors of the *Literary Digest* conducted a particularly large poll in advance of the presidential election. They randomly chose a sample of 10 million people from various lists, including names in telephone books and rosters of country clubs. They mailed a postcard "ballot" to each of these 10 million people. About 2.4 million people returned the postcard ballots. Based on the returned ballots, the editors of the *Literary Digest* predicted that Alf Landon would win the presidency by a margin of 57% to 43% over Franklin Roosevelt. Instead, Roosevelt won with 62% of the popular vote. How did such a large survey go so wrong?

The sample suffered from both selection bias and participation bias. The selection bias arose because the *Literary Digest* chose its 10 million names in ways that favored affluent people. For example, selecting names from telephone books meant choosing only from those who could afford telephones back in 1936. Similarly, country club members are usually quite wealthy. The selection bias favored the Republican Landon because affluent voters of the 1930s tended to vote for Republican candidates.

The participation bias arose because return of the postcard ballots was voluntary, so people who felt strongly about the election were more likely to be among those who returned their ballots. This bias also tended to favor Landon because he was the challenger—people who did not like President Roosevelt could express their desire for change by returning the postcards. Together, the two forms of bias made the sample results useless, despite the large number of people surveyed.

By the Way ...

A young pollster named George Gallup conducted his own survey prior to the 1936 election. Sending postcards to only 3,000 randomly selected people, he correctly predicted not only the outcome of the election, but also the outcome of the *Literary Digest* poll, to within 1%. Gallup went on to establish a very successful polling organization.

By the Way ...

More than a third of all Americans routinely shut the door or hang up the phone when contacted for a survey, thereby making self-selection a problem for legitimate pollsters. One reason people hang up may be the proliferation of selling under the guise of market research (often called "sugging"), in which a telemarketer pretends you are part of a survey in order to try to get you to buy something.

EXAMPLE 5 Self-Selected Poll

The television show *Nightline* conducted a poll in which viewers were asked whether the United Nations headquarters should be kept in the United States. Viewers could respond to the poll by paying 50 cents to call a "900" phone number with their opinions. The poll drew 186,000 responses, of which 67% favored moving the United Nations out of the United States. Around the same time, a poll using simple random sampling of 500 people found that 72%

wanted the United Nations to *stay* in the United States. Which poll is more likely to be representative of the general opinions of Americans?

Solution The *Nightline* sample was severely biased. It had selection bias because its sample was drawn only from the show's viewers, rather than from all Americans. The poll itself was a self-selected survey in which viewers not only chose whether to respond, but also had to *pay* 50 cents to participate. This cost made it even more likely that respondents would be those who felt a need for change. Despite its large number of respondents, the *Nightline* survey was therefore unlikely to give meaningful results. In contrast, the simple random sample of 500 people is quite likely to be representative, so the finding of this small survey has a better chance of representing the true opinions of all Americans.

By the Way ...

NASA's *Kepler* mission, scheduled for launch in late 2008, is an orbiting telescope that should be able to detect planets as small as Earth around other stars. *Kepler* will look for slight dimming of a star's light each time an orbiting planet passes in front of it (a "transit"), which means it will be able to detect planets only for the small fraction of stars that happen to have their planetary systems aligned with our line of sight.

EXAMPLE 6 Planets Around Other Stars

Until the mid-1990s, astronomers had never found conclusive evidence for planets outside our own solar system. But improving technology made it possible to begin finding such planets, and more than 200 had been discovered by early 2007. The existing technology makes it easier to find large planets than small ones and easier to find planets that orbit close to their stars than planets that orbit far from their stars. According to the leading theory of solar system formation, large planets that orbit close to their stars should be very rare. But they are fairly common among the first 200 discoveries. Does this mean there is something wrong with the leading theory of solar system formation?

Solution Although the theory suggests that large planets in close orbits should be rare, the technology makes these rare cases the easiest ones to find. The finding of many of these planets may therefore represent a *selection effect* that biases the sample (of discovered planets) toward a rare type. In fact, this seems almost certainly to be the case, because many "normal" planets are now being discovered, and astronomers have found that with slight modifications the existing theory can account for the rare types found in early discoveries.

Guideline 4: Look for Problems in Defining or Measuring the Variables of Interest

Results of a statistical study may be difficult to interpret if the variables under study are difficult to define or measure. For example, imagine trying to conduct a study of how exercise affects resting heart rates. The variables of interest would be *amount of exercise* and *resting heart rate*. Both variables are difficult to define and measure. In the case of *amount of exercise*, it's not clear what the definition covers—does it include walking to class? Even if we specify the definition, how can we measure *amount of exercise* given that some forms of exercise are more vigorous than others?

TIME OUT TO THINK

How would you measure your resting heart rate? Describe some difficulties in defining and measuring resting heart rate.

EXAMPLE 7 Can Money Buy Love?

A Roper poll reported in *USA Today* involved a survey of the wealthiest 1% of Americans. The survey found that these people would pay an average of $487,000 for *true love*, $407,000 for *great intellect*, $285,000 for *talent*, and $259,000 for *eternal youth*. Analyze this result according to Guideline 4.

Solution The variables in this study are very difficult to define. How, for example, do you define *true love*? And does it mean true love for a day, a lifetime, or something else? Similarly, does the ability to balance a spoon on your nose constitute *talent*? Because the variables are so poorly defined, it's likely that different people interpreted them differently, making the results very difficult to interpret.

EXAMPLE 8 Illegal Drug Supply

A commonly quoted statistic is that law enforcement authorities succeed in stopping only about 10% to 20% of the illegal drugs entering the United States. Should you believe this statistic?

Solution There are essentially two variables in a study of illegal drug interception: *quantity of illegal drugs intercepted* and *quantity of illegal drugs NOT intercepted.* It should be relatively easy to measure the quantity of illegal drugs that law enforcement officials intercept. However, because the drugs are illegal, it's unlikely that anyone is reporting the quantity of drugs that are *not* intercepted. How, then, can anyone know that the intercepted drugs are 10% to 20% of the total? In a *New York Times* analysis, a police officer was quoted as saying that his colleagues refer to this type of statistic as "PFA" for "pulled from the air."

Guideline 5: Watch Out for Confounding Variables

Often, variables that are *not intended* to be part of the study can make it difficult to interpret results properly. As discussed in Section 1.3, it's not always easy to discover these *confounding variables*. Sometimes they are discovered only years after a study is completed, and other times they are not discovered at all, in which case a study's conclusion may be accepted even though it's not correct. Fortunately, confounding variables are sometimes more obvious and can be discovered simply by thinking hard about factors that may have influenced a study's results.

EXAMPLE 9 Radon and Lung Cancer

Radon is a radioactive gas produced by natural processes (the decay of uranium) in the ground. The gas can leach into buildings through the foundation and can accumulate to relatively high concentrations if doors and windows are closed. Imagine a (hypothetical) study that seeks to determine whether radon gas causes lung cancer by comparing the lung cancer rate in Colorado, where radon gas is fairly common, with the lung cancer rate in Hong Kong, where radon gas is less common. Suppose the study finds that the lung cancer rates are nearly the same. Would it be reasonable to conclude that radon is *not* a significant cause of lung cancer?

Solution The variables of interest are *amount of radon* (an explanatory variable in this case) and *lung cancer rate* (a response variable) However, radon gas is not the only possible cause of lung cancer. For example, smoking can cause lung cancer, so *smoking rate* may be a confounding variable in this study—especially because the smoking rate in Hong Kong is much higher than the smoking rate in Colorado. As a result, we cannot draw any conclusions about radon and lung cancer without taking the smoking rate into account (and perhaps other variables as well). In fact, careful studies have shown that radon gas *can* cause lung cancer, and the U.S. Environmental Protection Agency (EPA) recommends taking steps to prevent radon from building up indoors.

By the Way ...

Many hardware stores sell simple kits that you can use to test whether radon gas is accumulating in your home. If it is, the problem can be eliminated by installing an appropriate "radon mitigation" system, which usually consists of a fan that blows the radon out from under the house before it can get into the house.

Guideline 6: Consider the Setting and Wording in Surveys

Even when a survey is conducted with proper sampling and with clearly defined terms and questions, you should watch for problems in the setting or wording that might produce inaccurate or dishonest responses. Dishonest responses are particularly likely when the survey concerns sensitive subjects, such as personal habits or income. For example, the question "Do you cheat on your income taxes?" is unlikely to elicit honest answers from those who cheat, unless they are assured of complete confidentiality (and perhaps not even then).

In other cases, even honest answers may not really be accurate if the wording of questions invites bias. Sometimes just the order of words in a question can affect the outcome. A poll conducted in Germany asked the following two questions.

- Would you say that traffic contributes more or less to air pollution than industry?
- Would you say that industry contributes more or less to air pollution than traffic?

The only difference is the order of the words *traffic* and *industry*, but this difference dramatically changed the results: With the first question, 45% answered traffic and 32% answered industry. With the second question, only 24% answered traffic while 57% answered industry.

By the Way ...

People are more likely to choose the item that comes first in a survey because of what psychologists call the *availability error*—the tendency to make judgments based on what is *available* in the mind. Professional polling organizations take great care to avoid this problem; for example, they may pose a question with two choices in one order to half the people in the sample and in the opposite order to the other half.

EXAMPLE 10 Do You Want a Tax Cut?

At a time when the U.S. government was running annual budget surpluses, Republicans in Congress proposed a tax cut and the Republican National Committee commissioned a poll to find out whether Americans supported the proposal. Asked "Do you favor a tax cut?," 67% of respondents answered *yes*. Should we conclude that Americans supported the proposal?

Solution A question like "Do you favor a tax cut?" is biased because it does not give other options. In fact, an independent poll conducted at the same time gave respondents a list of options for using surplus revenues. This poll found that 31% wanted the money devoted to Social Security, 26% wanted it used to reduce the national debt, and only 18% favored using it for a tax cut. (The remaining 25% of respondents chose a variety of other options.)

EXAMPLE 11 Sensitive Survey

Two surveys asked Catholics in the Boston area whether contraceptives should be made available to unmarried women. The first survey involved in-person interviews, and 44% of the respondents answered *yes*. The second survey was conducted by mail and telephone, and 75% of the respondents answered *yes*. Which survey was more likely to be accurate?

Solution Contraceptives are a sensitive topic, particularly among Catholics (because the Catholic Church officially opposes contraceptives). The first survey, with in-person interviews, may have encouraged dishonest responses. The second survey made responses seem more private and therefore was more likely to reflect the respondents' true opinions.

Guideline 7: Check That Results Are Fairly Represented in Graphics or Concluding Statements

Even when a statistical study is done well, it may be misrepresented in graphics or concluding statements. Researchers occasionally misinterpret the results of their own studies or jump to conclusions that are not supported by the results, particularly when they have personal biases. News reporters may misinterpret a survey or jump to unwarranted conclusions that make a story seem more spectacular. Misleading graphics are especially common (we will devote much of Chapter 3 to this topic). You should always look for inconsistencies between the interpretation of a study (in pictures and in words) and any actual data given along with it.

EXAMPLE 12 Does the School Board Need a Statistics Lesson?

The school board in Boulder, Colorado, created a hubbub when it announced that 28% of Boulder school children were reading "below grade level" and hence concluded that methods of teaching reading needed to be changed. The announcement was based on reading tests on which 28% of Boulder school children scored below the national average for their grade. Do these data support the board's conclusion?

Solution The fact that 28% of Boulder children scored below the national average for their grade implies that 72% scored at or above the national average. Thus, the school board's ominous statement about students reading "below grade level" makes sense only if "grade level" means the national average score for a particular grade. This interpretation of "grade level" is curious because it would imply that half the students in the nation are always below grade level—no matter how high the scores. It may still be the case that teaching methods needed to be improved, but these data did not justify that conclusion.

Guideline 8: Stand Back and Consider the Conclusions

Finally, even if a study seems reasonable according to all the previous guidelines, you should stand back and consider the conclusions. Ask yourself questions such at these:

- Did the study achieve its goals?
- Do the conclusions make sense?
- Can you rule out alternative explanations for the results?
- If the conclusions make sense, do they have any practical significance?

EXAMPLE 13 Extraordinary Claims

Suppose a (hypothetical) study concludes that wearing a gold chain increases your chances of surviving a car accident by 10%. The claim is based on a statistical analysis of data about survival rates and what people were wearing. Careful analysis of the research shows that it was conducted properly and carefully. Should you start wearing a gold chain whenever you drive a car?

Solution Despite the care that went into the study, the claim that a gold chain can save your life in a collision is difficult to believe. After all, how could a thin chain help in a high-speed collision? It's certainly *possible* that some unknown effect of gold chains makes the conclusion correct, but it seems far more likely that the results were either a fluke or due to an unidentified confounding variable (for example, perhaps those who wear gold chains are wealthier and drive newer cars with more advanced safety features, lowering their fatality rate).

EXAMPLE 14 Practical Significance

An experiment is conducted in which the weight losses of people who try a new "Fast Diet Supplement" are compared to the weight losses of a control group of people who try to lose weight in other ways. After eight weeks, the results show that the treatment group lost an average of one-half pound more than the control group. Assuming that it has no dangerous side effects, does this study suggest that the Fast Diet Supplement is a good treatment for people wanting to lose weight?

Solution Compared to the average person's body weight, a weight loss of one-half pound hardly matters at all. So while loss results may be interesting, they don't seem to have much practical significance.

Extraordinary claims require extraordinary evidence.

—Carl Sagan, astronomer and Pulitzer Prize winner

Section 1.4 Exercises

Statistical Literacy and Critical Thinking

1. **Peer Review.** What is peer review? How is it useful?
2. **Selection Bias and Participation Bias.** Describe and contrast selection bias and participation bias in sampling.
3. **Self-Selected Surveys.** Why are self-selected surveys always prone to participation bias?
4. **Confounding Variables.** What are confounding variables, and what problems can they cause?

Does It Make Sense? For Exercises 5–8, decide whether the statement makes sense (or is clearly true) or does not make sense (or is clearly false). Explain clearly; not all of these statements have definitive answers, so your explanation is more important than your chosen answer.

5. **Survey Validity.** The TV survey got more than 1 million phone-in responses, so it is clearly more valid than the survey by the professional pollsters, which involved interviews with only a few hundred people.
6. **Survey Location.** The survey of religious beliefs suffered from selection bias because the questionnaires were handed out only at Catholic churches.
7. **Vitamin C and Colds.** My experiment proved beyond a doubt that vitamin C can reduce the severity of colds, because I controlled the experiment carefully for every possible confounding variable.
8. **Faster Jogging.** Everyone who jogs for exercise should try the new training regimen, because careful studies suggest it can increase your speed by 1%.

Concepts and Applications

Applying Guidelines. In Exercises 9–16, determine which of the eight guidelines for evaluating statistical studies appears to be most relevant. Explain your reasoning.

9. **Quality Control.** The public relations department of Telektronics conducted a study of the defect rate of the circuit breakers manufactured by the company.
10. **Gender Selection.** After a test of a gender selection method designed to increase the likelihood of a baby girl, it was found that 23 babies were girls and 21 were boys, so a company spokesperson stated that the method is effective most of the time.
11. **Ethics.** In a study of 250 attorneys, each was asked whether he or she has good ethics.
12. **Agriculture.** Researchers conclude that an irrigation system used to grow tomatoes in California is more effective than a competing system used in Arizona.
13. **Prisoner Survey.** Prisoners attending classes at the college of one of the authors were mailed surveys about recidivism, and 10% of them responded.
14. **Election Poll.** Under the headline "Turner predicted to win in a landslide," it was reported that Turner received 55% of the votes in a pre-election poll, compared with 45% for her opponent.
15. **Nuclear Energy Poll.** Randomly selected adults were asked this question: "Do you agree or disagree with increasing the production of nuclear energy that could potentially kill thousands of innocent people?"
16. **Counterfeit Goods.** A consortium of manufacturers plans a study designed to compare the value of counterfeit goods produced in the United States in the year 2000 to the value of those produced in the current year.

Bias. In Exercises 17–20, identify and explain at least one source of bias in each study described. Then suggest how the bias might have been avoided.

17. **Healthy Chocolate.** The *New York Times* published an article that included these statements: "At long last, chocolate moves toward its rightful place in the food pyramid, somewhere in the high-tone neighborhood of red wine, fruits and vegetables, and green tea. Several studies, reported in the *Journal of Nutrition*, showed that after eating chocolate, test subjects had increased levels of antioxidants in their blood. Chocolate contains flavonoids, antioxidants that have been associated with decreased risk of heart disease and stroke. Mars Inc., the candy company, and the Chocolate Manufacturers Association financed much of the research."
18. **Famous Book.** When author Shere Hite wrote *Women and Love: A Cultural Revolution in Progress*, she based conclusions about the general population of all women on 4,500 replies that she received after mailing 100,000 questionnaires to various women's groups.
19. **Napster.** *Newsweek* magazine ran a survey about the Napster Web site for downloading music. Readers could register their responses on *Newsweek's* Web site.
20. **Survey Method.** You plan to conduct a survey to find the percentage of people in your state who can name the Lieutenant Governor (who plans to run for the United States Senate). You obtain addresses from a list of property

owners in the state and you mail a survey to 850 randomly selected people from the list.

21. **It's All in the Wording.** Princeton Survey Research Associates did a study for *Newsweek* magazine illustrating the effects of wording in a survey. Two questions were asked:

- Do you personally believe that abortion is wrong?
- Whatever your own personal view of abortion, do you favor or oppose a woman in this country having the choice to have an abortion with the advice of her doctor?

To the first question, 57% of the respondents replied *yes*, while 36% responded *no*. In response to the second question, 69% of the respondents favored the choice, while 24% opposed the choice. Discuss why the two questions produced seemingly contradictory results. How could the results of the questions be used selectively by various groups?

22. **Tax or Spend?** A Gallup poll asked the following two questions:

- Do you favor a tax cut or "increased spending on other government programs"? *Result*: 75% for the tax cut.
- Do you favor a tax cut or "spending to fund new retirement savings accounts, as well as increased spending on education, defense, Medicare and other programs"? *Result*: 60% for the spending.

Discuss why the two questions produced seemingly contradictory results. How could the results of the questions be used selectively by various groups?

What Do You Want to Know? Exercises 23–26 pose two related questions that might form the basis of a statistical study. Briefly discuss how the two questions differ and how these differences would affect the goal of a study and the design of the study.

23. **Internet Dating**

First question: What percentage of Internet dates lead to marriage?

Second question: What percentage of marriages begin with Internet dates?

24. **Full-Time Faculty**

First question: What percentage of introductory classes on campus are taught by full-time faculty members?

Second question: What percentage of full-time faculty members teach introductory classes?

25. **Binge Drinking**

First question: How often do college students binge drink?

Second question: How often is binge drinking done by college students?

26. **Statistics Courses**

First question: What is the proportion of college graduates who have taken a statistics course?

Second question: What proportion of all statistics courses are taken by college students?

Accurate Headlines? Exercises 27 and 28 give a headline and a brief description of the statistical news story that accompanied the headline. In each case, discuss whether the headline accurately represents the story.

27. Headline: "Drugs shown in 98 percent of movies"

Story summary: A "government study" claims that drug use, drinking, or smoking was depicted in 98% of the top movie rentals (Associated Press).

28. Headline: "Sex more important than jobs"

Story summary: A survey found that 82% of 500 people interviewed by phone ranked a satisfying sex life as important or very important, while 79% ranked job satisfaction as important or very important (Associated Press).

Stat-Bytes. Politicians commonly believe that they must make their political statements (often called sound-bytes) very short because the attention span of listeners is so short. A similar effect occurs in reporting statistical news. Major statistical studies are often reduced to one or two sentences. The summaries of statistical reports in Exercises 29–32 are taken from various news sources. Describe what crucial information is missing in the given statement and what more you would want to know before you acted on the report.

29. **Confidence in Military.** *USA Today* reports on a Harris poll claiming that the percentage of adults with a "great deal of confidence" in military leaders stands at 54% (up from 37% in 1997).

30. **Top Restaurants.** CNN reports on a Zagat Survey of America's Top Restaurants which found that "only nine restaurants achieved a rare 29 out of a possible 30 rating and none of those restaurants are in the Big Apple."

31. **Forecasting Weather.** A *USA Today* headline reported that "More companies try to bet on forecasting weather." The article gave examples of companies believing that long-range forecasts are reliable, and four companies were cited.

32. **Births in China.** A *USA Today* headline reported that "China thrown off balance as boys outnumber girls," and the accompanying graph showed that for every 100 girls born in China, 116.9 boys are born.

Projects for the Web and Beyond

For useful links, select "Links for Internet Projects" for Chapter 1 at www.aw.com/bbt.

33. Analyzing a Statistical Study. Find a detailed report on some recent statistical study of interest to you. Write a short report applying each of the eight guidelines given in this section. (Some of the guidelines may not apply to the particular study you are analyzing; in that case, explain why the guideline is not applicable.)

34. Harper's Index. Go to the Web site for Harper's Index and examine a few of the recently quoted statistics. If possible, obtain the sources for the statistics. Choose three statistics that you find particularly interesting and discuss whether you believe them, based on the guidelines given in this section.

35. Twin Studies. Researchers doing statistical studies in biology, psychology, and sociology are grateful for the existence of twins. Twins can be used to study whether certain traits are inherited from parents (nature) or acquired from one's surroundings during upbringing (nurture). Identical twins are formed from the same egg in the mother and have the same genetic material. Fraternal twins are formed from two separate eggs and share roughly half of the same genetic material. Find a published report of a twin study. Discuss how identical and fraternal twins are used to form case and control groups. Apply Guidelines 1–8 to the study and comment on whether you find the conclusions of the report convincing.

36. American Demographics. Consult an issue of *American Demographics,* a nontechnical magazine that specializes in summarizing statistical studies involving Americans and their lifestyles. Select one specific article and use the ideas of this section to summarize and evaluate the study. Be sure to cite information that you believe is missing and should be provided for you to make a complete analysis.

IN THE NEWS

37. Applying the Guidelines. Find a recent newspaper article or television report about a statistical study on a topic that you find interesting. Write a short report applying each of the eight guidelines given in this section. (Some of the guidelines may not apply to the particular study you are analyzing; in that case, explain why the guideline is not applicable.)

38. Believable Results. Find a recent news report about a statistical study whose results you believe are meaningful and important. In one page or less, summarize the study and explain why you find it believable.

39. Unbelievable Results. Find a recent news report about a statistical study whose results you *don't* believe are meaningful and important. In one page or less, summarize the study and explain why you don't believe its claims.

40. Self-Selected Survey. Find an example of a recent survey in which the sample was self-selected. Describe the makeup of the sample and how you think the self-selection affected the results of the survey.

Chapter Review Exercises

1. **Guns in Homes.** A Gallup poll involved a survey of adults selected in the United States. They were asked this question: "Do you have a gun in your home?" Among the 1,012 responses received, 38% were *yes*. The margin of error was reported as 3 percentage points.

   a. Interpret the margin of error by identifying the range of values likely to contain the proportion of all households with guns.

   b. Identify the population.

   c. Is this study an experiment or an observational study? Explain. Identify the variable of interest.

   d. Is the reported value of 38% a population parameter or a sample statistic? Why?

   e. If you learned that survey subjects responded to a magazine article asking readers to phone in their responses, would you consider the survey results to be valid? Why or why not?

   f. Describe a procedure for selecting the survey subjects using a simple random sample.

   g. Describe a procedure for selecting similar survey subjects using stratified sampling.

   h. Describe a procedure for selecting similar survey subjects using cluster sampling.

   i. Describe a procedure for selecting similar survey subjects using systematic sampling.

   j. Describe a procedure for selecting similar survey subjects using a convenience sample.

2. **Simple Random Sample.** An important element of this chapter is the concept of a simple random sample.

   a. What is a simple random sample?

   b. When the Bureau of Labor Statistics conducts a survey, it begins by partitioning the United States adult population into 2,007 groups called *primary sampling units*. Assume that these primary sampling units all contain the same number of adults. If you randomly select one adult from each primary sampling unit, is the result a simple random sample? Why or why not?

   c. Refer to the primary sampling units described in part b and describe a sampling plan that results in a simple random sample.

3. **Testing Zocor.** In clinical trials of the drug Zocor, used to treat high cholesterol levels, 1,583 Zocor users were observed for adverse reactions. It was found that 3.5% of the Zocor users experienced headaches.

   a. Based on the given information, can you conclude that in some cases Zocor causes headaches? Why or why not?

   b. While 1,583 subjects were treated with Zocor, another 157 subjects were given a placebo, and 5.1% of the placebo group experienced headaches. What does this additional information suggest about headaches as an adverse reaction to the use of Zocor?

   c. In this clinical trial, what is blinding and why is it important in testing the effects of Zocor?

   d. Is this clinical trial an observational study or an experiment? Explain.

   e. What is an experimenter effect, and how might this effect be minimized?

4. **Wording of a Survey Question.** In *The Superpollsters*, David W. Moore describes an experiment in which different subjects were asked if they agree with the following statements:

   - Too little money is being spent on welfare.
   - Too little money is being spent on assistance to the poor.

   Even though it is the poor who receive welfare, only 19% agreed when the word *welfare* was used, but 63% agreed with *assistance to the poor.*

   a. Which of the two questions should be used in a survey? Why?

   b. If you were working on a campaign for a conservative candidate for Congress and you wanted to emphasize opposition to the use of federal funds for assistance to the poor, which of the two questions would you use? Why?

   c. Is it ethical to deliberately word a survey question so as to influence responses? Why or why not?

Chapter Quiz

Choose the best answer to each of the following questions. Explain your reasoning with one or more complete sentences.

1. You conduct a poll in which you randomly select 1,000 registered voters from Texas and ask if they approve of the job their governor is doing. The population for this study is
   a. all registered voters in the state of Texas.
   b. the 1,000 people that you interview.
   c. the Governor of Texas.
2. For the poll described in Exercise 1, your results would most likely suffer from bias if you chose the participants from
   a. all registered voters in Texas.
   b. all people with a Texas drivers license.
   c. people who donated money to the Governor's campaign.
3. When we say that a sample is *representative* of the population, we mean that
   a. the results found for the sample are similar to those we would find for the entire population.
   b. the sample is very large.
   c. the sample was chosen in the best possible way.
4. Consider an experiment designed to see whether cash incentives can improve school attendance. The researcher chooses two groups of 100 high school students. She offers one group $10 for every week of perfect attendance. She tells the other group that they are part of an experiment but does not give them any incentive. The students who do not receive an incentive represent
   a. the treatment group.
   b. the control group.
   c. the observation group.
5. The experiment described in Exercise 4 is
   a. single-blind.
   b. double-blind.
   c. not blind.
6. The purpose of a placebo is
   a. to prevent participants from knowing whether they belong to the treatment group or the control group.
   b. to distinguish between the cases and the controls in a case-control study.
   c. to determine whether diseases can be cured without any treatment.
7. If we see a placebo effect in an experiment to test a new treatment designed to cure warts,
   a. the experiment was not properly double-blind.
   b. the experimental groups were too small.
   c. warts were cured among members of the control group.
8. An experiment is single-blind if
   a. it lacks a treatment group.
   b. it lacks a control group.
   c. the participants do not know whether they belong to the treatment or the control group.
9. Poll X predicts that Powell will receive 49% of the vote, while Poll Y predicts that she will receive 53% of the vote. Both polls have a margin of error of 3 percentage points. What can you conclude?
   a. One of the two polls must have been conducted poorly.
   b. The two polls are consistent with each another.
   c. Powell will receive 51% of the vote.
10. A survey reveals that 12% of Americans believe Elvis is still alive, with a margin of error of 4 percentage points. The confidence interval for this poll is
    a. from 10% to 14%.
    b. from 8% to 16%.
    c. from 4% to 20%.
11. A study conducted by Exxon Mobil shows that there was no lasting damage from a large oil spill in Alaska. This conclusion
    a. is definitely invalid, because the study was biased.
    b. may be correct, but the potential for bias means you should look very closely at how the conclusion was reached.
    c. could be correct if it falls within the confidence interval of the study.

12. The television show *American Idol* selects winners from votes cast by anyone who wants to vote. This means the winner

a. is the person most Americans want to win.

b. may or may not be the person most Americans want to win, because the voting is subject to participation bias.

c. may or may not be the person most Americans want to win, because the voting should have been double-blind.

13. Consider an experiment in which you measure the weights of 6-year-olds. The variable of interest in this study is

a. the size of the sample.

b. the weights of 6-year-olds.

c. the ages of the children under study.

14. Imagine a survey of randomly selected people in which it is found that people who use sunscreen were *more* likely to have been sunburned in the past year. Which explanation for this result seems most likely?

a. Sunscreen is useless.

b. The people in this study all used sunscreen that had passed its expiration date.

c. People who use sunscreen are more likely to spend time in the sun.

15. If a statistical study is carefully conducted in every possible way, then

a. its results must be correct.

b. we can have confidence in its results, but it is still possible that they are not correct.

c. we say that the study is perfectly biased.

FOCUS ON SOCIOLOGY

Does Daycare Breed Bullies?

Decades ago, it was almost a given that mothers would stay at home to take care of their young children. But with more women in the workforce, more children are spending more of their waking hours in some type of daycare. In 1975, fewer than 40% of women with children under age 6 worked full- or part-time outside the home. Today, more than 60% of such mothers do. This dramatic increase in the number of children in daycare led psychologists and sociologists to study the effects of daycare. In 1991, the National Institute of Child Health and Human Development (NICHD) started the Early Child Care study, which tracked more than 1,300 children enrolled in daycare at 10 different research sites around the United States and continued to track them through age 15. Notice that this set-up makes it an observational, *prospective* study (see Section 1.3).

The study has been conducted carefully by outstanding researchers, but some results have still provoked controversy. As an example, consider this result reported in 2001: Young children who spend more time in daycare are significantly more likely to be aggressive bullies as they get older. Working mothers everywhere felt pangs of guilt as they worried about the effects of daycare on their children, and the news media reported widely that daycare breeds bullies. But should we believe this claim? There are many ways to evaluate the claim, but for practice, let's use the eight guidelines given in Section 1.4 (p. 34).

Guideline 1: The goal of the study seems clear (to learn about effects of daycare), as does the population under study (all children in daycare). However, the type of study raises a potential problem. The study is observational, rather than a randomized experiment, because it would be unethical to tell parents whether or not to use daycare. Unfortunately, while an observational study can help establish a link (or *correlation*) between variables, it cannot by itself establish cause and effect (see Chapter 7). In this case, the study provides evidence of a link between daycare and bullying, but does not tell us whether one is the cause of the other. The problem is that, because parents *chose* whether their children were in daycare or at home, we cannot be sure that the sample and control groups are truly comparable. For example, it may be that mothers who must juggle children and work tend to be more stressed than nonworking mothers, which in turn might make daycare children more stressed than their stay-at-home counterparts. Or, it could be that mothers with rambunctious children are more likely to put their children in daycare. Thus, regardless of the validity of the link between daycare and bullying, the observational nature of the study makes it impossible to support a claim that daycare *causes* the bullying.

Guideline 2: The study has been conducted with care, and we have no reason to suspect any particular bias on the part of the researchers. However, daycare is an emotionally charged issue and researchers may have strong opinions about it, raising the potential for bias. Indeed, different researchers have interpreted the study results differently, which suggests that bias may play a role in the interpretations, if not in the study itself.

Guideline 3: Researchers used a type of stratified sampling to recruit participants for the study, because they wanted to match national demographics for factors such as socioeconomic background, race, and family structure. This type of sampling seems appropriate to the study. However, the families that were recruited had a choice about whether to participate in the study. This self-selection automatically raises the potential for participation bias, which may affect the study's results.

Guideline 4: The claim about daycare and bullying essentially involves two variables: *the amount of time a child spends in daycare* and *the child's level of aggressiveness*. Both are very difficult to define. For example, the Early Child Care study defines *daycare* as care provided by anyone other than the mother, which means that care by fathers or grandparents counts as daycare. Many people object to this definition. Defining aggression is also difficult; for example, it may be difficult to distinguish between active play and true aggression.

Guideline 5: There are many potentially confounding variables in the Early Child Care study. As noted earlier, the study does not provide data about whether mothers of children in daycare are subject to more stress than stay-at-home moms, which could confound the results. Perhaps of greater concern, there is no accepted way to measure or control for the quality of daycare. As a result, even some of the researchers involved in the study have suggested that the observed bullying is a result of poor or mediocre daycare rather than of daycare in general.

Guideline 6: The researchers used interviews to determine the type and amount of daycare and the level of aggression for each child. But because aggressiveness is difficult to define and may be thought of differently by different people, there is no clear way to ensure that all interviewers ask questions in precisely the same way (not only in wording, but also in facial expression and intonation) and that all respondents interpret the questions similarly.

Guideline 7: The reports produced by the Early Child Care researchers acknowledge potential flaws in the study and emphasize that the study shows only a *possible* link between daycare and bullying. So where does the claim that daycare breeds bullies come from? It comes from the way the findings are presented in the media. Because the actual results do not support this interpretation, the news reports clearly reflect either bias or misunderstanding. For example, journalists who have not studied statistics may not be aware of the difference between finding a link and finding a cause, and their misunderstanding is then reflected in their news reports. Alternatively, the media might deliberately sensationalize the results in order to attract more attention. And in some cases, people who believe that women should stay home with their children might misrepresent the results so as to claim that the study lends scientific support to their views.

Guideline 8: The question of practical significance is very important with this study. On an individual level, many working women don't believe they have a choice about daycare; they must work for either financial or professional reasons. They might therefore use daycare even if there were proof that home care was better. On the societal level, the findings might be practically significant if they led to new policies that could improve child care generally. But as long as the results do not yet prove cause and effect, they cannot give clear guidance to policy makers.

In summary, we've found that a provocative headline about daycare and bullying falls apart upon close scrutiny. The research might still be quite valuable, especially because the researchers themselves are aware of potential flaws in the study. The results may lead to future studies with more conclusive results—information that would surely be appreciated by working parents everywhere. But until then, we need to be careful about how we interpret the Early Child Care study results, and even more careful about how we react to any sensational headlines the study may produce.

QUESTIONS FOR DISCUSSION

1. A third potential confounding variable is *amount of time children spend watching television and playing video games*, which has also been linked to aggressive behavior. Do you think this variable might affect the Early Child Care study results and, if so, how? What other confounding variables might affect its results? Explain your answers clearly.

2. Suppose you were designing a follow-up study intended to give more conclusive results. Based on what has been learned from the Early Child Care study, what would you do differently?

3. What is your personal opinion about whether it is better for a parent to stay at home or to work? Would your opinion be swayed if the Early Child Care study produced any definitive results about daycare? Why or why not?

4. Overall, do you think the Early Child Care study was a worthwhile effort in terms of the time and expense it required? Defend your opinion.

SUGGESTED READING

Read more about the Study of Early Child Care (SECC) on its Web site at http://secc.rti.org.

Carey, Benedict. "Study Finds Rise in Behavior Problems After Significant Time in Day Care." *New York Times*, March 26, 2007.

Stolberg, Sheryl Gay. "Science, Studies, and Motherhood." *New York Times*, April 22, 2001.

Sweeney, Jennifer Foote. "The Day-Care Scare, Again." Salon.com, April 20, 2001.

FOCUS ON PUBLIC HEALTH

Is Your Lifestyle Healthy?

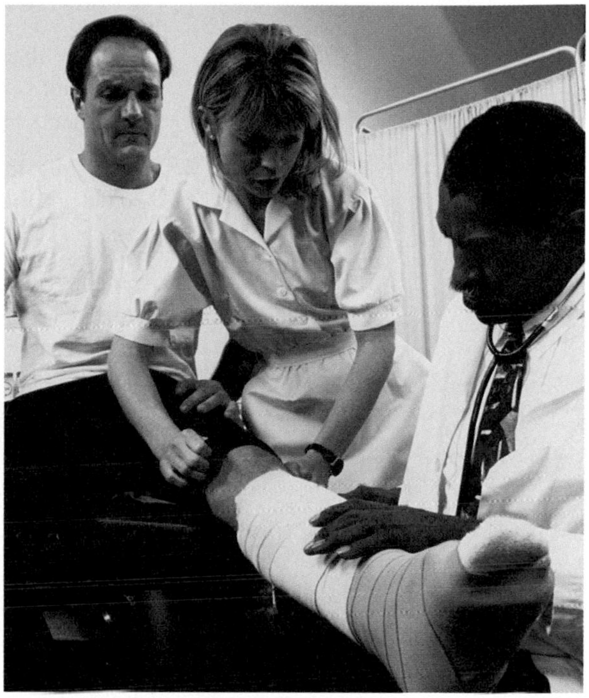

Consider the following findings from statistical studies:

- Smoking increases the risk of heart disease.
- Eating margarine can increase the risk of heart disease.
- One glass of wine per day can protect against heart disease, but increases the risk of breast cancer.

You are probably familiar with some of these findings, and perhaps you've even altered your lifestyle as a result of them. But where do they come from? Remarkably, these and hundreds of other important findings on public health come from a huge prospective study that has provided data for hundreds of smaller statistical studies. Known as the Harvard Nurses' Health Study, it is the most enduring public health study ever undertaken. If it has not already changed the way you live, it almost certainly will in the future.

The Harvard Nurses' Health Study began in 1976 when Dr. Frank E. Speizer, a professor at Harvard Medical School, decided to study the long-term effects of oral contraceptives. He mailed questionnaires to approximately 370,000 registered nurses and received more than 120,000 responses. He chose to survey nurses because he believed that their medical training would make their responses more reliable than those of the general public.

As Dr. Speizer and his colleagues sifted through the data in the returned questionnaires, they realized that the study could be expanded to include more than just the effects of contraceptives. Today, this research team continues to follow some 90% of the original 120,000 respondents. Hundreds of other researchers also use the data to study public health.

Annual questionnaires are still a vital part of the study, allowing researchers to gather data about what the nurses eat; what medicines and vitamins they take; whether and how much they exercise, drink, and smoke; and what illnesses they have contracted. Some of the nurses also provide blood samples, which are used to measure such things as cholesterol level, hormone levels, genetic variations, and residues from pesticides and environmental pollutants. Dr. Speizer's faith in nurses has proven justified, as they reliably complete surveys and almost always provide properly drawn and labeled blood samples upon request.

After more than 30 years of correspondence, both the researchers and the nurses say they feel a sense of closeness. Many of the nurses look forward to hearing from the researchers and say that the study has helped them to pay more attention to how they live their lives. Researchers feel deep sorrow when they must record the death of one of the nurses.

The sorrow of death will play an increasing role in the study, as many of the nurses are now entering their 70s and 80s. But this sorrow will also yield a wealth of valuable data about factors that influence longevity and health in old age. Researchers hope that the data will point the way toward definitive understanding about what constitutes a healthy diet, how pollution and chemical exposure influence health, and how we might prevent debilitating diseases like osteoporosis and Alzheimer's. In death, the 120,000 women of the Harvard Nurses' Health Study may give the gift of a better life to future generations.

QUESTIONS FOR DISCUSSION

1. Consider some of the results that are likely to come from the Harvard Nurses' Health Study over the next 10–20 years. What types of results do you think will be most important? Do you think the findings will alter the way you live your life?

2. Explain why the Harvard Nurses' Health Study is an observational study. Critics sometimes say that the results would be more valid if obtained by experiments rather than observations. Discuss whether it would be possible to gather similar data by carrying out experiments in a practical and ethical way.

3. In principle, the Harvard Nurses' Health Study is subject to participation bias because only 120,000 of the original 370,000 questionnaires were returned. Should the researchers be concerned about this bias? Why or why not?

4. Another potential pitfall comes from the fact that the questionnaires often deal with sensitive issues of personal health, and researchers have no way to confirm that the nurses answer honestly. Do you think that dishonesty could be leading researchers to incorrect conclusions? Defend your opinion.

5. All of the participants in the Harvard Nurses' Health Study are women. Do you think that the results also are of use to men? Why or why not?

6. Do a Web search for news articles that discuss results from the Harvard Nurses' Health Study. Choose one recent result that interests you, and discuss what it means and how it may affect public health or your own health in the future.

SUGGESTED READING

Read more about the Harvard Nurses' Health Study (headquartered at Harvard's Channing Laboratory) on its Web site at http://www.nurseshealthstudy.org.

Brophy, Beth. "Doing It for Science." *U.S. News & World Report*, Vol. 126, March 21, 1999, p. 67.

Conwell, Vikki. "Forever Friends: Ties Can Last a Lifetime and Sustain Body and Soul." *Atlanta Journal-Constitution*, January 2, 2007.

Parker-Pope, Tara. "Drinking Has Hidden Health Risks for Women." *Wall Street Journal*, December 26, 2006.

Yoon, Carol Kaesuk. "In Nurses' Lives, a Treasure Trove of Health Data." *New York Times*, September 15, 1998.

2

Practically no one knows what they're talking about when it comes to numbers in the newspapers. And that's because we're always quoting other people who don't know what they're talking about, like politicians and stock-market analysts.

—Molly Ivins (1944–2007), newspaper columnist

Measurement in Statistics

WE ALL KNOW HOW TO MEASURE QUANTITIES SUCH AS height, weight, and temperature. However, in statistical studies there are many other kinds of measurements, and we must be sure that they are defined, obtained, and reported carefully. In this chapter, we will discuss a few important concepts associated with measurements and statistics. As you'll see, these concepts are very useful to understanding the statistical reports you encounter in your daily life.

LEARNING GOALS

2.1 Data Types and Levels of Measurement

Be able to identify data as qualitative or quantitative, to identify quantitative data as discrete or continuous, and to assign data a level of measurement (nominal, ordinal, interval, or ratio).

2.2 Dealing with Errors

Understand the difference between random and systematic errors, be able to describe errors by their absolute and relative sizes, and know the difference between accuracy and precision in measurements.

2.3 Uses of Percentages in Statistics

Understand how percentages are used to report statistical results and recognize ways in which they are sometimes misused.

2.4 Index Numbers

Understand the concept of an index number; in particular, understand how the Consumer Price Index (CPI) is used to measure inflation.

2.1 Data Types and Levels of Measurement

One of the challenges in statistics is deciding how best to summarize and display data. Different types of data call for different types of summaries. In this section, we'll discuss how data are categorized, an idea that will help us when we consider data summaries and displays in later chapters.

Data Types

Data come in two basic types: qualitative and quantitative. **Qualitative data** have values that can be placed into *nonnumerical categories*. (For this reason, qualitative data are sometimes called *categorical* data.) For example, eye color data are qualitative because they are categorized by colors such as blue, brown, and hazel. Other examples of qualitative data include flavors of ice cream, names of employers, genders of animals, and "star ratings" of movies or restaurants. Note that star ratings are qualitative even though they involve a *number* of stars (such as three stars or four stars), because the numbers are not necessary and could not be used for computations; we could rate movies equally well with four nonnumerical categories such as bad, average, good, and excellent.

Quantitative data have numerical values representing counts or measurements. The times of runners in a race, the incomes of college graduates, and the numbers of students in different classes are all examples of quantitative data.

> **Data Types**
>
> **Qualitative** (or **categorical**) **data** consist of values that can be placed into nonnumerical categories.
>
> **Quantitative data** consist of values representing counts or measurements.

EXAMPLE 1 Data Types

Classify each of the following sets of data as qualitative or quantitative.

a. Brand names of shoes in a consumer survey

b. Scores on a multiple-choice exam

c. Letter grades on an essay assignment

d. Numbers on uniforms that identify players on a basketball team

Solution

a. Brand names are categories and therefore represent qualitative data.

b. Scores on a multiple-choice exam are quantitative because they are represent a count of the number of correct answers.

c. Letter grades on an essay assignment are qualitative because they represent different categories of performance (failing through excellent).

d. The players' uniform numbers are qualitative because they do not represent a count or measurement; they are used solely for identification. You can tell that these numbers are qualitative rather than quantitative because you could not use them for computations. For example, it would make no sense to add or subtract the uniform numbers of different players.

Discrete versus Continuous Data

Quantitative data can be further classified as continuous or discrete. Data are **continuous** if they can take on *any* value in a given interval. For example, a person's weight can be anything between 0 and a few hundred pounds, so data that consist of weights are continuous. Data are **discrete** if they can take on only particular values and not other values in between. For example, the number of students in your class is discrete because it must be a whole number, and shoe sizes are discrete because they take on only integer and half-integer values such as 7, $7\frac{1}{2}$, 8, and $8\frac{1}{2}$.

> **Discrete versus Continuous Data**
>
> **Continuous** data can take on *any* value in a given interval.
>
> **Discrete** data can take on only particular, distinct values and not other values in between.

EXAMPLE 2 Discrete or Continuous?

For each data set, indicate whether the data are discrete or continuous.

a. Measurements of the time it takes to walk a mile

b. The numbers of calendar years (such as 2007, 2008, 2009)

c. The numbers of dairy cows on different farms

d. The amounts of milk produced by dairy cows on a farm

Solution

a. Time can take on any value, so measurements of time are continuous.

b. The numbers of calendar years are discrete because they cannot have fractional values. For example, on New Year's Eve of 2009, the year will change from 2009 to 2010; we'll never say the year is $2009\frac{1}{2}$.

c. Each farm has a whole number of cows that we can count, so these data are discrete. (You cannot have fractional cows, for example.)

d. The amount of milk that a cow produces can take on any value in some range, so the milk production data are continuous.

Levels of Measurement

Another way to classify data is by their *level of measurement*. The simplest level of measurement applies to variables such as eye color, ice cream flavors, or gender of animals. These variables can be described solely by names, labels, or categories. We say that such data are at a **nominal level of measurement**. (The word *nominal* refers to *names* for categories.) The nominal level of measurement does not involve any ranking or ordering of the data. For example, we could not say that blue eyes come before brown eyes or that vanilla ranks higher than chocolate.

When we describe data with a ranking or ordering scheme, such as star ratings of movies or restaurants, we are using an **ordinal level of measurement**. (The word *ordinal* refers to *order*.) Such data generally cannot be used in any meaningful way for computations. For example, it doesn't make sense to add star ratings watching three one-star movies is not equivalent to watching one three-star movie.

TIME OUT TO THINK

Consider a survey that asks "What's your favorite flavor of ice cream?" We've said that ice cream flavors represent data at the nominal level of measurement. But suppose that, for convenience, the researchers enter the survey data into a computer by assigning numbers to the different flavors. For example, they assign 1 = vanilla, 2 = chocolate, 3 = cookies and cream, and so on. Does this change the ice cream flavor data from nominal to ordinal? Why or why not?

The ordinal level of measurement provides a ranking system, but it does not allow us to determine precise differences between measurements. For example, there is no way to determine the exact difference between a three-star movie and a two-star movie. In contrast, a temperature of 81°F is hotter than 80°F by the same amount that 28°F is hotter than 27°F. Temperature data are at a higher level of measurement, because the *intervals* (differences) between units on a temperature scale always mean the same definite amount. However, while intervals (which involve subtraction) between Fahrenheit temperatures are meaningful, *ratios* (which involve division) are not. For example, it is *not true* that 20°F is twice as hot as 10°F or that −40°F is twice as cold as −20°F. The reason ratios are meaningless on the Fahrenheit scale is that its *zero point is arbitrary* and does not represent a state of "no heat." If intervals are meaningful but ratios are not, as is the case with Fahrenheit temperatures, we say that the data are at the **interval level of measurement**.

When both intervals and ratios are meaningful, we say that data are at the **ratio level of measurement**. For example, data consisting of distances are at the ratio level of measurement because a distance of 10 kilometers really is twice as far as a distance of 5 kilometers. In general, the ratio level of measurement applies to any scale with a *true zero*, which is a value that means *none* of whatever is being measured. In the case of distances, a distance of zero means "no distance." Other examples of data at the ratio level of measurement include weights, speeds, and incomes.

Note that data at the nominal or ordinal level of measurement are always qualitative, while data at the interval or ratio level are always quantitative (and can therefore be either continuous or discrete). Figure 2.1 summarizes the possible data types and levels of measurement.

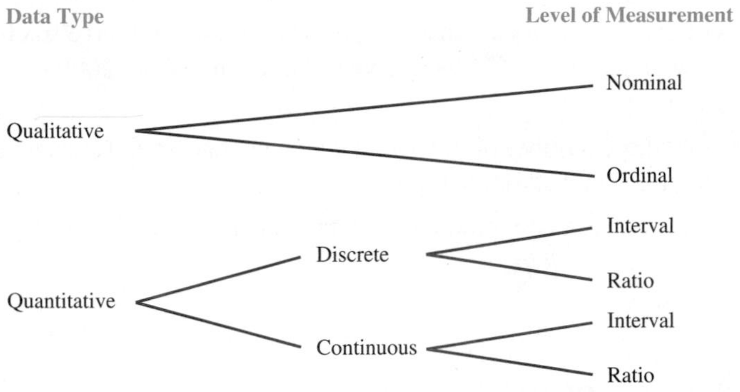

Figure 2.1 Data types and levels of measurement.

Levels of Measurement

The **nominal level of measurement** is characterized by data that consist of names, labels, or categories only. The data are qualitative and cannot be ranked or ordered.

The **ordinal level of measurement** applies to qualitative data that can be arranged in some order (such as low to high). It generally does not make sense to do computations with data at the ordinal level of measurement.

The **interval level of measurement** applies to quantitative data in which intervals are meaningful, but ratios are not. Data at this level have an arbitrary zero point.

The **ratio level of measurement** applies to quantitative data in which both intervals and ratios are meaningful. Data at this level have a true zero point.

EXAMPLE 3 Levels of Measurement

Identify the level of measurement (nominal, ordinal, interval, ratio) for each of the following sets of data.

a. Numbers on uniforms that identify players on a basketball team

b. Student rankings of cafeteria food as excellent, good, fair, or poor

c. Calendar years of historic events, such as 1776, 1945, or 2001

d. Temperatures on the Celsius scale

e. Runners' times in the Boston Marathon

Solution

a. As discussed in Example 1, numbers on uniforms don't count or measure anything. They are at the nominal level of measurement because they are labels and do not imply any kind of ordering.

b. A set of rankings represents data at the ordinal level of measurement because the categories (excellent, good, fair, or poor) have a definite order.

c. An interval of one calendar year always has the same meaning. But ratios of calendar years do not make sense because the choice of the year 0 is arbitrary and does not mean "the beginning of time." Calendar years are therefore at the interval level of measurement.

d. Like Fahrenheit temperatures, Celsius temperatures are at the interval level of measurement. An interval of 1°C always has the same meaning, but the zero point (0°C = freezing point of water) is arbitrary and does not mean "no heat."

e. Marathon times have meaningful ratios—for example, a time of 6 hours really is twice as long as a time of 3 hours—because they have a true zero point at a time of 0 hours.

By the Way ...

Scientists often measure temperatures on the Kelvin scale. Data on the Kelvin scale are at the ratio level of measurement, because the Kelvin scale has a true zero. A temperature of 0 Kelvin really is the coldest possible temperature. Called *absolute zero*, 0 K is equivalent to about −273.15°C or −459.67°F. (The degree symbol is not used for Kelvin temperatures.)

Temperature Scale

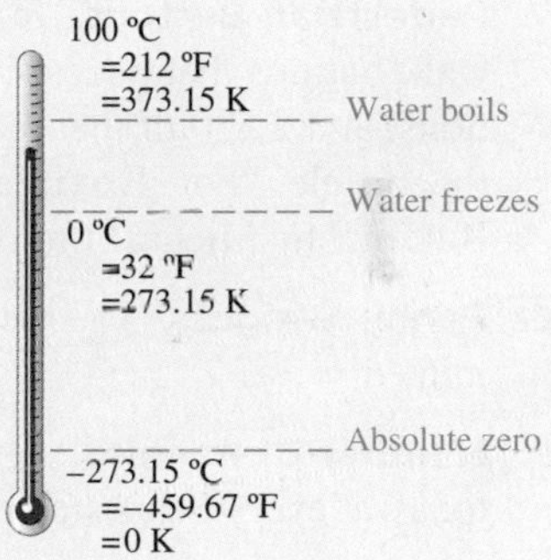

Section 2.1 Exercises

Statistical Literacy and Critical Thinking

1. **Qualitative vs. Quantitative Data.** What is the difference between qualitative data and quantitative data?

2. **Quantitative vs. Qualitative Data.** A football player is taking a statistics course and states that the names of the players on his team are qualitative, but they can be made quantitative by using the numbers on the jerseys of their uniforms. Is he correct? Why or why not?

3. **Qualitative vs. Quantitative Data.** Is a researcher correct when she argues that all data are either qualitative or quantitative? Explain.

4. **ZIP Codes.** A researcher argues that zip codes are quantitative data because they measure location, with low numbers in the east and high numbers in the west. Is she correct? Why or why not?

Concepts and Applications

Qualitative vs. Quantitative Data. In Exercises 5–16, determine whether the data described are qualitative or quantitative and explain why.

5. **Blood Groups.** The blood groups of A, B, AB, and O

6. **Weights.** The weights of subjects in a clinical trial of a new drug

7. **Heights.** The heights of subjects in a clinical trial of a new drug

8. **Movies.** The types of movies (drama, comedy, etc.)

9. **Movie Lengths.** The lengths (in minutes) of movies

10. **Survey Responses.** The responses (*yes*, *no*) from survey subjects when asked a question

11. **Nielsen Ratings.** The television shows being watched by households surveyed by Nielsen Media Research

12. **Nielsen Ratings.** The number of households with a television in use when surveyed by Nielsen Media Research

13. **Survey Refusals.** The number of people who refused to answer questions when surveyed

14. **Shoe Sizes.** The shoe sizes (such as 8 or $10\frac{1}{2}$) of test subjects

15. **Salary.** The salaries of all state governors

16. **Area Codes.** The area codes (such as 617) of the telephones of survey subjects

Discrete vs. Continuous Data. In Exercises 17–28, state whether the actual data are discrete or continuous and explain why.

17. **Pedestrian Buttons.** In New York City, there are 3,250 walk buttons that pedestrians can press at traffic intersections, and 2,500 of them do not work (based on data from the article "For Exercise in New York Futility, Push Button," by Michael Luo, *New York Times*).

18. **Penny Weights.** The weights of pennies currently being minted

19. **Number of Pennies.** The numbers of pennies collected today at every Sears store in the United States

20. **Gallup Poll.** In Gallup polls taken at different times, subjects were asked if they owned a gun, and the number of *yes* responses was recorded for each time period.

21. **Chemistry.** An experiment in chemistry was repeated, and the times required for a reaction to occur were recorded.

22. **Test Times.** The times required by students to complete a statistics test

23. **Test Scores.** The numerical scores on a statistics test, which are the numbers of correct responses from 25 multiple-choice questions

24. **Traffic Count.** The number of cars crossing the Golden Gate Bridge each hour

25. **Car Speeds.** The speeds of cars as they pass the center of the Golden Gate Bridge

26. **Movie Ratings.** The movie ratings by a critic: 0 stars, $\frac{1}{2}$ star, 1 star, and so on

27. **Stars.** The number of stars in each galaxy in the universe

28. **Calories.** The numbers of calories consumed by football players during one game

Levels of Measurement. For the data described in Exercises 29–40, identify the level of measurement as nominal, ordinal, interval, or ratio.

29. **Heights.** Heights of all players on the LA Lakers basketball team

30. **Movie Ratings.** A critic's movie recommendations of "must see," "good," "fair," "poor," "avoid at all costs"

31. **Movie Types.** Types of movies (drama, comedy, etc.)

32. **Temperatures.** Body temperatures of all students in a statistics class

33. **Cars.** Classifications of cars as subcompact, compact, intermediate, or full-size

34. **Clinical Trial.** Results from a clinical trial consisting of "true positive," "false positive," "true negative," "false negative"

35. **Grades.** Final course grades of A, B, C, D, F

36. **Weights.** Weights of all packages handled by UPS today

37. **SSN.** Social Security numbers

38. **Years.** Years in which Democrats won the U.S. presidential election

39. **Airline Baggage.** Weights of all checked baggage on United Flight 15

40. **Car Safety Ratings.** *Consumer Reports* safety ratings of cars, from 0 = unsafe up to 3 = safest

Meaningful Ratios? In Exercises 41–48, determine whether the given statement represents a meaningful ratio, so the ratio level of measurement applies. Explain.

41. **Movie Rating.** A movie with a 4-star rating is twice as good as one with a 2-star rating.

42. **Marathons.** A 10-kilometer race is twice as far as one that is 5 kilometers.

43. **Aircraft Speeds.** One aircraft moving at a speed of 450 miles per hour is going three times as fast as another aircraft going 150 miles per hour.

44. **Temperatures.** On August 6, it was 80°F in New York City, so it was twice as hot as on December 7, when it was 40°F.

45. **Weights of Fish.** One fish is found to weigh 2 pounds while another fish is found to weigh 4 pounds, so the heavier fish weighs twice as much as the lighter fish.

46. **Carbon Dating.** Through carbon dating, one sample of wood is found to be twice as old as another, because the first sample is found to be 2000 years old while the other is 1000 years old.

47. **Salary.** One employee has a salary of $150,000, which is twice the $75,000 salary of another employee.

48. **SAT Scores.** A person with an SAT score of 2200 is twice as qualified for college as a person with a score of 1100.

Complete Classification. In Exercises 49–56, determine whether the data described are qualitative or quantitative and give their level of measurement. If the data are quantitative, state whether they are continuous or discrete. Give a brief explanation.

49. Marathon Times. Finish times in the New York City marathon

50. Marathon Order. Finishing positions (such as 1st, 2nd, 3rd) of runners in a marathon

51. Employee ID Numbers. The randomly generated six-digit identification numbers assigned to employees of the Telektronics Corporation

52. Employee Service Times. The lengths of time that have passed since each employee at the Telektronics Corporation was first hired

53. Employee Hiring Years. The years in which employees were hired (such as 2003, 1998, 2007)

54. Political Survey. In a survey of voter preferences, the political parties of respondents are recorded as coded numbers 1, 2, 3, 4, or 5 (where 1 = Democrat, 2 = Republican, 3 = Liberal, 4 = Conservative, 5 = other).

55. Product Ratings. *Consumer Reports* magazine lists ratings of "best buy," "recommended," or "not recommended" for each of several different computers.

56. Quality Control. The Newport Electronics Company tests each of its manufactured radios and labels each as acceptable or defective.

2.2 Dealing with Errors

Now that you understand the different levels and types of measurement, we turn to the issue of how to deal with errors in measurement. First we will consider the various types of errors that can occur, and then we'll discuss how to account for possible errors when we state results. Note that while we will phrase most of this discussion in terms of measurements, it applies equally well to estimates or projections, such as population estimates or projected revenues for a corporation.

Mistakes are the portals of discovery.

—James Joyce

Types of Error: Random and Systematic

Broadly speaking, measurement errors fall into two categories: random errors and systematic errors. An example will illustrate the difference.

Suppose you work in a pediatric office and use a digital scale to weigh babies. If you've ever worked with babies, you know that they usually aren't very happy about being put on a scale. Their thrashing and crying tends to shake the scale, making the readout jump around. For the case shown in Figure 2.2a, you could conceivably record the baby's weight as anything between about 14.5 and 15.0 pounds. We say that the shaking of the scale introduces a **random error** because any particular measurement may be either too high or too low.

Now suppose you have been measuring weights of babies all day long with the scale shown in Figure 2.2b. At the end of the day, you notice that the scale reads 1.2 pounds when there is nothing on it. In that case, every measurement you made was high by 1.2 pounds. This type of error is called a **systematic error** because it is caused by an error in the measurement *system*—an error that consistently (systematically) affects all measurements.

By the Way ...

A systematic error in which a scale's measurements differ consistently from the true values is called a *calibration error*. You can test the calibration of a scale by putting known weights on it, such as 0-, 5-, 10-, and 20-pound weights, and making sure that the scale gives the expected readings.

Two Types of Measurement Error

Random errors occur because of random and inherently unpredictable events in the measurement process.

Systematic errors occur when there is a problem in the measurement system that affects all measurements in the same way.

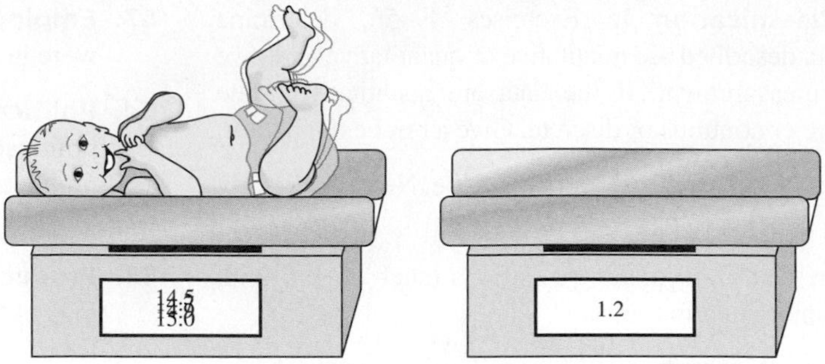

(a) (b)

Figure 2.2 (a) The baby's motion introduces random errors. (b) The scale reads 1.2 pounds when empty, introducing a systematic error that makes all measurements 1.2 pounds too high.

A systematic error affects all measurements in the same way, such as making them all too high or all too low. If you discover a systematic error, you can go back and adjust the affected measurements. In contrast, the unpredictable nature of random errors makes it impossible to correct for them. However, you can minimize the effects of random errors by making many measurements and averaging them. For example, if you measure the baby's weight ten times, your measurements will probably be too high in some cases and too low in others. You can therefore get a better value by averaging the ten individual measurements.

By the Way ...

The fact that urban areas tend to be warmer than they would be in the absence of human activity is often called the *urban heat island effect*. Major causes of this effect include heat released by burning fuel in automobiles, homes, and industry and the fact that pavement and large masonry buildings tend to retain heat from sunlight.

TIME OUT TO THINK

Go to a Web site (such as www.time.gov) that gives the current time. How far off is your clock or watch? Describe possible sources of random and systematic errors in your timekeeping.

EXAMPLE 1 Errors in Global Warming Data

Scientists studying global warming need to know how the average temperature of the entire Earth, or the *global average temperature*, has changed with time. Consider two difficulties in trying to interpret historical temperature data from the early 20th century: (1) Temperatures were measured with simple thermometers and the data were recorded by hand, and (2) most temperature measurements were recorded in or near urban areas, which tend to be warmer than surrounding rural areas because of heat released by human activity. Discuss whether each of these two difficulties produces random or systematic errors, and consider the implications of these errors.

Solution The first difficulty involves *random errors* because people undoubtedly made occasional errors in reading the thermometers, in calibrating the thermometers, and in recording temperature readings. There is no way to predict whether any individual reading is correct, too high, or too low. However, if there are several readings for the same region on the same day, averaging these readings can minimize the effects of the random errors.

The second difficulty involves a *systematic error* because the excess heat in urban areas always causes the temperature reading to be higher than it would be otherwise. If the researchers can estimate how much this systematic error affected the temperature readings, they can correct the data for this problem.

CASE STUDY **The Census**

The Constitution of the United States mandates a census of the population every 10 years. The U.S. Census Bureau conducts the census (and also does many other demographic studies).

In attempting to count the population, the Census Bureau relies largely on a survey that is supposed to be delivered to and returned by every household in the United States. However, many *random errors* occur in this survey process. For example, some people fill out their forms incorrectly and some responses are recorded incorrectly by Census Bureau employees.

The census also is subject to several types of *systematic error*. For example, it is very difficult to deliver surveys to the homeless, and it is difficult to count undocumented aliens, who may be reluctant to reveal their presence in the United States. These systematic errors lead to undercounts in the population. An example of a systematic error leading to an overcount is the double counting of some college students, who are counted by their parents and again at their school residence.

The 2000 census, for example, originally counted about 281.4 million people. However, follow-up statistical studies suggested that the census had missed about 7.6 million people and counted about 4.3 million people twice. If these statistical results are correct, then the census undercounted the population by more than 3 million people.

A little inaccuracy sometimes saves a ton of explanation.

—H. H. Munro (Saki)

TIME OUT TO THINK

The question of whether the Census Bureau should be allowed to adjust its "official" count on the basis of statistical surveys is very controversial. The Constitution calls for an "actual enumeration" of the population (Article 1, Section 2, Subsection 2). Do you believe that this wording precludes or allows the use of statistical surveys in the official count? Defend your opinion. Also discuss reasons why Democrats tend to favor the use of sampling methods while Republicans tend to oppose it.

Size of Errors: Absolute versus Relative

Besides wanting to know whether an error is random or systematic, we often want to know something about the overall size of the error. For example, is the error big enough to be of concern or small enough to be unimportant? Let's explore the concept of error size.

Suppose you go to the store and buy what you think is 6 pounds of hamburger, but because the store's scale is poorly calibrated you actually get only 4 pounds. You'd probably be upset by this 2-pound error. Now suppose you are buying hamburger for a huge town barbeque and you order 3,000 pounds of hamburger, but you actually receive only 2,998 pounds. You are short by the same 2 pounds as before, but in this case the error probably doesn't seem very important. This simple example shows that the size of an error can depend on how you look at it.

In more technical language, the 2-pound error in both cases is an **absolute error**—it describes how far the claimed or measured value lies from the true value. A **relative error** compares the size of the absolute error to the true value. The relative error for the first case is fairly large because the absolute error of 2 pounds is half the true weight of 4 pounds: We therefore say that the relative error is 2/4, or 50%. In contrast, the relative error for second case is quite small: It is the absolute error of 2 pounds divided by the true hamburger weight of 2,998 pounds, which means a relative error of only $2/2998 \approx 0.00067$, or 0.067%.

Absolute and Relative Error

The **absolute error** describes how far a claimed or measured value lies from the true value:

$$\text{absolute error} = \text{claimed or measured value} - \text{true value}$$

The **relative error** compares the size of the absolute error to the true value. It is often expressed as a percentage:

$$\text{relative error} = \frac{\text{absolute error}}{\text{true value}} \times 100\%$$

Note that the absolute and relative errors are *positive* when the claimed or measured value is greater than the true value, and *negative* when the claimed or measured value is less than the true value.

EXAMPLE 2 Absolute and Relative Error

Find the absolute and relative error in each case.

a. Your true weight is 100 pounds, but a scale says you weigh 105 pounds.

b. The government claims that a program costs \$99.0 billion, but an audit shows that the true cost is \$100.0 billion.

Solution

a. The measured value is the scale reading of 105 pounds and the true value is 100 pounds:

$$\begin{aligned}\text{absolute error} &= \text{measured value} - \text{true value}\\ &= 105\text{ lb} - 100\text{ lb}\\ &= 5\text{ lb}\end{aligned}$$

$$\begin{aligned}\text{relative error} &= \frac{\text{absolute error}}{\text{true value}} \times 100\%\\ &= \frac{5\text{ lb}}{100\text{ lb}} \times 100\%\\ &= 5\%\end{aligned}$$

The measured weight is too high by 5 pounds, or by 5%.

A billion here, a billion there; soon you're talking real money.

—Senator Everett Dirksen

b. The claimed cost is \$99.0 billion and the true cost is \$100.0 billion:

$$\begin{aligned}\text{absolute error} &= \text{claimed value} - \text{true value}\\ &= \$99.0\text{ billion} - \$100.0\text{ billion}\\ &= -\$1.0\text{ billion}\end{aligned}$$

$$\begin{aligned}\text{relative error} &= \frac{\text{absolute error}}{\text{true value}} \times 100\%\\ &= \frac{-\$1.0\text{ billion}}{\$100.0\text{ billion}} \times 100\%\\ &= -1\%\end{aligned}$$

The claimed cost is too low by \$1.0 billion, or by 1%.

Describing Results: Accuracy and Precision

Once a measurement is reported, we should evaluate it to see whether it is believable in light of any potential errors. In particular, we should consider two key ideas about any reported value: its *accuracy* and its *precision*. These terms are often used interchangeably in English, but in science they refer to two different concepts.

The goal of any measurement is to obtain a value that is as close as possible to the *true value*. **Accuracy** describes how close the measured value lies to the true value. **Precision** describes the amount of detail in the measurement. For example, suppose a census says that the population of your hometown is 72,453 but the true population is 96,000. The census value of 72,453 is quite precise because it seems to tell us the exact count, but it is not very accurate because it is nearly 25% smaller than the actual population of 96,000. Note that accuracy is usually defined by relative error rather than absolute error. For example, if a company projects sales of $7.30 billion and true sales turn out to be $7.32 billion, we say the projection was quite accurate because it was off by less than 1%, even though the error of $0.02 billion represents $20 million.

Definitions

Accuracy describes how closely a measurement approximates a true value. An accurate measurement is close to the true value. (*Close* is generally defined as a small *relative* error, rather than a small absolute error.)

Precision describes the amount of detail in a measurement.

By the Way ...

In 1999, NASA lost the $160 million Mars Climate Orbiter when engineers sent it very precise computer instructions in English units (pounds) but the spacecraft software interpreted them in metric units (kilograms). In other words, the loss occurred because the very precise instructions were actually quite inaccurate! NASA learned its lesson and has since sent four spacecraft to Mars successfully, with another (called *Phoenix*) scheduled to land on Mars in 2008.

We generally assume that the precision with which a number is reported reflects what was actually measured. If you say that you weigh 132 pounds, we assume that you measured your weight only to the nearest pound. In that case, a more precise measurement might find that you weighed, say, 132.3 pounds or 131.6 pounds (both of these would round to 132). In contrast, if you say that you weigh 132.0 pounds, we assume that you measured your weight to the nearest tenth of a pound. This assumption about precision means that numbers reported with unjustified precision are dishonest. For example, if you actually measured your weight only to the nearest pound, it would be dishonest to say that you weighed 132.0 pounds because that would imply you made a measurement to the nearest tenth of a pound.

EXAMPLE 3 Accuracy and Precision in Your Weight

Suppose that your true weight is 102.4 pounds. The scale at the doctor's office, which can be read only to the nearest quarter pound, says that you weigh $102\frac{1}{4}$ pounds. The scale at the gym, which gives a digital readout to the nearest 0.1 pound, says that you weigh 100.7 pounds. Which scale is more *precise*? Which is more *accurate*?

Solution The scale at the gym is more *precise* because it gives your weight to the nearest tenth of a pound, whereas the doctor's scale gives your weight only to the nearest quarter pound. However, the scale at the doctor's office is more *accurate* because its value is closer to your true weight.

TIME OUT TO THINK

In Example 3, we need to know your true weight to determine which scale is more accurate. But how would you know your true weight? Can you ever be sure that you know it? Explain.

CASE STUDY Does the Census Measure the True Population?

Upon completing the 2000 census, the U.S. Census Bureau reported a population of 281,421,906 (on April 1, 2000), thereby implying an exact count of everyone living in the United States. Unfortunately, such a precise count could not possibly be as accurate as it seems to imply.

Even in principle, the only way to get an exact count of the number of people living in the United States would be to count everyone *instantaneously*. Otherwise, the count would be off because the number of people changes very rapidly. An average of about eight births and four deaths occur every minute in the United States, and a new immigrant enters the United States about every three minutes on average. In fact, the Census Bureau spends many months collecting the data that lead to its count. Moreover, random and systematic errors can easily make the census inaccurate by a few million people (see the Case Study on p. 61).

In reality, then, the Census Bureau could not possibly know the exact population on a particular day, and the uncertainty in the actual population is at least a few million people. Given these facts, it is dishonest to report the population as an exact count. A more honest report would use much less precision—for example, stating the population as "about 280 million." In fairness to the Census Bureau, their detailed reports explained the uncertainties in the population count, but these uncertainties were rarely mentioned by the press.

By the Way ...

The digits in a number that were actually measured are called *significant digits*. All the digits in a number are significant except for zeros that are included so that the decimal point can be properly located. For example, 0.001234 has four significant digits; the zeros are required for proper placement of the decimal point. The number 1,234,000,000 has four significant digits for the same reason. The number 132.0 has four significant digits; the zero is significant because it is not required for the proper placement of the decimal point and therefore implies an actual measurement of zero tenths.

CASE STUDY Trillions Disappear in Unjustified Budget Precision

In early 2001, politicians heralded a projected U.S. federal budget *surplus* of \$5.6 trillion over the next 10 years. By 2007, rather than having gained trillions in surplus money, the government had instead gone another \$3 trillion into debt. In other words, with four years left to go in the 10-year projection, the projection had already proven to be in error by nearly \$9 trillion (going from *positive* \$5.6 trillion to *negative* \$3 trillion).

You've probably heard or read news reports about this multi-trillion-dollar turn in government fortunes, and people of different political persuasions may feel quite differently about its significance. But for our purposes here, the most remarkable thing about it was the unjustified precision of the original projection. After all, the projection of \$5.6 trillion was quoted precise to the nearest *tenth* of a trillion dollars (which means to the nearest hundred billion dollars), implying that the actual value would be between about \$5.5 and \$5.7 trillion. Now that's what you might call an "Ooops!"

In fairness to the economists responsible for the projection, they were well aware of the associated uncertainties. Moreover, changes that occurred after the projection was made—including the terrorist attacks of 9/11, wars in Afghanistan and Iraq, and substantial cuts in taxes—make it unsurprising that the projection turned out so far off base. But the lesson is clear: Next time you hear politicians or the media talking about future federal budgets with great precision, remember it is quite likely that their numbers will prove to be abysmally inaccurate.

Summary: Dealing with Errors

The ideas we've covered in this section are a bit technical, but very important to understanding measurements and errors. Let's briefly summarize how the ideas relate to one another.

- Errors can occur in many ways, but generally can be classified into one of two basic types: random errors or systematic errors.
- Whatever the source of an error, its size can be described in two different ways: as an absolute error or as a relative error.
- Once a measurement is reported, we can evaluate it in terms of its accuracy and its precision.

Section 2.2 Exercises

Statistical Literacy and Critical Thinking

1. **Error Type.** When recording driver reaction times in an experiment, a technician incorrectly records one of the actual times. Is this a random error or a systematic error? Explain.

2. **Standard Weight.** A standard weight defined to represent exactly 1 kilogram is kept by the National Institute of Standards and Technology. If you put this 1-kilogram true weight on a scale and the scale says it weighs 1.002 kilograms, what are the absolute error and the relative error of the measurement?

3. **Height Measurement.** Students in a statistics class measure the height of their instructor and record it as 174.0123668 centimeters. The instructor's true height is closer to 176 centimeters. Describe the accuracy and precision of the recorded height. Explain.

4. **Population Milestone.** The media reported that the United States population reached 300 million at 7:46 a.m. (Eastern time) on October 17, 2006. Describe the accuracy and precision of that reported time.

Does It Make Sense? For Exercises 5–8, decide whether the statement makes sense (or is clearly true) or does not make sense (or is clearly false). Explain clearly; not all of these statements have definitive answers, so your explanation is more important than your chosen answer.

5. **Species of Fish.** There are 24,627 species of fish on Earth.

6. **Relative Error.** The relative error that a microbiologist makes in measuring a cell must be less than the relative error that an astronomer makes in measuring a galaxy, because cells are smaller than galaxies.

7. **Scanner Error.** The Jenkins supermarket manager claims that the scanning errors on purchased items are random, and about half of the errors are in favor of the supermarket.

8. **Car Mileage.** When a car is purchased, the mileage is reported as 22.3655 miles per gallon. We can have much confidence in that value because it is very precise.

Concepts and Applications

9. **Tax Audit.** A tax auditor reviewing a tax return looks for several kinds of problems, including these two: (1) mistakes made in entering or calculating numbers on the tax return and (2) places where the taxpayer reported income dishonestly. Discuss whether each problem involves random or systematic errors.

10. **Safe Air Travel.** Before taking off, a pilot is supposed to set the aircraft altimeter to the elevation of the airport. A pilot leaves from Denver (altitude 5,280 feet) with her altimeter set to 2,500 feet. Explain how this affects the altimeter readings throughout the flight. What kind of error is this?

11. **Technical Specifications.** The Newport Manufacturing Company receives an order for 10,000 sheet metal screws that are supposed to be 25 millimeters long. After the screws are manufactured, a quality control analyst makes careful measurements and finds that the mean length of the screws is 25 millimeters, but about half of the screws are longer than 25 millimeters and half are shorter than 25 millimeters. What type of measurement error are we dealing with here? Explain.

12. **Suicide Data.** Concerning the collection of data on suicides, the *New York Times Almanac* claims, "Most experts believe that suicide statistics are grimmer than reported. They contend that numerous suicides are categorized as accidents or other deaths to spare families." If this is the case, what kind of error is introduced in suicide data and how does it affect the values of the data?

Sources of Errors. For each measurement described in Exercises 13–20, identify at least one likely source of random errors and also identify at least one likely source of systematic errors.

13. **Survey.** The annual incomes of 200 people surveyed by telephone

14. **Tax Returns.** The annual incomes of 200 people are obtained from their tax returns.

15. **Passenger Weights.** The Federal Aviation Administration required airlines to actually weigh a sample of passengers, and those passengers were weighed with their clothes and carry-on baggage.

16. **M&Ms.** The weights of individual M&M plain candies were obtained by placing each candy in a paper cup and then obtaining the weight without accounting for the weight of the cup.

17. **Radar Speeds.** Speeds of cars are recorded by a police officer who uses a radar gun.

18. **Counterfeit Products.** The police commissioner in New York City estimates the annual value of counterfeit goods sold in the city.

19. Cigarette Sales. The health commissioner of California estimates the number of cigarettes smoked in her state by finding the number of packs of cigarettes sold based on the tax stamps put on each pack.

20. Measuring Length. A groundskeeper measures the length and width of a football field by using a ruler that is one foot long.

Absolute and Relative Errors. In Exercises 21–24, find the values of the absolute and relative errors.

21. Credit Card Bill. One of the authors received a Visa credit card bill for $2,995, but it included a charge of $1,750 that was not valid.

22. Steak Weight. A student of one of the authors weighed a steak and found it to be 18 ounces, but the menu stated that it was 20 ounces.

23. Car Mileage. The sticker on a car states that it gets 22 miles per gallon on the highway, but measurements show that its highway mileage is actually 24 miles per gallon.

24. Baker's Dozen. The bakery menu shows that there are 12 doughnuts in a bag, but the baker routinely puts 13 doughnuts in each bag.

25. Minimizing Errors. Twenty-five people, including yourself, are to measure the length of a room to the nearest tenth of a millimeter. Assume that everyone uses the same well-calibrated measuring device, such as a tape measure.

a. All 25 measurements are not likely to be exactly the same; thus, the measurements will contain some sources of error. Are these errors systematic or random? Explain.

b. If you want to minimize the effect of random errors in determining the length of the room, which is the better choice: to report your own personal measurement as the length of the room or to report the average of all 25 measurements? Explain.

c. Describe any possible sources of systematic errors in the measurement of the room length.

d. Can the process of averaging all 25 measurements help reduce any systematic errors? Why or why not?

26. Minimizing Errors. The measuring instrument for weighing a model 22F car battery is very precise, and the weight is obtained 10 consecutive times.

a. All 10 measurements are not likely to be exactly the same; thus, the measurements will contain some errors. Are these errors systematic or random? Explain.

b. If you want to minimize the effect of random errors in determining the true weight of the battery, which is the better choice: to choose one of the 10 measurements at random or to report the average of all 10 measurements? Explain.

c. Describe any possible sources of systematic errors in the 10 measurements.

d. Can the process of averaging all 10 weights help reduce any systematic errors? Why or why not?

27. Accuracy and Precision in Weight of a Corvette. A new Corvette weights 3,273 pounds. A manufacturer's scale that is accurate to the nearest 10 pounds gives the weight as 3,250 pounds, while the U.S. Department of Transportation uses a scale that is accurate to the nearest 0.1 pound and obtains a weight of 3,298.2 pounds. Which measurement is more *precise*? Which is more *accurate*? Explain.

28. Accuracy and Precision in Height. Assume that President George W. Bush is exactly 71 inches tall. Assume that his height is measured with a tape measure that can be read to the nearest $\frac{1}{8}$ inch and the result is found to be $71\frac{1}{8}$ inches. With a new laser device at the doctor's office that gives readings to the nearest 0.05 inch, his height is found to be 71.15 inches. Which measurement is more *precise*? Which is more *accurate*? Explain.

29. Accuracy and Precision in Weight. Suppose your weight is 52.55 kilograms. A scale at a health clinic that gives weight measurements to the nearest half kilogram gives your weight as 53 kilograms. A digital scale at the gym that gives readings to the nearest 0.01 kilogram gives your weight as 52.88 kilograms. Which measurement is more *precise*? Which is more *accurate*? Explain.

30. Accuracy and Precision in Weight. Suppose your weight is 52.55 kilograms. A scale at a health clinic that gives weight measurements to the nearest half kilogram gives your weight as $52\frac{1}{2}$ kilograms. A digital scale at the gym that gives readings to the nearest 0.01 kilogram gives your weight as 51.48 kilograms. Which measurement is more *precise*? Which is more *accurate*? Explain.

Believable Facts? Exercises 31–38 give statements of "fact" coming from statistical measurements. For each statement, briefly discuss possible sources of error in the measurement. Then, in light of the precision with which the measurement is given, discuss whether you think the fact is believable.

31. Population. The population of the United States in 1860 was 31,443,321.

32. Motor Vehicle Deaths. Last year, there were 44,758 deaths in the United States due to motor vehicle crashes.

33. Population of China. Last year, the population of China was 1,372,557,236 people.

34. Tallest Building. The Tapei 101 building is 1,671 feet tall, making it the world's tallest building.

35. Gateway Arch. The St. Louis Gateway Arch is 630.2377599694 feet tall.

36. Cell Phones. InfoWorld reported that the number of cell phone users is currently 210 million.

37. College Students. There are currently 18,000,000 college students in the United States.

38. Threatened Species. According to the U.S. government, there are now 1,879 endangered or threatened species of animals and plants.

Projects for the Internet and Beyond

For useful links, select "Links for Internet Projects" for Chapter 2 at www.aw.com/bbt.

39. The 2010 Census. Go to the Web site for the U.S. Census Bureau and learn about plans for the 2010 census. How and when will data be collected? Are any significant changes in the collection process planned (compared to that used in 2000)? Overall, does it seem likely that the 2010 census will be any more accurate than the 2000 census? Defend your opinion.

40. Census Controversies. Use the Library of Congress's "Thomas" Web site to find out about any pending legislation concerning the collection or use of census data. If you find more than one legislative bill pending, choose one to study in depth. Summarize the proposed legislation, and briefly discuss arguments both for and against it.

41. Wristwatch Errors. Use a Web site that gives you the local time (such as www.time.gov) to set a watch to the nearest second. Then compare the time on your watch to the times on friends' watches. Record the errors with positive signs for watches that are ahead of the true time and negative signs for watches that are behind the true time. Use the concepts of this section to describe the accuracy of the wristwatches in your sample.

IN THE NEWS

42. Random and Systematic Errors. Find a recent news report that gives a quantity that was measured statistically (for example, a report of population, average income, or number of homeless people). Write a short description of how the quantity was measured, and briefly describe any likely sources of either random or systematic errors. Overall, do you think that the reported measurement was accurate? Why or why not?

43. Absolute and Relative Errors. Find a recent news report that describes some mistake in a measured, estimated, or projected number (for example, a budget projection that turned out to be incorrect). In words, describe the size of the error in terms of both absolute error and relative error.

44. Accuracy and Precision. Find a recent news article that causes you to question accuracy or precision. For example, the article might report a figure with more precision than you think is justified, or it might cite a figure that you know is inaccurate. Write a one-page summary of the report and explain why you question its accuracy or precision (or both).

2.3 Uses of Percentages in Statistics

Statistical results are often stated with percentages. A percentage is simply a way of expressing a fraction; the words *per cent* literally mean "divided by 100." However, percentages are often used in subtle ways. Consider a statement that appeared in a front-page article in the *New York Times*:

> *The rate [of smoking] among 10th graders jumped 45 percent, to 18.3 percent, and the rate for 8th graders is up 44 percent, to 10.4 percent.*

All the percentages in this statement are used correctly, but they are very confusing. For example, what does it mean to say "up 44%, to 10.4%"? In this section, we will investigate some of the subtle uses and abuses of percentages. Before we begin, you should review the following basic rules regarding conversions between fractions and percentages.

Conversions Between Fractions and Percentages

To convert a percentage to a common fraction: Replace the % symbol with division by 100; simplify the fraction if necessary.

Example: $25\% = \frac{25}{100} = \frac{1}{4}$

To convert a percentage to a decimal: Drop the % symbol and move the decimal point two places to the left (that is, divide by 100).

Example: $25\% = 0.25$

To convert a decimal to a percentage: Move the decimal point two places to the right (that is, multiply by 100) and add the % symbol.

Example: $0.43 = 43\%$

To convert a common fraction to a percentage: First convert the common fraction to a decimal; then convert the decimal to a percentage.

Example: $\frac{1}{5} = 0.2 = 20\%$

EXAMPLE 1 Newspaper Survey

A newspaper reports that 44% of 1,069 people surveyed said that the President is doing a good job. How many people said that the President is doing a good job?

Solution The 44% represents the fraction of respondents who said the President is doing a good job. Because "of" usually indicates multiplication, we multiply:

$$44\% \times 1{,}069 = 0.44 \times 1{,}069 = 470.36 \approx 470$$

About 470 out of the 1,069 people said the President is doing a good job. Note that we round the answer to 470 to obtain a whole number of people. (The symbol ≈ means "approximately equal to.")

Using Percentages to Describe Change

In statistics, percentages are often used to describe how data change with time. As an example, suppose the population of a town was 10,000 in 1970 and 15,000 in 2000. We can express the change in population in two basic ways:

- Because the population rose by 5,000 people (from 10,000 to 15,000), we say that the **absolute change** in the population was 5,000 people.
- Because the increase of 5,000 people was 50% of the starting population of 10,000, we say that the **relative change** in the population was 50%.

In general, calculating an absolute or relative change always involves two numbers: a starting number, or **reference value**, and a **new value**. Once we identify these two values, we can calculate the absolute and relative change with the following formulas. Note that the absolute and relative changes are positive if the new value is greater than the reference value and negative if the new value is less than the reference value.

Absolute and Relative Change

The **absolute change** describes the actual increase or decrease from a reference value to a new value:

$$\text{absolute change} = \text{new value} - \text{reference value}$$

The **relative change** describes the size of the absolute change in comparison to the reference value and can be expressed as a percentage:

$$\text{relative change} = \frac{\text{new value} - \text{reference value}}{\text{reference value}} \times 100\%$$

TIME OUT TO THINK

Compare the formulas for absolute and relative change to the formulas for absolute and relative error, given in Section 2.2. Briefly describe the similarities you notice.

EXAMPLE 2 World Population Growth

World population in 1950 was 2.6 billion. By the beginning of 2000, it had reached 6.0 billion. Describe the absolute and relative change in world population from 1950 to 2000.

Solution The reference value is the 1950 population of 2.6 billion and the new value is the 2000 population of 6.0 billion.

$$\begin{aligned} \text{absolute change} &= \text{new value} - \text{reference value} \\ &= 6.0 \text{ billion} - 2.6 \text{ billion} \\ &= 3.4 \text{ billion} \end{aligned}$$

$$\begin{aligned} \text{relative change} &= \frac{\text{new value} - \text{reference value}}{\text{reference value}} \times 100\% \\ &= \frac{6.0 \text{ billion} - 2.6 \text{ billion}}{2.6 \text{ billion}} \times 100\% \\ &= 130.7\% \end{aligned}$$

World population increased by 3.4 billion people, or by about 131%, from 1950 to 2000.

By the Way ...

According to United Nations and U.S. Census Bureau estimates, world population passed 6 billion in late 1999—only 12 years after passing the 5-billion mark. World population reached 6.5 billion in early 2006 and will probably reach the 7-billion mark by about 2012.

Using Percentages for Comparisons

Percentages are also commonly used to compare two numbers. In this case, the two numbers are the reference value and the compared value.

- The **reference value** is the number that we are using as the basis for a comparison.
- The **compared value** is the other number, which we compare to the reference value.

We can then express the absolute or relative difference between these two values with formulas very similar to those for absolute and relative change. Note that the absolute and relative differences are positive if the compared value is greater than the reference value and negative if the compared value is less than the reference value.

Absolute and Relative Difference

The **absolute difference** is the difference between the compared value and the reference value:

$$\text{absolute difference} = \text{compared value} - \text{reference value}$$

The **relative difference** describes the size of the absolute difference in comparison to the reference value and can be expressed as a percentage:

$$\text{relative difference} = \frac{\text{compared value} - \text{reference value}}{\text{reference value}} \times 100\%$$

By the Way ...

No one knows all the reasons for the low life expectancy of Russian men, but one contributing factor certainly is alcoholism, which is much more common in Russia than in America.

EXAMPLE 3 Russian and American Life Expectancy

Life expectancy for American men is about 75 years, while life expectancy for Russian men is about 59 years. Compare the life expectancy of American men to that of Russian men in absolute and relative terms. (See Section 6.4 for a discussion of the meaning of life expectancy.)

Solution We are told to compare the American male life expectancy to the Russian male life expectancy, which means that we use the Russian male life expectancy as the reference value and the American male life expectancy as the compared value:

$$\begin{aligned}\text{absolute difference} &= \text{compared value} - \text{reference value}\\ &= 75 \text{ years} - 59 \text{ years}\\ &= 16 \text{ years}\end{aligned}$$

$$\begin{aligned}\text{relative difference} &= \frac{\text{compared value} - \text{reference value}}{\text{reference value}} \times 100\%\\ &= \frac{75 \text{ years} - 59 \text{ years}}{59 \text{ years}} \times 100\%\\ &= 27\%\end{aligned}$$

The life expectancy of American men is 16 years greater in absolute terms and 27% greater in relative terms than the life expectancy of Russian men.

Of versus *More Than*

A subtlety in dealing with percentage statements comes from the way they are worded. Consider a population that *triples* in size from 200 to 600. There are two equivalent ways to state this change with percentages:

- Using *more than*: The new population is 200% *more than* the original population. Here we are looking at the relative change in the population:

$$\begin{aligned}\text{relative change} &= \frac{\text{new value} - \text{reference value}}{\text{reference value}} \times 100\%\\ &= \frac{600 - 200}{200} \times 100\%\\ &= 200\%\end{aligned}$$

- Using *of*: The new population is 300% *of* the original population, which means it is three times the original population. Here we are looking at the *ratio* of the new population to the original population:

$$\frac{\text{new population}}{\text{original population}} = \frac{600}{200} = 3.00 = 300\%$$

Notice that the percentages in the "more than" and "of" statements are related by 300% = 100% + 200%. This leads to the following general relationship.

Of versus *More Than* (or *Less Than*)

- If the new or compared value is *P% more than* the **reference value,** then it is (100 + *P*)% *of* the reference value.
- If the new or compared value is *P% less than* the reference value, then it is (100 – *P*)% *of* the reference value.

For example, 40% *more than* the reference value is 140% *of* the reference value, and 40% *less than* the reference value is 60% *of* the reference value. When you hear statistics quoted with percentages, it is very important to listen carefully for the key words *of* and *more than* (or *less than*)—and hope that the speaker knows the difference.

EXAMPLE 4 World Population

In Example 2, we found that world population in 2000 was about 131% more than world population in 1950. Express this change with an "of" statement.

Solution World population in 2000 was 131% more than world population in 1950. Because (100 + 131)% = 231%, the 2000 population was 231% *of* the 1950 population. This means that the 2000 population was 2.31 times the 1950 population.

By the Way ...

World population is currently growing by about 75 million people each year, which means that in just four years the world adds roughly as many people as the entire population of the United States.

EXAMPLE 5 Sale!

A store is having a "25% off" sale. How does a sale price compare to an original price?

Solution The "25% off" means that a sale price is 25% *less than* the original price, which means it is (100 – 25)% = 75% *of* the original price. For example, if an item's original price was $100, its sale price is $75.

TIME OUT TO THINK

One store advertises "1/3 off everything!" Another store advertises "Sale prices just 1/3 of original prices!" Which store is having the bigger sale? Explain.

Percentages of Percentages

Percentage changes and percentage differences can be particularly confusing when the values *themselves* are percentages. Suppose your bank increases the interest rate on your savings account from 3% to 4%. It's tempting to say that the interest rate increases by 1%, but that statement is ambiguous at best. The interest rate increases by 1 *percentage point*, but the relative change in the interest rate is 33%:

$$\frac{4\% - 3\%}{3\%} \times 100\% = 0.33 \times 100\% = 33\%$$

You can therefore say that the bank raised your interest rate by 33%, even though the actual rate increased by only 1 percentage point (from 3% to 4%).

Percentage Points versus %

When you see a change or difference expressed in *percentage points*, you can assume it is an *absolute* change or difference. If it is expressed as a percentage, it probably is a *relative* change or difference.

EXAMPLE 6 Margin of Error

Based on interviews with a sample of students at your school, you conclude that the percentage of all students who are vegetarians is probably between 20% and 30%. Should you report your result as "25% with a margin of error of 5%" or as "25% with a margin of error of 5 percentage points"? Explain. (See Section 1.1 to review the meaning of margin of error.)

Solution The range of 20% to 30% comes from subtracting and adding an *absolute difference* of 5 percentage points to 25%. That is,

$$20\% = (25 - 5)\% \quad \text{and} \quad 30\% = (25 + 5)\%$$

Therefore, the correct statement is "25% with a margin of error of 5 percentage points." If you instead said "25% with a margin of error of 5%," you would imply that the error was 5% of 25%, which is only 1.25%.

EXAMPLE 7 Care in Wording

If you can't convince them, confuse them.

—Harry S Truman

Assume that 40% of the registered voters in Carson City are Republicans. Read the following questions carefully and give the most appropriate answers.

a. The percentage of voters registered as Republicans is 25% higher in Freetown than in Carson City. What percentage of the registered voters in Freetown are Republicans?

b. The percentage of voters registered as Republicans is 25 percentage points higher in Freetown than in Carson City. What percentage of the registered voters in Freetown are Republicans?

Solution

a. We interpret the "25%" as a relative difference, and 25% of 40% is 10% (because $0.25 \times 0.40 = 0.10$). Therefore, the percentage of registered Republicans in Freetown is $40\% + 10\% = 50\%$.

b. In this case, we interpret the "25 percentage points" as an absolute difference, so we simply add this value to the percentage of Republicans in Carson City. Therefore, the percentage of registered Republicans in Freetown is $40\% + 25\% = 65\%$.

Section 2.3 Exercises

Statistical Literacy and Critical Thinking

1. **Percentage Interpretation.** Consider the quote from the beginning of this section: "The rate [of smoking] among 10th graders jumped 45 percent, to 18.3 percent, and the rate for 8th graders is up 44 percent, to 10.4 percent." Briefly explain the meaning of each of the percentages in this statement.

2. **Percentage.** A *New York Times* editorial criticized a chart caption that described a dental rinse as one that "reduces plaque on teeth by over 300%." If the dental rinse removes all of the plaque, what percentage is removed? Is it possible to reduce plaque by over 300%?

3. **Percentage Points.** One Gallup poll involved 1,236 adults, 5% of whom believed that bad luck follows breaking a mirror. That percentage has a margin of error of 1.2 *percentage points*. Why is it misleading to state that the percentage is 5% with a margin of error of 1.2%?

4. ***Of* and *More Than*.** A poll involves 1,400 respondents. What is 5% of 1400? What is 5% more than 1400?

Does It Make Sense? For Exercises 5–8, decide whether the statement makes sense (or is clearly true) or does not make sense (or is clearly false). Explain clearly; not all of these statements have definitive answers, so your explanation is more important than your chosen answer.

5. **Cell Phones.** The percentage of people with cell phones increased by 1.2 million people.

6. **Percentages.** The CEO of Telektronics has an annual salary that is 50% more than that of the CFO, so the salary of the CFO is 50% less than that of the CEO.

7. **Interest Rate.** The Jefferson Valley Bank increased its new-car loan rate by 100%.

8. **Interest Rate.** The Jefferson Valley Bank increased its new-car loan rate (annual) by 100 percentage points.

Concepts and Applications

9. **Fractions, Decimals, Percentages.** Express each of the following numbers in all three forms: fraction, decimal, and percentage.

   a. 2/5 **b.** 1.50 **c.** 0.25 **d.** 30%

10. **Fractions, Decimals, Percentages.** Express the following numbers in all three forms: fraction, decimal, and percentage.

   a. 225% **b.** 0.375 **c.** 1/20 **d.** −0.12

11. **Percentage Practice.** A study was conducted of pleas made by 1,028 criminals. Among those criminals, 956 pled guilty, and 392 of them were sentenced to prison. Among the 72 other criminals, who pled not guilty, 58 were sent to prison (based on data from "Does It Pay to Plead Guilty?" by Brereton and Casper, *Law and Society Review*, Vol. 16, No. 1).

   a. What percentage of the criminals pled guilty?

   b. What percentage of the criminals were sent to prison?

   c. Among those who pled guilty, what is the percentage who were sent to prison?

   d. Among those who pled not guilty, what is the percentage who were sent to prison?

12. **Percentage Practice.** A study was conducted to determine whether flipping a penny or spinning a penny has an effect on the proportion of heads. Among 49,437 trials, 29,015 involved flipping pennies, and 14,709 of those pennies turned up heads. The other 20,422 trials involved spinning pennies, and 9,197 of those pennies turned up heads (based on data from Robin Lock as given in *Chance News*).

   a. What percentage of the trials involved flipping pennies?

   b. What percentage of the trials involved spinning pennies?

   c. Among the pennies that were flipped, what is the percentage that turned up heads?

   d. Among the pennies that were spun, what is the percentage that turned up heads?

Relative Change. Exercises 13–20 each provide two values. For each pair of values, use a percentage to express the relative change or difference. Use the second given value as the reference value, and express results to the nearest percentage point. Also, write a statement describing the result.

13. **Newspapers.** The number of daily newspapers in the United States is now 1,456, and it was 2,226 in 1900.

14. **Cars.** There are now 136,651,000 registered passenger cars, and in 1980 there were 121,601,000.

15. **Aircraft.** There are now 8,206 commercial aircraft in the United States, and in 1980 there were 3,808.

16. **Bankruptcies.** There were 1,725,300 bankruptcy cases filed last year, and in the year 2000 there were 1,276,900 bankruptcy cases filed.

17. **Newspapers.** The daily circulation of the *Wall Street Journal* is about 1.75 million (the largest in the country). The daily circulation of the *New York Times* is about 1.09 million (the third largest in the country).

18. Car Sales. In a recent year, 13,525 Chrysler 300C cars were sold and 45,782 Honda hybrid cars were sold.

19. Airports. Chicago's O'Hare Airport handled 34 million passengers last year. The busiest airport in the world, Atlanta's Hartsfield Airport, handled 42 million passengers last year.

20. Tourists. In a recent year, France ranked as the number one tourist destination, with 75.1 million international arrivals. The United States ranked third, with 46.1 million international arrivals.

Surveys. Some important analyses of survey results require that you know the actual number of subjects whose responses fall into a particular category. In Exercises 21–24, find the actual number of respondents corresponding to the given percentage.

21. Personal Calls. In an At-A-Glance survey of 1,385 office workers, 4.8% said that they do not make personal phone calls.

22. Interview Mistakes. In an Accountemps survey of 150 executives, 47% said that the most common interview mistake is to have little or no knowledge of the company.

23. Televisions. In a Frank N. Magid Associates survey of 1,005 adults, 83% reported having more than one television at home.

24. Cell Phones. In a Frank N. Magid Associates survey of 1,109 consumers over the age of 11, 81% reported having at least one cell phone in their household.

***Of* vs. *More Than*.** Fill in the blanks in Exercises 25–28. Briefly explain your reasoning in each case.

25. Weights. If a truck weighs 40% more than a car, then the truck's weight is _____% of the car's weight.

26. Areas. If the area of Norway is 24% more than the area of Colorado, then Norway's area is _____% of Colorado's area.

27. Population. If the population of Montana is 20% less than the population of New Hampshire, then Montana's population is _____% of New Hampshire's population.

28. Salary. Jay Leno's salary is currently 18% less than David Letterman's salary, so Jay Leno's salary is _____% of David Letterman's salary.

29. Margin of Error. A Gallup poll of 1,012 adults showed that 89% of Americans say that human cloning should not be allowed. The margin of error was 3 percentage points. Would it matter if a newspaper reported the margin of error as 3%? Explain.

30. Margin of Error. A Pew Research Center survey of 3,002 adults showed that the percentage who listen to National Public Radio is probably between 14% and 18%. How should a newspaper report the margin of error? Explain.

Percentages of Percentages. Exercises 31–34 describe changes in which the measurements themselves are percentages. Express each change in two ways: (1) as an absolute difference in terms of percentage points and (2) as a relative difference in terms of percent.

31. The percentage of high school seniors using alcohol decreased from 68.2% in 1975 to 52.7% now.

32. The percentage of the world's population living in developed countries decreased from 27.1% in 1970 to 19.5% now.

33. The five-year survival rate for Caucasians for all forms of cancer increased from 39% in the 1960s to 61% now.

34. The five-year survival rate for African Americans for all forms of cancer increased from 27% in the 1960s to 48% now.

Projects for the Internet and Beyond

For useful links, select "Links for Internet Projects" for Chapter 2 at www.aw.com/bbt.

35. World Population. Find the current estimate of world population on the U.S. Census Bureau's world population clock. Describe the percentage change in population since the 6-billion mark was passed during 1999. Also find how the population clock estimates are made, and discuss the uncertainties in estimating world population.

36. Drug Use Statistics. Go to the Web site for the National Center on Addiction and Substance Abuse (CASA) and find a recent report giving statistics on substance abuse. Write a one-page summary of the new research, giving at least some of the conclusions in terms of percentages.

IN THE NEWS

37. Percentages. Find three recent news reports in which percentages are used to describe statistical results. In each case, describe the meaning of the percentage.

38. Percentage Change. Find a recent news report in which percentages are used to express the change in a statistical result from one time to another (such as an increase in population or in the number of children who smoke). Describe the meaning of the change. Be sure to watch for key words such as *of* or *more than*.

2.4 Index Numbers

If you listen to the nightly business report, you've probably heard about **index numbers**, such as the Consumer Price Index, the Producer Price Index, or the Consumer Confidence Index. Index numbers are very common in statistics because they provide a simple way to compare measurements made at different times or in different places. In this section, we'll investigate the meaning and use of index numbers, focusing on the Consumer Price Index (CPI). Let's start with an example using gasoline prices.

Table 2.1 shows the average price of gasoline in the United States for selected years from 1955 to 2005. (These are real prices from those years; that is, they have *not* been adjusted for inflation.) Suppose that, instead of the prices themselves, we want to know how the price of gasoline in different years compares to the 1975 price. One way to compare the prices would be to express each year's price as a percentage of the 1975 price. For example, by dividing the 1965 price by the 1975 price, we find that the 1965 price was 55.0% of the 1975 price:

$$\frac{1965 \text{ price}}{1975 \text{ price}} = \frac{31.2¢}{56.7¢} = 0.550 = 55.0\%$$

Proceeding similarly for each of the other years, we can calculate all the prices as percentages of the 1975 price. The third column of Table 2.1 shows the results. Note that the percentage for 1975 is 100%, because we chose the 1975 price as the reference value.

Table 2.1 Average Gasoline Prices (per gallon)

Year	Price	Price as a percentage of 1975 price	Price index (1975 = 100)
1955	29.1¢	51.3%	51.3
1965	31.2¢	55.0%	55.0
1975	56.7¢	100.0%	100.0
1985	119.6¢	210.9%	210.9
1995	120.5¢	212.5%	212.5
2000	155.0¢	273.4%	273.4
2005	231.0¢	407.4%	407.4

Source: U.S. Department of Energy.

By the Way ...

The prices in Table 2.1 are year-long averages that do not show how much prices varied. During 2005, for example, gas prices peaked shortly after Hurricane Katrina at a high of $3.13 per gallon.

Now look at the last column of Table 2.1. It is identical to the third column, except we dropped the % signs. This simple change converts the numbers from percentages to a *price index*, which is one type of index number. The statement "1975 = 100" in the column heading shows that the reference value is the 1975 price. In this case, there's really no difference between stating the comparisons as percentages and as index numbers—it's a matter of choice and convenience. However, as we'll see shortly, it's traditional to use index numbers rather than percentages in cases where many factors are being considered simultaneously.

Index Numbers

An **index number** provides a simple way to compare measurements made at different times or in different places. The value at one particular time (or place) must be chosen as the *reference value* (or *base value*). The index number for any other time (or place) is

$$\text{index number} = \frac{\text{value}}{\text{reference value}} \times 100$$

EXAMPLE 1 Finding an Index Number

Suppose the cost of gasoline today is \$3.20 per gallon. Using the 1975 price as the reference value, find the price index number for gasoline today.

Solution Table 2.1 shows that the price of gas was 56.7¢, or \$0.567, per gallon in 1975. If we use the 1975 price as the reference value and the price today is \$3.20, the index number for gasoline today is

$$\text{index number} = \frac{\text{current price}}{\text{1975 price}} \times 100 = \frac{\$3.20}{\$0.567} \times 100 = 564.4$$

This index number for the current price is 564.4, which means the current gasoline price is 564.4% of the 1975 price.

By the Way ...

The term *index* is commonly used for almost any kind of number that provides a useful comparison, even when the numbers are not standard index numbers. For example, body mass index (BMI) provides a way of comparing people by height and weight, but is defined without any reference value. Specifically, body mass index is defined as weight (in kilograms) divided by height (in meters) squared.

TIME OUT TO THINK

Find the actual price of gasoline today at a nearby gas station. What is the gasoline price index for today's price, with the 1975 price as the reference value?

Making Comparisons with Index Numbers

The primary purpose of index numbers is to facilitate comparisons. For example, suppose we want to know how much more expensive gas was in 2000 than in 1975. We can get the answer easily from Table 2.1, which uses the 1975 price as the reference value. This table shows that the price index for 2000 was 273.4, which means that the price of gasoline in 2000 was 273.4% of the 1975 price. Equivalently, we can say that the 2000 price was 2.734 times the 1975 price.

We can also do comparisons when neither value is the reference value. For example, suppose we want to know how much more expensive gas was in 1995 than in 1965. We find the answer by dividing the index numbers for the two years:

$$\frac{\text{index number for 1995}}{\text{index number for 1965}} = \frac{212.5}{55.0} = 3.86$$

The 1995 price was 3.86 times the 1965 price, or 386% of the 1965 price. In other words, the same amount of gas that cost \$1.00 in 1965 would have cost \$3.86 in 1995.

A study of economics usually reveals that the best time to buy anything is last year.

—Marty Allen

EXAMPLE 2 Using the Gas Price Index

Use Table 2.1 to answer the following questions.

a. Suppose that it cost \$7.00 to fill your gas tank in 1975. How much did it cost to buy the same amount of gas in 2005?

b. Suppose that it cost \$20.00 to fill your gas tank in 1995. How much did it cost to buy the same amount of gas in 1955?

Solution

a. Table 2.1 shows that the price index (1975 = 100) for 2005 was 407.4, which means that the price of gasoline in 2005 was 407.4% of the 1975 price. So the 1995 price of gas that cost \$7.00 in 1975 was

$$407.4\% \times \$7.00 = 4.074 \times \$7.00 = \$28.52$$

b. Table 2.1 shows that the price index (1975 = 100) for 1995 was 212.5 and the index for 1955 was 51.3. So the cost of gasoline in 1955 compared to the cost in 1995 was

$$\frac{\text{index number for 1955}}{\text{index number for 1995}} = \frac{51.3}{212.5} = 0.2414$$

Gas that cost \$20.00 in 1995 cost 0.2414 × \$20.00 = \$4.83 in 1955.

The Consumer Price Index

We've seen that the price of gas has risen substantially with time. Most other prices and wages have also risen, a phenomenon we call **inflation**. (Prices and wages occasionally decline with time, which is *deflation*.) Changes in the actual gasoline price therefore are not very meaningful unless we compare them to the overall rate of inflation, which is measured by the **Consumer Price Index (CPI)**.

The Consumer Price Index is computed and reported monthly by the U.S. Bureau of Labor Statistics. It represents an average of prices for a sample of goods, services, and housing. The monthly sample consists of more than 60,000 items. The details of the data collection and index calculation are fairly complex, but the CPI itself is a simple index number. Table 2.2 shows the average annual CPI over a 30-year period. Currently, the reference value for the CPI is an average of prices during the period 1982–1984, which is why the table says "1982–1984 = 100."

The Consumer Price Index

The **Consumer Price Index (CPI),** which is computed and reported monthly, is based on prices in a sample of more than 60,000 goods, services, and housing costs.

Table 2.2 Average Annual Consumer Price Index (1982–1984 = 100)

Year	CPI	Year	CPI	Year	CPI
1977	60.6	1987	113.6	1997	160.5
1978	65.2	1988	118.3	1998	163.0
1979	72.6	1989	124.0	1999	166.6
1980	82.4	1990	130.7	2000	172.2
1981	90.9	1991	136.2	2001	177.1
1982	96.5	1992	140.3	2002	179.9
1983	99.6	1993	144.5	2003	184.0
1984	103.9	1994	148.2	2004	188.9
1985	107.6	1995	152.4	2005	195.3
1986	109.6	1996	156.9	2006	201.6

TECHNICAL NOTE

The government measures two consumer price indices: The CPI-U is based on products thought to reflect the purchasing habits of all urban consumers, whereas the CPI-W is based on the purchasing habits of only wage earners. (Table 2.2 shows the CPI-U.)

The CPI allows us to compare overall prices at different times. For example, to find out how much higher typical prices were in 2005 than in 1995, we divide the CPIs for the two years:

$$\frac{\text{CPI for 2005}}{\text{CPI for 1995}} = \frac{195.3}{152.4} = 1.28$$

Based on the CPI, typical prices in 2005 were 1.28 times those in 1995. For example, a typical item that cost \$1,000 in 1995 would, on average, have cost \$1,280 in 2005. Of course, individual items may have price changes that are different from the average. For example, computer prices for equivalent computing power *fell* significantly from 1995 to 2005, meaning you could buy a much more powerful computer in 2005 for the same or less money. In contrast, the average price of health insurance more than doubled during the same period, so we say that health insurance prices rose much faster than the overall rate of inflation.

EXAMPLE 3 CPI Changes

Suppose you needed \$30,000 to maintain a particular standard of living in 2000. How much would you have needed in 2006 to maintain the same living standard? Assume that the average price of your typical purchases has risen at the same rate as the Consumer Price Index (CPI).

Solution We compare the Consumer Price Indices for 2006 and 2000:

$$\frac{\text{CPI for 2006}}{\text{CPI for 2000}} = \frac{201.6}{172.2} = 1.17$$

That is, typical prices in 2006 were about 1.17 times those in 2000. So if you needed \$30,000 in 2000, you would have needed $1.17 \times \$30{,}000 = \$35{,}100$ to have the same standard of living in 2006.

By the Way ...

Salaries of professional athletes were once kept low because the players were not allowed to offer their skills in the free market ("free agency"). That changed after star baseball player Curt Flood filed suit against Major League Baseball in 1970. Flood ultimately lost when the Supreme Court ruled in favor of baseball in 1972, but the process he set in motion (toward free agency) was unstoppable.

Adjusting Prices for Inflation

In mid-2000, gasoline prices suddenly jumped, reaching a peak of about \$1.87 per gallon. Although that might sound cheap by today's standards, at the time it caused consumer outrage as news reports trumpeted the "highest gasoline prices in history."

In terms of actual prices (as posted at the pump), the \$1.87 price per gallon did indeed shatter the previous record high of \$1.47 from 1981. But were gas prices *really* at a record high? If we want to compare prices fairly, we must take into account the effects of inflation. To do that, we must first see how typical prices changed between 1981 and 2000, which we can do by dividing the CPIs for those years, using data from Table 2.2:

$$\frac{\text{CPI for 2000}}{\text{CPI for 1981}} = \frac{172.2}{90.9} = 1.89$$

Because the CPI rose by a factor of 1.89, the 1981 gasoline price of \$1.47 was equivalent to a 2000 price of $\$1.47 \times 1.89 = \$2.78$. In the language of economics, we say that \$1.47 in "1981 dollars" was equivalent to \$2.78 in "2000 dollars." Because the actual price of gasoline in 2000 was much less than \$2.78, the 2000 price was *not* as high as the 1981 price in "real terms," meaning prices adjusted for inflation. Figure 2.3 shows more than 50 years of annual price data for both actual gasoline prices and prices adjusted to 2006 dollars. Notice that, in real terms, the 1981 average annual price remained the record through 2006.

EXAMPLE 4 Baseball Salaries

In 1987, the mean salary for major league baseball players was \$412,000. In 2006, it was \$2,867,000. Compare the increase in mean baseball salaries to the overall rate of inflation as measured by the Consumer Price Index.

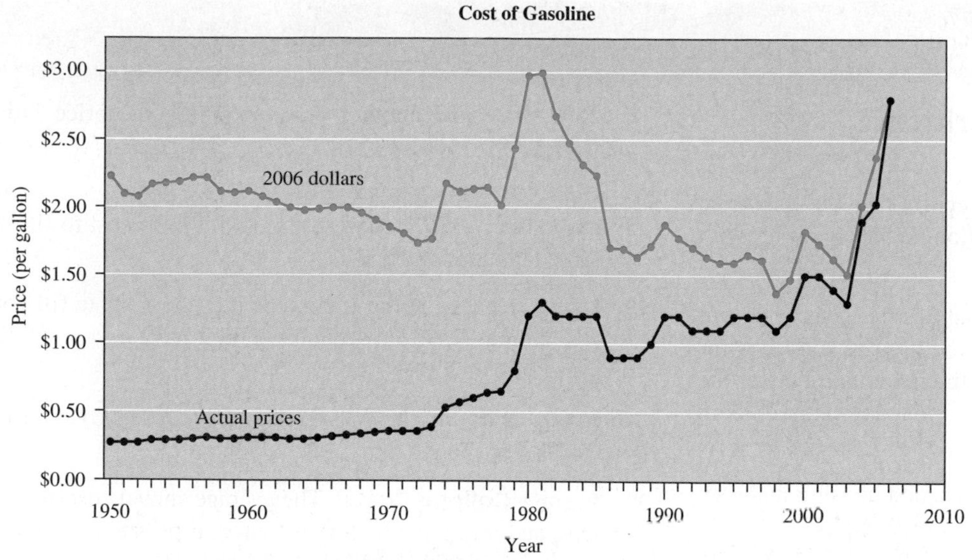

Figure 2.3 Gasoline prices (annual average), 1950–2006. Note that, because we use 2006 dollars for the inflation-adjusted prices, the actual and adjusted prices are the same for 2006. *Source:* American Petroleum Institute.

Solution First, we compare the Consumer Price Indices for 2006 and 1987:

$$\frac{\text{CPI for 2006}}{\text{CPI for 1987}} = \frac{201.6}{113.6} = 1.77$$

Next, we compare the average baseball salaries for those two years:

$$\frac{\text{average baseball salary for 2006}}{\text{average baseball salary for 1987}} = \frac{\$2,867,000}{\$412,000} = 6.96$$

During the same period of time that average prices (as measured by the CPI) rose about 77%, the mean baseball salary rose by almost 600%. In other words, the salaries of major league baseball players rose about $600/77 \approx 8$ times as much as the overall rate of inflation.

Other Index Numbers

The Consumer Price Index is only one of many index numbers that you'll see in news reports. Some are also price indices, such as the Producer Price Index (PPI), which measures the prices that producers (manufacturers) pay for the goods they purchase (rather than the prices that consumers pay). Other indices attempt to measure more qualitative variables. For example, the Consumer Confidence Index is based on a survey designed to measure consumer attitudes so that businesses can gauge whether people are likely to be spending or saving. New indices are created frequently by groups attempting to provide simple comparisons.

By the Way ...

Thinking of becoming a comedian? Then you'll probably want to check the Cost of Laughing Index (there really is one!), which tracks costs of items such as rubber chickens, Groucho Marx glasses, and admission to comedy clubs.

Section 2.4 Exercises

Statistical Literacy and Critical Thinking

1. **Index Number.** A newspaper reports that the gas price index in 2007 was \$2.27 per gallon. What is wrong with that statement?

2. **Index Number.** If computer costs in the year 2000 are set equal to 100 so that they can be used as the basis for determining index numbers, and the index number for the year 2005 is 20, what do we know about computer costs in 2005 compared to computer costs in 2000?

3. **CPI.** If the prices of goods, services, and housing increase, must the Consumer Price Index increase? Explain.

4. **CPI.** If the Consumer Price Index increases, must wages also increase? Explain.

Concepts and Applications

Gasoline Price Index. In Exercises 5–8, use the gasoline price index from Table 2.1. Briefly explain your reasoning in each case.

5. **Current Data.** Suppose the cost of gasoline today is \$3.25 per gallon. What is the price index number for gasoline today, with the 1975 price as the reference value?

6. **2006 Index.** The average price of a gallon of gas was \$2.62 in 2006. What is the price index for gasoline in 2006, with the 1975 price as the reference value?

7. **1998 Price.** With the 1975 price as the reference value, the gasoline price index for 1998 is 197.5. What was the cost of a gallon of gasoline in 1998?

8. **1978 Price.** With the 1975 price as the reference value, the gasoline price index for 1978 is 114.6. What was the cost of a gallon of gasoline in 1978?

9. **Reconstructing the Gasoline Price Index.** Identify the seven price index numbers in Table 2.1 that would result from using the price from 1965 as the reference value. (Hint: Create a column for price as a percentage of 1965 price and another column giving the price index with 1965 = 100.)

10. **Reconstructing the Gasoline Price Index.** Identify the seven price index numbers in Table 2.1 that would result from using the price from 2000 as the reference value. (Hint: Create a column for price as a percentage of 2000 price and another column giving the price index with 2000 = 100.)

11. **Using the Gas Price Index.** If it cost \$16.74 to fill your gas tank in 1985, how much would it have cost to fill the same tank in 2000?

12. **Using the Gas Price Index.** If it cost \$23.25 to fill your gas tank in 2000, how much would it have cost to fill the same tank in 2005?

Consumer Price Index. In Exercises 13–16, use the Consumer Price Index from Table 2.2.

13. **Private College Costs.** The average annual cost (tuition, fees, and room and board) at four-year private universities rose from \$5,900 in 1980 to \$27,516 in 2004. Calculate the percentage rise in cost from 1980 to 2004, and compare it to the overall rate of inflation as measured by the Consumer Price Index.

14. **Public College Costs.** The average annual cost (tuition, fees, and room and board) at four-year public universities rose from \$2,490 in 1980 to \$11,354 in 2004. Calculate the percentage rise in cost from 1980 to 2004, and compare it to the overall rate of inflation as measured by the Consumer Price Index.

15. **Home Prices—South.** The typical (median) price of a new single-family home in the southern part of the United States rose from \$75,300 in 1990 to \$155,500 in 2004. Calculate the percentage rise in cost of a home from 1990 to 2004, and compare it to the overall rate of inflation as measured by the Consumer Price Index.

16. **Home Prices—West.** The typical (median) price of a new single-family home in the western part of the United States rose from \$129,600 in 1990 to \$241,300 in 2004. Calculate the percentage rise in cost of a home from 1990 to 2004, and compare it to the overall rate of inflation as measured by the Consumer Price Index.

Housing Price Index. Realtors use an index to compare housing prices in major cities throughout the country. The index numbers for several cities are given in the table below. If you know the price of a particular house in your town, you can use the index to find the price of a comparable house in another town:

$$\underset{\text{(other town)}}{\text{price}} = \underset{\text{(your town)}}{\text{price}} \times \frac{\text{index in other town}}{\text{index in your town}}$$

Use the housing price index in Exercises 17–20.

City	Index	City	Index
Denver	100	Boston	358
Miami	194	Las Vegas	101
Phoenix	86	Dallas	81
Atlanta	90	Cheyenne	60
Baltimore	150	San Francisco	382

17. **Housing Prices.** For a house valued at $300,000 in Denver, find the price of a comparable house in Miami and Cheyenne.

18. **Housing Prices.** For a house valued at $500,000 in Boston, find the price of a comparable house in Baltimore and Phoenix.

19. **Housing Prices.** For a house valued at $250,000 in Cheyenne, find the price of a comparable house in San Francisco and Boston.

20. **Housing Prices.** For a house valued at $1,000,000 in Boston, find the price of a comparable house in San Francisco and Cheyenne.

Projects for the Internet and Beyond

For useful links, select "Links for Internet Projects" for Chapter 2 at www.aw.com/bbt.

21. **Consumer Price Index.** Go to the Consumer Price Index home page and find the latest news release with updated figures for the CPI. Summarize the news release and any important trends in the CPI.

22. **Producer Price Index.** Go to the Producer Price Index (PPI) home page. Read the overview and recent news releases. Write a short summary describing the purpose of the PPI and how it is different from the CPI. Also summarize any important recent trends in the PPI.

23. **Consumer Confidence Index.** Use a search engine to find recent news about the Consumer Confidence Index. After studying the news, write a short summary of what the Consumer Confidence Index is trying to measure and describe any recent trends in the Consumer Confidence Index.

24. **Human Development Index.** The United Nations Development Programme regularly releases its Human Development Report. A closely watched finding of this report is the Human Development Index (HDI), which measures the overall achievements in a country in three basic dimensions of human development: life expectancy, educational attainment, and adjusted income. Find the most recent copy of this report and investigate exactly how the HDI is defined and computed.

25. **Convenience Store Index.** Go to a local supermarket and find the prices of a few staples, such as bread, milk, juice, and coffee. Compute the total cost of those items. Then go to a few smaller convenience stores and find the prices of the same items. Using the supermarket total as the reference value, compute the index numbers for the convenience stores.

26. **Gasoline Prices.** Collect data, make a convincing graph, and write a persuasive argument to either defend or refute the statement that gasoline prices have *not* increased over the past 30 years relative to the cost of living.

IN THE NEWS

27. **Consumer Price Index.** Find a recent news report that includes a reference to the Consumer Price Index. Briefly describe how the Consumer Price Index is important in the story.

28. **Index Numbers.** Find a recent news report that includes an index number other than the Consumer Price Index. Describe the index number and its meaning, and discuss how the index is important in the story.

Chapter Review Exercises

1. **Titanic.** Of the 2,223 people who were aboard the *Titanic*, 31.76% survived when the ship sank on Monday, April 15, 1912.

   a. Find the true number of people who survived when the *Titanic* sank.

   b. Is the number of survivors a value from a discrete data set or a continuous data set? Explain.

   c. Of the people who were aboard the *Titanic*, 531 were women or children. What percentage of *Titanic* passengers were women or children?

   d. There were 45 girls aboard the *Titanic*, and there were 42% more boys than girls. How many boys were aboard?

   e. If we compile the ages of all passengers aboard the *Titanic*, what is the level of measurement (nominal, ordinal, interval, ratio) of those ages?

   f. If we list the passengers according to the categories of men, women, boys, and girls, what is the level of measurement (nominal, ordinal, interval, ratio) of this data set?

2. **AOL Poll.** In an America OnLine poll of 3,309 people, 66% said that they sometimes eat at KFC, 26% said that they never eat at KFC, and 8% said that they eat frequently at KFC.

   a. What is the number of respondents who said that they never eat at KFC?

   b. If 1,224 of the respondents said that KFC would neither lose nor gain business after the elimination of trans fats, what is the percentage of respondents who made that statement?

   c. Given that the possible responses are *always*, *sometimes*, and *never*, is the level of measurement of those responses nominal, ordinal, interval, or ratio?

   d. Given that the poll was conducted by asking America OnLine users to respond to a question that was posted on the Web site, what do you conclude about the poll results? Is it likely that the results reflect the opinion of the general population?

3. **Health Care Spending.** Total spending on health care in the United States rose from $80 billion in 1973 to $1.8 trillion in 2004. The Consumer Price Index was 44.4 in 1973, and it was 188.9 in 2004 (with 1982–1984 = 100). Compare the change in health care spending from 1973 to 2004 to the overall rate of inflation as measured by the Consumer Price Index.

4. **Minimum Wage.** The accompanying table lists the federal hourly minimum wage over the past 60 years, in both actual dollars at the time and 1996 dollars. Except for 2006, the table entries correspond to years in which the minimum wage changed (based on data from the Department of Labor).

Year	Actual dollars	1996 dollars
1938	0.25	2.78
1939	0.30	3.39
1945	0.40	3.49
1950	0.75	4.88
1956	1.00	5.77
1961	1.25	6.41
1967	1.40	6.58
1968	1.60	7.21
1974	2.00	6.37
1976	2.30	6.34
1978	2.65	6.38
1979	2.90	6.27
1981	3.35	5.78
1990	3.50	4.56
1991	4.25	4.90
1996	4.75	4.75
1997	5.15	5.03
2006	5.15	4.01

   a. According to the table, how much is $0.25 in 1938 dollars worth in 1996 dollars?

   b. According to the table, how much is $1.00 in 1956 dollars worth in 1996 dollars?

   c. Why is the minimum wage for 2006 in actual dollars greater than the minimum wage for 2006 in 1996 dollars?

Chapter Quiz

1. The eye colors of randomly selected subjects are recorded as part of a national health study. Is the level of measurement of those eye colors nominal, ordinal, interval, or ratio?

2. Head circumferences of randomly selected subjects are recorded as part of a national health study. Are those values continuous or discrete?

3. Is the level of measurement of the head circumferences described in Exercise 2 nominal, ordinal, interval, or ratio?

4. A researcher measures the head circumference of a subject and records a value of 45.4 centimeters, but the subject's actual head circumference is 55.4 centimeters. What is the absolute error?

5. A researcher measures the head circumference of a subject and records a value of 45.4 centimeters, but the subject's actual head circumference is 55.4 centimeters. What is the relative error?

6. In a Gallup poll of 1,038 adults, 540 said that second-hand smoke is very harmful. What is the percentage of adults who said that second-hand smoke is very harmful?

7. In a Gallup poll of 1,038 adults, 5% of the respondents said that second-hand smoke is not at all harmful. How many respondents said that second-hand smoke is not at all harmful?

8. Two different students measure the height of an instructor who is actually 178.44 centimeters tall. The first student obtains a measurement of 178 centimeters and the second student obtains a measurement of 179.18 centimeters. Which measurement is more accurate? Which measurement is more precise?

9. The Telektronics Company has been in business for 5 years, and the table lists the net profits in each of those years. Using the first year as a reference, find the index number for the net profit in the second year.

Year	Net profit
1	$12,335
2	$15,257
3	$23,444
4	$31,898
5	$47,296

10. Refer to the table in Exercise 9. If the net profit in the sixth year is projected to be 12% more than in the fifth year, what is the projected net profit for the sixth year?

FOCUS ON POLITICS

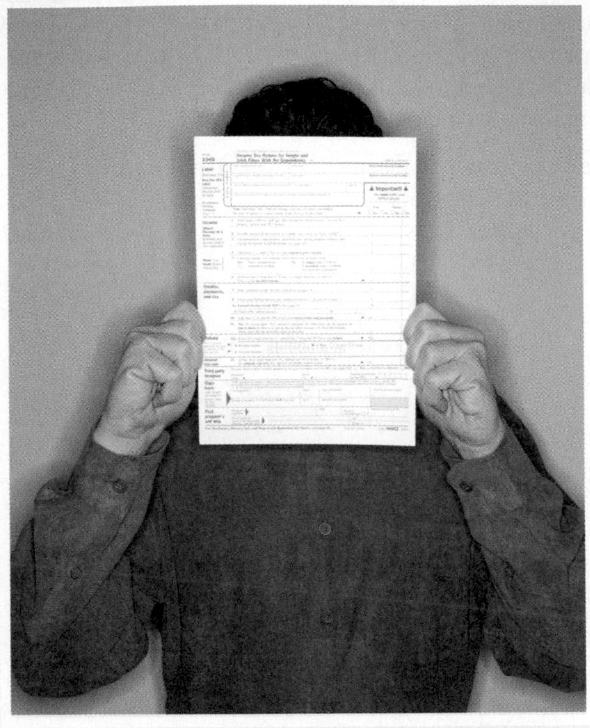

Who Benefits from a Tax Cut?

Politicians have a remarkable capacity to cast numbers in whatever light best supports their beliefs. Consider the two charts shown in Figure 2.4. Both purport to show the effects of the *same* proposed tax cut, yet they appear to support radically different conclusions. (This particular tax cut did not become law.)

The chart in Figure 2.4a reflects numbers supplied by Republicans, who supported the tax cut. It suggests that the tax cut would benefit families of all incomes similarly in percentage terms, with slightly greater benefits going to middle-income families than to the poor or the rich. Figure 2.4b reflects numbers supplied by Democrats, who opposed the tax cut. This chart suggests that the benefits would go disproportionately to the rich. If we assume that neither side is lying outright, how can they make such remarkably different claims? The answer lies in how each side chose to calculate the "benefits" of the tax cut.

The Republicans calculated the *average* tax cut that would be received by families in each group. For example, the last bar in Figure 2.4a shows that families with incomes over $200,000 would get an average tax cut of 2.9%. Of course, a 2.9% tax cut means many more actual dollars to someone paying high taxes than to someone paying a little. For example, someone who pays $100,000 in taxes would save $2,900 with a 2.9% tax cut, while someone paying only $1,000 in taxes would save only $29.

The Democrats calculated the percentage of *total* benefits that would go to each group. For example, the last bar in Figure 2.4b shows that families with incomes over $200,000 would receive 28.1% of the total benefits from the tax cut. But this does *not* mean that these families are getting a 28.1% tax cut. To see why, consider the effects of an across-the-board tax cut of

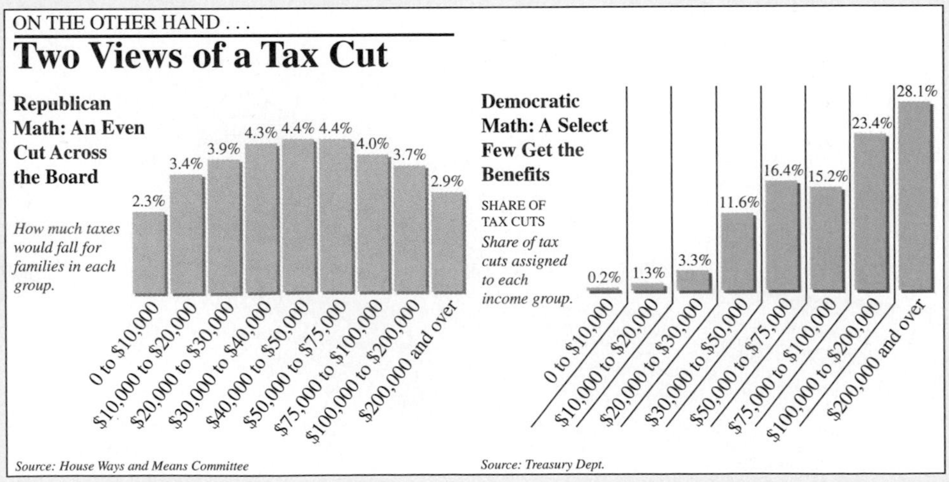

(a) (b)

Figure 2.4 *Source:* Adapted from the *New York Times.*

the same percentage for everyone. Because families with incomes above $200,000 pay more than one-fourth of the total income taxes collected by the U.S. government, these families would receive about one-fourth of the total benefit of *any* across-the-board cut. For example, if the government collected $1 trillion in taxes, a 10% across-the-board cut would mean savings of $100 billion for taxpayers as a whole. In that case, because one-fourth of the $1 trillion was paid by families with incomes over $200,000, one-fourth of the $100 billion in savings would go to these same families. These families would therefore get 25% of the *benefits* of the tax cut, even though their actual tax cut would be only 10%.

Which side was being more fair? Neither, really. The Republicans chose to neglect the fact that most of the total tax savings would go to the wealthy, while the Democrats chose to neglect the fact that the wealthy already pay most of the taxes. Unfortunately, this type of "selective truth" is very common when it comes to numbers, especially those tied up in politics.

Politicians and government officials usually abuse numbers and logic in the most elementary ways. They simply cook figures to suit their purpose, use obscure measures of economic performance, and indulge in horrendous examples of chart abuse, all in the name of disguising unpalatable truths.

—A. K. Dewdney, *200% of Nothing*

QUESTIONS FOR DISCUSSION

1. The percentages in Figure 2.4a represent the *relative* tax savings to each income group. How do these relative savings compare to the *absolute* savings for each group? (Hint: In estimating the absolute tax savings, remember that the amount of tax that any family pays is some percentage of its income, and that lower-income families generally pay a smaller percentage of their income in taxes.)
2. A secondary reason for the differences in the two graphs is that the two sides defined *income* differently. For example, the Democrats chose to allocate corporate profits to the individual incomes of stockholders, which means that a person holding stock was judged to have a higher income by the Democrats than by the Republicans. How did this difference in defining income affect the two graphs?
3. Do you think that either of the charts in Figure 2.4 accurately portrays the overall "fairness" of the proposed tax cut? If so, which one and why? If not, how do you think the numbers could be portrayed more fairly?
4. Have any tax cuts or tax increases been proposed by the U.S. Congress or the President this year? Discuss the fairness of current proposals.

SUGGESTED READING

"Two Views of a Tax Cut." *New York Times*, April 7, 1995.

Andrews, Edward. "Your Taxes: Cracking the Tax Code." *New York Times*, February 11, 2006.

Andrews, Edward. "Tax Cuts Offer Most for Very Rich, Study Says." *New York Times*, January 8, 2007.

FOCUS ON ECONOMICS

Is Our Standard of Living Improving?

Most of us would like our standard of living to keep improving year after year, and in principle it is easy to tell if that is the case: If your income is rising faster than the rate of inflation, then your standard of living will rise; if your income does not keep up with inflation, then your standard of living will fall.

In reality, however, the situation is much more complex. For example, even if your income rises substantially, your standard of living may fall if you face new expenses, such as a health care crisis that your insurance won't cover, or costs for yourself or a child to go to college, or an adjustable-rate mortgage payment that adjusts to a higher monthly cost.

Similar complexities affect the question of the national standard of living. Economists almost universally agree that the living standard of the average American rose substantially from the time after World War II through the early 1970s, and then again during the late 1990s. But, unless you are among the top few percent of earners (whose income has continued to rise; see "Focus on Economics: Are the Rich Getting Richer?" at the end of Chapter 4), there is considerable debate over living standards during the 1970s and 1980s and during the first few years of the new millennium.

The key question in the debate is whether *real income*—income adjusted for inflation—is rising or falling. This question, in turn, hinges on how we measure inflation. As discussed in Section 2.4, inflation is commonly measured with the Consumer Price Index (CPI). Figure 2.5 shows how the real wages of most Americans have changed since the 1980s, if we use the CPI as the measure of inflation. For example, a change in real wages of 0% means that actual wages and the CPI changed by precisely the same amount in a particular year. The change in real wages is positive when actual wages increased by more than the CPI and negative when actual wages increased by less than the CPI.

Notice that, according to this measure, real wages have declined (negative changes) in most years since the late 1970s. Indeed, if you calculate the total change over the period shown (and back a few years further), this picture indicates that real wages in 2006 were *lower* than they had been some three decades earlier. In the 2006 elections, these data prompted Democrats running for Congress to claim that wages—and therefore the standard of living—for most Americans had been stagnating for more than 30 years.

But is the picture of our changing standard of living really this bleak? Or is it even bleaker? The answer depends on whom you ask, because different economists have different opinions as to whether the CPI overstates, understates, or accurately measures inflation.

Let's start with those who argue that the CPI *overstates* inflation. If that is the case, then the real inflation rate is *lower* than the CPI indicates, which would mean that the zero line is too high up in Figure 2.5. For example, if the CPI overstates the rate of inflation by 1 percentage point each year, then the zero line should be at the level of –1 in Figure 2.5, rather than at 0. That change would make the majority of the years shown on the graph positive rather than negative, which would mean that rather than stagnating, the average standard of living has actually risen significantly over the past three decades.

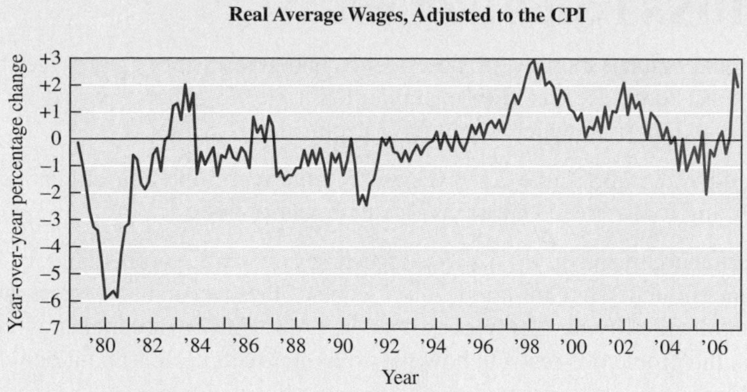

Figure 2.5 The change in wages, adjusted to the CPI, for non-management workers in the private sector. These workers represent about 80% of the work force, not counting government employees. Negative changes represent a decline in real wages and positive changes represent an increase in real wages. Adapted from the *New York Times*, January 3, 2007. *Sources:* U.S. Department of Labor; University of Michigan; National Bureau of Economic Research.

The thrust of this argument, documented most clearly in the 1996 "Boskin Report" (see the suggested reading list), is that the standard calculation of the CPI contains systematic errors that overstate the increase in the cost of living. The report pointed to two major systematic errors that seem to have this effect. First, from one month to the next, the CPI is based on changes in the prices of particular items at particular stores. In reality, however, if the price of an item rises at one store, consumers often buy it more cheaply at another store. And if the price of a particular item rises too much, consumers may substitute a similar but lower-priced item, which may mean as simple a change as switching brands. Consumers therefore don't find their actual costs rising as much as the CPI indicates. Second, the CPI tracks prices of items purchased by "typical" consumers at any given time. These items change with time, especially when it comes to technology, but the CPI does not account for how such changes affect our standard of living. For example, 20 years ago no one owned a DVD recorder, an iPod, a high-definition TV, or a computer with high-speed Internet access. Most economists would say that these items have improved our standard of living, but in the CPI they count as nothing more than items that "typical" consumers now buy. Based on these types of systematic errors, the Boskin report concluded that the CPI overstates the actual rate of inflation by between 0.8 and 1.6 percentage points per year.

If I had to populate an asylum with people certified insane, I'd just pick 'em from all those who claim to understand inflation.

—Will Rogers

While most economists acknowledge the systematic problems identified by the Boskin report, some argue that other systematic errors work in the opposite direction. For example, many middle-age workers now have substantial expenses associated with the care of their aging parents, something that was not true a few decades ago, and on average we all pay a much higher percentage of income for health insurance and health care than we did in the past. These types of expenditures leave less income for the types of items that we generally think of as improving our standard of living. Indeed, some economists argue that these changes have been so significant that the CPI actually *understates* the true rate of inflation, in which case our living standards have fallen more than the wage picture in Figure 2.5 indicates.

The lesson in this discussion should be clear. The statistics regarding the CPI and changes in actual wages are all quite clear, yet people of different political or economic belief systems can still draw very different conclusions from these data. That, after all, is why people can disagree in good faith about such questions as whether our standard of living is rising or falling. It is also why you should think carefully about statistical claims of all types, rather than just taking the latest political slogans at face value.

QUESTIONS FOR DISCUSSION

1. Based on the graph in Figure 2.5, describe in general terms how real wages have changed over the past three decades. For example, describe the general period in which wages have risen and those in which wages have declined.
2. Overall, do you think Figure 2.5 gives an accurate representation of changes in the standard of living for average Americans? Defend your opinion.
3. Consider the arguments in the article suggesting systematic errors that may cause the CPI to overstate or understate inflation. Choose one of these systematic errors and investigate it in more detail. Giving plenty of examples, draw your own conclusion about whether you think the problem is real and how you think it affects claims about our changing standard of living.
4. In Figure 2.5, notice the dramatic increase in real wages shown for 2006. Do some Web research to find out what has happened to wages since that time. Has the increase continued, or was it temporary?

SUGGESTED READING

Boskin, Michael J., Ellen R. Dulberger, Robert J. Gordon, Zvi Griliches, and Dale W. Jorgenson. "Toward a More Accurate Measure of the Cost of Living." Final Report to the Senate Finance Committee, from the Advisory Commission to Study the Consumer Price Index. Washington, DC, Senate Finance Committee, 1996.

Greenhouse, Steven. "Falling Fortunes of Wage Earners." *New York Times*, April 12, 2005.

Gross, Jane. "Elder-Care Costs Deplete Savings of a Generation." *New York Times*, December 30, 2006.

Leonhar, David. "It's the Year to Keep an Eye on Paychecks." *New York Times*, January 3, 2007.

3

The greatest value of a picture is when it forces us to notice what we never expected to see.
—John Tukey

Visual Displays of Data

WHETHER YOU LOOK AT A NEWSPAPER, A CORPORATE annual report, or a government study, you are almost sure to see tables and graphs of statistical data. Some of these tables and graphs are very simple; others can be quite complex. Some make it easy to understand the data; others may be confusing or even misleading. In this chapter, we'll study some of the many ways in which statistical data are commonly displayed in tables and graphs. Because the ability to convey concepts through graphs is so valuable in today's data-driven society, the skills developed in this chapter are crucial for success in nearly every profession.

LEARNING GOALS

3.1 Frequency Tables
Be able to create and interpret frequency tables.

3.2 Picturing Distributions of Data
Be able to create and interpret basic bar graphs, dotplots, pie charts, histograms, stem-and-leaf plots, line charts, and time-series diagrams.

3.3 Graphics in the Media
Understand how to interpret the many types of more complex graphics that are commonly found in news media.

3.4 A Few Cautions About Graphics
Critically evaluate graphics and identify common ways in which graphics can be misleading.

3.1 Frequency Tables

Table 3.1 Frequency Table for a Set of Essay Grades	
Grade	**Frequency**
A	4
B	7
C	9
D	3
F	2
Total	**25**

Professor Delaney records the following list of the grades she gave to her 25 students on a set of essays:

A C C B C D C C F D C C C
B B A B D B A A B F C B

This list contains all of the grades, but it is not very easy to read. A much easier way to display these data is by constructing a table in which we record the number of times, or **frequency**, that each grade appears. The result, shown in Table 3.1, is called a **frequency table**. The five possible grades (A, B, C, D, F) are called the **categories** (or classes) for the table.

Definition

A basic **frequency table** has two columns:

- One column lists all the **categories** of data.
- The other column lists the **frequency** of each category, which is the number of data values in the category.

EXAMPLE 1 Taste Test

The Rocky Mountain Beverage Company wants feedback on its new product, Coral Cola, and sets up a taste test with 20 people. Each individual is asked to rate the taste of the cola on a 5-point scale:

(bad taste) 1 2 3 4 5 (excellent taste)

The 20 ratings are as follows:

1 3 3 2 3 3 4 3 2 4
2 3 5 3 4 5 3 4 3 1

Construct a frequency table for these data.

Table 3.2 Taste Test Ratings	
Taste scale	**Frequency**
1	2
2	3
3	9
4	4
5	2
Total	**20**

Solution The variable of interest is *taste*, and this variable can take on five values: the taste categories 1 through 5. (Note that the data are qualitative and at the ordinal level of measurement.) We construct a table with these five categories in the left column and their frequencies in the right column, as shown in Table 3.2.

Binning Data

Consider the data in Table 3.3, showing the average annual energy use per person, in millions of BTUs, in each of the 50 states. The 50 numbers in this data set (not counting the U.S. average) range from 212 million BTUs per person (Rhode Island) to 1,175 million BTUs per person (Alaska), and most of the numbers appear only once. How can we efficiently make a frequency table of these data?

The answer is to create categories that each span some range of data values. For example, we could create a category for all data values between 200 and 299 million BTUs, then create a second

Table 3.3 Average Annual Energy Use per Person, by State, in Millions of BTUs

State	Millions of BTUs per person	State	Millions of BTUs per person	State	Millions of BTUs per person
Alabama	447	Louisiana	822	Ohio	349
Alaska	1,175	Maine	366	Oklahoma	425
Arizona	246	Maryland	281	Oregon	295
Arkansas	415	Massachusetts	248	Pennsylvania	321
California	229	Michigan	313	Rhode Island	212
Colorado	297	Minnesota	355	South Carolina	389
Connecticut	255	Mississippi	411	South Dakota	345
Delaware	383	Missouri	322	Tennessee	388
Florida	252	Montana	410	Texas	560
Georgia	343	Nebraska	372	Utah	296
Hawaii	248	Nevada	292	Vermont	252
Idaho	341	New Hampshire	254	Virginia	329
Illinois	310	New Jersey	298	Washington	316
Indiana	470	New Mexico	353	West Virginia	433
Iowa	400	New York	220	Wisconsin	335
Kansas	410	North Carolina	314	Wyoming	919
Kentucky	456	North Dakota	624	**U.S. Average**	**339**

Note: Data include all energy uses, including residential, commercial, industrial, and transportation.
Source: United States Energy Information Administration (2007 tables, showing 2003 data).

By the Way ...

A BTU, or British thermal unit, is the energy required to raise the temperature of one pound of water by 1°F. It is equivalent to 252 calories or 1,055 joules.

category for data values between 300 and 399 million BTUs, and so on. We then count the frequency (number of data values) in each category, generating the frequency table shown as Table 3.4. This process is called **binning** the data because each category acts like a separate bin into which we can pour some of the data values.

Definition

When it is impossible or impractical to have a category for every value in a data set, we **bin** (or group) the data into categories (*bins*), each covering a range of possible data values.

Table 3.4 Frequency Table for the Energy Use Data in Table 3.3

Annual energy use per person (millions of BTUs)	Frequency (number of states)
200–299	16
300–399	19
400–499	10
500–599	1
600–699	1
700–799	0
800–899	1
900–999	1
1,000–1,099	0
1,100–1,199	1
Total	**50**

Table 3.5 Data for Stocks in the Dow Jones Industrial Average, 2006

Company	Revenue (billions)	Return*	Rank**	Company	Revenue (billions)	Return*	Rank**
Wal-Mart	351.1	0.1%	1	Johnson & Johnson	53.3	12.4%	36
Exxon Mobil	347.3	39.2%	2	Pfizer	52.4	15.3%	39
General Motors	207.3	64.2%	3	United Technologies	47.8	13.7%	42
General Electric	168.3	26.5%	5	Microsoft	44.3	15.8%	49
Citigroup	146.8	19.4%	8	Caterpillar	41.6	7.9%	55
AIG	113.2	6.0%	10	Intel	35.4	–17.2%	62
J.P. Morgan Chase	100.0	25.5%	11	Walt Disney	34.3	44.5%	64
Verizon	93.2	34.6%	13	Honeywell	31.4	24.2%	69
Hewlett-Packard	91.7	45.3%	14	ALCOA	30.9	3.5%	71
IBM	91.4	19.8%	15	DuPont	29.0	18.7%	74
Home Depot	90.8	0.9%	17	American Express	27.1	19.1%	79
Altria	70.3	19.9%	23	Coca-Cola	24.1	23.1%	94
Procter & Gamble	68.2	13.3%	25	3M	22.9	3.0%	97
AT&T	63.1	52.9%	27	Merck	22.6	42.9%	99
Boeing	61.5	28.4%	28	McDonald's	21.6	34.6%	108

*Total return to investors, 2006, including both dividends and change in stock price.
**Rank based on revenue among Fortune 500 companies.
Source: Fortune.com.

By the Way ...

The 30 stocks that make up the Dow list are chosen by the editors of the *Wall Street Journal*. On occasion, the stocks on the list are changed. For example, in April 2004, AT&T, Eastman Kodak, and International Paper were removed from the Dow and replaced by AIG, Pfizer, and Verizon. AT&T soon returned to the index when another company on the list (SBC Communications) merged with it and took the AT&T name.

EXAMPLE 2 The Dow Stocks

For the 30 stocks of the Dow Jones Industrial Average, Table 3.5 shows the annual revenue (in billions of dollars), the one-year total return, and the rank on the Fortune 500 list of largest U.S. companies. Create a frequency table for the revenue. Discuss the pros and cons of the binning choices.

Solution The revenue data range from \$21.6 billion (McDonald's) to \$351.1 billion (Wal-Mart). There are many possible ways to bin data for this range; here's one good way and the reasons for it:

- We create bins spanning a range from \$0 to \$400 billion. This covers the full range of the data, with extra room below the lowest data value and above the highest data value.
- We give each bin a width of \$50 billion so that we can span the \$0 to \$400 billion range with eight bins. Also, the width of \$50 billion is a convenient number that helps make the table easy to read.
- Because the data values are given to the nearest tenth (of a billion dollars), we also define the bins to the nearest tenth so that they do not overlap. That is, bins go from \$0 to \$49.9 billion, from \$50.0 to \$99.9 billion, and so on.

Table 3.6 shows the resulting frequency table.

Table 3.6 Frequency Table for the Annual Revenue Data in Table 3.5

Annual revenue (billions of dollars)	Frequency (number of companies)
0–49.9	13
50–99.9	10
100–149.9	3
150–199.9	1
200–249.9	1
250–299.9	0
300–349.9	1
350–399.9	1
Total	**30**

TIME OUT TO THINK

Consider three other possible ways of binning the data in Table 3.6: 4 bins spanning the range \$0 to \$400 billion, 11 bins spanning the range \$0 to \$375 billion, and 36 bins spanning the range \$0 to \$360 billion. Briefly discuss the pros and cons of each of these choices.

Relative Frequency

Let's reconsider the essay grades listed in Table 3.1. We might want to know not only the number of students who received each grade, but also the fraction or percentage of students who received those grades. We call these fractions (or proportions or percentages) for each category the **relative frequencies**. For example, 4 of the 25 students received A grades, so the relative frequency of A grades is 4/25, or 0.16, or 16%. Table 3.7 repeats the data from Table 3.1, but this time with an added column for the relative frequency.

Table 3.7 Relative Frequency Table

Grade	Frequency	Relative frequency
A	4	4/25 = 0.16
B	7	7/25 = 0.28
C	9	9/25 = 0.36
D	3	3/25 = 0.12
F	2	2/25 = 0.08
Total	**25**	**1**

"Data! Data! Data!" he cried impatiently. "I can't make bricks without clay."

—Sherlock Holmes in Sir Arthur Conan Doyle's *The Adventure of the Copper Beeches*

The sum of the relative frequencies must equal 1 (or 100%), because each individual relative frequency is a fraction of the total frequency. (Sometimes, rounding causes the total to be slightly different from 1.)

Definition

The **relative frequency** of any category is the proportion or percentage of the data values that fall in that category:

$$\text{relative frequency} = \frac{\text{frequency in category}}{\text{total frequency}}$$

Cumulative Frequency

Look one more time at the essay grades in Table 3.1. What if we want to know how many students got a grade of C or better? We could, of course, add the frequencies of A, B, and C grades to find that 20 students got a C or better. Often, however, tables do this arithmetic for us by showing the **cumulative frequencies**, or the number of data values in a particular category *and all preceding* categories. Table 3.8 repeats the data from Table 3.1, but this time with an added column for the cumulative frequency. Note that the cumulative frequency for the last category must always equal the total number of data values, which is the total frequency.

TECHNICAL NOTE

Most frequency tables start with the lowest category, but tables of grades commonly start with the highest category (A). Category order does not affect frequencies or relative frequencies, but *does* affect cumulative frequency. For example, if the table to the right had the categories in reverse order, the cumulative frequency for C would be the number of grades of C or lower (rather than C or higher).

Table 3.8 Cumulative Frequency Table

Grade	Frequency	Cumulative frequency
A	4	4
B	7	7 + 4 = 11
C	9	9 + 7 + 4 = 20
D	3	3 + 9 + 7 + 4 = 23
F	2	2 + 3 + 9 + 7 + 4 = 25
Total	**25**	**25**

Definition

The **cumulative frequency** of any category is the number of data values in that category *and all preceding* categories.

Keep in mind that cumulative frequencies make sense only for data categories that have a clear order. That is, we can use cumulative frequencies for data at the ordinal, interval, and ratio levels of measurement, but not for data at the nominal level of measurement.

EXAMPLE 3 More on the Taste Test

Using the taste test data from Example 1, create a frequency table with columns for the relative and cumulative frequencies. What percentage of the respondents gave the cola the highest rating? What percentage gave the cola one of the three lowest ratings?

Solution We find the relative frequencies by dividing the frequency in each category by the total frequency of 20. We find the cumulative frequencies by adding the frequency in each category to the sum of the frequencies in all preceding categories. Table 3.9 shows the results. The relative frequency column shows that 0.10, or 10%, of the respondents gave the cola the highest rating. The cumulative frequency column shows that 14 out of 20 people, or 70%, gave the cola a rating of 3 or lower.

TECHNICAL NOTE

A cumulative frequency divided by the total frequency is called a *relative cumulative frequency*. For example, in Table 3.9 the relative cumulative frequency of 3 or lower is 14/20 = 0.70, meaning that 70% of the data values are in the categories 3, 2, or 1.

Table 3.9 Relative and Cumulative Frequencies

Taste scale	Frequency	Relative frequency	Cumulative frequency
1	2	2/20 = 0.10	2
2	3	3/20 = 0.15	3 + 2 = 5
3	9	9/20 = 0.45	9 + 3 + 2 = 14
4	4	4/20 = 0.20	4 + 9 + 3 + 2 = 18
5	2	2/20 = 0.10	2 + 4 + 9 + 3 + 2 = 20
Total	**20**	**1**	**20**

EXAMPLE 4 Energy Data

To the frequency table for energy data shown as Table 3.4, add columns for the relative and cumulative frequencies. Discuss any trends that seem particularly revealing or surprising.

Solution We find the relative frequencies by dividing the frequency in each category by the total frequency, which is 50 in this case. We find the cumulative frequencies by adding the frequency in each category to the sum of the frequencies in all preceding categories. Table 3.10 shows the results.

The table reveals many interesting facts about annual energy use per person in the different states. For example, the relative frequency column tells us that 38% (or 0.38) of the states fall into the category of 300–399 million BTUs per person per year. The cumulative frequency column shows that 35 of the 50 states fall into the first two categories, or energy use below 400 million BTUs per person per year. The three states with the highest annual energy use per person use more than twice as much energy per person as any of the lowest 35 states. (Using terminology that we'll discuss in Chapter 4, we say that these three states are outliers because their values differ so much from those of other states.)

Table 3.10 Binned Energy Data

Annual energy use per person (millions of BTUs)	Frequency	Relative frequency	Cumulative frequency
200–299	16	0.32	16
300–399	19	0.38	35
400–499	10	0.20	45
500–599	1	0.02	46
600–699	1	0.02	47
700–799	0	0.00	47
800–899	1	0.02	48
900–999	1	0.02	49
1,000–1,099	0	0.00	49
1,100–1,199	1	0.02	50
Total	**50**	**1**	**50**

By the Way ...

The portions of total energy going to residential, commercial, industrial, and transportation uses vary widely among the states. Industrial and transportation uses are particularly high in the three states with the highest total energy use (Alaska, Wyoming, and Louisiana).

TIME OUT TO THINK

Be careful in interpreting Table 3.10 in Example 4. For example, the 0.32 relative frequency for 200–299 million BTUs means that 32% of the 50 states have per capita energy use in this range. Does this also mean that 32% of all Americans use between 200 and 299 million BTUs of energy each year? Why or why not? (Hint: Look back at Table 3.3 and consider the populations of these states.)

Section 3.1 Exercises

Statistical Literacy and Critical Thinking

1. **Frequency Table.** What is a frequency table? Explain what we mean by the categories (or classes) and frequencies.

2. **Relative Frequency.** A frequency table has five classes with frequencies of 2, 9, 14, 12, and 3. What are the relative frequencies of the five classes?

3. **Cumulative Frequency.** A frequency table has five classes with frequencies of 2, 9, 14, 12, and 3. What are the cumulative frequencies of the five classes?

4. **Frequency Table.** The first class in a frequency table shows a frequency of 12 corresponding to the range of values from 15.0 to 15.9. Using only this information about the frequency table, is it possible to identify the original 12 sample values that are summarized by this class? Explain.

Does It Make Sense? For Exercises 5–8, decide whether the statement makes sense (or is clearly true) or does not make sense (or is clearly false). Explain clearly. Not all of these statements have definitive answers, so your explanation is more important than your chosen answer.

5. **Frequency Table.** A friend tells you that her frequency table has two columns labeled *State* and *Median Income*.

6. **Relative Frequency.** The relative frequency of category A in a table is 1.4.

7. **Cumulative Frequency.** The cumulative frequency of a category in a table is 1.4.

8. **Bins.** For a given data set, as the width of the bins decreases, the number of bins increases.

Concepts and Applications

9. **Frequency Table Practice.** Professor Diaz records the following final grades in one of her courses:

A	A	A	A	B	B	B	B	B	B	B	C
C	C	C	C	C	C	C	D	D	D	F	F

Construct a frequency table for these grades. Include columns for relative frequency and cumulative frequency. Briefly explain the meaning of each column.

10. **Frequency Table Practice.** A guidebook for New York City lists 5 five-star restaurants (the highest rating), 10 four-star restaurants, 20 three-star restaurants, 15 two-star restaurants, and 5 one-star restaurants. Make a frequency table for these ratings. Include columns for relative frequency and cumulative frequency. Briefly explain the meaning of each column.

11. **Weights of Coke.** Construct a frequency table for the weights (in pounds) given below of 36 cans of regular Coke. Start the first bin at 0.7900 pound and use a bin width of 0.0050 pound. Discuss your findings.

0.8192	0.8150	0.8163	0.8211	0.8181	0.8247
0.8062	0.8128	0.8172	0.8110	0.8251	0.8264
0.7901	0.8244	0.8073	0.8079	0.8044	0.8170
0.8161	0.8194	0.8189	0.8194	0.8176	0.8284
0.8165	0.8143	0.8229	0.8150	0.8152	0.8244
0.8207	0.8152	0.8126	0.8295	0.8161	0.8192

12. **Weights of Diet Coke.** Construct a frequency table for the weights (in pounds) given below of 36 cans of Diet Coke. Start the first bin at 0.7750 pound and use a bin width of 0.0050 pound. Discuss your findings.

0.7773	0.7758	0.7896	0.7868	0.7844	0.7861
0.7806	0.7830	0.7852	0.7879	0.7881	0.7826
0.7923	0.7852	0.7872	0.7813	0.7885	0.7760
0.7822	0.7874	0.7822	0.7839	0.7802	0.7892
0.7874	0.7907	0.7771	0.7870	0.7833	0.7822
0.7837	0.7910	0.7879	0.7923	0.7859	0.7811

13. **Oscar-Winning Actors.** The following data show the ages of recent Academy Award–winning male actors at the time when they won their award. Make a frequency table for the data, using bins of 20–29, 30–39, and so on. Discuss your findings.

32	37	36	32	51	53	33	61	35	45
55	39	76	37	42	40	32	60	38	56
48	48	40	43	62	43	42	44	41	56
39	46	31	47						

14. **Body Temperatures.** The following data show the body temperatures (°F) of randomly selected subjects. Construct a frequency table with seven classes: 96.9–97.2, 97.3–97.6, 97.7–98.0, and so on.

98.6	98.6	98.0	98.0	99.0	98.4	98.4	98.4
98.4	98.6	98.6	98.8	98.6	97.0	97.0	98.8
97.6	97.7	98.8	98.0	98.0	98.3	98.5	97.3
98.7	97.4	98.9	98.6	99.5	97.5	97.3	97.6
98.2	99.6	98.7	99.4	98.2	98.0	98.6	98.6
97.2	98.4	98.6	98.2	98.0	97.8	98.0	98.4
98.6	98.6						

15. **Missing Information.** The following table shows grades for a term paper in an English class. The table is incomplete. Use the information given to fill in the missing entries and complete the table.

Category	Frequency	Relative frequency
A	?	?
B	?	18%
C	?	24%
D	11	?
F	6	?
Total	**50**	**?**

16. **Missing Information.** The following table shows grades for performances in a drama class. The table is incomplete. Use the information given to fill in the missing entries and complete the table.

Category	Frequency	Cumulative frequency
A	?	1
B	6	?
C	7	?
D	?	23
F	?	25
Total	**?**	**?**

17. **Loaded Die.** One of the authors drilled a hole in a die, filled it with a lead weight, and then proceeded to roll it. The results are given in the following frequency table.

a. According to the data, how many times was the die rolled?

b. How many times was the outcome greater than 2?

c. What percentage of outcomes were 6?

d. List the relative frequencies that correspond to the given frequencies.

e. List the cumulative frequencies that correspond to the given frequencies.

Outcome	Frequency
1	27
2	31
3	42
4	40
5	28
6	32

18. Interpreting Family Data. Consider the following frequency table for the number of children in American families.

a. According to the data, how many families are there in America?

b. How many families have two or fewer children?

c. What percentage of American families have no children?

d. What percentage of American families have three or more children?

Number of children	Number of families (millions)
0	35.54
1	14.32
2	13.28
3	5.13
4 or more	1.97

19. Computer Keyboards. The traditional keyboard configuration is called a *Qwerty* keyboard because of the positioning of the letters QWERTY on the top row of letters. Developed in 1872, the Qwerty configuration supposedly forced people to type slower so that the early typewriters would not jam. Developed in 1936, the Dvorak keyboard supposedly provides a more efficient arrangement by positioning the most used keys on the middle row (or "home" row), where they are more accessible.

A *Discover* magazine article suggested that you can measure the ease of typing by using this point rating system: Count each letter on the home row as 0, count each letter on the top row as 1, and count each letter on the bottom row as 2. For example, the word *statistics* would result in a rating of 7 on the Qwerty keyboard and 1 on the Dvorak keyboard, as shown below.

	S	T	A	T	I	S	T	I	C	S	
Qwerty keyboard	0	1	0	1	1	0	1	1	2	0	(sum = 7)
Dvorak keyboard	0	0	0	0	0	0	0	0	1	0	(sum = 1)

Using this rating system with each of the 52 words in the Preamble to the Constitution, we get the rating values below.

Qwerty Keyboard Word Ratings:

2	2	5	1	2	6	3	3	4	2	4	0	5
7	7	5	6	6	8	10	7	2	2	10	5	8
2	5	4	2	6	2	6	1	7	2	7	2	3
8	1	5	2	5	2	14	2	2	6	3	1	7

Dvorak Keyboard Word Ratings:

2	0	3	1	0	0	0	0	2	0	4	0	3
4	0	3	3	1	3	5	4	2	0	5	1	4
0	3	5	0	2	0	4	1	5	0	4	0	1
3	0	1	0	3	0	1	2	0	0	0	1	4

a. Create a frequency table for the Qwerty word ratings data. Use bins of 0–2, 3–5, 6–8, 9–11, and 12–14. Include a column for relative frequency.

b. Create a frequency table for the Dvorak word ratings data, using the same bins as in part a. Include a column for relative frequency.

c. Based on your results from parts a and b, which keyboard arrangement is easier for typing? Explain.

20. Double Binning. The students in a statistics class conduct a transportation survey of students in their high school. Among other data, they record the age and mode of transportation between home and school for each student. The following table gives some of the data that were collected. For age: 1 = 14 years, 2 = 15 years, 3 = 16 years, 4 = 17 years, 5 = 18 years. For transportation: 1 = walk, 2 = school bus, 3 = public bus, 4 = drive, 5 = other.

Student	Age	Transportation	Student	Age	Transportation
1	1	1	11	3	5
2	5	1	12	5	5
3	2	2	13	1	2
4	3	5	14	5	5
5	4	3	15	5	5
6	1	1	16	4	4
7	5	2	17	2	2
8	2	1	18	3	1
9	3	4	19	3	3
10	1	3	20	1	4

a. Classify the two variables, *age* and *transportation*, as qualitative or quantitative, and give the level of measurement for each.

b. In order to be analyzed or displayed, the data must be binned with respect to both variables. Count the number of students in each of the 25 age/transportation categories and fill in the blank cells in the following table.

		Transportation				
		1	2	3	4	5
Age	1					
	2					
	3					
	4					
	5					

Projects for the Internet and Beyond

For useful links, select "Links for Internet Projects" for Chapter 3 at www.aw.com/bbt.

21. **Energy Table.** The U.S. Energy Information Administration (EIA) Web site offers dozens of tables relating to energy use, energy prices, and pollution. Explore the selection of tables. Find a table of raw data that is of interest to you and convert it to an appropriate frequency table. Briefly discuss what you can learn from the frequency table that is less obvious in the raw data table.

22. **Endangered Species.** The Web site for the World Conservation Monitoring Centre in Great Britain provides data on extinct, endangered, and threatened animal species. Explore these data and summarize some of your more interesting findings with frequency tables.

23. **Navel Data.** The *navel ratio* is defined to be a person's height divided by the height (from the floor) of his or her navel. An old theory says that, on average, the navel ratio of humans is the golden ratio: $(1 + \sqrt{5})/2$. Measure the navel ratio of each person in your class. What percentage of students have a navel ratio within 5% of the golden ratio? What percentage of students have a navel ratio within 10% of the golden ratio? Does the old theory seem reliable?

24. **Your Own Frequency Table (Unbinned).** Collect your own frequency data for some set of categories that will *not* require binning. (For example, you might collect data by asking friends to do a taste test on some brand of cookie.) State how you collected your data, and make a list of all your raw data. Then summarize the data in a frequency table. Include a column for relative frequency, and also include a column for cumulative frequency if it is appropriate. (Cumulative frequency is appropriate for all data except those at the nominal level of measurement.)

25. **Your Own Frequency Table (Binned).** Collect your own frequency data for some set of categories that *will* require binning (for example, weights of your friends or scores on a recent exam). State how you collected your data, and make a list of all your raw data. Then summarize the data in a frequency table. Include columns for relative frequency and cumulative frequency.

IN THE NEWS

26. **Frequency Tables.** Find a recent news article that includes some type of frequency table. Briefly describe the table and how it is useful to the news report. Do you think the table was constructed in the best possible way for the article? If so, why? If not, what would you have done differently?

27. **Relative Frequencies.** Find a recent news article that gives at least some data in the form of relative frequencies. Briefly describe the data, and discuss why relative frequencies were useful in this case.

28. **Cumulative Frequencies.** Find a recent news article that gives at least some data in the form of cumulative frequencies. Briefly describe the data, and discuss why cumulative frequencies were useful in this case.

29. **Temperature Data.** Look in a newspaper for a weather report that lists the expected high temperatures in many American cities. (Make a photocopy of the data so that your instructor can see them.) Choosing appropriate bins, make a frequency table for the high temperature data. Include columns for relative frequency and cumulative frequency. Briefly describe how and why you chose your bins.

3.2 Picturing Distributions of Data

A frequency table shows us how a variable is distributed over chosen categories, so we say that it summarizes the **distribution** of data. While tables can be extraordinarily useful, we often gain deeper insight into a distribution with a picture or graph. In this section, we'll study some of the most common methods for displaying distributions of data.

Definition

The **distribution** of a variable refers to the way its values are spread over all possible values. We can summarize a distribution in a table or show a distribution visually with a graph.

Bar Graphs, Dotplots, and Pareto Charts

A **bar graph** is one of the simplest ways to picture a distribution. Bar graphs are commonly used for qualitative data. Each bar represents the frequency (or relative frequency) of one category: the higher the frequency, the longer the bar. The bars can be either vertical or horizontal.

Let's create a vertical bar graph from the essay grade data in Table 3.1. We need five bars, one for each of the five categories (the grades A, B, C, D, F). The height of each bar should correspond to the frequency for its category. Figure 3.1 shows the result. Note the following key features of the graph's construction:

- Because the highest frequency is 9 (the frequency for C grades), we chose to make the vertical scale run from 0 to 10. This ensures that even the tallest bar does not quite touch the top of the graph.
- The graph should not be too short or too tall. In this case, it looks about right to choose a total height of 5 centimeters, which is convenient because it means that each centimeter of height corresponds to a frequency of 2.

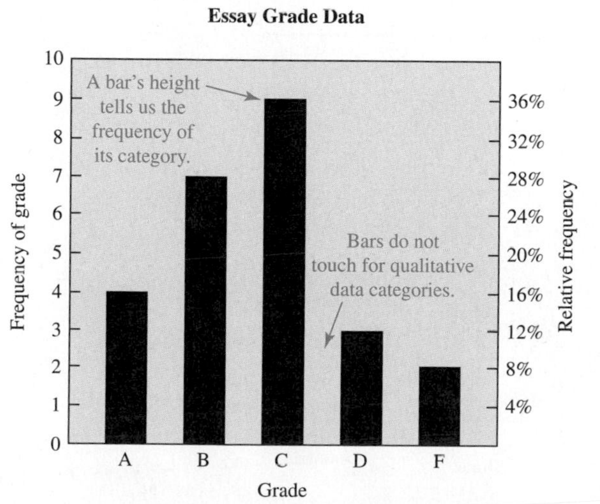

Figure 3.1 Bar graph for the essay grade data in Table 3.1.

Table 3.1 (repeated)

Grade	Frequency
A	4
B	7
C	9
D	3
F	2
Total	**25**

- The height of each bar should be proportional to its frequency. For example, because each centimeter of height corresponds to a frequency of 2, the bar representing a frequency of 4 should have a height of 2 centimeters.
- Because the data are qualitative, the widths of the bars have no special meaning, and there is no reason for them to touch each other. We therefore draw them with uniform widths.

Labeling graphs is extremely important. Without proper labels, a graph is meaningless. The following summary gives the important labels for almost any graph. Notice how these rules were applied in Figure 3.1.

Important Labels for Graphs

Title/caption: The graph should have a title or caption (or both) that explains what is being shown and, if applicable, lists the source of the data.

Vertical scale and label: Numbers along the vertical axis should clearly indicate the scale. The numbers should line up with the *tick marks*—the marks along the axis that precisely locate the numerical values. Include a label that describes the variable shown on the vertical axis.

Horizontal scale and label: The categories should be clearly indicated along the horizontal axis. (Tick marks may not be necessary for qualitative data, but should be included for quantitative data.) Include a label that describes the variable shown on the horizontal axis.

Legend: If multiple data sets are displayed on a single graph, include a legend or key to identify the individual data sets.

A **dotplot** is a variation on a bar graph in which we use dots rather than bars to represent the frequencies. Each dot represents one data value; for example, a stack of 4 dots means a frequency of 4. Figure 3.2 shows a dotplot for the essay data set. Dotplots are convenient when making graphs of raw data, because you can tally the data by making a dot for each data value. You may then choose to convert the graph to a bar chart for a formal report.

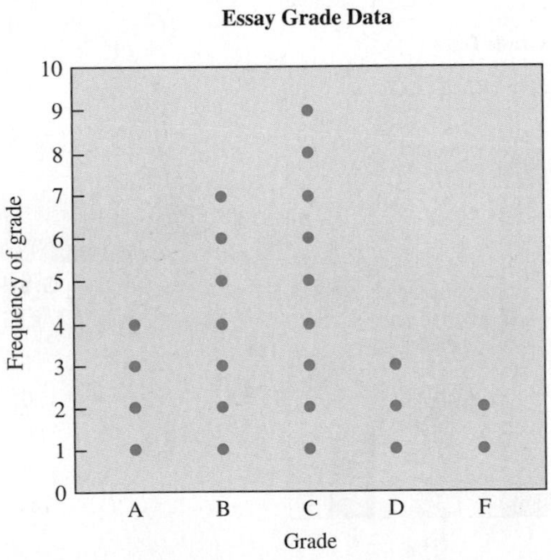

Figure 3.2 Dotplot for the essay grade data in Table 3.1.

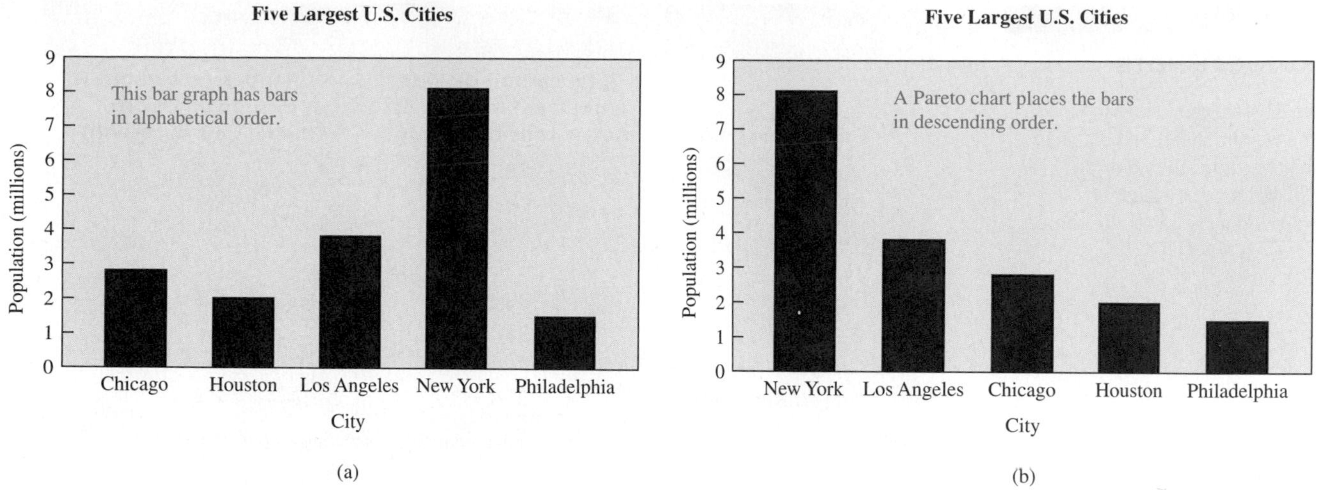

Figure 3.3 (a) Bar graph showing populations for the five largest cities in the United States (2005). (b) Pareto chart for the same data. *Source:* U.S. Census Bureau.

Figure 3.3a shows a bar graph of populations for the five largest cities in the United States. In this case, the five cities are the categories and their populations are the frequencies; the cities are listed in alphabetical order. Figure 3.3b shows the same data, but with the bars arranged in descending order. A bar graph in which the bars are arranged in frequency order is often called a **Pareto chart**. Note that rearranging the order of the bars makes sense only if the data are qualitative at the nominal level of measurement, as they are when the categories are cities. For example, it would not make sense to make a Pareto chart for the ordinal-level grade data in Figure 3.1, because putting the bars in frequency order would put the grades in the order C, B, A, D, F.

By the Way ...

Pareto charts were invented by Italian economist Vilfredo Pareto (1848–1923). Pareto is best known for developing methods of analyzing income distributions, but his most important contributions probably were in developing new ways of applying mathematics and statistics to economic analysis.

TIME OUT TO THINK

Would it be practical to make a dotplot for the population data in Figure 3.3? Would it make sense to make a Pareto chart for data concerning SAT scores? Explain.

Definitions

A **bar graph** consists of bars representing frequencies (or relative frequencies) for particular categories. The bar lengths are proportional to the frequencies.

A **dotplot** is similar to a bar graph, except each individual data value is represented with a dot.

A **Pareto chart** is a bar graph with the bars arranged in frequency order. Pareto charts make sense only for data at the nominal level of measurement.

EXAMPLE 1 Carbon Dioxide Emissions

Carbon dioxide is released into the atmosphere primarily by the combustion of fossil fuels (oil, coal, natural gas). Table 3.11 lists the eight countries that emit the most carbon dioxide each year. Construct Pareto charts for the total emissions and for the average emissions per person. Discuss why the two charts look so different.

TECHNICAL NOTE

Table 3.11 gives emissions in terms of the weight of carbon contained in the emitted carbon dioxide (CO_2); it does not include the weight of the oxygen in the carbon dioxide.

Table 3.11 The World's Eight Leading Emitters of Carbon Dioxide

Country/region	Total carbon dioxide emissions (millions of metric tons of carbon)	Per person carbon dioxide emissions (metric tons of carbon)
United States	1,582	5.4
China	966	0.7
Russia	438	3.0
Japan	329	2.6
India	280	0.3
Germany	230	2.8
Canada	164	5.2
United Kingdom	154	2.6

Source: U.S. Department of Energy. Data released in 2007, showing emissions for 2003.

Solution The categories are the countries and the frequencies are the data values. Because the range of data values for total carbon dioxide emissions goes from 154 to 1,582, a range of 0 to 1,600 makes a good choice for the vertical scale. To make the Pareto chart, we put the bars in descending order of size. Each bar's height corresponds to its data value, and we label the category (country) under the bar. Figure 3.4a shows the resulting graph for total carbon dioxide emissions. Figure 3.4b shows the Pareto chart for the per person emissions; a vertical scale range of 0 to 6 encompasses all the data.

Notice that the two Pareto charts have the countries in different orders. This tells us that the biggest total emitters of carbon dioxide are not necessarily the biggest per person emitters. For example, China ranks as the second largest total emitter, but its per person emissions are far below those of the United States and other more developed nations.

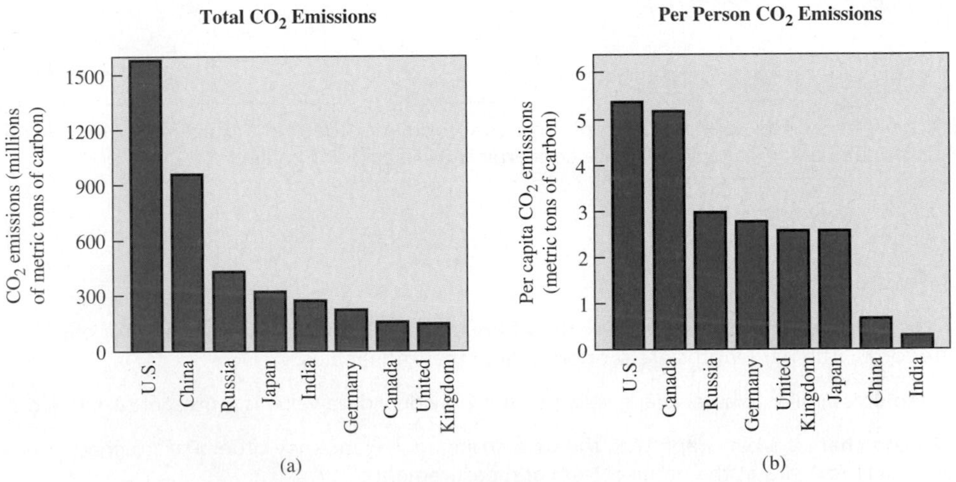

Figure 3.4 Pareto charts for (a) carbon dioxide emissions by country and (b) per person carbon dioxide emissions.

TIME OUT TO THINK

The combined population of China and India is more than eight times the U.S. population, yet U.S. carbon dioxide emissions are larger than those of China and India combined. Why do you think this is the case? What consequences might there be for the world if China and India had the same per person carbon dioxide emissions as the United States?

Pie Charts

Pie charts are usually used to show relative frequency distributions. A circular pie represents the total relative frequency of 100%, and the sizes of the individual slices, or wedges, represent the relative frequencies of different categories. Pie charts are used almost exclusively for qualitative data.

As a simple example, consider the registered voters in Rochester County: 25% are Democrats, 25% are Republicans, and 50% are independents. We can show these party affiliations in a pie chart. Because Democrats and Republicans each represent 25% of the voters, the wedges for Republicans and Democrats each occupy 25%, or one-fourth, of the pie. Independents represent half of the voters, so their wedge occupies the remaining half of the pie. Figure 3.5 shows the result. As always, note the importance of clear labeling.

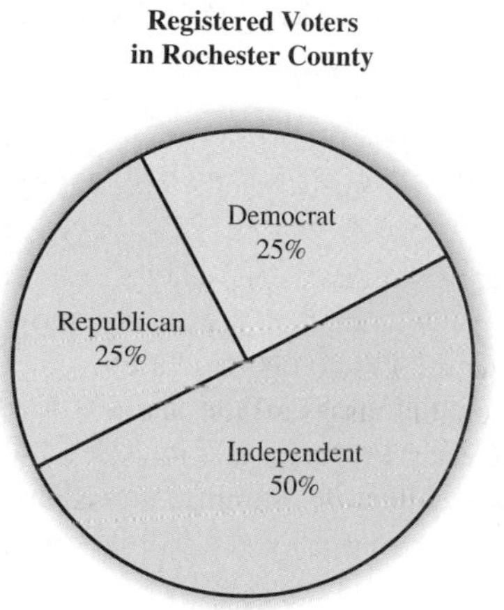

Figure 3.5 Party affiliations of registered voters in Rochester County.

By the Way ...

Nationally, as of the 2006 elections, 37% of voting-age Americans identified themselves as Democrats, 31% identified themselves as Republicans, and 32% identified themselves as independents or with other parties (*Statistical Abstract of the United States*).

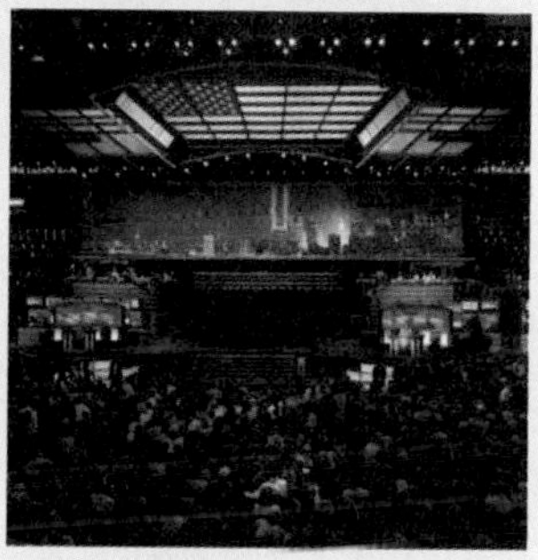

Definition

A **pie chart** is a circle divided so that each wedge represents the *relative frequency* of a particular category. The wedge size is proportional to the relative frequency. The entire pie represents the total relative frequency of 100%.

A pie chart like the one in Figure 3.5 is easy to create because the wedge sizes represent very simple fractions. For more complex pie charts, you must either make careful angle measurements or use software that makes the measurements. And while pie charts can be useful for simple data sets, the following example shows that complex pie charts may not always be the best way to present data.

EXAMPLE 2 Student Majors

Figure 3.6 is a pie chart showing planned major areas for first-year college students. Construct a Pareto chart showing the same data. What are the three most popular major areas? Comment on the relative ease with which this question can be answered with the pie chart and the Pareto chart.

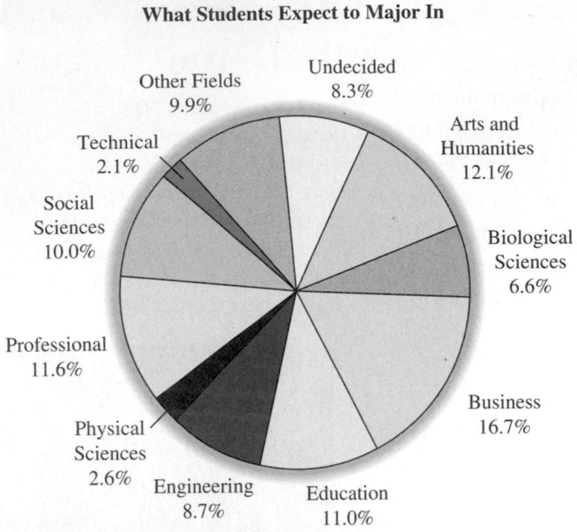

Figure 3.6 Planned major areas for first-year college students. *Source: The Chronicle of Higher Education.*

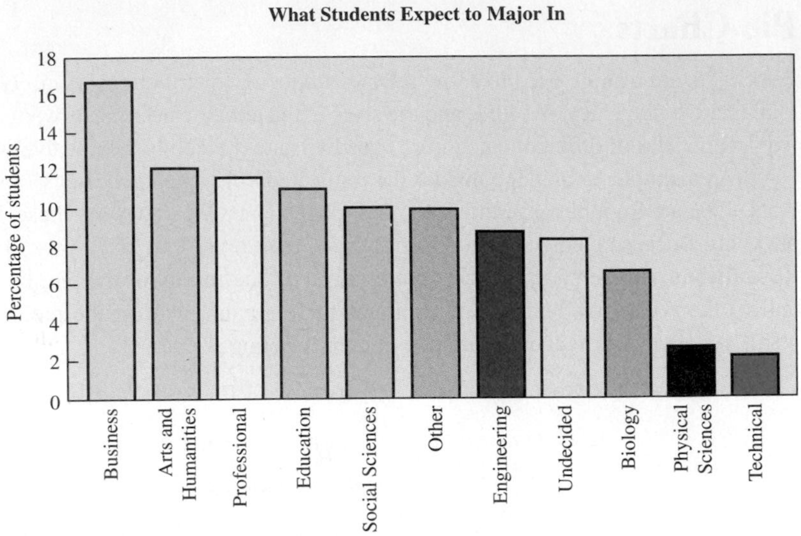

Figure 3.7 Pareto chart for the data in Figure 3.6.

Solution Figure 3.7 shows the Pareto chart for the data. This chart makes it immediately obvious that the three most popular major areas are business (16.7%), arts and humanities (12.1%), and professional (11.6%). ("Professional" includes students majoring in the fields with professional licensing, such as architecture, nursing, and pharmacology.) In contrast, it takes a fair amount of study of the pie chart before we can list the three most popular major areas.

TIME OUT TO THINK

Example 2 discussed an advantage of a Pareto chart over a pie chart for showing the data concerning major areas. Do you think the pie chart has any advantages over the Pareto chart? If so, what are they?

TECHNICAL NOTE

Different books define the terms *histogram* and *bar graph* differently, and there are no universally accepted definitions. In this book, a bar graph is any graph that uses bars, and histograms are the types of bar graphs used for quantitative data at the interval or ratio level of measurement.

Histograms and Stem-and-Leaf Plots

Figure 3.8 shows a bar graph for the energy use data of Table 3.4. The horizontal axis is marked with the categories of energy use per person per year (in millions of BTUs), and the vertical axis is marked with the frequencies (number of states) that correspond to each category. As in all bar graphs, the lengths of the bars are proportional to the frequencies. However, unlike the bars on the graphs we made earlier with qualitative data, the bars on this graph fall into a natural order based on the category values. In addition, the widths of the bars in Figure 3.8 have a specific meaning—in this case, telling us the range of values in each of the energy use categories. This type of bar graph, in which the bars have a natural order and the bar widths have specific meaning, is called a **histogram**. The bars in a histogram touch each other because there are no gaps between the categories.

The energy use histogram clearly reveals the trends listed in a frequency table like Table 3.4. However, neither the frequency table nor the histogram provides all the details of the original data set in Table 3.3. For example, the histogram tells us that one state has energy use in the category 800–899 million BTUs per person, but it does not tell us which one or the precise

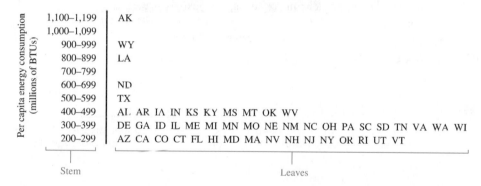

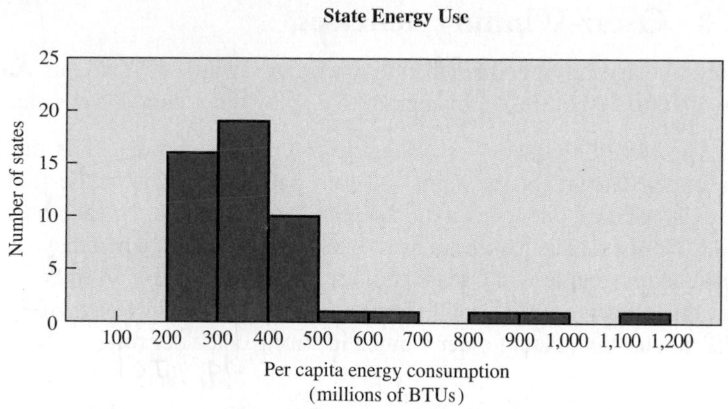

Figure 3.8 Histogram for the data in Table 3.4.

energy use value for this state. The **stem-and-leaf plot** (or *stemplot*) in Figure 3.9 gives a more detailed look at the data. It looks somewhat like a histogram turned sideways, except in place of bars we see a listing of data—in this case, the states—for each category. We can now see, for example, that the state in the category 800–899 million BTUs per person is Louisiana.

Per capita energy consumption (millions of BTUs) — Stem	Leaves
1,100–1,199	AK
1,000–1,099	
900–999	WY
800–899	LA
700–799	
600–699	ND
500–599	TX
400–499	AL AR IA IN KS KY MS MT OK WV
300–399	DE GA ID IL ME MI MN MO NE NM NC OH PA SC SD TN VA WA WI
200–299	AZ CA CO CT FL HI MD MA NV NH NJ NY OR RI UT VT

Figure 3.9 Stem-and-leaf plot for the energy use data from Table 3.3.

Another type of stem-and-leaf plot lists the individual data *values*. For example, Figure 3.10 shows a stem-and-leaf plot for the per person carbon dioxide emissions data in Table 3.11. In this case, each data value is represented with a stem consisting of the first digit (and the decimal point) and a leaf consisting of the remaining digit. We can read the data values directly from this plot. For example, the first row shows the data values 0.3 and 0.7.

Stem	Leaves
0.	3 7
1.	
2.	6 6 8
3.	0
4.	
5.	2 4

Figure 3.10 Stem-and-leaf plot showing numerical data—in this case, the per person carbon dioxide emissions from Table 3.11.

Definitions

A **histogram** is a bar graph showing a distribution for quantitative data (at the interval or ratio level of measurement); the bars have a natural order and the bar widths have specific meaning.

A **stem-and-leaf plot** (or *stemplot*) is somewhat like a histogram turned sideways, except in place of bars we see a listing of data.

Table 3.12 Ages of Actresses at Time of Academy Award, 1970–2007

Age	Number of actresses
20–29	8
30–39	18
40–49	8
50–59	0
60–69	2
70–79	1
80–89	1

EXAMPLE 3 Oscar-Winning Actresses

Table 3.12 shows the ages (at the time when they won the award) of Academy Award–winning actresses from 1970 to 2007. Make a histogram to display these data. Discuss the results.

Solution The largest frequency is 18 (actresses), so we choose 20 as the height of the histogram. As in any bar graph, the height of each bar corresponds to the frequency for the category. The width of each bar spans a full 10 years, so the bars touch each other. Figure 3.11 shows the result. We see that actresses are most likely to win Oscars when they are fairly young. In contrast, male actors tend to win Oscars at older ages (see Exercise 13 in Section 3.1). Many actresses believe these facts reflect a subtle form of discrimination, in which Hollywood producers rarely make movies that feature older women in strong character roles.

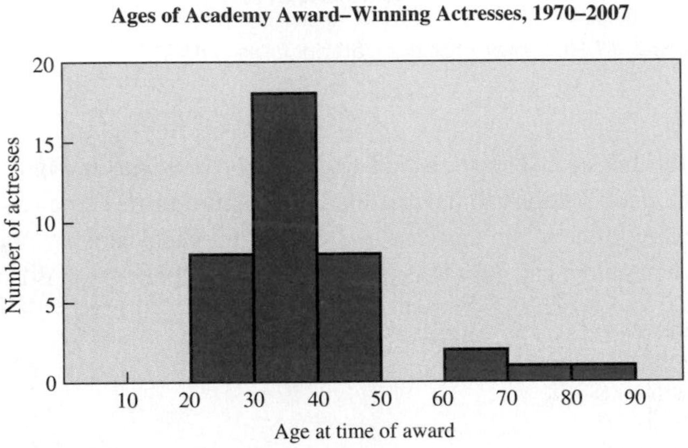

Figure 3.11 Histogram for ages of Academy Award–winning actresses.

By the Way ...

The oldest Oscar-winning actress was Jessica Tandy, who won in 1990 for the film *Driving Miss Daisy*.

TIME OUT TO THINK

What additional information would you need to create a stem-and-leaf plot for the ages of actresses when they won Academy Awards? What would the stem-and-leaf plot look like?

Line Charts

Like a histogram, a **line chart** shows a distribution of quantitative data. However, instead of using bars, a line chart connects a series of dots. The vertical position of each dot represents a frequency; the dot goes where the top of a bar would go on a histogram. Figure 3.12 shows a line chart for the energy data in Table 3.4. For comparison, it is overlaid by the histogram shown earlier (see Figure 3.8).

Note one important subtlety in interpreting a line chart: The horizontal positions of the dots correspond to the *centers* of the bins. For example, the dot representing the bin for 300–399 million BTUs is located at 349.5 million BTUs. Therefore, if you didn't know better, you might think the dot meant that 19 states had an energy use of *exactly* 349.5 million BTUs per person, when actually 19 states are in the *range* of 300–399 million BTUs per person.

TECHNICAL NOTE

A line chart created from a set of frequency data is often called a *frequency polygon* because it consists of many straight line segments that take the shape of a many-sided figure.

Definition

A **line chart** shows a distribution of quantitative data as a series of dots connected by lines. For each dot, the horizontal position is the *center* of the bin it represents and the vertical position is the frequency value for the bin.

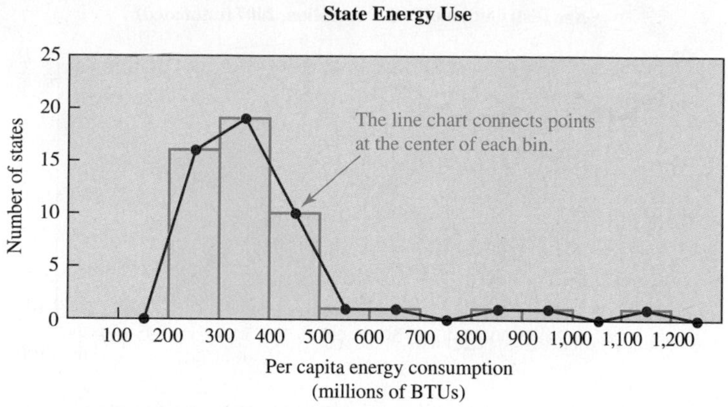

Figure 3.12 Line chart for the energy use data, with a histogram overlaid for comparison.

EXAMPLE 4 Age Distribution

Table 3.13 shows the distribution of the U.S. population according to age categories. Make a histogram and a line chart for these data. If you made a similar graph for age data 20 years from now, how would you expect it to be different?

Table 3.13 Estimated Age Distribution of U.S. Population, 2007

Age category	Population (millions)	Age category	Population (millions)	Age category	Population (millions)
0–4	21.0	35–39	19.0	70–74	9.0
5–9	19.4	40–44	20.4	75–79	7.2
10–14	19.9	45–49	22.2	80–84	5.6
15–19	21.7	50–54	22.0	85–89	3.5
20–24	21.2	55–59	19.2	90–94	1.6
25–29	19.8	60–64	16.3	95–99	0.6
30–34	19.0	65–69	12.2	100 or more	0.1

Source: U.S. Census Bureau.

Solution The data are already binned in 5-year intervals (except for the last category of "100 or more"). We make the histogram by representing the frequency (population) in each category with a bar. Figure 3.13 shows the result. Superimposed on this histogram is a line chart in which a dot is placed at the center of each category. For example, the dot representing 25–29-year-olds is located at 27.5 on the horizontal axis. For consistency, we show the "100 or more" category as if it were 100–104, but the label on the horizontal axis shows what it really means. Many frequency tables include open-ended categories such as "100 or more" when a small number of extreme data values are spread over a wide range.

If we made a similar graph in 20 years, we would expect to see two major differences. First, all the frequencies would be higher because of overall population growth. Second, there would be more growth in the higher age categories than in the lower age categories because advances in medical technology are allowing more people to live longer.

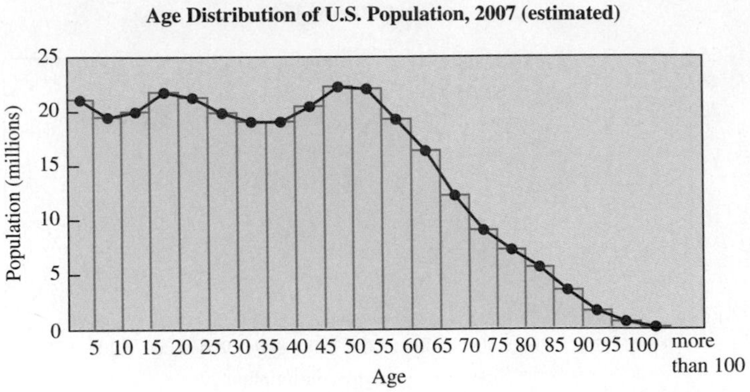

Figure 3.13 Histogram and line chart for the age data in Table 3.13.

Time-Series Diagrams

Table 3.14 shows how the homicide rate in the United States has changed with time. The categories are the years and the data are the homicide rates (measured in deaths per 100,000 people). We can represent these data with either a histogram or a line chart; Figure 3.14 shows a line chart. Because the horizontal axis represents time in this case, we say that this graph is a **time-series diagram**.

Table 3.14 U.S. Homicide Rate per 100,000 People Since 1960

Year	Homicides (per 100,000 people)	Year	Homicides (per 100,000 people)	Year	Homicides (per 100,000 people)
1960	5.1	1976	8.8	1992	9.3
1961	4.8	1977	8.8	1993	9.5
1962	4.6	1978	9.0	1994	9.0
1963	4.6	1979	9.7	1995	8.2
1964	4.9	1980	10.2	1996	7.4
1965	5.1	1981	9.8	1997	6.8
1966	5.6	1982	9.1	1998	6.3
1967	6.2	1983	8.3	1999	5.7
1968	6.9	1984	7.9	2000	5.5
1969	7.3	1985	7.9	2001	5.6
1970	7.9	1986	8.6	2002	5.6
1971	8.6	1987	8.3	2003	5.7
1972	9.0	1988	8.4	2004	5.5
1973	9.4	1989	8.7	2005	5.8
1974	9.8	1990	9.4		
1975	9.6	1991	9.8		

Source: United States Bureau of Justice Statistics.

Definition

A histogram or line chart in which the horizontal axis represents *time* is called a **time-series diagram**.

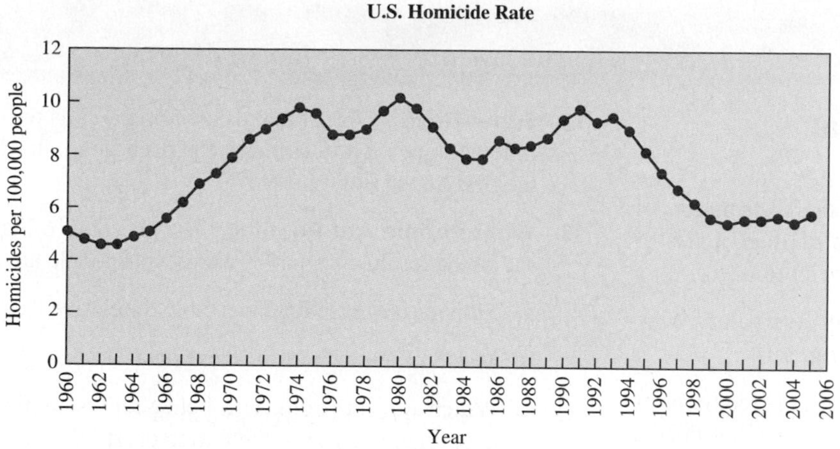

Figure 3.14 Time-series diagram for the homicide rate data of Table 3.14.

TIME OUT TO THINK

Look for data to extend Table 3.14 and Figure 3.14 beyond 2005. Do you see any trend in the homicide rate?

EXAMPLE 5 Declining Death Rates

Figure 3.15 shows a time-series diagram for the death rate (deaths per 1,000 people) in the United States since 1900. (For example, the 1905 death rate of 15 means that, for each 1,000 people living at the beginning of 1905, 15 people died during the year.) Discuss the general trend and reasons for the trend. Also consider the spike in 1919: If someone told you that this spike was due to battlefield deaths in World War I, would you believe it? Explain.

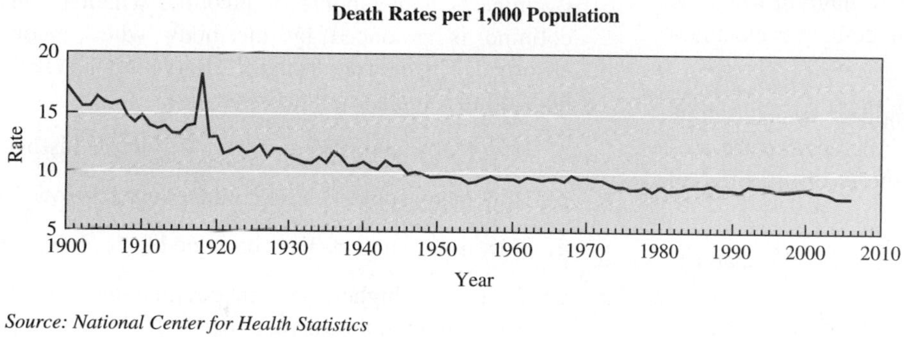

Figure 3.15 Historical U.S. death rates per 1,000 people.

Solution The general trend in death rates is clearly downward, presumably because of improvements in medical science. For example, bacterial diseases such as pneumonia were major killers in the early 1900s, but are largely curable with antibiotics today. The spike in 1919 *does* coincide with the end of World War I. However, if the spike were due to battlefield casualties, we might expect it to span the several years of World War I and we might expect to see a similar spike during World War II. The absence of these features suggests that the 1919 spike may not be due to World War I. In fact, the reason for the spike was a deadly epidemic of influenza.

By the Way ...

The influenza epidemic of 1919 killed 850,000 people in the United States and an estimated 20 million people worldwide. Most deaths were caused by secondary bacterial infections that would be curable today with antibiotics.

Section 3.2 Exercises

Statistical Literacy and Critical Thinking

1. **Distribution.** What do we mean by the distribution of data? How is a histogram more helpful than a list of sample values for understanding the distribution?

2. **Small Data Set.** If a data set has only five values, why should we not bother to construct a histogram?

3. **Time-Series Diagram.** What type of data is required for the construction of a time-series graph, and what does a time-series graph reveal about the data?

4. **Histogram and Stem-and-Leaf Plot.** Assume that a data set is used to construct a histogram and a stem-and-leaf plot. Using only the histogram, is it possible to re-create the original list of data values? Using only the stem-and-leaf plot, is it possible to re-create the original list of data values? What is an advantage of a stem-and-leaf plot over a histogram?

Does It Make Sense? For Exercises 5–8, decide whether the statement makes sense (or is clearly true) or does not make sense (or is clearly false). Explain clearly. Not all of these statements have definitive answers, so your explanation is more important than your chosen answer.

5. **Histogram.** A histogram is used to depict a sample of heights categorized according to the eye color of the subject.

6. **Time-Series Diagram.** A time-series diagram would be used to show the net profit of different computer manufacturers in 2007.

7. **Pareto Chart.** A quality control engineer wants to draw attention to the most serious causes of defects, so she uses a Pareto chart to illustrate the frequencies of the different causes of defects.

8. **Pie Chart.** An important disadvantage of pie charts is that they lack an appropriate scale.

Concepts and Applications

Most Appropriate Display. Exercises 9–12 describe data sets, but do not give actual data. For each data set, state the type of graphic that you believe would be most appropriate for displaying the data, if they were available. Explain your choice.

9. **SAT Scores.** SAT scores of college statistics students

10. **Eye Color.** Eye colors of actresses

11. **Causes of Death.** Causes of deaths of people who died last year

12. **Super Bowl.** Cost of television commercials during halftime at Super Bowl contests for each year, starting with the first Super Bowl in 1967

13. **What People Are Reading.** The pie chart in Figure 3.16 shows the results of a survey about what people are reading.

    a. Summarize these data in a table of relative frequencies.

    b. Make a Pareto chart for these data.

    c. Which do you think is a better representation of the data: the pie chart or the Pareto chart? Why?

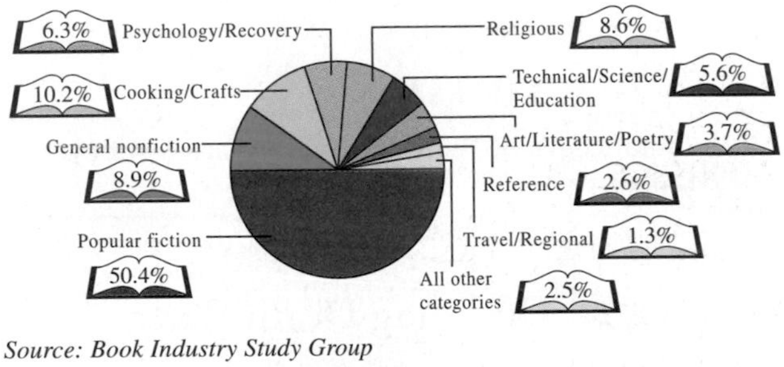

Figure 3.16 *Source: Wall Street Journal Almanac.*

14. **Histogram.** The STATDISK-generated histogram in Figure 3.17 depicts cotinine levels (in milligrams per milliliter) of a sample of subjects who smoke cigarettes. Cotinine is a metabolite of nicotine, which means that cotinine is produced by the body when nicotine is absorbed. The data are from the Third National Health and Nutrition Examination Survey.

    a. How many subjects are represented in the histogram?

    b. How many subjects have cotinine levels below 400?

    c. How many subjects have cotinine levels above 150?

    d. What is the highest possible cotinine level of a subject represented in this histogram?

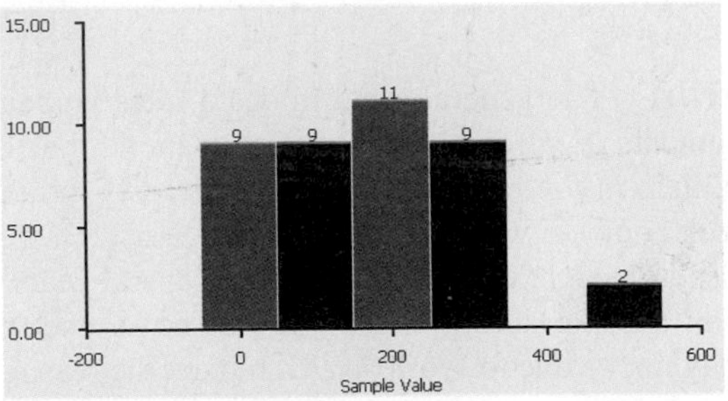

Figure 3.17

15. Weights of Coke. Exercise 11 in Section 3.1 required the construction of a frequency table from the weights (in pounds) of 36 cans of regular Coke. Use that frequency table to construct the corresponding histogram.

16. Weights of Diet Coke. Exercise 12 in Section 3.1 required the construction of a frequency table for the weights (in pounds) of 36 cans of Diet Coke. Use that frequency table to construct the corresponding histogram.

17. Oscar-Winning Actors. Exercise 13 in Section 3.1 required the construction of a frequency table for the ages of recent Academy Award–winning male actors at the time when they won their award. Use that frequency table to construct the corresponding histogram.

18. Body Temperatures. Exercise 14 in Section 3.1 required the construction of a frequency table for a list of body temperatures (°F) of randomly selected subjects. Use that frequency table to construct the corresponding histogram.

19. Job Hunting. A study was conducted to determine how people get jobs. The table below lists data from 400 randomly selected subjects. The data are based on results from the National Center for Career Strategies. Construct a Pareto chart that corresponds to the given data. If someone would like to get a job, what seems to be the most effective approach?

Job sources of survey respondents	Frequency
Help-wanted ads	56
Executive search firms	44
Networking	280
Mass mailing	20

20. Job Sources. Refer to the data given in Exercise 19, and construct a pie chart. Compare the pie chart to the Pareto chart. Can you determine which graph is more effective in showing the relative importance of job sources?

21. Causes of Train Derailments. An analysis of train derailment incidents showed that 23 derailments were caused by bad track, 9 were due to faulty equipment, 12 were attributable to human error, and 6 had other causes (based on data from the Federal Railroad Administration). Construct a pie chart representing the given data.

22. Analyzing Causes of Train Derailments. Refer to the data given in Exercise 21, and construct a Pareto chart. Compare the Pareto chart to the pie chart. Can you determine which graph is more effective in showing the relative importance of the causes of train derailments?

23. Dotplot. Refer to the QWERTY data in Exercise 19 in Section 3.1 and construct a dotplot.

24. Dotplot. Refer to the Dvorak data in Exercise 19 in Section 3.1 and construct a dotplot. Compare the result to the dotplot in Exercise 23. Based on the results, does either keyboard configuration appear to be better? Explain.

25. Time Series for DJIA. The following values are high values of the Dow Jones Industrial Average (DJIA) for years beginning with 1980. The data are arranged in order by row. Construct a time-series diagram and then determine whether there appears to be a trend. How might an investor profit from this trend?

1,000	1,024	1,071	1,287	1,287	1,553
1,956	2,722	2,184	2,791	3,000	3,169
3,413	3,794	3,978	5,216	6,561	8,259
9,374	11,568	11,401	11,350	10,635	10,454
10,855	10,941	12,464			

26. Time Series for Motor Vehicle Deaths. The following values are numbers of motor vehicle deaths in the United States for years beginning with 1980. The data are arranged in order by row. Construct a time-series graph and then determine whether there appears to be a trend. If so, provide a possible explanation.

51,091	49,301	43,945	42,589	44,257	43,825
46,087	46,390	47,087	45,582	44,599	41,508
39,250	40,150	40,716	41,817	42,065	42,013
41,501	41,717	41,945	42,196	43,005	42,884
42,836	43,443				

27. Stem-and-Leaf Plot. Construct a stem-and-leaf plot of these test scores: 67, 72, 85, 75, 89, 89, 88, 90, 99, 100. How does the stem-and-leaf plot show the distribution of these data?

28. Stem-and-Leaf Plot. Listed below are the lengths (in minutes) of animated children's movies. Construct a stem-and-leaf plot. Does the stem-and-leaf plot show the distribution of the data? If so, how?

83	88	120	64	69	71	76	74	75	76	75
75	79	80	78	78	83	77	71	83	80	73
72	82	74	84	90	89	81	81	90	79	92
82	89	82	74	86	76	81	75	75	77	70
75	64	73	74	71	94					

Projects for the Internet and Beyond

For useful links, select "Links for Internet Projects" for Chapter 3 at www.aw.com/bbt.

29. CO_2 Emissions. Look for updated data on international carbon dioxide emissions at the Web site for the *Inter-*

national Energy Annual, published by the U.S. Energy Information Administration (EIA). Create a graph that includes the latest data. Discuss any important features of your graph.

30. Energy Table. Explore the full set of energy tables at the U.S. Energy Information Administration (EIA) Web site. Choose a table that you find interesting and make a graph of its data. You may choose any of the graph types discussed in this section. Explain how you made your graph, and briefly discuss what can be learned from it.

31. Statistical Abstract. Go to the Web site for the *Statistical Abstract of the United States*. Explore the selection of "frequently requested tables." Choose one table of interest to you and make a graph from its data. You may choose any of the graph types discussed in this section. Explain how you made your graph and briefly discuss what can be learned from it.

32. Navel Data. Create an appropriate display of the navel data collected in Exercise 23 of Section 3.1. Discuss any special properties of this distribution.

IN THE NEWS

33. Bar Graphs. Find a recent news article that includes a bar graph with qualitative data categories.

a. Briefly explain what the bar graph shows, and discuss whether it helps make the point of the news article. Are the labels clear?

b. Briefly discuss whether the bar graph could be recast as a dotplot.

c. Is the bar graph already a Pareto chart? If so, explain why you think it was drawn this way. If not, do you think it would be clearer if the bars were rearranged to make a Pareto chart? Explain.

34. Pie Charts. Find a recent news article that includes a pie chart. Briefly discuss the effectiveness of the pie chart. For example, would it be better if the data were displayed in a bar graph rather than a pie chart? Could the pie chart be improved in other ways?

35. Histograms. Find a recent news article that includes a histogram. Briefly explain what the histogram shows, and discuss whether it helps make the point of the news article. Are the labels clear? Is the histogram a time-series diagram? Explain.

36. Line Charts. Find a recent news article that includes a line chart. Briefly explain what the line chart shows, and discuss whether it helps make the point of the news article. Are the labels clear? Is the line chart a time-series diagram? Explain.

3.3 Graphics in the Media

The basic graphs we have studied so far are only the beginning of the many ways to depict data visually. In this section, we will explore some of the more complex types of graphics that are common in the media.

Multiple Bar Graphs and Line Charts

A **multiple bar graph** is a simple extension of a regular bar graph: It has two or more sets of bars that allow comparison between two or more data sets. All the data sets must have the same categories so that they can be displayed on the same graph. Figure 3.18 is a multiple bar graph showing the amount of money that children receive from their parents. The categories are age ranges of children. The three sets of bars represent weekly income from three different parental sources: allowance, handouts, and payments for chores. (Note: The median, which we will discuss in Section 4.1, is the middle of a set of values; for example, the median allowance of \$3.00 for children ages 9–10 means that half these children received more than \$3.00 and half received less than \$3.00.)

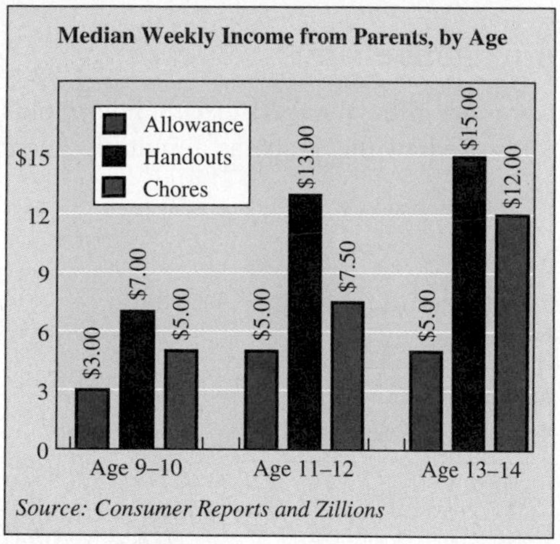

Figure 3.18 A multiple bar graph. *Source: Wall Street Journal Almanac.*

A **multiple line chart** follows the same basic idea as a multiple bar chart, but shows the related data sets with lines rather than bars. Figure 3.19 shows a multiple line chart of stock, bond, and gold prices over a 12-week period. This particular chart is also a time-series diagram, because it shows data over time.

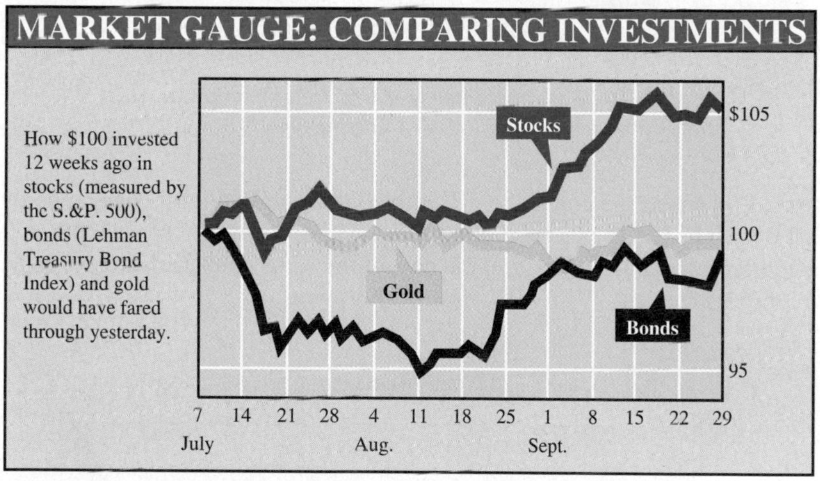

Figure 3.19 A multiple line chart. *Source: New York Times.*

By the Way ...

Gold was once considered to be a solid investment and an important part of any investment portfolio. However, gold prices have languished in recent decades. At the end of 2006, gold was worth only about $630 per ounce—much less than its inflation-adjusted value of more than $2,000 per ounce in 1980.

EXAMPLE 1 Reading the Investment Graph

Consider Figure 3.19. Suppose that, on July 7, you had invested $100 in a stock fund that tracks the S&P 500, $100 in a bond fund that follows the Lehman Index, and $100 in gold. If you sold all three funds on September 15, how much would you have gained or lost?

Solution The graph shows that the $100 in the stock fund would have been worth about $105 on September 15. The $100 bond investment would have declined in value to about $99. The gold investment would have held its initial value of $100. On September 15, your complete portfolio would have been worth

$$\$105 + \$99 + \$100 = \$304$$

You would have gained $4 on your total investment of $300.

By the Way ...

In 1993, only 3 million people worldwide were connected to the Internet. By 2007, Internet access was shared by more than 210 million Americans and more than 1 billion people worldwide.

EXAMPLE 2 Graphic Conversion

Figure 3.20 is a multiple bar graph of the numbers of U.S. households with computers and the number of on-line households. Redraw this graph as a multiple line chart. Briefly discuss the trends shown on the graphs.

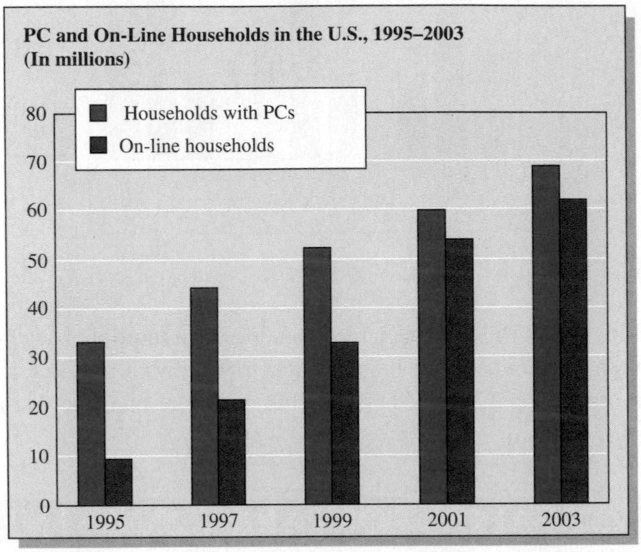

Figure 3.20 A multiple bar graph of trends in home computing. *Source: Statistical Abstract of the United States.*

Solution To convert the graphic to a multiple line chart, we must change the two sets of bars to a set of two lines. We place a dot corresponding to the height of each bar in the *center* of each category (year) and then connect the dots of the same color with a color-coded line, as shown in Figure 3.21.

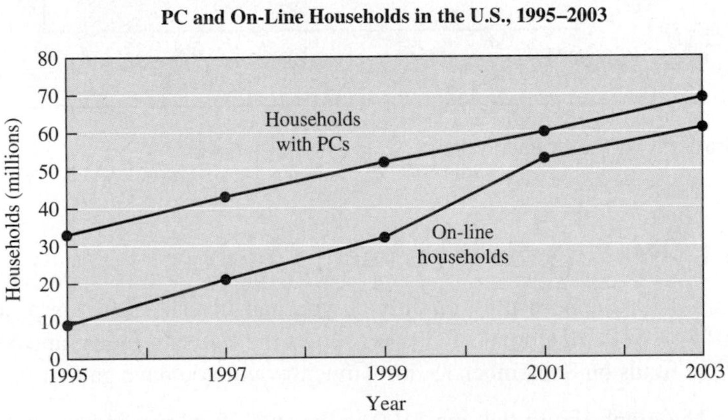

Figure 3.21 Multiple line chart showing the data from Figure 3.20.

The most obvious trend is that both data sets show an increase with time. We see a second trend by comparing the bars within each year. In 1995, the number of on-line homes (about 10 million) was less than one-third the number of homes with computers (about 33 million). By 2003, the number of on-line homes (about 62 million) was about 90% of the number of homes with computers (about 70 million). This tells us that a higher percentage of computer users are going on-line. If we project the trends into the future, it seems likely that the number of on-line households will approach the number of households with computers, and both will approach the total number of households in the United States.

Stack Plots

Another way to show two or more related data sets simultaneously is with a **stack plot**, which shows different data sets in a vertical stack. Although data can be stacked in both bar charts and line charts, the latter are much more common.

EXAMPLE 3 Stacked Line Chart

Figure 3.22 shows death rates (deaths per 100,000 people) for four diseases since 1900. Based on this graph, what was the death rate for cardiovascular disease in 1980? Discuss the general trends visible on this graph.

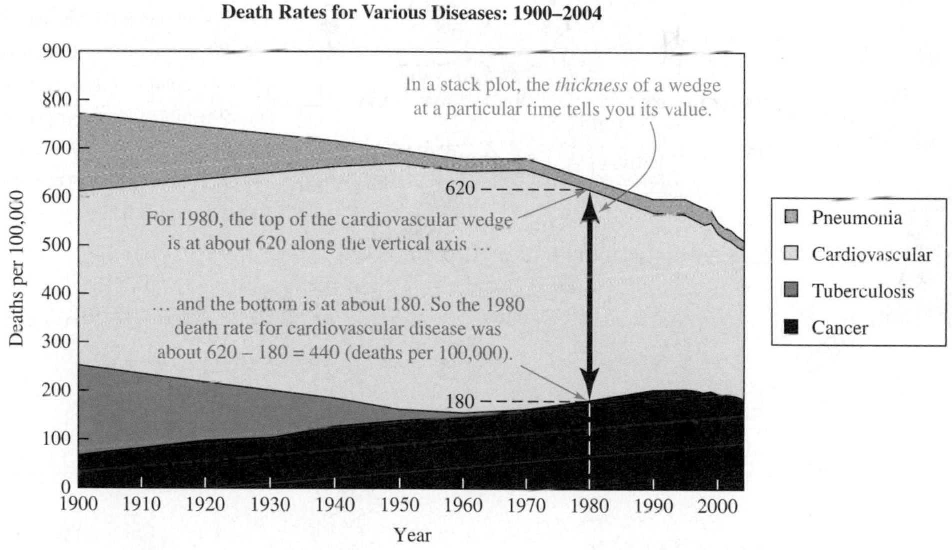

Figure 3.22 A stack plot using stacked wedges. *Sources:* National Center for Health Statistics, American Cancer Society.

Solution Each disease has its own color-coded region, or wedge; note the importance of the legend. The *thickness* of a wedge at a particular time tells us its value at that time. For 1980, the cardiovascular wedge extends from about 180 to 620 on the vertical axis, so its thickness is about 440. This tells us that the death rate in 1980 for cardiovascular disease was about 440 deaths per 100,000 people. The graph shows several important trends. First, the downward slope of the top wedge shows that the overall death rate from these four diseases decreased substantially, from nearly 800 deaths per 100,000 in 1900 to about 525 in 2003. The drastic decline in the thickness of the tuberculosis wedge shows that this disease was once a major killer, but has been nearly wiped out since 1950. Meanwhile, the cancer wedge shows that the death rate from cancer rose steadily until the mid-1990s, but has dropped somewhat since then.

By the Way ...

The death rate from cancer rose steadily through most of the 20th century, reaching a peak in 1995 at about 205 deaths per 100,000 people. Since that time, however, the cancer death rate has declined about 10%, and in 2006 the actual number of cancer deaths (not just the rate per 100,000) decreased for the first time. Scientists attribute the decline primarily to a decrease in the number of people who smoke (since smoking is estimated to cause about 1/3 of all cancers) and improvements in early detection and treatment of cancer.

Geographical Data

The energy use data in Table 3.3 are an example of **geographical data**, because the raw data correspond to different geographical locations. We used these data earlier to make a frequency table (Table 3.4), a histogram (Figure 3.8), and a stem-and-leaf plot (Figure 3.9). However, these displays do not give us a sense of any geographical patterns in the data. For example, we are unable to see whether states in the northeast all have similar levels of energy use. Figure 3.23 shows one way to display the geographical trends. The categories are color-coded according to the key, and each state is colored appropriately.

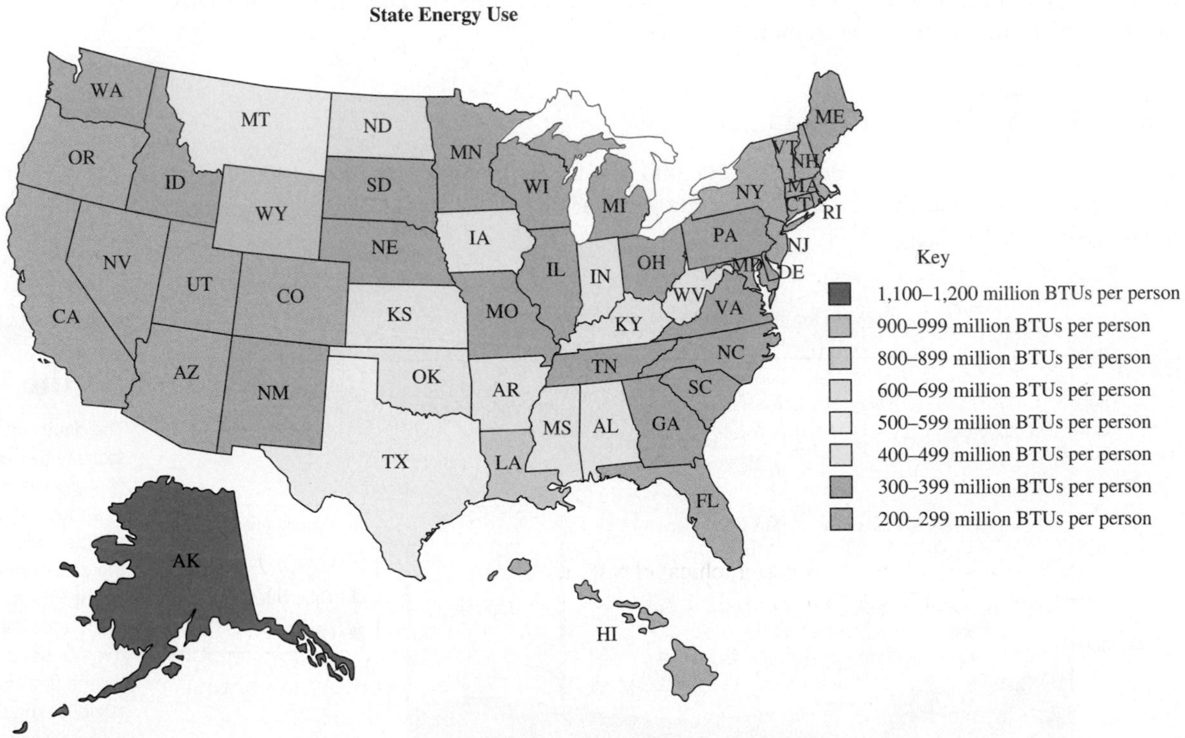

Figure 3.23 Geographical data can be displayed with a color-coded map.

TIME OUT TO THINK

What can you learn from the histogram in Figure 3.8 that you cannot learn easily from the geographical display in Figure 3.23? What can you learn from the geographical display that you cannot learn from the histogram? Do you see any surprising geographical trends in Figure 3.23? Explain.

Figure 3.23 works well for the energy data set because each state is associated with a unique number. For data that vary continuously across geographical areas, a **contour map** is more convenient. Figure 3.24 shows a contour map of temperature over the United States at a particular time. Each of the *contours* (lines) connects locations with the same temperature. For example, the temperature is 50°F everywhere along the contour labeled 50° and 60°F everywhere along the contour labeled 60°. Between these two contours, the temperature is between 50°F and 60°F. Note that greater temperature differences mean more tightly spaced contours. For example, the closely packed contours in the northeast indicate that the temperature varies substantially over small distances. To make the graph easier to read, the regions between adjacent contours are color-coded.

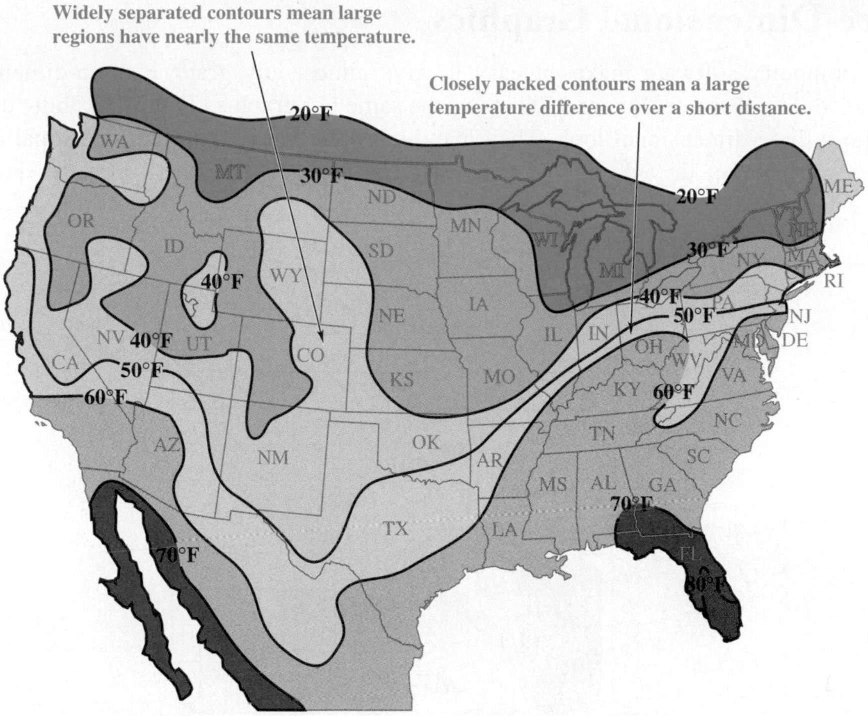

Figure 3.24 Geographical data that vary continuously, such as temperatures, can be displayed with a contour map.

EXAMPLE 4 A Contour Elevation Map

Contour plots are also used to show geographical elevations. Figure 3.25 shows elevation contours around Boulder, Colorado. Discuss a few of the key features shown on the map.

Solution The labels on the figure indicate key features. The contours are widely spaced in the east, where the terrain is relatively flat and the elevations are fairly constant. Westward, the contours become closely spaced where the mountains rise up from the plains. The concentric closed contours in the center of the map surround peaks.

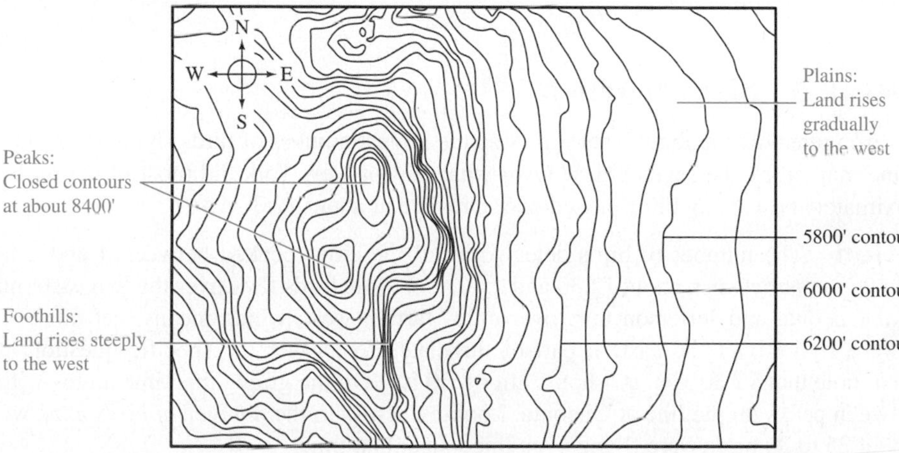

Figure 3.25 A contour elevation map for the region around Boulder, CO.

Three-Dimensional Graphics

Today, computer software makes it easy to give almost any graph a three-dimensional appearance. For example, Figure 3.26 shows the same bar graph as Figure 3.1, but "dressed up" with a three-dimensional look. They may look nice, but the three-dimensional effects are purely cosmetic; they do not provide any information that wasn't already shown in Figure 3.1.

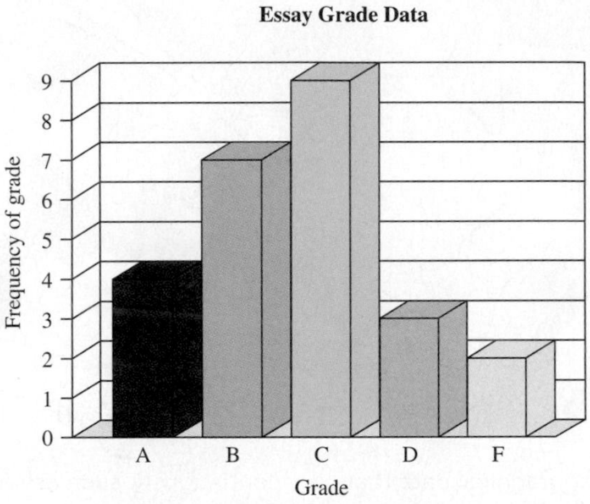

Figure 3.26 This graph has a three-dimensional appearance, but it shows only two-dimensional data.

In contrast, each of the three axes in Figure 3.27 carries distinct information, making the graph a true three-dimensional graph. Researchers studying migration patterns of a bird species (the *Bobolink*) counted the number of birds flying over seven New York cities throughout the night. As shown on the inset map, the cities were aligned east-west so that the researchers would learn what parts of the state the birds flew over, and at what times of night, as they headed south for the winter. Notice that the three axes measure *number of birds, time of night,* and *east-west location.*

EXAMPLE 5 Bird Migration

Based on Figure 3.27, at about what time was the largest number of birds flying over the east-west line marked by the seven cities? Over what part of New York did most of the birds fly? Approximately how many birds passed over Oneonta at about 12:30 A.M.?

Solution The number of birds detected in all the cities peaked between 4 and 6 hours after 8:30 P.M., or between about 12:30 and 2:30 A.M. More birds flew over the two easternmost cities of Oneonta and Jefferson than over cities farther west, which means that most of the birds were flying over the eastern part of the state. To answer the specific question about Oneonta, note that 12:30 A.M. is 4 hours after 8:30 P.M. On the graph, this time aligns with the dip between peaks on the line at Oneonta. Looking across to the *number of birds* axis, we see that about 25 to 30 birds were flying over Oneonta at that time.

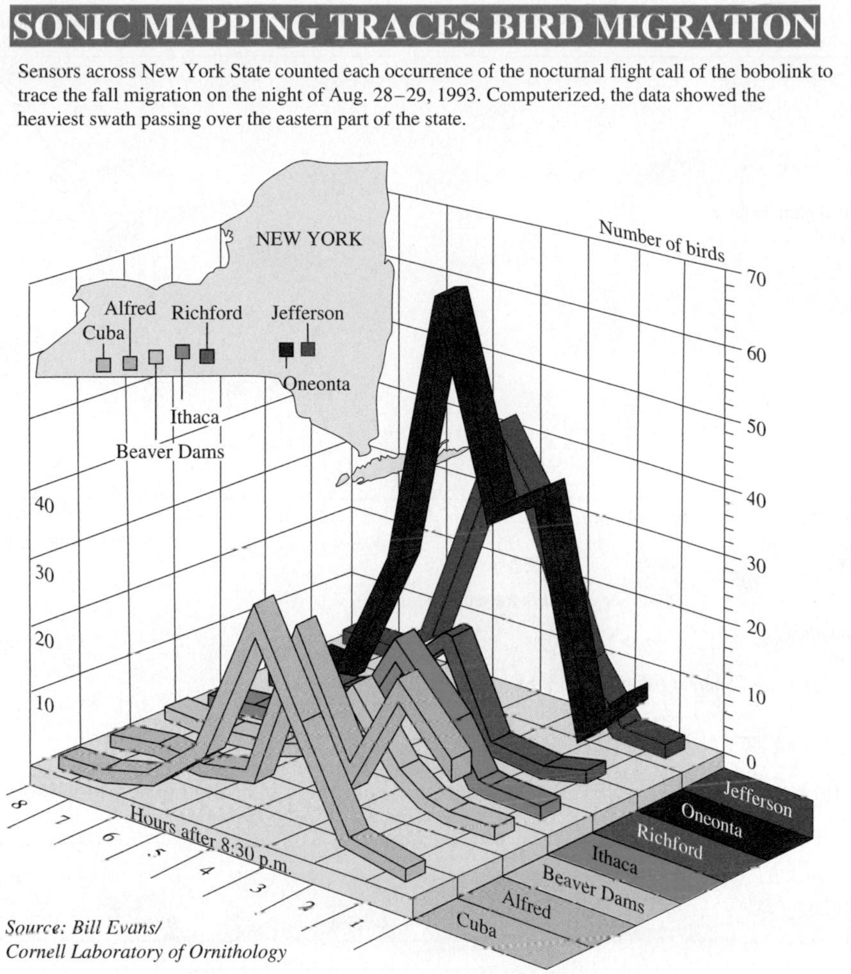

Figure 3.27 This graph shows true three-dimensional data. *Source: New York Times*.

Combination Graphics

All of the graphic types we have studied so far are common and fairly easy to create. But the media today are often filled with many varieties of even more complex graphics. For example, Figure 3.28 shows a graphic concerning the participation of women in the summer Olympics. This single graphic combines a line chart, many pie charts, and numerical data. It is certainly a case of a picture being worth far more than a thousand words.

EXAMPLE 6 Olympic Women

Describe three trends shown in Figure 3.28, on page 120.

Solution The line chart shows that the total number of women competing in the summer Olympics has risen fairly steadily, especially since the 1960s, reaching nearly 5,000 in the 2004 games. The pie charts show that the percentage of women among all competitors has also increased, reaching 44% in the 2004 games. The bold red numbers at the bottom show that the number of events in which women compete has also increased dramatically, reaching 135 in the 2004 games.

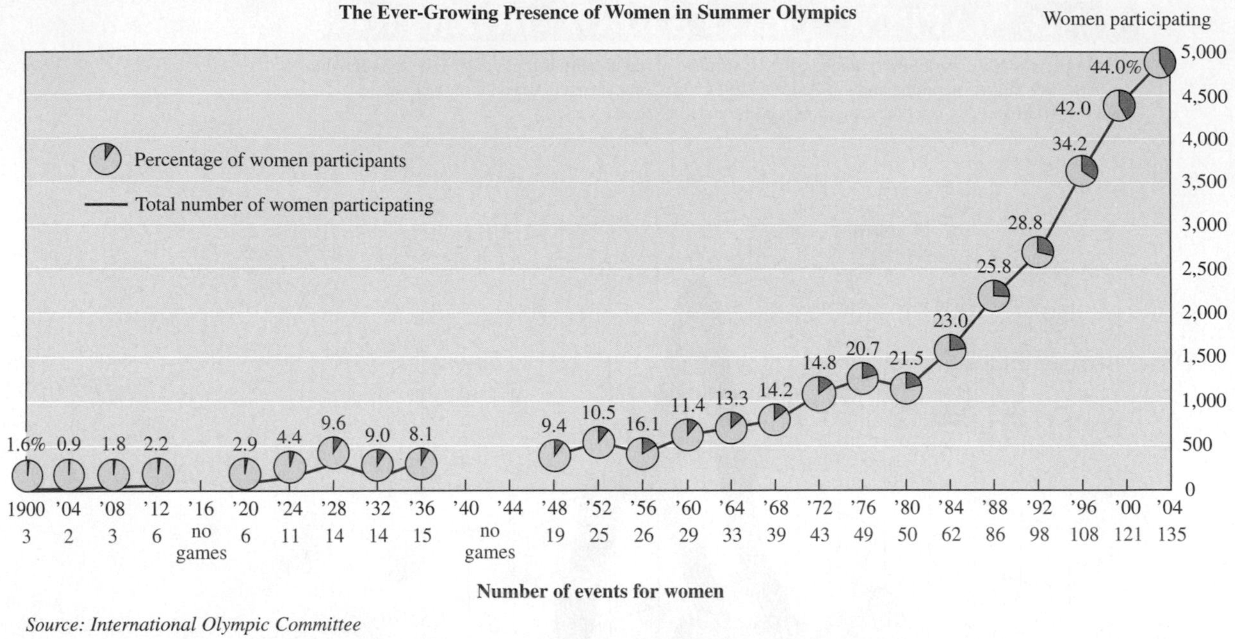

Figure 3.28 Women in the Olympics. *Source:* Adapted from the *New York Times.*

TIME OUT TO THINK

Which of the trends shown in Figure 3.28 are likely to continue over the next few Olympic games? Which are not? Explain.

Section 3.3 Exercises

Statistical Literacy and Critical Thinking

1. **Three-Dimensional Graphic.** One magazine used three-dimensional oil barrels of different sizes to depict the oil production from different countries. The oil barrels were clearly three-dimensional objects, but are oil production amounts from different countries an example of three-dimensional data? Are the oil barrels appropriate for these data? If not, how should the data be represented?

2. **Stack Plots.** What are stack plots, and how are they helpful?

3. **Geographical Data.** What are geographical data? Identify at least two ways to display geographical data.

4. **Contour Map.** What is a contour on a contour map? What does it mean when contours are close together? What does it mean when they are far apart?

Does It Make Sense? For Exercises 5–8, decide whether the statement makes sense (or is clearly true) or does not make sense (or is clearly false). Explain clearly. Not all of these statements have definitive answers, so your explanation is more important than your chosen answer.

5. **Three-Dimensional Graph.** Sue claims that she needs a three-dimensional graph to display the high temperature in Sacramento on each day of the past year.

6. **Stacked Line Chart.** A stacked line chart could be used to show the numbers of males and females at your college for each of the past 10 years.

7. **Contour Map.** A contour map could be used to display the population in each major city of the United States.

8. **Geographic Data.** A graphic artist for a magazine is depicting the populations of the major U.S. cities by using bars of different heights, with the bars positioned on the locations of the cities on a map of the United States.

Concepts and Applications

9. **Genders of Students.** The stack plot in Figure 3.29 shows the numbers of male and female higher education students for different years. Projections are from the U.S. National Center for Education Statistics.

   a. In words, discuss the trends revealed on this graphic.

b. Redraw the graph as a multiple line chart. Briefly discuss the advantages and disadvantages of the two different representations of this particular data set.

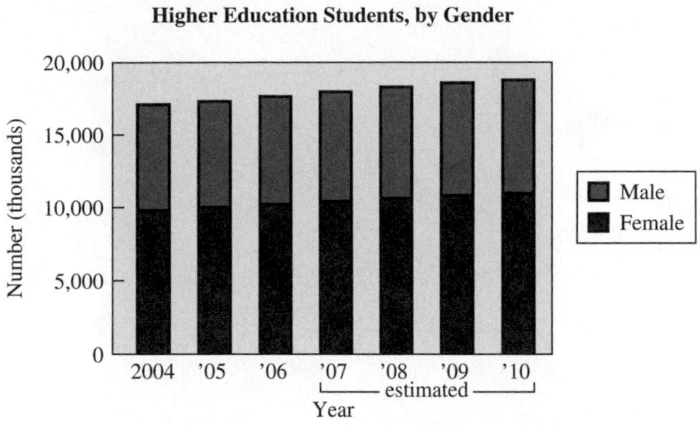

Figure 3.29

10. Home Prices by Region. The graph in Figure 3.30 shows home prices in different regions of the United States. Note that the data have *not* been adjusted for the effects of inflation.

a. In words, describe the general trends that apply to the home price data for all regions.

b. In words, describe any differences that you notice among the different regions.

c. In words, describe how this graph would look different if the data were adjusted for the effects of inflation.

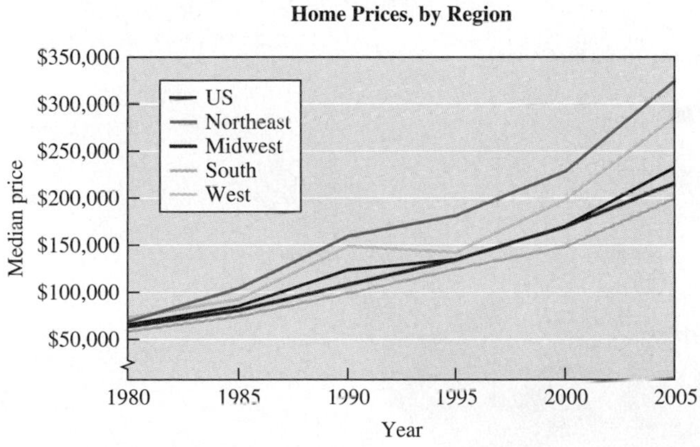

Figure 3.30

11. Gender and Income. Consider the display in Figure 3.31 of median incomes of males and females in different years (U.S. Census Bureau).

a. What story does the graph convey? Does the ratio of male incomes to female incomes appear to be decreasing? Why or why not?

b. Redraw the graph as a multiple (two) line chart. Briefly discuss the advantages and disadvantages of the two different representations of this particular data set.

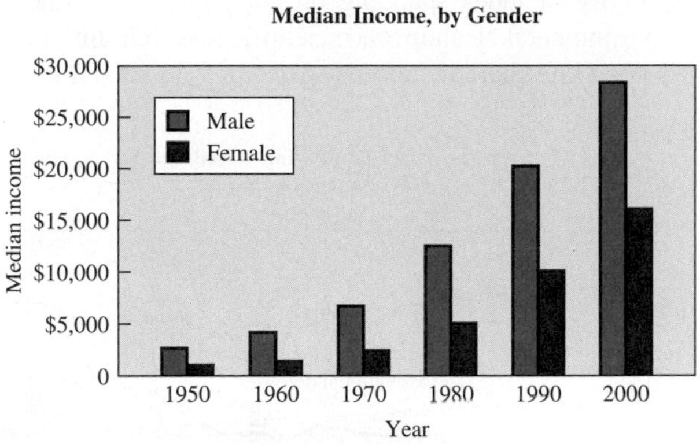

Figure 3.31

12. Marriage and Divorce Rates. The graph in Figure 3.32 depicts the U.S. marriage and divorce rates for selected years since 1900. Both rates are given in units of marriages/divorces per 1,000 people in the population (Department of Health and Human Services).

a. Why do these data consist of marriage and divorce *rates* rather than total numbers of marriages and divorces? Comment on any trends that you observe in these rates, and give plausible historical and sociological explanations for these trends.

b. Construct a stack plot of the marriage and divorce rate data. For each bar, place the divorce rate above the marriage rate. Which graph makes the comparisons easier: the multiple bar graph shown here or the stack plot? Explain.

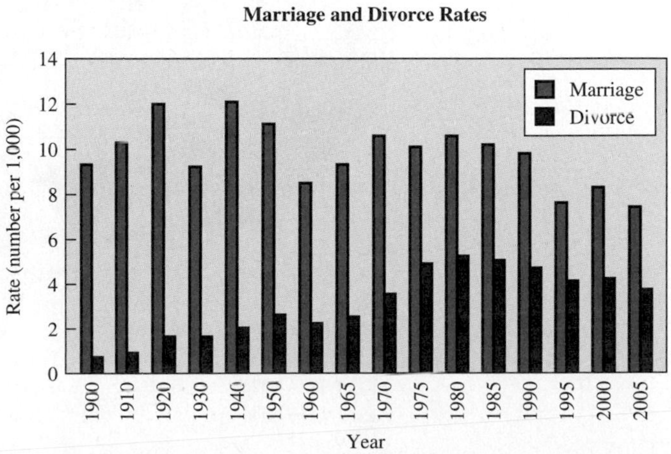

Figure 3.32

13. Federal Spending. The stacked line chart in Figure 3.33 shows the changes in major spending categories of the federal budget. Note that the "net interest" category represents interest payments on the national debt, and the "all other" category includes spending on such things as education, environmental cleanup, and scientific research. Interpret the stacked line chart and discuss some of the trends it reveals.

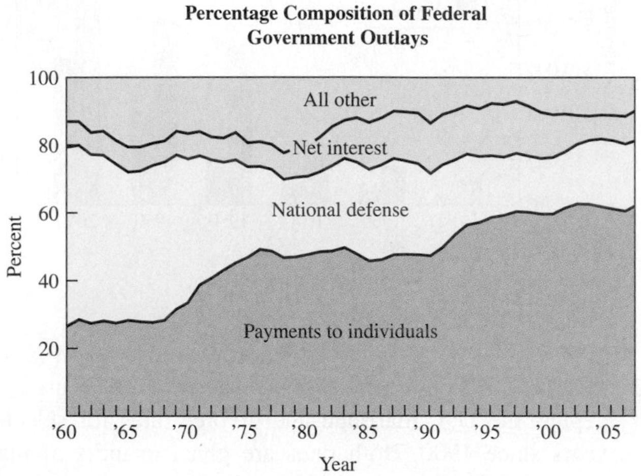

Figure 3.33 *Source:* Office of Management and Budget.

14. College Degrees. The stacked line chart in Figure 3.34 shows the numbers of college degrees awarded to men and women over time.

a. Estimate the numbers of college degrees awarded to men and to women (separately) in 1930 and in 2000.

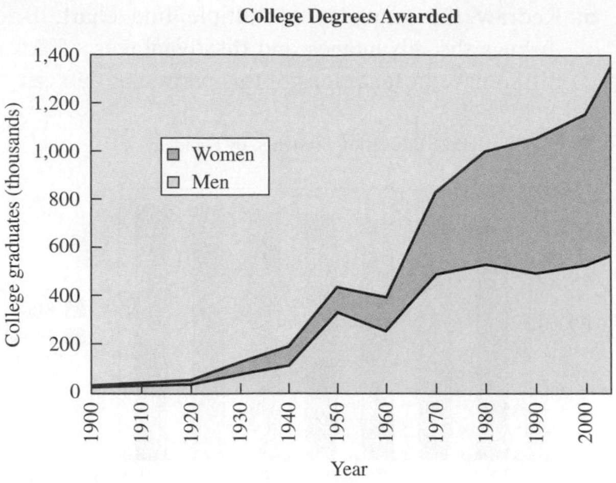

Figure 3.34

b. Compare the numbers of degrees awarded to men and to women (separately) in 1980 and 2000.

c. During what decade did the *total* number of degrees awarded increase the most?

d. Compare the *total* numbers of degrees awarded in 1950 and 2000.

e. Do you think the stacked line chart is an effective way to display these data? Briefly discuss other ways that might have been used instead.

15. Melanoma Mortality. Figure 3.35 shows how the mortality rate from *melanoma* (a form of skin cancer) varies on a county-by-county basis across the United States. The

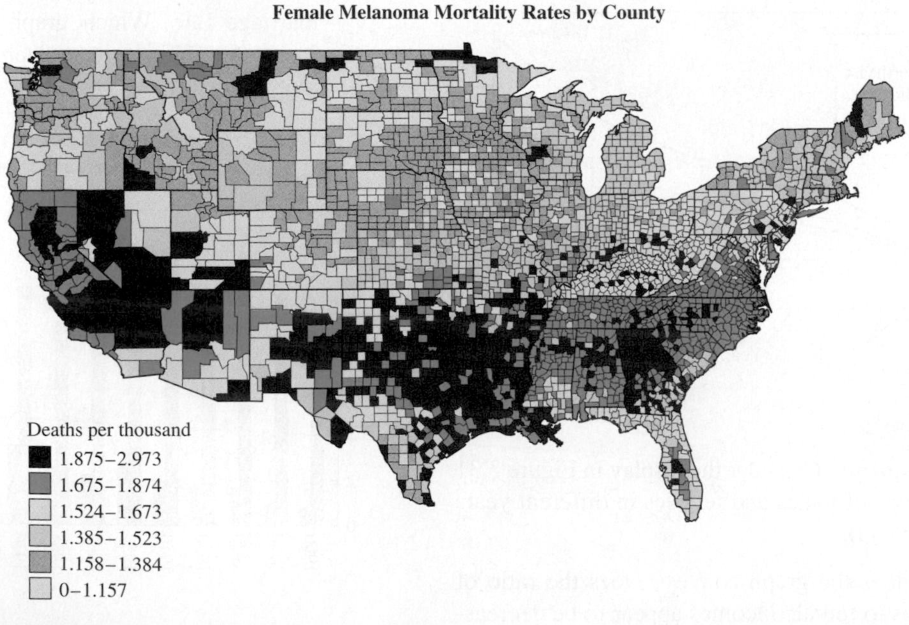

Figure 3.35 *Source:* Professor Karen Kafadar, Mathematics Department, University of Colorado at Denver.

Probability That a Black Student Would Have White Classmates

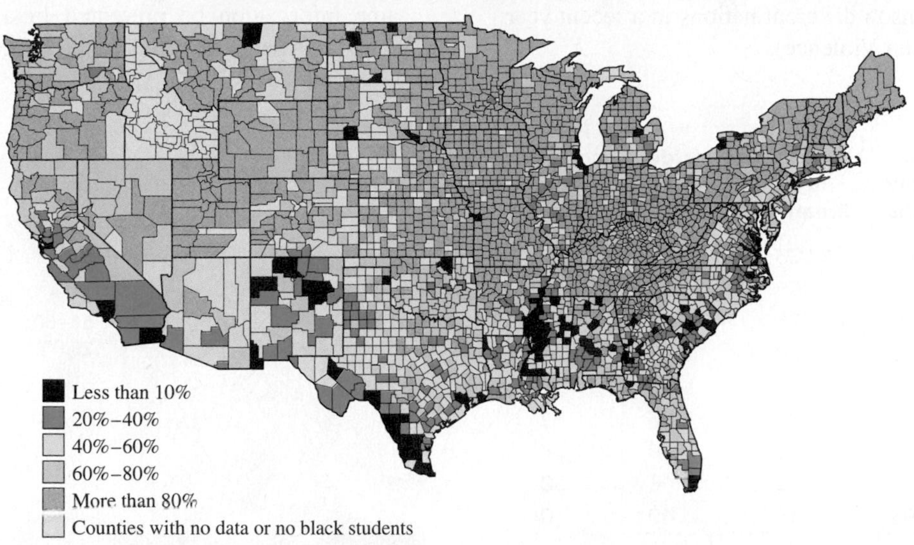

Figure 3.36 *Source: New York Times.*

legend shows that the darker the shading in a county, the higher the mortality rate. Discuss a few of the trends revealed in the figure. If you were researching skin cancer, which regions might warrant special study? Why?

16. **School Segregation.** One way of measuring segregation is the likelihood that a black student will have classmates who are white. A recent *New York Times* study found that, by this measure, segregation increased significantly in the 1990s. Figure 3.36 shows the probability that a black student would have white classmates, by county, during a recent academic year. Do there appear to be any significant regional differences? Can you pick out any differences between urban and rural areas? Discuss possible explanations for a few of the trends that you see in the figure.

Creating Graphics. Exercises 17–20 give tables of real data. For each table, make a graphical display of the data. You may choose any graphic type that you feel is appropriate to the data set. In addition to making the display, write a few sentences explaining why you chose this type of display and a few sentences describing interesting patterns in the data.

17. **Alcohol on the Road.** The following table gives the number of automobile fatalities in which (a) no alcohol was involved, (b) moderate alcohol was involved (blood alcohol content between 0.01 and 0.09), and (c) a high level of alcohol was involved (blood alcohol content above 0.1). All figures are in thousands of deaths (National Highway Traffic Safety Administration).

Year	No alcohol	Moderate alcohol	High alcohol
1982	18.7	4.8	20.4
1984	20.5	4.8	19.0
1986	22.0	5.1	18.9
1988	23.5	4.9	18.7
1990	22.5	4.4	17.7
1992	21.4	3.6	14.2
1994	24.1	3.5	13.1
1996	24.8	3.8	13.4
1998	25.5	3.7	12.2
2000 (est.)	25.1	3.9	12.8

18. **Daily Newspapers.** The following table gives the number of daily newspapers and their total circulation (in millions) for selected years since 1920 (*Editor & Publisher*).

Year	Number of daily newspapers	Circulation (millions)
1920	2,042	27.8
1930	1,942	39.6
1940	1,878	41.1
1950	1,772	53.9
1960	1,763	58.8
1970	1,748	62.1
1980	1,747	62.2
1990	1,611	62.3
2000	1,485	56.1
2003	1,456	55.2

19. Firearm Fatalities. The following table summarizes deaths due to firearms in different nations in a recent year (Coalition to Stop Gun Violence).

Country	Total firearms deaths	Homicides by firearms	Suicides by firearms	Fatal accidents by firearms
United States	35,563	15,835	18,503	1,225
Germany	1,197	168	1,004	25
Canada	1,189	176	975	38
Australia	536	96	420	20
Spain	396	76	219	101
United Kingdom	277	72	193	12
Sweden	200	27	169	4
Vietnam	131	85	16	30
Japan	93	34	49	10

20. Mothers in the Labor Force. The following table lists percentages of mothers in the labor force who have children in two different age brackets (based on data from the Bureau of Labor Statistics).

Year	6 to 17 years	Under 6 years
1955	38.4	18.2
1965	45.7	25.3
1975	54.9	39.0
1985	69.9	53.5
1995	76.4	62.3
2005	77.5	62.2

Projects for the Internet and Beyond

For useful links, select "Links for Internet Projects" for Chapter 3 at www.aw.com/bbt.

21. Weather Maps. Many Web sites offer contour maps with current weather data. For example, you can use the Yahoo! Weather site to generate many different contour weather maps. Generate at least two contour weather maps and discuss what they show.

22. The Federal Budget. Go to the Web site for the U.S. Office of Management and Budget (OMB) and look for some of its charts related to the federal budget. Pick two charts of particular interest to you and discuss the data they show.

23. Graphic Deception. Refer to Figure 3.37, in which percentages are represented by volumes of portions of someone's head. Are the data presented in a format that makes them easy to understand and compare? Are the data presented in a way that does not mislead? Could the same information be presented in a better way? If so, construct your own graph that better depicts the given information.

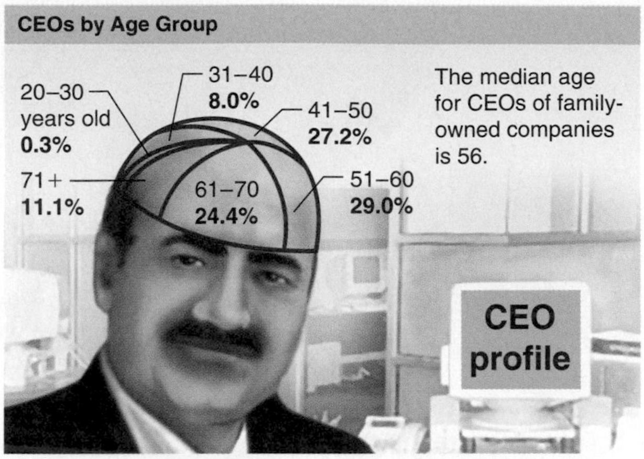

Figure 3.37 *Source:* Data from Arthur Anderson/Mass Mutual Family Business Survey '97.

IN THE NEWS

24. Multiple Bar Graphs. Find an example of a multiple bar graph or multiple line chart in a recent news report. Comment on the effectiveness of the display. Could another display have been used to depict the same data?

25. Stack Plots. Find an example of a stack plot in a recent news report. Comment on the effectiveness of the display. Could another display have been used to depict the same data?

26. Geographical Data. Find an example of a graph of geographical data in a recent news report. Comment on the effectiveness of the display. Could another display have been used to depict the same data?

27. Three-Dimensional Displays. Find an example of a three-dimensional display in a recent news report. Are three dimensions needed, or are they included for cosmetic reasons? Comment on the effectiveness of the display. Could another display have been used to depict the same data?

28. Fancy News Graphics. Find an example in the news of a graphic that combines two or more of the basic graphic types. Briefly explain what the graphic is showing, and discuss the effectiveness of the graphic.

3.4 A Few Cautions About Graphics

As we have seen, graphics can offer clear and meaningful summaries of statistical data. However, even well-made graphics can be misleading if we are not careful in interpreting them, and poorly made graphics are almost always misleading. Moreover, some people use graphics in deliberately misleading ways. In this section, we discuss a few of the more common ways in which graphics can lead us astray.

By the Way ...

German researchers in the latter part of the 19th century studied many types of graphics. The type of distortion shown in Figures 3.38 and 3.39 was so common that they gave it its own name, which translates roughly as "the old goosing up the effect by squaring the eyeball trick."

Perceptual Distortions

Many graphics are drawn in a way that distorts our perception of them. Figure 3.38 shows one of the most common types of distortion. Dollar-shaped bars are used to show the declining value of the dollar over time. The *lengths* of the bars represent the data, but our eyes tend to focus on the *areas* of the bars. For example, the right bar is supposed to show that a dollar in 2006 was worth 41% as much as a dollar in 1980. Its length is indeed 41% of that of the left bar, but its area is much smaller in comparison (about 17% of the area of the left bar). This gives the perception that the value of the dollar shrank even more than it really did.

Figure 3.38 The lengths of the dollars are proportional to their spending power, but our eyes are drawn to the areas, which decline more than the lengths.

Even greater distortion occurs when a graphic shows volume where length is the important measure. Figure 3.39 uses television sets to represent the number of houses with cable TV in 1980 and 2005. This number increased by a factor of about 4 during that period, from 18 million homes to 73 million homes, as shown by the *heights* of the TVs in the figure. However, our eyes are drawn to the *volumes* of the TVs, which differ by the much greater factor of about $4^3 = 64$ and therefore make the increase look much larger than it really was.

By the Way ...

The growth of cable TV has slowed in recent years because of competition from satellite TV services. Competition may increase further as telephone lines and the Internet become capable of delivering television programming.

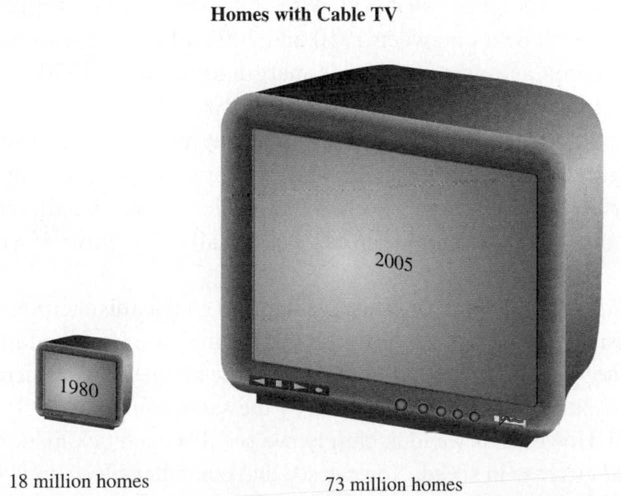

Figure 3.39 The heights of the TVs are the important measure in this figure, but our eyes are drawn to their volumes.

Home, Sweet Home

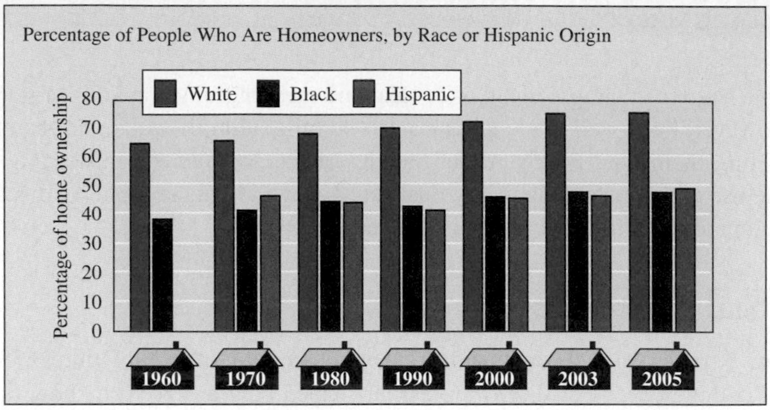

Source: U.S. Census Bureau

Figure 3.40 Figure 3.40 shows a multiple bar graph concerning home ownership in the United States. At first glance, you might think you can see some trends in the data. But look at the time (horizontal) scale more closely. The first five categories represent years that are a decade apart, while the last two categories represent the years 2003 and 2005. The fact that the time scale is not uniform can lead to misleading conclusions unless you are very careful and you examine the data. *Source:* U.S. Department of Commerce, U.S. Census Bureau; adapted from the *Wall Street Journal Almanac*.

Watch the Scales

Figure 3.40 shows a multiple bar graph concerning home ownership in the United States. At first glance, it appears that the percentage of people owning their own homes rose much faster during the period 1960–2000 than it did in recent years. But look at the horizontal scale more closely: The first five categories represent years that are a decade apart, while the last two categories represent the years 2003 and 2005. If the small increase from 2003 to 2005 were repeated each year for an entire decade, the increase would not look much different from the trends in other decades. This graph is misleading on first impression because it does not use a uniform scale for the horizontal axis.

By the Way ...

By 2005, the overall percentage of homes in the United States occupied by their owners had reached 69.0%—the highest percentage in history.

TIME OUT TO THINK

Examine Figure 3.40. Could you use this graph to predict home ownership rates in 2020? Why or why not? More generally, can you think of any reasons why a graph would be made with a non-uniform time axis, and do you think such graphs are legitimate?

Similar scaling problems can occur with the vertical axis. Figure 3.41a shows the percentage of college students between 1910 and 2005 who were women. At first glance, it appears that this percentage grew by a huge margin after about 1950. But the vertical axis scale does not begin at zero and does not end at 100%. The increase is still substantial but looks far less dramatic if we redraw the graph with the vertical axis covering the full range of 0 to 100% (Figure 3.41b). From a mathematical point of view, leaving out the zero point on a scale is perfectly honest and can make it easier to see small-scale trends in data. Nevertheless, as this example shows, it can be visually deceptive if you don't study the scale carefully.

Sometimes the scale may not be deceptive, but still can be misinterpreted unless you examine it closely. Consider Figure 3.42a, which shows how the speeds of the fastest computers have increased with time. At first glance, it appears that speeds have been increasing linearly. For example, it might look as if the speed increased by the same amount from 1990 to 2000 as it did from 1950 to 1960. However, if we look closely, we see that each tick mark on the vertical scale represents a *tenfold* increase in speed. Now we see that computer speed grew from about 1 to 100 calculations per second from 1950 to 1960 and from about 100 million to 10 billion calculations per second between 1990 and 2000. This type of scale is called an *exponential scale* (or

The easiest person to deceive is one's own self.

—Edward Bulwer-Lytton

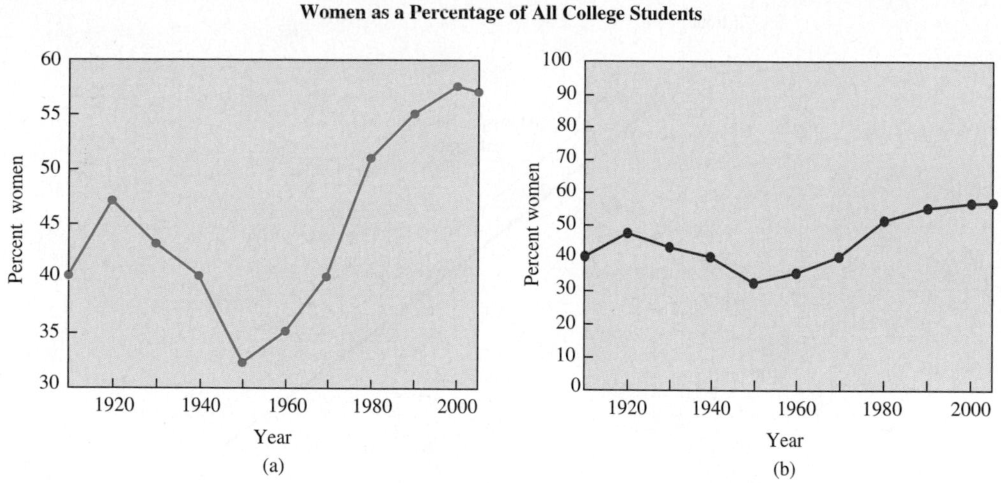

Figure 3.41 Both graphs show the same data, but they look very different because their vertical scales have different ranges.

logarithmic scale) because it grows by powers of 10 and powers of 10 are *exponents*. (For example, 3 is the exponent in 10^3.) Exponential scales are useful for displaying data that vary over a huge range of values, as you can see by comparing the exponential graph in Figure 3.42a to the ordinary graph in Figure 3.42b. Because the speeds have grown so rapidly, the ordinary scale makes it impossible to see any detail in the early years shown on the graph.

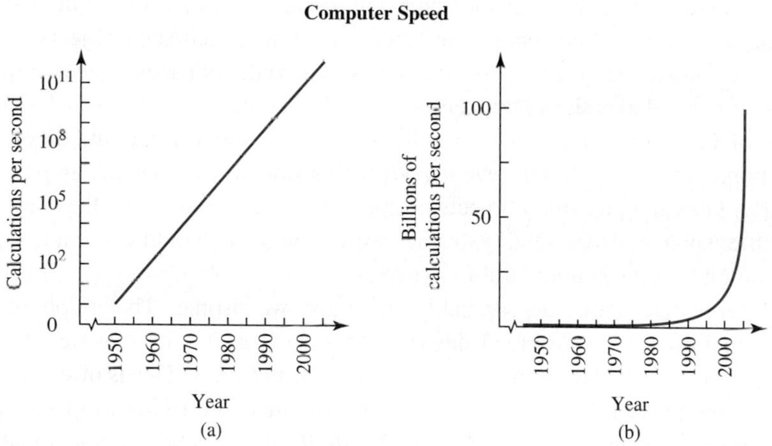

Figure 3.42 Both graphs show the same data, but the graph on the left uses an exponential scale.

By the Way ...

In 1965, Intel founder Gordon E. Moore predicted that advances in technology would allow computer chips to double in power roughly every two years. This idea is now called *Moore's law*, and it has held fairly true ever since Moore first stated it.

TIME OUT TO THINK

Based on Figure 3.42a, can you predict the speed of the fastest computers in 2015? Could you make the same prediction with Figure 3.42b? Explain.

CASE STUDY Asteroid Threat

Asteroids and comets occasionally hit the Earth. Small ones tend to burn up in the atmosphere or create small craters on impact. But larger ones can cause substantial devastation. About 65 million years ago, an asteroid about 10 kilometers in diameter hit the Earth, leaving a 200-kilometer-wide crater on

By the Way ...

More than 150 impact craters have been identified around the Earth. One of the best preserved is Meteor Crater in Arizona (photo below). It formed about 50,000 years ago when an asteroid about 50 meters across crashed to Earth, releasing energy equivalent to that of a 20-megaton hydrogen bomb and leaving a crater more than a kilometer across and 200 meters deep.

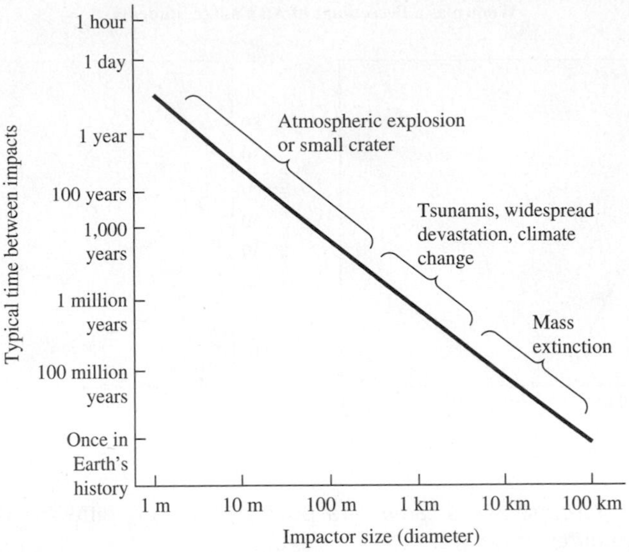

Figure 3.43 This graph shows how the frequency of impacts—and the magnitude of their effects—depends on the size of the impactor. Note that smaller impacts are much more frequent than larger ones.

the coast of the Yucatan peninsula in Mexico. Scientists estimate that this impact caused the extinction of about three-quarters of all species living on Earth at the time, including all the dinosaurs.

Clearly, a similar impact would be bad news for our civilization. We might therefore want to understand the likelihood of such an event. Figure 3.43 shows a graph relating the size of impacting asteroids and comets to the frequency with which such objects hit the Earth. Because of the wide range of sizes and time scales involved, *both* axes on this graph are exponential. The horizontal axis shows impactor (asteroid or comet) sizes, with each tick representing a power of 10. The vertical axis shows the frequency of impact; moving up on the vertical axis corresponds to more frequent events. With this double exponential graph, we can see trends clearly. For example, small objects of about 1 meter in size strike the Earth every day, but cause little damage. At the other extreme, objects large enough to cause a mass extinction hit only about once every hundred million years.

The intermediate cases are probably the most worrisome. The graph indicates that objects that could cause "widespread devastation"—such as wiping out the population of a large city—can be expected as often as once every thousand years. This is often enough to warrant at least some preventive action. Currently, astronomers are trying to make more precise predictions about when an object might hit the Earth. If they discover an object that will hit the Earth, scientists will need to find a way to deflect it to prevent the impact.

Percentage Change Graphs

Get your facts first, and then you can distort them as much as you please.

—Mark Twain

Is college getting more or less expensive? A quick look at Figure 3.44a might give the impression that the cost for private colleges has been holding fairly steady while the cost for public colleges fell steeply in 2006 after rising in prior years. But look more closely and you'll see that this is not the case at all. The vertical axis on this graph represents the *percentage increase* in costs. A flat graph means only that costs increased by the same percentage each year, not that costs held steady. Similarly, the drop in 2006 for public colleges means only that the cost rose by less in that year than in the preceding years.

In fact, the actual costs (not adjusted for inflation) for both public and private colleges have risen substantially with time, as shown in Figure 3.44b. Moreover, because the rate of

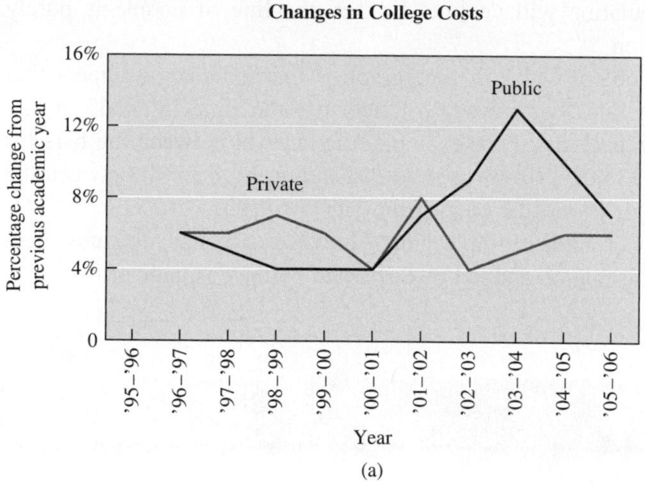

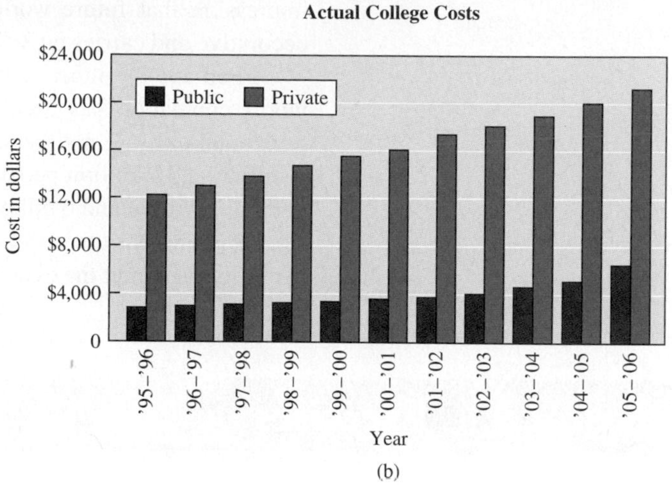

Figure 3.44 Trends in college costs: (a) annual percent change; (b) actual costs. *Source:* The College Board.

inflation (as measured by the CPI) has been less than the rate of increase in college costs, the real cost of public colleges has steadily risen. Graphs that show percentage change are very common, particularly with economic data. Although they are perfectly honest, you can be easily misled unless you interpret them with great care.

Pictographs

Pictographs are graphs embellished with additional artwork. The artwork may make the graph more appealing, but it can also distract or mislead. Figure 3.45 is a pictograph showing the rise in world population from 1804 to 2054 (numbers for future years are based on United Nations projections). The lengths of the bars correctly correspond to world population for the years listed. However, the artistic embellishments of this graph are deceptive in several ways. For example, your eye may be drawn to the figures of people lining the globe. Because this line of people rises from the left side of the pictograph to the center and then falls, it might give the

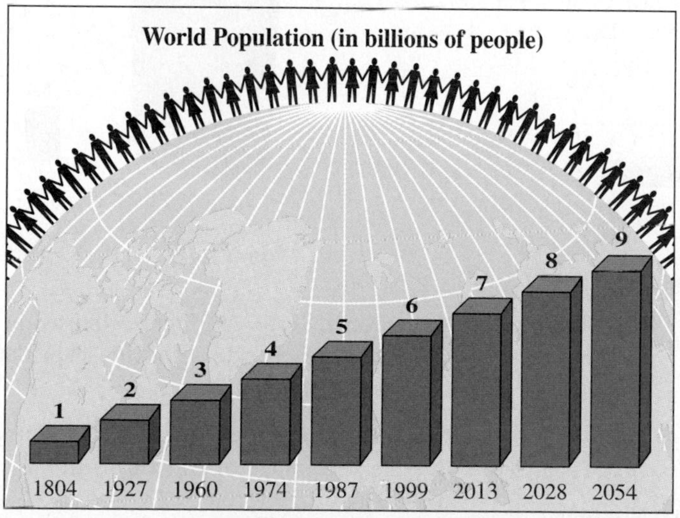

Figure 3.45 A pictograph. *Source:* Data from United Nations Population Division, World Population Prospects.

By the Way ...

Demographers often characterize population growth by a doubling time—the time it takes the population to double. During the late 20th century, the doubling time for the human population was about 40 years. If the population continued to double at this rate, world population would reach 34 billion by 2100 and 192 billion by 2200. By about 2650, the human population would be so large that it would not fit on the Earth, even if everyone stood elbow to elbow everywhere.

impression that future world population will decline. In fact, the line of people is purely decorative and carries no information.

Perhaps the most serious problem with this pictograph is that it makes it appear that world population has been rising linearly. However, notice that the time intervals on the horizontal axis are not the same in each case. For example, the interval between the bars for 1 billion and 2 billion people is 123 years (from 1804 to 1927), but the interval between the bars for 5 billion and 6 billion people is only 12 years (from 1987 to 1999).

Pictographs are very common. As this example shows, however, you have to study them carefully to extract the essential information and not be distracted by the cosmetic effects.

Section 3.4 Exercises

Statistical Literacy and Critical Thinking

1. **Vertical Scale.** How is a graph misleading when its vertical scale does not begin at zero?

2. **Pictograph.** What is a pictograph? Briefly identify an advantage and a disadvantage of a pictograph.

3. **Exponential Scale.** What is an exponential scale? When is it helpful to use an exponential scale in a graph?

4. **Percentage Change Graphs.** Explain how a graph that shows percentage change can show descending bars (or a descending line) even when the variable of interest is increasing.

Concepts and Applications

5. **Braking Distances.** Figure 3.46 shows the braking distance for different cars measured under the same conditions. Discuss the ways in which this display might be deceptive. How much greater is the braking distance of an Acura RL than the braking distance of a Volvo S80? Draw the display in a way that depicts the data more fairly.

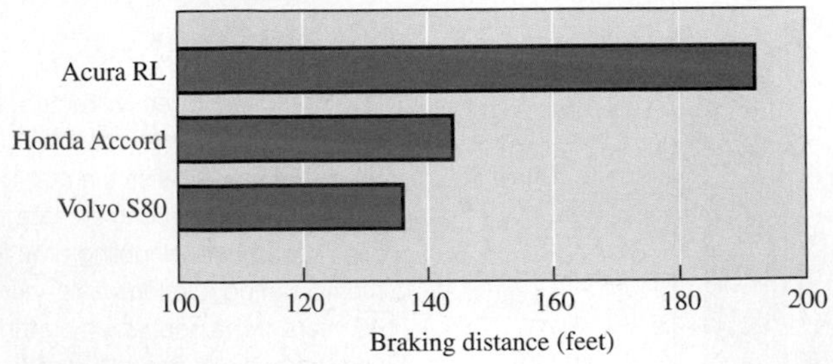

Figure 3.46 Braking distances.

6. **Comparing Earnings.** Consider the bar graph in Figure 3.47, which compares the median annual earnings of men and women. Identify any misleading aspects of the display. Draw the display in a fairer way.

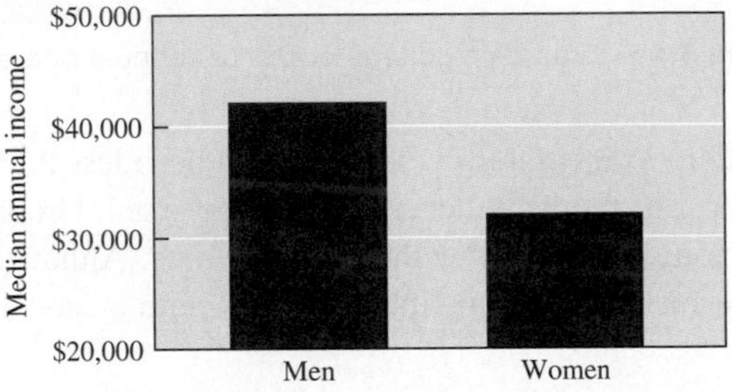

Figure 3.47 *Source:* U.S. Census Bureau.

7. **Pictograph.** Figure 3.48 depicts the amounts of daily oil consumption in the United States and Japan. Does the illustration accurately depict the data? Why or why not?

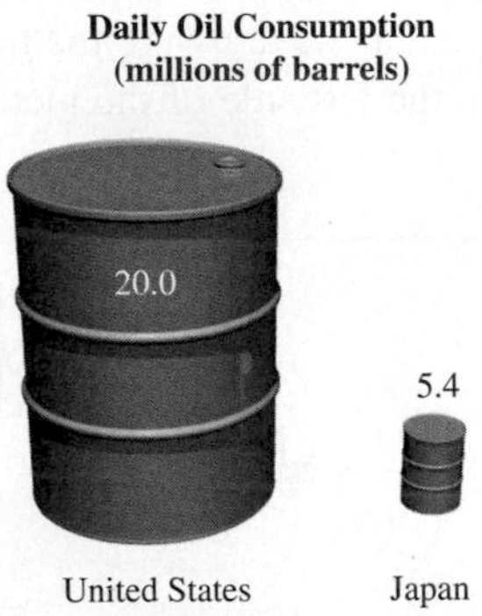

Figure 3.48

8. **Pictograph.** Refer to Figure 3.48 used in Exercise 7 and construct a bar chart to depict the same data in a way that is fair and objective.

9. **Three-Dimensional Pies.** The pie charts in Figure 3.49 give the percentage of Americans in three age categories in 1990 and 2050 (projected).

   a. Consider the 1990 age distribution. The actual percentages for the three categories for 1990 were 87.5%

(others), 11.3% (60–84), and 1.2% (85+). Does the pie chart show these values accurately? Explain.

b. Consider the 2050 age distribution. The actual percentages for the three categories for 2050 were 80.0% (others), 15.4% (60–84), and 4.6% (85+). Does the pie chart show these values accurately? Explain.

c. Using the actual percentages given in parts a and b, draw flat (two-dimensional) pie charts to display these data. Explain why these pie charts give a more accurate picture than the three-dimensional pies.

d. Comment on the general trends shown in the two pie charts.

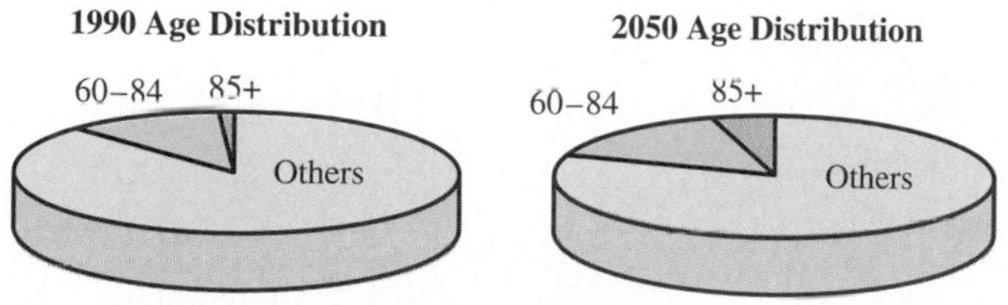

Figure 3.49 *Source:* U.S. Census Bureau.

10. Cigarette Pictograph. Consider the pictograph in Figure 3.50. Briefly describe what the graph is showing. Discuss whether it is effective in its purpose and whether it is deceptive in any way.

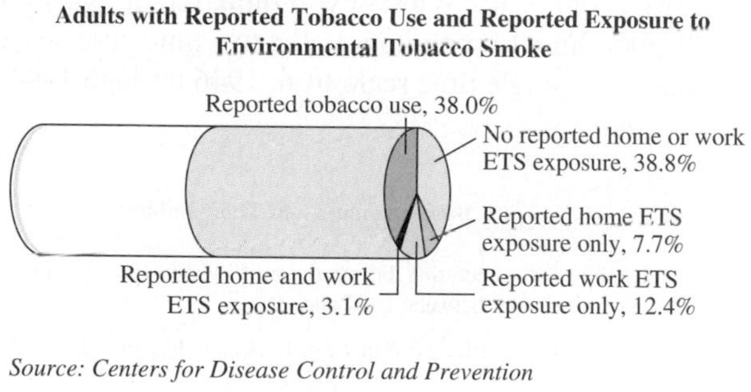

Figure 3.50 *Source: Wall Street Journal Almanac.*

11. Cell Phone Subscriptions. The following table shows the numbers of cell phone subscriptions in the United States.

Year	Number	Year	Number
1985	340	1997	55,312
1987	1,231	1999	86,047
1989	3,509	2001	128,375
1991	7,557	2003	158,722
1993	16,009	2005	207,900
1995	33,786		

a. Construct a time-series graph of these data, using a uniform scale on both axes.

b. Make an exponential graph of these data in which the subdivisions on the vertical axis are 1, 10, 100, 1000, 10,000, 100,000, and 1,000,000.

c. Compare the graphs in parts a and b.

12. Moore's Law. In 1965, Intel co-founder Gordon Moore initiated what has since become known as *Moore's law:* The number of transistors per square inch on integrated circuits will double approximately every 18 months. The table below lists the number of transistors (in thousands) for several different years.

Year	Transistors	Year	Transistors
1971	2.3	1993	3100
1974	5	1997	7500
1978	29	1999	24,000
1982	120	2000	42,000
1985	275	2002	220,000
1989	1180	2003	410,000

a. Construct a time-series graph of these data, using a uniform scale on both axes.

b. Make an exponential graph of these data in which the subdivisions on the vertical axis are 1, 10, 100, 1000, 10,000, 100,000, and 1,000,000.

c. Compare the graphs in parts a and b.

13. Percentage Change in the CPI. The graph in Figure 3.51 shows the percentage change in the CPI over 17 years. In what year (of the years displayed) was the change in the CPI the greatest? In what year was the change in the CPI the least? How do actual prices in 1991 compare to those in 1990? Based on this graph, what can you conclude about changes in prices during the period shown?

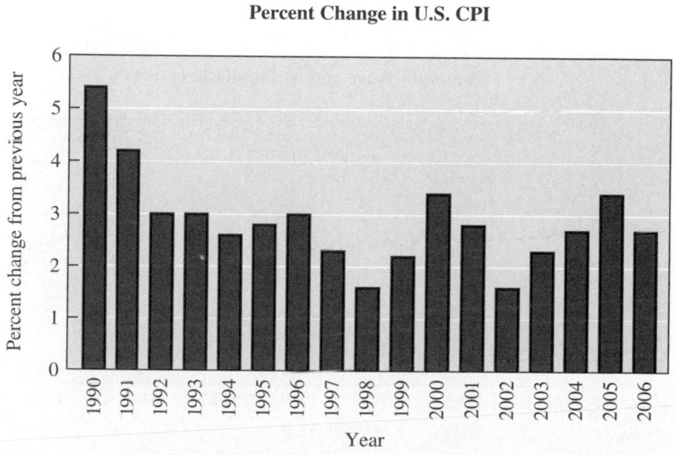

Figure 3.51 *Source:* U.S. Bureau of Labor Statistics.

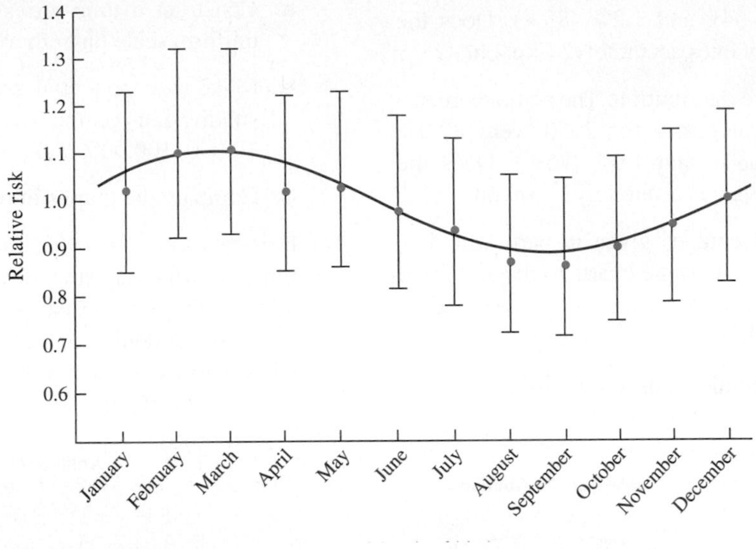

Figure 3.52 *Source: New England Journal of Medicine.*

14. Seasonal Effects on Schizophrenia? The graph in Figure 3.52 shows data regarding the relative risk of schizophrenia among people born in different months.

a. Note that the scale of the vertical axis does not include zero. Sketch the same risk curve using an axis that includes zero. Comment on the effect of this change.

b. Each value of the relative risk is shown with a dot at its most likely value and with an "error bar" indicating the range in which the data value probably lies. The study concludes that "the risk was also significantly associated with the season of birth." Given the size of the error bars, does this claim appear justified? (Is it possible to draw a flat line that passes through all of the error bars?)

15. Constant Dollars. The graph in Figure 3.53 shows the minimum wage in the United States, together with its purchasing power, which is adjusted for inflation with 1996 used as the reference year. The graph represents the years from 1955 to 2006. Summarize what the graph shows.

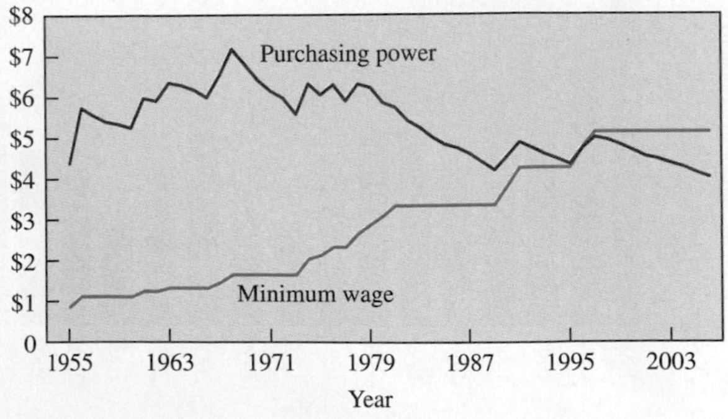

Figure 3.53 *Source:* U.S. Department of Labor.

16. Double Horizontal Scale. The graph in Figure 3.54 shows *simultaneously* the number of births in this country during two time periods: 1946–1964 and 1977–1994. When did the first baby boom peak? When did the second baby boom peak? Why do you think the designers of this display chose to superimpose the two time intervals, rather than use a single time scale from 1946 through 1994?

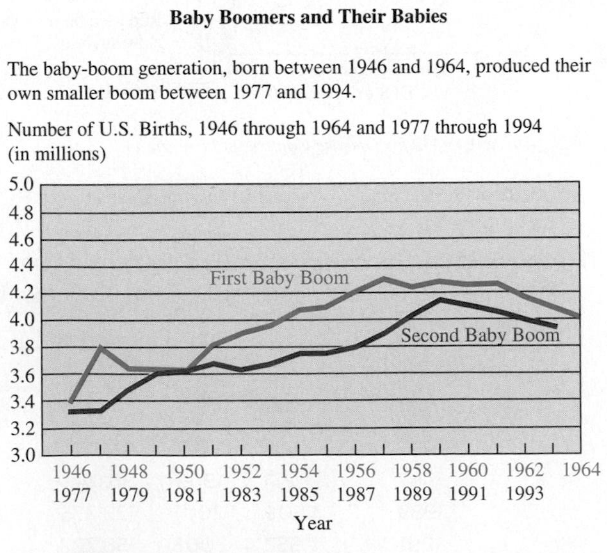

Figure 3.54 *Source: Wall Street Journal Almanac.*

Projects for the Internet and Beyond

For useful links, select "Links for Internet Projects" for Chapter 3 at www.aw.com/bbt.

17. **USA Snapshot.** *USA Today* offers a daily pictograph for its "USA Snapshot." Obtain a snapshot from this week's issues of *USA Today.* Briefly discuss its purpose and effectiveness.

18. **Creating Displays.** Currently, women earn less than men for doing the same job. Find the current amount earned by women for each $1 earned by men. Draw a graph that depicts this information objectively, and then draw another graph that exaggerates the difference.

IN THE NEWS

19. **Distortions in the News.** Find an example in a recent news report of a graph that involves some type of perceptual distortion. Explain the effects of the distortion, and describe how the graph could have been drawn more honestly.

20. **Scale Problems in the News.** Find an example in a recent news report of a graph in which the vertical scale does not start at zero. Suggest why the graph was drawn that way and also discuss any ways in which the graph might be misleading as a result.

21. **Economic Graph in the News.** Find an example in a recent news report of a graph that shows economic data over time. Are the data adjusted for inflation? Discuss the meaning of the graph and any ways in which it might be deceptive.

22. **Pictograph in the News.** Find an example of a pictograph in a recent news report. Discuss what the pictograph attempts to show, and discuss whether the artistic embellishments help or hinder this purpose.

23. **Outstanding News Graph.** Find a graph from a recent news report that, in your opinion, is truly outstanding in displaying data visually. Discuss what the graph shows, and explain why you think it is so outstanding.

24. **Not-So-Outstanding News Graph.** Find a graph from a recent news report that, in your opinion, fails in its attempt to display data visually in a meaningful way. Discuss what the graph was trying to show, explain why it failed, and explain how it could have been done better.

Chapter Review Exercises

Listed below are measured weights (in pounds) of the contents in samples of cans of regular Pepsi and diet Pepsi. Use these data for Exercises 1–3.

Regular Pepsi:

0.8258	0.8156	0.8211	0.8170	0.8216	0.8302
0.8192	0.8192	0.8271	0.8251	0.8227	0.8256
0.8139	0.8260	0.8227	0.8388	0.8260	0.8317
0.8247	0.8200	0.8172	0.8227	0.8244	0.8244
0.8319	0.8247	0.8214	0.8291	0.8227	0.8211
0.8401	0.8233	0.8291	0.8172	0.8233	0.8211

Diet Pepsi:

0.7925	0.7868	0.7846	0.7938	0.7861	0.7844
0.7795	0.7883	0.7879	0.7850	0.7899	0.7877
0.7852	0.7756	0.7837	0.7879	0.7839	0.7817
0.7822	0.7742	0.7833	0.7835	0.7855	0.7859
0.7775	0.7833	0.7835	0.7826	0.7815	0.7791
0.7866	0.7855	0.7848	0.7806	0.7773	0.7775

1. a. Construct a frequency table for the weights of regular Pepsi. Use bins of

0.8130–0.8179
0.8180–0.8229
0.8230–0.8279
0.8280–0.8329
0.8330–0.8379
0.8380–0.8429

b. Construct a frequency table for the weights of diet Pepsi. Use bins of

0.7740–0.7779
0.7780–0.7819
0.7820–0.7859
0.7860–0.7899
0.7900–0.7939

c. Compare the frequency tables from parts a and b. What notable differences are there? How can those notable differences be explained?

2. a. Construct a relative frequency table for the weights of regular Pepsi. Use bins of

0.8130–0.8179
0.8180–0.8229
0.8230–0.8279
0.8280–0.8329
0.8330–0.8379
0.8380–0.8429

b. Construct a cumulative frequency table for the weights of regular Pepsi.

3. a. Use the result from Exercise 1a to construct a histogram for the weights of regular Pepsi.

b. Use the result from Exercise 1b to construct a histogram for the weights of diet Pepsi.

c. Compare the histograms from parts a and b. How are they similar and how are they different?

4. Pie Chart. In a recent year, 5,524 people were killed while working. Here is a breakdown of causes:

transportation, 2,375
contact with objects or equipment, 884
assaults or violent acts, 829
falls, 718
exposure to harmful substances or a harmful environment, 552
fires or explosions, 166

(The data are from the Bureau of Labor Statistics.) Construct a pie chart representing the given data.

5. Pareto Chart. Construct a Pareto chart from the data given in Exercise 4. Compare the Pareto chart to the pie chart. Which graph is more effective in showing the causes of death while working? Explain.

6. Bar Chart. Figure 3.55 shows the numbers of U.S. adoptions from China in the years 2005 and 2000. What is wrong with this graph? Draw a graph that depicts the data in a fair and objective way.

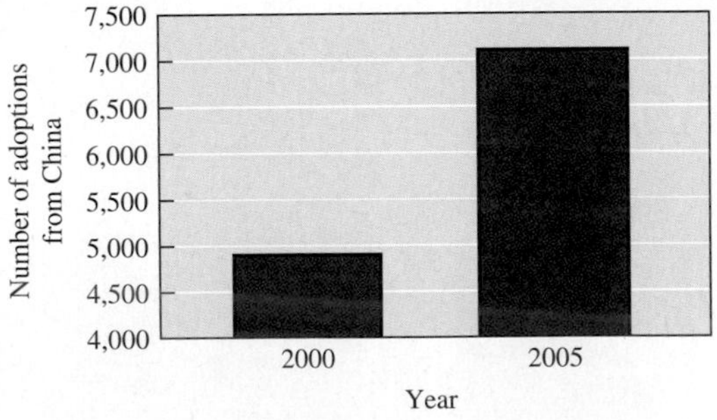

Figure 3.55

Chapter Quiz

1. The frequencies of different eye colors are obtained from a randomly selected sample of subjects who participate in a national health study. Which graph would be most appropriate for these data: a histogram, bar chart, multiple bar graph, or stack plot?

2. To investigate gender and eye color, the frequencies of different eye colors are obtained from a random sample of males and another random sample of females who participate in a national health study. Which graph would be most appropriate for these data: a histogram, bar chart, or multiple bar graph?

3. Hip breadths are measured for randomly selected subjects participating in a national health study. Which graph would be most appropriate for these data: a histogram, bar chart, multiple bar graph, or stack plot?

4. The first class in a frequency table is 50–59 and the corresponding frequency is 7. What does the value of 7 indicate?

5. The first class in a relative frequency table is 50–59 and the corresponding relative frequency is 0.2. What does the value of 0.2 indicate?

6. Identify the values represented in the following stem-and-leaf plot.

0	5
0	
0	
1	0
1	22
1	4

7. The bar chart in Figure 3.56 depicts the number of twin births in the United States in the years 2000 and 2004. In what way is this graph misleading?

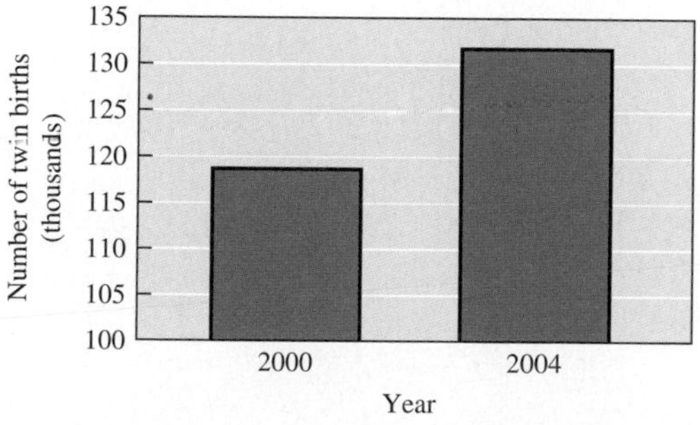

Figure 3.56

8. Identify the largest single data value represented in the dotplot in Figure 3.57.

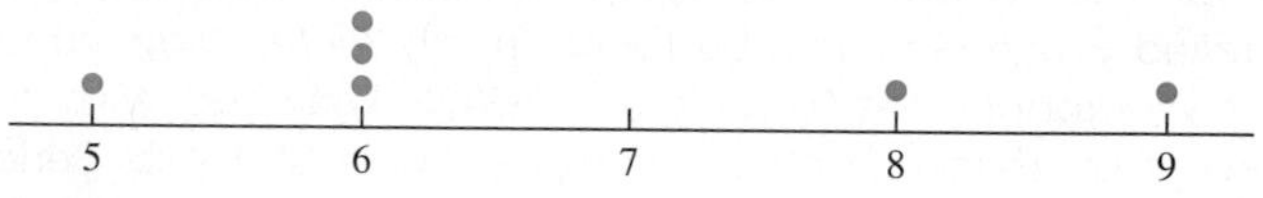

Figure 3.57

9. Figure 3.58 shows the number of automobile fatalities in the United States in which alcohol was involved for each year from 1982 to 2000. How many alcohol-related fatalities were there in 1982? In 2000?

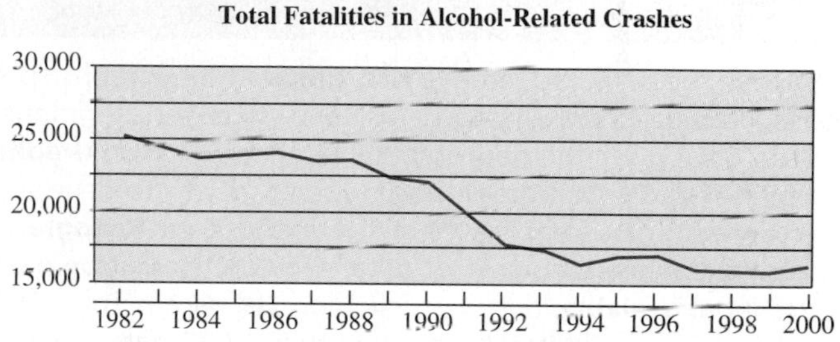

Source: National Highway Traffic Safety Administration

Figure 3.58

10. A newspaper reporter wants to construct a graph to depict the amount of gasoline consumed by cars each year in the United States for a recent sequence of years. Why does it not make sense to adjust the annual amounts by using a particular reference year, such as 1996?

Using Technology

Powerful software packages are now quite effective for generating a wide variety of impressive graphs. The detailed instructions can vary from extremely easy to extremely complex, so we provide some relevant comments below. More detailed procedures can be found in the *Technology Manual and Workbook* that is on the CD-ROM included with this book. Also, these supplements to the Triola Statistics Series, all available from Pearson Addison Wesley, may be helpful: *SPSS Student Laboratory Manual and Workbook* by James J. Ball, *Excel Student Laboratory Manual and Workbook* by Johanna Halsey and Ellena Reda, and *Statdisk Student Laboratory Manual and Workbook* by Mario F. Triola. Supplements are also available for Minitab, SAS, and the TI-83/84 Plus calculator.

SPSS

SPSS can be used to create bar graphs, line graphs, pie charts, Pareto charts, histograms, and time-series graphs. First enter or open the desired data set. Then click on **Graphs** and proceed to select the desired type of graph. To generate a histogram, for example, select the menu item **Histogram**, click on the desired column of data, click on the button next to the label **Variable**, and then click on **OK**. Note: The SPSS student version is limited to no more than 1,500 data values.

Excel

Although it can generate histograms, bar graphs, line charts, pie charts, stack plots, stacked line charts, and multiple bar graphs, Excel is typically difficult to use. The DDXL add-in can be installed in Excel, and it can be used to easily generate a histogram, bar chart, dotplot, or pie chart. The DDXL add-in is available at www.aw.com/bbt. After DDXL has been installed in Excel, proceed as follows.

First enter the sample data in column A; then click on the add-in **DDXL**, select **Charts and Plots**, click on the box labeled **function type**, and select the desired graph from the menu. To generate a histogram, for example, select **Histogram** from the menu; then click on the pencil icon and enter the range of cells containing the data, such as A1:A500 for 500 values in rows 1 through 500 of column A.

STATDISK

STATDISK is free to those who purchase new copies of this book, and it is available at www.aw.com/bbt. STATDISK easily generates histograms. Enter the data in the STATDISK data window, click on **Data**, click on **Histogram**, and then click on the **Plot** button. (If you prefer to enter your own class width and starting point, click on the **User defined** button before clicking on **Plot**.)

To generate a pie chart, first enter the category names in column 1 of the STATDISK data window and then enter the corresponding frequencies in column 2; select **Data**, click on **Pie Chart**, and click on **Chart**.

FOCUS ON HISTORY

Can War Be Described with a Graph?

Can a war be described with a graph? Figure 3.59 (on page 138), created by Charles Joseph Minard in 1869, does so remarkably well. This graph tells the story of Napoleon's ill-fated Russian campaign of 1812, sometimes called Napoleon's death march.

The underlying map on Minard's graph shows a roughly 500-mile strip of land extending from the Niemen River on the Polish-Russian border to Moscow. The blue strip depicts the forward march of Napoleon's army. On Minard's original drawing, each millimeter of width represented 6,000 men; this reproduction is shown at a smaller size than the original. The march begins at the far left, where the strip is widest. Here, an army of 422,000 men triumphantly began a march toward Moscow on June 24, 1812. At the time, it was the largest army ever mobilized.

The narrowing of the strip as it approaches Moscow represents the unfolding decimation of the army. (The offshoots represent battalions that were sent off in other directions along the way.) Napoleon had brought only minimal food supplies, and hot summer weather accompanied by heavy rains brought rampant disease. Starvation, disease, and combat losses killed thousands of men each day. By the time the army entered Moscow on September 14, it had shrunk to 100,000 men. The worst was yet to come.

To Napoleon's dismay, the Russians evacuated Moscow prior to the French army's arrival. Deprived of the opportunity to engage the Russian troops and feeling that his army's condition was too poor to continue on to the Russian capital of St. Petersburg, Napoleon took his troops southward out of Moscow. The lower part of the strip on the graph (shown in maroon) represents the retreat, and the dark blue line near the bottom of the figure shows the nighttime temperatures as winter approached. We see that freezing temperatures had already set in by October 18.

Temperatures plunged below 0°F in late November. The sudden narrowing of the lower strip around November 28 shows where 22,000 men perished on the banks of the Berezina River. Three-fourths of the survivors froze to death over the next few days, many on the bitter cold night of December 6. By the time the army reached Poland on December 14, only 10,000 of the original 422,000 remained.

In a famous analysis of graphical techniques, author Edward Tufte described Minard's graph as possibly "the best statistical graphic ever drawn." But a more dramatic statement came from a contemporary of Minard, E. J. Marey, who wrote that this graphic "brought tears to the eyes of all France."

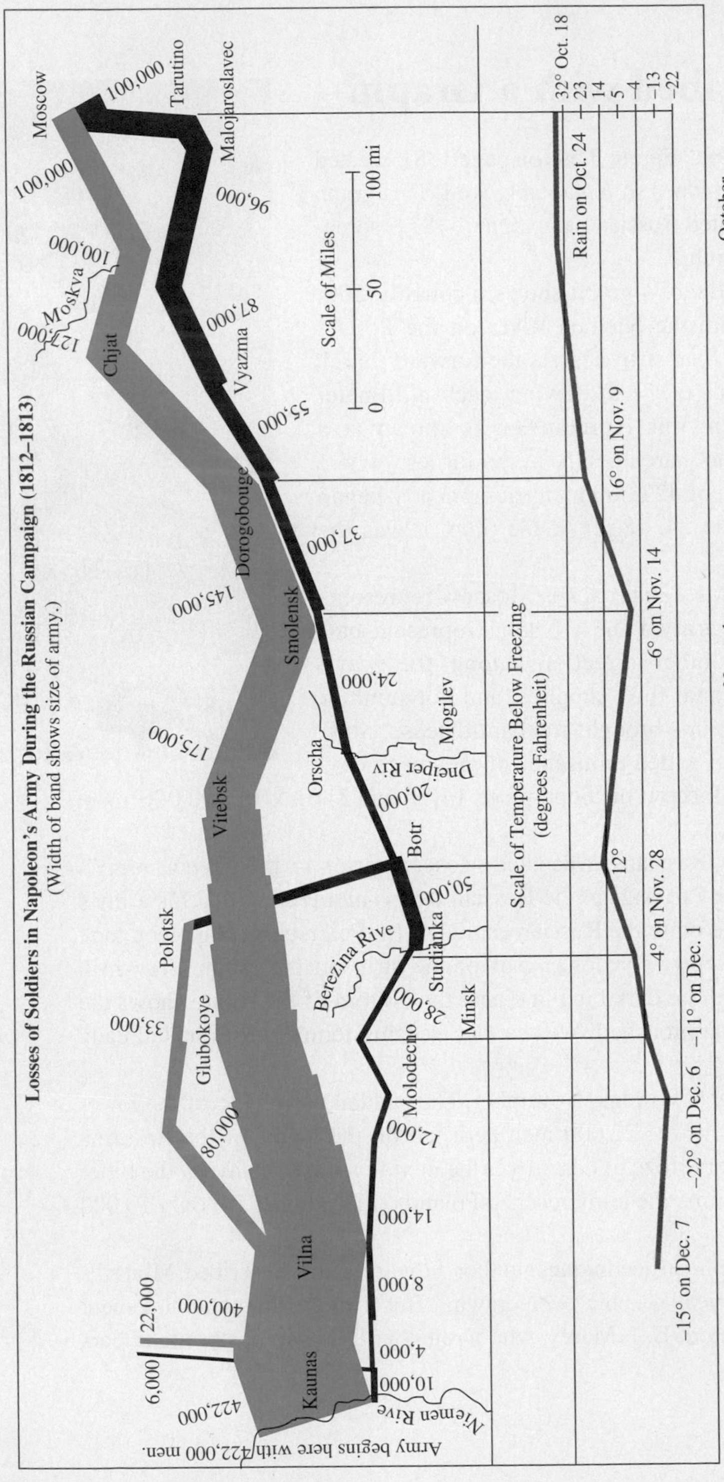

Figure 3.59 *Source:* Edward R. Tufte, *The Visual Display of Quantitative Information* (Cheshire, CT: Graphics Press, 1983). Reprinted with permission.

QUESTIONS FOR DISCUSSION

1. Discuss how this graph helps to overcome the impersonal nature of the many deaths in a war. What kind of impact does it have on you personally?
2. Note that this graph plots six variables: two variables of direction (north-south and east-west), the size of the army, the location of the army, the direction of the army's movement, and temperatures during the retreat. Do you think Minard could have gotten the point across with fewer variables? Why or why not?
3. Discuss how you might make a similar graph for some other historical or political event.

SUGGESTED READING

Tufte, Edward R. *The Visual Display of Quantitative Information*. Graphics Press, 1992.

Wainer, Howard. *Visual Revelations: Graphical Tales of Fate and Deception from Napoleon Bonaparte to Ross Perot*. Copernicus (New York), 1997.

FOCUS ON ENVIRONMENT

How Can We Visualize Global Warming?

Global warming is certainly one of the most discussed issues of our time. But how serious is the problem, and how can we decide whether the warming is caused by human activity or by natural cycles of Earth's climate? These questions can be addressed statistically, and, like many statistical ideas, they are best visualized with graphics.

The obvious starting point for the scientific study of global warming is to find out how much warming is actually occurring. You might at first guess that we could learn about temperature trends simply by comparing local temperatures from old newspapers with temperatures in the same places today. However, global warming refers to an increase in the *average* temperature of our whole planet, which means individual localities may warm more or less than this average. Indeed, we should expect that some places might actually get colder even as the planet as a whole warms up. If we want to learn whether our planet is warming up, we therefore need data that can tell us how the *global average temperature*—the average temperature of the entire Earth—is changing with time.

Today, orbiting satellites provide data that allow us to determine the global average temperature quite accurately, because they give us a view of our entire planet. We can validate these records with "ground truth" measurements recorded at more than 7,000 weather stations around the world, along with measurements of ocean temperatures generally obtained by measuring the temperature of water collected by ships' intake valves. As a result, we have reliable temperature data for the approximately four decades for which we have satellite observations of our planet. The data become somewhat less reliable for years prior to the satellite era, because we can look only at data for specific locations, and the number of locations is smaller as we look further back into the past couple of centuries. Moreover, most past temperature records were kept in cities, which tend to become warmer over time for reasons independent of global warming (see Section 2.2, Example 1). Scientists can often account for this "urban heat island" effect, but even then are left primarily with data for land temperatures and few records of temperatures over the oceans that cover three-fourths of Earth's surface. Scientists have devoted a lot of effort in the past few years to examining past temperature data in detail. Through statistical analysis, it is possible to reconstruct a fairly reliable temperature history for most of the past two centuries, though the uncertainties become larger as we look further back.

Figure 3.60 shows the reconstructed history of Earth's global average temperature since 1860. Despite the uncertainties, the overall conclusion is clear: Global average temperatures have risen by about 0.8°C (1.4°F) in the past century. Moreover, most of the warming (about 0.6°C) has occurred in just the last thirty years, the period for which the data are most reliable. The warming trend also seems to be accelerating. For example, nine of the ten hottest years on record occurred in the most recent 10-year period shown on the graph.

With the temperature data showing a clear warming trend, the next question is how much we might expect the temperature to increase in the future. As we'll discuss further in Chapter 7 (see the "Focus" on p. 328), the rise in temperature is clearly linked to an increase

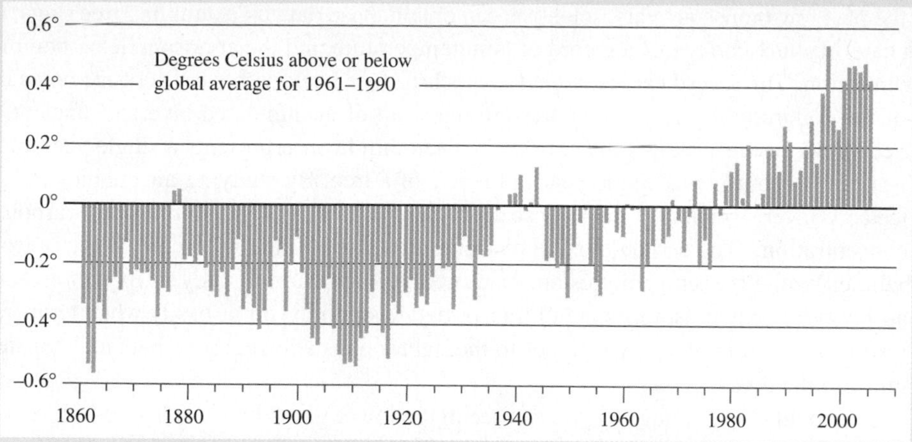

Figure 3.60 Global average temperatures since 1860, compared to the average temperature over the 30-year period from 1961 to 1990. Notice the clear global warming trend of the past few decades. *Source:* National Center for Atmospheric Research.

in the atmospheric concentration of carbon dioxide (CO_2) and other *greenhouse gases*. Therefore, if we wish to predict how much the temperature will rise in the future, we need data on the recent rise in the atmospheric CO_2 concentration, as well as past data concerning how the CO_2 concentration is linked to global temperature changes.

For the recent rise in the CO_2 concentration, we have an extraordinary data set collected since 1958 by scientists working at a station on Mauna Loa (Hawaii). These data, shown on the right side of Figure 3.61, represent direct measurements of the changing CO_2 concentration. Notice that the units of the concentration are *parts per million* (ppm); for example, 320 ppm means that there are 320 carbon dioxide molecules in every million molecules of air. The yearly wiggles on the graph show that the carbon dioxide concentration varies with the seasons. Despite these wiggles, the trend is clearly upward.

To learn how much we might expect the temperature to rise, we would like to have data showing the correlation between carbon dioxide concentration and temperature in the past.

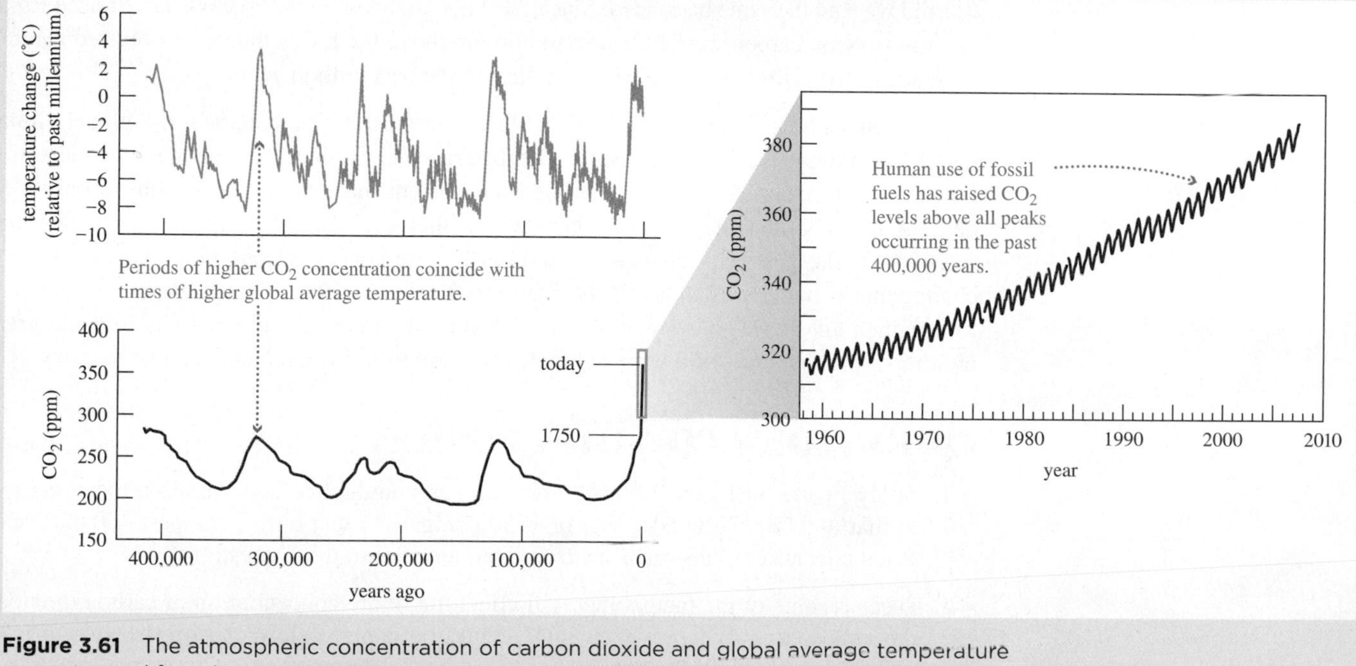

Figure 3.61 The atmospheric concentration of carbon dioxide and global average temperature reconstructed from ice cores and other data for the past 400,000 years. The recent CO_2 data (right) represent direct measurements taken at Mauna Loa (Hawaii).

For the past few thousand years, scientists can obtain these data by examining tree rings, which with careful study can reveal a record of both temperature and the atmospheric carbon dioxide concentration. The record can be extended much further back with studies of ice cores drilled out of the Antarctic ice sheet. Ice cores are made up of accumulated layers of ancient, compressed snow. Because snow falls seasonally, each thin layer represents a single year, much as each tree ring represents a single year in the life of a tree. By studying air bubbles trapped in the ice core layers, scientists can reconstruct the past history of temperatures and carbon dioxide concentration. (The carbon dioxide concentration can be measured fairly directly from the air bubbles, while the temperatures are measured through careful study of oxygen isotopes in the air bubbles: heavier isotopes aren't transported through the air as easily when temperatures are lower, so the ratio of heavy isotopes to the lighter ones allows researchers to estimate temperatures in the distant past.)

The results are striking. As you can see in the lower graph on the left side of Figure 3.61, the ice core data provide a record of temperatures and carbon dioxide concentration going back more than 400,000 years, a period during which Earth went through numerous ice ages and warm periods. At least three conclusions should jump out at you as you study the figure:

1. There is a correlation between temperature and carbon dioxide concentration: Periods of higher temperature tend to also be periods of higher carbon dioxide concentration. Although this does not prove that a rise in carbon dioxide concentration *causes* the higher temperature, it certainly makes it seem likely that the two go together—which means it makes sense that the recent rise in the carbon dioxide concentration has been accompanied by a corresponding rise in temperature.
2. Both the temperature and the carbon dioxide concentration vary substantially and naturally with time. Global average temperatures have risen or fallen by more than 10°C (about 18°F) several times during the past few hundred thousand years, and the carbon dioxide concentration has varied naturally between less than 200 and nearly 300 parts per million (ppm).
3. When we compare the recent carbon dioxide data on the right side of the graph to the past data on the left, we find we are in completely new territory: The current carbon dioxide concentration is far above anything that our planet has seen in the past 400,000 years. Indeed, recent data have preliminarily extended the ice core record back more than 1 million years, and we find the same basic idea: Since the dawn of the industrial age, we have raised Earth's atmospheric carbon dioxide concentration far above the levels that have occurred naturally during any of the warm periods or ice ages of the past million years.

Even without precise predictions of the expected future changes in the global climate, these data provide a sobering message. The difference between an ice age in which most of the United States is buried under glaciers and our current more balmy conditions is only a few degrees Celsius, and the ice core data tell us that even larger temperature changes have occurred in the past with changes in carbon dioxide concentrations much smaller than the change that is under way today. If the past trends are indicative of what we can expect in the future, then unless we act soon to stop the buildup of carbon dioxide, our children and grandchildren will inhabit a world with a climate quite different from the one we live in today.

QUESTIONS FOR DISCUSSION

1. Study Figure 3.61 carefully. How does the carbon dioxide concentration today compare to that of 1750? How does that of 1750 compare to that during the past 400,000 years? What conclusions can you draw from your answers to these questions?

2. Discuss some of the factors that will affect the future concentration of carbon dioxide in the atmosphere. What do you think should be done to slow or stop the growth in the carbon dioxide concentration?

3. What kinds of consequences might we expect from global warming? Discuss what you have heard from news reports, and consider which consequences seem likely and which ones seem overly optimistic or overly alarming.

4. Only a decade or so ago, there was still a fair amount of controversy over whether global warming was really occurring and whether it qualified as a serious problem. Today, that controversy is all but ended. Discuss the role that new data have played in clarifying the threat of global warming. How important are clear graphics to our understanding of these new data?

SUGGESTED READING

You can find numerous scientific Web sites with the latest updates to global warming data. Good places to start include the Web sites of the Intergovernmental Panel on Climate Change (IPCC), the National Center for Atmospheric Research (NCAR), and the U.S. Environmental Protection Agency (EPA).

Bennett, Jeffrey, and Shostak, Seth. *Life in the Universe*, 2nd ed. Addison Wesley, 2007 (see Section 10.5).

Flannery, Tim. *The Weather Makers*. Atlantic Monthly Press, 2005.

4

It's no use trying to sum people up. One must follow hints, not exactly what is said, nor yet entirely what is done.

—Virginia Woolf

Describing Data

LEARNING GOALS

4.1 What Is Average?
Understand the difference between a mean, median, and mode and how each is affected by outliers. Also understand how these different types of "average" can lead to confusion and when it is appropriate to use a weighted mean.

4.2 Shapes of Distributions
Be able to describe the general shape of a distribution in terms of its number of modes, skewness, and variation.

4.3 Measures of Variation
Understand and interpret these common measures of variation: range, the five-number summary, and standard deviation.

4.4 Statistical Paradoxes
Investigate a few common paradoxes that arise in statistics, such as how it is possible that most people who fail a "90% accurate" polygraph test may actually be telling the truth.

IN CHAPTER 3, WE DISCUSSED METHODS FOR DISPLAYING data distributions with tables and graphs. Now we are ready to study how we can summarize the characteristics of a distribution in terms of just a few properties and numbers. In particular, we'll discuss common methods for describing the center, shape, and variation of a collection of data. These methods are central to data analysis and, as you'll see, have applications to nearly every statistical study you encounter in the news. We'll conclude the chapter by studying a few surprises that occasionally turn up even when we look at data carefully.

4.1 What Is Average?

The term *average* comes up frequently in the news and other reports, but it does not always have the same meaning. As you will see in this section, the most appropriate definition of *average* depends on the situation.

Mean, Median, and Mode

Table 4.1 Five Science Fiction Movie Series

Series	Number of movies (as of 2007)
Alien	4
Back to the Future	3
Planet of the Apes	6
Star Trek	10
Star Wars	6

Table 4.1 shows the number of movies (original and sequels or prequels) in each of five popular science fiction series. What is the average number of films in these series? One way to answer this question is to compute the **mean**. (The formal term of *arithmetic mean* is commonly referred to simply as the *mean*.) We find the mean by dividing the total number of movies by five (because there are five series listed in the data set):

$$\text{mean} = \frac{4 + 3 + 6 + 10 + 6}{5} = \frac{29}{5} = 5.8$$

In other words, these five series have a mean of 5.8 movies. More generally, we find the mean of any data set by dividing the sum of all the data values by the number of data values. The mean is what most people think of as the average. In essence, it represents the balance point for a quantitative data distribution, as shown in Figure 4.1.

We could also describe the average number of films by computing the **median**, or middle value, of the data set. To find a median, we arrange the data values in ascending (or descending) order, repeating data values that appear more than once. If the number of values is odd, there is exactly one value in the middle of the list, and this value is the median. If the number of values is even, there are two values in the middle of the list, and the median is the number that lies halfway between them. Putting the data in Table 4.1 in ascending order gives the list 3, 4, 6, 6, 10. The median number of movies is 6 because 6 is the middle number in the list.

The **mode** is the most common value or group of values in a data set. In the case of the movies, the mode is 6 because this value occurs twice in the data set, while the other values occur only once. A data set may have one mode, more than one mode, or no mode. Sometimes the mode refers to a group of closely spaced values rather than a single value. The mode is used

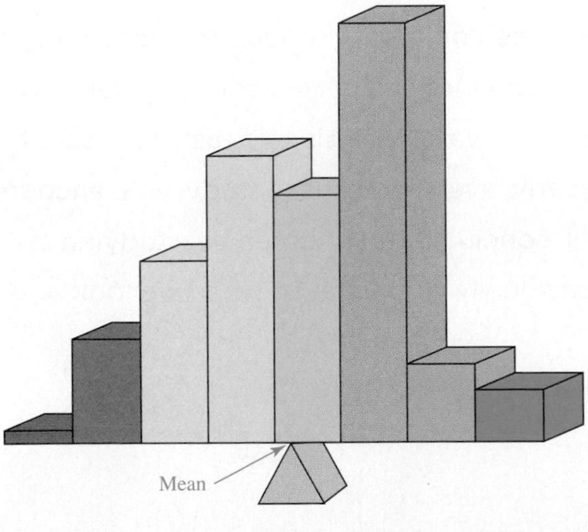

Figure 4.1 A histogram made from blocks would balance at the position of its mean.

more commonly for qualitative data than for quantitative data, as neither the mean nor the median can be used with qualitative data.

Definitions—Measures of Center in a Distribution

The **mean** is what we most commonly call the average value. It is found as follows:

$$\text{mean} = \frac{\text{sum of all values}}{\text{total number of values}}$$

The **median** is the middle value in the sorted data set (or halfway between the two middle values if the number of values is even).

The **mode** is the most common value (or group of values) in a data set.

When rounding, we will use the following rule for all the calculations discussed in this chapter.

Rounding Rule for Statistical Calculations

State your answers with *one more* decimal place of precision than is found in the raw data. Example: The mean of 2, 3, and 5 is 3.3333 . . . , which we round to 3.3. Because the raw data are whole numbers, we round to the nearest tenth. *As always, round only the final answer and not any intermediate values used in your calculations.*

Note that we applied this rule in our example of the movies. The data in Table 4.1 consist of whole numbers, but we stated the mean as 5.8.

TECHNICAL NOTE

If the measure of center has the same number of significant digits as the original data, you can either include an extra zero or use the exact result without the extra decimal place. For example, the mean of 2 and 4 can be expressed as 3 or 3.0.

EXAMPLE 1 Price Data

Eight grocery stores sell the PR energy bar for the following prices:

\$1.09 \$1.29 \$1.29 \$1.35 \$1.39 \$1.49 \$1.59 \$1.79

Find the mean, median, and mode for these prices.

Solution The *mean* price is \$1.41:

$$\text{mean} = \frac{\$1.09 + \$1.29 + \$1.29 + \$1.35 + \$1.39 + \$1.49 + \$1.59 + \$1.79}{8}$$

$$= \$1.41$$

To find the *median*, we first sort the data in ascending order:

$$\underbrace{\$1.09,\ \ \$1.29,\ \ \$1.29,}_{\text{3 values below}}\quad \underbrace{\$1.35,\ \ \$1.39,}_{\text{2 middle values}}\quad \underbrace{\$1.49,\ \ \$1.59,\ \ \$1.79}_{\text{3 values above}}$$

Because there are eight prices (an even number), there are two values in the middle of the list: \$1.35 and \$1.39. Therefore the median lies halfway between these two values, which we calculate by adding them and dividing by 2:

$$\text{median} = \frac{\$1.35 + \$1.39}{2} = \$1.37$$

Using the rounding rule, we could express the mean and median as \$1.410 and \$1.370, respectively.

The *mode* is \$1.29 because this price occurs more times than any other price.

By the Way...

A recent study at the University of Chicago discovered that the median allowance for teenagers in the United States is \$50 per week. The study estimated that teenagers receive over \$1 billion per week in allowances.

FRAZZ reprinted by permission of United Feature Syndicate, Inc.

EXAMPLE 2 Oceans and Seas

Table 4.2 lists areas, in square kilometers, of the world's oceans and seas. For these areas, find the mean, median, and mode.

Table 4.2 Areas of the World's Oceans and Seas

Ocean/Sea	Area (sq. km)	Ocean/Sea	Area (sq. km)	Ocean/Sea	Area (sq. km)
Pacific Ocean	165,760,000	South China Sea	2,319,000	Japan Sea	1,007,800
Atlantic Ocean	82,400,000	Bering Sea	2,291,900	Andaman Sea	797,700
Indian Ocean	65,527,000	Gulf of Mexico	1,592,800	North Sea	575,200
Arctic Ocean	14,090,000	Okhotsk Sea	1,589,700	Red Sea	438,000
Mediterranean Sea	2,965,800	East China Sea	1,249,200	Baltic Sea	422,200
Caribbean Sea	2,718,200	Hudson Bay	1,232,300		

Source: TIME Almanac.

Solution If you check, you'll find that the sum of the areas for the 17 oceans and seas listed is 346,976,800 square kilometers. We find the *mean* by dividing this sum by 17:

$$\text{mean} = \frac{346{,}976{,}800 \text{ sq. km}}{17} = 20{,}410{,}400 \text{ sq. km}$$

Table 4.2 is already arranged in descending order, which makes it easy to find the median. With 17 items on the list, the 9th value is the middle value because 8 values lie above it and 8 values lie below it. The 9th item on the list is the Gulf of Mexico, so the *median* is its area of 1,592,800 square kilometers.

These data have no *mode* because no value occurs more than once. However, if we look at groups of values, 5 of the 17 oceans and seas have areas between 1 and 2 million square kilometers. We might therefore say that the most common areas are those between about 1 and 2 million square kilometers.

Effects of Outliers

To explore the differences among the mean, median, and mode, imagine that the five graduating seniors on a college basketball team receive the following first-year contract offers to play in the National Basketball Association (zero indicates that the player did not receive a contract offer):

$$0 \quad 0 \quad 0 \quad 0 \quad \$3{,}500{,}000$$

The mean contract offer is

$$\text{mean} = \frac{0 + 0 + 0 + 0 + \$3{,}500{,}000}{5} = \$700{,}000$$

Is it therefore fair to say that the *average* senior on this basketball team received a $700,000 contract offer?

Not really. The problem is that the single player receiving the large offer makes the mean much larger than it would be otherwise. If we ignore this one player and look only at the other four, the mean contract offer is zero. Because this one value of $3,500,000 is so extreme compared to the others, we say that it is an **outlier** (or *outlying value*). As our example shows, an outlier can pull the mean significantly upward (or downward), thereby making the mean unrepresentative of the data set as a whole.

Definition

An **outlier** in a data set is a value that is much higher or much lower than almost all other values.

While the outlier pulls the mean contract offer upward, it has no effect on the median contract offer, which remains zero for the five players. In general, the value of an outlier has no effect on the median, because outliers don't lie in the middle of a data set. Outliers do not affect the mode either. (However, the median may change if we delete an outlier, because we are changing the number of values in the data set.) Table 4.3 summarizes the characteristics of the mean, median, and mode, including the effects of outliers on each measure.

TIME OUT TO THINK

Is it fair to use the median as the average contract offer for the five players? Why or why not?

By the Way ...

A survey once found that geography majors from the University of North Carolina had a far higher mean starting salary than geography majors from other schools. The reason for the high mean turned out to be a single outlier—the basketball superstar and geography major named Michael Jordan.

Table 4.3 Comparison of Mean, Median, and Mode

Measure	Definition	How common?	Existence	Takes every value into account?	Affected by outliers?	Advantages
Mean	$\frac{\text{sum of all values}}{\text{total number of values}}$	most familiar "average"	always exists	yes	yes	commonly understood; works well with many statistical methods
Median	middle value	common	always exists	no (aside from counting the total number of values)	no	when there are outliers, may be more representative of an "average" than the mean
Mode	most frequent value	sometimes used	may be no mode, one mode, or more than one mode	no	no	most appropriate for qualitative data (see Section 2.1)

Deciding how to deal with outliers is one of the more important issues in statistics. Sometimes, as in our basketball example, an outlier is a legitimate value that must be understood in order to interpret the mean and median properly. Other times, outliers may indicate mistakes in a data set. Deciding when outliers are important and when they may simply be mistakes can be very difficult.

EXAMPLE 3 Mistake?

A track coach wants to determine an appropriate heart rate for her athletes during their workouts. She chooses five of her best runners and asks them to wear heart monitors during a workout. In the middle of the workout, she reads the following heart rates for the five athletes: 130, 135, 140, 145, 325. Which is a better measure of the average in this case—the mean or the median? Why?

Solution Four of the five values are fairly close together and seem reasonable for mid-workout heart rates. The high value of 325 is an outlier. This outlier seems likely to be a mistake (perhaps caused by a faulty heart monitor), because anyone with such a high heart rate should be in cardiac arrest. If the coach uses the mean as the average, she will be including this outlier—which means she will be including any mistake made when it was recorded. If she uses the median as the average, she'll have a more reasonable value, because the median won't be affected by the outlier.

Confusion About "Average"

The different meanings of *average* can lead to confusion. Sometimes this confusion arises because we are not told whether the average is the mean or the median, and other times because we are not given enough information about how the average was computed. The following examples illustrate a few such situations.

EXAMPLE 4 Wage Dispute

A newspaper surveys wages for workers in regional high-tech companies and reports an average of \$22 per hour. The workers at one large firm immediately request a pay raise, claiming that they work as hard as employees at other companies but their average wage is only \$19. The management rejects their request, telling them that they are *overpaid* because their average wage, in fact, is \$23. Can both sides be right? Explain.

Solution Both sides can be right if they are using different definitions of *average*. In this case, the workers may be using the median while the management uses the mean. For example, imagine that there are only five workers at the company and their wages are \$19, \$19, \$19, \$19, and \$39. The median of these five wages is \$19 (as the workers claimed), but the mean is \$23 (as management claimed).

EXAMPLE 5 Which Mean?

All 100 first-year students at a small college take three courses in the Core Studies program. Two courses are taught in large lectures, with all 100 students in a single class. The third course is taught in 10 classes of 10 students each. Students and administrators get into an argument about whether classes are too large. The students claim that the mean size of their Core Studies classes is 70. The administrators claim that the mean class size is only 25. Can both sides be right? Explain.

Solution Both sides are right, but they are talking about different means. The students calculated the mean size of the classes in which each student is personally enrolled. Each student is taking two classes with enrollments of 100 and one class with an enrollment of 10, so the mean size of each student's classes is

$$\frac{\text{total enrollment in student's classes}}{\text{number of classes student is taking}} = \frac{100 + 100 + 10}{3} = 70$$

The administrators calculated the mean enrollment in all classes. There are two classes with 100 students and 10 classes with 10 students, making a total enrollment of 300 students in 12 classes. The mean enrollment per class is

$$\frac{\text{total enrollment}}{\text{number of classes}} = \frac{300}{12} = 25$$

The two claims about the mean are both correct, but very different because the students and administrators are talking about different means. The students calculated the mean *class size per student*, while the administrators calculated the mean number of *students per class*.

Figures won't lie, but liars will figure.

—Charles H. Grosvenor

TIME OUT TO THINK

In Example 5, could the administrators redistribute faculty assignments so that all classes have 25 students each? How? Discuss the advantages and disadvantages of such a change.

Weighted Mean

Suppose your course grade is based on four quizzes and one final exam. Each quiz counts as 15% of your final grade, and the final counts as 40%. Your quiz scores are 75, 80, 84, and 88, and your final exam score is 96. What is your overall score?

Because the final exam counts more than the quizzes, a simple mean of the five scores does not give your final score. Instead, we must assign a *weight* (indicating the relative importance) to each score. In this case, we assign weights of 15 (for the 15%) to each of the quizzes and 40 (for the 40%) to the final. We then find the **weighted mean** by adding the products of each score and its weight and then dividing by the sum of the weights:

$$\text{weighted mean} = \frac{(75 \times 15) + (80 \times 15) + (84 \times 15) + (88 \times 15) + (96 \times 40)}{15 + 15 + 15 + 15 + 40}$$

$$= \frac{8745}{100} = 87.45$$

The weighted mean of 87.45 properly accounts for the different weights of the quizzes and the exam. Following the rounding rule, we round this score to 87.5.

Weighted means are appropriate whenever the data values vary in their degree of importance. You can always find a weighted mean using the following formula.

Definition

A **weighted mean** accounts for variations in the relative importance of data values. Each data value is assigned a weight and the weighted mean is

$$\text{weighted mean} = \frac{\text{sum of (each data value} \times \text{its weight)}}{\text{sum of all weights}}$$

By the Way ...

Sports statistics that rate players or teams according to their performance in many different categories are usually weighted means. Examples include the earned run average (ERA) and slugging percentage in baseball, the quarterback rating in football, and computerized rankings of college teams.

TIME OUT TO THINK

Because the weights are percentages in the course grade example, we could think of the weights as 0.15 and 0.40 rather than 15 and 40. Calculate the weighted mean by using the weights of 0.15 and 0.40. Do you still find the same answer? Why or why not?

EXAMPLE 6 GPA

Randall has 38 credits with a grade of A, 22 credits with a grade of B, and 7 credits with a grade of C. What is his grade point average (GPA)? Base the GPA on values of 4.0 points for an A, 3.0 points for a B, and 2.0 points for a C.

Solution The grades of A, B, and C represent data values of 4.0, 3.0, and 2.0, respectively. The numbers of credits are the weights. The As represent a data value of 4 with a weight of 38, the Bs represent a data value of 3 with a weight of 22, and the Cs represent a data value of 2 with a weight of 7. The weighted mean is

$$\text{weighted mean} = \frac{(4 \times 38) + (3 \times 22) + (2 \times 7)}{38 + 22 + 7} = \frac{232}{67} = 3.46$$

Following our rounding rule, we round Randall's GPA from 3.46 to 3.5.

EXAMPLE 7 Stock Voting

Voting in corporate elections is usually weighted by the amount of stock owned by each voter. Suppose a company has five stockholders who vote on whether the company should embark on a new advertising campaign. The votes (Y = yes, N = no) are as follows:

Stockholder	Shares owned	Vote
A	225	Y
B	170	Y
C	275	Y
D	500	N
E	90	N

According to the company's bylaws, the measure needs 60% of the vote to pass. Does it pass?

Solution We can regard a yes vote as a value of 1 and a no vote as a value of 0. The number of shares is the weight for the vote of each stockholder, so Stockholder A's vote represents a value of 1 with a weight of 225, stockholder B's vote represents a value of 1 with a weight of 170, and so on. The weighted mean vote is

$$\text{weighted mean} = \frac{(1 \times 225) + (1 \times 170) + (1 \times 275) + (0 \times 500) + (0 \times 90)}{225 + 170 + 275 + 500 + 90}$$

$$= \frac{670}{1260} \approx 0.53$$

The weighted vote is 53% (or 0.53) in favor, which is short of the required 60%, so the measure does not pass.

Means with Summation Notation (Optional Section)

Many statistical formulas, including the formula for the mean, can be written compactly with a mathematical notation called *summation notation*. The symbol Σ (the Greek capital letter *sigma*) is called the *summation sign* and indicates that a set of numbers should be added. We use the symbol x to represent *each* value in a data set, so we write the sum of all the data values as

$$\text{sum of all values} = \Sigma x$$

For example, if a sample consists of 25 exam scores, Σx represents the sum of all 25 scores. Similarly, if a sample consists of the incomes of 10,000 families, Σx represents the total dollar value of all 10,000 incomes.

We use n to represent the total number of values in the sample. Thus, the general formula for the mean is

$$\bar{x} = \text{sample mean} = \frac{\text{sum of all values}}{\text{total number of values}} = \frac{\Sigma x}{n}$$

The symbol $\bar{x}$ is the standard symbol for the mean of a sample. When dealing with the mean of a population rather than a sample, statisticians instead use the Greek letter μ (*mu*).

Summation notation also makes it easy to express a general formula for the weighted mean. Again we use the symbol x to represent each data value, and we let w represent the weight of each data value. The sum of the products of each data value and its corresponding weight is $\Sigma(x \times w)$. The sum of the weights is Σw. Therefore, the formula for the weighted mean is

$$\text{weighted mean} = \frac{\Sigma(x \times w)}{\Sigma w}$$

TECHNICAL NOTE

Summations are often written with the use of an *index* that specifies how to step through the sum. For example, the symbol x_i indicates the *i*th data value in the set; the letter *i* is the index. We then write the sum of all values as

$$\sum_{i=1}^{n} x_i$$

We read this expression as "the sum of the x_i values, starting with $i = 1$ and continuing to $i = n$, where n is the total number of data values in the set." With this notation, the mean is written

$$\bar{x} = \frac{1}{n}\sum_{i=1}^{n} x_i$$

Means and Medians with Binned Data (Optional Section)

The ideas of this section can be extended to binned data simply by assuming that the middle value in the bin represents all the data values in the bin. For example, consider the following table of 50 binned data values:

Bin	Frequency
0–6	10
7–13	10
14–20	10
21–27	20

The middle value of the first bin is 3, so we assume that the value of 3 occurs 10 times. Continuing this way, we have for the total of the 50 values in the table

$$(3 \times 10) + (10 \times 10) + (17 \times 10) + (24 \times 20) = 780$$

Thus, the mean is 780/50 = 15.6. With 50 values, the median is between the 25th and 26th values. These values fall within the bin 14–20, so we call this bin the **median class** for the data. The mode is the bin with the highest frequency—the bin 21–27 in this case.

Section 4.1 Exercises

Statistical Literacy and Critical Thinking

1. **Outlier.** What is an outlier? Are outliers defined in an exact way so that they can all be clearly and objectively identified? Explain.

2. **Mean and Median.** A statistics class consists of 24 students, all of whom are unemployed or are employed in low-paying part-time jobs. The class also includes a professor who is paid an enormous salary. Which does a better job of describing the income of a typical person in the class of 25 people (including the professor): the mean or the median? Why?

3. **Mean Commuting Time.** A sociologist wants to find the mean commuting time for all working U.S. residents. She knows that it is not practical to survey each of the millions of working people, so she conducts an Internet search and finds the mean commuting time for each of the 50 states. She adds the 50 commuting times and divides by 50. Is the result likely to be a good estimate of the mean commuting time for all workers? Why or why not?

4. **Nominal Data.** In Chapter 2, it was noted that data are at the nominal level of measurement if they consist of names or labels only. A New England Patriots football fan records the number on the jersey of each Patriots player in a Super Bowl game. Does it make sense to calculate the mean of those numbers? Why or why not?

Does It Make Sense? For Exercises 5–8, decide whether the statement makes sense (or is clearly true) or does not make sense (or is clearly false). Explain clearly; not all of these statements have definitive answers, so your explanation is more important than your chosen answer.

5. **Mean.** A data set has means of 65.2 and 72.3.

6. **Mode.** A data set has modes of 65.2 and 72.3.

7. **Mean, Median, and Mode.** A researcher finds that a data set has the same value for the mean, median, and mode.

8. **Weighted Mean.** A researcher calculates the value of the mean for a data set and then constructs a frequency table and calculates the mean from the frequency table. The researcher concludes that an error was made because two different results are obtained.

Appropriate Average. Exercises 9–12 list "averages" that someone might want to know. In each case, state whether the mean or median would give a better description of the "average." Explain your reasoning.

9. **Income.** The average income of all adults in a large city

10. **Weight.** The average weight of the oranges in a large box

11. **Job Changes.** The average number of times that people change jobs during their careers

12. **Lost Baggage.** The average number of pieces of lost luggage per flight for each airline company

Concepts and Applications

Mean, Median, and Mode. Exercises 13–20 each list a set of numbers. In each case, find the mean, median, and mode of the listed numbers.

13. **Perception of Time.** Actual times (in seconds) recorded when statistics students participated in an experiment to test their ability to determine when one minute (60 seconds) had passed:

53	52	75	62	68	58	49	49

14. **Body Temperatures.** Body temperatures (in degrees Fahrenheit) of randomly selected normal and healthy adults:

98.6	98.6	98.0	98.0	99.0
98.4	98.4	98.4	98.4	98.6

15. **Blood Alcohol.** Blood alcohol concentrations of drivers involved in fatal crashes and then given jail sentences (based on data from the U.S. Department of Justice):

0.27	0.17	0.17	0.16	0.13	0.24
0.29	0.24	0.14	0.16	0.12	0.16

16. **Old Faithful Geyser.** Time intervals (in minutes) between eruptions of Old Faithful geyser in Yellowstone National Park:

98	92	95	87	96	90
65	92	95	93	98	94

17. **Fruit Flies.** Thorax lengths (in millimeters) of a sample of male fruit flies:

0.72	0.90	0.84	0.68	0.84	0.90
0.92	0.84	0.64	0.84	0.76	

18. **Ages of Presidents.** Ages of selected U.S. Presidents at the time of inauguration:

57	61	57	57	58	57	61	54
68	51	49	64	50	48	65	

19. Weights of M&Ms. Weights (in grams) of randomly selected M&M plain candies:

0.957 0.912 0.842 0.925 0.939 0.886
0.914 0.913 0.958 0.947 0.920

20. Quarters. Weights (in grams) of quarters in circulation:

5.60 5.63 5.58 5.56 5.66 5.58 5.57 5.59
5.67 5.61 5.84 5.73 5.53 5.58 5.52 5.65
5.57 5.71 5.59 5.53 5.63 5.68

21. Alphabetic States. The following table gives the total area in square miles (land and water) of the seven states with names beginning with the letters A through C.

State	Area
Alabama	52,200
Alaska	615,200
Arizona	114,000
Arkansas	53,200
California	158,900
Colorado	104,100
Connecticut	5,500

a. Find the mean area and median area for these states.

b. Which state is an outlier on the high end? If you eliminate this state, what are the new mean and median areas for this data set?

c. Which state is an outlier on the low end? If you eliminate this state, what are the new mean and median areas for this data set?

22. Outlier Coke. Cans of regular Coca-Cola vary slightly in weight. Here are the measured weights of seven cans, in pounds:

0.8161 0.8194 0.8165 0.8176
0.7901 0.8143 0.8126

a. Find the mean and median of these weights.

b. Which, if any, of these weights would you consider to be an outlier? Explain.

c. What are the mean and median weights if the outlier is excluded?

23. Raising Your Grade. Suppose you have scores of 70, 75, 80, and 70 on quizzes in a mathematics class.

a. What is the mean of these scores?

b. What score would you need on the next quiz to have an overall mean of 75?

c. If the maximum score on a quiz is 100, is it possible to have a mean of 80 after the fifth quiz? Explain.

24. Raising Your Grade. Suppose you have scores of 60, 70, 65, 85, and 85 on exams in a sociology class.

a. What is the mean of these scores?

b. What score would you need on the next exam to have an overall mean of 75?

c. If the maximum score on an exam is 100, what is the maximum mean score that you could possibly have after the next exam? Explain.

25. New Mean. Suppose that after six quizzes you have a mean score of 80 (out of a possible 100). If you get a score of 90 on the next quiz, what is your new mean? What is the maximum mean score that you could have after the next quiz? What is the minimum mean score that you could have after the next quiz?

26. New Batting Average. Suppose that after 30 at-bats a baseball player has a batting average of 0.300. If she gets a hit on the next at-bat, what will her new batting average be? (The batting average is the number of hits divided by the number of at-bats.)

27. Comparing Averages. Suppose that school district officials claim that the average reading score for fourth-graders in the district is 73 (out of a possible 100). As a principal, you know that your fourth-graders had the following scores: 55, 60, 68, 70, 87, 88, 95. Would you be justified in claiming that your students scored above the district average? Explain.

28. Comparing Averages. Suppose the National Basketball Association (NBA) reports that the average height of basketball players is 6′8″. As a coach, you know that the players on your starting lineup have heights of 6′5″, 6′6″, 6′6″, 7′0″, and 7′2″. Would you be justified in claiming that your starting lineup has above average height for the NBA? Explain.

29. Average Peaches. A grocer has three baskets of peaches. One holds 50 peaches and weighs 18 pounds, one holds 55 peaches and weighs 22 pounds, and the third holds 60 peaches and weighs 24 pounds. What is the mean weight of all of the peaches combined? Explain.

30. Average Confusion. An instructor has a first-period class with 25 students, and they had a mean score of 86% on the midterm exam. The second-period class has 30 students, and they had a mean score of 84% on the same exam. Does it follow that the mean score for both classes combined is 85%? Explain.

31. Which Mean? All of the 300 students at a high school take the same four courses. Three of the courses are taught in 15 classes of 20 students each. The fourth course is taught in 3 classes of 100 students each. Explain how the school's principal could claim that the mean class size is 25 students, while upset parents could claim that the mean class size is 40 students. Which mean do you think is a fairer description of class size?

32. Final Grade. Your course grade is based on one midterm that counts as 15% of your final grade, one class project that counts as 20% of your final grade, a set of homework assignments that counts as 40% of your final grade, and a final exam that counts as 25% of your final grade. Your midterm score is 75, your project score is 90, your homework score is 85, and your final exam score is 72. What is your overall final score?

33. Batting Average. Recall that a batting average in baseball is determined by dividing the total number of hits by the total number of at-bats (neglecting walks, errors, and a few other special cases). A player goes 2 for 4 (2 hits in 4 at-bats) in the first game, 0 for 3 in the second game, and 3 for 5 in the third game. What is his batting average? In what way is this number an "average"?

34. Averaging Averages. Suppose a player has a batting average over many games of 0.400. In his next game, he goes 2 for 4, which is a batting average of 0.500 for the game. Does it follow that his new batting average is $(0.400 + 0.500)/2 = 0.450$? Explain.

35. Egg Percentages. A farm inspector finds that 8% of the eggs tested at one farm contain salmonella and 12% of the eggs at another farm contain salmonella. Does it follow that between the two farms, a total of 10% of the eggs tested contain salmonella? Why or why not?

36. Slugging Average. In addition to the batting average, another measure of hitting performance in baseball is called the slugging average. In finding a slugging average, a single is worth 1 point, a double is worth 2 points, a triple is worth 3 points, and a home run is worth 4 points. A player's slugging average is the total number of points divided by the total number of at-bats (neglecting walks, errors, and a few other special cases). A player has three singles in five at-bats in the first game, a triple and a single in four at-bats in the second game, and a double and a home run in five at-bats in the third game.

- **a.** What is his batting average?
- **b.** What is his slugging average?
- **c.** Is it possible for a slugging average to be more than 1? Explain.

37. Stockholder Voting. A small company has four stockholders. One stockholder has 400 shares, a second stockholder has 300 shares, a third stockholder has 200 shares, and the fourth stockholder has 100 shares. In a vote on a new advertising campaign, the first stockholder votes yes, and the other three stockholders vote no. Explain how the outcome of the vote can be expressed as a weighted mean. What is the outcome of the vote?

38. GPA. One common system for computing a grade point average (GPA) assigns 4 points to an A, 3 points to a B, 2 points to a C, and 1 point to a D. What is the GPA of a student who gets an A in a 5-credit course and a B, a C, and a D, respectively, in each of three 3-credit courses?

39. Phenotypes of Peas. An experiment was conducted to determine whether a deficiency of carbon dioxide in soil affects the phenotypes of peas. Listed below are the phenotype codes, where 1 = smooth yellow, 2 = smooth green, 3 = wrinkled yellow, and 4 = wrinkled green. Can the measures of center be obtained for these values? Do the results make sense?

2	1	1	1	1	1	1	4	1	2	2	1	2
3	3	2	3	1	3	1	3	1	3	2	2	

40. U.S. Population Center. Imagine taking a huge flat map of the United States and placing weights on it to represent where people live. The point at which the map would balance is called the mean center of population. Figure 4.2 shows how the location of the mean center of population has shifted from 1790 to 1990. Briefly explain the pattern shown on this map.

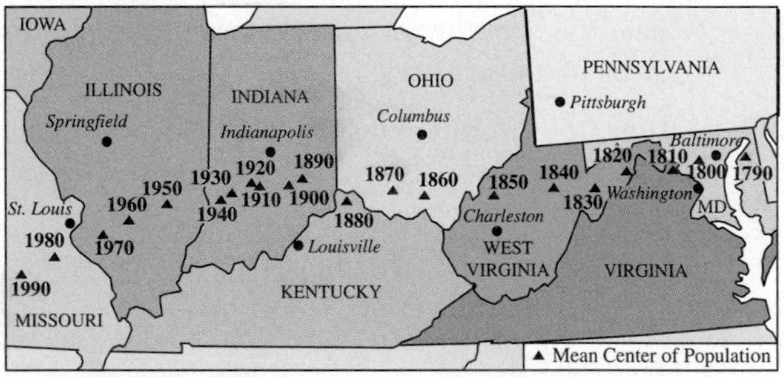

Figure 4.2 Mean center of population. *Source: Statistical Abstract of the United States.*

Projects for the Internet and Beyond

For useful links, select "Links for Internet Projects" for Chapter 4 at www.aw.com/bbt.

41. Salary Data. Many Web sites offer data on salaries in different careers. Find salary data for a career you are considering. What are the mean and median salaries for this career? How do these salaries compare to those of other careers that interest you?

42. Is the Median the Message? Read the article "The Median Isn't the Message," by Stephen Jay Gould, which is posted on the Web. Write a few paragraphs in which you describe the message that Gould is trying to get across. How is this message important to other patients diagnosed with cancer?

43. Navel Data. Bin the data collected in Exercise 23 of Section 3.1. Then make a frequency table, and draw a histogram of the distribution. What is the mean of the distribution? What is the median of the distribution? An old theory says that, on average, the navel ratio of humans is the golden ratio: $(1 + \sqrt{5})/2$. Does this theory seem accurate based on your observations?

IN THE NEWS

44. Daily Averages. Cite three examples of averages that you deal with in your own life (such as grade point average or batting average). In each case, explain whether the average is a mean, a median, or some other type of average. Briefly describe how the average is useful to you.

45. Averages in the News. Find three recent news articles that refer to some type of average. In each case, explain whether the average is a mean, a median, or some other type of average.

4.2 Shapes of Distributions

So far in this chapter we have discussed how to describe the center of a quantitative data distribution with measures such as the mean and median. We now turn our attention to describing the overall *shape* of a distribution. We can *see* the complete shape of a distribution on a graph. Our current goal is to describe the general shape in words. Although such descriptions carry less information than the complete graph, they are still useful. We will focus on three characteristics of a distribution: its number of modes, its symmetry or skewness, and its variation.

Because we are interested primarily in the *general* shapes of distributions, it's often easier to examine graphs that show smooth curves rather than the original data sets. Figure 4.3 shows three examples of this idea, two in which the distributions are shown as histograms and one in which the distribution is shown as a line chart. The smooth curves approximate these distributions, but do not show all of their details.

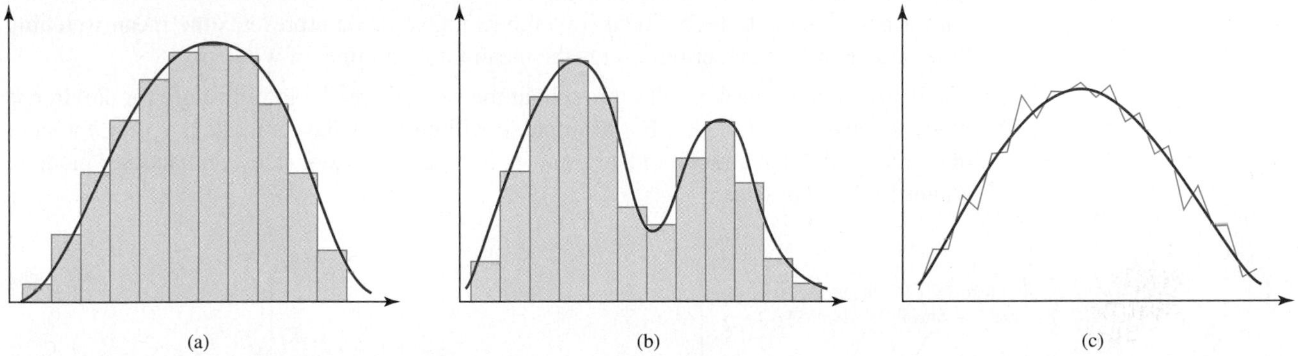

Figure 4.3 The smooth curves approximate the shapes of the distributions.

Number of Modes

One simple way to describe the shape of a distribution is by its number of peaks, or modes. Figure 4.4a shows a distribution, called a **uniform distribution**, that has no mode because all data values have the same frequency. Figure 4.4b shows a distribution with a single peak as its

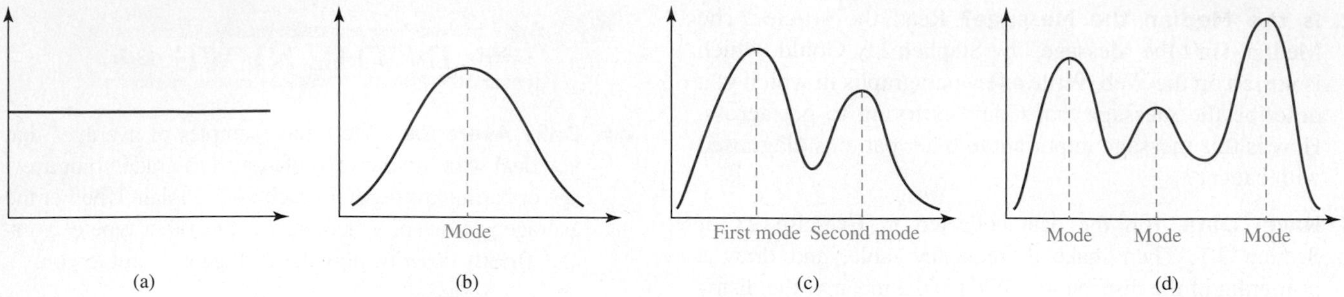

Figure 4.4 (a) A uniform distribution has no mode. (b) A single-peaked distribution has one mode. (c) A bimodal distribution has two modes. (d) A trimodal distribution has three modes.

mode. It is called a **single-peaked**, or **unimodal**, distribution. By convention, any peak in a distribution is considered a mode, even if not all peaks have the same height. For example, the distribution in Figure 4.4c is said to have two modes—even though the second peak is lower than the first; it is a *bimodal* distribution. Similarly, the distribution in Figure 4.4d is said to have three modes; it is a *trimodal* distribution.

EXAMPLE 1 Number of Modes

How many modes would you expect for each of the following distributions? Why? Make a rough sketch for each distribution, with clearly labeled axes.

a. Heights of 1,000 randomly selected adult women

b. Hours spent watching football on TV in January for 1,000 randomly selected adult Americans

c. Weekly sales throughout the year at a retail clothing store for children

d. The number of people with particular last digits (0 through 9) in their Social Security numbers

Solution Figure 4.5 shows sketches of the distributions.

a. The distribution of heights of women is single-peaked because many women are at or near the mean height, with fewer and fewer women at heights much greater or less than the mean.

b. The distribution of times spent watching football on TV for 1,000 randomly selected adult Americans is likely to be bimodal (two modes). One mode represents the mean watching time of men, and the other represents the mean watching time of women.

c. The distribution of weekly sales throughout the year at a retail clothing store for children is likely to have several modes. For example, it will probably have a mode in spring for sales of summer clothing, a mode in late summer for back-to-school sales, and another mode in winter for holiday sales.

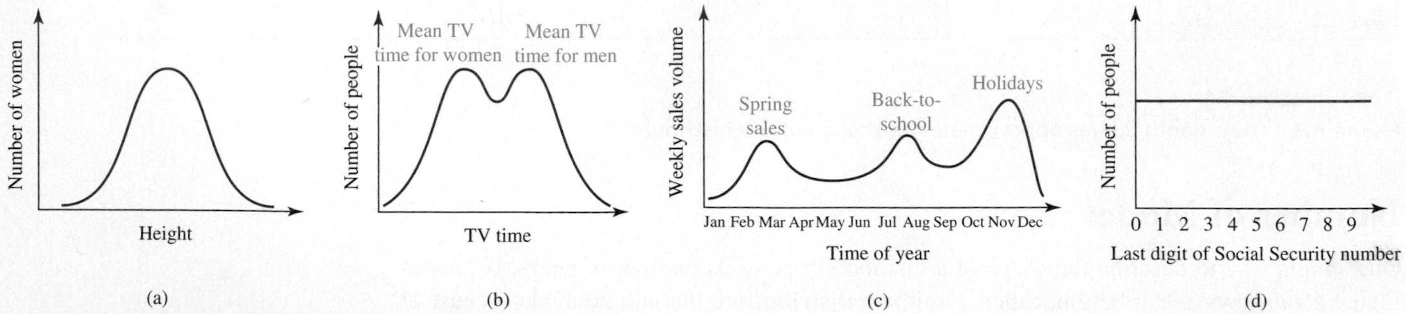

Figure 4.5 Sketches for Example 1.

d. The last digits of Social Security numbers are essentially random, so the number of people with each different last digit (0 through 9) should be about the same. That is, about 10% of all Social Security numbers end in 0, 10% end in 1, and so on. It is therefore a uniform distribution with no mode.

Symmetry or Skewness

A second simple way to describe the shape of a distribution is in terms of its symmetry or skewness. A distribution is **symmetric** if its left half is a mirror image of its right half. The distributions in Figure 4.6 are all symmetric. The symmetric distribution in Figure 4.6a, with a single peak and a characteristic bell shape, is known in statistics as a *normal distribution*; it is so important that we will devote Chapter 5 to its study.

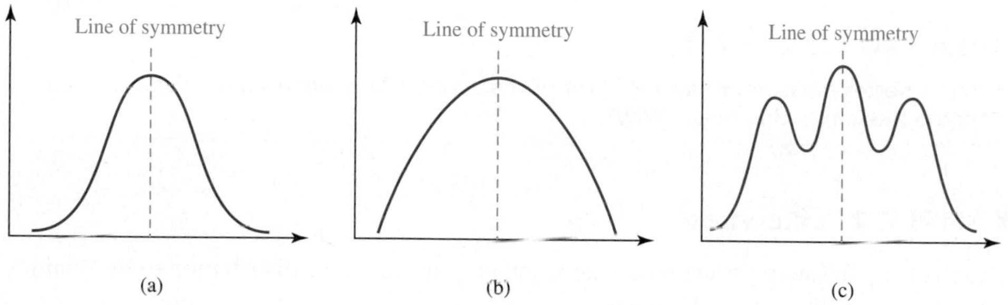

Figure 4.6 These distributions are all symmetric because their left halves are mirror images of their right halves. Note that (a) and (b) are single-peaked (unimodal), whereas (c) is triple-peaked (trimodal).

A distribution that is not symmetric must have values that tend to be more spread out on one side than on the other. In this case, we say that the distribution is **skewed**. Figure 4.7a shows a distribution in which the values are more spread out on the left, meaning that some values are outliers at low values. We say that such a distribution is **left-skewed** (or *negatively* skewed). It is helpful to think of such a distribution as having a tail that has been pulled toward the left. Figure 4.7b shows a distribution in which the values are more spread out on the right, making it **right-skewed** (or *positively* skewed). It looks as if it has a tail pulled toward the right.

Figure 4.7 also shows how skewness affects the relative positions of the mean, median, and mode. By definition, the mode is at the peak in a single-peaked distribution. A left-skewed

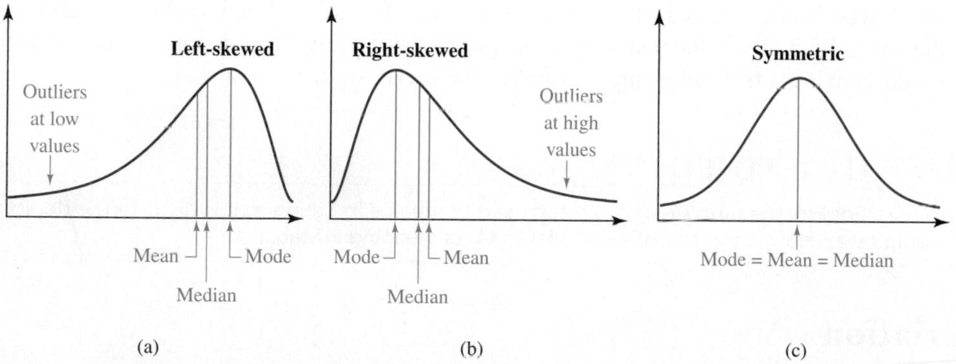

Figure 4.7 (a) Skewed to the left (left-skewed): The mean and median are less than the mode. (b) Skewed to the right (right-skewed): The mean and median are greater than the mode. (c) Symmetric distribution: The mean, median, and mode are the same.

distribution pulls the mean and median to the left—that is, to values less than the mode. Moreover, outliers at the low end of the data set tend to make the mean less than the median (see Table 4.3 on page 149). Similarly, a right-skewed distribution pulls the mean and median to the right—that is, to values greater than the mode. In such cases, the outliers at the high end of the data set tend to make the mean greater than the median. When the distribution is symmetric and single-peaked, both the mean and the median are equal to the mode.

TECHNICAL NOTE

A left-skewed distribution is also called *negatively skewed*, and a right-skewed distribution is also called *positively skewed*. A symmetric distribution has *zero skewness*.

Definitions

A distribution is **symmetric** if its left half is a mirror image of its right half.

A distribution is **left-skewed** if its values are more spread out on the left side.

A distribution is **right-skewed** if its values are more spread out on the right side.

TIME OUT TO THINK

Which is a better measure of "average" (or of the *center* of the distribution) for a skewed distribution: the median or the mean? Why?

EXAMPLE 2 Skewness

For each of the following situations, state whether you expect the distribution to be symmetric, left-skewed, or right-skewed. Explain.

a. Heights of a sample of 100 women
b. Family income in the United States
c. Speeds of cars on a road where a visible patrol car is using radar to detect speeders

Solution

a. The distribution of heights of women is symmetric because roughly equal numbers of women are shorter and taller than the mean, and extremes of height are rare on either side of the mean.
b. The distribution of family income is right-skewed. Most families are middle-class, so the mode of this distribution is a middle-class income (somewhere around $50,000 in the United States). But a few very high-income families pull the mean to a considerably higher value, stretching the distribution to the right (high-income) side.
c. Drivers usually slow down when they are aware of a patrol car looking for speeders. Few if any drivers will be exceeding the speed limit, but some drivers tend to slow to well below the speed limit. Thus, the distribution of speeds will be left-skewed, with a mode near the speed limit but a few cars going well below the speed limit.

By the Way ...

Speed kills. On average, in the United States, someone is killed in an auto accident about every 12 minutes. About one-third of these fatalities involve a speeding driver.

TIME OUT TO THINK

In ordinary English, the term *skewed* is often used to mean something that is distorted or depicted in an unfair way. How is this use of *skew* related to its meaning in statistics?

Variation

A third way to describe a distribution is by its **variation**, which is a measure of how much the data values are spread out. A distribution in which most data are clustered together has a low variation. As shown in Figure 4.8a, such a distribution has a fairly sharp peak. The variation is higher when the data are distributed more widely around the center, which makes the peak broader. Figure 4.8b shows a distribution with a moderate variation and Figure 4.8c shows a distribution with a high variation. In the next section, we'll discuss methods for describing the variation quantitatively.

Definition

Variation describes how widely data are spread out about the center of a data set.

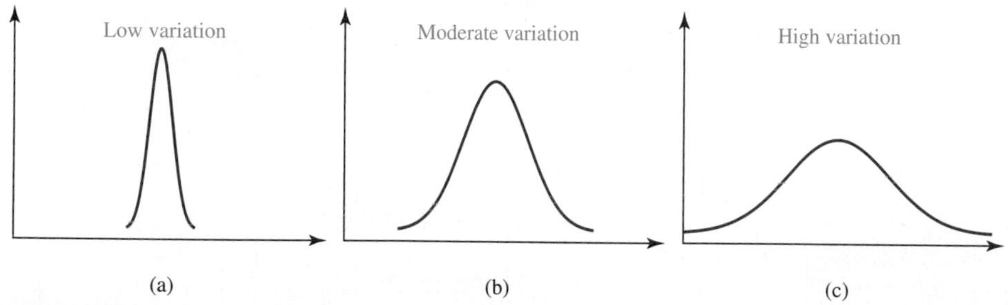

Figure 4.8 From left to right, these three distributions have increasing variation.

EXAMPLE 3 Variation in Marathon Times

How would you expect the variation to differ between times in the Olympic marathon and times in the New York City marathon? Explain.

Solution The Olympic marathon invites only elite runners, whose times are likely to be clustered relatively near world-record times. The New York City marathon allows runners of all abilities, whose times are spread over a very wide range (from near the world record of just over two hours to many hours). Therefore, the variation among the times should be greater in the New York City marathon than in the Olympic marathon.

By the Way ...

In 1997, doctors told 25-year-old Lance Armstrong that, according to the averages, he had less than a 50-50 chance of surviving his testicular cancer, which had already spread to his brain. He not only survived, but went on to win the Tour de France (bicycle race) a record seven consecutive times. His story shows that, on a personal level, variation can be much more important than any average.

Section 4.2 Exercises

Statistical Literacy and Critical Thinking

1. **Symmetry.** What do we mean when we say that a graph is symmetric?

2. **Distribution.** The voltage supplied to a particular home varies between 123.0 volts and 125.0 volts, and all of the voltage levels are equally likely. Which term best describes this distribution: skewed, bimodal, uniform, or unimodal? Explain.

3. **IQ Scores.** Consider the IQ scores of college students in a statistics class compared to the IQ scores of randomly selected adults. Which of these two sets of IQ scores has less variation? What effect does the lower variation have on a graph of the distribution of those IQ scores?

4. **Skewness.** What is skewness in a graph?

Does It Make Sense? For Exercises 5–8, decide whether the statement makes sense (or is clearly true) or does not make sense (or is clearly false). Explain clearly; not all of these statements have definitive answers, so your explanation is more important than your chosen answer.

5. **Symmetry.** Because a data set has two modes, it cannot be symmetric.

6. **Symmetry.** Examination of the data set reveals that it is symmetric with a mean of 75.0 and a median of 80.0.

7. **Uniform Distribution.** Examination of a data set reveals that it is uniform and unimodal.

8. **Distribution.** Examination of a data set reveals that its distribution is left-skewed and unimodal.

Concepts and Applications

9. **Old Faithful.** The histogram in Figure 4.9 shows the times between eruptions of Old Faithful geyser in Yellowstone National Park for a sample of 300 eruptions (which means 299 times between eruptions). Over the histogram, draw a smooth curve that captures its general features. Then classify the distribution according to its number of modes and its symmetry or skewness. In words, summarize the meaning of your results.

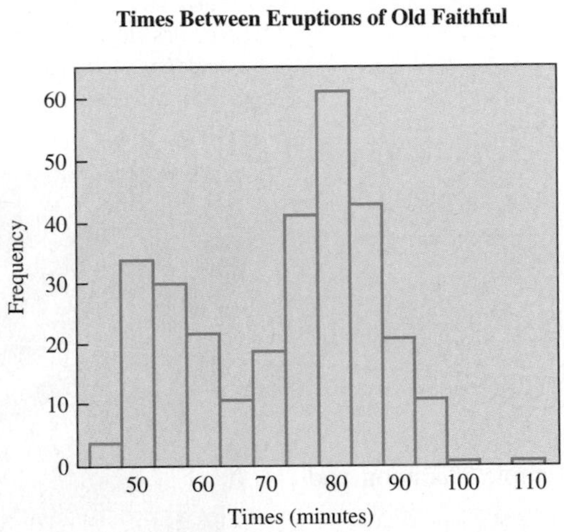

Figure 4.9 *Source:* Hand et al., *Handbook of Small Data Sets.*

10. **Chip Failures.** The histogram in Figure 4.10 shows the time until failure for a sample of 108 computer chips. Over the histogram, draw a smooth curve that captures its general features. Then classify the distribution according to its number of modes and its symmetry or skewness. In words, summarize the meaning of your results.

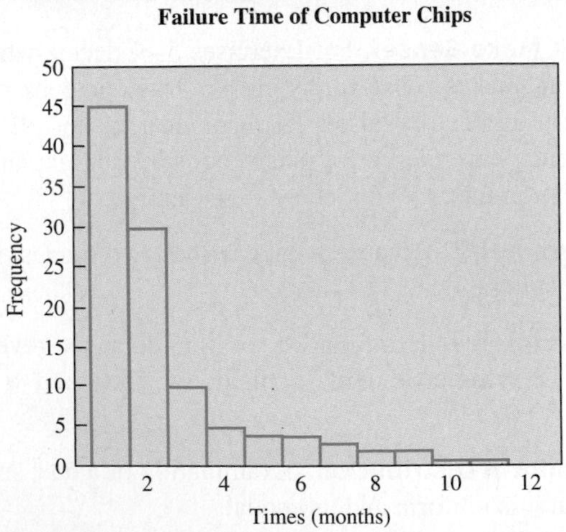

Figure 4.10 *Source:* Hand et al., *Handbook of Small Data Sets.*

11. **Rugby Weights.** The histogram in Figure 4.11 shows the weights of a sample of 391 rugby players. Over the histogram, draw a smooth curve that captures its general features. Then classify the distribution according to its number of modes and its symmetry or skewness. In words, summarize the meaning of your results.

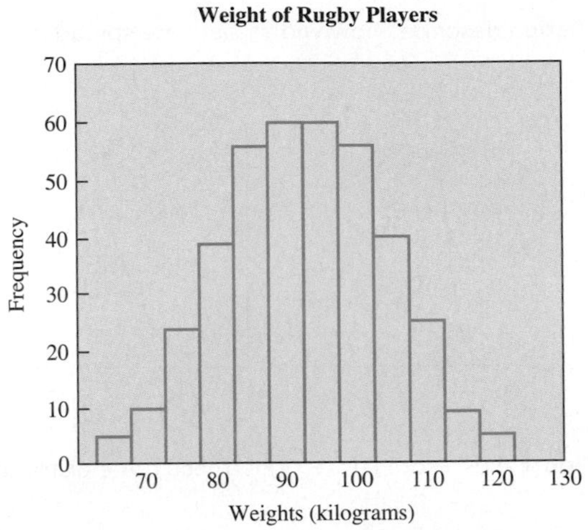

Figure 4.11 *Source:* Hand et al., *Handbook of Small Data Sets.*

12. **Penny Weights.** The histogram in Figure 4.12 shows the weights (in grams) of 72 pennies. Over the histogram, draw a smooth curve that captures its general features. Then classify the distribution according to its number of modes and its symmetry or skewness. What feature of the graph reflects the fact that 35 of the pennies were made before 1983 and consist of 97% copper and 3% zinc whereas the other 37 pennies were made after 1983 and are 3% copper and 97% zinc?

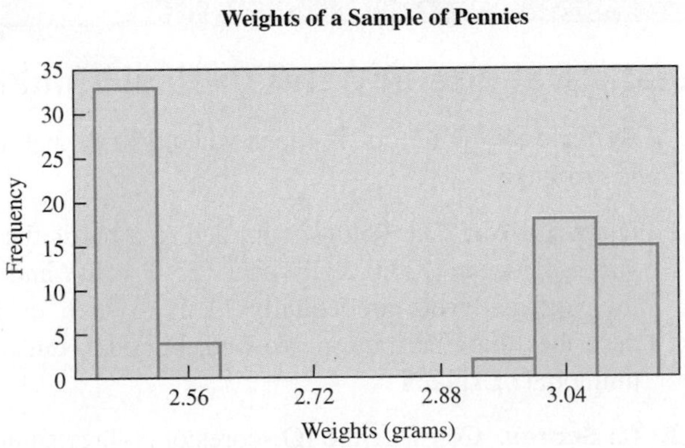

Figure 4.12 *Source:* Measurements by Mario F. Triola.

13. **Family Income.** Suppose you study family income in a random sample of 300 families. Your results can be summarized as follows:

- The mean family income was $41,000.
- The median family income was $35,000.
- The highest and lowest incomes were $250,000 and $2,400, respectively.

a. Draw a rough sketch of the income distribution, with clearly labeled axes. Describe the distribution as symmetric, left-skewed, or right-skewed.

b. How many families in the sample earned less than $35,000? Explain how you know.

c. Based on the given data, can you determine how many families earned more than $41,000? Why or why not?

14. Boston Rainfall. The daily rainfall amounts (in inches) for Boston in a recent year consist of 365 values with these properties:

- The mean daily rainfall amount is 0.083 inch.
- The median of the daily rainfall amounts is 0 inches.
- The minimum daily rainfall amount is 0 inches and the maximum is 1.48 inches.

a. How is it possible that the minimum of the 365 values is 0 inches and the median is also 0 inches?

b. Describe the distribution as symmetric, left-skewed, or right-skewed.

c. Can you determine the exact number of days that it rained? Can you conclude anything about the number of days that it rained? Explain.

Describing Distributions. For each distribution described in Exercises 15–30, answer the following questions:

a. How many modes would you expect for the distribution? Explain.

b. Would you expect the distribution to be symmetric, left-skewed, or right-skewed? Explain.

15. Incomes. The annual incomes of all those in a statistics class, including the instructor

16. Test Scores. The test scores of all 200 students who took a fair test with a mean score of 75%

17. Student Weights. The weights of 100 eighth-grade students

18. Football Player Weights. The weights of 100 professional football players

19. Weights of Ice Skaters. The weights of people who use an ice rink that is open only to professional figure skaters in the morning and to professional hockey players in the afternoon and evening

20. Vehicle Weights. The weights of cars in the new car lot at a car dealership in which about half the inventory consists of compact cars and the other half consists of sport utility vehicles

21. Departure Delays. The delays in minutes of scheduled flights from a large airport

22. Speeds. The speeds of drivers as they pass through a school zone

23. Patron Ages. The ages of people who visit an art museum (not including school groups)

24. Patron Ages. The ages of people who visit an amusement park

25. Tennis Players. The number of players in each doubles match of a women's tennis tournament

26. Sales. The monthly sales of swimming suits over a one-year period at a store in San Diego

27. Incomes. The incomes of people sitting in luxury boxes at the Super Bowl

28. Incomes. The incomes of people watching the Super Bowl on TV

29. Salaries. The salaries of major league baseball players

30. Batting Averages. The batting averages of major league baseball players

Projects for the Internet and Beyond

For useful links, select "Links for Internet Projects" for Chapter 4 at www.aw.com/bbt.

31. New York Marathon. The Web site for the New York City marathon gives frequency data for finish times in the most recent marathon. Study the data, make a rough sketch of the distribution, and describe the shape of the distribution in words.

32. Tax Stats. The IRS Web site provides statistics collected from tax returns on income, refunds, and much more. Choose a set of statistics from this Web site and study the distribution. Describe the distribution in words, and discuss anything you learn that is relevant to national tax policies.

33. Social Security Data. Survey a sample of fellow students, asking each to indicate the last digit of her or his Social Security number. Also ask each participant to indicate the fifth digit. Draw one graph showing the distribution of the last digits and another graph showing the distribution of the fifth digits. Compare the two graphs. What notable difference becomes apparent?

IN THE NEWS

34. Distributions in the News. Find three recent examples in the news of distributions shown as histograms or line charts. Over each distribution, draw a smooth curve that captures its general features. Then classify the distribution according to its number of modes, symmetry or skewness, and variation.

35. Trimodal Distribution. Give an example of a real distribution that you expect to have *three* modes. Make a rough sketch of the distribution; be sure to label the axes on your sketch.

36. Skewed Distribution. Give an example of a real distribution that you would expect to be either right- or left-skewed. Make a rough sketch of the distribution; be sure to label the axes on your sketch.

4.3 Measures of Variation

In Section 4.2, we saw how to describe variation qualitatively. Here, we turn to quantitative measures of variation.

Why Variation Matters

We mortals cross the ocean
of this world,
Each in his average cabin
of a life.

—Robert Browning

Imagine that you observe customers waiting in line for tellers at two different banks. Customers at Big Bank can enter any one of three different lines leading to three different tellers. Best Bank also has three tellers, but all customers wait in a single line and are called to the next available teller. The following values are waiting times, in minutes, for 11 customers at each bank. The times are arranged in ascending order.

Big Bank (three lines):	4.1	5.2	5.6	6.2	6.7	7.2	7.7	7.7	8.5	9.3	11.0
Best Bank (one line):	6.6	6.7	6.7	6.9	7.1	7.2	7.3	7.4	7.7	7.8	7.8

You'll probably find more unhappy customers at Big Bank than at Best Bank, but this is *not* because the average wait is any longer. In fact, you should verify for yourself that the mean and median waiting times are 7.2 minutes at both banks. The difference in customer satisfaction comes from the *variation* at the two banks. The waiting times at Big Bank vary over a fairly wide range, so a few customers have long waits and are likely to become annoyed. In contrast, the variation of the waiting times at Best Bank is small, so all customers feel they are being treated roughly equally. Figure 4.13 shows the difference in the two variations with histograms in which the data values are binned to the nearest minute.

By the Way ...

The idea of waiting in line (or *queuing*) is important not only for people but also for data, particularly for data streaming through the Internet. Major corporations often employ statisticians to help them make sure that data move smoothly and without bottlenecks through their servers and Web pages.

TIME OUT TO THINK

Explain *why* Big Bank, with three separate lines, should have a greater variation in waiting times than Best Bank. Then consider several places where you commonly wait in lines, such as a grocery store, a bank, a concert ticket outlet, or a fast food restaurant. Do these places use a single customer line that feeds multiple clerks or multiple lines? If a place uses multiple lines, do you think a single line would be better? Explain.

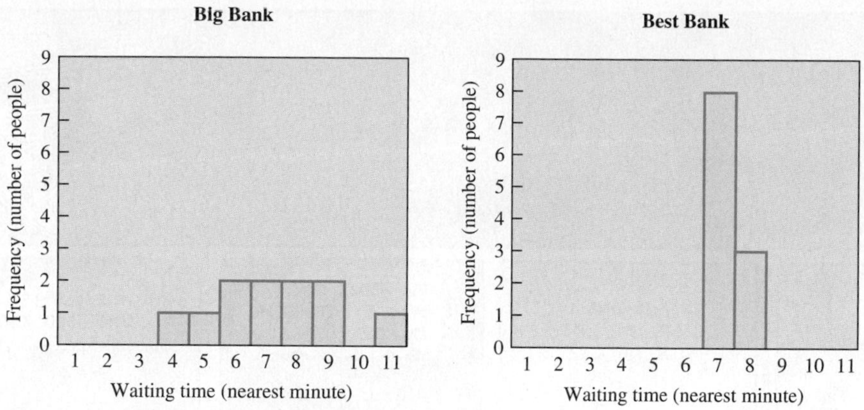

Figure 4.13 Histograms for the waiting times at Big Bank and Best Bank, shown with data binned to the nearest minute.

Range

The simplest (but not necessarily the best) way to describe the variation of a data set is to compute its **range**, defined as the difference between the lowest (minimum) and highest (maximum) values. For the example of the two banks, the waiting times for Big Bank vary from 4.1 to 11.0 minutes, so the range is $11.0 - 4.1 = 6.9$ minutes. The waiting times for Best Bank vary from 6.6 to 7.8 minutes, so the range is $7.8 - 6.6 = 1.2$ minutes. The range for Big Bank is much larger, reflecting its greater variation.

Definition

The **range** of a set of data values is the difference between its highest and lowest data values:

$$\text{range} = \text{highest value (max)} - \text{lowest value (min)}$$

Although the range is easy to compute and can be useful, it occasionally can be misleading, as the next example shows.

EXAMPLE 1 Misleading Range

Consider the following two sets of quiz scores for nine students. Which set has the greater range? Would you also say that this set has the greater variation?

Quiz 1:	1	10	10	10	10	10	10	10	10
Quiz 2:	2	3	4	5	6	7	8	9	10

Solution The range for Quiz 1 is $10 - 1 = 9$ points and the range for Quiz 2 is $10 - 2 = 8$ points. Thus, the range is greater for Quiz 1. However, aside from a single low score (an outlier), Quiz 1 has no variation at all because every other student got a 10. In contrast, no two students got the same score on Quiz 2, and the scores are spread throughout the list of possible scores. Quiz 2 therefore has greater variation even though Quiz 1 has greater range.

Quartiles and the Five-Number Summary

A better way to describe variation is to consider a few intermediate data values in addition to the high and low values. A common way involves looking at the **quartiles**, or values that divide the data distribution into quarters. The following list repeats the waiting times at the two banks, with the quartiles shown in bold. Note that the middle quartile, which divides the data set in half, is simply the median.

			Lower quartile (Q_1) ↓			Median (Q_2) ↓			Upper quartile (Q_3) ↓		
Big Bank:	4.1	5.2	**5.6**	6.2	6.7	**7.2**	7.7	7.7	**8.5**	9.3	11.0
Best Bank:	6.6	6.7	**6.7**	6.9	7.1	**7.2**	7.3	7.4	**7.7**	7.8	7.8

TECHNICAL NOTE

Statisticians do not universally agree on the procedure for calculating quartiles, and different procedures can result in slightly different values.

Definitions

The **lower quartile** (or **first quartile** or $\mathbf{Q_1}$) divides the lowest fourth of a data set from the upper three-fourths. It is the median of the data values in the *lower half* of a data set. (Exclude the middle value in the data set if the number of data points is odd.)

The **middle quartile** (or **second quartile** or $\mathbf{Q_2}$) is the overall median.

The **upper quartile** (or **third quartile** or $\mathbf{Q_3}$) divides the lowest three-fourths of a data set from the upper fourth. It is the median of the data values in the *upper half* of a data set. (Exclude the middle value in the data set if the number of data points is odd.)

Once we know the quartiles, we can describe a distribution with a **five-number summary**, consisting of the low value, the lower quartile, the median, the upper quartile, and the high value. For the waiting times at the two banks, the five-number summaries are as follows:

Big Bank:		*Best Bank:*	
low	= 4.1	low	= 6.6
lower quartile	= 5.6	lower quartile	= 6.7
median	= 7.2	median	= 7.2
upper quartile	= 8.5	upper quartile	= 7.7
high	= 11.0	high	= 7.8

The Five-Number Summary

The **five-number summary** for a data distribution consists of the following five numbers:

low value lower quartile median upper quartile high value

We can display the five-number summary with a graph called a **boxplot** (sometimes called a *box-and-whisker plot*). Using a number line for reference, we enclose the values from the lower to the upper quartiles in a box. We then draw a line through the box at the median and add two "whiskers," extending from the box to the low and high values. Figure 4.14 shows boxplots for the bank waiting times. Both the box and the whiskers for Big Bank are broader than those for Best Bank, indicating that the waiting times have greater variation at Big Bank.

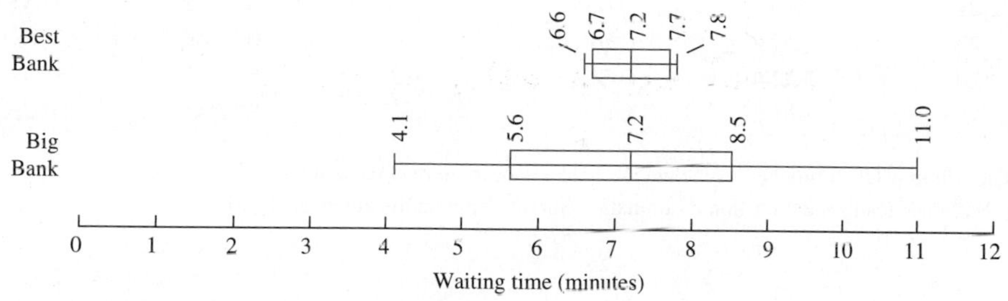

Figure 4.14 Boxplots show that the variation of the waiting times is greater at Big Bank than at Best Bank.

Drawing a Boxplot

Step 1. Draw a number line that spans all the values in the data set.

Step 2. Enclose the values from the lower to the upper quartile in a box. (The thickness of the box has no meaning.)

Step 3. Draw a line through the box at the median.

Step 4. Add "whiskers" extending to the low and high values.

TECHNICAL NOTE

The boxplots shown in this book are called *skeletal boxplots.* Some boxplots are drawn with outliers marked by an asterisk (*) and the whiskers extending only to the smallest and largest *non*outliers; these types of boxplots are called *modified boxplots.*

EXAMPLE 2 Passive and Active Smoke

One way to study exposure to cigarette smoke is by measuring blood levels of *serum cotinine*, a metabolic product of nicotine that the body absorbs from cigarette smoke. Table 4.4 lists serum cotinine levels from samples of 50 smokers ("active smoke") and 50 nonsmokers

Table 4.4 Serum Cotinine Levels (nanograms per milliliter of blood) in Samples of 50 Smokers and 50 Nonsmokers Exposed to Passive Smoke, with Data Values Listed in Ascending Order

Order number	Smokers	Nonsmokers	Order number	Smokers	Nonsmokers
1	0.08	0.03	26	34.21	0.82
2	0.14	0.07	27	36.73	0.97
3	0.27	0.08	28	37.73	1.12
4	0.44	0.08	29	39.48	1.23
5	0.51	0.09	30	48.58	1.37
6	1.78	0.09	31	51.21	1.40
7	2.55	0.10	32	56.74	1.67
8	3.03	0.11	33	58.69	1.98
9	3.44	0.12	34	72.37	2.33
10	4.98	0.12	35	104.54	2.42
11	6.87	0.14	36	114.49	2.66
12	11.12	0.17	37	145.43	2.87
13	12.58	0.20	38	187.34	3.13
14	13.73	0.23	39	226.82	3.54
15	14.42	0.27	40	267.83	3.76
16	18.22	0.28	41	328.46	4.58
17	19.28	0.30	42	388.74	5.31
18	20.16	0.33	43	405.28	6.20
19	23.67	0.37	44	415.38	7.14
20	25.00	0.38	45	417.82	7.25
21	25.39	0.44	46	539.62	10.23
22	29.41	0.49	47	592.79	10.83
23	30.71	0.51	48	688.36	17.11
24	32.54	0.51	49	692.51	37.44
25	32.56	0.68	50	983.41	61.33

Note: The column "Order number" is included to make it easier to read the table.
Source: National Health and Nutrition Examination Survey, National Institutes of Health.

By the Way ...

Passive smoke appears to be particularly harmful to young children. Apparently, the toxins in cigarette smoke have a greater effect on developing bodies than on full-grown adults. A similar effect is found for most other toxins, which is why it is especially important to limit children's exposure to toxic chemicals.

who are exposed to cigarette smoke at home or at work ("passive smoke"). Compare the two data sets (smokers and nonsmokers) with five-number summaries and boxplots, and discuss your results.

Solution The two data sets are already in ascending order, making it easy to construct the five-number summary. Each has 50 data points, so the median lies halfway between the 25th and 26th values. For the smokers, the 25th and 26th values are 32.56 and 34.21, respectively, so the median is

$$\frac{32.56 + 34.21}{2} = 33.385$$

For the nonsmokers, the 25th and 26th values are 0.68 and 0.82, respectively, so the median is

$$\frac{0.68 + 0.82}{2} = 0.75$$

The lower quartile is the median of the *lower half* of the values, which is the 13th value in each set. The upper quartile is the median of the *upper half* of the values, which is the 38th value in each set. The five-number summaries for the two data sets are as follows:

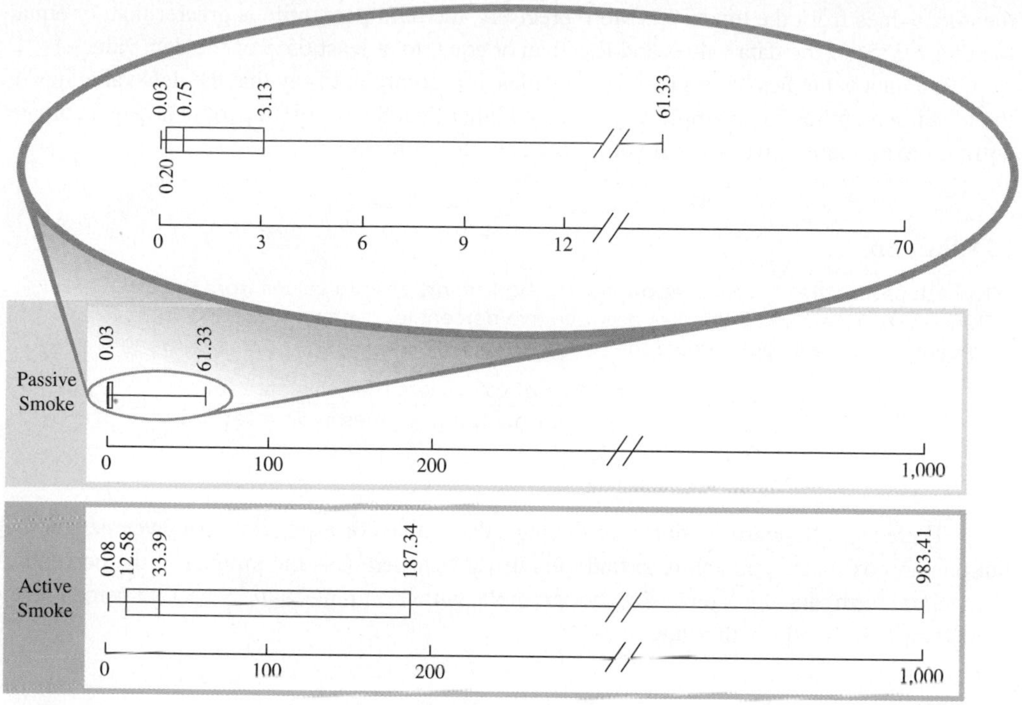

Figure 4.15 Boxplots for the data in Table 4.4.

Active smoke:		*Passive smoke:*	
low	= 0.08 ng/ml	low	= 0.03 ng/ml
lower quartile	= 12.58 ng/ml	lower quartile	= 0.20 ng/ml
median	= 33.385 ng/ml	median	= 0.75 ng/ml
upper quartile	= 187.34 ng/ml	upper quartile	= 3.13 ng/ml
high	= 983.41 ng/ml	high	= 61.33 ng/ml

Figure 4.15 shows boxplots for the two data sets. The boxplots make it easy to see some key features of the data sets. For example, it is immediately clear that the active smokers have a higher median level of serum cotinine, as well as a greater variation in levels. We conclude that smokers absorb considerably more nicotine than do nonsmokers exposed to passive smoke. Nevertheless, the levels in the passive smokers are much higher than those found in people who had no exposure to cigarette smoke (as demonstrated by other data, not shown here). Indeed, the nonsmoker with the high value for passive smoke has absorbed more nicotine than the median smoker. We conclude that passive smoke can expose nonsmokers to significant amounts of nicotine. Given the known dangers of cigarette smoke, these results give us reason to be concerned about possible health effects from passive smoke.

Percentiles

Quartiles divide a data set into 4 segments. It is possible to divide a data set even more. For example, *quintiles* divide a data set into 5 segments, and *deciles* divide a data set into 10 segments. It is particularly common to divide data sets into 100 segments using **percentiles**. Roughly speaking, the 35th percentile, for example, is a value that separates the bottom 35% of

TECHNICAL NOTE

As with quartiles, statisticians and various statistics software packages may use slightly different procedures to calculate percentiles, resulting in slightly different values.

the data values from the top 65%. (More precisely, the 35th percentile is greater than or equal to at least 35% of the data values and less than or equal to at least 65% of the data values.)

If a data value lies between two percentiles, it is common to say that the data value lies *in* the lower percentile. For example, if you score higher than 84.7% of all people taking a college entrance examination, we say that your score is in the 84th percentile.

Definition

The ***n*th percentile** of a data set divides the bottom *n*% of data values from the top (100 − *n*)%. A data value that lies between two percentiles is often said to lie *in* the lower percentile. You can approximate the percentile of any data value with the following formula:

$$\text{percentile of data value} = \frac{\text{number of values less than this data value}}{\text{total number of values in data set}} \times 100$$

There are different procedures for finding a data value corresponding to a given percentile, but one approximate approach is to find the *L*th value, where *L* is the product of the percentile (in decimal form) and the sample size. For example, with 50 sample values, the 12th percentile is around the $0.12 \times 50 = 6$th value.

EXAMPLE 3 Smoke Exposure Percentiles

Answer the following questions concerning the data in Table 4.4.

a. What is the percentile for the data value of 104.54 ng/ml for smokers?

b. What is the percentile for the data value of 61.33 ng/ml for nonsmokers?

c. What data value marks the 36th percentile for the smokers? For the nonsmokers?

Solution The following results are approximate.

a. The data value of 104.54 ng/ml for smokers is the 35th data value in the set, which means that 34 data values lie below it. Thus, its percentile is

$$\frac{\text{number of values less than 104.54 ng/ml}}{\text{total number of values in data set}} \times 100 = \frac{34}{50} \times 100 = 68$$

In other words, the 35th data value marks the 68th percentile.

b. The data value of 61.33 ng/ml for nonsmokers is the 50th and highest data value in the set, which means that 49 data values lie below it. Thus, its percentile is

$$\frac{\text{number of values less than 61.33 ng/ml}}{\text{total number of values in data set}} \times 100 = \frac{49}{50} \times 100 = 98$$

In other words, the highest data value in this set lies in the 98th percentile.

c. Because there are 50 data values in the set, the 36th percentile is around the $0.36 \times 50 = 18$th value. For smokers this value is 20.16 ng/ml, and for nonsmokers it is 0.33 ng/ml.

Standard Deviation

The five-number summary characterizes variation well, but statisticians often prefer to describe variation with a single number. The single number most commonly used to describe variation is called the **standard deviation**.

The standard deviation is a measure of how widely data values are spread around the mean of a data set. To calculate a standard deviation, we first find the mean and then find how much each data value "deviates" from the mean. Consider our bank data sets, in which the mean waiting time was 7.2 minutes for both Big Bank and Best Bank. A waiting time of 8.2 minutes has a **deviation from the mean** (or just **deviation**) of 8.2 minutes − 7.2 minutes = 1.0 minute, meaning that it is 1.0 minute greater than the mean. A waiting time of 5.2 minutes has a deviation from the mean of 5.2 minutes − 7.2 minutes = −2 minutes (*negative* 2 minutes), because it is 2.0 minutes *less* than the mean.

The standard deviation is a measure of the average of all the deviations from the mean. However, the actual mean of the deviations is always zero, because the positive deviations exactly balance the negative deviations. To avoid always coming up with a value of zero, we calculate the standard deviation by first finding a mean of the *squares* of the deviations (because squares are always positive) and taking a square root in the end. (For technical reasons, we divide the sum of the squares by the total number of data values *minus* 1.)

TECHNICAL NOTE

In finding the standard deviation when dealing with data from a *sample*, one part of the calculation involves dividing the sum of the squared deviations by the total number of data values *minus* 1. When dealing with an *entire population*, we do not subtract the 1. In this book, we will use only the formula for a sample.

Calculating the Standard Deviation

To calculate the standard deviation for any data set:

Step 1. Compute the mean of the data set. Then find the deviation from the mean for every data value by subtracting the mean from the data value. That is, for every data value,

$$\text{deviation from mean} = \text{data value} - \text{mean}$$

Step 2. Find the squares (second power) of all the deviations from the mean.

Step 3. Add all the squares of the deviations from the mean.

Step 4. Divide this sum by the total number of data values *minus* 1.

Step 5. The standard deviation is the square root of this quotient.

Overall, these steps produce the standard deviation formula:

$$\text{standard deviation} = \sqrt{\frac{\text{sum of (deviations from the mean)}^2}{\text{total number of data values} - 1}}$$

(This formula is shown in summation notation on page 174.)

TECHNICAL NOTE

The result of Step 4 is called the *variance* of the distribution. In other words, the standard deviation is the square root of the variance. Although the variance is used in many advanced statistical computations, we will not use it in this book.

Note that, because we square the deviations in Step 3 and then take the square root in Step 5, the units of the standard deviation are the same as the units of the data values. For example, if the data values have units of minutes, the standard deviation also has units of minutes.

The standard deviation formula is relatively easy to use in principle, but the actual calculations become quite tedious for all but the smallest data sets. As a result, it is usually calculated with the aid of a calculator or computer (see the Using Technology section at the end of this chapter). Nevertheless, you'll find the standard deviation formula easier to understand if you try a few examples in which you work through the calculations in detail.

EXAMPLE 4 Calculating Standard Deviation

Calculate the standard deviations for the waiting times at Big Bank and Best Bank.

Solution We follow the five steps to calculate the standard deviations. Table 4.5 shows how to organize the work in the first three steps. The first column for each bank lists the waiting times (in minutes), the second column lists the deviations from the mean (Step 1), and the third

Table 4.5 Calculating Standard Deviation					
Big Bank			**Best Bank**		
Time	**Deviation (Time − Mean)**	**(Deviation)^2**	**Time**	**Deviation (Time − Mean)**	**(Deviation)^2**
4.1	4.1 − 7.2 = −3.1	$(-3.1)^2 = 9.61$	6.6	6.6 − 7.2 = −0.6	$(-0.6)^2 = 0.36$
5.2	5.2 − 7.2 = −2.0	$(-2.0)^2 = 4.00$	6.7	6.7 − 7.2 = −0.5	$(-0.5)^2 = 0.25$
5.6	5.6 − 7.2 = −1.6	$(-1.6)^2 = 2.56$	6.7	6.7 − 7.2 = −0.5	$(-0.5)^2 = 0.25$
6.2	6.2 − 7.2 = −1.0	$(-1.0)^2 = 1.00$	6.9	6.9 − 7.2 = −0.3	$(-0.3)^2 = 0.09$
6.7	6.7 − 7.2 = −0.5	$(-0.5)^2 = 0.25$	7.1	7.1 − 7.2 = −0.1	$(-0.1)^2 = 0.01$
7.2	7.2 − 7.2 = 0.0	$(0.0)^2 = 0.0$	7.2	7.2 − 7.2 = 0.0	$(0.0)^2 = 0.0$
7.7	7.7 − 7.2 = 0.5	$(0.5)^2 = 0.25$	7.3	7.3 − 7.2 = 0.1	$(0.1)^2 = 0.01$
7.7	7.7 − 7.2 = 0.5	$(0.5)^2 = 0.25$	7.4	7.4 − 7.2 = 0.2	$(0.2)^2 = 0.04$
8.5	8.5 − 7.2 = 1.3	$(1.3)^2 = 1.69$	7.7	7.7 − 7.2 = 0.5	$(0.5)^2 = 0.25$
9.3	9.3 − 7.2 = 2.1	$(2.1)^2 = 4.41$	7.8	7.8 − 7.2 = 0.6	$(0.6)^2 = 0.36$
11.0	11.0 − 7.2 = 3.8	$(3.8)^2 = 14.44$	7.8	7.8 − 7.2 = 0.6	$(0.6)^2 = 0.36$
		Sum = 38.46			**Sum = 1.98**

column lists the squares of the deviations (Step 2). We add all the squared deviations to find the sum at the bottom of the third column (Step 3). Note that we can calculate the deviations because we already know that the mean waiting time is 7.2 minutes for both banks. For Step 4, we divide the sums from Step 3 by the total number of data values *minus* 1. Because there are 11 data values, we divide by 10:

$$\textit{Big Bank:} \quad \frac{38.46}{10} = 3.846$$

$$\textit{Best Bank:} \quad \frac{1.98}{10} = 0.198$$

Finally, Step 5 tells us that the standard deviations are the square roots of the numbers from Step 4:

$$\textit{Big Bank:} \quad \text{standard deviation} = \sqrt{3.846} = 1.96 \text{ minutes}$$

$$\textit{Best Bank:} \quad \text{standard deviation} = \sqrt{0.198} = 0.44 \text{ minutes}$$

In other words, while both banks had the same mean waiting time of 7.2 minutes, typical waiting times tended to be about 1.96 minutes away from this mean at Big Bank but only 0.44 minute away from this mean at Best Bank. Again, we see that the waiting times showed greater variation at Big Bank, explaining why the lines at Big Bank annoyed more customers than did those at Best Bank.

TIME OUT TO THINK

Look closely at the individual deviations in Table 4.5 in Example 4. Do the standard deviations for the two data sets seem like reasonable "averages" for the deviations? Explain.

Interpreting the Standard Deviation

A good way to develop a deeper understanding of the standard deviation is to consider an approximation called the **range rule of thumb**, summarized in the following box.

The Range Rule of Thumb

The standard deviation is *approximately* related to the range of a distribution by the **range rule of thumb**:

$$\text{standard deviation} \approx \frac{\text{range}}{4}$$

If we know the range of a distribution (range = high − low), we can use this rule to estimate the standard deviation. Alternatively, if we know the standard deviation, we can use this rule to estimate the low and high values as follows:

$$\text{low value} \approx \text{mean} - (2 \times \text{standard deviation})$$
$$\text{high value} \approx \text{mean} + (2 \times \text{standard deviation})$$

The range rule of thumb does not work well when the high or low values are outliers.

TECHNICAL NOTE

Another way of interpreting the standard deviation uses a mathematical rule called *Chebyshev's Theorem*. It states that, for any data distribution, at least 75% of all data values lie within two standard deviations of the mean, and at least 89% of all data values lie within three deviations of the mean. Although we will not use this theorem in this book, you may encounter it if you take another statistics course.

The range rule of thumb works reasonably well for data sets in which values are distributed fairly evenly. It does not work well when the high or low values are extreme outliers. You must therefore use judgment in deciding whether the range rule of thumb is applicable in a particular case, and in all cases remember that the range rule of thumb yields rough approximations, not exact results.

EXAMPLE 5 Using the Range Rule of Thumb

Use the range rule of thumb to estimate the standard deviations for the waiting times at Big Bank and Best Bank. Compare the estimates to the actual values found in Example 4.

Solution The waiting times for Big Bank vary from 4.1 to 11.0 minutes, which means a range of $11.0 - 4.1 = 6.9$ minutes. The waiting times for Best Bank vary from 6.6 to 7.8 minutes, for a range of $7.8 - 6.6 = 1.2$ minutes. Thus, the range rule of thumb gives the following estimates for the standard deviations:

Big Bank: $$\text{standard deviation} \approx \frac{6.9}{4} = 1.7$$

Best Bank: $$\text{standard deviation} \approx \frac{1.2}{4} = 0.3$$

The actual standard deviations calculated in Example 4 are 1.96 and 0.44, respectively. For these two cases, the estimates from the range rule of thumb slightly underestimate the actual standard deviations. Nevertheless, the estimates put us in the right ballpark, showing that the rule is useful.

EXAMPLE 6 Estimating a Range

Studies of the gas mileage of a BMW under varying driving conditions show that it gets a mean of 22 miles per gallon with a standard deviation of 3 miles per gallon. Estimate the minimum and maximum typical gas mileage amounts that you can expect under ordinary driving conditions.

Solution From the range rule of thumb, the low and high values for gas mileage are approximately

$$\text{low value} \approx \text{mean} - (2 \times \text{standard deviation}) = 22 - (2 \times 3) = 16$$
$$\text{high value} \approx \text{mean} + (2 \times \text{standard deviation}) = 22 + (2 \times 3) = 28$$

The range of gas mileage for the car is roughly from a minimum of 16 miles per gallon to a maximum of 28 miles per gallon.

By the Way ...

Technologies such as catalytic converters have helped reduce the amounts of many pollutants emitted by cars (per mile driven), but burning less gasoline is the only way to reduce carbon dioxide emissions that cause global warming. This is a major reason why auto manufacturers are developing high-mileage hybrid vehicles and zero-emission vehicles that run on electricity or fuel cells.

Standard Deviation with Summation Notation (Optional Section)

The summation notation introduced earlier makes it easy to write the standard deviation formula in a compact form. Recall that x represents the individual values in a data set and $\bar{x}$ represents the mean of the data set. We can therefore write the deviation from the mean for any data value as

$$\text{deviation} = \text{data value} - \text{mean} = x - \bar{x}$$

We can now write the sum of all squared deviations as

$$\text{sum of all squared deviations} = \Sigma(x - \bar{x})^2$$

The remaining steps in the calculation of the standard deviation are to divide this sum by $n - 1$ and then take the square root. You should confirm that the following formula summarizes the five steps in the earlier box:

$$s = \text{standard deviation} = \sqrt{\frac{\Sigma(x - \bar{x})^2}{n - 1}}$$

The symbol s is the conventional symbol for the standard deviation of a sample. For the standard deviation of a population, statisticians use the Greek letter σ (*sigma*), and the term $n - 1$ in the formula is replaced by n. Consequently, you will get slightly different results for the standard deviation depending on whether you assume the data represent a sample or a population.

TECHNICAL NOTE

The formula for the *variance* is

$$s^2 = \frac{\Sigma(x - \bar{x})^2}{n - 1}$$

The standard symbol for the variance, s^2, reflects the fact that it is the square of the standard deviation.

Section 4.3 Exercises

Statistical Literacy and Critical Thinking

1. **Variation.** Why is the standard deviation considered a measure of variation? In your own words, describe the characteristic of a data set that is measured by the standard deviation.
2. **Comparing Variation.** Which do you think has more variation: the IQ scores of 30 students in a physics class or the IQ scores of 30 patrons in a movie theater? Why?
3. **Correct Statement?** In the book *How to Lie with Charts,* the author writes that "the standard deviation is usually shown as plus or minus the difference between the high and the mean, and the low and the mean. For example, if the mean is 1, the high is 3, and the low is −1, the standard deviation is ±2." Is that statement correct? Why or why not?
4. **Quartiles.** For a large data set, the lower quartile is found to be 93.2. What do we mean when we say that 93.2 is the lower quartile?

Does It Make Sense? For Exercises 5–8, decide whether the statement makes sense (or is clearly true) or does not make sense (or is clearly false). Explain clearly; not all of these statements have definitive answers, so your explanation is more important than your chosen answer.

5. **SAT Score.** An SAT score was at the median value and was in the 60th percentile.
6. **Family Income.** A report in *USA Today* stated that for a recent year the median annual income of a family is \$43,200 and the mean is \$70,700.
7. **Incomes.** The standard deviation of the annual incomes of 50 full-time statistics instructors is less than the standard deviation of the annual incomes of 50 physicians.
8. **Sample Sizes.** The standard deviation of the annual incomes of 50 randomly selected full-time statistics instructors is much less than the standard deviation of the annual incomes of 200 other randomly selected full-time statistics instructors.

Concepts and Applications

Range and Standard Deviation. Exercises 9–16 each list a set of numbers. In each case, find the range and standard deviation. (The same sets of numbers were used in Exercises 13–20 in Section 4.1.)

9. **Perception of Time.** Actual times (in seconds) recorded when statistics students participated in an experiment to test

their ability to determine when one minute (60 seconds) had passed:

53 52 75 62 68 58 49 49

10. Body Temperatures. Body temperatures (in degrees Fahrenheit) of randomly selected normal and healthy adults:

98.6	98.6	98.0	98.0	99.0
98.4	98.4	98.4	98.4	98.6

11. Blood Alcohol. Blood alcohol concentrations of drivers involved in fatal crashes and then given jail sentences (based on data from the U.S. Department of Justice):

0.27	0.17	0.17	0.16	0.13	0.24
0.29	0.24	0.14	0.16	0.12	0.16

12. Old Faithful Geyser. Time intervals (in minutes) between eruptions of Old Faithful geyser in Yellowstone National Park:

98	92	95	87	96	90
65	92	95	93	98	94

13. Fruit Flies. Thorax lengths (in millimeters) of a sample of male fruit flies:

0.72	0.90	0.84	0.68	0.84	0.90
0.92	0.84	0.64	0.84	0.76	

14. Ages of Presidents. Ages of selected U.S. Presidents at the time of inauguration:

57	61	57	57	58	57	61	54
68	51	49	64	50	48	65	

15. Weights of M&Ms. Weights (in grams) of randomly selected M&M plain candies:

0.957	0.912	0.842	0.925	0.939	0.886
0.914	0.913	0.958	0.947	0.920	

16. Quarters. Weights (in grams) of quarters in circulation:

5.60	5.63	5.58	5.56	5.66	5.58	5.57	5.59
5.67	5.61	5.84	5.73	5.53	5.58	5.52	5.65
5.57	5.71	5.59	5.53	5.63	5.68		

Comparing Variation. In Exercises 17–20, find the range and standard deviation for each of the two samples and then compare the two sets of results.

17. It's Raining Cats. Statistics are sometimes used to compare or identify authors of different works. The lengths of the first 20 words in the foreword by Tennessee Williams in *Cat on a Hot Tin Roof* are listed along with the lengths of the first 20 words in *The Cat in the Hat* by Dr. Seuss. Does there appear to be a difference in variation?

Cat on a Hot Tin Roof:

2	6	2	2	1	4	4	2	4	2
3	8	4	2	2	7	7	2	3	11

The Cat in the Hat:

3	3	3	3	5	2	3	3	3	2
4	2	2	3	2	3	5	3	4	4

18. Ages of Stowaways. The *Queen Mary* sailed between England and the United States, and stowaways were sometimes found on board. The ages (in years) of stowaways from eastbound crossings and westbound crossings are given below (data from the Cunard Steamship Co., Ltd.). Compare the variation in the two data sets.

Eastbound:

24 24 34 15 19 22 18 20 20 17

Westbound:

41 24 32 26 39 45 24 21 22 21

19. Weather Forecast Accuracy. In an analysis of the accuracy of weather forecasts, the actual high temperatures are compared to the high temperatures predicted one day earlier and the high temperatures predicted five days earlier. Listed below are the errors between the predicted temperatures and the actual high temperatures for consecutive days in Dutchess County, New York. Do the standard deviations suggest that the temperatures predicted one day in advance are more accurate than those predicted five days in advance, as we might expect?

(*actual high*) – (*high predicted one day earlier*):

2	2	0	0	−3	−2	1
−2	8	1	0	−1	0	1

(*actual high*) – (*high predicted five days earlier*):

0	−3	2	5	−6	−9	4
−1	6	−2	−2	−1	6	−4

20. Treatment Effect. Researchers at Pennsylvania State University conducted experiments with poplar trees. Listed below are weights (in kilograms) of poplar trees given no treatment and poplar trees treated with fertilizer and irrigation. Does there appear to be a difference between the two standard deviations?

No treatment:

0.15 0.02 0.16 0.37 0.22

Fertilizer and irrigation:

2.03 0.27 0.92 1.07 2.38

21. Calculating Percentiles. One of the authors with too much time on his hands weighed each M&M candy in a bag of 465 plain M&M candies.

a. One of the M&Ms weighed 0.776 gram and it was heavier than 25 of the other M&Ms. What is the percentile of this particular value?

b. One of the M&Ms weighed 0.876 gram and it was heavier than 322 of the other M&Ms. What is the percentile of this particular value?

c. One of the M&Ms weighed 0.856 gram and it was heavier than 224 of the other M&Ms. What is the percentile of this particular value?

22. Calculating Percentiles. A data set consists of the 76 ages of women at the time that they won an Oscar in the category of best actress.

a. One of the actresses was 34 years of age, and she was older than 38 of the other actresses at the time that they won Oscars. What is the percentile of the age of 34?

b. One of the actresses was 29 years of age, and she was older than 20 of the other actresses at the time that they won Oscars. What is the percentile of the age of 29?

c. One of the actresses was 60 years of age, and she was older than 71 of the other actresses at the time that they won Oscars. What is the percentile of the age of 60?

23. Understanding Standard Deviation. The following four sets of 7 numbers all have a mean of 9.

{9, 9, 9, 9, 9, 9, 9,} {8, 8, 9, 9, 9, 10, 10}
{8, 8, 8, 9, 10, 10, 10} {6, 6, 6, 9, 12, 12, 12}

a. Make a histogram for each set.

b. Give the five-number summary and draw a boxplot for each set.

c. Compute the standard deviation for each set.

d. Based on your results, briefly explain how the standard deviation provides a useful single-number summary of the variation in these data sets.

24. Understanding Standard Deviation. The following four sets of 7 numbers all have a mean of 6.

{6, 6, 6, 6, 6, 6, 6}, {5, 5, 6, 6, 6, 7, 7}
{5, 5, 5, 6, 7, 7, 7}, {3, 3, 3, 6, 9, 9, 9}

a. Make a histogram for each set.

b. Give the five-number summary and draw a boxplot for each set.

c. Compute the standard deviation for each set.

d. Based on your results, briefly explain how the standard deviation provides a useful single-number summary of the variation in these data sets.

Comparing Variations. For each of Exercises 25–28, do the following:

a. Find the mean, median, and range for each of the two data sets.

b. Give the five-number summary and draw a boxplot for each of the two data sets.

c. Find the standard deviation for each of the two data sets.

d. Apply the range rule of thumb to estimate the standard deviation of each of the two data sets. How well does the rule work in each case? Briefly discuss why it does or does not work well.

e. Based on all your results, compare and discuss the two data sets in terms of their center and variation.

25. The following data sets give the ages in years of a sample of cars in a faculty parking lot and a student parking lot at the College of Portland.

Faculty:
2 3 1 0 1 2 4 3 3 2 1

Student:
5 6 8 2 7 10 1 4 6 10 9

26. The following data sets give the driving speeds in miles per hour of the first nine cars to pass through a school zone and the first nine cars to pass through a downtown intersection.

School:
20 18 23 21 19 18 17 24 25

Downtown:
29 31 35 24 31 26 36 31 28

27. The following data sets show the ages of the first seven U.S. Presidents (Washington through Jackson) and seven recent U.S. Presidents (Nixon through G. W. Bush) at the time of inauguration.

First 7:
57 61 57 57 58 57 61

Last 7:
56 61 52 69 64 46 54

28. The following data sets give the approximate lengths of Beethoven's nine symphonies and Mahler's nine symphonies (in minutes).

Beethoven:
28 36 50 33 30 40 38 26 68

Mahler:
52 85 94 50 72 72 80 90 80

29. Pizza Deliveries. After recording pizza delivery times for two different pizza shops, you conclude that one pizza shop has a mean delivery time of 45 minutes with a standard deviation of 3 minutes and the other shop has a mean delivery time of 42 minutes with a standard deviation of 20 minutes. Interpret these figures. If you liked the pizzas from both shops equally well, which one would you order from? Why?

30. Managing Complaints. You manage a small ice cream shop in which your employees scoop the ice cream by hand. Each night, you total your sales and the total volume of ice cream sold. You find that on nights when an employee named Sam is working, the mean price of the ice cream sold is \$1.75 per pint with a standard deviation of \$0.05. On nights when an employee named Kevin is working, the mean price of the ice cream sold is \$1.70 per pint with a standard deviation of \$0.35. Which employee is more likely to be generating complaints of "too small" servings? Explain.

31. Portfolio Standard Deviation. The book *Investments*, by Zvi Bodie, Alex Kane, and Alan Marcus, claims that the returns for investment portfolios with a single stock have a standard deviation of 0.55, while the returns for portfolios with 32 stocks have a standard deviation of 0.325. Explain how the standard deviation measures the risk in these two types of portfolios.

32. Batting Standard Deviation. For the last 100 years, the mean batting average in the major leagues has remained fairly constant at about 0.260. However, the standard deviation of batting averages has decreased from about 0.049 in the 1870s to 0.031 in the present. What does this tell us about the batting averages of players? Based on these facts, would you expect batting averages above 0.350 to be more or less common today than in the past? Explain.

Projects for the Internet and Beyond

For useful links, select "Links for Internet Projects" for Chapter 4 at www.aw.com/bbt.

33. Secondhand Smoke. At the Web sites of the American Lung Association and the U.S. Environmental Protection Agency, find statistical data concerning the health effects of secondhand (passive) smoke. Write a short summary of your findings and your opinions about whether and how this health issue should be addressed by government.

34. Kids and the Media. A recent study by the Kaiser Family Foundation looked at the role of media (for example, television, books, computers) in the lives of children. The report, which is on the Kaiser Family Foundation Web site, gives many data distributions concerning, for example, how much time children spend daily with each medium. Study at least three of the distributions in the report that you find particularly interesting. Summarize each distribution in words, and discuss your opinions of the social consequences of the findings.

35. Measuring Variation. The range and standard deviation use different approaches to measure variation in a data set. Construct two different data sets configured so that the range of the first set is *greater than* the range of the second set (suggesting that the first set has more variation) but the standard deviation of the first set is *less than* the standard deviation of the second set (suggesting that the first set has less variation).

IN THE NEWS

36. Ranges in the News. Find two examples of data distributions in recent news reports; they may be given either as tables or as graphs. In each case, state the range of the distribution and explain its meaning in the context of the news report. Estimate the standard deviation by applying the range rule of thumb.

37. Summarizing a News Data Set. Find an example of a data distribution given in the form of a table in a recent news report. Make a five-number summary and a boxplot for the distribution.

4.4 Statistical Paradoxes

The government administers polygraph tests ("lie detectors") to new applicants for sensitive security jobs. The polygraph tests are reputed to be 90% accurate; that is, they catch 90% of the people who are lying and validate 90% of the people who are truthful. But how many people who fail a polygraph test are actually telling the truth? Most people guess that only about 10% of those who fail a test are falsely identified. In fact, the actual percentage of false accusations can be much higher. How can this be?

We'll discuss the answer soon, but the moral of this story should already be clear: Even when we describe data carefully according to the principles discussed in the first three sections of this chapter, we may still be led to very surprising conclusions. Before we get to the polygraph issue, let's start with a couple of other statistical surprises.

Better in Each Case, But Worse Overall

Suppose a pharmaceutical company creates a new treatment for acne. To decide whether the new treatment is better than an older treatment, the company gives the old treatment to 90 patients and gives the new treatment to 110 patients. Some patients had mild acne and others had severe acne. Table 4.6 summarizes the results after four weeks of treatment, broken down by which treatment was given and whether the patient's acne was mild or severe. If you study the table carefully, you will notice these key facts:

- Among patients with *mild* acne:
  10 received the old treatment and 2 were cured, for a 20% cure rate.
  90 received the new treatment and 30 were cured, for a 33% cure rate.
- Among patients with *severe* acne:
  80 received the old treatment and 40 were cured, for a 50% cure rate.
  20 received the new treatment and 12 were cured, for a 60% cure rate.

Table 4.6 Results of Acne Treatments

	Mild acne		Severe acne	
	Cured	**Not cured**	**Cured**	**Not cured**
Old treatment	2	8	40	40
New treatment	30	60	12	8

By the Way ...

The general case in which a set of data gives different results for each of several group comparisons than it does when the groups are taken together is known as *Simpson's paradox*, so named because it was described by Edward Simpson in 1951. However, the same idea was actually described around 1900 by Scottish statistician George Yule.

Notice that the new treatment had a higher cure rate *both* for patients with mild acne (33% for the new treatment vs. 20% for the old) and for patients with severe acne (60% for the new treatment vs. 50% for the old). Is it therefore fair for the company to claim that their new treatment is better than the old treatment?

At first, this might seem to make sense. But instead of looking at the data for the mild and severe acne patients separately, let's look at the *overall* results:

- A total of 90 patients received the old treatment and 42 were cured (2 out of 10 with mild acne and 40 out of 80 with severe acne), for an overall cure rate of $42/90 = 46.7\%$.
- A total of 110 patients received the new treatment and 42 were cured (30 out of 90 with mild acne and 12 out of 20 with severe acne), for an overall cure rate of $42/110 = 38.2\%$.

Overall, the *old* treatment had the higher cure rate, despite the fact that the new treatment had a higher rate for both mild and severe acne cases.

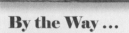

This example illustrates that it is possible for something to appear better in each of two or more group comparisons but actually be worse overall. If you look carefully, you'll see that this occurs because of the way in which the overall results are divided into unequally sized groups (in this case, mild acne patients and severe acne patients).

EXAMPLE 1 Who Played Better?

Table 4.7 gives the shooting performance of two players in each half of a basketball game. Shaq had a higher shooting percentage in both the first half (40% to 25%) and the second half (75% to 70%). Can Shaq claim that he had the better game?

Table 4.7 Basketball Shots

Player	First half			Second half		
	Baskets	Attempts	Percent	Baskets	Attempts	Percent
Shaq	4	10	40%	3	4	75%
Vince	1	4	25%	7	10	70%

Solution No, and we can see why by looking at the overall game statistics. Shaq made a total of 7 baskets (4 in the first half and 3 in the second half) on 14 shots (10 in the first half and 4 in the second half), for an overall shooting percentage of 7/14 = 50%. Vince made a total of 8 baskets on 14 shots, for an overall shooting percentage of 8/14 = 57.1%. Surprisingly, even though Shaq had a higher shooting percentage in both halves, Vince had a better overall shooting percentage for the game.

Does a Positive Mammogram Mean Cancer?

We often associate tumors with cancers, but most tumors are not cancers. Medically, any kind of abnormal swelling or tissue growth is considered a tumor. A tumor caused by cancer is said to be *malignant* (or *cancerous*); all others are said to be *benign.*

Imagine you are a doctor or nurse treating a patient who has a breast tumor. The patient will be understandably nervous, but you can give her some comfort by telling her that only about 1 in 100 breast tumors turns out to be malignant. But, just to be safe, you order a mammogram to determine whether her tumor is one of the 1% that are malignant.

Now, suppose the mammogram comes back positive, suggesting that the tumor is malignant. Mammograms are not perfect, so the positive result does not necessarily mean that your patient has breast cancer. More specifically, let's assume that the mammogram screening is 85% accurate: It will correctly identify 85% of malignant tumors as malignant and 85% of benign tumors as benign. When you tell your patient that her mammogram was positive, what should you tell her about the chance that she actually has cancer?

Because the mammogram screening is 85% accurate, most people guess that the positive result means that the patient probably has cancer. Studies have shown that most doctors also believe this to be the case, and would tell the patient to be prepared for cancer treatment. But a more careful analysis shows otherwise. In fact, the chance that the patient has cancer is still quite small—about 5%. We can see why by analyzing some numbers.

Consider a study in which mammograms are given to 10,000 women with breast tumors. Assuming that 1% of tumors are malignant, 1% $\times$10,000 = 100 of the women actually have cancer; the remaining 9,900 women have benign tumors. Table 4.8 summarizes the mammogram results. Notice the following:

- The mammogram screening correctly identifies 85% of the 100 malignant tumors as malignant. Thus, it gives positive (malignant) results for 85 of the malignant tumors; these cases

By the Way . . .

This mammogram example and the polygraph example that follows illustrate cases in which conditional probabilities (discussed in Section 6.5) lead to confusion. The proper way of handling conditional probabilities was discovered by the Reverend Thomas Bayes (1702–1761) and is often called *Bayes rule.*

Table 4.8 Summary of Results for 10,000 Mammograms (when in fact 100 tumors are malignant and 9,900 are benign)

	Tumor is malignant	Tumor is benign	Total
Positive mammogram	85 true positives	1,485 false positives	**1,570**
Negative mammogram	15 false negatives	8,415 true negatives	**8,430**
Total	**100**	**9,900**	**10,000**

By the Way ...

The accuracy of breast cancer screening is rapidly improving; newer technologies, including digital mammograms and ultrasounds, appear to achieve accuracies near 98%. The most definitive test for cancer is a biopsy, though even biopsies can miss cancers if they are not taken with sufficient care. If you have negative tests but are still concerned about an abnormality, ask for a second opinion. It may save your life.

are called **true positives**. In the other 15 malignant cases, the result is negative, even though the women actually have cancer; these cases are **false negatives**.

- The mammogram screening correctly identifies 85% of the 9,900 benign tumors as benign. Thus, it gives negative (benign) results for 85% × 9,900 = 8,415 of the benign tumors; these cases are **true negatives**. The remaining 9,900 − 8,415 = 1,485 women get positive results in which the mammogram incorrectly identifies their tumors as malignant; these cases are **false positives**.

Overall, the mammogram screening gives positive results to 85 women who actually have cancer and to 1,485 women who do *not* have cancer. The total number of positive results is 85 + 1,485 = 1,570. Because only 85 of these are true positives (the rest are false positives), the chance that a positive result really means cancer is only 85/1,570 = 0.054, or 5.4%. Therefore, when your patient's mammogram comes back positive, you should reassure her that there's still only a small chance that she has cancer.

EXAMPLE 2 False Negatives

Suppose you are a doctor seeing a patient with a breast tumor. Her mammogram comes back negative. Based on the numbers in Table 4.8, what is the chance that she has cancer?

Solution For the 10,000 cases summarized in Table 4.8, the mammograms are negative for 15 women with cancer and for 8,415 women with benign tumors. The total number of negative results is 15 + 8,415 = 8,430. Thus, the fraction of women with cancer who have false negatives is 15/8,430 = 0.0018, or slightly less than 2 in 1,000. In other words, the chance that a woman with a negative mammogram has cancer is only about 2 in 1,000.

TIME OUT TO THINK

While the chance of cancer with a negative mammogram is small, it is not zero. Therefore, it might seem like a good idea to biopsy all tumors, just to be sure. However, biopsies involve surgery, which means they can be painful and expensive, among other things. Given these facts, do you think that biopsies should be routine for all tumors? Should they be routine for cases of positive mammograms? Defend your opinion.

Polygraphs and Drug Tests

We're now ready to return to the question asked at the beginning of this section, about how a 90% accurate polygraph test can lead to a surprising number of false accusations. The explanation is very similar to that used in the case of the mammograms.

Suppose the government gives the polygraph test to 1,000 applicants for sensitive security jobs. Further suppose that 990 of these 1,000 people tell the truth on their polygraph test, while only 10 people lie. For a test that is 90% accurate, we find the following results:

- Of the 10 people who lie, the polygraph correctly identifies 90%, meaning that 9 fail the test (they are identified as liars) and 1 passes.
- Of the 990 people who tell the truth, the polygraph correctly identifies 90%, meaning that 90% × 990 = 891 truthful people pass the test and the other 10% × 990 = 99 truthful people fail the test.

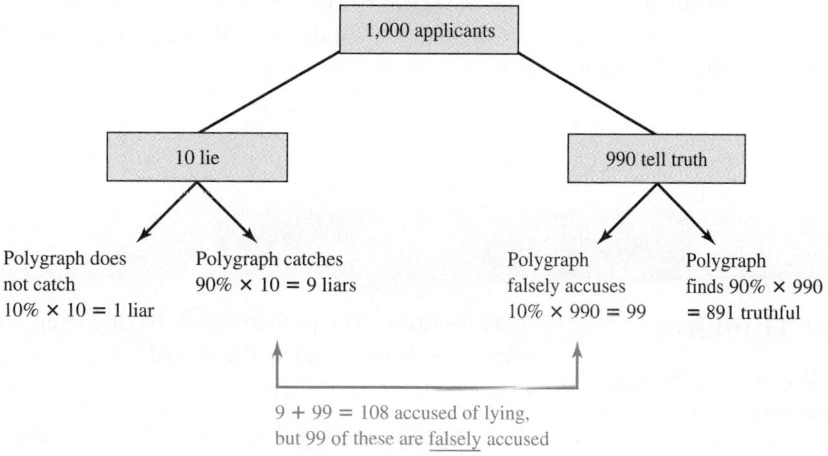

Figure 4.16 A tree diagram summarizes results of a 90% accurate polygraph test for 1,000 people, of whom only 10 are lying.

Figure 4.16 summarizes these results. The total number of people who fail the test is 9 + 99 = 108. Of these, only 9 were actually liars; the other 99 were falsely accused of lying. That is, 99 out of 108, or 99/108 = 91.7%, of the people who fail the test were actually telling the truth.

The percentage of people who are falsely accused in any real situation depends on both the accuracy of the test and the proportion of people who are lying. Nevertheless, for the numbers given here, we have an astounding result: Assuming the government rejects applicants who fail the polygraph test, then almost 92% of the rejected applicants were actually being truthful and may have been highly qualified for the jobs.

By the Way ...

A polygraph, often called a "lie detector," measures a variety of bodily functions including heart rate, skin temperature, and blood pressure. Polygraph operators look for subtle changes in these functions that typically occur when people lie. However, polygraph results have never been allowed as evidence in criminal proceedings. First of all, 90% accuracy is far too low for justice. In addition, studies show that polygraphs are easily fooled by people who train to beat them.

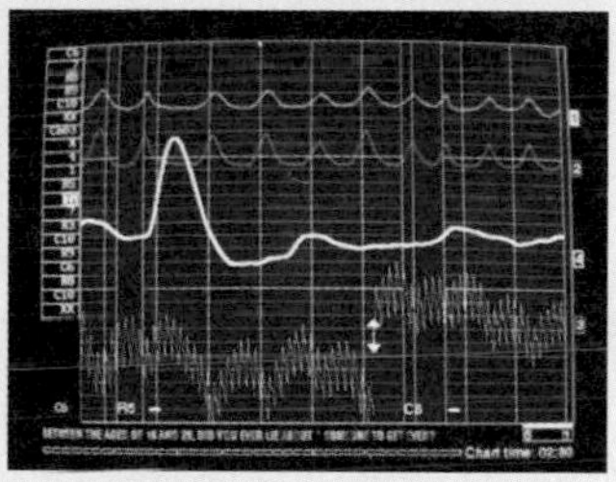

TIME OUT TO THINK

Imagine that you are falsely accused of a crime. The police suggest that, if you are truly innocent, you should agree to take a polygraph test. Would you do it? Why or why not?

EXAMPLE 3 High School Drug Testing

All athletes participating in a regional high school track and field championship must provide a urine sample for a drug test. Those who fail are eliminated from the meet and suspended from competition for the following year. Studies show that, at the laboratory selected, the drug tests are 95% accurate. Assume that 4% of the athletes actually use drugs. What fraction of the athletes who fail the test are falsely accused and therefore suspended without cause?

Solution The easiest way to answer this question is by using some sample numbers. Suppose there are 1,000 athletes in the meet. Then 4%, or 40 athletes, actually use drugs; the remaining 960 athletes do not use drugs. In that case, the 95% accurate drug test should return the following results:

- 95% of the 40 athletes who use drugs, or $0.95 \times 40 = 38$ athletes, fail the test. The other 2 athletes who use drugs pass the test.
- 95% of the 960 athletes who do not use drugs pass the test, but 5% of these 960, or $0.05 \times 960 = 48$ athletes, fail.

The total number of athletes who fail the test is $38 + 48 = 86$. But 48 of these athletes who fail the test, or $48/86 = 56\%$, are actually nonusers. Despite the 95% accuracy of the drug test, more than half of the suspended students are innocent of drug use.

Section 4.4 Exercises

Statistical Literacy and Critical Thinking

1. **False Positive and False Negative.** When someone is given a test for drug use, what is a false positive? What is a false negative?

2. **Positive Test Result.** Briefly explain why a positive result on a cancer test such as a mammogram does not necessarily mean that a patient has cancer.

3. **Test Accuracy.** How it is possible that a polygraph test can be very accurate while resulting in a large proportion of false accusations?

4. **Better in Each Half, Worse Overall.** When two football teams play, can one of the quarterbacks have a higher passing percentage in each half while having a lower passing percentage for the entire game?

Does It Make Sense? For Exercises 5–8, decide whether the statement makes sense (or is clearly true) or does not make sense (or is clearly false). Explain clearly; not all of these statements have definitive answers, so your explanation is more important than your chosen answer.

5. **Course Average.** Ann and Bret are taking the same statistics course, in which the final grade is determined by assignments and exams. Ann's average on the assignments is higher than Bret's, and Ann's average on the exams is higher than Bret's. It follows that Ann's overall average in the course is higher than Bret's.

6. **Batting Average.** Ann's batting average for the first half of the season is higher than Bret's, and Ann's batting average for the second half of the season is higher than Bret's. It follows that Ann's batting average for the entire season is higher than Bret's.

7. **Test Results.** The probability of having strep if your test is positive is the same as the probability of having a positive test if you have strep.

8. **Test Accuracy.** If a drug test is 90% accurate, 90% of those who test positive are actual drug users.

Concepts and Applications

9. **Batting Percentages.** The table below shows the batting records of two baseball players in the first half (first 81 games) and last half of a season.

	First half		
Player	**Hits**	**At-bats**	**Batting average**
Josh	50	150	.333
Jude	10	50	.200
	Second half		
Player	**Hits**	**At-bats**	**Batting average**
Josh	35	70	.500
Jude	70	150	.467

Who had the higher batting average in the first half of the season? Who had the higher batting average in the second half? Who had the higher overall batting average? Explain how these results illustrate Simpson's paradox.

10. **Passing Percentages.** The table below shows the passing records of two rival quarterbacks in the first half and second half of a football game.

	First half		
Player	**Completions**	**Attempts**	**Percent**
Allan	8	20	40%
Abner	2	6	33%
	Second half		
Player	**Completions**	**Attempts**	**Percent**
Allan	3	6	50%
Abner	12	25	48%

Who had the higher completion percentage in the first half? Who had the higher completion percentage in the second half? Who had the higher overall completion percentage? Explain how these results illustrate Simpson's paradox.

11. Test Scores. The table below shows eighth-grade mathematics test scores in Nebraska and New Jersey. The scores are separated according to the race of the student. Also shown are the state averages for all races.

	White	Nonwhite	Average for all races
Nebraska	281	250	277
New Jersey	283	252	272

Source: National Assessment of Educational Progress, from *Chance* magazine.

a. Which state had the higher scores in both racial categories? Which state had the higher overall average across both racial categories?

b. Explain how a state could score lower in both categories and still have a higher overall average.

c. Now consider the table below, which gives the percentages of whites and nonwhites in each state. Use these percentages to verify that the overall average test score in Nebraska is 277, as claimed in the first table.

	White	Nonwhite
Nebraska	87%	13%
New Jersey	66%	34%

d. Use the racial percentages to verify that the overall average test score in New Jersey is 272, as claimed in the first table.

e. Explain briefly, in your own words, how Simpson's paradox appeared in this case.

12. Test Scores. Consider the following table comparing the grade point averages (GPAs) and mathematics SAT scores of high school students in 1988 and 1998.

	% students		SAT score		
GPA	**1988**	**1998**	**1988**	**1998**	**Change**
A+	4	7	632	629	−3
A	11	15	586	582	−4
A−	13	16	556	554	−2
B	53	48	490	487	−3
C	19	14	431	428	−3
Overall average			504	514	+10

Source: Cited in *Chance*, Vol. 12, No. 2, 1999, from data in *New York Times*, September 2, 1999.

a. In general terms, how did the SAT scores of the students in the five grade categories change between 1988 and 1998?

b. How did the overall average SAT score change between 1988 and 1998?

c. How is this an example of Simpson's paradox?

13. Tuberculosis Deaths. The following table shows deaths due to tuberculosis (TB) in New York City and Richmond, Virginia, in 1910.

	New York	
Race	**Population**	**TB deaths**
White	4,675,000	8400
Nonwhite	92,000	500
Total	4,767,000	8900
	Richmond	
Race	**Population**	**TB deaths**
White	81,000	130
Nonwhite	47,000	160
Total	128,000	290

Source: Cohen and Nagel, *An Introduction to Logic and Scientific Method*, Harcourt, Brace and World, 1934.

a. Compute the death rates for whites, nonwhites, and all residents in New York City.

b. Compute the death rates for whites, nonwhites, and all residents in Richmond.

c. Explain why this is an example of Simpson's paradox and explain how the paradox arises.

14. Weight Training. Two cross-country running teams participated in a (hypothetical) study in which a fraction of each team used weight training to supplement a running workout. The remaining runners did not use weight training. At the end of the season, the mean improvement in race times (in seconds) was recorded in the table below.

	Mean improvement (seconds)		
	Weight training	No weight training	Team average
Gazelles	10	2	6.0
Cheetahs	9	1	6.2

Describe how Simpson's paradox arises in this table. Resolve the paradox by finding the percentage of each team that used weight training.

15. Basketball Records. Consider the following (hypothetical) basketball records for Spelman and Morehouse Colleges.

	Spelman College	Morehouse College
Home games	10 wins, 19 losses	9 wins, 19 losses
Away games	12 wins, 4 losses	56 wins, 20 losses

a. Give numerical evidence to support the claim that Spelman College has a better team than Morehouse College.

b. Give numerical evidence to support the claim that Morehouse College has a better team than Spelman College.

c. Which claim do you think makes more sense? Why?

16. Better Drug. Two drugs, A and B, were tested on a total of 2,000 patients, half of whom were women and half of whom were men. Drug A was given to 900 patients and Drug B to 1,100 patients. The results appear in the table below.

	Women	Men
Drug A	5 of 100 cured	400 of 800 cured
Drug B	101 of 900 cured	196 of 200 cured

a. Give numerical evidence to support the claim that Drug B is more effective than Drug A.

b. Give numerical evidence to support the claim that Drug A is more effective than Drug B.

c. Which claim do you think makes more sense? Why?

17. Polygraph Test. Suppose that a polygraph is 90% accurate (it will correctly detect 90% of people who are lying and it will correctly detect 90% of people who are telling the truth). The 2,000 employees of a company are given a polygraph test during which they are asked whether they use drugs. All of them deny drug use when, in fact, 1% of the employees actually use drugs. Assume that anyone whom the polygraph operator finds untruthful is accused of lying.

a. Verify that the entries in the table below follow from the given information. Explain each entry.

	Users	Nonusers	Total
Test finds employee lying	18	198	216
Test finds employee truthful	2	1,782	1,784
Total	20	1,980	2,000

b. How many employees are accused of lying? Of these, how many were actually lying and how many were telling the truth? What percentage of those accused of lying were falsely accused?

c. How many employees are found truthful? Of these, how many were actually truthful? What percentage of those found truthful really were truthful?

18. Disease Test. Suppose a test for a disease is 80% accurate for those who have the disease (true positives) and 80% accurate for those who do not have the disease (true negatives). Within a sample of 4,000 patients, the incidence rate of the disease matches the national average, which is 1.5%.

a. Verify that the entries in the table below follow from the information given and that the overall incidence rate is 1.5%. Explain.

	Disease	No disease	Total
Test positive	48	788	836
Test negative	12	3,152	3,164
Total	60	3,940	4,000

b. Of those with the disease, what percentage test positive?

c. Of those who test positive, what percentage have the disease? Compare this result to the one in part b and explain why they are different.

d. Suppose a patient tests positive for the disease. As a doctor using this table, how would you describe the patient's chance of actually having the disease? Compare this figure to the overall incidence rate of the disease.

Further Applications

19. Hiring Statistics. (This problem is based on an example in "Ask Marilyn," *Parade Magazine*, April 28, 1996.) A company decided to expand, so it opened a factory, generating 455 jobs. For the 70 white-collar positions, 200 males and 200 females applied. Of the females who applied, 20% were hired, while only 15% of the males were hired. Of the 400 males applying for the blue-collar positions, 75% were hired, while 85% of the 100 females who applied were hired. How does looking at the white-collar and blue-collar positions separately suggest a hiring preference for women? Do the overall data support the idea that the company hires women preferentially? Explain why this is an example of Simpson's paradox and how the paradox can be resolved.

20. Drug Trials. (This problem is based on an example in "Ask Marilyn," *Parade Magazine*, April 28, 1996.) A company runs two trials of two treatments for an illness. In the first trial, Treatment A cures 20% of the cases (40 out of 200) and Treatment B cures 15% of the cases (30 out of 200). In the second trial, Treatment A cures 85% of the cases (85 out of 100) and Treatment B cures 75% of the cases (300 out of 400). Which treatment had the better cure rate in the two trials individually? Which treatment had the better overall cure rate? Explain why this is an example of Simpson's paradox and how the paradox can be resolved.

21. HIV Risks. The New York State Department of Health estimates a 10% rate of HIV for the at-risk population and a 0.3% rate for the general population. Tests for HIV are 95% accurate in detecting both true negatives and true positives. Random selection and testing of 5,000 at-risk people and 20,000 people from the general population results in the following table.

	At-risk population	
	Test positive	**Test negative**
Infected	475	25
Not infected	225	4,275
	General population	
	Test positive	**Test negative**
Infected	57	3
Not infected	997	18,943

a. Verify that incidence rates for the general and at-risk populations are 0.3% and 10%, respectively. Also verify that detection rates for the general and at-risk populations are 95%.

b. Consider the at-risk population. Of those with HIV, what percentage test positive? Of those who test positive, what percentage have HIV? Explain why these two percentages are different.

c. Suppose a patient in the at-risk category tests positive for the disease. As a doctor using this table, how would you describe the patient's chance of actually having the disease? Compare this figure to the overall incidence rate of the disease.

d. Consider the general population. Of those with HIV, what percentage test positive? Of those who test positive, what percentage have HIV? Explain why these two percentages are different.

e. Suppose a patient in the general population tests positive for the disease. As a doctor using this table, how would you describe the patient's chance of actually having the disease? Compare this figure to the overall incidence rate of the disease.

Projects for the Internet and Beyond

For useful links, select "Links for Internet Projects" for Chapter 4 at www.aw.com/bbt.

22. Polygraph Arguments. Visit Web sites devoted to either opposing or supporting the use of polygraph tests. Summarize the arguments on both sides, specifically noting the role that false negative rates play in the discussion.

23. Drug Testing. Explore the issue of drug testing either in the workplace or in athletic competitions. Discuss the legality of drug testing in these settings and the accuracy of the tests that are commonly conducted.

24. Cancer Screening. Investigate recommendations concerning routine screening for some type of cancer (for example, breast cancer, prostate cancer, or colon cancer). Explain how the accuracy of the screening test is measured. How is the test useful? How can its results be misleading?

IN THE NEWS

25. Polygraphs. Find a recent article in which someone or some group proposes a polygraph test to determine whether a person is being truthful. In light of what you know about polygraph tests, do you think the results will be meaningful? Why or why not?

26. Drug Testing and Athletes. Find a news report concerning drug testing of athletes. Summarize how the testing is being used, and discuss whether the testing is reliable.

Chapter Review Exercises

1. Refer to the weights (in grams) of randomly selected plain M&Ms.

Red M&Ms:

0.751	0.841	0.856	0.799	0.966	0.859	0.857
0.942	0.873	0.809	0.890	0.878	0.905	

Green M&Ms:

0.925	0.914	0.881	0.865	0.865	1.015	0.876
0.809	0.865	0.848	0.940	0.833	0.845	0.852
0.778	0.814	0.791	0.810	0.881		

a. Find the mean and median for each of the two data sets.

b. Find the range and standard deviation for each of the two data sets.

c. Give the five-number summary and construct a boxplot for each of the two data sets.

d. Apply the range rule of thumb to estimate the standard deviation of each of the two data sets. How well does the rule work in each case? Briefly discuss why it does or does not work well.

e. Based on all your results, compare and discuss the two data sets in terms of their center and variation. Does there appear to be a difference between the weights of red M&Ms and the weights of green M&Ms?

2. Combine the two samples of weights from Review Exercise 1 and find the following:

a. The percentile for the weight of 0.845 gram

b. The mode

3. a. What is the standard deviation for a sample of 50 values, all of which are the same?

b. Which of the following two car batteries would you prefer to buy, and why?

- One taken from a population with a mean life of 48 months and a standard deviation of 2 months
- One taken from a population with a mean life of 48 months and a standard deviation of 6 months

c. If an outlier is included with a sample of 50 values, what is the effect of the outlier on the mean?

d. If an outlier is included with a sample of 50 values, what is the effect of the outlier on the median?

e. If an outlier is included with a sample of 50 values, what is the effect of the outlier on the range?

f. If an outlier is included with a sample of 50 values, what is the effect of the outlier on the standard deviation?

Chapter Quiz

1. When you add the values 3, 5, 8, 12, and 20 and then divide by the number of values, the result is 9.6. Which term best describes this value: average, mean, median, mode, or standard deviation?

2. To measure variation in a sample, which of the following statistics should be used: average, mean, median, mode, or standard deviation?

3. When the range rule of thumb is used to find the value of the standard deviation of a sample of data, is the result always the actual value of the standard deviation?

4. A data set consists of a set of numerical values. Which, if any, of the following statements could be correct?

a. There is no mode.

b. There are two modes.

c. There are three modes.

5. Indicate whether the given statement could apply to a data set consisting of 1,000 values that are all different.

a. The 20th percentile is greater than the 30th percentile.

b. The median is greater than the first quartile.

c. The third quartile is greater than the first quartile.

d. The mean is equal to the median.

e. The range is zero.

6. A standard test for anxiety is designed so that the mean score is 50 and the standard deviation is 10. Based on the range rule of thumb, what are the likely low and high values?

7. Use the range rule of thumb to estimate the standard deviation of the values 2, 4, 5, 8, and 10.

8. What is the standard deviation of the values 5.8, 5.8, 5.8, 5.8, and 5.8?

9. What is the range of the values 2.0, 3.7, 4.9, 5.0, 5.7, 6.7, 8.5, and 9.0?

10. Identify the components that constitute the five-number summary for a data set.

Using Technology

Many computer software programs allow you to enter a data set and use one operation to get several different sample statistics, referred to as *descriptive statistics*. Such descriptive statistics typically include the mean, median, range, and standard deviation. Here are some of the procedures for obtaining such displays.

Important notes about the following procedures:

1. SPSS and Excel provide "modified" boxplots that show special points representing outliers, which are data values far from almost all of the other data values.
2. There is not universal agreement on a single procedure for calculating quartiles, and different computer programs often yield different results. For example, if you use the data set 1, 3, 6, 10, 15, 21, 28, and 36, you will get these results:

	Q_1	Q_2	Q_3
SPSS	3.75	12.5	26.25
Excel	5.25	12.5	22.75
STATDISK	4.5	12.5	24.5
Minitab	3.75	12.5	26.25
SAS	4.5	12.5	24.5
TI-83/84 Plus	4.5	12.5	24.5

SPSS

Descriptive Statistics: Begin by entering a data set or opening an existing data set. Click on **Analyze**, select **Descriptive Statistics**, and then select **Frequency**. Double click on the variable or column of data at the left. Now click on the **Statistics** button and proceed to click on the boxes corresponding to the statistics that you want. You can select quartiles, standard deviation, range, minimum, maximum, mean, median, and others. Click on the **Continue** button; then click on the **OK** button to obtain the values of the selected statistics.

Boxplot: Enter or open a data set. Click on **Graphs** and select **Boxplot**. In the boxplot window, select the option **Summaries of separate variables** and **Simple**, and then click on the **Define** button. Now click on the variable or column at the left, click on the button in the middle of the box, and then click on **OK**.

Excel

Descriptive Statistics: Excel's Data Analysis add-in must be installed. If the following procedure does not produce the data analysis window, you must install the Data Analysis add-in. If you are using Excel 2003, click on **Help**; then enter **Data Analysis ToolPak** and press **ENTER**. Select the help option **Load or unload add-in programs**, and the instructions will be displayed. If you are using Excel 2007, select the help feature by clicking on the question mark icon. Then enter **Data Analysis** and select the help item **Load the Analysis ToolPak**, and the instructions will be displayed.

Enter the sample data in column A. Select **Tools** and then **Data Analysis**. If you are using Excel 2007, click on **Data** and then click on **Data Analysis**. In the displayed window, select **Descriptive Statistics** and click on **OK**. In the dialog box, enter the input range (such as A1:A76 for 76 values in column A), click on **Summary Statistics**, and then click on **OK**. The descriptive statistics include the mean, median, standard deviation, range, minimum, and maximum.

Boxplot: Although Excel is not designed to generate boxplots, they can be generated using the DDXL add-in that is a supplement to this book.

First enter the data in column A. Select **Charts and Plots**. For the function type, select the option **Boxplot**. In the dialog box, click on the pencil icon and enter the range of data, such as A1:A76 if you have 76 values listed in column A. Click on **OK**. The result will be a modified boxplot with mild outliers and extreme outliers. The values of the 5-number summary will also be displayed.

STATDISK

Descriptive Statistics: Enter the data in the data window. Click on **Data**, and select **Descriptive Statistics**. Now click on **Evaluate** to get the various descriptive statistics, including the mean, median, range, standard deviation, and the values that constitute the five-number summary (minimum, first quartile, median or second quartile, third quartile, and maximum).

Boxplot: Enter the data in the data window; then click on **Data** and then **Boxplot**. Click on the columns that you want to include, and then click on **Plot**.

FOCUS ON THE STOCK MARKET

What's Average About the Dow?

As "averages" go, this one is extraordinary. You can't watch the news without hearing what happened to it, and many people spend hours tracking it each day. It is by far the most famous indicator of stock market performance. We are talking, of course, about the Dow Jones Industrial Average, or DJIA for short. But what exactly is it?

The easiest way to understand the DJIA is by looking at its history. As the modern industrial era got under way in the late 19th century, most people considered stocks to be dangerous and highly speculative investments. One reason was a lack of regulation that made it easy for wealthy speculators, unscrupulous managers, and corporate raiders to manipulate stock prices. But another reason was that, given the complexities of daily stock trading, even Wall Street professionals had a hard time figuring out whether stocks in general were going up (a "bull market") or down (a "bear market"). Charles H. Dow, the founder (along with Edward D. Jones) and first editor of the *Wall Street Journal*, believed he could rectify this problem by creating an "average" for the stock market as a whole. If the average was up, the market was up, and if the average was down, the market was down.

To keep the average simple, Dow chose 12 large corporations to include in his average. On May 26, 1896, he added the stock prices of these 12 companies and divided by 12, finding a mean stock price of $40.94. This was the first value for the DJIA. As Dow had hoped, it suddenly became easy for the public to follow the market's direction just by comparing his average from day to day, month to month, or year to year.

The basic idea behind the DJIA is still the same, although the list now includes 30 stocks rather than 12 (see Table 3.5). However, the DJIA is no longer the mean price of its 30 stocks. Instead, it is calculated by adding the prices of its 30 stocks and dividing by a special divisor. Because of this divisor, we now think of the DJIA as an index that helps us keep track of stock values, rather than as an actual average of stock prices.

The divisor is designed to preserve continuity in the underlying value represented by the DJIA, and it therefore must change whenever the list of 30 stocks changes and whenever a company on the list has a stock split. A simple example shows why the divisor must change when the list changes. Suppose the DJIA consisted of only 2 stocks (rather than 30): Stock A with a price of $100 and Stock B with a price of $50. The mean price of these two stocks is ($100 + $50)/2 = $75. Now, suppose that we change the list by replacing Stock B with Stock C and that Stock C's price is $200. The new mean is ($100 + $200)/2 = $150, so merely replacing one stock on the list would raise the mean price from $75 to $150. Therefore, to keep the "value" of the DJIA constant when we change this list, we must divide the new mean of $150 by 2. In this way, the DJIA remains 75 both before and after the list change, but we can no longer think of this 75 as a mean price in dollars.

To see why a stock split changes the divisor, again suppose the index consists of just two stocks: Stock X at $100 and Stock Y at $50, for a mean price of $75. Now, suppose Stock X undergoes a 2-for-1 stock split, so that its new price is $50. With both stocks now priced at $50, the mean price after the stock split would also be $50. In other words, even though a stock split does not affect a company's total value (it only changes the number and prices of its shares), we'd find a drop in the mean price from \$75 to \$50. In this case, we can preserve continuity by dividing the new mean of 50 by 2/3 (which is equivalent to multiplying by 3/2) so that the DJIA holds at 75 both before and after the stock split.

Just as in these simple examples, the real divisor changes with every list change or stock split, so it has changed many times since Charles Dow first calculated the DJIA as an actual mean. The current value of the divisor is published daily in the *Wall Street Journal*.

Given that there are now well over 10,000 actively traded stocks, it might seem remarkable that a sample of only 30 could reflect overall market activity. But today, when computers make it easy to calculate stock market "averages" in many other ways, we can look at historical data and see that the DJIA has indeed been a reliable indicator of overall market performance. Figure 4.17 shows the historical performance of the DJIA.

If you study Figure 4.17 carefully, you may be tempted to think that you can see patterns that would allow you to forecast precise values of the market in the future. Unfortunately, no one has ever found a way to make reliable forecasts, and most economists now believe that such forecasts are impossible.

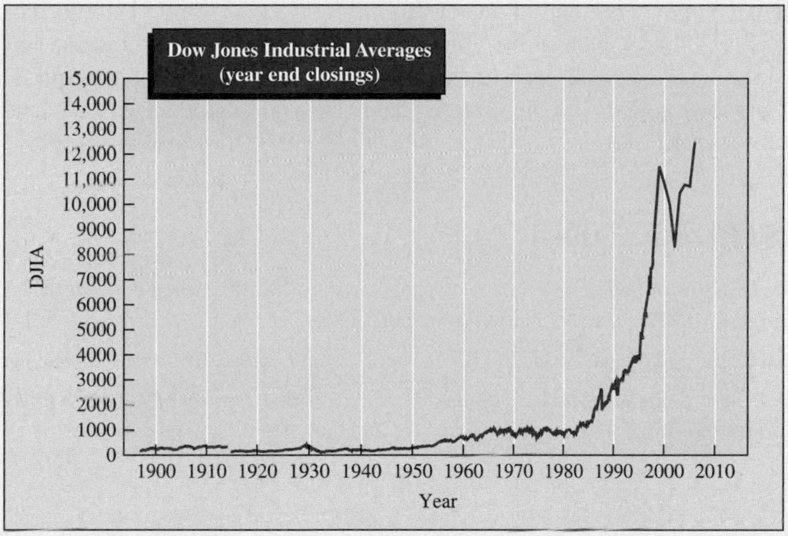

Figure 4.17 Year-end closing values of the DJIA, through 2006.

The futility of trying to forecast the market is illustrated by the story of the esteemed Professor Benjamin Graham, often called the father of "value investing." In the spring of 1951, one of his students came to him for some investment advice. Professor Graham noted that the DJIA then stood at 250, but that it had fallen below 200 at least once during every year since its inception in 1896. Because it had not yet fallen below 200 in 1951, Professor Graham advised his student to hold off on buying until it did. Professor Graham presumably followed his own advice, but the student did not. Instead, the student invested his "about 10 thousand bucks" in the market right away. As it turned out, the market never did fall below 200 in 1951 or any time thereafter. And the student, named Warren Buffet, became a billionaire many times over.

QUESTIONS FOR DISCUSSION

1. The stock market is still considered a riskier investment than, say, bank savings accounts or bonds. Nevertheless, financial advisors almost universally recommend holding at least some stocks, which is quite different from the situation that prevailed a century ago. What role do you think the DJIA played in building investors' confidence in the stock market?

2. The DJIA is only one of many different stock market indices in wide use today. Briefly look up a few other indices, such as the S&P 500, the Russell 2000, and the NASDAQ. How do these indices differ from the DJIA? Do you think that any of them should be considered more reliable indicators of the overall market than the DJIA? Why or why not?

3. The 30 stocks in the DJIA represent a sample of the more than 10,000 actively traded stocks. However, it is not a random sample because it is chosen by particular editors for particular reasons that sometimes include personal biases. Suppose that you chose a random sample of 30 stocks and tracked their prices. Do you think that such a random sample would track the market as well as the stocks in the DJIA? Why or why not?

4. Create your own "portfolio" of 10 stocks that you'd like to own, and assume you own 100 shares of each. Calculate the total value of your portfolio today, and track price changes over the next month. At the end of the month, calculate the percent change in the value of your portfolio. How did the performance of your portfolio compare to the performance of the DJIA during the month? If you really owned these stocks, would you continue to hold them or would you sell? Explain.

5. Figure 4.17 shows the end-of-year DJIA through 2006. Find year-end data for years since 2006 and add them to the graph. Then find data for the highs and lows for each year since 1995 and plot each of those as a separate curve on the graph. How do the highs and lows compare to the year-end data? Do you think the year-end data are a valid way of tracking historical changes in the DJIA?

SUGGESTED READING

You can find information about the DJIA regularly in the *Wall Street Journal, New York Times, Money, Business Week*, and numerous other periodicals.

Morris, K. M., and Siegel, A. M. *The Wall Street Journal Guide to Personal Finance*. Fireside, 2000.

Prestbo, John (ed.). *The Market's Measure: An Illustrated History of America Told Through the Dow Jones Industrial Average*. Dow Jones & Company, 1999.

FOCUS ON ECONOMICS

Are the Rich Getting Richer?

At the height of the "dot com bubble" in 1999, when technology stock prices soared to levels never seen before and not seen since, Microsoft founder Bill Gates briefly became the world's first person with a net worth of over $100 billion. His wealth thereby exceeded the gross national products of all but the 18 richest countries in the world. Gates's wealth has actually declined since then, both because of a decline in Microsoft's stock price and because he has given a large portion of his net worth to charity. Nevertheless, media reports continually highlight stories about the super-rich, making it seem that the rich keep getting richer while the rest of us are left behind. But is it true?

If we wish to draw general conclusions about how the average person is faring compared to the rich, we must look at the overall income distribution. Economists have developed a number, called the Gini Index, that is used to describe the level of equality or inequality in the income distribution. The Gini Index is defined so that it can range only between 0 and 1. A Gini Index of 0 indicates perfect income equality, in which every person has precisely the same income. A Gini Index of 1 indicates perfect inequality, in which a single person has all the income and no one else has anything. Figure 4.18 shows the Gini Index in the United States since 1947. Note that the Gini Index fell from 1947 to 1968, indicating that the income distribution became more uniform during this period. The Gini Index has generally risen ever since, indicating that the rich are, indeed, getting richer.

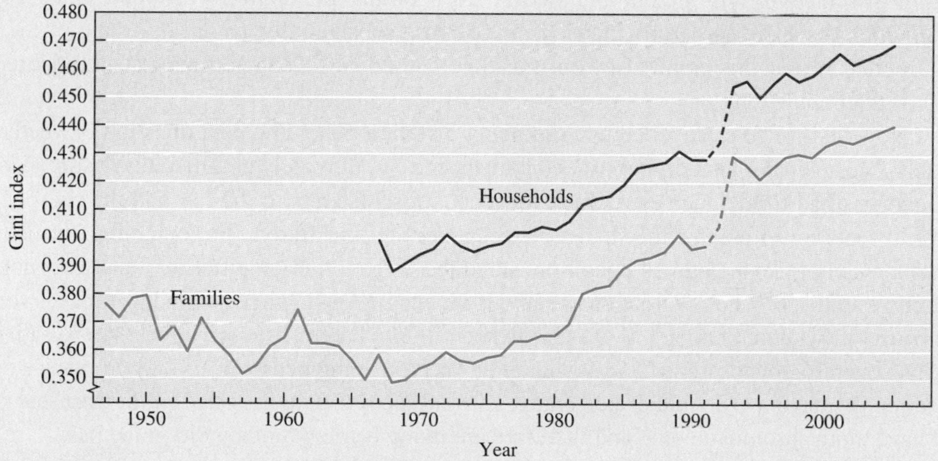

Figure 4.18 Gini Index for families and households, 1947–2005. Household data, which include single people and households in which the members are not part of the same family, have been taken only since 1967. The dashed segments in 1993 indicate a change in the methodology for data collection, so the corresponding rise in the Gini Index may be partially or wholly due to this change rather than a real change in income inequality. *Source:* U.S. Census Bureau.

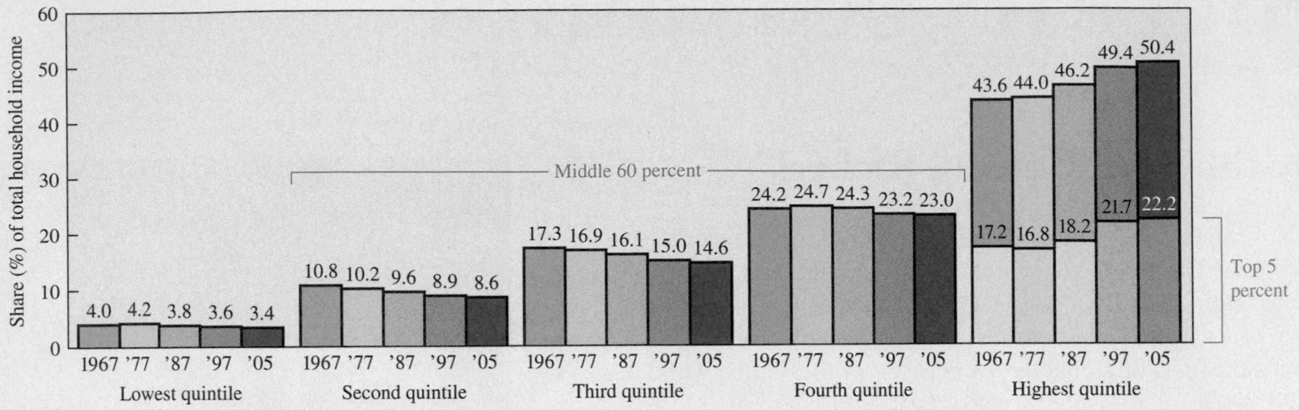

Figure 4.19 Share of total household income by quintile (and top 5%): 1967, 1977, 1987, 1997, and 2005. *Source:* U.S. Census Bureau.

Although the Gini Index provides a simple single-number summary of income inequality, the number itself is fairly difficult to interpret (and to calculate). An alternative way to look at the income distribution is to study income quintiles, which divide the population into fifths by income. Often, the highest quintile is further broken down to show how the top 5% of income earners compares to others.

Figure 4.19 shows the share of total income received by each quintile and the top 5% in the United States in different decades. The height of each bar (the number on top of it) represents the share of total income. For example, the 3.4 on the bar for the lowest quintile in 2005 means that the poorest 20% of the population received only 3.4% of the total income in the United States. Similarly, the 50.4 on the bar for the top quintile in 2005 means that the richest 20% of the population received 50.4% of the income. Note also that the richest 5% received 22.2% of the income—nearly double the total income of the poorest 40% of the population. If you study this graph carefully, you'll see that the share of income earned by the first four quintiles—which means all but the richest 20% of the population—dropped since 1967. Meanwhile, the share earned by the richest 20% rose substantially, as did the share of the top 5%. In other words, this graph also confirms that the rich have been getting richer compared to most of the population.

Now that we've established that the rich are getting richer, the next question is whether it matters. Most people, including most economists, have traditionally assumed that rising income inequality is bad for democracies. But a few economists from both the left and the right of the political spectrum argue that the change in recent decades is different. For one thing, the change meets a widely accepted ethical condition called the Pareto criterion, after the Italian economist Vilfredo Pareto (for whom Pareto charts are also named): Any change is good if it makes someone better off without making anyone else worse off. The Pareto criterion appears to be satisfied because overall growth in the U.S. economy has been helping nearly everyone. In other words, most people may have a smaller percentage of total income than they had in the past, but they still have more absolute income and therefore are living better than they did in the past.

Secondly, today's rich differ from the rich in the past. For example, as recently as 1980, 60% of the "Forbes 400" (the richest 400 people) had inherited most of their wealth. By 2005, less than 20% of the Forbes 400 represented old money. The implication is that while you had to be born rich in the past, today you can *become* rich by getting educated and working hard. Surely, it is a good thing to encourage education and hard work.

Finally, while overall income inequality has increased, the income inequality among different races and between men and women has decreased. In other words, it is now easier than it was in the past for African Americans, Hispanics, and women to earn as much as white males. Again, this is surely a good thing for our democratic values, even if we still have a long way to go before the inequalities are completely eliminated.

Of course, even if the recent increase in income inequality has been good for the United States, it may not be good if it continues. Unfortunately, no one knows precisely what causes income inequality to increase, and therefore no one knows how to reverse the current trend.

QUESTIONS FOR DISCUSSION

1. Compare several different ways of looking at the data shown in Figure 4.18 and Figure 4.19. For example, does one seem to indicate a larger change in income inequality than the other? Can you think of other possible ways to display income data that might give a different picture than those shown here?
2. Do you agree that the Pareto criterion is a good way to evaluate the ethics of economic change? Why or why not?
3. Overall, do you think the increase in income inequality has been a good or bad thing for the United States? Will it be good if the trend continues? Defend your opinion.
4. Although economic data suggest that the vast majority of Americans are better off today than they were a few decades ago, the poorest Americans still live in very bad economic conditions. What do you think can or should be done to help improve the lives of the poor? Can your suggestion be implemented without harming the overall economy? Explain.

SUGGESTED READING

You can find extensive data and reports on the distribution of income by going to the "Income" area of the U.S. Census Bureau Web site at www.census.gov.

Arrow, Kenneth, et al. (eds.). *Meritocracy and Economic Inequality*. Princeton University Press, 2000.

Firebaugh, Glenn. *The New Geography of Global Income Inequality*. Harvard University Press, 2006.

Nasar, Sylvia. "Is the U.S. Income Gap Really a Big Problem?" *New York Times*. April 4, 1999.

5

Nothing in life is to be feared. It is only to be understood.
—Marie Curie

A Normal World

WHEN YOU WALK INTO A STORE, HOW DO YOU KNOW IF a sale price is really a good price? When you exercise and your heart rate rises, how do you know if it has risen enough, but not too much, for a good workout? If your 12-year-old daughter runs a mile in 5 minutes, is she a future Olympic hopeful? These questions seem very different, but from a statistical standpoint they are very similar: Each one asks whether a particular number (price, heart rate, running time) is somehow unusual. In this chapter, we will discuss how we can answer such questions with the aid of the bell-shaped *normal distribution.*

LEARNING GOALS

5.1 What Is Normal?
Understand what is meant by a normal distribution and be able to identify situations in which a normal distribution is likely to arise.

5.2 Properties of the Normal Distribution
Know how to interpret the normal distribution in terms of the 68-95-99.7 rule, standard scores, and percentiles.

5.3 The Central Limit Theorem
Understand the basic idea behind the Central Limit Theorem and its important role in statistics.

5.1 What Is Normal?

Suppose a friend is pregnant and due to give birth on June 30. Would you advise her to schedule an important business meeting for June 16, two weeks before the due date? Answering this question requires knowing whether the baby is likely to arrive more than 14 days before the due date. For that, we need to examine data concerning due dates and actual birth dates.

Figure 5.1 is a histogram for a distribution of 300 natural births at Providence Memorial Hospital; the data are hypothetical, but based on how births would be distributed without medical intervention. The horizontal axis shows how many days before or after the due date a baby was born: Negative numbers represent births *prior* to the due date, zero represents a birth on the due date, and positive numbers represent births *after* the due date. The left vertical axis shows the number of births for each 4-day bin. For example, the frequency of 35 for the highest bar corresponds to the bin from −2 days to 2 days; it shows that out of the 300 total births in the sample, 35 births occurred within 2 days of the due date.

To answer our question about whether a birth is likely to occur more than 14 days early, it is more useful to look at the *relative frequencies*. Recall that the relative frequency of any data value is its frequency divided by the total number of data values (see Section 3.1). Figure 5.1 shows relative frequencies on the right vertical axis. For example, the bin for −14 days to −10 days (shaded dark blue) has a relative frequency of about 0.07, or 7%. That is, about 7% of the 300 births occurred between 14 days and 10 days before the due date.

Now we can find the proportion of births that occurred more than 14 days before the due date: We simply add the relative frequencies for the bins to the left of −14; you can measure the graph to confirm that these bins have a total relative frequency of about 0.21, which says that about 21% of the births in this data set occurred more than 14 days before the due date. We could say that your friend has about a 1 in 5 chance of her baby being born on or before the date of the business meeting. If the meeting is important, it might be good to schedule it earlier.

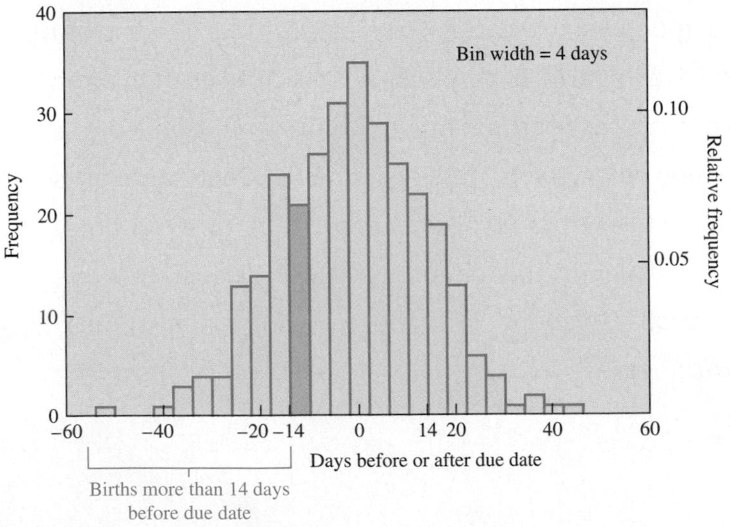

Figure 5.1 Histogram of frequencies (left axis) and relative frequencies (right axis) for birth dates relative to due date. (These data are hypothetical.) Negative numbers refer to births before the due date; positive numbers refer to births after the due date. The width of each bin is 4 days. For example, the bin shaded dark blue represents births occurring between 10 and 14 days early.

TIME OUT TO THINK

Suppose the friend plans to take a three-month maternity leave after the birth. Based on the data in Figure 5.1 and assuming a due date of June 30, should she promise to be at work on October 10?

The Normal Shape

The distribution of the birth data has a fairly distinctive shape, which is easier to see if we overlay the histogram with a smooth curve (Figure 5.2). For our present purposes, the shape of this smooth distribution has three very important characteristics:

- The distribution is *single-peaked.* Its mode, or most common birth date, is the due date.
- The distribution is *symmetric* around its single peak; therefore, its median and mean are the same as its mode. The median is the due date because equal numbers of births occur before and after this date. The mean is also the due date because, for every birth before the due date, there is a birth the same number of days after the due date.
- The distribution is spread out in a way that makes it resemble the shape of a bell, so we call it a "bell-shaped" distribution.

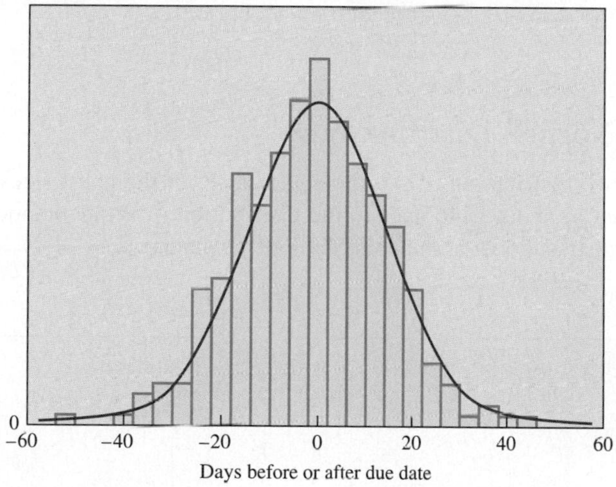

Figure 5.2 A smooth normal distribution curve is drawn over the histogram of Figure 5.1.

TIME OUT TO THINK

The histogram in Figure 5.2, which is based on natural births, is fairly symmetric. Today, doctors usually induce birth if a woman goes too far past her due date. How would the shape of the histogram change if it included induced labor births?

The smooth distribution in Figure 5.2, with these three characteristics, is called a **normal distribution**. (Note that the given birth data are not exactly normal.) All normal distributions have the same characteristic bell shape, but they can differ in their mean and in their variation. Figure 5.3 shows two different normal distributions. Both have the same mean, but distribution (a) has greater variation. As we'll discuss in the next section, knowing the *standard deviation* (see Section 4.3) of a normal distribution tells us everything we need to know about its variation. Therefore, a normal distribution can be fully described with just two numbers: its mean and its standard deviation.

TECHNICAL NOTE

A normal distribution with mean μ and standard deviation σ is given by the formula

$$y = \frac{e^{-\frac{1}{2}[(x-\mu)/\sigma]^2}}{\sigma\sqrt{2\pi}}$$

This formula is not used in this book, but it does algebraically describe the shape of the normal distribution.

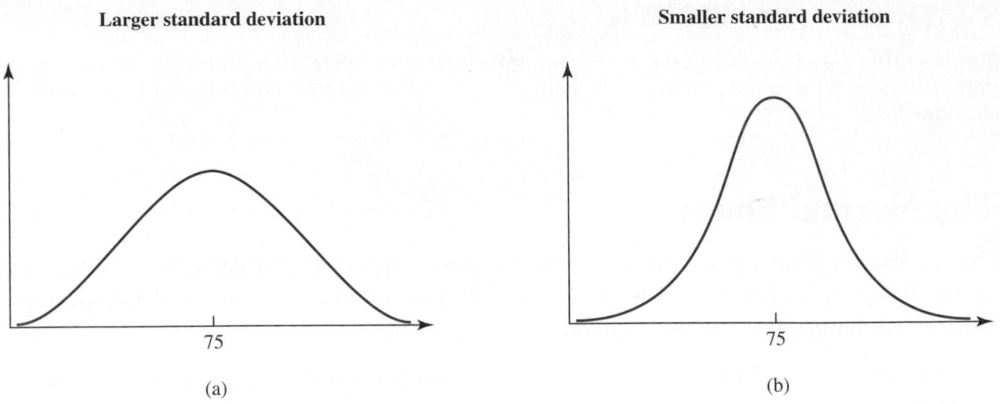

Figure 5.3 Both distributions are normal and have the same mean of 75, but the distribution on the left has a larger standard deviation.

Definition

The **normal distribution** is a symmetric, bell-shaped distribution with a single peak. Its peak corresponds to the mean, median, and mode of the distribution. Its variation can be characterized by the standard deviation of the distribution.

EXAMPLE 1 Normal Distributions?

Figure 5.4 shows two distributions: (a) a famous data set of the chest sizes of 5,738 Scottish militiamen collected in about 1846 and (b) the distribution of the population densities of the 50 states. Is either distribution a normal distribution? Explain.

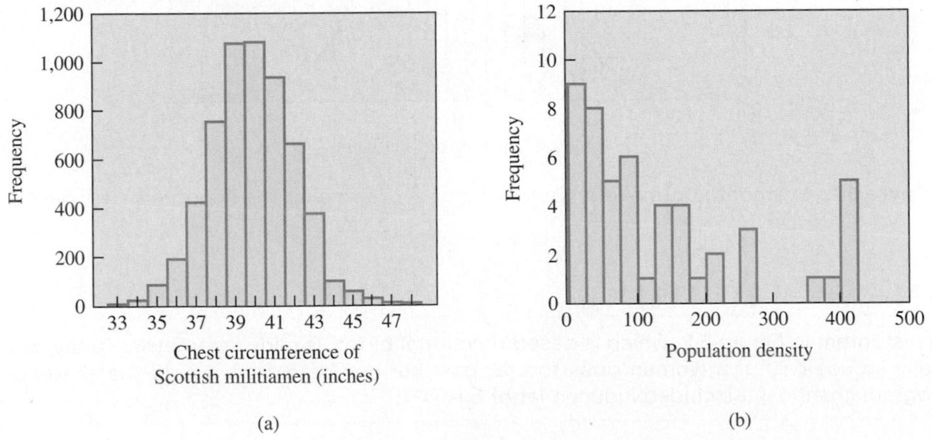

Figure 5.4 *Source of* (a): Adolphe Quetelet, *Lettres à S. A. R. le Duc Régnant de Saxe-Cobourg et Gotha,* 1846.

Solution The distribution in Figure 5.4a is nearly symmetric, with a mean between 39 and 40 inches; it is nearly normal. The distribution in Figure 5.4b shows that most states have low population densities, but a few have much higher densities. This fact makes the distribution right-skewed, so it is not a normal distribution.

By the Way ...

Using data taken from French and Scottish soldiers, the Belgian social scientist Adolphe Quetelet realized in the 1830s that human characteristics such as height and chest circumference are normally distributed. This observation led him to coin the term "the average man." Quetelet was the first foreign member of the American Statistical Association.

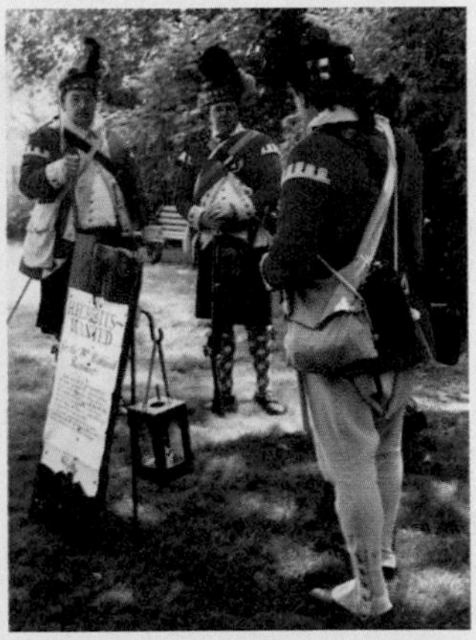

The Normal Distribution and Relative Frequencies

Recall that the total relative frequency for any data set must be 1 (see Section 3.1). Now consider the smooth curve for the normal distribution in Figure 5.2. Although we no longer have individual bars, we can still associate the height of the normal curve with the relative frequency. The fact that the relative frequencies must sum to 1 becomes the condition that the area under the normal curve must be 1.

Figure 5.5 gives an example of how we interpret the normal curve. The key idea is this: *The relative frequency for any range of data values is the area under the curve covering that range of values.* For example, a precise calculation shows that the shaded region to the left of –14 days comprises about 18% of the total area under the curve. We therefore conclude that the relative frequency is about 0.18 for data values less than –14 days, which means that about 18% of births are more than 14 days early. Similarly, the shaded region to the right of 18 days comprises about 12% of the total area under the curve. We therefore conclude that the relative frequency is about 0.12 for data values greater than 18 days, which means that about 12% of births are more than 18 days late. Altogether, we see that 18% + 12% = 30% of all births are either more than 14 days early or more than 18 days late. (Note that these relative frequencies are correct for the smooth normal distribution in Figure 5.5, but are only approximate for the original data in the histogram of Figure 5.1.)

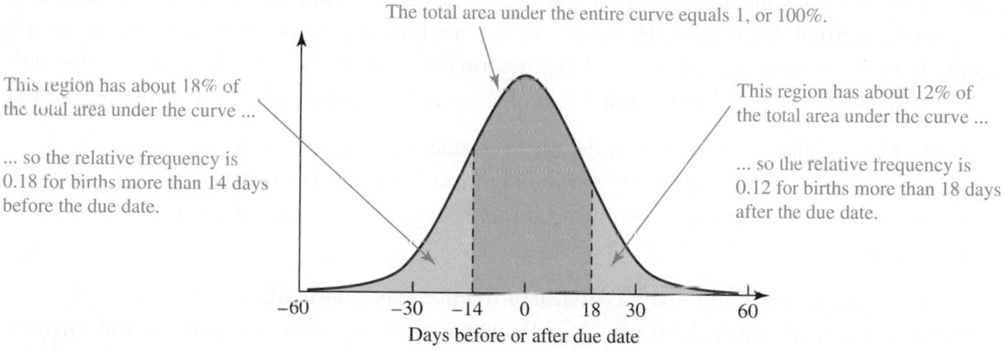

Figure 5.5 The percentage of the total area in any region under the normal curve tells us the relative frequency of data values in that region.

Relative Frequencies and the Normal Distribution

- The area that lies under the normal distribution curve corresponding to a range of values on the horizontal axis is the relative frequency of those values.
- Because the total relative frequency must be 1, the total area under the normal distribution curve must equal 1, or 100%.

TIME OUT TO THINK

According to Figure 5.5, what percentage of births occur between 14 days early and 18 days late? Explain. (Hint: Remember that the total area under the curve is 100%.)

By the Way ...

The Scottish politician John Sinclair (1754–1835) was one of the first collectors of economic, demographic, and agricultural data. He is credited with introducing the words *statistics* and *statistical* into the English language, having heard them used in Germany to refer to matters of state.

EXAMPLE 2 Estimating Areas

Look again at the normal distribution in Figure 5.5.

a. Estimate the percentage of births occurring between 0 and 60 days after the due date.

b. Estimate the percentage of births occurring between 14 days before and 14 days after the due date.

Solution

a. About half of the total area under the curve lies in the region between 0 days and 60 days. This means that about 50% of the births in the sample occur between 0 and 60 days after the due date.

b. Figure 5.5 shows that about 18% of the births occur more than 14 days before the due date. Because the distribution is symmetric, about 18% must also occur more than 14 days after the due date. Therefore, a total of about 18% + 18% = 36% of births occur either more than 14 days before or more than 14 days after the due date. The question asked about the remaining region, which means *between* 14 days before and 14 days after the due date, so this region must represent 100% − 36% = 64% of the births.

When Can We Expect a Normal Distribution?

The normal distribution is a good approximation to the distribution of many variables of practical interest. Physical characteristics such as weight, height, blood pressure, and reflex times generally follow a normal distribution. Standardized test scores, such as those for SATs or IQ tests, are normally distributed. Sports statistics, such as batting averages, times in a swimming event, or high jump results in a track meet, also tend to follow a normal distribution. Indeed, much of statistics is based on the normal distribution. However, not all variables follow a normal distribution, so it is important to understand when the normal distribution can be used and when it is not appropriate.

By studying graphs of normal distributions, we can see what makes a distribution normal.

- It must have values clustered near the mean so that it is single-peaked, or unimodal.
- The values must be spread evenly around the mean so that it is symmetric.
- Large deviations from the mean must be increasingly rare so that it has the characteristic bell shape.

On a deeper level, any quantity that is influenced by many different factors tends to be normally distributed. Physical traits are influenced by many different genetic and environmental effects. Standardized test scores reflect the performance of many individuals working on many different test questions. Sports statistics involve many players with different skills performing under different conditions.

By the Way ...

The normal distribution curve is often called a *Gaussian curve* in honor of the 19th-century German mathematician Carl Friedrich Gauss. The American logician Charles Peirce introduced the term *normal distribution* in about 1870.

Conditions for a Normal Distribution

A data set that satisfies the following four criteria is likely to have a nearly normal distribution:

1. Most data values are clustered near the mean, giving the distribution a well-defined single peak.
2. Data values are spread evenly around the mean, making the distribution symmetric.
3. Larger deviations from the mean become increasingly rare, producing the tapering tails of the distribution.
4. Individual data values result from a combination of many different factors, such as genetic and environmental factors.

EXAMPLE 3 Is It a Normal Distribution?

Which of the following variables would you expect to have a normal or nearly normal distribution?

a. Scores on a very easy test

b. Heights of a random sample of adult women

c. The number of Macintosh apples in each of 100 full bushel baskets

Solution

a. Tests have a maximum possible score (100%) that limits the size of data values. If the test is easy, the mean will be high and many scores will be close to the maximum possible. The few lower scores may be spread out well below the mean. We therefore expect the distribution of scores to be left-skewed and non-normal.

b. Height is determined by a combination of many factors (the genetic makeup of both parents and possibly environmental or nutritional factors). We expect the mean height for the sample to be close to the mode (most common height). We also expect there to be roughly equal numbers of women above and below the mean, and extremely large and small heights should be rare. That is why height is nearly normally distributed.

c. The number of apples in a bushel basket varies with the size of the apples. We expect that in the distribution there will be a single mode that should be close to the mean number of apples per basket. The number of baskets with more than the mean number of apples should be close to the number of baskets with fewer than the mean number of apples. We therefore expect the number of apples per basket to have a nearly normal distribution.

TIME OUT TO THINK

Would you expect scores on a moderately difficult exam to have a normal distribution? Suggest two more quantities that you would expect to be normally distributed.

Section 5.1 Exercises

Statistical Literacy and Critical Thinking

1. **Normal Distribution.** When we refer to a "normal" distribution, does the word *normal* have the same meaning as in ordinary language, or does it have a special meaning in statistics? What exactly is a normal distribution?

2. **Normal Distribution.** A normal distribution is informally and loosely described as a probability distribution that is "bell-shaped" when graphed. What is the "bell shape"?

3. **Random Digits.** Computers are often used to randomly generate digits of telephone numbers to be called when a survey is conducted. Is the distribution of those digits a normal distribution? Why or why not?

4. **Areas.** If you determine that, on the graph of a normal distribution, the area to the left of a particular value is 0.2, what is the area to the right of that value?

Does It Make Sense? For Exercises 5–8, decide whether the statement makes sense (or is clearly true) or does not make sense (or is clearly false). Explain clearly; not all of these statements have definitive answers, so your explanation is more important than your chosen answer.

5. **SAT Scores.** A random sample of SAT scores has a normal distribution with a mean of 1518 and a median of 1490.

6. **Weights.** As part of a national health study, each person in a random sample of subjects is weighed, and those weights have a normal distribution with two modes.

7. **IQ Scores.** The mean of a normally distributed set of IQ scores is 100, and 60% of the scores are over 105.

8. **Volumes of Pepsi.** A quality control analyst measures the contents of 100 cans of regular Pepsi and predicts that the distribution of volumes will be a normal distribution.

Concepts and Applications

9. **What Is Normal?** Identify the distribution in Figure 5.6 that is not normal. Of the two normal distributions, which has the larger standard deviation?

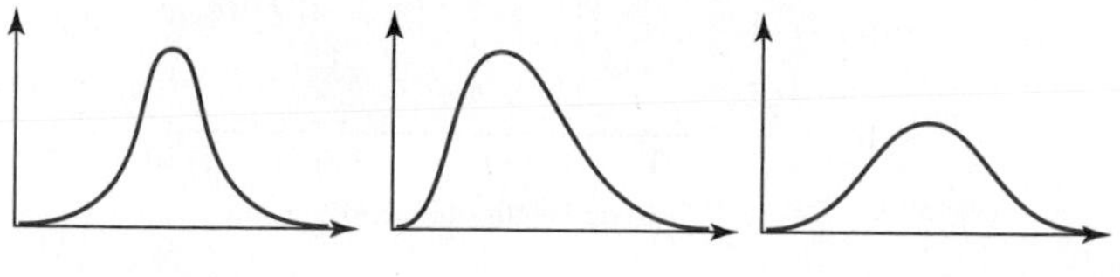

(a) (b) (c)

Figure 5.6

10. **What Is Normal?** Identify the distribution in Figure 5.7 (on next page) that is not normal. Of the two normal distributions, which has the larger standard deviation?

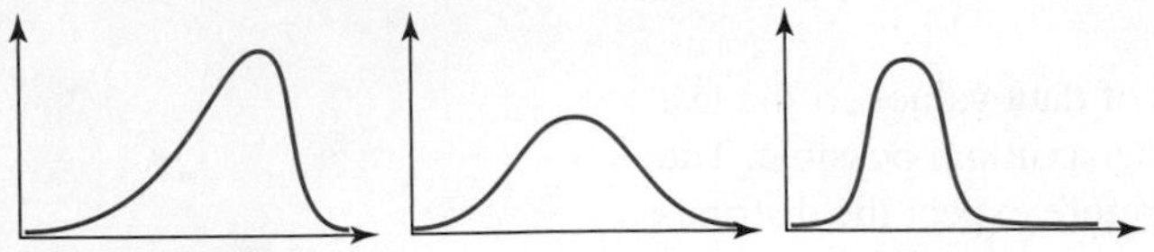

Figure 5.7

Normal Variables. For each of the data sets in Exercises 11–18, state whether you would expect it to be normally distributed. Explain your reasoning.

11. Weights of CDs. The exact weights of a random sample of Sony CDs

12. Incomes. The incomes of randomly selected adults in the United States

13. Casting a Die. The results from 500 tosses of a fair die

14. SAT Scores. All of the SAT scores from last year

15. Grip Strength. The measures of grip strength for a random sample of males

16. Flight Delays. The lengths of time that commercial aircraft are delayed before departing

17. Waiting Times. The waiting times at a bus stop if the bus comes once every ten minutes and you arrive at random times

18. Parking Ticket Fines. The amounts of the fines from parking tickets found on a random sample of 1,000 cars

19. Movie Lengths. Figure 5.8 shows a histogram for the lengths of 60 movies. The mean movie length is 110.5 minutes. Is this distribution close to normal? Should this variable have a normal distribution? Why or why not?

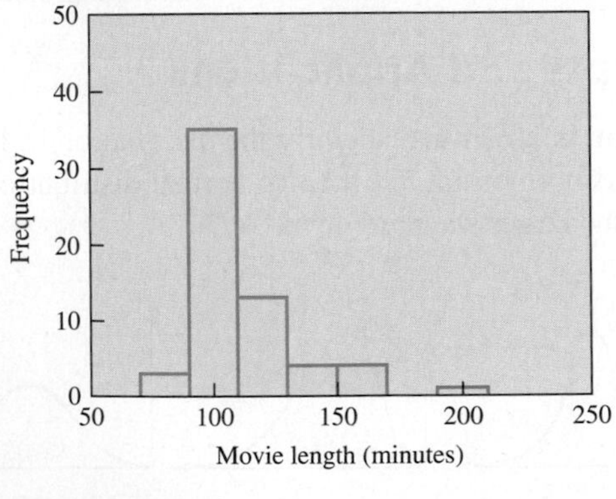

Figure 5.8

20. Heart Rates. Figure 5.9 shows a histogram for the heart rates of 98 students. The mean heart rate is 71.2 beats per minute. Is this distribution close to normal? Should this variable have a normal distribution? Why or why not?

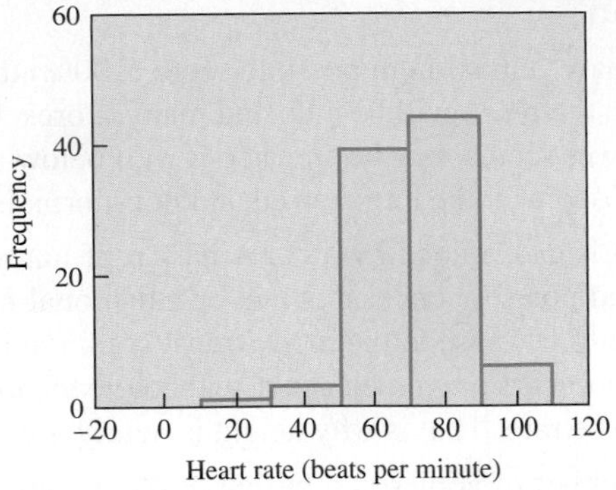

Figure 5.9

21. Quarter Weights. Figure 5.10 shows a histogram for the weights of 50 randomly selected quarters. The mean weight is 5.62 grams. Is this distribution close to normal? Should this variable have a normal distribution? Why or why not?

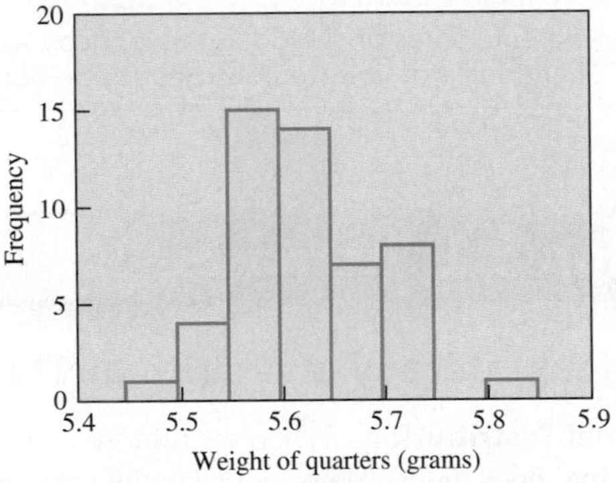

Figure 5.10

22. Aspirin Weights. Figure 5.11 shows a histogram for the weights of 30 randomly selected aspirin tablets. The mean weight is 665.4 milligrams. Is this distribution close to normal? Should this variable have a normal distribution? Why or why not?

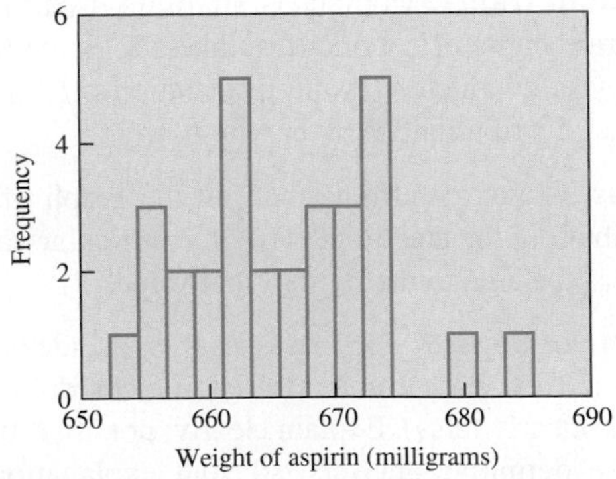

Figure 5.11

23. Areas and Relative Frequencies. Consider the normal curve in Figure 5.12, which gives relative frequencies in a distribution of men's heights. The distribution has a mean of 69.6 inches and a standard deviation of 2.8 inches.

a. What is the total area under the curve?

b. Estimate (using area) the relative frequency of values less than 67.

c. Estimate the relative frequency of values greater than 67.

d. Estimate the relative frequency of values between 67 and 70.

e. Estimate the relative frequency of values greater than 70.

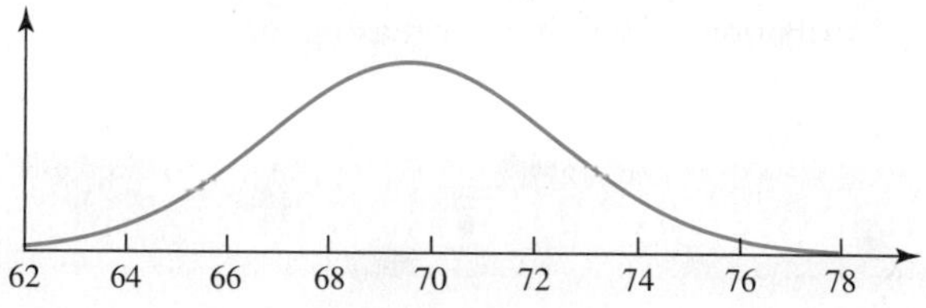

Figure 5.12

24. Areas and Relative Frequencies. Consider the normal curve in Figure 5.13, which shows the relative frequencies in a distribution of IQ scores. The distribution has a mean of 100 and a standard deviation of 16.

a. What is the total area under the curve?

b. Estimate (using area) the relative frequency of values less than 100.

c. Estimate the relative frequency of values greater than 110.

d. Estimate the relative frequency of values less than 110.

e. Estimate the relative frequency of values between 100 and 110.

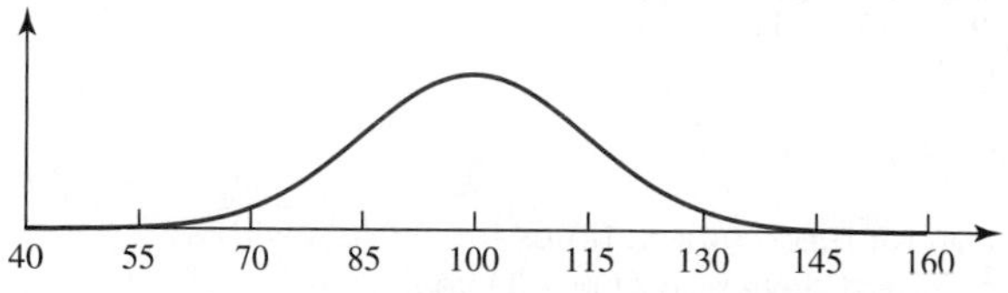

Figure 5.13

25. Estimating Areas. Consider the normal curve in Figure 5.14, which illustrates the relative frequencies in a distribution of systolic blood pressures for a sample of female students. The distribution has a standard deviation of 14.

a. What is the mean of the distribution?

b. Estimate (using area) the percentage of students whose blood pressure is less than 100.

c. Estimate the percentage of students whose blood pressure is between 110 and 130.

d. Estimate the percentage of students whose blood pressure is greater than 130.

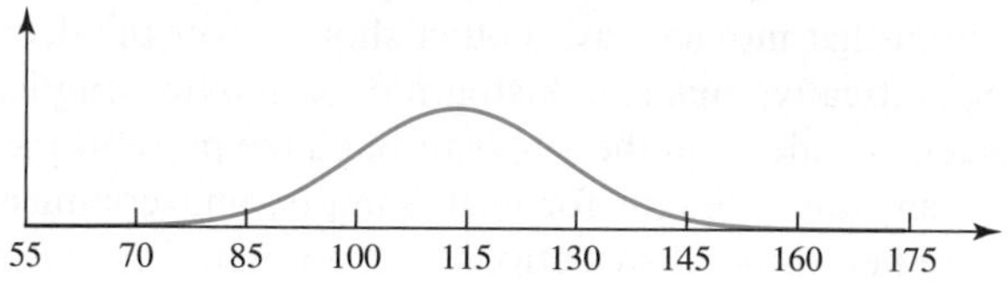

Figure 5.14

26. Estimating Areas. Consider the normal curve in Figure 5.15, which gives the relative frequencies in a distribution of body weights for a sample of male students.

a. What is the mean of the distribution?

b. Estimate (using area) the percentage of students whose weight is less than 140.

c. Estimate the percentage of students whose weight is greater than 170.

d. Estimate the percentage of students whose weight is between 140 and 160.

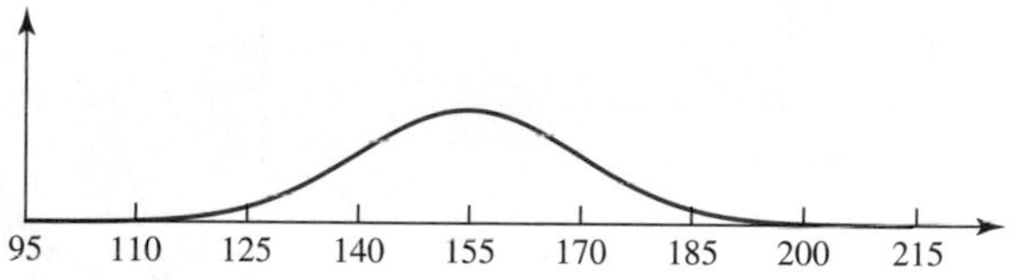

Figure 5.15

Projects for the Internet and Beyond

For useful links, select "Links for Internet Projects" for Chapter 5 at www.aw.com/bbt.

27. SAT Score Distributions. The College Board Web site gives the distribution of SAT scores (usually in 50-point bins). Collect these data and construct a histogram for each part of the test. Discuss the truth of the claim that SAT scores are normally distributed.

28. Finding Normal Distributions. Using the guidelines given in the text, choose a variable that you think should be nearly normally distributed. Collect at least 30 data values for the variable and make a histogram. Comment on how closely the distribution fits a normal distribution. In what ways does it differ from a normal distribution? Try to explain the differences.

29. **Movie Lengths.** Collect data to support or refute the claim that movies have gotten shorter over the decades. Specifically, make a histogram of movie lengths for each decade from the 1940s through the present, find the mean movie length for each sample, and comment on whether these distributions are normal. Discuss your results and give plausible reasons for any trends that you observe.

IN THE NEWS

30. **Normal Distributions.** Rarely does a news article refer to the actual distribution of a variable or state that a variable is normally distributed. Nevertheless, variables mentioned in news reports must have some distribution. Find two variables in news reports that you suspect have nearly normal distributions. Explain your reasoning.

31. **Non-normal Distributions.** Find two variables in news reports that you suspect *do not* have nearly normal distributions. Explain your reasoning.

5.2 Properties of the Normal Distribution

Consider a *Consumer Reports* survey in which participants were asked how long they owned their last TV set before they replaced it. The variable of interest in this survey is *replacement time for television sets.* Based on the survey, the distribution of replacement times has a mean of about 8.2 years, which we denote as μ (the Greek letter *mu*). The standard deviation of the distribution is about 1.1 years, which we denote as σ (the Greek letter *sigma*). Making the reasonable assumption that the distribution of TV replacement times is approximately normal, we can picture it as shown in Figure 5.16.

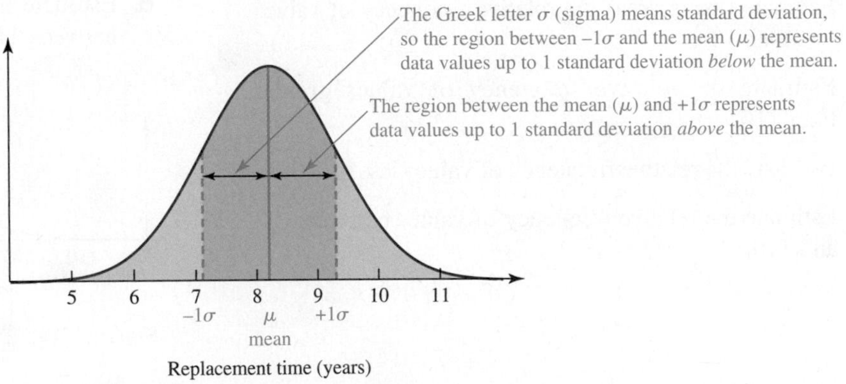
The Greek letter σ (sigma) means standard deviation, so the region between −1σ and the mean (μ) represents data values up to 1 standard deviation below the mean.
The region between the mean (μ) and +1σ represents data values up to 1 standard deviation above the mean.
5
6
7
−1σ
8
μ
mean
9
+1σ
10
11
Replacement time (years)

Figure 5.16 Normal distribution for replacement times for TV sets with a mean of μ = 8.2 years and a standard deviation of σ = 1.1 years.

TIME OUT TO THINK

Apply the four criteria for a normal distribution (see Section 5.1) to explain why the distribution of TV replacement times should be approximately normal.

TECHNICAL NOTE

A normal distribution can have any value for the mean and any positive value for the standard deviation. The term *standard normal distribution* specifically refers to a normal distribution with a mean of 0 and a standard deviation of 1.

Because all normal distributions have the same bell shape, knowing the mean and standard deviation of a distribution allows us to say a lot about where the data values lie. For example, if we measure areas under the curve in Figure 5.16, we find that about two-thirds of the area lies within 1 standard deviation of the mean, which in this case is between $8.2 - 1.1 = 7.1$ years and $8.2 + 1.1 = 9.3$ years. Therefore, the TV replacement time is between 7.1 and 9.3 years for about two-thirds of the people surveyed. Similarly, about 95% of the area lies within 2 standard deviations of the mean, which in this case is between 8.2 – 2.2 = 6.0 years and 8.2 + 2.2 = 10.4 years. We conclude that the TV replacement time is between 6.0 and 10.4 years for about 95% of the people surveyed.

A simple rule, called the **68-95-99.7 rule**, gives precise guidelines for the percentage of data values that lie within 1, 2, and 3 standard deviations of the mean for any normal distribution. The following box states the rule in words, and Figure 5.17 shows it visually.

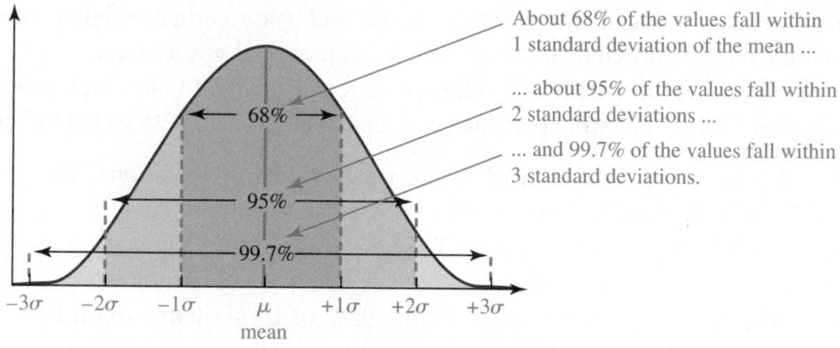
About 68% of the values fall within 1 standard deviation of the mean ...
... about 95% of the values fall within 2 standard deviations ...
... and 99.7% of the values fall within 3 standard deviations.
68%
95%
99.7%
-3σ -2σ -1σ μ $+1\sigma$ $+2\sigma$ $+3\sigma$
mean

Figure 5.17 Normal distribution illustrating the 68-95-99.7 rule.

The 68-95-99.7 Rule for a Normal Distribution

- About 68% (more precisely, 68.3%), or just over two-thirds, of the data points fall within 1 standard deviation of the mean.
- About 95% (more precisely, 95.4%) of the data points fall within 2 standard deviations of the mean.
- About 99.7% of the data points fall within 3 standard deviations of the mean.

EXAMPLE 1 SAT Scores

The tests that make up the verbal (critical reading) and mathematics SAT (and the GRE, LSAT, and GMAT) are designed so that their scores are normally distributed with a mean of $\mu = 500$ and a standard deviation of $\sigma = 100$. Interpret this statement.

Solution From the 68-95-99.7 rule, about 68% of students have scores within 1 standard deviation (100 points) of the mean of 500 points; that is, about 68% of students score between 400 and 600. About 95% of students score within 2 standard deviations (200 points) of the mean, or between 300 and 700. And about 99.7% of students score within 3 standard deviations (300 points) of the mean, or between 200 and 800. Figure 5.18 shows this

TECHNICAL NOTE

As discussed in the "Focus on Psychology" section at the end of the chapter, the mean score on a particular SAT may differ from 500 depending on the test and the year it is given.

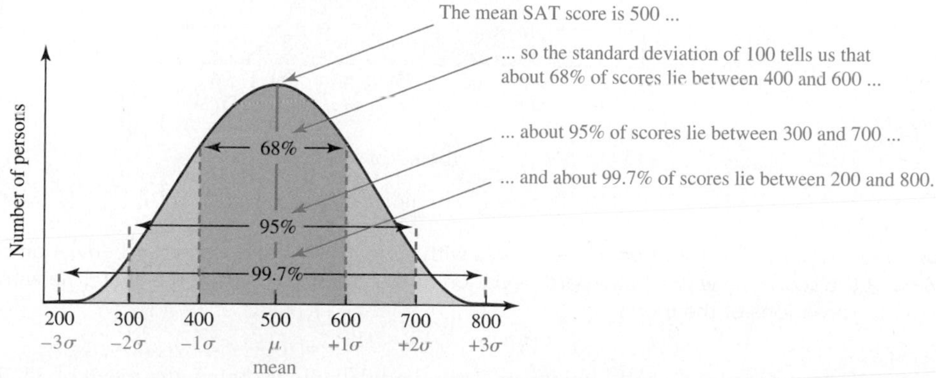
The mean SAT score is 500 ...
... so the standard deviation of 100 tells us that about 68% of scores lie between 400 and 600 ...
... about 95% of scores lie between 300 and 700 ...
... and about 99.7% of scores lie between 200 and 800.
Number of persons
68%
95%
99.7%
200 300 400 500 600 700 800
-3σ -2σ -1σ μ $+1\sigma$ $+2\sigma$ $+3\sigma$
mean

Figure 5.18 Normal distribution for SAT scores, showing the percentages associated with 1, 2, and 3 standard deviations.

interpretation graphically; note that the horizontal axis shows both actual scores and distance from the mean in standard deviations.

EXAMPLE 2 Detecting Counterfeits

Vending machines can be adjusted to reject coins above and below certain weights. The weights of legal U.S. quarters have a normal distribution with a mean of 5.67 grams and a standard deviation of 0.0700 gram. If a vending machine is adjusted to reject quarters that weigh more than 5.81 grams and less than 5.53 grams, what percentage of legal quarters will be rejected by the machine?

Solution A weight of 5.81 is 0.14 gram, or 2 standard deviations, above the mean. A weight of 5.53 is 0.14 gram, or 2 standard deviations, below the mean. Therefore, by accepting only quarters within the weight range 5.53 to 5.81 grams, the machine accepts quarters that are within 2 standard deviations of the mean and rejects those that are more than 2 standard deviations from the mean. By the 68-95-99.7 rule, 95% of legal quarters will be accepted and 5% of legal quarters will be rejected.

By the Way ...

The U.S. Mint makes almost a billion pennies per month. And yet, according to a report of the U.S. General Accounting Office, the circulation rate for pennies is estimated to be about 34% (compared to 88% for quarters). This means that about two-thirds of pennies produced make a one-way trip from the U.S. Mint to piggy banks or penny jars.

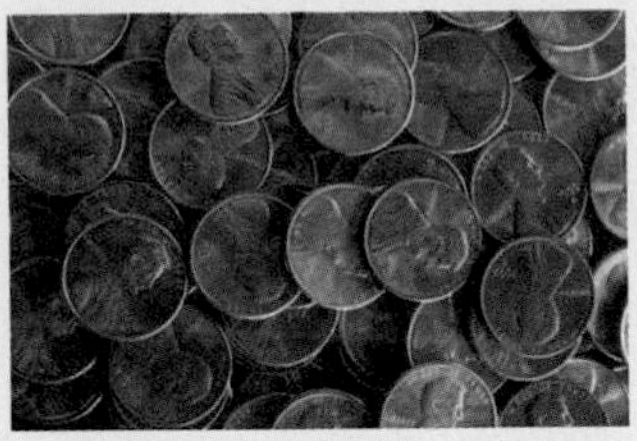

Applying the 68-95-99.7 Rule

We can apply the 68-95-99.7 rule to determine when data values lie 1, 2, or 3 standard deviations from the mean. For example, suppose that 1,000 students take an exam and the scores are normally distributed with a mean of $\mu = 75$ and a standard deviation of $\sigma = 7$.

A score of 82 is 7 points, or 1 standard deviation, above the mean of 75. The 68-95-99.7 rule tells us that about 68% of the scores are *within* 1 standard deviation of the mean. It follows that about 100% − 68% = 32% of the scores are *more than* 1 standard deviation from the mean. Half of this 32%, or 16%, of the scores are more than 1 standard deviation *below* the mean; the other 16% of the scores are more than 1 standard deviation *above* the mean (Figure 5.19a). Thus, about 16% of 1,000 students, or 160 students, scored above 82.

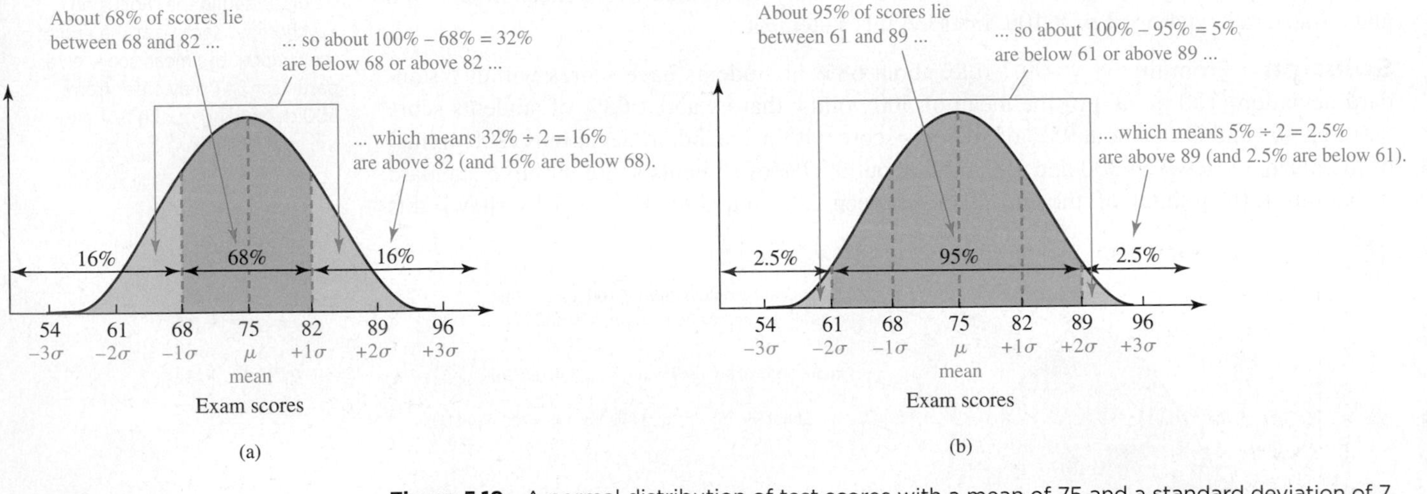

Figure 5.19 A normal distribution of test scores with a mean of 75 and a standard deviation of 7. (a) 68% of the scores lie within 1 standard deviation of the mean. (b) 95% of the scores lie within 2 standard deviations of the mean.

Similarly, a score of 61 is 14 points, or 2 standard deviations, below the mean of 75. The 68-95-99.7 rule tells us that about 95% of the scores are *within* 2 standard deviations of the mean, so about 5% of the scores are *more than* 2 standard deviations from the mean. Half of this 5%, or 2.5%, of the scores are more than 2 standard deviations *below* the mean (Figure 5.19b). Thus, about 2.5% of 1,000 students, or 25 students, scored below 61.

Because 95% of the scores fall between 61 and 89, we sometimes say the scores outside this range are unusual, since they are relatively rare.

Identifying Unusual Results

In statistics, we often need to distinguish values that are typical, or "usual," from values that are "unusual." By applying the 68-95-99.7 rule, we find that about 95% of all values from a normal distribution lie within 2 standard deviations of the mean. This implies that, among all values, 5% lie more than 2 standard deviations away from the mean. We can use this property to identify values that are relatively "unusual": **Unusual values** are values that are more than 2 standard deviations away from the mean.

EXAMPLE 3 Traveling and Pregnancy

Consider again the question of whether you should advise a pregnant friend to schedule an important business meeting two weeks before her due date. Actual data suggest that the number of days between the birth date and the due date is normally distributed with a mean of $\mu = 0$ days and a standard deviation of $\sigma = 15$ days. How would you help your friend make the decision? Would a birth two weeks before the due date be considered "unusual"?

Solution Your friend is assuming that she will *not* have given birth 14 days, or roughly 1 standard deviation, before her due date. But because this outcome is well within 2 standard deviations of the mean, it is *not* unusual. By the 68-95-99.7 rule, the day of birth for 68% of pregnancies is within 1 standard deviation of the mean, or between –15 days and 15 days from the due date. This means that about 100% – 68% = 32% of births occur either more than 15 days early or more than 15 days late. Therefore, half of this 32%, or 16%, of all births are more than 15 days early (see Figure 5.20). You should tell your friend that about 16% of all births occur more than 15 days before the due date. If your friend likes to think in terms of probability, you could say that there is a 0.16 (about 1 in 6) chance that she will give birth on or before her meeting date.

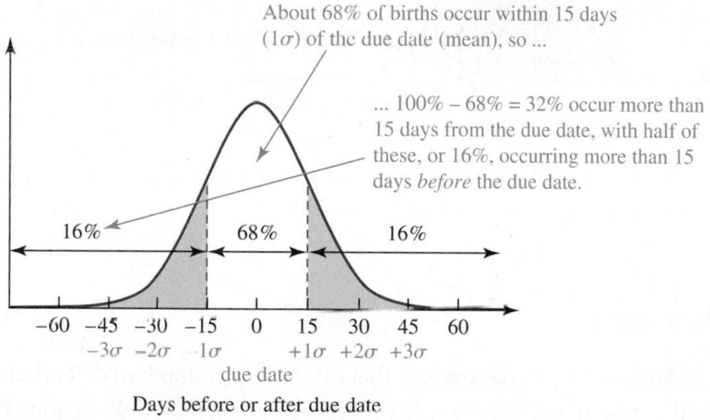

Figure 5.20 About 16% of births occur more than 15 days before the due date.

EXAMPLE 4 Normal Heart Rate

You measure your resting heart rate at noon every day for a year and record the data. You discover that the data have a normal distribution with a mean of 66 and a standard deviation of 4. On how many days was your heart rate below 58 beats per minute?

Solution A heart rate of 58 is 8 (or 2 standard deviations) below the mean. According to the 68-95-99.7 rule, about 95% of the data points are within 2 standard deviations of the mean. Therefore, 2.5% of the data points are more than 2 standard deviations *below* the mean, and 2.5% of the data points are more than 2 standard deviations *above* the mean. On 2.5% of 365 days, or about 9 days, your measured heart rate was below 58 beats per minute.

TIME OUT TO THINK

As Example 4 suggests, many measurements of the resting heart rate of a *single* individual are normally distributed. Would you expect the average resting heart rates of *many* individuals to be normally distributed? Which distribution would you expect to have the larger standard deviation? Why?

EXAMPLE 5 Finding a Percentile

On a visit to the doctor's office, your fourth-grade daughter is told that her height is 1 standard deviation above the mean for her age and sex. What is her percentile for height? Assume that heights of fourth-grade girls are normally distributed.

Solution Recall that a data value lies in the *n*th percentile of a distribution if $n\%$ of the data values are *less than or equal to* it (see Section 4.3). According to the 68-95-99.7 rule, 68% of the heights are within 1 standard deviation of the mean. Therefore, 34% of the heights (half of 68%) are between 0 and 1 standard deviation *above* the mean. We also know that, because the distribution is symmetric, 50% of all heights are below the mean. Therefore, $50\% + 34\% = 84\%$ of all heights are less than 1 standard deviation above the mean (Figure 5.21). Your daughter is in the 84th percentile for heights among fourth-grade girls.

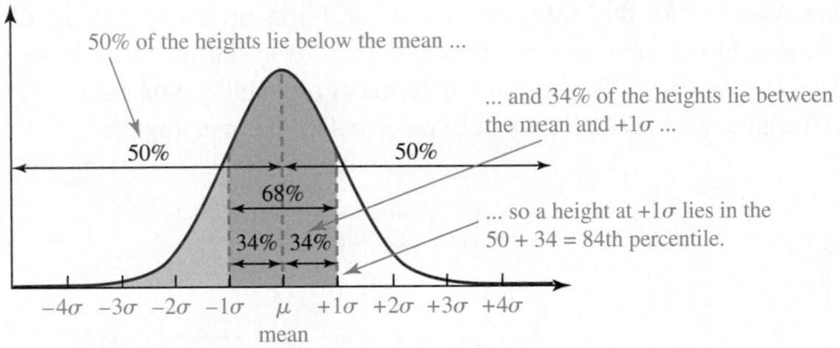

Figure 5.21 Normal distribution curve showing 84% of scores less than 1 standard deviation above the mean.

Standard Scores

The 68-95-99.7 rule applies only to data values that are 1, 2, or 3 standard deviations from the mean. We can generalize this rule if we know precisely how many standard deviations from the mean a particular data value lies. The number of standard deviations a data value lies above or below the mean is called its **standard score** (or *z-score*), often abbreviated by the letter z. For example:

- The standard score of the mean is $z = 0$, because it is 0 standard deviations from the mean.
- The standard score of a data value 1.5 standard deviations *above* the mean is $z = 1.5$.
- The standard score of a data value 2.4 standard deviations *below* the mean is $z = -2.4$.

The following box summarizes the computation of standard scores.

Computing Standard Scores

The number of standard deviations a data value lies above or below the mean is called its standard score (or z-score), defined by

$$z = \text{standard score} = \frac{\text{data value} - \text{mean}}{\text{standard deviation}}$$

The standard score is positive for data values above the mean and negative for data values below the mean.

EXAMPLE 6 Finding Standard Scores

The Stanford-Binet IQ test is scaled so that scores have a mean of 100 and a standard deviation of 16. Find the standard scores for IQs of 85, 100, and 125.

Solution We calculate the standard scores for these IQs by using the standard score formula with a mean of 100 and standard deviation of 16.

$$\textit{standard score for 85:} \quad z = \frac{85 - 100}{16} = -0.94$$

$$\textit{standard score for 100:} \quad z = \frac{100 - 100}{16} = 0.00$$

$$\textit{standard score for 125:} \quad z = \frac{125 - 100}{16} = 1.56$$

We can interpret these standard scores as follows: 85 is 0.94 standard deviation *below* the mean, 100 is equal to the mean, and 125 is 1.56 standard deviations *above* the mean. Figure 5.22 shows these values on the distribution of IQ scores.

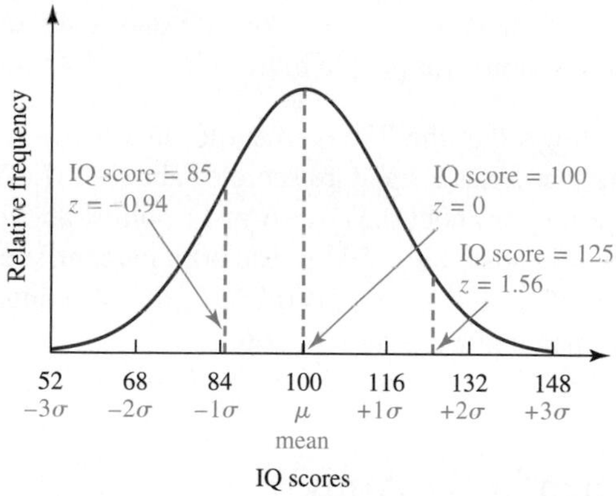

Figure 5.22 Standard scores for IQ scores of 85, 100, and 125.

Standard Scores and Percentiles

Once we know the standard score of a data value, the properties of the normal distribution allow us to find its *percentile* in the distribution. This is usually done with a *standard score table*, such as Table 5.1. (Appendix A has a more detailed standard score table.) For each of many standard scores in a normal distribution, the table gives the percentage of values in the distribution less than that value. For example, the table shows that 55.96% of the values in a normal distribution have a standard score less than 0.15. In other words, a data value with a standard score of 0.15 lies in the 55th percentile.

EXAMPLE 7 Cholesterol Levels

Cholesterol levels in men 18 to 24 years of age are normally distributed with a mean of 178 and a standard deviation of 41.

a. What is the percentile for a 20-year-old man with a cholesterol level of 190?

b. What cholesterol level corresponds to the 90th percentile, the level at which treatment may be necessary?

Solution

a. The *standard score* for a cholesterol level of 190 is

$$z = \text{standard score} = \frac{\text{data value} - \text{mean}}{\text{standard deviation}} = \frac{190 - 178}{41} \approx 0.29$$

Table 5.1 shows that a standard score of 0.29 corresponds to about the 61st percentile.

b. Table 5.1 shows that 90.32% of all data values have a standard score less than 1.3. Thus, the 90th percentile is about 1.3 standard deviations above the mean. Given the mean cholesterol level of 178 and the standard deviation of 41, a cholesterol level 1.3 standard deviations above the mean is

$$\underbrace{178}_{\text{mean}} + \underbrace{(1.3 \times 41)}_{\text{1.3 standard deviations}} = 231.3$$

A cholesterol level of about 231 corresponds to the 90th percentile.

EXAMPLE 8 IQ Scores

IQ scores are normally distributed with a mean of 100 and a standard deviation of 16 (see Example 6). What are the IQ scores for people in the 75th and 40th percentiles on IQ tests?

Solution Table 5.1 shows that the 75th percentile falls *between* standard scores of 0.65 and 0.70; we can estimate that it has a standard score of about 0.67. This corresponds to an IQ that is 0.67 standard deviation, or about $0.67 \times 16 = 11$ points, above the mean of 100. Thus, a person in the 75th percentile has an IQ of 111. The 40th percentile corresponds to a standard score of approximately −0.25, or a score that is $0.25 \times 16 = 4$ points *below* the mean of 100. Thus, a person in the 40th percentile has an IQ of 96.

EXAMPLE 9 Women in the Army

The heights of American women aged 18 to 24 are normally distributed with a mean of 65 inches and a standard deviation of 2.5 inches. In order to serve in the U.S. Army, women must be between 58 inches and 80 inches tall. What percentage of women are ineligible to serve based on their height?

Solution The standard scores for the army's minimum and maximum heights of 58 inches and 80 inches are

$$\textit{For 58 inches:} \quad z = \frac{58 - 65}{2.5} = -2.8$$

$$\textit{For 80 inches:} \quad z = \frac{80 - 65}{2.5} = 6.0$$

Table 5.1 Standard Scores and Percentiles for a Normal Distribution (cumulative values from the *left*)

Standard score	%	Standard score	%	Standard score	%	Standard score	%
−3.5	0.02	−1.0	15.87	0.0	50.00	1.1	86.43
−3.0	0.13	−0.95	17.11	0.05	51.99	1.2	88.49
−2.9	0.19	−0.90	18.41	0.10	53.98	1.3	90.32
−2.8	0.26	−0.85	19.77	0.15	55.96	1.4	91.92
−2.7	0.35	−0.80	21.19	0.20	57.93	1.5	93.32
−2.6	0.47	−0.75	22.66	0.25	59.87	1.6	94.52
−2.5	0.62	−0.70	24.20	0.30	61.79	1.7	95.54
−2.4	0.82	−0.65	25.78	0.35	63.68	1.8	96.41
−2.3	1.07	−0.60	27.43	0.40	65.54	1.9	97.13
−2.2	1.39	−0.55	29.12	0.45	67.36	2.0	97.72
−2.1	1.79	−0.50	30.85	0.50	69.15	2.1	98.21
−2.0	2.28	−0.45	32.64	0.55	70.88	2.2	98.61
−1.9	2.87	−0.40	34.46	0.60	72.57	2.3	98.93
−1.8	3.59	−0.35	36.32	0.65	74.22	2.4	99.18
−1.7	4.46	−0.30	38.21	0.70	75.80	2.5	99.38
−1.6	5.48	−0.25	40.13	0.75	77.34	2.6	99.53
−1.5	6.68	−0.20	42.07	0.80	78.81	2.7	99.65
−1.4	8.08	−0.15	44.04	0.85	80.23	2.8	99.74
−1.3	9.68	−0.10	46.02	0.90	81.59	2.9	99.81
−1.2	11.51	−0.05	48.01	0.95	82.89	3.0	99.87
−1.1	13.57	0.0	50.00	1.0	84.13	3.5	99.98

Note: The table shows percentiles for standard scores between −3.5 and +3.5, though much lower and higher standard scores are possible. (Appendix A has a more detailed standard score table.) The % column gives the percentage of values in the distribution less than the corresponding standard score.

Table 5.1 shows that a standard score of –2.8 corresponds to the 0.26 percentile. A standard score of 6.0 does not appear in Table 5.1, which means it is above the 99.98th percentile (the highest percentile shown in the table). Thus, 0.26% of all women are too short to serve in the army and fewer than 0.02% of all women are too tall to serve in the army. Altogether, fewer than about 0.28% of all women, or about 1 out of every 400 women, are ineligible to serve in the army based on their height.

Toward Probability

Suppose you pick a baby at random and ask whether the baby was born more than 15 days prior to his or her due date. Because births are normally distributed around the due date with a standard deviation of 15 days, we know that 16% of all births occur more than 15 days prior to the due date (see Example 3). For an individual baby chosen at random, we can therefore say that there's a 0.16 chance (about 1 in 6) that the baby was born more than 15 days early. In other words, the properties of the normal distribution allow us to make a *probability statement* about an individual. In this case, our statement is that the probability of a birth occurring more than 15 days early is 0.16.

This example shows that the properties of the normal distribution can be restated in terms of ideas of probability. In fact, much of the work we will do throughout the rest of this text is closely tied to ideas of probability. For this reason, we will devote the next chapter to studying fundamental ideas of probability. But first, in the next section, we will use the basic ideas we have discussed so far to introduce one of the most important concepts in statistics.

Section 5.2 Exercises

Statistical Literacy and Critical Thinking

1. **Standard Score.** What is the standard z-score for the mean in a normally distributed set of values?

2. **Standard Score.** The total percentage of values to the left of a particular z-score is 75%. What is the percentage of values to the right of that same z-score?

3. **Distributions.** For rolling a die, the mean outcome is 3.5. Can we apply the 68-95-99.7 rule and conclude that 95% of all outcomes fall within 2 standard deviations of 3.5? Why or why not?

4. ***z*-Scores and Percentages.** Table 5.1 includes standard scores and percentiles. Can a z-score be a negative value? Can a percentile be a negative value? Explain.

Does It Make Sense? For Exercises 5–8, decide whether the statement makes sense (or is clearly true) or does not make sense (or is clearly false). Explain clearly; not all of these statements have definitive answers, so your explanation is more important than your chosen answer.

5. **Psychology Test.** Scores on a psychology test are all positive, and they are normally distributed with a mean of 55 and a standard deviation of 65.

6. **Birth Weights.** Birth weights in the United States are normally distributed with a mean of 3,420 grams and a standard deviation of 495 grams.

7. **SAT Scores.** SAT scores are normally distributed with a mean of 1518 and a standard deviation of 325.

8. **SAT Scores.** SAT scores are normally distributed with a mean of 1518 and a standard deviation of 325, and Marc's score of 1840 is unusually high.

Concepts and Applications

9. **Using the 68-95-99.7 Rule.** Assume that a set of test scores is normally distributed with a mean of 100 and a standard deviation of 20. Use the 68-95-99.7 rule to find the following quantities.

   a. Percentage of scores less than 100

   b. Relative frequency of scores less than 120

   c. Percentage of scores less than 140

   d. Percentage of scores less than 80

   e. Relative frequency of scores less than 60

   f. Percentage of scores greater than 120

   g. Percentage of scores greater than 140

   h. Relative frequency of scores greater than 80

   i. Percentage of scores between 80 and 120

   j. Percentage of scores between 80 and 140

10. **Using the 68-95-99.7 Rule.** Assume the resting heart rates for a sample of individuals are normally distributed with a mean of 70 and a standard deviation of 15. Use the 68-95-99.7 rule to find the following quantities.

    a. Percentage of rates less than 70

    b. Percentage of rates less than 55

    c. Relative frequency of rates less than 40

    d. Percentage of rates less than 85

    e. Relative frequency of rates less than 100

    f. Percentage of rates greater than 85

    g. Percentage of rates greater than 55

    h. Relative frequency of rates greater than 40

    i. Percentage of rates between 55 and 85

    j. Percentage of rates between 70 and 100

11. **Applying the 68-95-99.7 Rule.** In a study of facial behavior, people in a control group are timed for eye contact in a 5-minute period. Their times are normally distributed with a mean of 184.0 seconds and a standard deviation of 55.0 seconds (based on data from "Ethological Study of Facial Behavior in Nonparanoid and Paranoid Schizophrenic Patients," by Pittman, Olk, Orr, and Singh, *Psychiatry*, Vol. 144, No. 1). Use the 68-95-99.7 rule to find the indicated quantity.

    a. Find the percentage of times within 55.0 seconds of the mean of 184.0 seconds.

    b. Find the percentage of times within 110.0 seconds of the mean of 184.0 seconds.

    c. Find the percentage of times within 165.0 seconds of the mean of 184.0 seconds.

    d. Find the percentage of times between 184.0 seconds and 294 seconds.

12. **Applying the 68-95-99.7 Rule.** When designing the placement of a CD player in a new model car, engineers must consider the forward grip reach of the driver. Women have forward grip reaches that are normally distributed with a mean of 27.0 inches and a standard deviation of

1.3 inches (based on anthropometric survey data from Gordon, Churchill, et al.). Use the 68-95-99.7 rule to find the indicated quantity.

a. Find the percentage of women with forward grip reaches between 24.4 inches and 29.6 inches

b. Find the percentage of women with forward grip reaches less than 30.9 inches

c. Find the percentage of women with forward grip reaches between 27.0 inches and 28.3 inches

IQ Scores. For Exercises 13–24, use the normal distribution of IQ scores, which has a mean of 100 and a standard deviation of 16. Use Table 5.1 (on page 211) to find the indicated quantities. Note: Table 5.1 shows standard scores from −3.5 to +3.5. In these problems, for standard scores above 3.5, use a percentile of 99.99%; for all standard scores below −3.5, use a percentile of 0.01%.

13. Percentage of scores less than 100

14. Percentage of scores less than 84

15. Percentage of scores greater than 116

16. Percentage of scores less than 76

17. Percentage of scores greater than 132

18. Percentage of scores less than 80

19. Percentage of scores less than 65

20. Percentage of scores less than 129

21. Percentage of scores between 84 and 116

22. Percentage of scores between 76 and 124

23. Percentage of scores between 80 and 120

24. Percentage of scores between 92 and 115

Heights of Women. For Exercises 25–36, use the normal distribution of heights of adult women, which has a mean of 162 centimeters and a standard deviation of 6 centimeters. Use Table 5.1 to find the indicated quantities. Note: Table 5.1 shows standard scores from −3.5 to +3.5. In these problems, for standard scores above 3.5, use a percentile of 99.99%; for all standard scores below −3.5, use a percentile of 0.01%.

25. The percentage of heights greater than 162 centimeters

26. The percentage of heights less than 168 centimeters

27. The percentage of heights greater than 156 centimeters

28. The percentage of heights greater than 171 centimeters

29. The percentage of heights less than 177 centimeters

30. The percentage of heights less than 147 centimeters

31. The percentage of heights less than 144 centimeters

32. The percentage of heights greater than 179 centimeters

33. The percentage of heights between 156 centimeters and 168 centimeters

34. The percentage of heights between 159 centimeters and 165 centimeters

35. The percentage of heights between 148 centimeters and 170 centimeters

36. The percentage of heights between 146 centimeters and 156 centimeters

37. Coin Weights. Consider the following table, showing the official mean weight and estimated standard deviation for five U.S. coins. Suppose a vending machine is designed to reject all coins with weights more than 2 standard deviations above or below the mean. For each coin, find the range of weights that are acceptable to the vending machine. In each case, what percentage of legal coins are rejected by the machine?

Coin	Weight (grams)	Estimated standard deviation (grams)
Cent	2.500	0.03
Nickel	5.000	0.06
Dime	2.268	0.03
Quarter	5.670	0.07
Half dollar	11.340	0.14

38. Pregnancy Lengths. Lengths of pregnancies are normally distributed with a mean of 268 days and a standard deviation of 15 days.

a. What is the percentage of pregnancies that last less than 250 days?

b. What is the percentage of pregnancies that last more than 300 days?

c. If a birth is considered premature if the pregnancy lasts less than 238 days, what is the percentage of premature births?

39. SAT Scores. Based on data from the College Board, SAT scores are normally distributed with a mean of 1518 and a standard deviation of 325.

a. Find the percentage of SAT scores greater than 2000.

b. Find the percentage of SAT scores less than 1500.

c. Find the percentage of SAT scores between 1600 and 2100.

40. GRE Scores. Assume that the scores on the Graduate Record Exam (GRE) are normally distributed with a mean of 497 and a standard deviation of 115.

a. A graduate school requires a GRE score of 650 for admission. To what percentile does this correspond?

b. A graduate school requires a GRE score in the 95th percentile for admission. To what actual score does this correspond?

41. Calibrating Barometers. Researchers for a manufacturer of barometers (devices to measure atmospheric pressure) read each of 50 barometers at the same time of day. The mean of the readings is 30.4 (inches of mercury) with a standard deviation of 0.23 inch, and the readings appear to be normally distributed.

a. What percentage of the barometers read over 31?

b. What percentage of the barometers read less than 30?

c. The company decides to reject barometers that read more than 1.5 standard deviations above or below the mean. What is the critical reading below which barometers will be rejected? What is the critical reading above which barometers will be rejected?

d. What would you take as the actual atmospheric pressure at the time the barometers were read? Explain.

42. Spelling Bee Scores. At the district spelling bee, the 60 girls have a mean score of 71 points with a standard deviation of 6, while the 50 boys have a mean score of 66 points with a standard deviation of 5 points. Those students with a score greater than 75 are eligible to go to the state spelling bee. What percentage of those going to the state bee will be girls?

43. Being a Marine. According to data from the National Health Survey, the heights of adult men are normally distributed with a mean of 69.0 inches and a standard deviation of 2.8 inches. The U.S. Marine Corps requires that men have heights between 64 inches and 78 inches. What percentage of American men are eligible for the Marines based on height?

44. Movie Lengths. Based on a random sample of movie lengths, the mean length is 110.5 minutes with a standard deviation of 22.4 minutes. Assume that movie lengths are normally distributed.

a. What fraction of movies are more than 2 hours long?

b. What fraction of movies are less than $1\frac{1}{2}$ hours long?

c. What is the probability that a randomly selected movie will be less than $2\frac{1}{2}$ hours long?

Projects for the Internet and Beyond

For useful links, select "Links for Internet Projects" for Chapter 5 at www.aw.com/bbt.

45. Data and Story Library. Visit the Data and Story Library Web site and find one example (of many) of a normal distribution. Write a one-page account of your findings. Include a careful description of the variable under consideration, details of the distribution, and an explanation of why you expect this variable to have an approximately normal distribution.

46. State Vital Statistics. The National Center for Health Statistics provides vital statistics by state (and county) under State Tabulated Data. Find a variable (for example, births, deaths, or infant deaths). Make a histogram showing this variable for each state. Comment on whether the variable is normally distributed across the states.

47. Normal Distribution Demonstrations on the Web. Do an Internet search on the keywords "normal distribution" and find an animated demonstration of the normal distribution. Describe how the demonstration works and the useful features that you observed.

48. Estimating a Minute. Ask survey subjects to estimate 1 minute without looking at a watch or clock. Each subject should say "go" at the beginning of the minute and then "stop" when he or she thinks that 1 minute has passed. (Alternatively, you could repeatedly time yourself by looking away from any watch or clock and then noting the correct time when you think that 1 minute has passed.) Use a watch to record the actual times. Construct a graph of the estimates. Is the graph approximately normal? Does the mean appear to be close to 1 minute?

5.3 The Central Limit Theorem

A high school English teacher has 100 seniors taking a college placement test. The test is designed to have a mean score of 500 and a standard deviation of 100. Using the methods of this chapter, she can determine the percentage of individual scores that are likely to be above, say, 600. But can she predict anything about the performance of her *group* of 100 students? For example, what is the likelihood that the mean score of the group will be above 600? This type of question, in which we ask about the mean score for a group or sample drawn from a much larger population, can be answered with the *Central Limit Theorem*.

Before we get to the theorem itself, we can develop some insight by thinking about dice rolling. Suppose we roll *one* die 1,000 times and record the outcome of each roll, which can be the number 1, 2, 3, 4, 5, or 6. Figure 5.23 shows a histogram of outcomes. All six outcomes have roughly the same relative frequency, because the die is equally likely to land in each of the six possible ways. That is, the histogram shows a (nearly) *uniform distribution* (see Section 4.2). Using the methods described in Chapter 4, we can compute the mean and standard deviation for this distribution. It turns out that the distribution in Figure 5.23 has a mean of 3.41 and a standard deviation of 1.73.

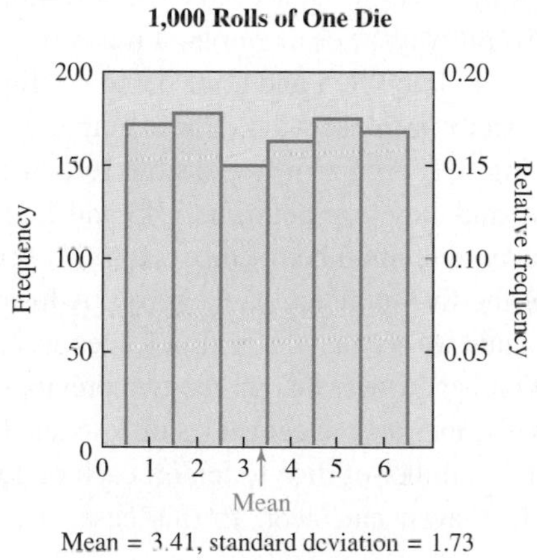

Figure 5.23 Frequency and relative frequency distribution of outcomes from rolling one die 1,000 times.

Now suppose we roll *two* dice 1,000 times and record the *mean* of the two numbers that appear on each roll (Figure 5.24). To find the mean for a single roll, we add the two numbers and divide by 2. For example, if the two dice come up 3 and 5, the mean for the roll is $(3 + 5)/2 = 4$. The possible values of the mean on a roll of two dice are 1.0, 1.5, 2.0, . . . , 5.0, 5.5, 6.0.

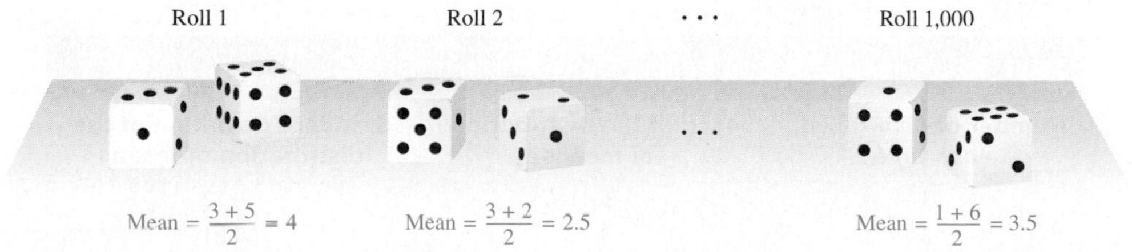

Figure 5.24 This diagram represents the idea of rolling two dice 1,000 times and recording the *mean* on each roll. The mean of the values on two dice is their sum divided by 2. This mean can range from $(1 + 1)/2 = 1$ to $(6 + 6)/2 = 6$.

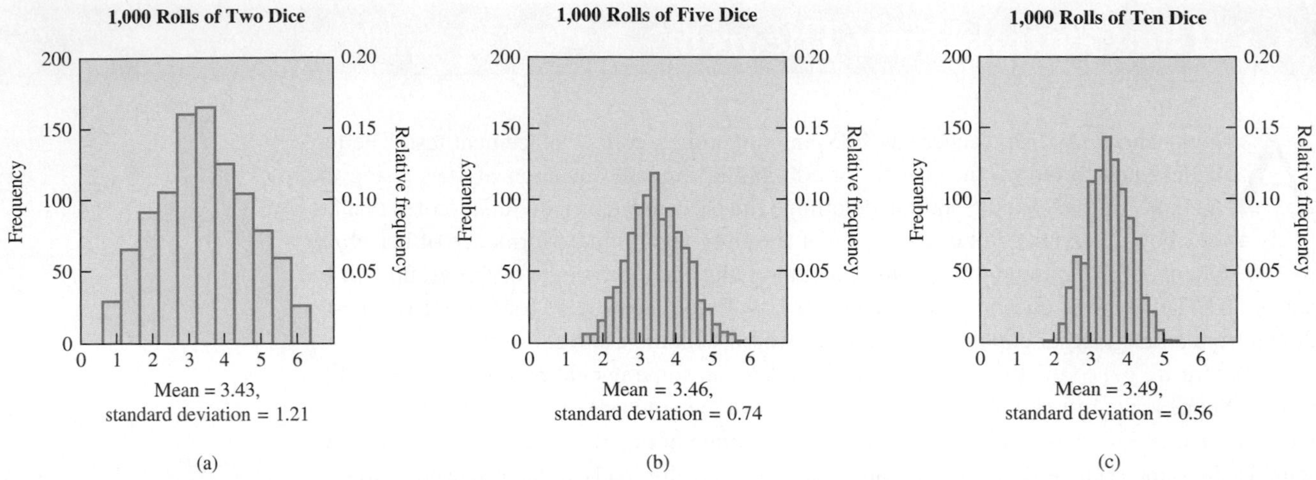

Figure 5.25 Frequency and relative frequency distributions of sample means from rolling (a) two dice 1,000 times, (b) five dice 1,000 times, and (c) ten dice 1,000 times.

Figure 5.25a shows a typical result of rolling two dice 1,000 times. The most common values in this distribution are the central values 3.0, 3.5, and 4.0. These values are common because they can occur in several ways. For example, a mean of 3.5 can occur if the two dice land as 1 and 6, 2 and 5, 3 and 4, 4 and 3, 5 and 2, or 6 and 1. High and low values occur less frequently because they can occur in fewer ways. For example, a roll can have a mean of 1.0 only if both dice land showing a 1. Again, it is possible to compute the mean and standard deviation for this distribution, and they turn out to be 3.43 and 1.21, respectively.

What happens if we increase the number of dice we roll? Suppose we roll five dice 1,000 times and record the mean of the five numbers on each roll. A histogram for this experiment is shown in Figure 5.25b. Once again we see that the central values around 3.5 occur most frequently, but the spread of the distribution is narrower than in the two previous cases. Computing the mean and standard deviation of this distribution, we get values of 3.46 and 0.74, respectively.

If we further increase the number of dice to ten on each of 1,000 rolls, we find the histogram in Figure 5.25c, which is even narrower. In this case, the mean is 3.49 and standard deviation is 0.56.

Table 5.2 summarizes the four experiments we've described. Columns 2 and 3 of the table refer to a *distribution of means* because, in each of the dice rolling experiments, we recorded the mean of each of 1,000 rolls (that is, the mean of one die on each roll, of two dice on each roll, of five dice on each roll, or of ten dice on each roll). Thus, the mean for all 1,000 rolls in an experiment is *a mean of the distribution of means* (Column 2). Similarly, the standard deviation for all 1,000 rolls in an experiment is a *standard deviation of the distribution of means* (Column 3).

Table 5.2 Summary of Dice Rolling Experiments

Number of dice rolled each time	Mean of the distribution of means	Standard deviation of the distribution of means
1	3.41	1.73
2	3.43	1.21
5	3.46	0.74
10	3.49	0.56

A remarkable insight emerges from these four experiments. Rolling $n = 1$ die 1,000 times can be regarded as taking 1,000 samples of size $n = 1$ from the population of all possible dice rolls. Rolling $n = 2$ dice 1,000 times can be viewed as taking 1,000 samples of size $n = 2$. Likewise, rolling $n = 5$ and $n = 10$ dice 1,000 times is like taking 1,000 samples of size $n = 5$ and $n = 10$, respectively. Table 5.2 shows that as the sample size increases, the mean of the distribution of means approaches the value 3.5 and the standard deviation becomes smaller (making the distribution narrower). More important, the distribution looks more and more like a normal distribution as the sample size increases. This latter fact may seem surprising because we have taken samples from a *uniform* distribution (the outcomes of rolling a single die shown in Figure 5.23), *not* from a normal distribution. Nevertheless, the distribution of means clearly approaches a normal distribution for large sample sizes. This fact is a consequence of the **Central Limit Theorem**.

The Central Limit Theorem

Suppose we take many random samples of size n for a variable with any distribution (not necessarily a normal distribution) and record the distribution of the *means* of each sample. Then,

1. The distribution of means will be approximately a normal distribution for large sample sizes.
2. The mean of the distribution of means approaches the population mean, μ, for large sample sizes.
3. The standard deviation of the distribution of means approaches $\sigma/\sqrt{n}$ for large sample sizes, where σ is the standard deviation of the population.

TECHNICAL NOTE

(1) For practical purposes, the distribution of means will be nearly normal if the sample size is larger than 30. (2) If the original population is normally distributed, then the sample means will be normally distributed for *any* sample size n. (3) In the ideal case, where the distribution of means is formed from *all* possible samples, the mean of the distribution of means *equals* μ and the standard deviation of the distribution of means *equals* $\sigma/\sqrt{n}$.

Be sure to note the very important adjustment, described by item 3 above, that must be made when working with samples or groups instead of individuals:

The standard deviation of the distribution of sample means is not the standard deviation of the population, σ, but rather $\sigma/\sqrt{n}$, where n is the size of the samples.

TIME OUT TO THINK

Confirm that the standard deviations of the distributions of means given in Table 5.2 for $n = 2, 5, 10$ agree with the prediction of the Central Limit Theorem, given that $\sigma = 1.73$ (the population standard deviation found in Figure 5.23). For example, with $n = 2$, $\sigma/\sqrt{2} = 1.22 \approx 1.21$.

Let's summarize the ingredients of the Central Limit Theorem. We always start with a particular variable, such as the outcomes of rolling a die or the weights of people, that varies randomly over a population. The variable has a certain mean, μ, and standard deviation, σ, which we may or may not know. This variable can have *any* sort of distribution, not necessarily normal. Now we take many samples of that variable, with n items in each sample, and find the mean of each sample (such as the mean value of n dice or the mean weight of a sample of n people). If we then make a histogram of the means from the many samples, we will see a distribution that is close to a normal distribution. The larger the sample size, n, the more closely the distribution of means approximates a normal distribution. Careful study of Figure 5.26 should help solidify these important ideas.

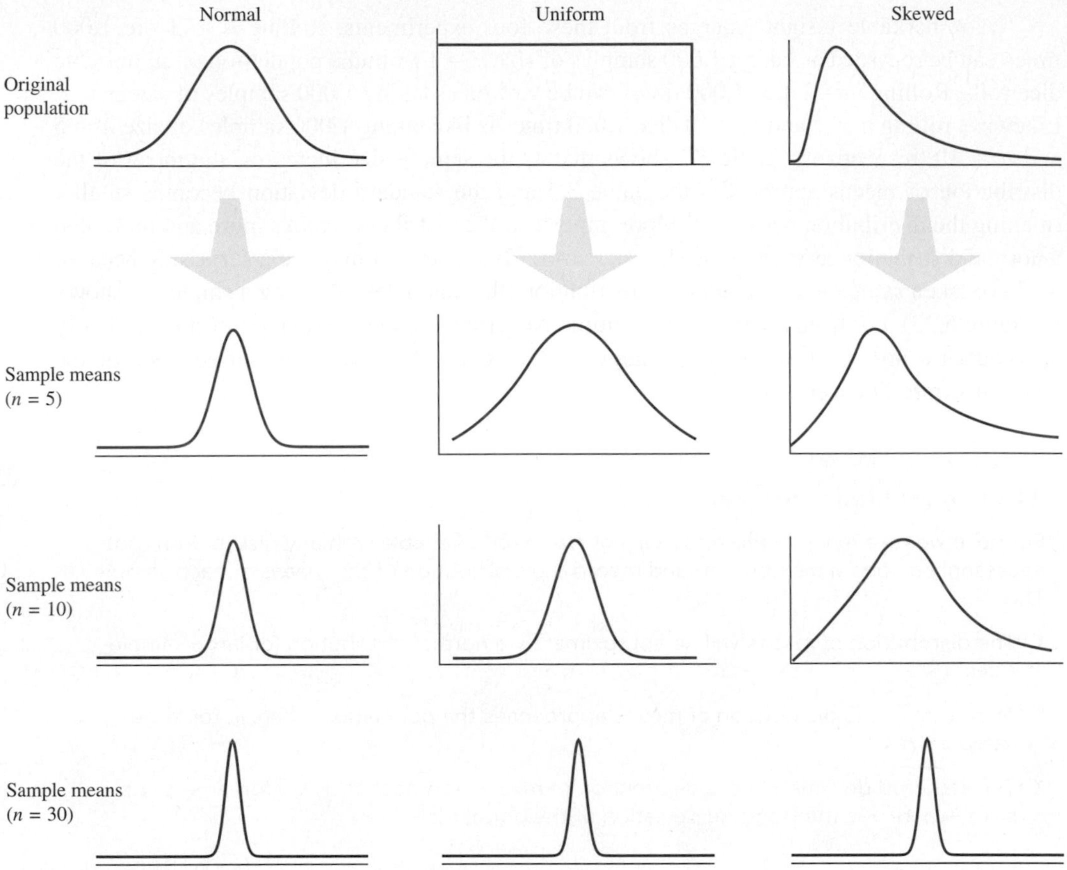

Figure 5.26 As the sample size increases (n = 5, 10, 30), the distribution of sample means approaches a normal distribution, regardless of the shape of the original distribution. The larger the sample size, the smaller is the standard deviation of the distribution of sample means.

EXAMPLE 1 Predicting Test Scores

You are a middle school principal and your 100 eighth-graders are about to take a national standardized test. The test is designed so that the mean score is $\mu = 400$ with a standard deviation of $\sigma = 70$. Assume the scores are normally distributed.

a. What is the likelihood that *one* of your eighth-graders, selected at random, will score below 375 on the exam?

b. Your performance as a principal depends on how well your entire *group* of eighth-graders scores on the exam. What is the likelihood that your group of 100 eighth-graders will have a *mean* score below 375?

Solution

a. In dealing with an individual score, we use the method of standard scores discussed in Section 5.2. Given the mean of 400 and standard deviation of 70, a score of 375 has a standard score of

$$z = \frac{\text{data value} - \text{mean}}{\text{standard deviation}} = \frac{375 - 400}{70} = -0.36$$

According to Table 5.1, a standard score of –0.36 corresponds to about the 36th percentile—that is, 36% of all students can be expected to score below 375. Thus, there is about a

0.36 chance that a randomly selected student will score below 375. Notice that we need to know that the scores have a normal distribution in order to make this calculation, because the table of standard scores applies only to normal distributions.

b. The question about the mean of a *group* of students must be handled with the Central Limit Theorem. According to this theorem, if we take random samples of size $n = 100$ students and compute the mean test score of each group, the distribution of means is approximately normal. Moreover, the mean of this distribution is $\mu = 400$ and its standard deviation is $\sigma/\sqrt{n} = 70/\sqrt{100} = 7$. With these values for the mean and standard deviation, the standard score for a mean test score of 375 is

$$z = \frac{\text{data value} - \text{mean}}{\text{standard deviation}} = \frac{375 - 400}{7} = -3.57$$

Table 5.1 shows that a standard score of −3.5 corresponds to the 0.02th percentile, and the standard score in this case is even lower. In other words, fewer than 0.02% of all random samples of 100 students will have a mean score of less than 375. Therefore, the chance that a randomly selected group of 100 students will have a mean score below 375 is less than 0.0002, or about 1 in 5,000. Notice that this calculation regarding the group mean did *not* depend on the individual scores' having a normal distribution.

This example has an important lesson. The likelihood of an *individual* scoring below 375 is more than 1 in 3 (36%), but the likelihood of a *group* of 100 students having a mean score below 375 is less than 1 in 5,000 (0.02%). In other words, there is much more variation in the scores of individuals than in the means of groups of individuals.

When you are listening to corn pop, are you hearing the Central Limit Theorem?

—William A. Massey

EXAMPLE 2 Salary Equity

The mean salary of the 9,000 employees at Holley.com is $\mu = \$26{,}400$ with a standard deviation of $\sigma = \$2{,}420$. A pollster samples 400 randomly selected employees and finds that the mean salary of the sample is \$26,650. Is it likely that the pollster would get these results by chance, or does the discrepancy suggest that the pollster's results are suspect?

Solution The question deals with the mean of a *group* of 400 individuals, which is a case for the Central Limit Theorem. The theorem tells us that if we select many groups of 400 individuals and compute the mean of each group, the distribution of means will be close to normal with a mean of $\mu = \$26{,}400$ and a standard deviation of $\sigma/\sqrt{n} = \$2{,}420/\sqrt{400} = \$121$. Within the distribution of means, a mean salary of \$26,650 has a *standard score* of

$$z = \frac{\text{data value} - \text{mean}}{\text{standard deviation}} = \frac{\$26{,}650 - \$26{,}400}{\$121} = 2.07$$

In other words, if we assume that the sample is randomly selected, its mean salary is more than 2 standard deviations above the mean salary of the entire company. According to Table 5.1, a standard score of 2.07 lies near the 98th percentile. Thus, the mean salary of *this* sample is greater than the mean salary we would find in 98% of the possible samples of 400 workers. That is, the likelihood of selecting a group of 400 workers with a mean salary above \$26,650 is about 2%, or 0.02. The mean salary of the sample is surprisingly high; perhaps the survey was flawed.

TIME OUT TO THINK

Would a salary of \$26,650 for an *individual* worker lie above or below the 98th percentile? Explain.

The Value of the Central Limit Theorem

The Central Limit Theorem allows us to say something about the mean of a group if we know the mean, μ, and the standard deviation, σ, of the entire population. This can be useful, but it turns out that the opposite application is far more important.

Two major activities of statistics are making estimates of population means and testing claims about population means. Suppose we do *not* know the mean of a variable for the entire population. Is it possible to make a good estimate of the population mean (such as the mean income of all Internet users) knowing only the mean of a much smaller sample? As you can probably guess, being able to answer this type of question lies at the heart of statistical sampling, especially in polls and surveys. The Central Limit Theorem provides the key to answering such questions. We will return to this topic in Chapter 8.

Section 5.3 Exercises

Statistical Literacy and Critical Thinking

1. **Convenience Sample.** Because a statistics student waited until the last minute to do a project, she has only enough time to collect heights from female friends and female relatives. She then calculates the mean height of the females in her sample. Assuming that females have heights that are normally distributed with a mean of 63.6 inches and a standard deviation of 2.5 inches, can she use the Central Limit Theorem when analyzing the mean height of her sample?

2. **Notation.** In this section, it was noted that the standard deviation of sample means is $\sigma/\sqrt{n}$. In that expression, what does σ represent and what does n represent?

3. **Central Limit Theorem.** A process consists of repeating this operation: Randomly select two values from some population and then find the mean of the two values. Does the Central Limit Theorem suggest that the sample means will be normally distributed? Why or why not?

4. **Central Limit Theorem.** A process consists of repeating this operation: Randomly select four values from a population with a normal distribution and a standard deviation σ. What is the standard deviation of the sample means?

Concepts and Applications

5. **IQ Scores and the Central Limit Theorem.** IQ scores are normally distributed with a mean of 100 and a standard deviation of 16. Assume that many samples of size n are taken from a large population of people and the mean IQ score is computed for each sample.

   a. If the sample size is $n = 64$, find the mean and standard deviation of the distribution of sample means.

   b. If the sample size is $n = 100$, find the mean and standard deviation of the distribution of sample means.

   c. Why is the standard deviation in part a different from the standard deviation in part b?

6. **SAT Scores and the Central Limit Theorem.** Based on data from the College Board, assume that SAT scores are normally distributed with a mean of 1518 and a standard deviation of 325. Assume that many samples of size n are taken from a large population of students and the mean SAT score is computed for each sample.

   a. If the sample size is $n = 100$, find the mean and standard deviation of the distribution of sample means.

   b. If the sample size is $n = 2{,}500$, find the mean and standard deviation of the distribution of sample means.

   c. Why is the standard deviation in part a different from the standard deviation in part b?

7. **Twelve-Sided Dice and the Central Limit Theorem.** Rolling a fair *twelve-sided* die produces a uniformly distributed set of numbers between 1 and 12 with a mean of 6.5 and a standard deviation of 3.452. Assume that n twelve-sided dice are rolled many times and the mean of the n outcomes is computed each time.

   a. Find the mean and standard deviation of the resulting distribution of sample means for $n = 81$.

   b. Find the mean and standard deviation of the resulting distribution of sample means for $n = 100$.

   c. Why is the standard deviation in part a different from the standard deviation in part b?

8. **Ten-Sided Dice and the Central Limit Theorem.** Rolling a fair *ten-sided* die produces a uniformly distributed set of numbers between 1 and 10 with a mean of 5.5 and a standard deviation of 2.872. Assume that n ten-sided dice are rolled many times and the mean of the n outcomes is computed each time.

   a. Find the mean and standard deviation of the resulting distribution of sample means for $n = 49$.

   b. Find the mean and standard deviation of the resulting distribution of sample means for $n = 400$.

   c. Why is the standard deviation in part a different from the standard deviation in part b?

Aircraft Ages. In Exercises 9–12, assume that the ages of commercial aircraft are normally distributed with a mean of 13.0 years and a standard deviation of 7.9 years (data from the Aviation Data Services).

9. What percentage of individual aircraft have ages greater than 15 years? Assume that a random sample of 49 aircraft is selected and the mean age of the sample is computed. What percentage of sample means have ages greater than 15 years?

10. What percentage of individual aircraft have ages less than 10 years? Assume that a random sample of 84 aircraft is selected and the mean age of the sample is computed. What percentage of sample means have ages less than 10 years?

11. What percentage of individual aircraft have ages between 10 years and 16 years? Assume that a random sample of 81 aircraft is selected and the mean age of the sample is computed. What percentage of sample means are between 10 years and 16 years?

12. What percentage of individual aircraft have ages between 12.5 years and 13.5 years? Assume that a random sample of 400 aircraft is selected and the mean age of the sample is computed. What percentage of sample means are between 12.5 years and 13.5 years?

13. **Amounts of Cola.** Assume that cans of cola are filled so that the actual amounts are normally distributed with a mean of 12.00 ounces and a standard deviation of 0.11 ounce.

   a. What is the likelihood that a sample of 36 cans will have a mean amount of at least 12.05 ounces?

   b. Given the result in part a, is it reasonable to believe that the cans are actually filled with a mean of 12.00 ounces? If the mean is not 12.00 ounces, are consumers being cheated?

14. **Designing Strobe Lights.** An aircraft strobe light is designed so that the times between flashes are normally distributed with a mean of 3.00 seconds and a standard deviation of 0.40 second.

   a. What is the likelihood that an individual time is greater than 4.00 seconds?

   b. What is the likelihood that the mean for 60 randomly selected times is greater than 4.00 seconds?

   c. Given that the strobe light is intended to help other pilots see an aircraft, which result is more relevant for assessing the safety of the strobe light: the result in part a or the result in part b? Why?

15. **Designing Motorcycle Helmets.** Engineers must consider the breadths of male heads when designing motorcycle helmets for men. Men have head breadths that are normally distributed with a mean of 6.0 inches and a standard deviation of 1.0 inch (based on anthropometric survey data from Gordon, Churchill, et al.).

   a. If one male is randomly selected, what is the likelihood that his head breadth is less than 6.2 inches?

   b. The Safeguard Helmet company plans an initial production run of 100 helmets. How likely is it that 100 randomly selected men have a mean head breadth of less than 6.2 inches?

   c. The production manager sees the result in part b and reasons that all helmets should be made for men with head breadths of less than 6.2 inches, because they would fit all but a few men. What is wrong with that reasoning?

16. **Staying Out of Hot Water.** In planning for hot water requirements, the manager of the Luxurion Hotel finds that guests spend a mean of 11.4 minutes each day in the shower (based on data from the Opinion Research Corporation). Assume that the shower times are normally distributed with a standard deviation of 2.7 minutes.

   a. Find the percentage of guests who shower for more than 12 minutes.

   b. The hotel has installed a system that can provide enough hot water provided that the mean shower time for 84 guests is less than 12 minutes. If the hotel currently has 84 guests, how likely is it that there will not be enough hot water? Does the current system appear to be effective?

17. **Redesign of Ejection Seats.** When women were allowed to become pilots of fighter jets, engineers needed to redesign the ejection seats because they had been designed for men only. The ACES-II ejection seats were designed for men weighing between 140 pounds and 211 pounds. The population of women has normally

distributed weights with a mean of 143 pounds and a standard deviation of 29 pounds (based on data from the National Health Survey).

a. What percentage of women have weights between 140 pounds and 211 pounds?

b. If 36 women are randomly selected, how likely is it that their mean weight is between 140 pounds and 211 pounds?

c. For redesigning the fighter jet ejection seats to better accommodate women, which probability is more relevant: the result in part a or the result in part b? Why?

18. Labeling of M&M Packages. M&M plain candies have weights that are normally distributed with a mean weight of 0.8565 gram and a standard deviation of 0.0518 gram (based on measurements from one of the authors). A random sample of 100 M&M candies is obtained from a package containing 465 candies; the package label states that the net weight is 396.9 grams. (If every package has 465 candies, the mean weight of the candies must exceed 396.9/465 = 0.8535 for the net contents to weigh at least 396.9 grams.)

a. If 1 M&M plain candy is randomly selected, how likely is it that it weighs more than 0.8535 gram?

b. If 465 M&M plain candies are randomly selected, how likely is it that their mean weight is at least 0.8535 gram?

c. Given these results, does it seem that the Mars Company is providing M&M consumers with the amount claimed on the label?

19. Vending Machines. Currently, quarters have weights that are normally distributed with a mean of 5.670 grams and a standard deviation of 0.062 gram. A vending machine is configured to accept only those quarters with weights between 5.550 grams and 5.790 grams.

a. If 280 different quarters are inserted into the vending machine, what is the expected number of rejected quarters?

b. If 280 different quarters are inserted into the vending machine, how likely is it that the mean falls between the limits of 5.550 grams and 5.790 grams?

c. If you owned the vending machine, which result would concern you more: the result in part a or the result in part b? Why?

20. Aircraft Safety Standards. Under old Federal Aviation Administration rules, airlines had to estimate the weight of a passenger as 185 pounds. (That amount is for an adult traveling in winter, and it includes 20 pounds of carry-on baggage.) Current rules require an estimate of 195 pounds. Men have weights that are normally distributed with a mean of 172 pounds and a standard deviation of 29 pounds.

a. If one adult male is randomly selected and is assumed to have 20 pounds of carry-on baggage, how likely is it that his total weight is greater than 195 pounds?

b. If a Boeing 767-300 aircraft is full of 213 adult male passengers and each is assumed to have 20 pounds of carry-on baggage, how likely is it that the mean passenger weight (including carry-on baggage) is greater than 195 pounds? Does a pilot have to be concerned about exceeding this weight limit?

21. Generic Variable. Suppose a variable, such as height or blood pressure, is normally distributed across a large population with a mean of μ and a standard deviation of σ.

a. Suppose we take many samples of size $n = 100$ and compute the mean of the variable for each sample. How does the standard deviation of the distribution of sample means compare to σ? Explain why.

b. Suppose we take many samples of size $n = 1{,}000$ and compute the mean of the variable for each sample. How does the standard deviation of this distribution of sample means compare to the standard deviation of the distribution of sample means in part a? How does the standard deviation of this distribution of sample means compare to σ? Explain your answers.

c. How does the standard deviation of the distribution of sample means change as we take larger and larger sample sizes?

22. Blood Pressure in Women. Systolic blood pressure for women between the ages of 18 and 24 is normally distributed with a mean of 114.8 (millimeters of mercury) and a standard deviation of 13.1.

a. What is the likelihood that an individual woman has a blood pressure above 125?

b. Suppose a random sample of $n = 300$ women is selected and the mean blood pressure for the sample is computed. What is the likelihood that the mean blood pressure for the sample will be above 125?

c. Suppose a random sample of $n = 300$ women is selected and the mean blood pressure for the sample is computed. What is the likelihood that the mean blood pressure for the sample will be below 114?

Projects for the Internet and Beyond

For useful links, select "Links for Internet Projects" for Chapter 5 at www.aw.com/bbt.

23. Central Limit Theorem on the Internet. Doing an Internet search on "central limit theorem" will uncover many sites devoted to this subject. Find a site that has animated demonstrations of the Central Limit Theorem. Describe in your own words what you observed and how it illustrates the Central Limit Theorem.

24. The Quincunx on the Internet. Do an Internet search on "central limit theorem" or "quincunx" and find a site that has an animated demonstration of the quincunx (or Galton's board). Describe the quincunx and explain how it illustrates the Central Limit Theorem.

25. Dice Rolling. Demonstrate the Central Limit Theorem using dice, as discussed in this section. Give each person in your class as many dice as possible. Begin by rolling one die and making a histogram of the outcomes. Then let every person roll two dice and make a histogram of the mean for each roll. Increase the number of dice in each roll as long as dice and time allow. Comment on the appearance of the histogram at each stage.

Chapter Review Exercises

1. For each of the following situations, state whether the distribution of values is likely to be a normal distribution. Give a brief explanation justifying your choice.

a. Numbers resulting from spins of a roulette wheel. (There are 38 equally likely slots with numbers 0, 00, 1, 2, 3, . . . , 36.)

b. Weights of adult Golden Retriever dogs

c. Measured braking reaction times of 18-year-old drivers

2. Test scores on the ACT exam are normally distributed with a mean of 21.1 and a standard deviation of 4.8.

a. Using the 68-95-99.7 rule, find the percentage of ACT scores within 9.6 of the mean of 21.1.

b. Using the 68-95-99.7 rule, find the percentage of ACT scores within 14.4 of the mean of 21.1.

c. Is a score of 35.6 unusual? Why or why not?

3. Assume that body temperatures of healthy adults are normally distributed with a mean of 98.20°F and a standard deviation of 0.62°F (based on data from University of Maryland researchers).

a. If you have a body temperature of 99.00°F, what is your percentile score?

b. Convert 99.00°F to a standard score (or *z*-score).

c. Is a body temperature of 99.00°F "unusual"? Why or why not?

d. Fifty adults are randomly selected. What is the likelihood that the mean of their body temperatures is 97.98°F or lower?

e. A person's body temperature is found to be 101.00°F. Is this result "unusual"? Why or why not? What should you conclude?

f. What body temperature is the 95th percentile?

g. What body temperature is the 5th percentile?

h. Bellevue Hospital in New York City uses 100.6°F as the lowest temperature considered to indicate a fever. What percentage of normal and healthy adults would be considered to have a fever? Does this percentage suggest that a cutoff of 100.6°F is appropriate?

i. If, instead of assuming that the mean body temperature is 98.20°F, we assume that the mean is 98.60°F (as many people believe), what is the chance of randomly selecting 106 people and getting a mean of 98.20°F or lower? (Continue to assume that the standard deviation is 0.62°F.) University of Maryland researchers did get such a result. What should we conclude?

Chapter Quiz

1. Which of the following statements are correct?

a. A normal distribution is any distribution that is not unusual.

b. The graph of a normal distribution is bell-shaped.

c. If a population has a normal distribution, the mean and median are not equal.

d. In a normal distribution, it is possible that the standard deviation is 0.

e. The graph of a normal distribution is symmetric.

2. A test of perception is designed so that scores are normally distributed with a mean of 50 and a standard deviation of 10. Using the 68-95-99.7 rule, find the percentage of scores within 20 points of the mean of 50.

3. A population is normally distributed with a mean of 50, and samples of size 100 are randomly selected. What is the mean of the sample means?

4. A population is normally distributed with a standard deviation of 5, and samples of size 100 are randomly selected. What is the standard deviation of the sample means?

5. A population has a normal distribution with a mean of 50 and a standard deviation of 10. What is the standard *z*-score corresponding to 70?

6. A population has a normal distribution with a mean of 50 and a standard deviation of 10. What is the standard *z*-score corresponding to 40?

7. A population of test scores has a normal distribution with a mean of 50 and a standard deviation of 10. What percentage of scores are greater than 50?

8. A population of test scores has a normal distribution with a mean of 50 and a standard deviation of 10. If 97.72% of the scores are less than 70, what percentage of scores are greater than 70?

9. A population of test scores has a normal distribution with a mean of 50 and a standard deviation of 10. If 97.72% of the scores are less than 70, what percentage of scores are less than 30?

10. Which of the following is likely to have a distribution that is closest to a normal distribution?

a. The outcomes that occur when a single die is rolled many times

b. The outcomes that occur when two dice are rolled many times and the mean is computed each time

c. The outcomes that occur when five dice are rolled many times and the mean is computed each time

Using Technology

Table 5.1 includes percentiles corresponding to limited values of standard *z*-scores, but statistics software packages typically allow the use of any standard *z*-scores. Table 5.1 lists percentiles based on cumulative values from the left; statistics software packages typically have that same format, but the software packages typically use areas instead of percentiles. For example, Table 5.1 shows the cumulative left percentile of 97.72 for $z = 2.0$, but software packages will yield a cumulative left *area* (or probability) of 0.9772.

Among the software packages discussed below, STATDISK is by far the easiest to use, and it is available for free on the CD that is included with this book. STATDISK or any other suitable software package replaces Table 5.1 and gives you a much wider range of values that can be used.

SPSS

- To find the cumulative area (or probability) to the left of a particular value, follow this procedure:
  1. Enter a value of *x* in the SPSS data editor window, and then press **ENTER**.
  2. Click on the menu item **Transform** in the SPSS data editor window.
  3. Click on the item **Compute**.
  4. In the "Compute Variable" window, make these two entries:
     a. Enter **prob** in the "Target Variable" box.
     b. Enter **CDF.NORMAL(*x*, mean, standard deviation)** in the "Numeric Expression" box, where specific values are used for each of the three entries. (If using standard *z*-scores, enter a mean of 0 and a standard deviation of 1.)
  5. Click on **OK**.
  6. The probability (or area) corresponding to the region to the left of the *x* value will be displayed. Click on that value to see more decimal places.
- To find the value of *x* that corresponds to a cumulative area from the left, follow this procedure:
  1. In the SPSS data editor window, enter the value for the cumulative area (or probability) corresponding to the region to the left of the desired value *x*, and then press **ENTER**.
  2. Click on the menu item **Transform** in the SPSS data editor window.
  3. Click on the item **Compute**.
  4. In the "Compute Variable" window, make these two entries:
     a. Enter ***x*** in the "Target Variable" box.
     b. Enter **IDF.NORMAL(*p*, mean, standard deviation)** in the "Numeric Expression" box, where specific values are used for each of the three entries. The value of *p* should be the cumulative area to the left of the desired value. (If using standard *z*-scores, enter a mean of 0 and a standard deviation of 1.)
  5. Click on **OK**.
  6. The value of *x* will be displayed. Click on that value to see more decimal places.

Excel

- To find the cumulative area to the left of a value (such as 2), click on ***fx*** and then select **Statistical, NORMDIST.** In the dialog box, enter the value (such as 2) for *x*, enter the mean and standard deviation, and enter 1 in the "cumulative" space. (If using standard *z*-scores, enter a mean of 0 and a standard deviation of 1, and enter 1 in the "cumulative" space.)
- To find a value corresponding to a known area (such as 0.95), select ***fx*, Statistical, NORMINV**, and proceed to make the entries in the dialog box. (If using standard scores, enter a mean of 0 and a standard deviation of 1.) When entering the probability value, enter the total area to the left of the given value.

STATDISK

Select **Analysis, Probability Distributions, Normal Distribution**. Either enter the standard *z*-score to find corresponding areas or enter the cumulative area from the left to find the *z*-score. After entering a value, click on **Evaluate**.

FOCUS ON EDUCATION

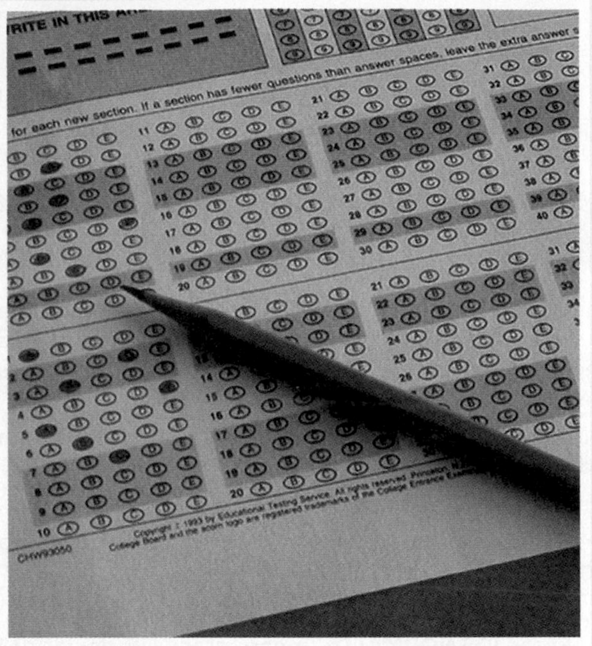

What Can We Learn from SAT Trends?

The Scholastic Aptitude Test (SAT) has been taken by college-bound high school students since 1941, when 11,000 students took the first test. Today it is taken by approximately 1.5 million high school students each year, and it is taken at some time by nearly half of all high school graduates. Until 2006, there were two parts to the SAT: the verbal test and the mathematics test. In 2006, the verbal test was changed to become the "critical reading" test and a third test was added, on writing. Changes were also made to the mathematics test.

The scores on each part of the SAT were originally scaled to have a mean of 500 and a standard deviation of 100. The maximum and minimum scores on each part are 800 and 200, respectively, which are 3 standard deviations above and below the mean. (Scores more than 3 standard deviations above or below the mean just get the maximum 800 or minimum 200.)

The mean does not stay at 500 from year to year, however. Each year's test shares some common questions with prior years' tests, and the College Board (which designs the test) uses these common questions to compare each year's test results to those from prior years. The College Board therefore claims that a score of 500 represents the same level of achievement on the test no matter what year the test is taken, so variations in the mean from one year to another represent real changes in student performance on the test. For example, if the mean score in a particular year is higher than 500, it implies students performed better on average than did students at the time that the average was originally set to 500.

Trends in SAT scores have been widely used to assess the general state of American education. Figure 5.27 shows the average scores on the verbal/critical reading and math parts of the SAT between 1972 and 2006; the label "verbal/critical reading" is used because it refers to the verbal test prior to 2006 and to the critical reading test starting in 2006. Notice that the scores for both parts generally declined until the early 1990s, but have recovered in recent years. Results for the writing test are not shown, because there are not yet enough data to look for trends over time.

The trends show that, if the scores from year to year are truly comparable, then students taking the SAT in recent years have poorer verbal skills than students of a few decades ago, and combined scores are down somewhat as well. But before we accept this conclusion, we must answer two important questions:

1. Are trends among the *sample* of high school students who take the SAT representative of trends for the *population* of all high school students?
2. Aside from the small number of common questions that link one year to the next, the test changes every year. Can test scores from one year legitimately be compared to scores in other years?

Unfortunately, neither question can be answered with a clear yes. For example, many people argue that much of the long-term decline in SAT scores is the result of changes in the sample of students who take the test. Only about a third of high school graduates took the SAT in the 1970s, while nearly 50% of high school graduates took the SAT in 2006. If these sam-

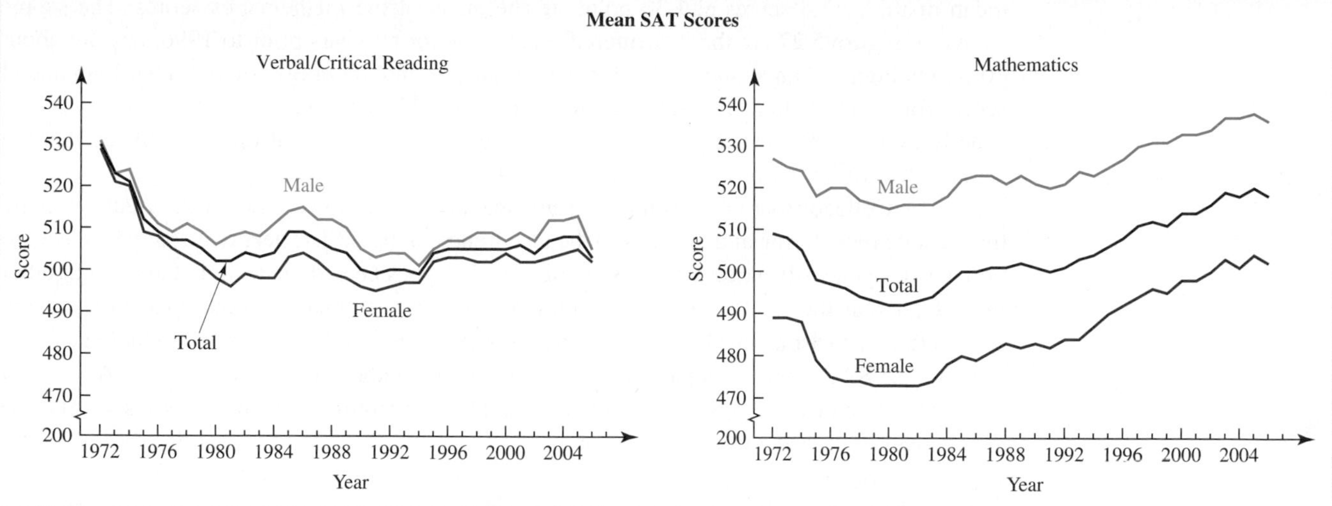

Figure 5.27 Verbal/critical reading and mathematics SAT scores, 1972–2006. Scores for years before 1996 are shown at their "recentered" values, rather than their original values. *Source:* The College Board.

ples represent the top tier of high school students, the decline in score averages might simply reflect the fact that a greater range of student abilities is represented in later samples. Support for this idea comes from state-by-state SAT data. In general, the states with the highest mean SAT scores are also those with the lowest proportions of their population taking the test. The presumed explanation for this trend—an explanation that is also supported by more detailed statistical analysis—is that when a smaller proportion of a state's students take the test, those students tend to be among the best high school students in the state.

The second question is even more difficult. Even if some test questions stay the same from one year to the next, different material may be emphasized in high schools, thereby changing the likelihood that students will answer the same questions correctly from one year to the next.

Further complications arise when the test undergoes major changes. For example, in 1994 the use of calculators was allowed for the first time on the mathematics section, and additional time was allotted for the entire test (the verbal test also underwent significant changes that year). Is it a coincidence that mathematics scores began an upward trend starting that year, or could it be that the test became "easier" as a result of the changes? Similarly, in 2006 the combined critical reading and mathematics scores suffered their largest decline in more than 30 years. Critics attributed the decline to the fact that the SAT became longer and harder as a result of adding the writing test, while the College Board attributed the decline to a decrease in the number of repeat test-takers. (Repeat test-takers tend to raise the average score, because individual students make significant gains in their scores when they take the test a second or third time.) Clearly, changes in the structure of the test make it difficult to determine whether changes in average scores reflect real changes in education.

Perhaps even more significantly, the calibration of the SAT test underwent a major change in 1996. By that time, scores had declined significantly from the originally planned mean of 500: The mean verbal score had fallen to about 420, while the mean mathematics score had fallen to about 470. Because the range of possible scores goes only from 200 to 800, these declines meant that the minimum and maximum scores no longer represented the same number of standard deviations from the mean, which created difficulties with statistical analysis. The College Board therefore decided to "recenter" all scores to a mean of 500 in 1996. The recentering affected different percentiles differently, but it effectively added 80 points to the mean of the verbal scores and 30 points to the mean of the mathematics scores. The scores shown in Figure 5.27 are the "recentered" scores, so for all years prior to 1996 they are about 80 points higher than actual scores for the verbal test and about 30 points higher than actual scores for the mathematics test. For example, Figure 5.27 shows a mean verbal score of almost exactly 500 in 1994, but if you look back at news reports from that time, you'll see that the actual mean verbal score was about 420 in 1994.

The College Board argues that its statistical analysis has been done with great care, and that despite recentering and changes to the test, trends in the SAT reflect real changes in education. Critics generally agree that the statistical analysis has been done well, but argue that other factors, such as the change in the test-taking population and changes in our education system, make the comparisons invalid. For students, however, the key fact is that many colleges continue to use SAT scores as a part of their decision on whether to admit applicants. As long as that remains the case, the SAT will remain one of the nation's most important tests—and the debate over its merits will surely continue.

QUESTIONS FOR DISCUSSION

1. Do you think that comparisons of SAT scores over two years are meaningful? Over ten years? Do you agree that the long-term trends indicate that students today have poorer verbal skills but better mathematical skills than those of a few decades ago? Defend your opinions.
2. The downward trend in verbal skills was a major reason why the College Board added the writing test in 2006. The hope was that adding the new test would cause students to focus more on writing in high school and therefore improve both in writing and in overall verbal abilities. What do you think of this rationale? Do you think it will be successful?
3. Notice that scores for males have been consistently higher than scores for females. Why do you think this is the case? Do you think that changes in our education system could eliminate this gap? Defend your opinions.
4. Discuss personal experiences with the SAT among your classmates. Based on these personal experiences, what do you think the SAT is measuring? Do you think the test is a reasonable way to predict students' performance in college? Why or why not?

SUGGESTED READING

Arenson, Karen W. "Scores on Reading and Math Portions of SAT Show Significant Decline." *New York Times,* August 30, 2006.

College Board. "The Effects of SAT Scale Recentering on Percentiles." Research Summary RS-05, 1999, available at http://www.collegeboard.com/repository/rs05_3962.pdf.

Crouse, James, and Trusheim, Dale. *The Case Against the SAT.* University of Chicago Press, 1988.

FOCUS ON PSYCHOLOGY

Are We Smarter than Our Parents?

Most kids tend to think that they're smarter than their parents, but is it possible that they're right? If you believe the results of IQ tests, not only are we smarter than our parents, on average, but our parents are smarter than our grandparents. In fact, almost all of us would have ranked as geniuses if we'd lived a hundred years ago. Of course, before any of us start touting our Einstein-like abilities, it would be good to investigate what lies behind this startling claim.

The idea of an IQ, which stands for *intelligence quotient*, was invented by French psychologist Alfred Binet (1857–1911). Binet created a test that he hoped would identify children in need of special help in school. He gave his test to many children, and then calculated each child's IQ by dividing the child's "mental age" by his or her physical age (and multiplying by 100). For example, a 5-year-old child who scored as well as an average 6-year-old was said to have a mental age of 6, and therefore an IQ of $(6 \div 5) \times 100$, or 120. Note that, by this definition, IQ tests make sense only for children. However, later researchers, especially psychologists for the U.S. Army, extended the idea of IQ so that it could be applied to adults as well.

Today, IQ is defined by a normal distribution with a mean of 100 and a standard deviation of 16. According to the 68-95-99.7 rule, 68% of the people taking an IQ test get scores between 84 and 116, 95% get scores between 68 and 132, and 99.7% get scores between 52 and 148. Traditionally, psychologists classified people with an IQ below 70 (about 2 standard deviations below the mean of 100) as "intellectually deficient," and people who score above 130 (about 2 standard deviations above the mean) as "intellectually superior."

You're probably aware of the controversy that surrounds IQ tests, which boils down to two key issues:

- Do IQ tests measure intelligence or something else?
- If they do measure intelligence, is it something that is innate and determined by heredity or something that can be molded by environment and education?

A full discussion of these issues is too involved to cover here, but a recently discovered trend in IQ scores sheds light on these issues in a surprising way. As we'll see shortly, the trend is quite pronounced, but it was long hidden because of the way IQ tests are scored. There are several different, competing versions of IQ tests, and most of them are regularly changed and updated. But in all cases, the scores are adjusted to fit a normal distribution with a mean of 100 and standard deviation of 16. In other words, the scoring of an IQ test is essentially done in the same way that an instructor might grade an exam "on a curve." Because of this adjustment, the mean on IQ tests is *always* 100, which makes it impossible for measured IQ scores to rise and fall with time.

However, a few IQ tests have not been changed and updated over time, including some given by the military. In other cases, tests that have been updated sometimes still repeat old questions. In the early 1980s, a political science professor named Dr. James Flynn began to look at the raw, unadjusted scores on unchanged tests and questions. The results were astounding.

Dr. Flynn found that raw scores have been steadily rising, although the precise amount of the rise varies somewhat with the type of IQ test. The highest rates of increase are found on

By the Way ...

Binet himself assumed that intelligence could be molded and warned against taking his tests as a measure of any innate or inherited abilities. However, many later psychologists concluded that IQ tests could measure innate intelligence, which led to their being used for separating school children, military recruits, and many other groups of people according to supposed intellectual ability.

tests that purport to measure abstract reasoning abilities (such as the "Raven's" tests). For these tests, Dr. Flynn found that the unadjusted IQ scores of people in industrialized countries have been rising at a rate of about 6 points per decade. In other words, a person who scored 100 on a test given in 2000 would have scored about 106 on a test in 1990, 112 on a test in 1980, and so on. Over a hundred years, this would imply a rise of some 60 points, suggesting that someone who scores an "intellectually deficient" IQ of 70 today would have rated an "intellectually superior" IQ of 130 a century ago.

This long-term trend toward rising scores on IQ tests is now called the *Flynn effect*. It is present for all types of IQ tests, though not always to the same degree as with abstract reasoning tests. For example, Figure 5.28 shows how results changed on one of the most widely used IQ tests (the Stanford-Binet test) between 1932 and 1997. Note that, in terms of unadjusted scores, the mean rose by 20 points during that time period. In other words, if people who scored an IQ of 100 on a 1997 test were instead scored on a 1932 test, they would rate an IQ of 120. As the figure shows, about one-fourth of the 1997 test-takers would have rated "intellectually superior" on the 1932 test. There is some evidence that the rise in scores may have begun to slow or halt in recent years, though the data are still subject to debate.

Many other scientists have investigated the Flynn effect, and there seems to be little doubt that the long-term trend is real. The implication is clear: Whatever IQ tests measure, people today really *do* have more of it than people just a few decades ago. If IQ tests measure intelligence, then it means we really are smarter than our parents (on average), who in turn are smarter than our grandparents (on average).

Of course, if IQ tests don't measure "intelligence" but only measure some type of skill, then the rise in scores may indicate only that today's children have more practice at that skill than past children. The fact that the greatest rise is seen on tests of abstract thinking lends some support to this idea. These tests often involve such problems as solving puzzles and looking for patterns among sets of shapes, and these types of problems are now much more common in games than they were in the past.

While the Flynn effect does not answer the question of whether IQ tests measure intelligence, it may tell us one important thing: If IQs really have been rising as the Flynn effect suggests, then IQ must *not* be a primarily inherited trait, because inherited traits cannot change

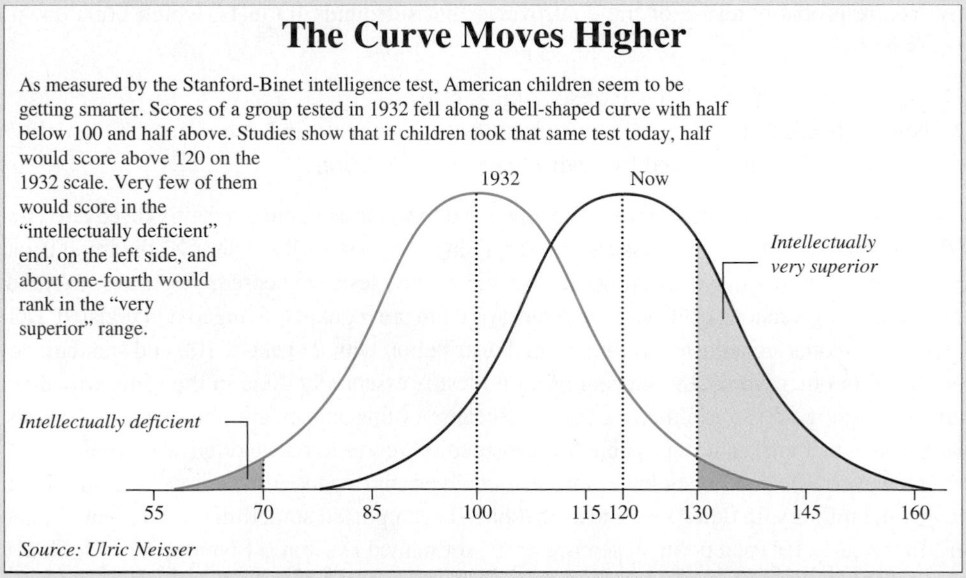

Figure 5.28 *Source: New York Times.*

that much in just a few decades. Thus, if IQ tests are measuring intelligence, then intelligence can be molded by environmental factors. Conversely, if intelligence is hereditary, then IQ tests are not measuring it.

Dr. Flynn's discovery has already changed the way psychologists look at IQ tests, and it is sure to be an active topic of research in coming decades. Moreover, given the many uses to which modern society has put IQ tests, the Flynn effect is likely to have profound social and political consequences as well. So back to our starting question: Are we smarter than our parents? We really can't say, but we can certainly hope so, because it will take a lot of brainpower to solve the problems of the future.

QUESTIONS FOR DISCUSSION

1. Which explanation do you favor for the Flynn effect: that people are getting smarter or that people are merely getting more practice at the skills measured on IQ tests? Defend your opinion.
2. The rise in performance on IQ tests contrasts sharply with a steady decline in performance over the past few decades on many tests that measure factual knowledge, such as the SAT. Think of several possible ways to explain these contrasting results, and form an opinion as to the most likely explanation.
3. Results on IQ tests tend to differ among different ethnic groups. Some people have used this fact to argue that some ethnic groups tend to be intellectually superior to others. Can such an argument still be supported in light of the Flynn effect? Defend your opinion.
4. Discuss some of the common uses of IQ tests. Do you think that IQ tests *should* be used for these purposes? Does the Flynn effect alter your thoughts about the uses of IQ tests? Explain.

SUGGESTED READING

Hall, Trish. "I.Q. Scores Are Up, and Psychologists Wonder Why." *New York Times*, February 24, 1998.

Neisser, Ulric (ed.). *The Rising Curve: Long-Term Gains in IQ and Related Measures*. American Psychological Association, 1998.

6

You can take as understood
That your luck changes only
if it is good.

—Ogden Nash,
Roulette Us Be Gay

Probability in Statistics

MOST STATISTICAL STUDIES SEEK TO LEARN SOMETHING about a *population* from a much smaller *sample*. Therefore, a key question in any statistical study is whether it is valid to generalize from a sample to the population. To answer this question, we must understand the likelihood, or probability, that what we've learned about the sample also applies to the population. In this chapter we will focus on a few basic ideas of probability that are commonly used in statistics. As you will see, these ideas of probability also have many applications in their own right.

LEARNING GOALS

6.1 The Role of Probability in Statistics: Statistical Significance
Understand the concept of statistical significance and the essential role that probability plays in defining it.

6.2 Basics of Probability
Know how to find probabilities using theoretical and relative frequency methods and understand how to construct basic probability distributions.

6.3 Probabilities with Large Numbers
Understand the law of large numbers, use this law to understand and calculate expected values, and recognize how misunderstanding of the law of large numbers leads to the gambler's fallacy.

6.4 Ideas of Risk and Life Expectancy
Compute and interpret various measures of risk as they apply to travel, disease, and life expectancy.

6.5 Combining Probabilities (Supplementary Section)
Distinguish between independent and dependent events and between overlapping and non-overlapping events, and be able to calculate *and* and *either/or* probabilities.

6.1 The Role of Probability in Statistics: Statistical Significance

To see why probability is so important in statistics, let's begin with a simple example of a coin toss. Suppose you are trying to test whether a coin is fair—that is, whether it is equally likely to land on heads or tails. If you toss the coin 100 times and get 52 heads and 48 tails, should you conclude that the coin is unfair? No. While we should expect to see *roughly* 50 heads and 50 tails in every 100 tosses of a fair coin, we should also expect some variation from one sample of 100 tosses to another. A small deviation from a perfect 50-50 split between heads and tails does not necessarily mean that the coin is unfair, because we expect small deviations to occur *by chance*.

Now, suppose you toss a coin 100 times and the results are 20 heads and 80 tails. This is a substantial deviation from a 50–50 split, making it seem much less likely that it was just a chance set of tosses with a fair coin. In other words, while it's *possible* that you observed a rare set of 100 tosses, it's more likely that the coin is unfair. When the difference between what is observed and what is expected seems unlikely to be explained by chance alone, we say the difference is **statistically significant**.

Definition

A set of measurements or observations in a statistical study is said to be **statistically significant** if it is unlikely to have occurred by chance.

EXAMPLE 1 Likely Stories?

a. *A detective in Detroit finds that 25 of the 62 guns used in crimes during the past week were sold by the same gun shop*. This finding is statistically significant. Because there are many gun shops in the Detroit area, having 25 out of 62 guns come from the same shop seems unlikely to have occurred by chance.

b. *In terms of the global average temperature, five of the years between 1990 and 1999 were the five hottest years in the 20th century*. Having the five hottest years in 1990–1999 is statistically significant. By chance alone, any particular year in a century would have a 5 in 100, or 1 in 20, chance of being one of the five hottest years. Having five of those years come in the same decade is very unlikely to have occurred by chance alone. This statistical significance suggests that the world may be warming up.

c. *The team with the worst win-loss record in basketball wins one game against the defending league champions*. This one win is *not* statistically significant because although we expect a team with a poor win-loss record to lose most of its games, we also expect it to win occasionally, even against the defending league champions.

From Sample to Population

Let's look at the idea of statistical significance in an opinion poll. Suppose that in a poll of 1,000 randomly selected people, 51% support the President. A week later, in another poll with a different randomly selected sample of 1,000 people, only 49% support the President. Should you conclude that the opinions of Americans changed during the one week between the polls?

You can probably guess that the answer is *no.* The poll results are *sample statistics* (see Section 1.1): 51% of the people in the first sample support the President. We can use this result to estimate the *population parameter*, which is the percentage of *all* Americans who support the President. At best, if the first poll were conducted well, it would say that the percentage of Americans who support the President is *close* to 51%. Similarly, the 49% result in the second poll means that the percentage of Americans supporting the President is *close* to 49%. Because the two sample statistics differed only slightly (51% versus 49%), it's quite possible that the real percentage of Americans supporting the President did not change at all. Instead, the two polls reflect expected and reasonable differences between the two samples.

In contrast, suppose the first poll found that 75% of the sample supported the President, and the second poll, taken a week later, found that only 30% supported the President. Assuming the polls were carefully conducted, it's highly unlikely that two groups of 1,000 randomly chosen people could differ so much by chance alone. In this case, we would look for another explanation. Perhaps Americans' opinions about the President really did change in the week between the polls.

In terms of statistical significance, the change from 51% to 49% in the first set of polls is *not* statistically significant, because we can reasonably attribute this change to chance variations between the two samples. However, in the second set of polls, the change from 75% to 30% is statistically significant, because it is unlikely to have occurred by chance.

TECHNICAL NOTE

The difference between 49% and 51% is not statistically significant for typical polls, but it can be for a poll involving a very large sample size. In general, any difference can be significant if the sample size is large enough.

EXAMPLE 2 Statistical Significance in Experiments

A researcher conducts a double-blind experiment that tests whether a new herbal formula is effective in preventing colds. During a three-month period, the 100 randomly selected people in a treatment group take the herbal formula while the 100 randomly selected people in a control group take a placebo. The results show that 30 people in the treatment group get colds, compared to 32 people in the control group. Can we conclude that the herbal formula is effective in preventing colds?

Solution Whether a person gets a cold during any three-month period depends on many unpredictable factors. Therefore, we should not expect the number of people with colds in any two groups of 100 people to be exactly the same. In this case, the difference between 30 people getting colds in the treatment group and 32 people getting colds in the control group seems small enough to be explainable by chance. So the difference is not statistically significant, and we should not conclude that the treatment is effective.

Quantifying Statistical Significance

In Example 2, we said that the difference between 30 colds in the treatment group and 32 colds in the control group was not statistically significant. This conclusion was fairly obvious because the difference was so small. But suppose that 24 people in the treatment group had colds, compared to the 32 in the control group. Would the difference between 24 and 32 be large enough to be considered statistically significant? The definition of statistical significance that we've been using so far is too vague to answer this question. We need a way to quantify the idea of statistical significance.

He that leaves nothing to chance will do few things ill, but will do very few things.

—George Savile Halifax

In general, we determine statistical significance by using probability to quantify the likelihood that a result may have occurred by chance. We therefore ask a question like this one: *Is the probability that the observed difference occurred by chance less than or equal to 0.05 (or 1 in 20)?* If the answer is *yes* (the probability is less than or equal to 0.05), then we say that the difference is *statistically significant at the 0.05 level.* If the answer is *no*, the observed difference is reasonably likely to have occurred by chance, so we say that it is not statistically significant.

The choice of 0.05 is somewhat arbitrary, but it's a figure that statisticians frequently use. Nevertheless, other probabilities are sometimes used, such as 0.1 or 0.01. Statistical significance at the 0.01 level is stronger than significance at the 0.05 level, which is stronger than significance at the 0.1 level.

Quantifying Statistical Significance

- If the probability of an observed difference occurring by chance is 0.05 (or 1 in 20) or less, the difference is statistically significant at the 0.05 level.
- If the probability of an observed difference occurring by chance is 0.01 (or 1 in 100) or less, the difference is statistically significant at the 0.01 level.

You can probably see that caution is in order when working with statistical significance. We would expect roughly 1 in 20 trials to give results that are statistically significant at the 0.05 level even when the results actually occurred by chance. Thus, statistical significance at the 0.05 level—or at almost any level, for that matter—is *no guarantee* that an important effect or difference is present.

TIME OUT TO THINK

Suppose an experiment finds that people taking a new herbal remedy get fewer colds than people taking a placebo, and the results are statistically significant at the 0.01 level. Has the experiment *proven* that the herbal remedy works? Explain.

EXAMPLE 3 Polio Vaccine Significance

In the test of the Salk polio vaccine (see Section 1.1), 33 of the 200,000 children in the treatment group got paralytic polio, while 115 of the 200,000 in the control group got paralytic polio. Calculations show that the probability of this difference between the groups occurring by chance is less than 0.01. Describe the implications of this result.

Solution The results of the polio vaccine test are statistically significant at the 0.01 level, meaning that there is a 0.01 chance (or less) that the difference between the control and treatment groups occurred by chance. Therefore, we can be fairly confident that the vaccine really was responsible for the fewer cases of polio in the treatment group. (In fact, the probability of the Salk results occurring by chance is *much* less than 0.01, so researchers were quite convinced that the vaccine worked; as we'll discuss in Chapter 9, this probability is called a "*P*-value.")

Section 6.1 Exercises

Statistical Literacy and Critical Thinking

1. **Statistical Significance.** In an experiment testing a method of gender selection intended to increase the likelihood that a baby is a girl, 20 couples give birth to 11 girls and 9 boys. A company representative claims that this is evidence that the method is effective, because the percentage of girls exceeds the 50% rate that is normally expected. Do you agree with that claim? Why or why not?
2. **Statistical Significance.** Does the term *statistical significance* refer to results that are significant in the sense that they have great importance? Explain.
3. **Statistical Significance.** If a particular result is statistically significant at the 0.05 level, must it also be statistically significant at the 0.01 level? Why or why not?
4. **Statistical Significance.** If a particular result is statistically significant at the 0.01 level, must it also be statistically significant at the 0.05 level? Why or why not?

Does It Make Sense? For Exercises 5–8, decide whether the statement makes sense (or is clearly true) or does not make sense (or is clearly false). Explain clearly; not all of these statements have definitive answers, so your explanation is more important than your chosen answer.

5. Drunk Driving. The drunk driving fatality rate has statistical significance because it seriously affects so many people.

6. Statistical Significance. In an experiment testing a method of gender selection, 100 couples give birth to 80 girls and 20 boys. Because those results are statistically significant, they could not have occurred by chance.

7. Gender Selection. In a test of a technique of gender selection, the 100 babies born consist of at least 80 girls. Because there is about 1 chance in a billion of getting at least 80 girls among 100 babies, the results are statistically significant.

8. Clinical Trial. In a clinical trial of a treatment for reducing back pain, the difference between the treatment group and the control group (with no treatment) was found to be statistically significant. This means that the treatment will definitely ease back pain for everyone.

Concepts and Applications

Subjective Significance. For each event in Exercises 9–16, state whether the difference between what occurred and what you would have expected by chance is statistically significant. Discuss any implications of the statistical significance.

9. Coin Tosses. In 500 tosses of a coin, you observe 255 tails.

10. Coin Tosses. In 500 tosses of a coin, you observe 400 tails.

11. Rolls of a Die. In 60 rolls of a six-sided die, the outcome of 3 never appears.

12. Spins of a Roulette Wheel. A roulette wheel has 38 slots numbered 0, 00, 1, 2, 3, . . . , 36. In five spins of a roulette wheel, the outcome of the number 5 occurs on five consecutive spins.

13. Survey. In conducting a survey of adults in the United States, a pollster claims that he randomly selected 20 subjects and all of them were women.

14. Statistics Class. An instructor walks into her first statistics class and finds that all of the students are women who are at least 6 feet tall.

15. Jury Composition. For a trial on a charge of failure to pay child support, the jury consists of exactly 6 men and 6 women.

16. Clinical Trial. In a clinical trial of a new drug intended to treat allergies, 5 of the 80 subjects in the treatment group experienced headaches, and 8 of the 160 subjects in the control group experienced headaches.

17. Fuel Tests. Thirty identical cars are selected for a fuel test. Half of the cars are filled with regular gasoline, and the other half are filled with a new experimental fuel. The cars in the first group average 29.3 miles per gallon, while the cars in the second group average 35.5 miles per gallon. Discuss whether this difference seems statistically significant.

18. Carpal Tunnel Syndrome Treatments. An experiment was conducted to determine whether there is a difference between the success rates from treating carpal tunnel syndrome with surgery and with splinting. The success rate for 73 patients treated with surgery was 92%, and the success rate for 83 patients treated with splints was 72% (based on data from "Splinting vs. Surgery in the Treatment of Carpal Tunnel Syndrome" by Gerritsen et al., *Journal of the American Medical Association*, Vol. 288, No. 10). Discuss whether this difference appears to be statistically significant.

19. Gender Selection. The Genetics and IVF Institute conducted a clinical trial of its method for gender selection. At the time this book was written, 325 babies had been born to parents using the XSORT method to increase the probability of conceiving a girl, and 295 of those babies were girls. Discuss whether these results appear to be statistically significant.

20. Bednets and Malaria. In a randomized controlled trial in Kenya, insecticide-treated bednets were tested as a way to reduce malaria. Among 343 infants who used the bednets, 15 developed malaria. Among 294 infants not using bednets, 27 developed malaria (based on data from "Sustainability of Reductions in Malaria Transmission and Infant Mortality in Western Kenya with Use of Insecticide-Treated Bednets" by Lindblade et al., *Journal of the American Medical Association*, Vol. 291, No. 21). Assuming that the bednets have no effect, there is a probability of 0.015 of getting these results by chance. Do the results appear to have statistical significance? Do the bednets appear to be effective?

21. Human Body Temperature. In a study by researchers at the University of Maryland, the body temperatures of 106 individuals were measured; the mean for the sample was 98.20°F. The accepted value for human body temperature is 98.60°F. The difference between the sample mean and the accepted value is significant at the 0.05 level.

a. Discuss the meaning of the significance level in this case.

b. If we assume that the mean body temperature is actually 98.6°F, the probability of getting a sample with a mean of 98.20°F or less is 0.000000001. Interpret this probability value.

22. Seat Belts and Children. In a study of children injured in automobile crashes (*American Journal of Public Health*, Vol. 82, No. 3), those wearing seat belts had a mean stay of 0.83 day in an intensive care unit. Those not wearing seat belts had a mean stay of 1.39 days. The difference in means between the two groups is significant at the 0.0001 level. Interpret this result.

23. **SAT Preparation.** A study of 75 students who took an SAT preparation course (*American Education Research Journal*, Vol. 19, No. 3) concluded that the mean improvement on the SAT was 0.6 point. If we assume that the preparation course has no effect, the probability of getting a mean improvement of 0.6 point by chance is 0.08. Discuss whether this preparation course results in statistically significant improvement.

24. **Weight by Age.** A National Health Survey determined that the mean weight of a sample of 804 men aged 25 to 34 years was 176 pounds, while the mean weight of a sample of 1,657 men aged 65 to 74 years was 164 pounds. The difference is significant at the 0.01 level. Interpret this result.

Projects for the Internet and Beyond

For useful links, select "Links for Internet Projects" for Chapter 6 at www.aw.com/bbt.

25. **Significance in Vital Statistics.** Visit a Web site that has vital statistics (for example, the U.S. Census Bureau or the National Center for Health Statistics). Choose a question such as the following:
    - Are there significant differences in numbers of births among months?
    - Are there significant differences in numbers of natural deaths among days of the week?
    - Are there significant differences in infant mortality rates among selected states?
    - Are there significant differences in incidences of a particular disease among selected states?
    - Are there significant differences among the marriage rates in various states?

    Collect the relevant data and determine subjectively whether you think the observed differences are significant; that is, explain if they could occur by chance or provide some alternative explanations.

26. **Lengths of Rivers.** Using an almanac or the Internet, find the lengths of the principal rivers of the world. Construct a list of the leading digits only. Does any particular digit occur more often than the others? Does that digit occur significantly more often? Explain.

IN THE NEWS

27. **Statistical Significance.** Find a recent newspaper article on a statistical study in which the idea of statistical significance is used. Write a one-page summary of the study and the result that is considered to be statistically significant. Also include a brief discussion of whether you believe the result, given its statistical significance.

28. **Significant Experiment?** Find a recent news story about a statistical study that used an experiment to determine whether some new treatment was effective. Based on the available information, briefly discuss what you can conclude about the statistical significance of the results. Given this significance (or lack thereof), do you think the new treatment is useful? Explain.

29. **Personal Statistical Significance.** Describe an incident in your own life that did not meet your expectation, defied the odds, or seemed unlikely to have occurred by chance. Would you call this incident statistically significant? To what did you attribute the event?

6.2 Basics of Probability

In Section 6.1, we saw that ideas of probability are fundamental to statistics, in part through the concept of statistical significance. We will return to the topic of statistical significance in Chapters 7 through 10. First, however, we need to explore a few essential ideas of probability and their applications in everyday life.

Let's begin by considering a toss of two coins. Figure 6.1 shows that there are four different ways the coins could fall. We say that each of these four ways is a different **outcome** of the coin toss. The outcomes are the most basic possible results of the coin toss. But suppose we are interested only in the number of heads. Because the two middle outcomes in Figure 6.1 each have 1 head, we say that these two outcomes represent the same event. An **event** describes one or more possible outcomes that all have the same property of interest—in this case, the same

In this world, nothing is certain but death and taxes.

—Benjamin Franklin

Figure 6.1 The four possible outcomes for a toss of two coins. The middle outcomes both represent the same *event* of 1 head.

number of heads. Figure 6.1 shows that there are four possible outcomes for the two-coin toss, but only three possible events: 0 heads, 1 head, and 2 heads.

Definitions

Outcomes are the most basic possible results of observations or experiments.

An **event** is a collection of one or more outcomes that share a property of interest.

Mathematically, we express probabilities as numbers between 0 and 1. For example, the probability of a coin landing on heads is one-half, or 0.5. If an event is impossible, we assign it a probability of 0. For example, the probability of meeting a married bachelor is 0. At the other extreme, an event that is certain to occur is given a probability of 1. For example, according to the old saying from Benjamin Franklin, the probability of death and taxes is 1. We write *P*(event) to mean the probability of an event. We often denote events by letters or symbols. For example, using H to represent the event of a head on a coin toss, we write *P*(H) = 0.5.

Expressing Probability

The probability of an event, expressed as *P*(event), is always between 0 and 1 inclusive. A probability of 0 means that the event is impossible, and a probability of 1 means that the event is certain.

Figure 6.2 shows the scale of probability values, along with common expressions of likelihood. It's helpful to develop the sense that a probability value such as 0.95 indicates that an event is very likely to occur, but is not certain. It should occur about 95 times out of 100. In contrast, a probability value of 0.01 describes an event that is very unlikely to occur. The event is possible, but will occur only about once in 100 times.

With these basic definitions and notation, we are ready to discuss the three basic techniques for finding probabilities. They go by the names *theoretical method*, *relative frequency method*, and *subjective method*. We'll explore each in turn.

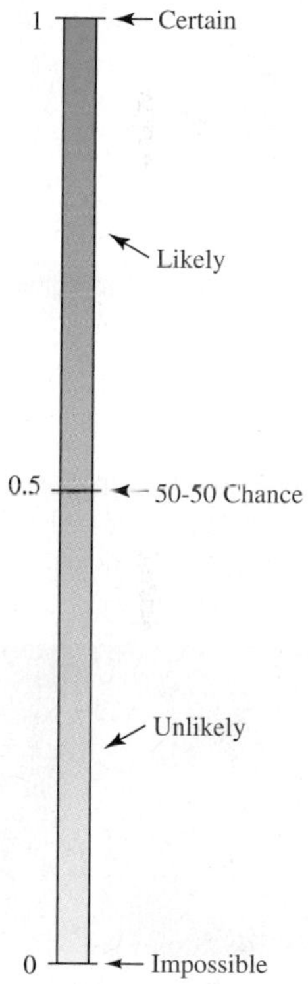

Figure 6.2 The scale shows various degrees of certainty as expressed by probabilities.

Theoretical Probabilities

When we say that the probability of heads on a coin toss is 1/2, we are *assuming* that the coin is fair and is equally likely to land on heads or tails. In essence, the probability is based on a theory of how the coin behaves, so we say that the probability of 1/2 comes from the theoretical method. As another example, consider rolling a single die. Because there are six equally likely outcomes (Figure 6.3), the theoretical probability for each outcome is 1/6.

As long as all outcomes are equally likely, we can use the following procedure to calculate theoretical probabilities.

By the Way ...

A sophisticated analysis by mathematician Persi Diaconis has shown that seven shuffles of a new deck of cards are needed to give it an ordering in which all card arrangements are equally likely.

By the Way ...

As early as 3600 BC, rounded bones called *astragali* were used much like dice for games of chance in the Middle East. Our familiar cubical dice appeared about 2000 BC in Egypt and in China. Playing cards were invented in China in the 10th century and brought to Europe in the 14th century.

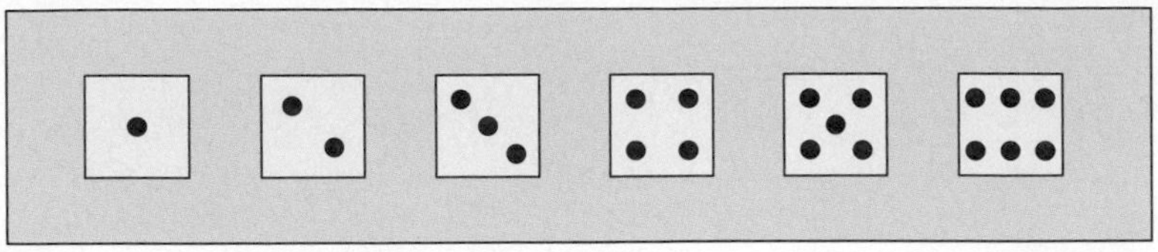

Figure 6.3 The six possible outcomes for a roll of one six-sided die.

Theoretical Method for Equally Likely Outcomes

Step 1. Count the total number of possible outcomes.

Step 2. Among all the possible outcomes, count the number of ways the event of interest, A, can occur.

Step 3. Determine the probability, $P(A)$, from

$$P(A) = \frac{\text{number of ways } A \text{ can occur}}{\text{total number of outcomes}}$$

EXAMPLE 1 Guessing Birthdays

Suppose you select a person at random from a large group at a conference. What is the probability that the person selected has a birthday in July? Assume 365 days in a year.

Solution If we assume that all birthdays are equally likely, we can use the three-step theoretical method.

Step 1. Each possible birthday represents an outcome, so there are 365 possible outcomes.

Step 2. July has 31 days, so 31 of the 365 possible outcomes represent the event of a July birthday.

Step 3. The probability that a randomly selected person has a birthday in July is

$$P(\text{July birthday}) = \frac{31}{365} \approx 0.0849$$

which is slightly more than 1 in 12.

By the Way ...

Birthdays are not randomly distributed throughout the year. After accounting for the different lengths of months, January has the lowest birth rate and June and July have the highest birth rates.

Counting Outcomes

Suppose we toss two coins and want to count the total number of outcomes. The toss of the first coin has two possible outcomes: heads (H) or tails (T). The toss of the second coin also has two possible outcomes. The two outcomes for the first coin can occur with either of the two outcomes for the second coin. So the total number of outcomes for two tosses is $2 \times 2 = 4$; they are HH, HT, TH, and TT, as shown in the tree diagram of Figure 6.4a.

TIME OUT TO THINK

Explain why the outcomes for tossing one coin twice in a row are the same as those for tossing two coins at the same time.

We can now extend this thinking. If we toss three coins, we have a total of $2 \times 2 \times 2 = 8$ possible outcomes: HHH, HHT, HTH, HTT, THH, THT, TTH, and TTT, as shown in Figure 6.4b. This idea is the basis for the following counting rule.

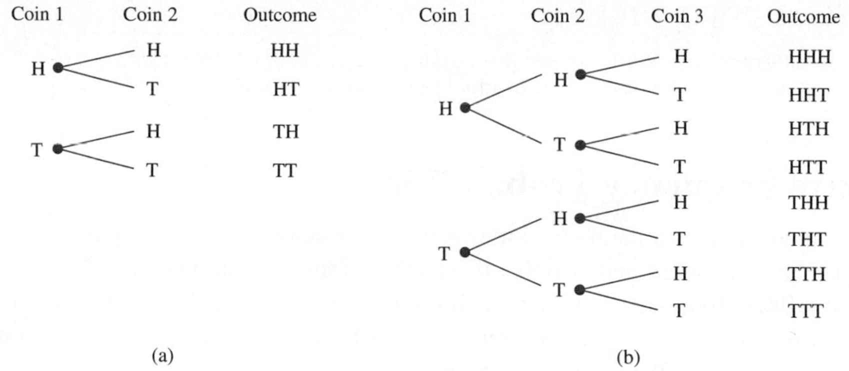

Figure 6.4 Tree diagrams showing the outcomes of tossing (a) two and (b) three coins.

Counting Outcomes

Suppose process A has *a* possible outcomes and process B has *b* possible outcomes. Assuming the outcomes of the processes do not affect each other, the number of different outcomes for the two processes combined is $a \times b$. This idea extends to any number of processes. For example, if a third process C has *c* possible outcomes, the number of possible outcomes for the three processes combined is $a \times b \times c$.

EXAMPLE 2 Some Counting

a. How many outcomes are there if you roll a fair die and toss a fair coin?

b. What is the probability of rolling two 1's (snake eyes) when two fair dice are rolled?

Solution

a. The first process, rolling a fair die, has six outcomes (1, 2, 3, 4, 5, 6). The second process, tossing a fair coin, has two outcomes (H, T). Therefore, there are $6 \times 2 = 12$ outcomes for the two processes together (1H, 1T, 2H, 2T, . . . , 6H, 6T).

b. Rolling a single die has six equally likely outcomes. Therefore, when two fair dice are rolled, there are $6 \times 6 = 36$ different outcomes. Of these 36 outcomes, only one is the event of interest (two 1's). So the probability of rolling two 1's is

$$P(\text{two 1's}) = \frac{\text{number of ways two 1's can occur}}{\text{total number of outcomes}} = \frac{1}{36} = 0.0278$$

EXAMPLE 3 Counting Children

What is the probability that, in a randomly selected family with three children, the oldest child is a boy, the second child is a girl, and the youngest child is a girl? Assume boys and girls are equally likely.

Solution There are two possible outcomes for each birth: boy or girl. For a family with three children, the total number of possible outcomes (birth orders) is $2 \times 2 \times 2 = 8$ (BBB, BBG, BGB, BGG, GBB, GBG, GGB, GGG). The question asks about one particular birth order (boy-girl-girl), so this is 1 of 8 possible outcomes. Therefore, this birth order has a probability of 1 in 8, or $1/8 = 0.125$.

By the Way ...

Births of boys and girls are *not* equally likely. Naturally, there are approximately 105 male births for every 100 female births. However, male death rates are higher than female death rates, so female adults outnumber male adults.

TIME OUT TO THINK

How many different four-child families are possible if birth order is taken into account? What is the probability of a couple having a four-child family with four girls?

Relative Frequency Probabilities

The second way to determine probabilities is to *approximate* the probability of an event A by making many observations and counting the number of times event A occurs. This approach is called the **relative frequency** (or **empirical) method**. For example, if we observe that it rains an average of 100 days per year, we might say that the probability of rain on a randomly selected day is 100/365. We can use a general rule for this method.

Relative Frequency Method

Step 1. Repeat or observe a process many times and count the number of times the event of interest, A, occurs.

Step 2. Estimate $P(A)$ by

$$P(A) = \frac{\text{number of times } A \text{ occurred}}{\text{total number of observations}}$$

EXAMPLE 4 500-Year Flood

Geological records indicate that a river has crested above a particular high flood level four times in the past 2,000 years. What is the relative frequency probability that the river will crest above the high flood level next year?

Solution Based on the data, the probability of the river cresting above this flood level in any single year is

$$\frac{\text{number of years with flood}}{\text{total number of years}} = \frac{4}{2{,}000} = \frac{1}{500}$$

Because a flood of this magnitude occurs on average once every 500 years, it is called a "500-year flood." The probability of having a flood of this magnitude in any given year is 1/500, or 0.002.

Subjective Probabilities

The third method for determining probabilities is to estimate a **subjective probability** using experience or intuition. For example, you could make a subjective estimate of the probability that a friend will be married in the next year or of the probability that a good grade in statistics will help you get the job you want.

Three Approaches to Finding Probability

A **theoretical probability** is based on assuming that all outcomes are equally likely. It is determined by dividing the number of ways an event can occur by the total number of possible outcomes.

A **relative frequency probability** is based on observations or experiments. It is the relative frequency of the event of interest.

A **subjective probability** is an estimate based on experience or intuition.

By the Way ...

Theoretical methods are also called *a priori* methods. The words *a priori* are Latin for "before the fact" or "before experience."

EXAMPLE 5 Which Method?

Identify the method that resulted in the following statements.

a. The chance that you'll get married in the next year is zero.

b. Based on government data, the chance of dying in an automobile accident is 1 in 7,000 (per year).

c. The chance of rolling a 7 with a twelve-sided die is 1/12.

Solution

a. This is a subjective probability because it is based on a feeling at the current moment.

b. This is a relative frequency probability because it is based on observed data on past automobile accidents.

c. This is a theoretical probability because it is based on assuming that a fair twelve-sided die is equally likely to land on any of its twelve sides.

By the Way ...

Another approach to finding probabilities, called the *Monte Carlo method*, uses computer simulations. This technique essentially finds relative frequency probabilities; in this case, observations are made with the computer.

EXAMPLE 6 Hurricane Probabilities

Figure 6.5 shows a map released when Hurricane Floyd approached the southeast coast of the United States. The map shows "strike probabilities" in three different regions along the path of the hurricane. Interpret the term "strike probability" and discuss the value of these probabilities and how they are estimated.

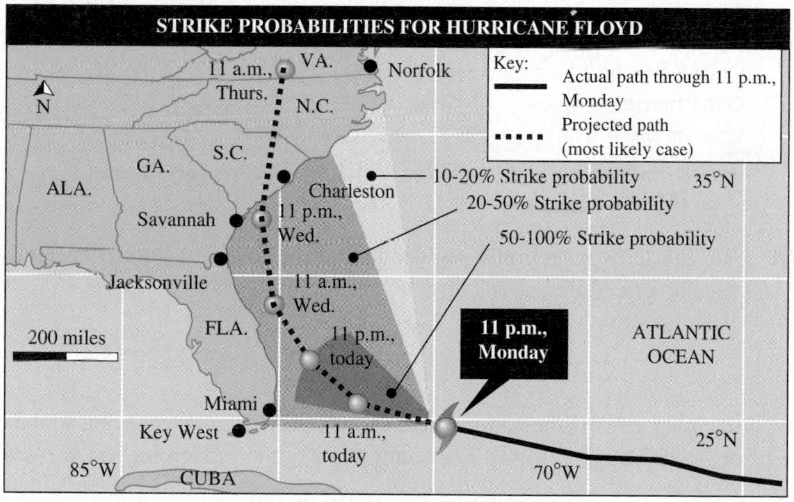

Figure 6.5 This map shows the actual path of Hurricane Floyd up to the time the map was made ("11 p.m., Monday"). It then shows the probabilities of various alternative paths, with the most likely path a dashed line. *Source: Virginia Pilot.*

Solution The strike probabilities tell how likely it is that the eye of the hurricane will pass over a particular location. For example, for a person in the 20–50% strike probability region, there is a 0.2 to 0.5 probability of being hit by the hurricane. Knowing these probabilities helps people make decisions about preparations and possible evacuations. The probabilities are calculated with computer models that predict hurricane behavior based on current weather patterns in the hurricane region, the physics of hurricanes, and observational data from past hurricanes.

Probability of an Event *Not* Occurring

Suppose we are interested in the probability that a particular event or outcome does *not* occur. For example, consider the probability of a wrong answer on a multiple-choice question with five possible answers. The probability of answering correctly with a random guess is 1/5, so the probability of *not* answering correctly is 4/5. Notice that the probability of either a right answer or a wrong answer must be 1. We can generalize this idea.

Probability of an Event *Not* Occurring

If the probability of an event A is $P(A)$, then the probability that event A does not occur is $P(\text{not } A)$. Because the event must either occur or not occur, we can write

$$P(A) + P(\text{not } A) = 1 \quad \text{or} \quad P(\text{not } A) = 1 - P(A)$$

Note: The event *not A* is called the **complement** of the event A; the "not" is often designated by a bar, so $\bar{A}$ means *not A*.

By the Way ...

A *lot* is a person's portion or allocation. The term came to be used for any object or marker that identified a person in a selection by chance. The casting of lots was used in the Trojan Wars to determine who would throw the first spear and in the Old Testament to divide conquered lands.

EXAMPLE 7 Is Scanner Accuracy the Same for Specials?

In a study of checkout scanning systems, samples of purchases were used to compare the scanned prices to the posted prices. Table 6.1 summarizes results for a sample of 819 items. Based on these data, what is the probability that a regular-priced item has a scanning error? What is the probability that an advertised-special item has a scanning error?

Table 6.1 Scanner Accuracy

	Regular-priced items	Advertised-special items
Undercharge	20	7
Overcharge	15	29
Correct price	384	364

Source: Ronald Goodstein, "UPC Scanner Pricing Systems: Are They Accurate?" *Journal of Marketing,* Vol. 58.

Solution We can let R represent a regular-priced item being scanned correctly. Because 384 of the 419 regular-priced items are correctly scanned,

$$P(R) = \frac{384}{419} = 0.916$$

The event of a scanning error is the complement of the event of a correct scan, so the probability of a regular-priced item being subject to a scanning error (either undercharged or overcharged) is

$$P(\text{not } R) = 1 - 0.916 = 0.084$$

Now let A represent an advertised-special item being scanned correctly. Of the 400 advertised-special items in the sample, 364 are scanned correctly. Therefore,

$$P(A) = \frac{364}{400} = 0.910$$

The probability of an advertised-special item being scanned incorrectly is

$$P(\text{not } A) = 1 - 0.910 = 0.090$$

The error rates are nearly equal. However, notice in Table 6.1 that most of the errors made with advertised-special items are not to the customer's advantage. With regular-priced items, over half of the errors are to the customer's advantage.

Probability Distributions

In Chapters 3 through 5, we worked with frequency and relative frequency distributions—for example, distributions of age or income. One of the most fundamental ideas in probability and statistics is that of a *probability distribution*. As the name suggests, a probability distribution is a distribution in which the variable of interest is associated with a probability.

Suppose you toss two coins simultaneously. The outcomes are the various combinations of a head and a tail on the two coins. Because *each* coin can land in two possible ways (heads or tails), the *two* coins can land in $2 \times 2 = 4$ different ways. Table 6.2 has a row for each of these four outcomes.

Table 6.2 Outcomes of Tossing Two Fair Coins

Coin 1	Coin 2	Outcome	Probability
H	H	HH	1/4
H	T	HT	1/4
T	H	TH	1/4
T	T	TT	1/4

Notice that the four outcomes represent only three different events: 2 heads (HH), 2 tails (TT), and 1 head and 1 tail (HT or TH). Therefore, the probability of two heads is $P(\text{HH}) = 1/4 = 0.25$; the probability of two tails is $P(\text{TT}) = 1/4 = 0.25$; and the probability of one head and one tail is $P(\text{H and T}) = 2/4 = 0.50$. These probabilities result in a **probability distribution** that can be displayed as a table (Table 6.3) or a histogram (Figure 6.6). Note that the sum of all the probabilities must be 1 (because exactly one of the possible results must occur).

Table 6.3 Tossing Two Coins

Result	Probability
2 heads, 0 tails	0.25
1 head, 1 tail	0.50
0 heads, 2 tails	0.25
Total	**1**

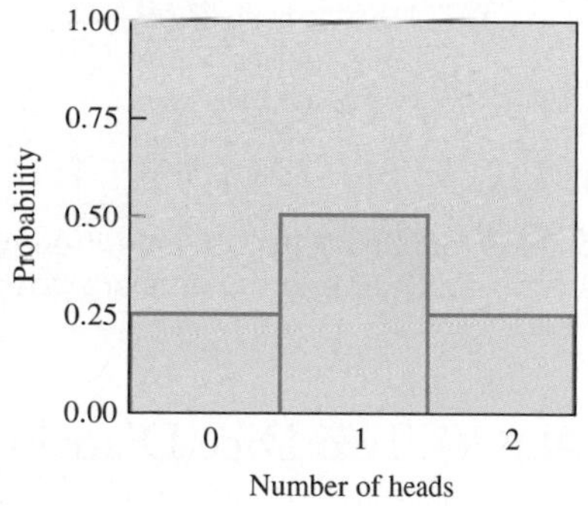

Figure 6.6 Histogram showing the probability distribution for the results of tossing two coins.

Making a Probability Distribution

A **probability distribution** represents the probabilities of all possible events. Do the following to make a display of a probability distribution:

Step 1. List all possible *outcomes*. Use a table or figure if it is helpful.

Step 2. Identify outcomes that represent the same *event*. Find the probability of each event.

Step 3. Make a table in which one column lists each event and another column lists each probability. The sum of all the probabilities must be 1.

By the Way ...

Another common way to express likelihood is to use odds. The odds *against* an event are the ratio of the probability that the event does not occur to the probability that it does occur. For example, the odds against rolling a 6 with a fair die are (5/6)/(1/6), or 5 to 1. The odds used in gambling are called *payoff odds*; they express your net gain on a winning bet. For example, suppose that the payoff odds on a particular horse at a horse race are 3 to 1. This means that for each $1 you bet on this horse, you will gain $3 if the horse wins (and get your original $1 back).

EXAMPLE 8 Tossing Three Coins

Make a probability distribution for the number of heads that occurs when three coins are tossed simultaneously.

Solution We apply the three-step process.

Step 1. The number of different outcomes when three coins are tossed is $2 \times 2 \times 2 = 8$. Figure 6.4b (page 241) shows how we find all eight possible outcomes, which are HHH, HHT, HTH, HTT, THH, THT, TTH, and TTT.

Step 2. There are four possible events: 0 heads, 1 head, 2 heads, and 3 heads. By looking at the eight possible outcomes, we find the following:

- Only one of the eight outcomes represents the event of 0 heads, so its probability is 1/8.
- Three of the eight outcomes represent the event of 1 head (and 2 tails): HTT, THT, and TTH. This event therefore has a probability of 3/8.
- Three of the eight outcomes represent the event of 2 heads (and 1 tail): HHT, HTH, and THH. This event also has a probability of 3/8.
- Only one of the eight outcomes represents the event of 3 heads, so its probability is 1/8.

Step 3. We make a table with the four events listed in the left column and their probabilities in the right column. Table 6.4 shows the result.

Table 6.4 Tossing Three Coins

Result	Probability
3 heads (0 tails)	1/8
2 heads (1 tail)	3/8
1 head (2 tails)	3/8
0 heads (3 tails)	1/8
Total	**1**

TIME OUT TO THINK

When you toss four coins, how many different outcomes are possible? If you record the number of heads, how many different events are possible?

EXAMPLE 9 Two Dice Distribution

Make a probability distribution for the sum of the dice when two dice are rolled. Express the distribution as a table and as a histogram.

Solution Because there are six ways for each die to land, there are $6 \times 6 = 36$ outcomes of rolling two dice. We enumerate all 36 outcomes in Table 6.5 by listing one die along the

Table 6.5 Outcomes and Sums for the Roll of Two Dice

	1	2	3	4	5	6
1	1 + 1 = 2	1 + 2 = 3	1 + 3 = 4	1 + 4 = 5	1 + 5 = 6	1 + 6 = 7
2	2 + 1 = 3	2 + 2 = 4	2 + 3 = 5	2 + 4 = 6	2 + 5 = 7	2 + 6 = 8
3	3 + 1 = 4	3 + 2 = 5	3 + 3 = 6	3 + 4 = 7	3 + 5 = 8	3 + 6 = 9
4	4 + 1 = 5	4 + 2 = 6	4 + 3 = 7	4 + 4 = 8	4 + 5 = 9	4 + 6 = 10
5	5 + 1 = 6	5 + 2 = 7	5 + 3 = 8	5 + 4 = 9	5 + 5 = 10	5 + 6 = 11
6	6 + 1 = 7	6 + 2 = 8	6 + 3 = 9	6 + 4 = 10	6 + 5 = 11	6 + 6 = 12

rows and the other along the columns. In each cell, we show the sum of the numbers on the two dice.

The possible *events* are the sums from 2 to 12. These are the *events* of interest in this problem. We find the probability of each event by counting all the outcomes for each sum and then dividing the number of outcomes by 36. For example, the five highlighted outcomes in the table have a sum of 8, so the probability of a sum of 8 is 5/36. Table 6.6 shows the complete probability distribution, and Figure 6.7 shows the distribution as a histogram.

Table 6.6 Probability Distribution for the Sum of Two Dice

Event (sum)	2	3	4	5	6	7	8	9	10	11	12	Total
Probability	$\frac{1}{36}$	$\frac{2}{36}$	$\frac{3}{36}$	$\frac{4}{36}$	$\frac{5}{36}$	$\frac{6}{36}$	$\frac{5}{36}$	$\frac{4}{36}$	$\frac{3}{36}$	$\frac{2}{36}$	$\frac{1}{36}$	1

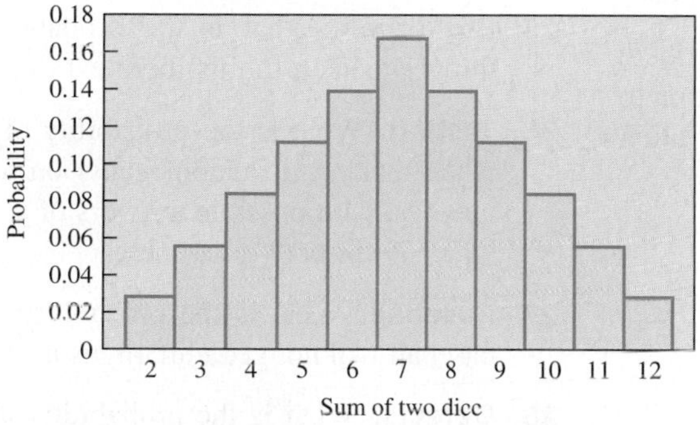

Figure 6.7 Histogram showing the probability distribution for the sum of two dice

Section 6.2 Exercises

Statistical Literacy and Critical Thinking

1. **Notation.** What does the notation $P(A)$ represent? What does $P(\text{not } A)$, also written $P(\bar{A})$, represent?

2. **Interpreting Probability.** What do we mean when we say that "the probability of winning the grand prize in the Illinois lottery is 1/20,358,520? Is such a win *unusual*? Why or why not?

3. **Probability of Rain.** When writing about the probability that it will rain in Boston on July 4 of next year, a newspaper reporter states that the probability is 1/2, because either it will rain or it will not. Is this reasoning correct? Why or why not?

4. **Subjective Probability.** Use subjective judgment to estimate the probability that the next time you ride an elevator, it gets stuck between floors.

Does It Make Sense? For Exercises 5–10, decide whether the statement makes sense (or is clearly true) or does not make sense (or is clearly false). Explain clearly; not all of these statements have definitive answers, so your explanation is more important than your chosen answer.

5. **Impossible Event.** Because it is impossible to get a total of 14 when two ordinary dice are rolled, the probability of 14 is 0.

6. **Car Crash.** An insurance company states that the probability that a particular car will be involved in a car crash this year is 0.14, and the probability that the car will not be involved in a car crash this year is 0.82.

7. **Opposites.** If $P(A) = 0.25$, then $P(\text{not } A) = 0.75$.

8. **Certain Event.** If $P(A) = 1$, then event A will definitely occur.

9. **Lightning.** Jack estimates that the subjective probability of his being struck by lightning sometime next year is 1/2.

10. **Lightning.** Jill estimates that the subjective probability of her being struck by lightning sometime next year is 1/1,000,000.

Concepts and Applications

11. Outcomes or Events. For each observation described, state whether there is one way for the given observation to occur or more than one way for the given observation to occur (in which case it is an event).

- **a.** Tossing a head with a fair coin
- **b.** Tossing two heads with three fair coins
- **c.** Rolling a 6 with a fair die
- **d.** Rolling an even number with a fair die
- **e.** Rolling a sum of 7 with two fair dice
- **f.** Drawing a pair of kings from a regular deck of cards

12. Outcomes or Events. For each observation described, state whether there is one way for the given observation to occur or more than one way for the given observation to occur (in which case it is an event).

- **a.** Tossing a tail with a fair coin
- **b.** Tossing one tail with two fair coins
- **c.** Rolling at least one 5 with two fair dice
- **d.** Rolling an odd number with a fair die
- **e.** Drawing a jack from a regular deck of cards
- **f.** Drawing three aces from a regular deck of cards

Theoretical Probabilities. For Exercises 13–20, use the theoretical method to determine the probability of the given outcome or event. State any assumptions that you need to make.

13. Die. Rolling a die and getting an outcome that is less than 3

14. Die. Rolling a die and getting an outcome that is greater than 6

15. Roulette. Getting an outcome of 0 or 00 when a roulette wheel is spun (A roulette wheel has slots of 0, 00, 1, 2, 3, . . . , 36.)

16. Birthday. Finding that the next president of the United States was born on Saturday

17. Birthday. Finding that the next person you meet has the same birthday as yours (Ignore leap years.)

18. Births. Finding that the next baby born in Alaska is a girl

19. Births. Finding that the next baby born to a couple is a girl, given that the couple already has two children and they are both boys

20. Testing. Randomly guessing the correct answer on a multiple-choice test question with possible answers of a, b, c, d, and e, one of which is correct

Opposite Events. For Exercises 21–28, determine the probability of the given opposite event. State any assumptions that you use.

21. Die. What is the probability of rolling a fair die and not getting an outcome less than 3?

22. Die. What is the probability of rolling a fair die and not getting an outcome that is greater than 6?

23. Roulette. What is the probability of spinning a roulette wheel and not getting an outcome of 0 or 00? (A roulette wheel has slots of 0, 00, 1, 2, 3, . . . , 36.)

24. Birthday. What is the probability of finding that the next president of the United States was not born on Saturday?

25. Basketball. What is the probability that a 55% free-throw shooter will miss her next free throw?

26. Testing. What is the probability of guessing incorrectly when making a random guess on a multiple-choice test question with possible answers of a, b, c, d, and e, one of which is correct?

27. Baseball. What is the probability that a 0.280 hitter in baseball will not get a hit on his next at-bat?

28. Defects. What is the probability of not getting a defective fuse when one fuse is randomly selected from an assembly line and 2% of the fuses are defective?

Theoretical Probabilities. For Exercises 29–32, use the theoretical method to determine the probability of the given outcome or event. State any assumptions that you need to make.

29. M&Ms. A bag contains 10 red M&Ms, 15 blue M&Ms, and 20 yellow M&Ms. What is the probability of drawing a red M&M? A blue M&M? A yellow M&M? Something besides a yellow M&M?

30. Test Questions. The New England College of Medicine uses an admissions test with multiple-choice questions, each with five possible answers, only one of which is correct. If you guess randomly on every question, what score might you expect to get (in percentage terms)?

31. Three-Child Family. Suppose you randomly select a family with three children. Assume that births of boys and girls are equally likely. What is the probability that the family has each of the following?

- **a.** Three girls
- **b.** Two boys and a girl
- **c.** A girl, a boy, and a boy, in that order
- **d.** At least one girl
- **e.** At least two boys

32. Four-Child Family. Suppose you randomly select a family with four children. Assume that births of boys and girls are equally likely.

a. How many birth orders are possible? List all of them.

b. What is the probability that the family has four boys? Four girls?

c. What is the probability that the family has a boy, a girl, a boy, and a girl, in that order?

d. What is the probability that the family has two girls and two boys in any order?

Relative Frequency Probabilities. Use the relative frequency method to estimate the probabilities in Exercises 33–36.

33. Weather Forecast. After recording the forecasts of your local weatherman for 30 days, you conclude that he gave a correct forecast 12 times. What is the probability that his next forecast will be correct?

34. Flood. What is the probability of a 100-year flood this year?

35. Basketball. Halfway through the season, a basketball player has hit 86% of her free throws. What is the probability that her next free throw will be successful?

36. Surgery. In a clinical trial of 73 carpal tunnel syndrome patients treated with surgery, 67 had successful treatments (based on data from "Splinting vs. Surgery in the Treatment of Carpal Tunnel Syndrome" by Gerritsen et al., *Journal of the American Medical Association*, Vol. 288, No. 10). What is the probability that the next surgery treatment will be successful?

37. Senior Citizen Probabilities. In the year 2000, there were 34.7 million people over 65 years of age out of a U.S. population of 281 million. In the year 2050, it is estimated that there will be 78.9 million people over 65 years of age out of a U.S. population of 394 million. Would your chances of meeting a person over 65 at random be greater in 2000 or in 2050? Explain.

38. Age at First Marriage. The following table gives percentages of women and men married for the first time in several age categories (U.S. Census Bureau).

	Under 20	20–24	25–29	30–34	35–44	45–64	Over 65
Women	16.6	40.8	27.2	10.1	4.5	0.7	0.1
Men	6.6	36.0	34.3	14.8	7.1	1.1	0.1

a. What is the probability that a randomly encountered married woman was married, for the first time, between the ages of 35 and 44?

b. What is the probability that a randomly encountered married man was married, for the first time, before he was 20 years old?

c. Construct a bar chart consisting of side-by-side bars representing men and women.

39. Four-Coin Probability Distribution

a. Construct a table similar to Table 6.2, showing all possible outcomes of tossing four coins at once.

b. Construct a table similar to Table 6.4, showing the probability distribution for the events 4 heads, 3 heads, 2 heads, 1 head, and 0 heads.

c. What is the probability of getting 2 heads and 2 tails when you toss four coins at once?

d. What is the probability of tossing anything except 4 heads when you toss four coins at once?

e. Which event is most likely to occur?

40. Colorado Lottery Distribution. The histogram in Figure 6.8 shows the distribution of 5,964 Colorado lottery numbers (possible values range from 1 to 42).

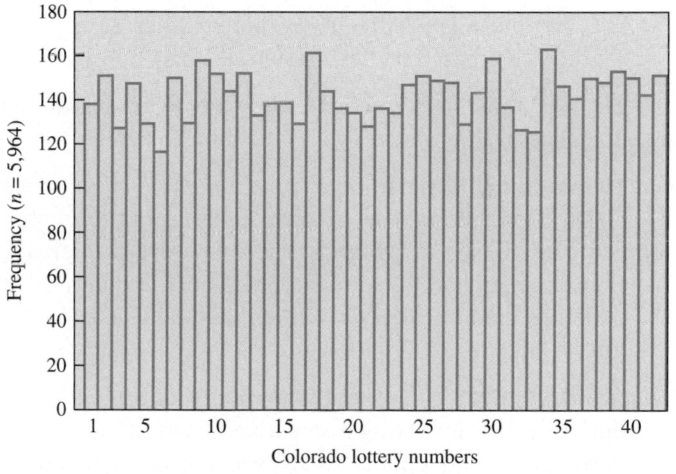

Figure 6.8 Colorado lottery distribution.

a. Assuming the lottery drawings are random, what would you expect the probability of any number to be?

b. Based on the histogram, what is the relative frequency probability of the most frequently appearing number?

c. Based on the histogram, what is the relative frequency probability of the least frequently appearing number?

d. Comment on the deviations of the empirical probabilities from the expected probabilities. Would you say these deviations are significant?

Projects for the Internet and Beyond

For useful links, select "Links for Internet Projects" for Chapter 6 at www.aw.com/bbt.

41. Blood Groups. The four major blood groups are designated A, B, AB, and O. Within each group there are two Rh types: positive and negative. Using library resources or the Internet, find data on the relative frequency of blood groups, including the Rh types. Construct a table showing the probability of meeting someone in each of the eight combinations of blood group and Rh type.

42. Age and Gender. The proportions of men and women in the population change with age. Using current data from a Web site, construct a table showing the probability of meeting a male or a female in each of these age categories: 0–5, 6–10, 11–20, 21–30, 31–40, 41–50, 51–60, 61–70, 71–80, over 80.

43. Thumb Tack Probabilities. Find a standard thumb tack and practice tossing it onto a flat surface. Notice that there are two different outcomes: The tack can land point down or point up.

a. Toss the tack 50 times and record the outcomes.

b. Give the relative frequency probabilities of the two outcomes based on these results.

c. If possible, ask several other people to repeat the process. How well do your probabilities agree?

44. Three-Coin Experiment. Toss three coins at once 50 times and record the outcomes in terms of the number of heads. Based on your observations, give the relative frequency probabilities of the outcomes. Do they agree with the theoretical probabilities? Explain and discuss your results.

45. Randomizing a Survey. Suppose you want to conduct a survey involving a sensitive question that not all participants may choose to answer honestly (for example, a question involving cheating on taxes or drug use). Here is a way to conduct the survey and protect the identity of respondents. We will assume that the sensitive question requires a *yes* or *no* answer. First, ask all respondents to toss a fair coin. Then give the following instructions:

- If you toss a head, then answer the decoy question (yes/no): Were you born on an even day of the month?
- If you toss a tail, then answer the real survey question (yes/no).

After all participants have answered *yes* or *no* to the question they were assigned, count the total numbers of *yes* and *no* responses.

a. Choose a question that may not produce totally honest responses and conduct a survey in your class using this technique.

b. Given only the total numbers of *yes* and *no* responses to both questions, explain how you can estimate the number of people who answered *yes* and *no* to the real question.

c. Will the results computed in part b be exact? Explain.

d. Suppose the decoy question was replaced by these instructions: If you toss a head, then answer *yes*. Can you still determine the number of people who answered *yes* and *no* to the real question?

IN THE NEWS

46. Theoretical Probabilities. Find a news article or research report that cites a theoretical probability. Provide a one-paragraph discussion.

47. Relative Frequency Probabilities. Find a news article or research report that makes use of a relative frequency (or empirical) probability. Provide a one-paragraph discussion.

48. Subjective Probabilities. Find a news article or research report that refers to a subjective probability. Provide a one-paragraph discussion.

49. Probability Distributions. Find a news article or research report that cites or makes use of a probability distribution. Provide a one-paragraph discussion.

6.3 Probabilities with Large Numbers

The false ideas prevalent among all classes of the community respecting chance and luck illustrate the truth that common consent argues almost of necessity of error.

—Richard Proctor, *Chance and Luck* (1887 textbook)

When we make observations or measurements that have random outcomes, it is impossible to predict any single outcome. We can state only the *probability* of a single outcome. However, when we make many observations or measurements, we expect the distribution of the outcomes to show some pattern or regularity. In this section, we use the ideas of probability to make useful statements about large samples. In the process, you will see one of the most important connections between probability and statistics.

The Law of Large Numbers

If you toss a coin once, you cannot predict exactly how it will land; you can state only that the probability of a head is 0.5. If you toss the coin 100 times, you still cannot predict precisely how many heads will occur. However, you can reasonably expect to get heads *close to* 50% of the time. If you toss the coin 1,000 times, you can expect the proportion of heads to be even closer to 50%. In general, the more times you toss the coin, the closer the percentage of heads will be to exactly 50%. The idea that large numbers of events may show some pattern even while individual events are unpredictable is called the **law of large numbers** (or the *law of averages*).

The Law of Large Numbers

The **law of large numbers** (or law of averages) applies to a process for which the probability of an event A is $P(A)$ and the results of repeated trials do not depend on results of earlier trials (they are *independent*). It states: *If the process is repeated through many trials, the proportion of the trials in which event A occurs will be close to the probability P(A). The larger the number of trials, the closer the proportion should be to P(A).*

We can illustrate the law of large numbers with a die-rolling experiment. The probability of a 1 on a single roll is $P(1) = 1/6 = 0.167$. To avoid the tedium of rolling the die many times, we can let a computer *simulate* random rolls of the die. Figure 6.9 shows the results of a computer simulation of rolling a single die 5,000 times. The horizontal axis gives the number of rolls, and the height of the curve gives the proportion of 1's. Although the curve bounces around when the number of rolls is small, for larger numbers of rolls the proportion of 1's approaches the probability of 0.167—just as predicted by the law of large numbers.

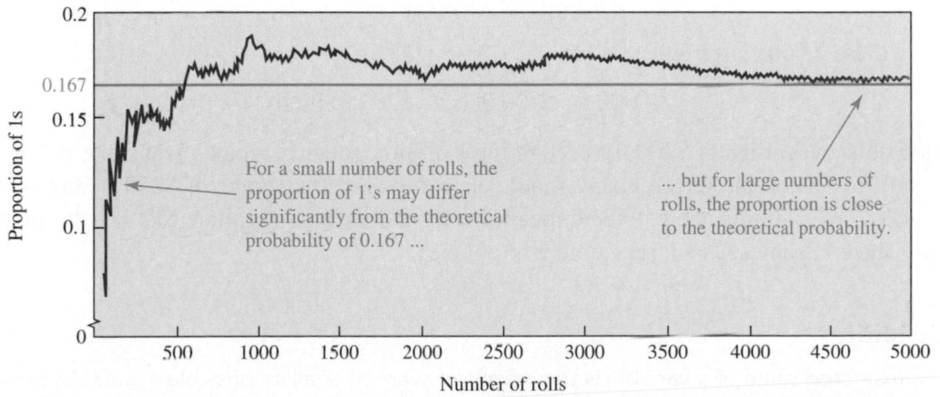
0.2
0.167
0.15
0.1
0
Proportion of 1s
500
1000
1500
2000
2500
3000
3500
4000
4500
5000
Number of rolls
For a small number of rolls, the proportion of 1's may differ significantly from the theoretical probability of 0.167 ...
... but for large numbers of rolls, the proportion is close to the theoretical probability.

Figure 6.9 Results of a computer simulation of rolling a die. As the number of rolls grows large, the proportion of 1's gets close to the theoretical probability of a 1 on a single roll, which is 0.167.

EXAMPLE 1 Roulette

A roulette wheel has 38 numbers: 18 black numbers, 18 red numbers, and the numbers 0 and 00 in green. (Assume that all outcomes—the 38 numbers—have equal probability.)

a. What is the probability of getting a red number on any spin?

b. If patrons in a casino spin the wheel 100,000 times, how many times should you expect a red number?

Solution

a. The theoretical probability of getting a red number on any spin is

$$P(A) = \frac{\text{number of ways red can occur}}{\text{total number of outcomes}} = \frac{18}{38} = 0.474$$

b. The law of large numbers tells us that as the game is played more and more times, the proportion of times that a red number appears should get closer to 0.474. In 100,000 tries, the wheel should come up red close to 47.4% of the time, or about 47,400 times.

Expected Value

Suppose the InsureAll Company sells a special type of insurance in which it promises to pay you \$100,000 in the event that you must quit your job because of serious illness. Based on data from past claims, the probability that a policyholder will be paid for loss of job is 1 in 500. Should the insurance company expect to earn a profit if it sells the policies for \$250 each?

If InsureAll sells only a few policies, the profit or loss is unpredictable. For example, selling 100 policies for \$250 each would generate revenue of $100 \times \$250 = \$25{,}000$. If none of the 100 policyholders files a claim, the company will make a tidy profit. On the other hand, if InsureAll must pay a \$100,000 claim to even one policyholder, it will face a huge loss.

In contrast, if InsureAll sells a large number of policies, the law of large numbers tells us that the proportion of policies for which claims will have to be paid should be very close to the 1 in 500 probability for a single policy. For example, if the company sells 1 million policies, it should expect that the number of policyholders collecting on a \$100,000 claim will be close to

$$\underbrace{1{,}000{,}000}_{\text{number of policies}} \times \underbrace{\frac{1}{500}}_{\text{probability of \$100,000 claim}} = 2{,}000$$

Paying these 2,000 claims will cost

$$2{,}000 \times \$100{,}000 = \$200 \text{ million}$$

This cost is an *average* of \$200 for each of the 1 million policies, which means that if the policies sell for \$250 each, the company should expect to earn an average of $\$250 - \$200 = \$50$ per policy. We call this average the **expected value** for each policy; note that it is "expected" only if the company sells a large number of policies.

By the Way ...

Hurricane Katrina (in 2005) was the costliest natural catastrophe in U.S. history, responsible for nearly 2,000 deaths and damages estimated at more than \$80 billion.

Definition

The **expected value** of a variable is the weighted average of all its possible events. Because it is an average, we should expect to find the "expected value" only when there are a large number of events, so that the law of large numbers comes into play.

We can find the same expected value with a more formal procedure. The insurance example involves two distinct events, each with a particular *probability* and *value* for the company:

1. In the event that a person buys a policy, the value to the company is the \$250 price of the policy. The probability of this event is 1 because everyone who buys a policy pays \$250.
2. In the event that a person is paid for a claim, the value to the company is −\$100,000; it is negative because the company loses \$100,000 in this case. The probability of this event is 1/500.

We now multiply the value of each event by its probability and add the results to find the expected value of each insurance policy:

$$\text{expected value} = \underbrace{\$250}_{\substack{\text{value of}\\ \text{policy sale}}} \times \underbrace{1}_{\substack{\text{probability of}\\ \text{earning \$250 on sale}}} + \underbrace{(-\$100{,}000)}_{\substack{\text{value of}\\ \text{claim}}} \times \underbrace{\frac{1}{500}}_{\substack{\text{probability of}\\ \text{paying claim}}}$$

$$= \$250 - \$200 = \$50$$

This expected profit of \$50 per policy is the same answer we found earlier. Note that it amounts to a profit of \$50 million on sales of 1 million policies.

TECHNICAL NOTE

There are alternative ways to calculate expected value. In the insurance example, we could instead define Event 1 as a policy with no claim and Event 2 as a policy on which the company pays a \$100,000 claim. Event 1 has a value to the company of \$250 (the sales price of a policy) and a probability of 499/500. Event 2 has a value to the company of \$250 (the policy price) *minus* the \$100,000 paid out with a claim, or −\$99,750, and a probability of 1/500. Notice that this alternative method still gives the same expected value of \$50:

$$\left(\$250 \times \frac{499}{500}\right) + \left(-\$99{,}750 \times \frac{1}{500}\right) = \$50$$

Some statisticians prefer this alternative method because the sum of the event probabilities is 1.

Calculating Expected Value

Consider two events, each with its own value and probability. The **expected value** is

$$\text{expected value} = \begin{pmatrix}\text{value of}\\ \text{event 1}\end{pmatrix} \times \begin{pmatrix}\text{probability of}\\ \text{event 1}\end{pmatrix} + \begin{pmatrix}\text{value of}\\ \text{event 2}\end{pmatrix} \times \begin{pmatrix}\text{probability of}\\ \text{event 2}\end{pmatrix}$$

This formula can be extended to any number of events by including more terms in the sum.

TIME OUT TO THINK

Should the insurance company expect to see a profit of \$50 on each individual policy? Should it expect a profit of \$50,000 on 1,000 policies? Explain.

EXAMPLE 2 Lottery Expectations

Suppose that \$1 lottery tickets have the following probabilities: 1 in 5 to win a free ticket (worth \$1), 1 in 100 to win \$5, 1 in 100,000 to win \$1,000, and 1 in 10 million to win \$1 million. What is the expected value of a lottery ticket? Discuss the implications. (Note: Winners do *not* get back the \$1 they spend on the ticket.)

Solution The easiest way to proceed is to make a table (below) of all the relevant events with their values and probabilities. We are calculating the expected value of a lottery ticket to *you;* thus, the ticket price has a negative value because it costs you money, while the values of the winnings are positive.

Event	Value	Probability	Value × probability
Ticket purchase	−\$1	1	$(-\$1) \times 1 = -\$1.00$
Win free ticket	\$1	$\frac{1}{5}$	$\$1 \times \frac{1}{5} = \$0.20$
Win \$5	\$5	$\frac{1}{100}$	$\$5 \times \frac{1}{100} = \$0.05$
Win \$1,000	\$1,000	$\frac{1}{100{,}000}$	$\$1{,}000 \times \frac{1}{100{,}000} = \$0.01$
Win \$1 million	\$1,000,000	$\frac{1}{10{,}000{,}000}$	$\$1{,}000{,}000 \times \frac{1}{10{,}000{,}000} = \$0.10$
			Sum of last column: −\$0.64

The expected value is the sum of all the products *value* × *probability*, which the final column of the table shows to be –$0.64. Thus, averaged over many tickets, you should expect to lose 64¢ for each lottery ticket that you buy. If you buy, say, 1,000 tickets, you should expect to *lose* about 1,000 × $0.64 = $640.

TIME OUT TO THINK

Many states use lotteries to finance worthy causes such as parks, recreation, and education. Lotteries also tend to keep state taxes at lower levels. On the other hand, research shows that lotteries are played by people with low incomes. Do you think lotteries are good social policy? Do you think lotteries are good economic policy?

The Gambler's Fallacy

Consider a simple game involving a coin toss: You win $1 if the coin lands heads, and you lose $1 if it lands tails. Suppose you toss the coin 100 times and get 45 heads and 55 tails, putting you $10 in the hole. Are you "due" for a streak of better luck?

You probably recognize that the answer is *no*: Your past bad luck has no bearing on your future chances. However, many gamblers—especially compulsive gamblers—guess just the opposite. They believe that when their luck has been bad, it's due for a change. This mistaken belief is often called the **gambler's fallacy** (or the *gambler's ruin*).

Definition

The **gambler's fallacy** is the mistaken belief that a streak of bad luck makes a person "due" for a streak of good luck.

Everyone who bets any part of his fortune, however small, on a mathematically unfair game of chance acts irrationally. . . . The imprudence of a gambler will be the greater the larger part of his fortune which he exposes to a game of chance.

—Daniel Bernoulli, 18th-century mathematician

One reason people succumb to the gambler's fallacy is a misunderstanding of the law of large numbers. In the coin-toss game, the law of large numbers tells us that the proportion of heads tends to be closer to 50% for larger numbers of tosses. But this does *not* mean that you are likely to recover early losses. To see why, study Table 6.7, which shows results from computer simulations of large numbers of coin tosses. Note that as the number of tosses increases, the percentage of heads gets closer to exactly 50%, just as the law of large numbers predicts. However, the last column shows that the *difference* between the number of heads and the number of tails continues to grow—meaning that in this particular set of simulations, the losses (the difference between the numbers of heads and tails) grow larger even as the proportion of heads approaches 50%.

Table 6.7 Outcomes of Coin Tossing Trials

Number of tosses	Number of heads	Percentage of heads	Difference between numbers of heads and tails
100	45	45%	10
1,000	470	47%	60
10,000	4,950	49.5%	100
100,000	49,900	49.9%	200

Figure 6.10 shows this result graphically for a computer simulation of a coin tossed 1,000 times. Although the percentage of heads approaches 50% over the course of 1,000 tosses, we see fairly large excursions away from equal numbers of heads and tails.

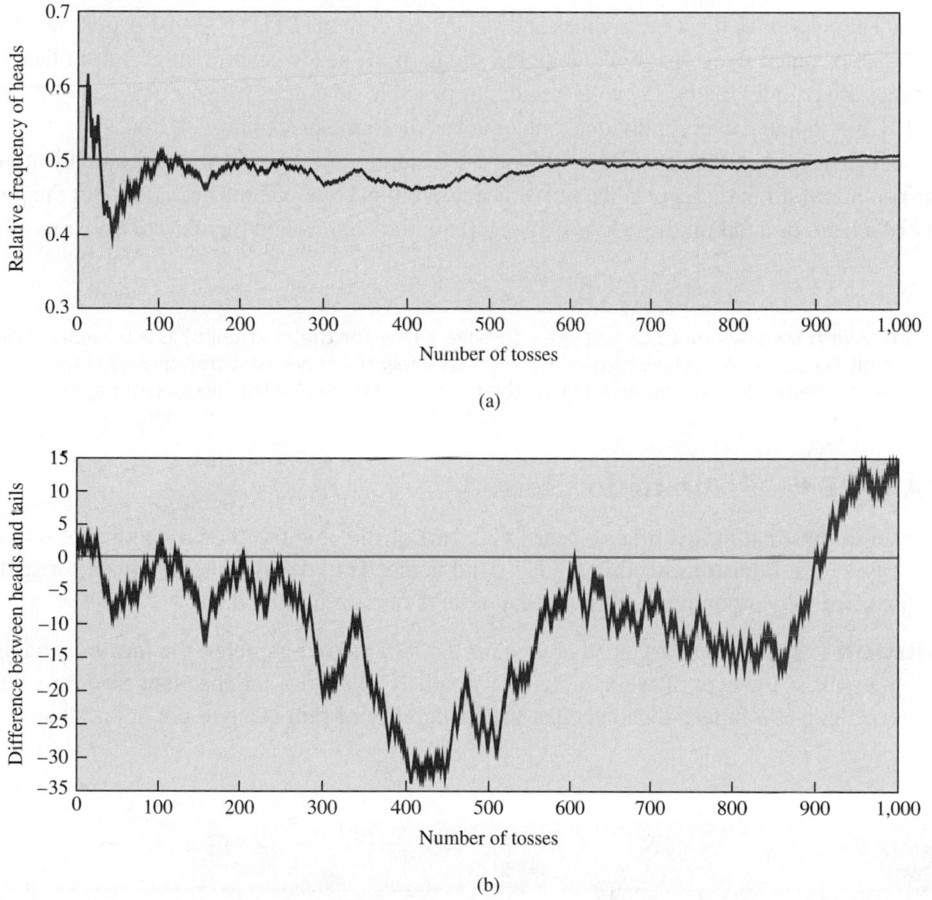

Figure 6.10 A computer simulation of 1,000 coin tosses. (a) This graph shows how the proportion of heads changes as the number of tosses increases. Note that the proportion approaches 0.5, as we expect from the law of large numbers. (b) This graph shows how the *difference* between the numbers of heads and tails changes as the number of tosses increases. Note that the difference can actually grow larger with more tosses, despite the fact that the proportion of heads approaches 0.5.

EXAMPLE 3 Continued Losses

You are playing the coin-toss game in which you win $1 for heads and lose $1 for tails. After 100 tosses, you are $10 in the hole because you have 45 heads and 55 tails. You continue playing until you've tossed the coin 1,000 times, at which point you've gotten 480 heads and

520 tails. Is this result consistent with what we expect from the law of large numbers? Have you gained back any of your losses? Explain.

Solution The proportion of heads in your first 100 tosses was 45%. After 1,000 tosses, the proportion of heads has increased to 480 out of 1,000, or 48%. Because the proportion of heads moved closer to 50%, the results are consistent with what we expect from the law of large numbers. However, you've now won \$480 (for the 480 heads) and lost \$520 (for the 520 tails), for a net loss of \$40. Thus, your losses *increased*, despite the fact that the proportion of heads grew closer to 50%.

Streaks

Another common misunderstanding that contributes to the gambler's fallacy involves expectations about streaks. Suppose you toss a coin six times and see the outcome HHHHHH (all heads). Then you toss it six more times and see the outcome HTTHTH. Most people would say that the latter outcome is "natural" while the streak of all heads is surprising. But, in fact, both outcomes are equally likely. The total number of possible outcomes for six coins is $2 \times 2 \times 2 \times 2 \times 2 \times 2 = 64$, and every individual outcome has the same probability of 1/64.

Moreover, suppose you just tossed six heads and had to bet on the outcome of the next toss. You might think that, given the run of heads, a tail is "due" on the next toss. But the probability of a head or a tail on the next toss is still 0.50; the coin has no memory of previous tosses.

By the Way ...

The same kind of thinking about streaks applies to selecting lottery numbers. There are no special combinations of lottery numbers that are more likely to be drawn than other combinations. However, people do pick special combinations (such as 1, 2, 3, 4, 5, 6) more often than ordinary combinations. Therefore, if you do win a lottery with an ordinary combination, you are less likely to split the prize with others, with the net effect being that your prize will be larger.

TIME OUT TO THINK

Is a family with six boys more or less likely to have a boy for the next child? Is a basketball player who has hit 20 consecutive free throws more or less likely to hit her next free throw? Is the weather on one day independent of the weather on the next (as assumed in the next example)? Explain.

EXAMPLE 4 Planning for Rain

A farmer knows that at this time of year in his part of the country, the probability of rain on a given day is 0.5. It hasn't rained in 10 days, and he needs to decide whether to start irrigating. Is he justified in postponing irrigation because he is due for a rainy day?

Solution The 10-day dry spell is unexpected, and, like a gambler, the farmer is having a "losing streak." However, if we assume that weather events are independent from one day to the next, then it is a fallacy to expect that the probability of rain is any more or less than 0.5.

Section 6.3 Exercises

Statistical Literacy and Critical Thinking

1. **Law of Large Numbers.** In your own words, describe the law of large numbers.

2. **Expected Value.** A geneticist computes the expected number of girls in 5 births and obtains the result of 2.5 girls. He rounds the result to 3 girls, reasoning that it is impossible to get 2.5 girls in 5 births. Is that reasoning correct? Why or why not?

3. **Gambling Strategy.** A professional gambler playing blackjack in the Venetian casino has lost each of his first 10 bets. He begins to place larger bets, reasoning that his current proportion of wins (which is 0) will increase to get closer to the average number of wins. Is his betting strategy sound? Is his reasoning correct? Explain.

4. **Gambler's Fallacy.** In your own words, describe the gambler's fallacy.

Does It Make Sense? For Exercises 5–8, decide whether the statement makes sense (or is clearly true) or does not make sense (or is clearly false). Explain clearly; not all of these statements have definitive answers, so your explanation is more important than your chosen answer.

5. **Roulette Strategy.** Steve has just lost $200 by placing 40 consecutive bets on the number 13 in roulette and not winning any of those bets. He reasons that, according to the law of large numbers, his streak of losses should end so that his proportion of wins gets closer to 1/38. Thus, he decides to recover his losses by placing larger bets on future spins of the roulette wheel.

6. **Lottery.** Kelly works in a convenience store and has access to a lottery machine. She reasons that if she buys a ticket for every possible combination of numbers in the state lottery she will definitely win the jackpot, so she plans to raise enough money to pay for all of those tickets.

7. **Lottery.** Kim purchases a state lottery ticket and selects the numbers of 1, 2, 3, 4, 5, and 6. She reasons that this combination has the same chance of winning as any other combination.

8. **Coin Tossing.** When a fair coin is tossed 10 times, the outcome of 10 heads will never occur.

Concepts and Applications

9. **Understanding the Law of Large Numbers.** Suppose you toss a fair coin 10,000 times. Should you expect to get exactly 5,000 heads? Why or why not? What does the law of large numbers tell you about the results you are likely to get?

10. **Speedy Driver.** A person who has a habit of driving fast has never had an accident or traffic citation. What does it mean to say that "the law of averages will catch up with him"? Is it true? Explain.

11. **Should You Play?** Suppose someone gives you 5 to 1 odds that you cannot roll two even numbers with the roll of two fair dice. This means you win $5 if you succeed and you lose $1 if you fail. What is the expected value of this game to you? Should you expect to win or lose the expected value in the first game? What can you expect if you play 100 times? Explain. (Table 6.5 on page 246 will be helpful in finding the required probabilities.)

12. **Kentucky's Pick 4 Lottery.** If you bet $1 in Kentucky's Pick 4 lottery game, you either lose $1 or gain $4,999. (The winning prize is $5,000, but your $1 bet is not returned, so the net gain is $4,999.) The game is played by selecting a four-digit number between 0000 and 9999. What is the probability of winning? If you bet $1 on 1234, what is the expected value of your gain or loss?

13. **Extra Points in Football.** Football teams have the option of trying to score either 1 or 2 extra points after a touchdown. They get 1 point by kicking the ball through the goal posts or 2 points by running or passing the ball across the goal line. For a recent year in the NFL, 1-point kicks were successful 94% of the time, while 2-point attempts were successful only 37% of the time. In either case, failure means zero points. Calculate the expected values of the 1-point and 2-point attempts. Based on these expected values, which option makes more sense in most cases? Can you think of any circumstances in which a team should make a decision different from what the expected values suggest? Explain.

14. **Insurance Claims.** An actuary at an insurance company estimates from existing data that on a $1,000 policy, an average of 1 in 100 policyholders will file a $20,000 claim, an average of 1 in 200 policyholders will file a $50,000 claim, and an average of 1 in 500 policyholders will file a $100,000 claim.

    a. What is the expected value to the company for each policy sold?

    b. If the company sells 100,000 policies, can it expect a profit? Explain the assumptions of this calculation.

15. **Expected Waiting Time.** Suppose that you arrive at a bus stop randomly, so all arrival times are equally likely. The bus arrives regularly every 30 minutes without delay (say, on the hour and on the half hour). What is the expected value of your waiting time? Explain how you got your answer.

16. **Powerball Lottery.** The multi-state Powerball lottery advertises the following prizes and probabilities of winning for a single $1 ticket. Assume the jackpot has a value of $30 million one week. Note that there is more than one way to win some of the monetary prizes (for example, two ways to win $100), so the table gives the probability for each way. What is the expected value of the winnings for a single lottery ticket? If you spend $365 per year on the lottery, how much can you expect to win or lose?

Prize	Probability
Jackpot	1 in 80,089,128
$100,000	1 in 1,953,393
$5,000	1 in 364,042
$100	1 in 8,879
$100	1 in 8,466
$7	1 in 207
$7	1 in 605
$4	1 in 188
$3	1 in 74

17. **Big Game.** The Multi-State Big Game lottery advertises the following prizes and probabilities of winning for a single $1 ticket. The jackpot is variable, but assume it has an average value of $3 million. Note that the same prize can

be given to two outcomes with different probabilities. What is the expected value of a single lottery ticket to you? If you spend $365 per year on the lottery, how much can you expect to win or lose?

Prize	Probability
Jackpot	1 in 76,275,360
$150,000	1 in 2,179,296
$5,000	1 in 339,002
$150	1 in 9,686
$100	1 in 7,705
$5	1 in 220
$5	1 in 538
$2	1 in 102
$1	1 in 62

18. **Expected Value in Roulette.** When you give the Venetian casino in Las Vegas $5 for a bet on the number 7 in roulette, you have a 37/38 probability of losing $5 and you have a 1/38 probability of making a net gain of $175. (The prize is $180, but your $5 bet is not returned, so the net gain is $175.) If you bet $5 that the outcome is an odd number, the probability of losing $5 is 20/38 and the probability of making a net gain of $5 is 18/38. (If you bet $5 on an odd number and win, you are given $10 that includes your bet, so the net gain is $5.)

    a. If you bet $5 on the number 7, what is your expected value?

    b. If you bet $5 that the outcome is an odd number, what is your expected value?

    c. Which of these options is best: bet on 7, bet on odd, or don't bet? Why?

19. **Expected Value in Casino Dice.** When you give a casino $5 for a bet on the "pass line" in a casino game of dice, there is a 251/495 probability that you will lose $5 and there is a 244/495 probability that you will make a net gain of $5. (If you win, the casino gives you $5 and you get to keep your $5 bet, so the net gain is $5.) What is your expected value? In the long run, how much do you lose for each dollar bet?

20. **Mean Household Size.** It is estimated that 57% of Americans live in households with 1 or 2 people, 32% live in households with 3 or 4 people, and 11% live in households with 5 or more people. Explain how you would find the expected number of people in an American household. How is this related to the mean household size?

21. **Psychology of Expected Values.** In 1953, a French economist named Maurice Allais conducted a survey of how people assess risk. Here are two scenarios that he used, each of which required people to choose between two options.

    Decision 1:

    Option A: 100% chance of gaining $1,000,000

    Option B: 10% chance of gaining $2,500,000; 89% chance of gaining $1,000,000; and 1% chance of gaining nothing

    Decision 2:

    Option A: 11% chance of gaining $1,000,000 and 89% chance of gaining nothing

    Option B: 10% chance of gaining $2,500,000 and 90% chance of gaining nothing

    Allais discovered that for decision 1, most people chose option A, while for decision 2, most people chose option B.

    a. For each decision, find the expected value of each option.

    b. Are the responses given in the survey consistent with the expected values?

    c. Give a possible explanation for the responses in Allais's survey.

22. **Expected Value for a Magazine Sweepstakes.** *Reader's Digest* ran a sweepstakes in which prizes were listed along with the chances of winning: $1,000,000 (1 chance in 90,000,000), $100,000 (1 chance in 110,000,000), $25,000 (1 chance in 110,000,000), $5,000 (1 chance in 36,667,000), and $2,500 (1 chance in 27,500,000).

    a. Assuming that there is no cost to enter the sweepstakes, find the expected value of the amount won for one entry.

    b. Find the expected value if the cost of entering this sweepstakes is the cost of a postage stamp. Is it worth entering this contest?

23. **Gambler's Fallacy and Dice.** Suppose you roll a die with a friend, with the following rules: For every even number you roll, you win $1 from your friend; for every odd number you roll, you pay $1 to your friend.

    a. What are the chances of rolling an even number on one roll of a fair die? An odd number?

    b. Suppose that on the first 100 rolls, you roll 45 even numbers. How much money have you won or lost?

    c. Suppose that on the second 100 rolls, your luck improves and you roll 47 even numbers. How much money have you won or lost over 200 rolls?

    d. Suppose that over the next 300 rolls, your luck again improves and you roll 148 even numbers. How much money have you won or lost over 500 rolls?

e. What was the percentage of even numbers after 100, 200, and 500 rolls? Explain why this game illustrates the gambler's fallacy.

f. How many even numbers would you have to roll in the next 100 rolls to break even?

24. Behind in Coin Tossing: Can You Catch Up? Suppose that you toss a fair coin 100 times, getting 38 heads and 62 tails, which is 24 more tails than heads.

a. Explain why, on your next toss, the *difference* in the numbers of heads and tails is as likely to grow to 25 as it is to shrink to 23.

b. Extend your explanation from part a to explain why, if you toss the coin 1,000 more times, the final difference in the numbers of heads and tails is as likely to be larger than 24 as it is to be smaller than 24.

c. Suppose that you continue tossing the coin. Explain why the following statement is true: If you stop at any random time, you always are more likely to have fewer heads than tails, in total.

d. Suppose that you are betting on heads with each coin toss. After the first 100 tosses, you are well on the losing side (having lost the bet 62 times while winning only 38 times). Explain why, if you continue to bet, you will most likely remain on the losing side. How is this answer related to the gambler's fallacy?

Projects for the Internet and Beyond

For useful links, select "Links for Internet Projects" for Chapter 6 at www.aw.com/bbt.

25. Analyzing Lotteries on the Web. Go to the Web site for all U.S. lotteries and study the summary of state and multi-state lottery odds and prizes. Pick five lotteries and determine the expected value for winnings in each case. Discuss your results.

26. Law of Large Numbers. Use a coin to simulate 100 births: Flip the coin 100 times, recording the results, and then convert the outcomes to genders of babies (tail = boy and head = girl). Use the results to fill in the following table. What happens to the proportion of girls as the sample size increases? How does this illustrate the law of large numbers?

Number of births	10	20	30	40	50	60	70	80	90	100
Proportion of girls										

IN THE NEWS

27. Personal Law of Large Numbers. Describe a situation in which you personally have made use of the law of large numbers, either correctly or incorrectly. Why did you use the law of large numbers in this situation? Was it helpful?

28. Gambler's Fallacy in Life. Describe a situation in which you or someone you know has fallen victim to the gambler's fallacy. How could the situation have been dealt with correctly?

29. Gambler's Ruin. Describe a situation you know of in which someone lost nearly everything through gambling. Did his or her strategy appear to be rational, or did it appear to be the result of a destructive addiction?

6.4 Ideas of Risk and Life Expectancy

A smooth-talking but honest salesman comes to you, offering a new product:

> *I can't reveal the details yet, but you will love this product! It will improve your life in more ways than you can count. Its only downside is that it will eventually kill everyone who uses it. Will you buy one?*

Not likely. After all, could any product be so great that you would die for it? A few weeks later, the salesman shows up again:

> *No one was buying, so we've made some improvements. Your chance of being killed by the product is now only 1 in 10. Ready to buy?*

Despite the improvement, most people still would send the salesman home and wait for his inevitable return:

> *Okay, this time we've really perfected it. We've made it so safe that it would take 20 years for it to kill as many people as live in San Francisco. And it can be yours for a mere $20,000.*

You may be surprised to realize that, if you are like most Americans, you'll jump at this offer. The product is, after all, the automobile. It does indeed improve our lives in many ways, and $20,000 is a typical price. And, given that roughly 40,000 Americans are killed in auto accidents each year, it kills the equivalent of the population of San Francisco (roughly 800,000) in about 20 years.

As this example shows, we frequently make tradeoffs between benefits and risks. In this section, we'll see how ideas of probability can help us quantify risk, thereby allowing us to make informed decisions about such tradeoffs.

By the Way ...

Roughly 115 people die in automobile accidents per day, or 1 every 13 minutes. Automobile accidents are the leading cause of death among people of ages 6 to 27 years. Vehicle occupants account for 85% of automobile fatalities, while pedestrians and cyclists account for 15%.

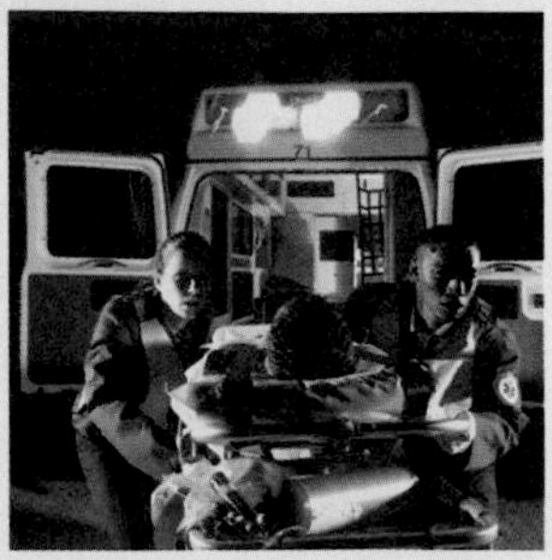

Risk and Travel

Are you safer in a small car or in a sport utility vehicle? Are cars today safer than those 30 years ago? If you need to travel across country, are you safer flying or driving? To answer these and many similar questions, we must quantify the risk involved in travel. We can then make decisions appropriate for our own personal circumstances.

Travel risk is often expressed in terms of an **accident rate** or **death rate**. For example, suppose an annual accident rate is 750 accidents per 100,000 people. This means that, within a group of 100,000 people, on average 750 will have an accident over the period of a year. The statement is in essence an expected value, which means it also represents a probability: It tells us that the probability of a person being involved in an accident (in one year) is 750 in 100,000, or 0.0075.

This concept of travel risk is straightforward, but we must still interpret the numbers with care. For example, travel risks are sometimes stated *per 100,000 people,* as above, but other times they are stated *per trip* or *per mile.* If we use death rates *per trip* to compare the risks of flying and driving, we neglect the fact that airplane trips are typically much longer than automobile trips. Similarly, if we use accident rates *per person,* we neglect the fact that most automobile accidents involve only minor injuries.

EXAMPLE 1 Is Driving Getting Safer?

Figure 6.11 shows the number of automobile fatalities and the total number of miles driven (among all Americans) for each year over a period of more than three decades. In terms of death rate per mile driven, how has the risk of driving changed?

Solution Figure 6.11a shows that the annual number of fatalities decreased from about 52,000 in 1970 to about 43,000 in 2004. Meanwhile, Figure 6.11b shows that the number of miles driven increased from about 1,000 billion (1×10^{12}) to about 2,900 billion (2.9×10^{12}). Therefore, the death rates per mile for the beginning and end of the period were

$$\textit{1970:} \quad \frac{52{,}000 \text{ deaths}}{1 \times 10^{12} \text{ miles}} \approx 5.2 \times 10^{-8} \text{ death per mile}$$

$$\textit{2004:} \quad \frac{43{,}000 \text{ deaths}}{2.9 \times 10^{12} \text{ miles}} \approx 1.5 \times 10^{-8} \text{ death per mile}$$

Note that, because $10^8 = 100$ million, 5.2×10^{-8} death per mile is equivalent to 5.2 deaths per 100 million miles. Thus, over 34 years, the death rate per 100 million miles dropped from 5.2 to 1.5. By this measure, driving became much safer over the period. Most researchers believe the improvements resulted from better automobile design and from safety features, such as shoulder belts and air bags, that are much more common today.

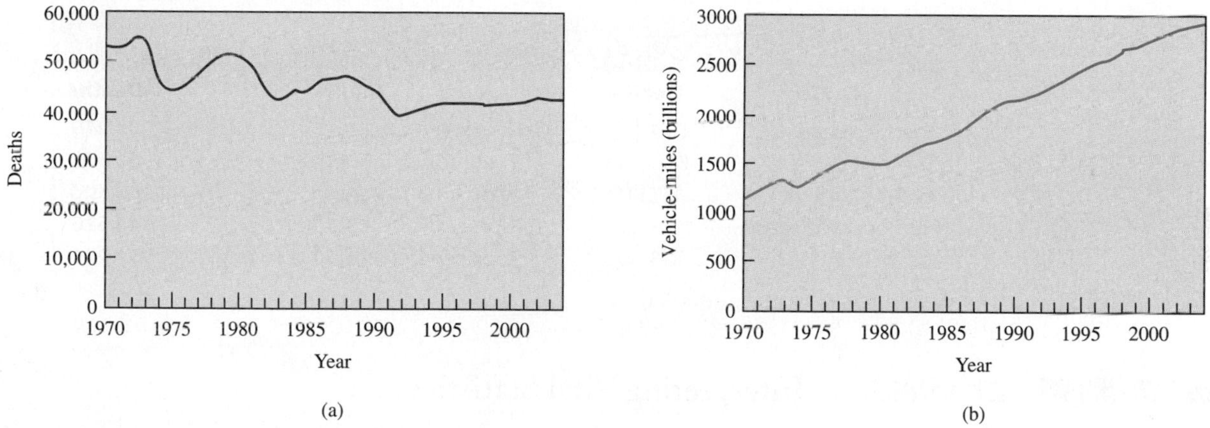

Figure 6.11 (a) Annual automobile fatalities. (b) Total miles driven annually. Both sets of data are for the United States only. *Source:* National Transportation Safety Board.

EXAMPLE 2 Which Is Safer: Flying or Driving?

Over the past 20 years in the United States, the average (mean) number of deaths in commercial airplane accidents has been roughly 100 per year. (The actual number varies significantly from year to year.) Currently, airplane passengers in the United States travel a total of about 8 billion miles per year. Use these numbers to calculate the death rate per mile of air travel. Compare the risk of flying to the risk of driving.

The cost of living is going up and the chance of living is going down.

—Flip Wilson, comedian

Solution Assuming 100 deaths and 8 billion miles in an average year, the risk of air travel is

$$\frac{100 \text{ deaths}}{8 \times 10^{9} \text{ miles}} \approx 1.3 \times 10^{-8} \text{ death per mile}$$

This risk is equivalent to 1.3 deaths per 100 million miles, or slightly lower than the risk of 1.5 deaths per 100 million miles for driving (see Example 1). Note that, because the average air trip covers a considerably longer distance than the average driving trip, the risk *per trip* is much higher for air travel, although the risk *per mile* is lower.

TIME OUT TO THINK

Suppose you need to make the 800-mile trip from Atlanta to Houston. Do you think it is safer to fly or to drive? Why?

Vital Statistics

Only those who risk going too far can possibly find out how far one can go.

—T. S. Eliot

Data concerning births and deaths of citizens, often called *vital statistics*, are very important to understanding risk-benefit tradeoffs. For example, insurance companies use vital statistics to assess risks and set rates. Health professionals study vital statistics to assess medical progress and decide where research resources should be concentrated. Demographers use birth and death rates to predict future population trends.

One important set of vital statistics, shown in Table 6.8, concerns causes of death. These data are extremely general; a more complete table would categorize data by age, sex, and race. Vital statistics are often expressed in terms of deaths per person or per 100,000 people, which makes it easier to compare the rates for different years and for different states or countries.

Table 6.8 Leading Causes of Death in the United States (in a single recent year)

Cause	Deaths	Cause	Deaths
Heart disease	684,462	Diabetes	73,249
Cancer	554,643	Pneumonia/Influenza	65,681
Stroke	157,803	Alzheimer's disease	63,343
Pulmonary disease	126,128	Kidney disease	42,536
Accidents	105,695	Septicemia (blood poisoning)	34,243

Source: Centers for Disease Control.

By the Way ...

Among college-age students, alcohol consumption presents one of the most serious health risks. The National Institutes of Health estimates that alcohol contributes to 1,400 deaths and 500,000 injuries among college students each year, as well as to 70,000 cases of sexual assault.

EXAMPLE 3 Interpreting Vital Statistics

Assuming a U.S. population of 300 million, find and compare risks per person and per 100,000 people for pneumonia (and influenza) and cancer.

Solution We find the risk per person by dividing the number of deaths by the total population of 300 million:

$$\textit{Pneumonia/influenza:} \quad \frac{65{,}681 \text{ deaths}}{300{,}000{,}000 \text{ people}} \approx 0.00022 \text{ death per person}$$

$$\textit{Cancer:} \quad \frac{554{,}643 \text{ deaths}}{300{,}000{,}000 \text{ people}} \approx 0.0018 \text{ death per person}$$

We can interpret these numbers as probabilities: The probability of death by pneumonia or influenza is about 2.2 in 10,000, while the probability of death by cancer is about 18 in 10,000. To put them in terms of deaths per 100,000 people, we simply multiply the per person rates by 100,000. We get a pneumonia/influenza death rate of 22 deaths per 100,000 people and a cancer death rate of 180 deaths per 100,000 people. The probability of death by cancer is more than eight times that of death by pneumonia or influenza.

TIME OUT TO THINK

Table 6.8 suggests that the probability of death by stroke is about 50% higher than the probability of death by accident, but these data include all age groups. How do you think the risks of stroke and accident would differ between young people and older people? Explain.

Life Expectancy

We now turn to the idea of *life expectancy*, which is often used to compare overall health at different times or in different countries. The idea will be clearer if we start by looking at death rates. Figure 6.12a shows the overall U.S. death rate (or mortality rate), in deaths per 1,000 people, for different age groups. Note that there is an elevated risk of death near birth, after which the death rate drops to very low levels. At about 15 years of age, the death rate begins a gradual rise.

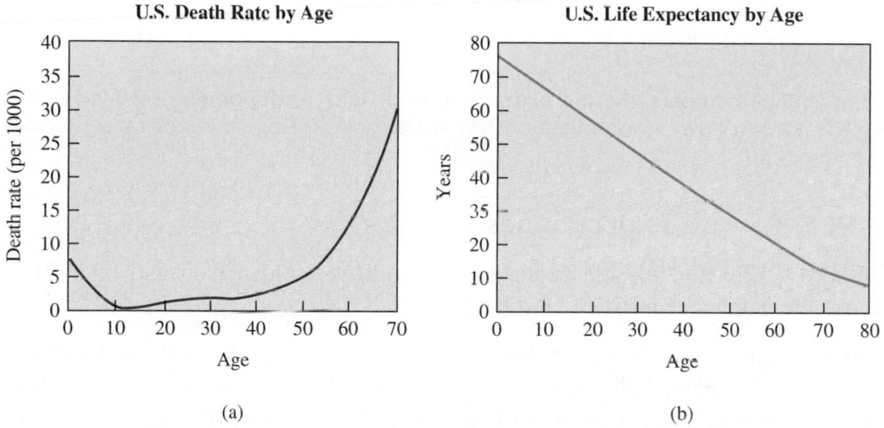

Figure 6.12 (a) The overall U.S. death rate (deaths per 1,000 people) for different ages. (b) Life expectancy for different ages. *Source:* U.S. National Center for Health Statistics.

Figure 6.12b shows the **life expectancy** of Americans of different ages, defined as the number of years a person of a given age can expect to live on average. As we would expect, life expectancy is higher for younger people because, on average, they have longer left to live. At birth, the life expectancy of Americans today is about 78 years (75 years for men and 81 years for women).

Definition

Life expectancy is the number of years a person with a given age today can expect to live on average.

The subtlety in interpreting life expectancy comes from changes in medical science and public health. Life expectancies are calculated by studying *current* death rates. For example, when we say that the life expectancy of infants born today is 78 years, we mean that the average baby will live to age 78 *if there are no future changes* in medical science or public health. Thus, while life expectancy provides a useful measure of current overall health, it should not be considered a *prediction* of future life spans.

In fact, because of advances in both medical science and public health, life expectancies increased dramatically during the past century, rising by about 60% (Figure 6.13). If this trend continues, today's infants are likely to live much longer than 78 years on average.

By the Way ...

Life expectancies vary widely around the world, and life expectancy in the United States (78 years) trails that of many other developed countries including Japan (81 years), Australia (80.5 years), and Canada (80.2 years). The lowest life expectancies are found in war- and disease-ravaged countries of sub-Saharan Africa, such as Angola (38 years) and Sierra Leone (40 years).

The reports of my death are greatly exaggerated.

—Mark Twain, from London, in a cable to the Associated Press

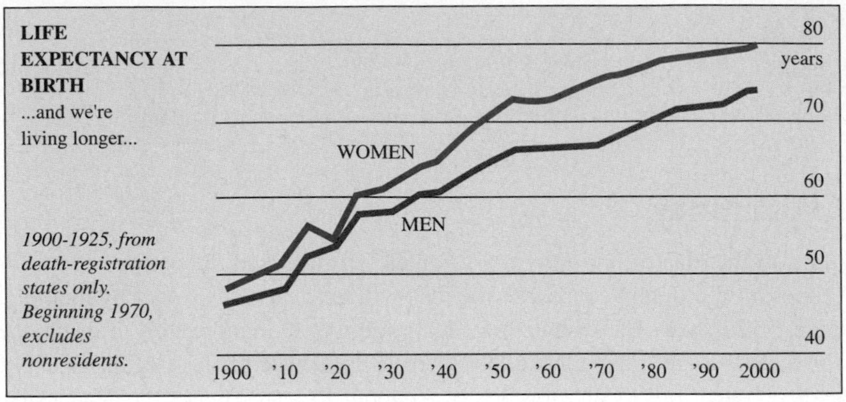

Figure 6.13 Changes in U.S. life expectancy during the 20th century. *Source: New York Times* and National Center for Health Science Statistics.

TIME OUT TO THINK

Using Figure 6.13, compare the life expectancies of men and women. Briefly discuss these differences. Do they have any implications for social policy? For insurance rates? Explain.

EXAMPLE 4 Life Expectancies

Using Figure 6.12b, find the life expectancy of a 20-year-old person and of a 60-year-old person. Are the numbers consistent? Explain.

Solution The graph shows that the life expectancy at age 20 is about 58 years and at age 60 is about 21 years. This means that an average 20-year-old can expect to live about 58 more years, to age 78. An average 60-year-old can expect to live about 21 more years, to age 81.

It might at first seem strange that 60-year-olds have a longer average life span than 20-year-olds (81 years versus 78 years). But remember that life expectancies are based on *current* data. If there were no changes in medicine or public health, a 60-year-old would have a greater probability of reaching age 81 than a 20-year-old simply because he or she has already made it to age 60. However, if medicine and public health continue to improve, today's 20-year-olds may live to older ages than today's 60-year-olds.

CASE STUDY Life Expectancy and Social Security

Because of the changing age makeup of the U.S. population, the number of retirees collecting Social Security benefits is expected to become much larger in the future, while the number of wage earners paying Social Security taxes is expected to remain relatively constant. As a result, one of the biggest challenges to the future of Social Security is finding a way to make sure there is enough money to pay benefits for future retirees.

Current projections show that, without significant changes, the Social Security program will be bankrupt and unable to pay full benefits after about 2040. (This may sound far off, but today's college students will be in their prime earning years at this time.) Social Security officials have proposed several different ways to solve this problem, including changes in the amounts of benefits paid, increases in the Social Security tax rate, changes in the retirement age, and partially or fully privatizing the Social Security program. Each of these proposals faces political obstacles. But, in addition, all of these proposals are based on assumptions about future life expectancy. Without corresponding changes in retirement age, longer lives mean more years of Social Security benefits.

More specifically, recent proposals by Social Security officials have assumed that life expectancy at birth will rise to 79.3 by 2030 and to 81.5 by 2070 (for both sexes combined). But these numbers may be far too pessimistic. For example, the Social Security projections assume that American women will not reach a life expectancy of 82 years until 2033—but women in many European and Asian nations have *already* achieved this life expectancy. Will it really take the United States decades to catch up? One expert panel suggested that more realistic estimates of future life expectancy would be 81 in 2030 and 85 in 2070, meaning nearly four more years of Social Security benefits for the average person in 2070. And, if medical science achieves any major breakthroughs in allowing people to live longer, the Social Security crisis could become far worse.

TIME OUT TO THINK

According to some biologists, there is a good chance that 21st-century advances in medical science will allow most people to live to ages of 100 or more. How would that affect programs like Social Security? What other effects would you expect it to have on society? Overall, do you think large increases in life expectancy would be good or bad for society? Defend your opinion.

Section 6.4 Exercises

Statistical Literacy and Critical Thinking

1. **Birth Rate.** The current U.S. birth rate is given as 14.1 per 1,000 population. When comparing population growth in different countries, why is it better to use birth *rates* instead of the actual numbers of births?

2. **Vital Statistics.** What are vital statistics?

3. **Life Expectancy.** What is life expectancy? Does a 30-year-old person have the same life expectancy as a 20-year-old person? Why or why not?

4. **Life Expectancy.** Based on recent data, a 20-year-old person in the United States has a life expectancy of 57.8 years. What does that mean?

Does It Make Sense? For Exercises 5–8, decide whether the statement makes sense (or is clearly true) or does not make sense (or is clearly false). Explain clearly. Not all of these statements have definitive answers, so your explanation is more important than your chosen answer.

5. **Plane Crash.** Plane crashes receive much attention in the media, even if the crash involves one or two people. Because plane crashes are given so much attention, the risk of dying in a plane crash must be greater than the risk of dying in a car crash.

6. **Life Expectancy.** A public health official states that your life expectancy decreases as you age.

7. **Expected Age at Death.** As you become older, your expected age at death increases.

8. **Risk of Death.** In a recent year, the total numbers of deaths in the United States due to accidents and pneumonia were approximately equal. Therefore, the risks of death by accident and pneumonia per 100,000 people are approximately equal.

Commercial Aviation Fatality Rates. For Exercises 9–12, use the following table, which summarizes data on commercial aviation flights in the United States for three separate years.

Year	Departures (thousands)	Fatalities	Passenger miles (billions)	Passengers (millions)
1995	8062	168	540.7	547.8
2000	9035	92	692.8	666.2
2004	11,182	14	731.9	697.8

9. For each of the three years, find the fatality rate in deaths per 1,000 departures. On the basis of those rates, which year was the safest? Why?

10. For each of the three years, find the fatality rate in deaths per 10 billion passenger miles. On the basis of those rates, which year was the safest? Why?

11. For each of the three years, find the fatality rate in deaths per 10 million passengers. On the basis of those rates, which year was the safest? Why?

12. For the year 2004, find the fatality rate in deaths per passenger mile. Why don't we report the fatality rate in units of deaths per passenger mile?

Life Table. For Exercises 13–16, use the data in the following table for people in the United States between the ages of 16 and 21 years.

Age interval	Probability of dying during the interval	Number surviving to the beginning of the interval	Number of deaths during the interval	Expected remaining lifetime (from the beginning of the interval)
16–17	0.000607	98,943	60	61.7
17–18	0.000706	98,883	70	60.7
18–19	0.000780	98,814	77	59.7
19–20	0.000833	98,736	82	58.8
20–21	0.000888	98,654	88	57.8

13. Expected Lifetime. How many years is a randomly selected 19-year-old expected to live beyond his or her 19th birthday?

14. Expected Lifetime. How many years is a randomly selected 17-year-old expected to live beyond his or her 17th birthday?

15. Death Rate. Life insurance companies must carefully monitor death rates. Before issuing a life insurance policy for a 19-year-old, the company needs to know the death rate for that age group. Find the death rate per 10,000 for people during their 19th year.

16. Death Rate. Find the death rate per 10,000 for people during their 16th year.

17. High/Low U.S. Birth Rates. The highest and lowest birth *rates* in the United States in 2003 were in Utah and Maine, respectively. Utah reported 49,870 births with a population of about 2.4 million people. Maine reported 13,861 births with a population of about 1.3 million people. Use these data to answer the following questions.

a. How many people were born per day in Utah? In Maine?

b. What was the birth rate in Utah in births per 1,000 people? In Maine?

18. High/Low U.S. Death Numbers. In a recent year, there were 235,000 deaths in California, the highest number in the United States. The state with the lowest number of deaths was Alaska, with 3,000 deaths. The populations of California and Alaska were approximately 35,463,000 and 648,000, respectively.

a. Compute the death *rates* for California and Alaska in deaths per 1,000.

b. Based on the fact that California and Alaska had the highest and lowest death numbers, respectively, does it follow that California and Alaska had the highest and lowest death *rates*? Why or why not?

19. U.S. Birth and Death Rates. In 2006, the U.S. population reached 300 million. The overall birth rate was estimated to be 14.1 births per 1,000, and the overall death rate was estimated to be 8.4 deaths per 1,000.

a. Approximately how many births were there in the United States?

b. About how many deaths were there in the United States?

c. Based on births and deaths alone (i.e., not counting immigration and emigration), about how much did the U.S. population rise during 2006?

d. Ignoring immigration and emigration, what is the 2006 rate of population growth of the United States? What is the population growth rate expressed as a percentage?

20. China Birth and Death Rates. These estimated 2006 values are for China: population = 1,313,973,713; birth rate = 13.25 per 1,000; death rate = 6.97 per 1,000.

a. Approximately how many births were there in China?

b. About how many deaths were there in China?

c. Based on births and deaths alone (i.e., not counting immigration and emigration), about how much did the population of China rise during 2006?

d. Ignoring immigration and emigration, what is the 2006 rate of population growth of China? What is the population growth rate expressed as a percentage?

Projects for the Internet and Beyond

For useful links, select "Links for Internet Projects" for Chapter 6 at www.aw.com/bbt.

21. U.S. vs. World Life Expectancy. You can find a great deal of data on the Web about life expectancies around the world. How does U.S. life expectancy compare to life expectancy in other developed countries? What might explain the differences you see? Based on your findings, discuss potential implications for social or government policy in the United States.

22. Male and Female Life Expectancies. Find data about how and why male and female life expectancies are changing with time. Why do women have longer life expectancies than men? Should we expect male life expectancies to catch up with female life expectancies in the future? Summarize your findings with a short report.

23. Uganda Case Study. Find data regarding life expectancies in Uganda over the past few decades. You will see that Ugandan life expectancy has risen dramatically in the past decade. What explains this rise, and what lessons might it have for improving life expectancy in other African nations?

24. Life Expectancy Calculations. You will find many life expectancy calculators available on the Internet; try a few of them. Do they seem to give accurate or realistic results? Explore the statistical techniques that are used to make life expectancy tables.

25. Richter Scale for Risk. The Royal Statistical Society has proposed a system of risk magnitudes and risk factors analogous to the Richter scale for measuring earthquakes. Go to the Internet to learn how these measures of risk are defined and computed. Using these measures, discuss the risks of various activities and events.

26. Understanding Risk. The book *Against the Gods: The Remarkable Story of Risk* by Peter Bernstein (John Wiley, 1996) is an award-winning account of the history of probability and risk assessment. Find the book in a library or bookstore (it's a worthwhile purchase) and identify a particular event that changed our understanding of risk. Write a two-page essay on both the history and the consequences of this particular event.

IN THE NEWS

27. Travel Safety. Find a recent news article discussing some aspect of travel safety (such as risk of accidents in automobiles or airplanes, the efficacy of child car seats, or the effects of driving while talking on a cell phone). Summarize any given statistics about risks, and give your overall opinion regarding the safety issue under discussion.

28. Vital Statistics. Find a recent news report that gives current data about vital statistics or life expectancy. Summarize the report and the statistics, and discuss any personal or social implications of the new data.

6.5 Combining Probabilities (Supplementary Section)

The ideas of probability that we have discussed to this point in the chapter will be sufficient for most of the work we will do in this book. However, probability has many more applications, both in statistics and in other areas of life. In this section, we investigate a few more ideas of probability and a few of their many applications.

And Probabilities

Chance favors only the prepared mind.

—Louis Pasteur, 19th-century scientist

Suppose you toss two fair dice and want to know the probability that *both* will come up 4. One way to find the probability is to consider the two tossed dice as a *single* toss of two dice. Then we can find the probability using the *theoretical* method (see Section 6.2). Because "double 4's" is 1 of 36 possible outcomes, its probability is 1/36.

Alternatively, we can consider the two dice individually. For each die, the probability of a 4 is 1/6. We find the probability that both dice show a 4 by multiplying the individual probabilities:

$$P(\text{double 4's}) = P(4) \times P(4) = \frac{1}{6} \times \frac{1}{6} = \frac{1}{36}$$

By either method, the probability of rolling double 4's is 1/36. In general, we call the probability of event *A and* event *B* occurring an ***and* probability** (or *joint probability*).

The advantage of the multiplication technique is that it can easily be extended to situations involving more than two events. For example, we might want to find the probability of getting 10 heads on 10 coin tosses or of having a baby *and* getting a pay raise in the same year.

By the Way

Whether events are independent is not always obvious. Consider a set of basketball free throws by a single player. Some people argue that the probability of a basket in the same for each free throw. Others suggest that the psychological impact of success or failure on one free throw affects the player's chances of making the next one. Despite many statistical studies of basketball free throws, no one has clearly established which view is correct.

However, there is an important distinction that must be made when working with *and* probabilities. We must distinguish between events that are *independent* and events that are *dependent* upon each other. Let's investigate each case.

Independent Events

The repeated roll of a single die produces **independent events** because the outcome of one roll does not affect the probabilities of the other rolls. Similarly, coin tosses are independent. Deciding whether events are independent is important, but the answer is not always obvious. For example, in analyzing a succession of free throws of a basketball player, should we assume that one free throw is independent of the others? Whenever events *are* independent, we can calculate the *and* probability of two or more events by multiplying.

And Probability for Independent Events

Two events are **independent** if the outcome of one event does not affect the probability of the other event. Consider two independent events A and B with probabilities $P(A)$ and $P(B)$. The probability that A *and* B occur together is

$$P(A \text{ and } B) = P(A) \times P(B)$$

This principle can be extended to any number of independent events. For example, the probability of A, B, and a third independent event C is

$$P(A \text{ and } B \text{ and } C) = P(A) \times P(B) \times P(C)$$

EXAMPLE 1 Three Coins

Suppose you toss three fair coins. What is the probability of getting three tails?

Solution Because coin tosses are independent, we multiply the probability of tails on each individual coin:

$$P(3 \text{ tails}) = \underbrace{P(\text{tails})}_{\text{coin 1}} \times \underbrace{P(\text{tails})}_{\text{coin 2}} \times \underbrace{P(\text{tails})}_{\text{coin 3}} = \frac{1}{2} \times \frac{1}{2} \times \frac{1}{2} = \frac{1}{8}$$

The probability that three tossed coins all land on tails is 1/8 (which we determined in Example 8 of Section 6.2 with much more work).

Dependent Events

A batch of 15 memory cards contains 5 defective cards. If you select a card at random from the batch, the probability of getting a defect is 5/15. Now, suppose that you select a defect on the first selection and put it in your pocket. What is the probability of getting a defect on the second selection?

Because you've removed one defective card from the batch, the batch now contains only 14 cards, of which 4 are defective. Thus, the probability of getting a defective card on the second draw is 4/14. This probability is less than the 5/15 probability on the first selection because the first selection changed the contents of the batch. Because the outcome of the first event affects the probability of the second event, these are **dependent events**.

Calculating the probability for dependent events still involves multiplying the individual probabilities, but we must take into account how prior events affect subsequent events. In the case of the batch of memory cards, we find the probability of getting two defective cards in a row by multiplying the 5/15 probability for the first selection by the 4/14 probability for the second selection.

$$P(\text{2 defectives}) = \underbrace{P(\text{defective})}_{\text{first selection}} \times \underbrace{P(\text{defective})}_{\substack{\text{second selection,} \\ \text{if first selection} \\ \text{defective}}} = \frac{5}{15} \times \frac{4}{14} = 0.0952$$

The probability of drawing two defective cards in a row is 0.0952, which is slightly less than $(5/15) \times (5/15) = 0.111$, the probability we get if we replace the first card before the second selection.

And Probability for Dependent Events

Two events are **dependent** if the outcome of one event affects the probability of the other event. The probability that dependent events *A* and *B* occur together is

$$P(A \text{ and } B) = P(A) \times P(B \text{ given } A)$$

where *P*(*B* given *A*) means the probability of event *B* given the occurrence of event *A*.

This principle can be extended to any number of individual events. For example, the probability of dependent events *A*, *B*, and *C* is

$$P(A \text{ and } B \text{ and } C) = P(A) \times P(B \text{ given } A) \times P(C \text{ given } A \text{ and } B)$$

TECHNICAL NOTE

P(*B* given *A*) is called a *conditional probability*. In some books, it is denoted $P(B|A)$.

EXAMPLE 2 Playing BINGO

The game of BINGO involves drawing labeled buttons from a bin at random, without replacement. There are 75 buttons, 15 for each of the letters B, I, N, G, and O. What is the probability of drawing two B buttons in the first two selections?

Solution BINGO involves dependent events because removing a button changes the contents of the bin. The probability of drawing a B on the first draw is 15/75. If this occurs, 74 buttons remain in the bin, of which 14 are Bs. Therefore, the probability of drawing a B button on the second draw is 14/74. The probability of drawing two B buttons in the first two selections is

$$P(\text{B and B}) = \underbrace{P(\text{B})}_{\text{first draw}} \times \underbrace{P(\text{B})}_{\substack{\text{second draw,} \\ \text{given B on} \\ \text{first draw}}} = \frac{15}{75} \times \frac{14}{74} = 0.0378$$

TIME OUT TO THINK

Without doing calculations, compare the probability of drawing three hearts in a row when the card is replaced and the deck is shuffled after each draw to the probability of drawing three hearts in a row without replacement. Which probability is larger and why?

EXAMPLE 3 When Can We Treat Dependent Events as Independent Events?

A polling organization has a list of 1,000 people for a telephone survey. The pollsters know that 433 people out of the 1,000 are members of the Democratic Party. Assuming that a person cannot be called more than once, what is the probability that the first two people called will be members of the Democratic Party?

Solution This problem involves an *and* probability for dependent events: Once a person is called, that person cannot be called again. The probability of calling a member of the Democratic Party on the first call is 433/1,000. With that person removed from the calling

The importance of probability can only be derived from the judgment that it is rational to be guided by it in action.

—John Maynard Keynes

pool, the probability of calling a member of the Democratic Party on the second call is 432/999. Therefore, the probability of calling two members of the Democratic Party on the first two calls is

$$\frac{433}{1{,}000} \times \frac{432}{999} = 0.1872$$

If we treat the two calls as independent events, the probability of calling a member of the Democratic Party is 433/1,000 on both calls. Then the probability of calling two members of the Democratic Party is

$$\frac{433}{1{,}000} \times \frac{433}{1{,}000} = 0.1875$$

which is nearly identical to the result assuming dependent events. In general, if relatively few items or people are selected from a large pool (in this case, 2 people out of 1,000), then dependent events can be treated as independent events with very little error. A common guideline is that independence can be assumed when the sample size is less than 5% of the population size. This practice is commonly used by polling organizations.

By the Way ...

In the late 1960s, the famed baby doctor Benjamin Spock was convicted by an all-male jury of encouraging draft resistance during the Vietnam War. His defense argued that a jury with women would have been more sympathetic.

EXAMPLE 4 Jury Selection

A nine-person jury is selected at random from a very large pool of people that has equal numbers of men and women. What is the probability of selecting an all-male jury?

Solution When a juror is selected, that juror is removed from the pool. However, because we are selecting a small number of jurors from a large pool, we can treat the selection of jurors as independent events. Assuming that equal numbers of men and women are available, the probability of selecting a single male juror is $P(\text{male}) = 0.5$. Thus, the probability of selecting 9 male jurors is

$$P(9 \text{ males}) = \underbrace{0.5 \times 0.5 \times \cdots \times 0.5}_{9 \text{ times}} = 0.5^9 = 0.00195$$

The probability of selecting an all-male jury by random selection is 0.00195, or roughly 2 in 1,000. (Note: Expressions of the form a^b, such as 0.5^9, can be evaluated on many calculators using the key marked x^y or y^x or $>$.)

Either/Or Probabilities

So far we have considered the probability that one event occurs in a first trial *and* another event occurs in a second trial. Suppose we want to know the probability that, when one trial is conducted, *either* of two events occurs. In that case, we are looking for an ***either/or* probability**, such as the probability of having either a blue-eyed *or* a green-eyed baby or the probability of losing your home either to a fire *or* to a hurricane.

Non-Overlapping Events

A coin can land on either heads *or* tails, but it can't land on both heads *and* tails at the same time. When two events cannot possibly occur at the same time, they are said to be **non-overlapping** (or *mutually exclusive*). We can represent non-overlapping events with a Venn diagram in which each circle represents an event. If the circles do not overlap, it means that the corresponding events cannot occur together. For example, we show the possibilities of heads and tails in a coin toss as two non-overlapping circles because a coin cannot land on both heads and tails at the same time (Figure 6.14).

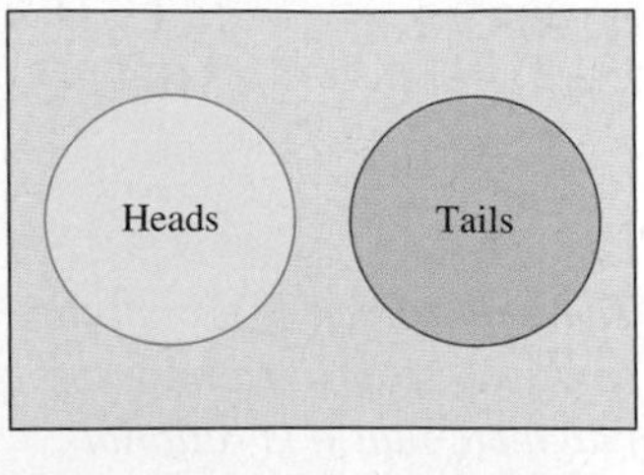

Figure 6.14 Venn diagram for non-overlapping events.

Suppose we roll a die once and want to find the probability of rolling a 1 *or* a 2. Because there are six equally likely outcomes and because two of those outcomes correspond to the events in question, we can use the theoretical method to conclude that the probability of rolling a 1 *or* a 2 is $P(1 \text{ or } 2) = 2/6 = 1/3$. Alternatively, we can find this probability by adding the individual probabilities $P(1) = 1/6$ and $P(2) = 1/6$. In general, we find the *either/or* probability of two non-overlapping events by adding the individual probabilities.

Either/Or Probability for Non-Overlapping Events

Two events are **non-overlapping** if they cannot occur at the same time. If A and B are non-overlapping events, the probability that either A or B occurs is

$$P(A \text{ or } B) = P(A) + P(B)$$

This principle can be extended to any number of non-overlapping events. For example, the probability that either event A, event B, or event C occurs is

$$P(A \text{ or } B \text{ or } C) = P(A) + P(B) + P(C)$$

provided that A, B, and C are all non-overlapping events.

EXAMPLE 5 *Either/Or* Dice Probability

Suppose you roll a single die. What is the probability of getting an even number?

Solution The even outcomes of 2, 4, and 6 are non-overlapping because a single die can yield only one result. We know that $P(2) = 1/6$, $P(4) = 1/6$, and $P(6) = 1/6$. Therefore, the combined probability is

$$P(2 \text{ or } 4 \text{ or } 6) = P(2) + P(4) + P(6) = \frac{1}{6} + \frac{1}{6} + \frac{1}{6} = \frac{1}{2}$$

The probability of rolling an even number is 1/2.

Overlapping Events

To improve tourism between France and the United States, the two governments form a committee consisting of 20 people: 2 American men, 4 French men, 6 American women, and 8 French women (as shown in Table 6.9). If you meet one of these people at random, what is the probability that the person will be *either* a woman *or* a French person?

Table 6.9 Tourism Committee

	Men	Women
American	2	6
French	4	8

Twelve of the 20 people are French, so the probability of meeting a French person is 12/20. Similarly, 14 of the 20 people are women, so the probability of meeting a woman is 14/20. The sum of these two probabilities is

$$\frac{12}{20} + \frac{14}{20} = \frac{26}{20}$$

This cannot be the correct probability of meeting either a woman or a French person, because probabilities cannot be greater than 1. The Venn diagram in Figure 6.15 shows why simple addition was wrong in this situation. The left circle contains the 12 French people, the right circle contains the 14 women, and the American men are in neither circle. We see that there are 18 people who are either French or women (or both). Because the total number of people in the room is 20, the probability of meeting a person who is *either* French *or* a woman is 18/20 = 9/10.

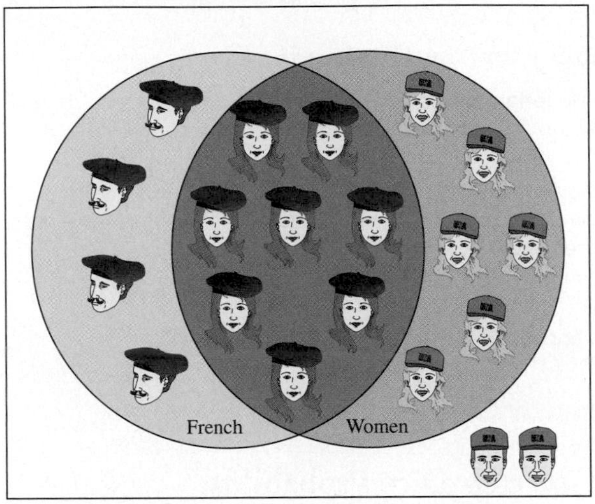

Figure 6.15 Venn diagram for overlapping events.

As the Venn diagram shows, simple addition was incorrect because the region in which the circles overlap contains 8 people who are *both* French *and* women. If we add the two individual probabilities, these 8 people get counted twice: once as women and once as French people. The probability of meeting one of these French women is 8/20. We can correct the double counting error by subtracting out this probability. Thus, the probability of meeting a person who is either French or a woman is

$$P(\text{woman or French}) = \underbrace{\frac{14}{20}}_{\substack{\text{probability}\\ \text{of a woman}}} + \underbrace{\frac{12}{20}}_{\substack{\text{probability of}\\ \text{a French person}}} - \underbrace{\frac{8}{20}}_{\substack{\text{probability of}\\ \text{a French woman}}} = \frac{18}{20} = \frac{9}{10}$$

which agrees with the result found by counting.

We say that meeting a woman and meeting a French person are **overlapping** (or *non–mutually exclusive*) events because both can occur at the same time. Generalizing the procedure we used in this example, we find the following rule.

Either/Or Probability for Overlapping Events

Two events *A* and *B* are **overlapping** if they can occur together. The probability that either *A* or *B* occurs is

$$P(A \text{ or } B) = P(A) + P(B) - P(A \text{ and } B)$$

The last term, $P(A \text{ and } B)$, corrects for the double counting of events in which *A* and *B* both occur together. Note that it is not necessary to use this formula. The correct probability can always be found by counting carefully and avoiding double counting.

TIME OUT TO THINK

Are the events of being born on a Wednesday or being born in Las Vegas overlapping? Are the events of being born on a Wednesday or being born in March overlapping? Are the events of being born on a Wednesday or being born on a Friday overlapping? Explain.

EXAMPLE 6 Minorities and Poverty

Pine Creek is an "average" American town: Of its 2,350 citizens, 1,950 are white, of whom 11%, or 215 people, live below the poverty level. Of the 400 minority citizens, 28%, or 112 people, live below the poverty level. If you visit Pine Creek, what is the probability of meeting (at random) a person who is *either* a minority *or* living below the poverty level? (This example is hypothetical but the percentages are consistent with national demographics.)

Solution Meeting a minority citizen and meeting a person living in poverty are overlapping events. It's useful to make a small table such as Table 6.10, showing how many citizens are in each of the four categories.

Table 6.10 Citizens in Pine Creek

	In poverty	Above poverty
White	215	1,735
Minority	112	288

You should check that the figures in the table are consistent with the given data and that the total in all four categories is 2,350. Because there are 400 minority citizens, the probability of (randomly) meeting a minority citizen is $400/2{,}350 = 0.170$. Because there are $215 + 112 = 327$ people living in poverty, the probability of meeting a citizen in poverty is $327/2{,}350 = 0.139$. The probability of meeting a person who is *both* a minority citizen and a person living in poverty is $112/2{,}350 = 0.0477$. According to the rule for overlapping events, the probability of meeting *either* a minority citizen *or* a person living in poverty is

$$P(\text{minority or poverty}) = 0.170 + 0.139 - 0.0477 = 0.261$$

The probability of meeting a citizen who is *either* a minority or a person living below the poverty level is about 1 in 4. Notice the importance of subtracting out the term that corresponds to meeting a person who is *both* a minority citizen and a person living in poverty.

Summary

Table 6.11 provides a summary of the formulas we've used in combining probabilities.

Table 6.11 Summary of Combining Probabilities

***And* probability: independent events**	***And* probability: dependent events**	***Either/or* probability: non-overlapping events**	***Either/or* probability: overlapping events**
$P(A \text{ and } B) = P(A) \times P(B)$	$P(A \text{ and } B) = P(A) \times P(B \text{ given } A)$	$P(A \text{ or } B) = P(A) + P(B)$	$P(A \text{ or } B) = P(A) + P(B) - P(A \text{ and } B)$

Section 6.5 Exercises

Statistical Literacy and Critical Thinking

1. **Independence.** In your own words, state what it means for two events to be independent.

2. **Non-overlapping Events.** In your own words, state what it means for two events to be non-overlapping.

3. **Sampling with Replacement?** The professor in a class of 25 students randomly selects a student and then randomly selects a second student. If all 25 students are available for the second selection, is this sampling with replacement or sampling without replacement? Is the second outcome independent of the first?

4. **Opposites.** If two events are opposites, such as A and *not* A, must those two events be non-overlapping? Why or why not?

Does It Make Sense? For Exercises 5–8, decide whether the statement makes sense (or is clearly true) or does not make sense (or is clearly false). Explain clearly; not all of these statements have definitive answers, so your explanation is more important than your chosen answer.

5. **Lottery.** The numbers 5, 17, 18, 27, 36, and 41 were drawn in the last lottery; they should not be bet on in the next lottery because they are now less likely to occur.

6. **Combining Probabilities.** The probability of flipping a coin and getting heads is 0.5. The probability of selecting a red card when one card is drawn from a shuffled deck is also 0.5. When flipping a coin and drawing a card, the probability of getting heads or a red card is $0.5 + 0.5 = 1$.

7. ***Either/Or* Probability.** $P(A) = 0.5$ and $P(A \text{ or } B) = 0.4$.

8. **Lottery.** When lottery numbers are drawn, the combination 1, 2, 3, 4, 5, and 6 is less likely to be drawn than other combinations.

Concepts and Applications

9. **Births.** Assuming that boys and girls are equally likely and that the gender of a child is not affected by the gender of any brothers or sisters, find the probability that a couple has three girls when they have three children.

10. **Guessing.** A quick quiz consists of a true/false question followed by a multiple-choice question with four possible answers (a, b, c, d). If both questions are answered with random guesses, find the probability that both responses are correct. Does guessing appear to be a good strategy on this quiz?

11. **Password.** A new computer owner creates a password consisting of two characters. She randomly selects a letter of the alphabet for the first character and a digit (0, 1, 2, 3, 4, 5, 6, 7, 8, 9) for the second character. What is the probability that her password is K9? Would this password be effective as a deterrent against someone trying to gain access to her computer?

12. **Wearing Hunter Orange.** A study of hunting injuries and the wearing of hunter orange clothing showed that among 123 hunters injured when mistaken for game, 6 were wearing orange (based on data from the Centers for Disease Control). If a follow-up study begins with the random selection of hunters from this sample of 123, find the probability that the first two hunters selected were both wearing orange.
    - **a.** Assume that the first hunter is replaced before the next one is selected.
    - **b.** Assume that the first hunter is not replaced before the second one is selected.
    - **c.** Which makes more sense in this situation: selecting with replacement or selecting without replacement? Why?

13. **Radio Tunes.** An MP3 player is loaded with 60 musical selections: 30 rock selections, 15 jazz selections, and 15 blues selections. The player is set on "random play," so selections are played randomly and can be repeated. What is the probability of each of the following events?
    - **a.** The first four selections are all jazz.
    - **b.** The first five selections are all blues.
    - **c.** The first selection is jazz and the second is rock.
    - **d.** Among the first four selections, none is rock.
    - **e.** The second selection is the same song as the first.

14. **Polling Calls.** A telephone pollster has names and telephone numbers for 45 voters, 20 of whom are registered Democrats and 25 of whom are registered Republicans. Calls are made in random order. Suppose you want to find the probability that the first two calls are to Republicans.
    - **a.** Are these independent or dependent events? Explain.
    - **b.** If you treat them as *dependent* events, what is the probability that the first two calls are to Republicans?
    - **c.** If you treat them as *independent* events, what is the probability that the first two calls are to Republicans?
    - **d.** Compare the results of parts b and c.

Probability and Court Decisions. The data in the following table show the outcomes of guilty and not-guilty pleas in 1,028 criminal court cases. Use the data to answer Exercises 15–20.

	Guilty plea	Not-guilty plea
Sent to prison	392	58
Not sent to prison	564	14

Source: Brereton and Casper, "Does It Pay to Plead Guilty? Differential Sentencing and the Functioning of the Criminal Courts," *Law and Society Review,* Vol. 16, No. 1.

15. What is the probability that a randomly selected defendant either pled guilty or was sent to prison?

16. What is the probability that a randomly selected defendant either pled not guilty or was not sent to prison?

17. If two different defendants are randomly selected, what is the probability that they both entered guilty pleas?

18. If two different defendants are randomly selected, what is the probability that they both were sentenced to prison?

19. If a defendant is randomly selected, what is the probability that the defendant entered a guilty plea and was sent to prison?

20. If a defendant is randomly selected, what is the probability that the defendant entered a guilty plea and was not sent to prison?

Pedestrian Deaths. For Exercises 21–26, use the the following table, which summarizes data on 985 pedestrian deaths that were caused by accidents (based on data from the National Highway Traffic Safety Administration).

		Pedestrian intoxicated?	
		Yes	No
Driver intoxicated?	Yes	59	79
	No	266	581

21. If one of the pedestrian deaths is randomly selected, find the probability that the pedestrian was intoxicated or the driver was intoxicated.

22. If one of the pedestrian deaths is randomly selected, find the probability that the pedestrian was not intoxicated or the driver was not intoxicated.

23. If one of the pedestrian deaths is randomly selected, find the probability that the pedestrian was intoxicated or the driver was not intoxicated.

24. If one of the pedestrian deaths is randomly selected, find the probability that the driver was intoxicated or the pedestrian was not intoxicated.

25. If two different pedestrian deaths are randomly selected, find the probability that they both involved intoxicated drivers.

26. If two different pedestrian deaths are randomly selected, find the probability that in both cases the pedestrians were intoxicated.

27. Drug Tests. An allergy drug is tested by giving 120 people the drug and 100 people a placebo. A control group consists of 80 people who were given no treatment. The number of people in each group who showed improvement appears in the table below.

	Allergy drug	Placebo	Control	Total
Improvement	65	42	31	**138**
No improvement	55	58	49	**162**
Total	**120**	**100**	**80**	**300**

a. What is the probability that a randomly selected person in the study was given either the drug or the placebo?

b. What is the probability that a randomly selected person either improved or did not improve?

c. What is the probability that a randomly selected person either was given the drug or improved?

d. What is the probability that a randomly selected person was given the drug and improved?

28. Political Party Affiliations. In a typical town (consistent with national demographics), the adults have the political affiliations listed in the table below. All figures are percentages.

	Republican	Democrat	Independent	Total
Men	17	15	18	**50**
Women	14	20	16	**50**
Total	**31**	**35**	**34**	**100**

a. What is the probability that a randomly selected person in the town is a Republican or a Democrat?

b. What is the probability that a randomly selected person in the town is a Republican or a woman?

c. What is the probability that a randomly selected person in the town is an Independent or a man?

d. What is the probability that a randomly selected person in the town is a Republican and a woman?

29. Probability Distributions and Genetics. Many traits are controlled by a dominant gene, **A**, and a recessive gene, **a**. Suppose that two parents carry these genes in the proportion 3:1; that is, the probability of either parent giving the **A** gene is 0.75, and the probability of either parent giving the **a** gene is 0.25. Assume that the genes are selected from each parent randomly. To answer the following questions, imagine 100 trial "births."

a. What is the probability that a child receives an **A** gene from both parents?

b. What is the probability that a child receives an **A** gene from one parent and an **a** gene from the other parent? Note that this can occur in two ways.

c. What is the probability that a child receives an **a** gene from both parents?

d. Make a table showing the probability distribution for all events.

e. If the combinations **AA** and **Aa** both result in the same dominant trait (say, brown hair) and **aa** results in the recessive trait (say, blond hair), what is the probability that a child will have the dominant trait?

30. BINGO. The game of BINGO involves drawing numbered and lettered buttons at random from a barrel. The B numbers are 1–15, the I numbers are 16–30, the N numbers are 31–45, the G numbers are 46–60, and the O numbers are 61–75. Buttons are not replaced after they have been selected. What is the probability of each of the following events on the initial selections?

a. Drawing a B button

b. Drawing two B buttons in a row

c. Drawing a B or an O

d. Drawing a B, then a G, then an N, in that order

e. Drawing anything but a B on five draws

At Least Once Problems. A common problem asks for the probability that an event occurs at least once in a given number of trials. Suppose the probability of a particular event is p (for example, the probability of drawing a heart from a deck of cards is 0.25). Then the probability that the event occurs at least once in N trials is

$$1 - (1 - p)^N$$

For example, the probability of drawing at least one heart in 10 draws (with replacement) is

$$1 - (1 - 0.25)^{10} = 0.944$$

Use this rule to solve Exercises 31 and 32.

31. The Bets of the Chevalier de Mère. It is said that probability theory was invented in the 17th century to explain the gambling of a nobleman named the Chevalier de Mère.

a. In his first game, the Chevalier bet on rolling at least one 6 with four rolls of a fair die. If played repeatedly, is this a game he should expect to win?

b. In his second game, the Chevalier bet on rolling at least one double-6 with 24 rolls of two fair dice. If played repeatedly, is this a game he should expect to win?

32. HIV among College Students. Suppose that 3% of the students at a particular college are known to carry HIV.

a. If a student has 6 sexual partners during the course of a year, what is the probability that at least one of them carries HIV?

b. If a student has 12 sexual partners during the course of a year, what is the probability that at least one of them carries HIV?

c. How many partners would a student need to have before the probability of an HIV encounter exceeded 50%?

Projects for the Internet and Beyond

For useful links, select "Links for Internet Projects" for Chapter 6 at www.aw.com/bbt.

33. Simulation. A classic probability problem involves a king who wants to increase the proportion of women in his kingdom. He decrees that after a mother gives birth to a son, she is prohibited from having any more children. The king reasons that some families will have just one boy whereas other families will have a few girls and one boy, so the proportion of girls will be increased. Use coin tossing to simulate a kingdom that abides by this decree: "After a mother gives birth to a son, she will not have any other children." If this decree is followed, does the proportion of girls increase?

IN THE NEWS

34. A columnist for the *New York Daily News* (Stephen Allensworth) provided tips for selecting numbers in New York State's lottery. He advocated a system based on the use of "cold digits," which are digits that hit once or not at all in a seven-day period. He made this statement: "That [system] produces the combos 5–8–9, 7–8–9, 6–8–9, 0–8–9 and 3–8–9. These five combos have an excellent chance of being drawn this week. Good luck to all." Can this system work? Why or why not?

Chapter Review Exercises

For Exercises 1–6, use the data below, obtained from a clinical trial of the Abbot blood test, a blood test for pregnancy (based on data from "Specificity and Detection Limit of Ten Pregnancy Tests," by Tiitinen and Stenman, *Scandinavian Journal of Clinical Laboratory Investigation*, Vol. 53, Supp. 216).

	Positive test result (pregnancy is indicated)	**Negative test result (pregnancy is not indicated)**
Subject is pregnant	80	5
Subject is not pregnant	3	11

1. If 1 of the 99 subjects is randomly selected, find the probability of getting a subject who is pregnant.

2. If 1 of the 99 subjects is randomly selected, find the probability of getting a subject who tested positive.

3. If 1 of the 99 subjects is randomly selected, find the probability of getting a subject who is pregnant or tested positive.

4. If 1 of the 99 subjects is randomly selected, find the probability of getting a subject who tested negative or is pregnant.

5. If two *different* subjects are randomly selected, find the probability that they both tested positive.

6. If two *different* subjects are randomly selected, find the probability that they both are pregnant.

7. The Binary Computer Company manufactures computer chips used in DVD players. Those chips are made with a 27% yield, meaning that 27% of them are good and the others are defective.

a. If one chip is randomly selected, find the probability that it is *not* good.

b. If two chips are randomly selected, find the probability that they are both good.

c. If five chips are randomly selected, what is the *expected number* of good chips?

d. If five chips are randomly selected, find the probability that they are all good. If you did get five good chips among the five selected, would you continue to believe that the yield was 27%? Why or why not?

For Exercises 8–10, consider an event to be "unusual" if its probability is less than or equal to 0.05.

8. a. An instructor obsessed with the metric system insists that all multiple-choice questions have 10 different possible answers, one of which is correct. What is the probability of answering such a question correctly if a random guess is made?

b. Is it "unusual" to answer a question correctly by guessing?

9. a. A roulette wheel has 38 slots: One slot is 0, another slot is 00, and the other slots are numbered 1 through 36. If you bet all of your textbook money on the number 13 for one spin, what is the probability that you will win?

b. Is it "unusual" to win when you bet on a single number in roulette?

10. a. A study of 400 randomly selected American Airlines flights showed that 344 arrived on time. What is the estimated probability of an American Airlines flight arriving late?

b. Is it "unusual" for an American Airlines flight to arrive late?

Chapter Quiz

1. A multiple-choice test question has answers of a, b, c, d, e, and f, and only one answer is correct. If you make a random guess, what is the probability that you are correct?

2. Three multiple-choice test questions each have answers of a, b, c, d, e, and f, and only one answer is correct in each case. If you make random guesses for the three answers, what is the probability that all three of your answers are correct?

3. If $P(A) = 0.45$, find $P(\bar{A})$.

4. If a coin is tossed 15 times, find the expected number of heads.

5. In a clinical trial, some subjects were treated with Nicorette and others were given a placebo. The results show a 0.279 probability that an adverse reaction is due to chance. Does chance appear to be a reasonable explanation for the results?

In Exercises 6–10, use the following data from a clinical trial of Nicorette, a chewing gum designed to help people stop smoking (based on data from Merrell Dow Pharmaceuticals, Inc.).

	Nicorette	Placebo
Mouth or throat soreness	43	35
No mouth or throat soreness	109	118

6. If one of the test subjects is randomly selected, find the probability of getting someone who was given a placebo.

7. If one of the subjects is randomly selected, find the probability of getting someone who used Nicorette and experienced mouth or throat soreness.

8. If one of the subjects is randomly selected, find the probability of getting someone who used Nicorette or experienced mouth or throat soreness.

9. If two different subjects are randomly selected, find the probability that they are both from the placebo group.

10. If three different subjects are randomly selected, find the probability that they are all from the Nicorette treatment group.

FOCUS ON SOCIAL SCIENCE

Are Lotteries Fair?

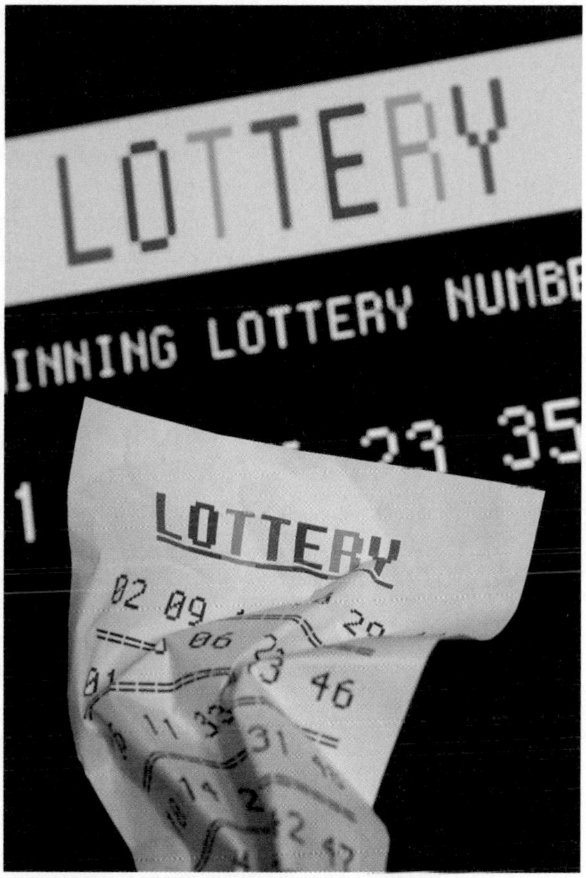

Lotteries have become part of the American way of life. Most states now have legal lotteries, including multi-state lotteries such as Powerball and the Big Game. National statistics show that per capita (average per person) lottery spending is approaching $200 per year. Since many people do not play lotteries at all, this means that active players tend to spend much more than $200 per year.

The mathematics of lottery odds involves counting the various combinations of numbers that are winners. While these calculations can become complex, the essential conclusion is always the same: The probability of winning a big prize is infinitesimally small. Advertisements may make lotteries sound like a good deal, but the expected value associated with a lottery is always negative. On average, those who play regularly can expect to lose about half of what they spend.

Lottery proponents point to several positive aspects. For example, lotteries produce billions of dollars of revenue that states use for education, recreation, and environmental initiatives. This revenue allows states to keep tax rates lower than they would be otherwise. Proponents also point out that lottery participation is voluntary and enjoyed by a representative cross-section of society. Indeed, a recent Gallup poll shows that three-fourths of Americans approve of state lotteries (two-thirds approve of legal gambling in general).

This favorable picture is part of the marketing and public relations of state lotteries. For example, Colorado state lottery officials offer statistics on the age, income, and education of lottery players compared to the general population (Figure 6.16). Within a few percentage points, the age of lottery players parallels that of the population as a whole. Similarly, the histogram for the income of lottery players gives the impression that lottery players as a whole are

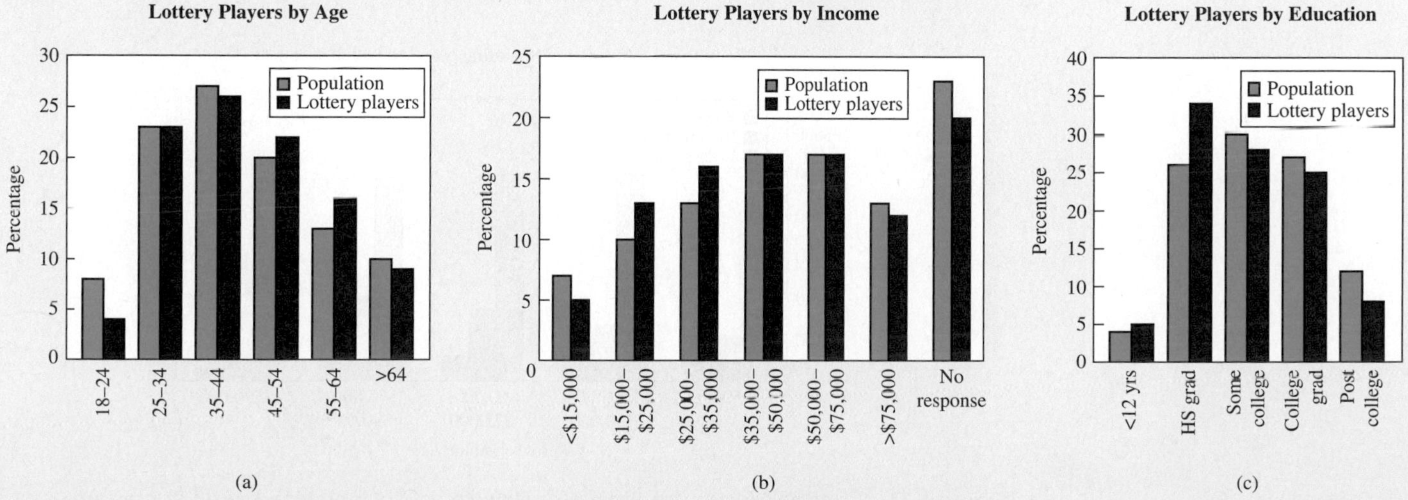

Figure 6.16 Three figures showing (a) age, (b) income, and (c) education of Colorado lottery players compared to population.

typical citizens—with the exception of the bars for incomes of $15,000–$25,000 and $25,000–$35,000, which show that the poor tend to play more than we would expect for their proportion of the population.

Despite the apparent benefits of lotteries, critics have long argued that lotteries are merely an unfair form of taxation. Some support for this view comes from a recent report by the National Gambling Impact Study Commission and a *New York Times* study of lotteries in New Jersey. Both of these studies focus on the *amount* of money spent on lotteries by individuals.

The *New York Times* study was based on data from 48,875 people who had won at least $600 in New Jersey lottery games. (In an ingenious bit of sampling, these winners were taken to be a random sample of all lottery players; after all, lottery winners are determined randomly. However, the sample is not really representative of all lottery players because winners tend to buy more than an average number of tickets.) By identifying the home zip codes of the lottery players, researchers were able to determine whether players came from areas with high or low income, high or low average education, and various demographic characteristics. The overwhelming conclusion of the *New York Times* study is that lottery spending has a much greater impact *in relative terms* on those players with lower incomes and lower educational backgrounds. For example, the following were among the specific findings:

- People in the state's lowest income areas spend five times as much of their income on lotteries as those in the state's highest income areas. Spending in the lowest income areas on one particular lottery game was $29 per $10,000 of annual income, compared to less than $5 per $10,000 of annual income in the highest income areas.
- The number of lottery sales outlets (where lottery tickets can be purchased) is nearly twice as high per 10,000 people in low-income areas as in high-income areas.
- People in areas with the lowest percentage of college education spent over five times as much per $10,000 of annual income as those in areas with the highest percentage of college education.
- Advertising and promotion of lotteries is focused in low-income areas.

Some of the results of the *New York Times* study are summarized in Figure 6.17. It suggests that while New Jersey has a progressive tax system (higher-income people pay a greater percentage of their income in taxes), the "lottery tax" is regressive. Moreover, the study found that the areas that generate the largest percentage of lottery revenues do not receive a proportional share of state funding.

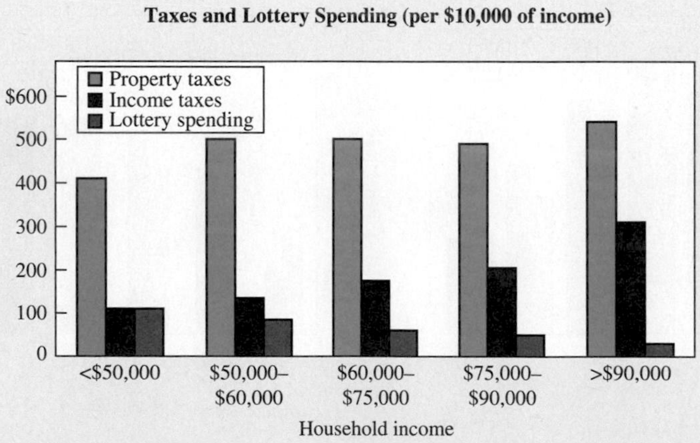

Figure 6.17 Taxes and lottery spending for New Jersey lottery winners (all figures are per $10,000 of income).

Similar studies reveal the same patterns in other states. The overall conclusions are inescapable: While lotteries provide many benefits to state governments, the revenue they produce comes disproportionately from poorer and less educated individuals. Indeed, a report by the National Gambling Impact Study Commission concluded that lotteries are "the most widespread form of gambling in the United States" and that state governments have "irresponsibly intruded gambling into society on a massive scale . . . through such measures as incessant advertising and the ubiquitous placement of lottery machines in neighborhood stores."

QUESTIONS FOR DISCUSSION

1. Study Figure 6.16. Do lottery players appear to be a typical cross-section of American society based on age? Based on income? Based on level of education? Explain. How does the "no response" category affect these conclusions?
2. Some lottery players use "systems" for choosing numbers. For example, they consult "experts" who tell them (sometimes for a fee) which numbers are popular or due to appear. Can such systems really improve your odds of winning the lottery? Why or why not? Do you think the use of such systems is related to educational background?
3. Considering all factors presented in this section and other facts that you can find, do you think lotteries are fair to poor or uneducated people? Should they remain legal? Should they be restricted in any way?
4. Find and study a particular lottery advertisement, and determine whether it is misleading in any way.
5. An anonymous quote circulated on the Internet read "Lotteries are a tax on people who are bad at math." Comment on the meaning and accuracy of this quote.

SUGGESTED READING

Pulley, Brett. "Living Off the Daily Dream of Winning a Lottery Prize." *New York Times,* May 22, 1999.

Safire, William. "Lotteries Are Losers." *New York Times,* June 21, 1999.

Schwartz, Nelson D., and Nixon, Ron. "Some States Consider Leasing Their Lotteries." *New York Times,* October 14, 2007.

Stodghill, Ron, and Nixon, Ron. "For Schools, No Lucky Numbers." *New York Times,* October 7, 2007.

Walsh, James. *True Odds*. Merritt Publishing, 1996.

FOCUS ON LAW

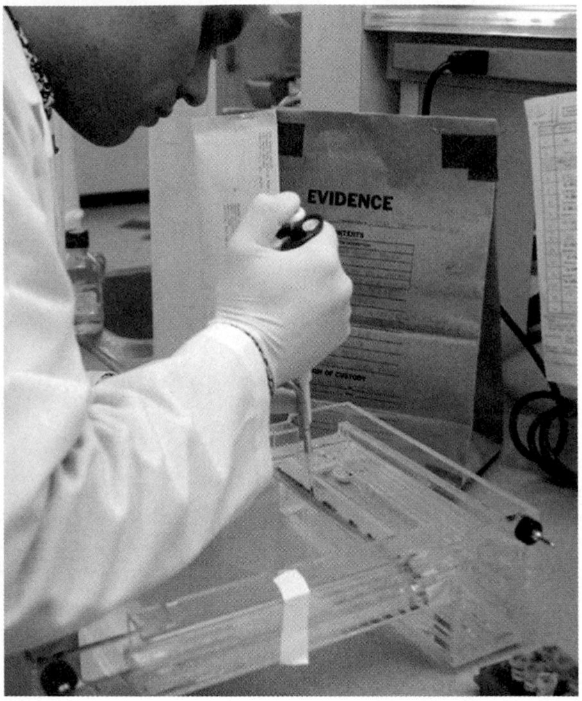

Is DNA Fingerprinting Reliable?

DNA fingerprinting (also called DNA profiling or DNA identification) has become a major tool of law enforcement. It is used in criminal cases, in paternity cases, and even in the identification of human remains. (DNA fingerprinting was the primary way by which remains of victims were identified after the 9/11 terrorist attacks on the World Trade Center in 2001.)

The scientific foundation for DNA identification has been in place for several decades. However, these ideas were first used in the courtroom only in 1986. The case involved a 17-year-old boy accused of the rape and murder of two schoolgirls in Narborough, in the Midlands of England. During his interrogation, the suspect asked for a blood test, which was sent to the laboratory of a noted geneticist, Alec Jeffreys, at the nearby University of Leicester. Using methods that he had already developed for paternity testing, Jeffreys compared the suspect's DNA to that found in samples from the victims. The tests showed that the rapes were committed by the same person, but not by the suspect in custody. The following year, after over 4,500 blood samples were collected, researchers made a positive identification of the murderer using Jeffreys's methods.

Word of the British case and Jeffreys's methods spread rapidly. The techniques were swiftly tested, commercialized, and promoted. Not surprisingly, the use of DNA identification also immediately met opposition and controversy.

To explore the essential roles that probability and statistics play in DNA identification, consider a simple eyewitness analogy. Suppose you are looking for a person who helped you out during a moment of need and you remember only three things about this person:

- The person was female.
- She had green eyes.
- She had long red hair.

If you find someone who matches this profile, can you conclude that this person is the woman who helped you? To answer, you need data telling you the probabilities that randomly selected indivivduals in the population have these characteristics. The probability that a person is female is about 1/2. Let's say that the probability of green eyes is about 0.06 (6% of the population has green eyes) and the probability of long red hair is 0.0075. If we assume that these characteristics occur independently of one another, then the probability that a randomly selected person has all three characteristics is

$$0.5 \times 0.06 \times 0.0075 = 0.000225$$

or about 2 in 10,000. This may seem relatively low, but it probably is not low enough to draw a definitive conclusion. For example, a profile matched by only 2 in 10,000 people will still be matched by some 200 people in a city with a population of 1 million. Clearly, if you want to be sure that you've found the right person, you'll need additional information for the profile.

DNA identification is based on a similar idea, but it is designed so that the probability of a profile match is much lower. The DNA of every individual is unique and is the same throughout the individual. A single physical trait is determined by a small piece of DNA called a *gene* at a specific *locus* (location) on a *chromosome*; humans have 23 chromosomes and over

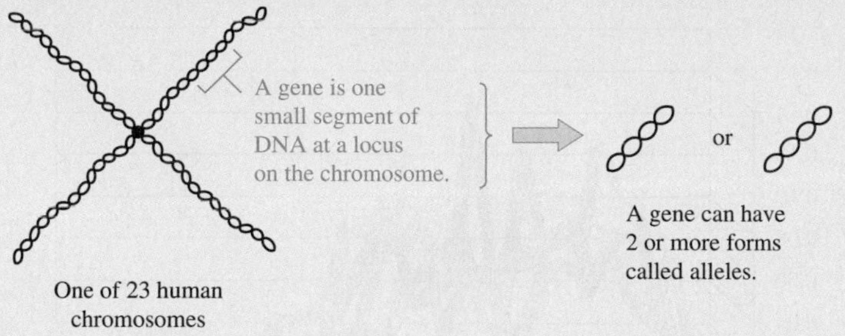

Figure 6.18 Diagram showing the relationship among chromosomes, alleles, genes, and loci.

30,000 genes (Figure 6.18). A gene can take two or more (often hundreds) of different forms, called *alleles* (pronounced a-leels). Different alleles give rise to variations of a trait (for example, different hair colors or different blood types). Not only can different alleles appear at a locus, but the corresponding piece of DNA can have different lengths in different people (called *variable number of tandem repeats* or VNTRs). The genetic evidence that is collected and analyzed in the lab consists of the allele lengths or allele types at five to eight different loci.

Collecting genetic evidence (from samples of blood, tissues, hair, semen, or even saliva on a postage stamp) and analyzing it are straightforward, at least in theory. Nevertheless, the process is subject to both controversy and sources of error. Suppose, in our analogy, a person is found with "reddish brown" hair instead of "red" hair. Because many characteristics are continuous (not discrete) variables, should you rule this person out or assume that reddish brown is close enough? For this reason, the issue of *binning* becomes extremely important (see Section 3.1). You might choose to include all people with hair that is any shade of red, or you could choose a narrower bin—say, bright red hair only—which would give a more discriminating test.

The same issue arises in genetic tests. When allele types or lengths are measured in the lab, there is enough variability or error in the measurements that these variables are continuous. Bin widths need to be chosen, and the choice is the source of debate. Bins with a small width give a more refined test, exclude more suspects, and ultimately provide stronger evidence against a defendant.

Another source of scientific controversy is the assumption that the characteristics in a genetic profile are independent. Expert witnesses (geneticists and molecular biologists) have testified at great lengths about this point without agreeing. There does seem to be agreement that the assumption of independence does not introduce significant errors in light of other sources of error. But it is important to see how a difficult scientific question affects the mathematics involved: If characteristics are independent, then the multiplication rule for independent events is justified; if not, a different sort of probability needs to be used.

A third point of controversy concerns the choice of a reference population. Some argue that if genetic data from an entire population are used, they might not fairly represent ethnic subpopulations. For example, a particular allele may have a very different frequency in ethnic Italians than in the entire U.S. population. Such a discrepancy would change the outcome of the probability calculation and could either strengthen or damage the case of a defendant.

Figure 6.19 shows an example of the reference population data used in genetic testing. The horizontal axis shows 30 bins for the different allele measurements, and the vertical axis shows the frequency for each bin. Of equal interest are the frequency curves for four different Asian subpopulations. The significant variation among the curves is the evidence for using specialized databases for subpopulations. However, until such databases are complete and there is supporting evidence within a case for using a specialized database, the choice is still to use the full population as a reference.

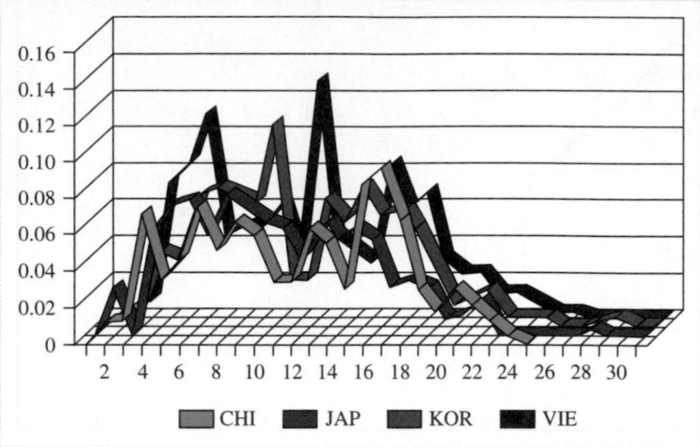

Figure 6.19 Binned frequency data for one allele, showing variability among four ethnic groups. *Source:* Kathryn Roeder, "DNA Fingerprinting: A Review of the Controversy," *Statistical Science,* Vol. 9, No. 2, pp. 222–247.

DNA technology has created new entwined pathways for scientists and legal scholars that are changing our society. Not surprisingly, statistical methods and thinking are an essential part of the change.

QUESTIONS FOR DISCUSSION

1. The result of a DNA test is considered physical (as opposed to circumstantial) evidence. Yet it is much more sophisticated and difficult to understand than a typical piece of physical evidence, such as a weapon or a piece of clothing. Some people have therefore argued that DNA evidence should not be used in a criminal trial in which jury members may not fully understand how the evidence is collected and analyzed. What do you think of this argument? Defend your opinion.
2. Suppose that an allele has a *greater* frequency in a subpopulation than in the full population. Explain how this would change the evidence for or against a defendant. Suppose that an allele has a *smaller* frequency in a subpopulation than in the full population. Explain how this would change the evidence for or against a defendant.
3. Evidence from blood tests can identify a suspect with a probability of about 1 in 200. Evidence from DNA tests often provides probabilities claimed to be on the order of 1 in 10 million. If you were a juror, would you accept such a probability as positive identification of a suspect?
4. The Innocence Project uses DNA to try to clear suspects wrongfully convicted of crimes. Is innocence easier to establish than guilt through DNA testing? Explain.
5. Discuss a few other ways DNA tests can be useful, such as in settling issues of paternity. Overall, how much do you think DNA evidence is likely to affect our society in the future?

SUGGESTED READING

DNA fingerprinting technology is constantly being refined and updated. You can find a great deal of information by searching on the Web.

Learn more about Project Innocence at http://www.innocenceproject.org.

7

The person who knows "how" will always have a job.
The person who knows "why" will always be his boss.
—Diane Ravitch

Correlation and Causality

DOES SMOKING CAUSE LUNG CANCER? DOES LOW unemployment lead to inflation? Does human use of fossil fuels cause global warming? A major goal of many statistical studies is to search for relationships among different variables so that researchers can then determine whether one factor *causes* another. Once a relationship is discovered, we can try to determine whether there is an underlying cause. In this chapter, we will study relationships known as correlations and explore how they are important to the more difficult task of searching for causality.

LEARNING GOALS

7.1 Seeking Correlation
Be able to define correlation, recognize positive and negative correlations on scatter diagrams, and understand the correlation coefficient as a measure of the strength of a correlation.

7.2 Interpreting Correlations
Be aware of important cautions concerning the interpretation of correlations, especially the effects of outliers, the effects of grouping data, and the crucial fact that correlation does not necessarily imply causality.

7.3 Best-Fit Lines and Prediction
Become familiar with the concept of a best-fit line for a correlation, recognize when such lines have predictive value and when they may not, understand how the square of the correlation coefficient is related to the quality of the fit, and qualitatively understand the use of multiple regression.

7.4 The Search for Causality
Understand the difficulty of establishing causality from correlation, and investigate guidelines that can be used to help establish confidence in causality.

7.1 Seeking Correlation

What does it mean when we say that smoking *causes* lung cancer? It certainly does *not* mean that you'll get lung cancer if you smoke a single cigarette. It does not even mean that you'll definitely get lung cancer if you smoke heavily for many years, since some heavy smokers do not get lung cancer. Rather, it is a *statistical* statement meaning that you are *much more likely* to get lung cancer if you smoke than if you don't smoke.

Smoking is one of the leading causes of statistics.

—Fletcher Knebel

How did researchers learn that smoking causes lung cancer? The process began with informal observations, as doctors noticed that a surprisingly high proportion of their patients with lung cancer were smokers. These observations led to carefully conducted studies in which researchers compared lung cancer rates among smokers and nonsmokers. These studies showed clearly that heavier smokers were more likely to get lung cancer. In more formal terms, we say that there is a **correlation** between the variables *amount of smoking* and *likelihood of lung cancer.* A correlation is a special type of relationship between variables, in which a rise or fall in one goes along with a corresponding rise or fall in the other.

Definition

A **correlation** exists between two variables when higher values of one variable consistently go with higher values of another variable or when higher values of one variable consistently go with lower values of another variable.

Here are a few other examples of correlations:

- There is a correlation between the variables *height* and *weight* for people; that is, taller people tend to weigh more than shorter people.
- There is a correlation between the variables *demand for apples* and *price of apples;* that is, demand tends to decrease as price increases.
- There is a correlation between *practice time* and *skill* among piano players; that is, those who practice more tend to be more skilled.

It's important to realize that establishing a correlation between two variables does *not* mean that a change in one variable *causes* a change in the other. The correlation between smoking and lung cancer did not by itself prove that smoking causes lung cancer. We could imagine, for example, that some gene predisposes a person both to smoking and to lung cancer. Nevertheless, identifying the correlation was the crucial first step in learning that smoking causes lung cancer. We will discuss the difficult task of establishing causality later in this chapter. For now, we concentrate on how we look for, identify, and interpret correlations.

By the Way ...

Smoking is linked to many serious diseases besides lung cancer, including heart disease and emphysema. Smoking is also linked with less lethal health conditions such as premature skin wrinkling and sexual impotence.

TIME OUT TO THINK

Suppose there really were a gene that made people prone to both smoking and lung cancer. Explain why we would still find a strong correlation between smoking and lung cancer in that case, but would not be able to say that smoking causes lung cancer.

Scatter Diagrams

Table 7.1 lists data for a sample of gem-store diamonds—their prices and several common measures that help determine their value. Because advertisements for diamonds often quote only their weights (in carats), we might suspect a correlation between the weights and the prices. We can look for such a correlation by making a **scatter diagram** (or *scatterplot*) showing the relationship between the variables *weight* and *price*.

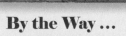

Table 7.1 Prices and Characteristics of a Sample of 23 Diamonds from Gem Dealers

Diamond	Price	Weight (carats)	Depth	Table	Color	Clarity
1	$6,958	1.00	60.5	65	3	4
2	$5,885	1.00	59.2	65	5	4
3	$6,333	1.01	62.3	55	4	4
4	$4,299	1.01	64.4	62	5	5
5	$9,589	1.02	63.9	58	2	3
6	$6,921	1.04	60.0	61	4	4
7	$4,426	1.04	62.0	62	5	5
8	$6,885	1.07	63.6	61	4	3
9	$5,826	1.07	61.6	62	5	5
10	$3,670	1.11	60.4	60	9	4
11	$7,176	1.12	60.2	65	2	3
12	$7,497	1.16	59.5	60	5	3
13	$5,170	1.20	62.6	61	6	4
14	$5,547	1.23	59.2	65	7	4
15	$7,521	1.29	59.6	59	6	2
16	$7,260	1.50	61.1	65	6	4
17	$8,139	1.51	63.0	60	6	4
18	$12,196	1.67	58.7	64	3	5
19	$14,998	1.72	58.5	61	4	3
20	$9,736	1.76	57.9	62	8	2
21	$9,859	1.80	59.6	63	5	5
22	$12,398	1.88	62.9	62	6	2
23	$11,008	2.03	62.0	63	8	3

Notes: Weight is measured in carats (1 carat = 0.2 gram). Depth is defined as 100 times the ratio of height to diameter. Table is the size of the upper flat surface. (Depth and table determine "cut.") Color and clarity are each measured on standard scales, where 1 is best. For color, 1 = colorless, and increasing numbers indicate more yellow. For clarity, 1 = flawless, and 6 indicates that defects can be seen by eye.

By the Way ...

The word *karats* (with a *k*) used to describe gold does not have the same meaning as the term *carats* (with a *c*) for diamonds and other gems. A carat is a measure of weight equal to 0.2 gram. Karats are a measure of the purity of gold: 24-karat gold is 100% pure gold; 18-karat gold is 75% pure (and 25% other metals); 12-karat gold is 50% pure (and 50% other metals); and so on.

Definition

A **scatter diagram** (or *scatterplot*) is a graph in which each point represents the values of two variables.

The following procedure describes how to make the scatter diagram in Figure 7.1.

1. We assign one variable to each axis and label the axis with values that comfortably fit all the data. Sometimes the axis selection is arbitrary, but if we suspect that one variable depends on the other then we plot the *explanatory variable* on the horizontal axis and the *response variable* on the vertical axis. In this case, we expect the diamond price to depend at least in part on its weight; we therefore say that *weight* is the explanatory variable (because it helps *explain* the price) and *price* is the response variable (because it *responds* to changes in the explanatory variable). We choose a range of 0 to 2.5 carats for the *weight* axis and $0 to $16,000 for the *price* axis.

2. For each diamond in Table 7.1, we plot a *single point* at the horizontal position corresponding to its weight and the vertical position corresponding to its price. For example, the point for Diamond 10 goes at a position of 1.11 carats on the horizontal axis and $3,670 on the vertical axis. The dashed lines on Figure 7.1 show how we locate this point.

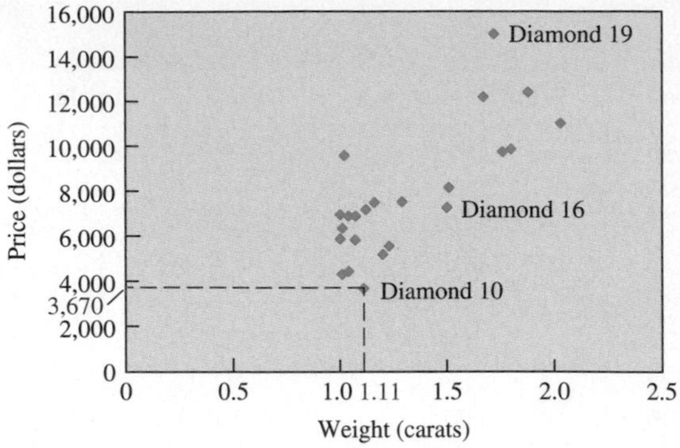

Figure 7.1 Scatter diagram showing the relationship between the variables *price* and *weight* for the diamonds in Table 7.1. The dashed lines show how we find the position of the point for Diamond 10.

3. (Optional) We can label some (or all) of the data points, as is done for Diamonds 10, 16, and 19 in Figure 7.1.

Scatter diagrams get their name because the way in which the points are scattered may reveal a relationship between the variables. In Figure 7.1, we see a general upward trend indicating that diamonds with greater weight tend to be more expensive. The correlation is not perfect. For example, the heaviest diamond is not the most expensive. But the overall trend seems fairly clear.

TIME OUT TO THINK

Identify the points in Figure 7.1 that represent Diamonds 3, 7, and 23.

EXAMPLE 1 Color and Price

Using the data in Table 7.1, create a scatter diagram to look for a correlation between a diamond's *color* and *price*. Comment on the correlation.

Solution We expect price to depend on color, so we plot the explanatory variable *color* on the horizontal axis and the response variable *price* on the vertical axis in Figure 7.2. (You should check a few of the points against the data in Table 7.1.) The points appear much more scattered than in Figure 7.1. Nevertheless, you may notice a weak trend diagonally downward from the upper left toward the lower right. This trend represents a weak correlation in which diamonds with more yellow color (higher numbers for color) are less expensive. This trend is

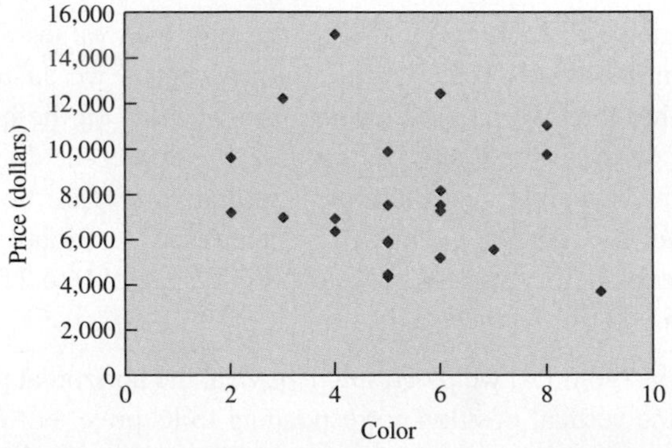

Figure 7.2 Scatter diagram for the color and price data in Table 7.1.

consistent with what we would expect, because colorless diamonds appear to sparkle more and are generally considered more desirable.

TIME OUT TO THINK

Thanks to a large bonus at work, you have a budget of $6,000 for a diamond ring. A dealer offers you the following two choices for that price. One diamond weighs 1.20 carats and has color = 4. The other weighs 1.18 carats and has color = 3. Assuming all other characteristics of the diamonds are equal, which would you choose? Why?

Types of Correlation

We have seen two examples of correlation. Figure 7.1 shows a fairly strong correlation between weight and price, while Figure 7.2 shows a weak correlation between color and price. We are now ready to generalize about types of correlation. Figure 7.3 shows eight

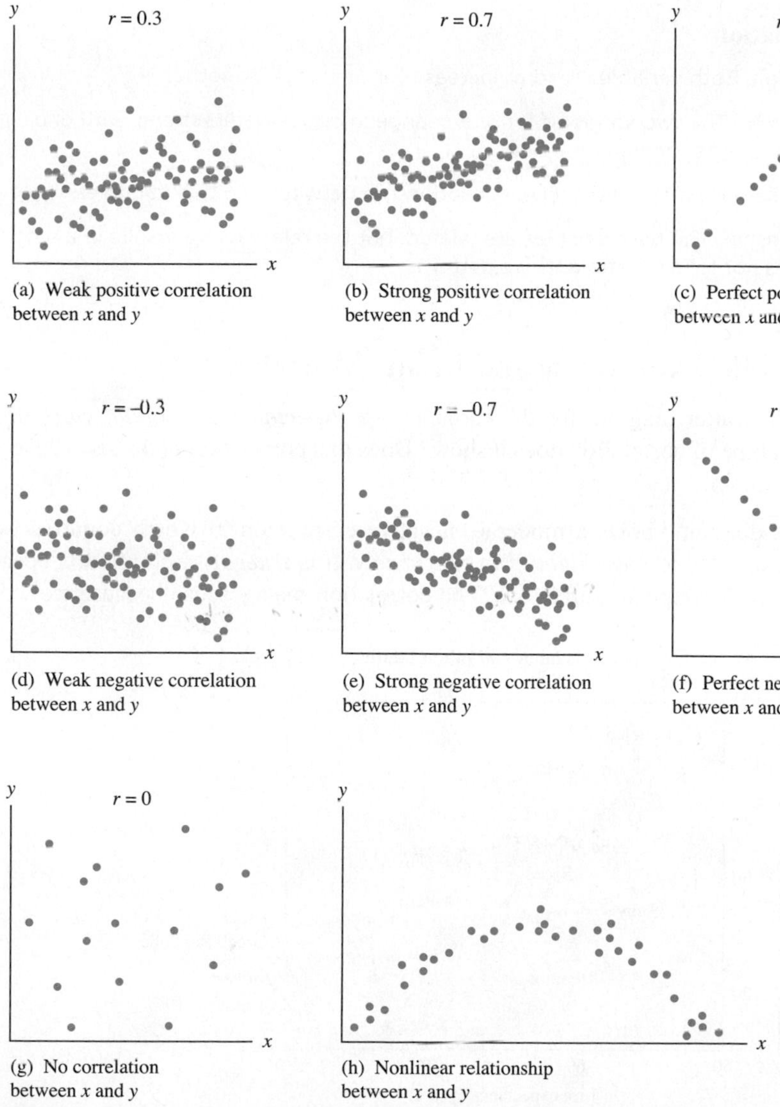

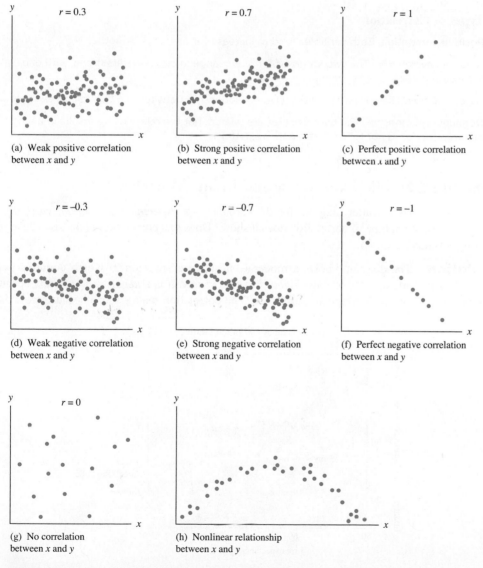

Figure 7.3 Types of correlation seen on scatter diagrams.

scatter diagrams for variables called x and y. Note the following key features of these diagrams:

- Parts a to c of Figure 7.3 show **positive correlations**, in which the values of y tend to increase with increasing values of x. The correlation becomes stronger as we proceed from a to c. In fact, c shows a perfect positive correlation, in which all the points fall along a straight line.
- Parts d to f of Figure 7.3 show **negative correlations**, in which the values of y tend to decrease with increasing values of x. The correlation becomes stronger as we proceed from d to f. In fact, f shows a perfect negative correlation, in which all the points fall along a straight line.
- Part g of Figure 7.3 shows **no correlation** between x and y. In other words, values of x do not appear to be linked to values of y in any way.
- Part h of Figure 7.3 shows a **nonlinear relationship**, in which x and y appear to be related but the relationship does not correspond to a straight line. (*Linear* means along a straight line, and *nonlinear* means *not* along a straight line.)

TECHNICAL NOTE

In this text we use the term *correlation* only for *linear* relationships. Some statisticians refer to nonlinear relationships as "nonlinear correlations." There are techniques for working with nonlinear relationships that are similar to those described in this book for linear relationships.

Types of Correlation

Positive correlation: Both variables tend to increase (or decrease) together.

Negative correlation: The two variables tend to change in opposite directions, with one increasing while the other decreases.

No correlation: There is no apparent (linear) relationship between the two variables.

Nonlinear relationship: The two variables are related, but the relationship results in a scatter diagram that does not follow a straight-line pattern.

EXAMPLE 2 Life Expectancy and Infant Mortality

Figure 7.4 shows a scatter diagram for the variables *life expectancy* and *infant mortality* in 16 countries. What type of correlation does it show? Does this correlation make sense? Does it imply causality? Explain.

Solution The diagram shows a moderate negative correlation in which countries with *lower* infant mortality tend to have *higher* life expectancy. It is a *negative* correlation because the two variables vary in opposite directions. The correlation makes sense because we would

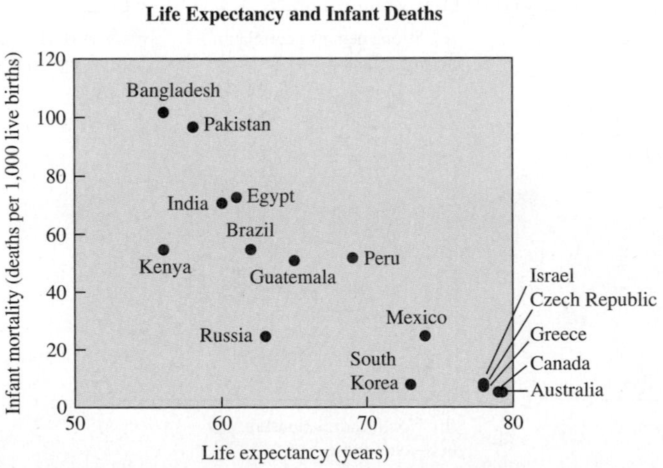

Figure 7.4 Scatter diagram for life expectancy and infant mortality data. *Source:* United Nations.

expect that countries with better health care would have both lower infant mortality and higher life expectancy. However, it does *not* imply causality between infant mortality and life expectancy: We would not expect that a concerted effort to reduce infant mortality would increase life expectancy significantly unless it was part of an overall effort to improve health care. (Reducing infant mortality will *slightly* increase life expectancy because having fewer infant deaths tends to raise the mean age of death for the population.)

Measuring the Strength of a Correlation

For most purposes, it is enough to state whether a correlation is strong, weak, or nonexistent. However, sometimes it is useful to describe the strength of a correlation in more precise terms. Statisticians measure the strength of a correlation with a number called the **correlation coefficient**, represented by the letter r.

Correlation coefficients are easy to interpret, although they are tedious to calculate unless you use a calculator or computer. Look again at Figure 7.3, and notice that it shows the value of the correlation coefficient r for each scatter diagram. The correlation coefficient is always between -1 and 1. When points in a scatter diagram lie close to an ascending straight line, the correlation coefficient is positive and close to 1. Similarly, points lying close to a descending straight line have a negative correlation coefficient with a value close to -1. Points that do not fit any type of straight-line pattern or that lie close to a *horizontal* straight line (indicating that the y values have no dependence on the x values) result in a correlation coefficient close to 0.

Properties of the Correlation Coefficient, r

- The correlation coefficient, r, is a measure of the strength of a correlation. Its value can range only from -1 to 1.
- If there is no correlation, the points do not follow any ascending or descending straight-line pattern, and the value of r is close to 0.
- If there is a positive correlation, the correlation coefficient is positive ($0 < r \leq 1$): Both variables increase together. A perfect positive correlation (in which all the points on a scatter diagram lie on an ascending straight line) has a correlation coefficient $r = 1$. Values of r close to 1 mean a strong positive correlation and positive values closer to 0 mean a weak positive correlation.
- If there is a negative correlation, the correlation coefficient is negative ($-1 \leq r < 0$): When one variable increases, the other decreases. A perfect negative correlation (in which all the points lie on a descending straight line) has a correlation coefficient $r = -1$. Values of r close to -1 mean a strong negative correlation and negative values closer to 0 mean a weak negative correlation.

TECHNICAL NOTE

For the methods of this section, there is a requirement that the two variables result in data having a "bivariate normal distribution." This basically means that for any fixed value of one variable, the corresponding values of the other variable have a normal distribution. This requirement is usually very difficult to check, so the check is often reduced to verifying that both variables result in data that are normally distributed.

EXAMPLE 3 U.S. Farm Size

Figure 7.5 shows a scatter diagram for the variables *number of farms* and *mean farm size* in the United States. Each dot represents data from a single year between 1950 and 2000; on this diagram, the earlier years generally are on the right and the later years on the left. Estimate the correlation coefficient by comparing this diagram to those in Figure 7.3 and discuss the underlying reasons for the correlation.

Solution The scatter diagram shows a strong negative correlation that most closely resembles the scatter diagram in Figure 7.3f, suggesting a correlation coefficient around $r = -0.9$. The correlation shows that as the number of farms decreases, the size of the remaining farms

By the Way ...

In 1900, over 40% of the U.S. population worked on farms; by 2000, less than 2% of the population worked on farms.

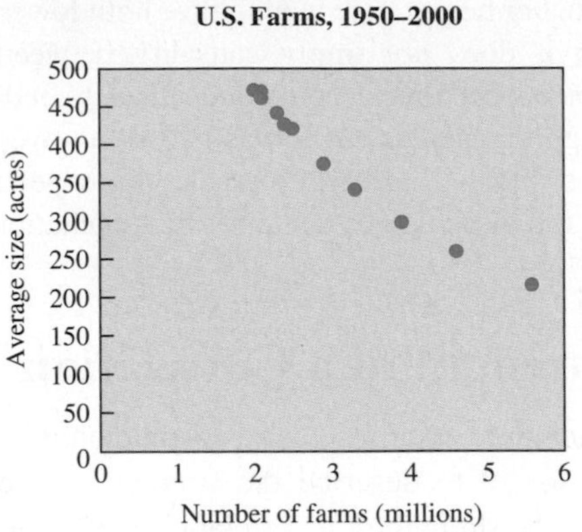

Figure 7.5 Scatter diagram for farm size data. *Source:* U.S. Department of Agriculture.

increases. This trend reflects a basic change in the nature of farming: Prior to 1950, most farms were small family farms. Over time, these small farms have been replaced by large farms owned by agribusiness corporations.

EXAMPLE 4 Accuracy of Weather Forecasts

The scatter diagrams in Figure 7.6 show two weeks of data comparing the actual high temperature for the day with the same-day forecast (part a) and the three-day forecast (part b). Estimate the correlation coefficient for each data set and discuss what these coefficients imply about weather forecasts.

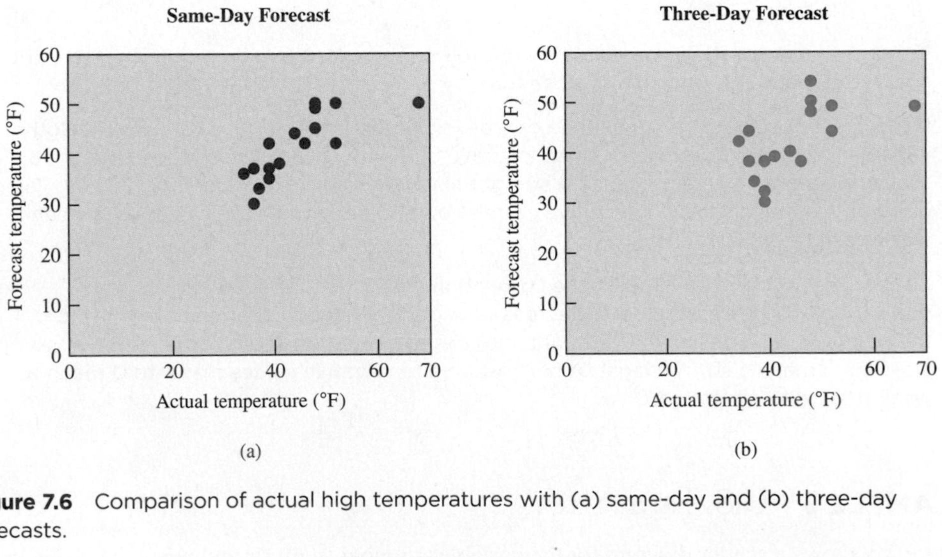

Figure 7.6 Comparison of actual high temperatures with (a) same-day and (b) three-day forecasts.

Solution If every forecast were perfect, each actual temperature would equal the corresponding forecasted temperature. This would result in all points lying on a straight line and a correlation coefficient of $r = 1$. In Figure 7.6a, in which the forecasts were made at the beginning of the same day, the points lie fairly close to a straight line, meaning that same-day forecasts are closely related to actual temperatures. By comparing this scatter diagram to the diagrams in Figure 7.3, we can reasonably estimate this correlation coefficient to be about $r = 0.8$. The correlation is weaker in Figure 7.6b, indicating that forecasts made three days in advance aren't as close to actual temperatures as same-day forecasts. This correlation coefficient is about $r = 0.6$. These results are unsurprising, since we expect longer-term forecasts to be less accurate.

EXAMPLE 5 Movie Success

Table 7.2 shows the production cost and gross receipts for the 15 biggest-budget science fiction and fantasy movies of all time (through 2006). Make a scatter diagram for the relationship between production cost and gross receipts. Estimate the correlation coefficient and discuss its meaning. ("Gross receipts" is the total amount of money collected in movie theater ticket sales.)

Table 7.2 Biggest-Budget Science Fiction and Fantasy Movies

Movie	Production cost (millions of dollars)	Gross receipts (millions of dollars)
King Kong (2005)	207	218
Spider-Man 2 (2004)	200	373
Chronicles of Narnia (2005)	180	292
Waterworld (1995)	175	88
Van Helsing (2004)	170	120
Polar Express (2004)	170	172
Terminator 3 (2003)	170	150
Poseidon (2006)	160	52
Batman Begins (2005)	150	205
Harry Potter/Goblet of Fire (2005)	150	290
Armageddon (1998)	140	201
Men in Black 2 (2002)	140	190
Spider-Man (2002)	139	403
Final Fantasy: The Spirits Within (2001)	137	32
Hulk (2003)	137	132

Note: Gross receipts are for United States only; worldwide receipts are often substantially higher. These figures are not adjusted for inflation.

By the Way ...

In percentage terms, the most profitable film of all time was the *Blair Witch Project*. Produced for $35,000 by two friends who came out of the University of Central Florida's film program, the movie grossed over $140 million—a gross of $4,000 for every $1 it cost to produce.

Solution Figure 7.7 shows the scatter diagram with production cost on the horizontal axis and gross receipts on the vertical axis. (You should check a few of the points against the data in Table 7.2.) The scatter diagram most closely resembles Figure 7.3g, which exhibits no

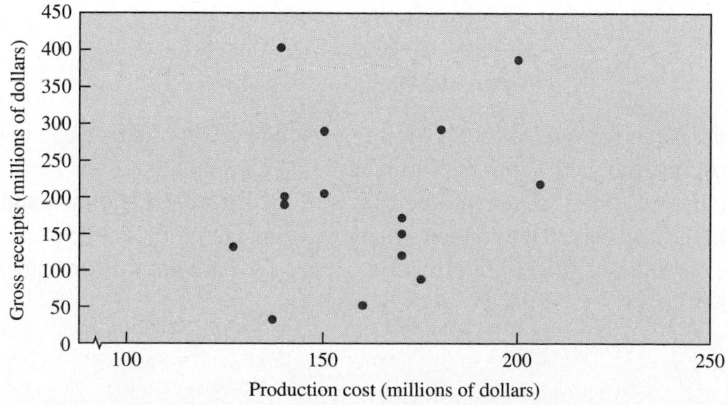

Figure 7.7 Scatter diagram for the data in Table 7.2.

correlation, so we conclude that the correlation coefficient is near $r = 0$. In other words, at least for these movies, there is no correlation between the amount of money spent producing the movie and the amount of money it earned in gross receipts.

TIME OUT TO THINK

For further practice, visually estimate the correlation coefficients for the data for diamond weight and price (Figure 7.1) and diamond color and price (Figure 7.2).

Calculating the Correlation Coefficient (Optional Section)

The formula for the (linear) correlation coefficient r can be expressed in several different ways that are all algebraically equivalent, which means that they produce the same value. The following expression has the advantage of relating more directly to the underlying rationale for r:

$$r = \frac{\Sigma\left[\frac{(x-\bar{x})}{s_x}\frac{(y-\bar{y})}{s_y}\right]}{n-1}$$

In the above expression, division by $n - 1$ (where n is the number of pairs of data) shows that r is a type of average, so it does not increase simply because more pairs of data values are included. The symbol s_x denotes the standard deviation of the x values (or the values of the first variable), and s_y denotes the standard deviation of the y values. The expression $(x - \bar{x})/s_x$ is in the same format as the *standard score* introduced in Section 5.2. By using the standard scores for x and y, we ensure that the value of r does not change simply because a different scale of values is used. The key to understanding the rationale for r is to focus on the product of the standard scores for x and the standard scores for y. Those products tend to be positive when there is a positive correlation, and they tend to be negative when there is a negative correlation. For data with no correlation, some of the products are positive and some are negative, with the net effect that the sum is relatively close to 0.

The following alternative formula for r has the advantage of simplifying calculations, so it is often used whenever manual calculations are necessary. The following formula is also easy to program into statistical software or calculators:

$$r = \frac{n \times \Sigma(x \times y) - (\Sigma x) \times (\Sigma y)}{\sqrt{n \times (\Sigma x^2) - (\Sigma x)^2} \times \sqrt{n \times (\Sigma y^2) - (\Sigma y)^2}}$$

This formula is straightforward to use, at least in principle: First calculate each of the required sums, then substitute the values into the formula. Be sure to note that (Σx^2) and $(\Sigma x)^2$ are *not* equal: (Σx^2) tells you to first square all the values of the variable x and then add them; $(\Sigma x)^2$ tells you to add the x values first and then square this sum. In other words, perform the operation within the parentheses first. Similarly, (Σy^2) and $(\Sigma y)^2$ are not the same.

Section 7.1 Exercises

Statistical Literacy and Critical Thinking

1. **Correlation.** What is a correlation? Does the term *correlation* have the same meaning in statistics as in ordinary usage?

2. **Correlation.** After computing the correlation coefficient from five pairs of data, you conclude that there is no correlation. Does it follow that there would be a conclusion of no correlation if 100 more pairs of similar data were obtained? Why or why not?

3. **Scatterplot.** If a correlation coefficient has been computed to be $r = 0.997$, describe the pattern of points in the corresponding scatter diagram.

4. **Coincidental Correlation.** A study showed a significant correlation between per capita wine consumption and the incomes of teachers. It's obviously silly to conclude that teachers spend their additional income on wine (isn't it?), so identify at least one other factor that might explain the correlation.

Does It Make Sense? For Exercises 5–8, decide whether the statement makes sense (or is clearly true) or does not make sense (or is clearly false). Explain clearly; not all of these statements have definitive answers, so your explanation is more important than your chosen answer.

5. **Births.** A study showed that for one town, as the stork population increased, the numbers of births in the town also increased. It therefore follows that the increase in the stork population caused the numbers of births to increase.

6. **Positive Effect.** A researcher plans to investigate the relationship between a decreasing infant mortality rate and per capita income in different nations. Because a decreasing infant mortality rate is a good, or "positive," effect, we know that there is a positive correlation between the infant mortality rate and the per capita income of different nations.

7. **Correlation.** Two studies both found a correlation between low birth weight and weakened immune systems. The second study had a much larger sample size, so the correlation it found must be stronger.

8. **Interpreting *r*.** A researcher uses 20 pairs of data to calculate the value of the linear correlation coefficient and he concludes that $r = 1.2$.

Concepts and Applications

Types of Correlation. Exercises 9–16 list pairs of variables. For each pair, state whether you believe the two variables are correlated. If you believe they are correlated, state whether the correlation is positive or negative. Explain your reasoning.

9. **Height/Weight.** The heights and weights of 50 randomly selected females between the ages of 1 and 21

10. **Studying/Grades.** The amount of time students spend studying for a history test and their grade on that test

11. **Weight/Fuel Consumption.** The weights of cars and their fuel consumption rates measured in miles per gallon

12. **IQ/Hat Size.** The IQ scores and hat sizes of randomly selected adults

13. **Marathon.** The time (in seconds) it takes to run a marathon and the order of finish (first, second, etc.)

14. **Altitude/Temperature.** The outside air temperature and the altitude of aircraft

15. **Height/SAT Score.** The heights and SAT scores of randomly selected subjects who take the SAT

16. **Golf Score/Prize Money.** Golf scores and prize money won by professional golfers

17. **World Meat and Grain Production.** The scatter diagram in Figure 7.8 shows the relationship between world production of meat and world production of grain, both measured in kilograms per person. The data points correspond to 11 different years between 1950 and 2000. Estimate the correlation coefficient and discuss the underlying reasons for the correlation.

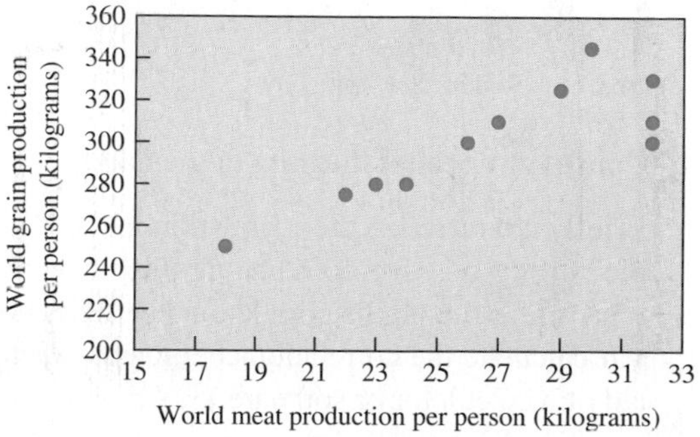

Figure 7.8 Scatter diagram for meat and grain production data for 11 different years. *Source:* U.S. Department of Agriculture.

18. **Two-Day Forecast.** Figure 7.9 shows a scatter diagram in which the actual high temperature for the day is compared with a forecast made two days in advance. Estimate the correlation coefficient and discuss what these data imply about weather forecasts. Do you think you would

get similar results if you made similar diagrams for other two-week periods? Why or why not?

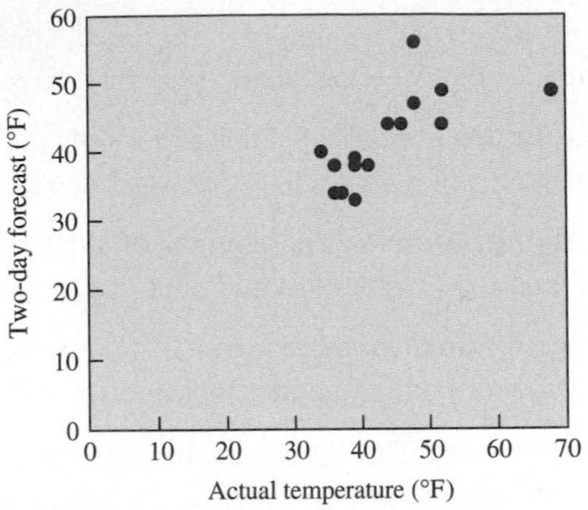

Figure 7.9

19. **Safe Speeds?** Consider the following table showing speed limits and death rates from automobile accidents in selected countries in the 1980s.

Country	Death rate (per 100 million vehicle-miles)	Speed limit (miles per hour)
Norway	3.0	55
United States	3.3	55
Finland	3.4	55
Britain	3.5	70
Denmark	4.1	55
Canada	4.3	60
Japan	4.7	55
Australia	4.9	65
Netherlands	5.1	60
Italy	6.1	75

Source: D. J. Rivkin, *New York Times.*

a. Construct a scatter diagram of the data.

b. Briefly characterize the correlation in words (for example, strong positive correlation, weak negative correlation) and estimate the correlation coefficient of the data. (Or calculate the correlation coefficient exactly with the aid of a calculator or software.)

c. In the newspaper, these data were presented in an article titled "Fifty-five mph speed limit is no safety guarantee." Based on the data, do you agree with this claim? Explain.

20. **Population Growth.** Consider the following table showing percentage change in population and birth rate (per 1,000 of population) for ten states over a period of ten years.

State	Percentage change in population	Birth rate
Nevada	50.1%	16.3
California	25.7%	16.9
New Hampshire	20.5%	12.5
Utah	17.9%	21.0
Colorado	14.0%	14.6
Minnesota	7.3%	13.7
Montana	1.6%	12.3
Illinois	0%	15.5
Iowa	−4.7%	13.0
West Virginia	−8.0%	11.4

Source: U.S. Census Bureau and Department of Health and Human Services.

a. Construct a scatter diagram for the data.

b. Briefly characterize the correlation in words and estimate the correlation coefficient.

c. Overall, does birth rate appear to be a good predictor of a state's population growth rate? If not, what other factor(s) may be affecting the growth rate?

21. **Most Valuable Players.** Consider the following table showing the number of home runs and the batting average for baseball's Most Valuable Players (NL = National League and AL = American League).

Player	Home runs	Batting average
Jeff Kent (2000 NL)	33	.334
Jason Giambi (2000 AL)	43	.333
Barry Bonds (2001 NL)	73	.328
Ichiro Suzuki (2001 AL)	8	.350
Barry Bonds (2002 NL)	46	.370
Miguel Tejada (2002 AL)	34	.308
Barry Bonds (2003 NL)	45	.341
Alex Rodriguez (2003 AL)	47	.298
Barry Bonds (2004 NL)	45	.362
Vladimir Guerrero (2004 AL)	39	.337
Albert Pujols (2005 NL)	41	.330
Alex Rodriguez (2005 AL)	48	.321
Ryan Howard (2006 NL)	58	.313
Justin Moreau (2006 AL)	34	.321

a. Construct a scatter diagram for the data.

b. Briefly characterize the correlation in words and estimate the correlation coefficient.

c. Do these data suggest that a high batting average is a good predictor of home runs? Explain.

22. Movie Data. Consider the following table showing total box office receipts and total attendance for all American films, 1990–2002.

Year	Total receipts (billions of dollars)	Total attendance (billions)
1990	5.0	1.18
1991	4.8	1.14
1992	4.9	1.17
1993	5.2	1.24
1994	5.4	1.29
1995	5.5	1.26
1996	5.9	1.34
1997	6.4	1.39
1998	7.0	1.48
1999	7.5	1.47
2000	7.7	1.42
2001	8.4	1.49
2002	9.5	1.64

Source: Motion Picture Association of America.

a. Make a scatter diagram for the data.

b. Briefly characterize the correlation in words and estimate the correlation coefficient.

c. How does the fact that the price of movies has increased since 1990 affect these data? If you were a movie executive, what general conclusions from these data would be important to you?

23. TV Time. Consider the following table showing the average hours of television watched in households in five categories of annual income.

Household income	Weekly TV hours
Less than \$30,000	56.3
\$30,000–\$40,000	51.0
\$40,000–\$50,000	50.5
\$50,000–\$60,000	49.7
More than \$60,000	48.7

Source: Nielsen Media Research.

a. Construct a scatter diagram for the data. To locate the dots, use the midpoint of each income category. Use a value of \$25,000 for the category "less than \$30,000," and use \$70,000 for "more than \$60,000."

b. Briefly characterize the correlation in words and estimate the correlation coefficient.

c. Suggest a reason why families with higher incomes watch less TV. Do you think these data imply that you can increase your income simply by watching less TV? Explain.

24. January Weather. Consider the following table showing January mean monthly precipitation and mean daily high temperature for ten Northern Hemisphere cities (National Oceanic and Atmospheric Administration).

City	Mean daily high temperature for January (°F)	Mean January precipitation (inches)
Athens	54	2.2
Bombay	88	0.1
Copenhagen	36	1.6
Jerusalem	55	5.1
London	44	2.0
Montreal	21	3.8
Oslo	30	1.7
Rome	54	3.3
Tokyo	47	1.9
Vienna	34	1.5

Source: The New York Times Almanac.

a. Construct a scatter diagram for the data.

b. Briefly characterize the correlation in words and estimate the correlation coefficient.

c. Can you draw any general conclusions about January temperatures and precipitation from these data? Explain.

25. Retail Sales. Consider the following table showing one year's total sales (revenue) and profits for eight large retailers in the United States.

Company	Total sales (billions of dollars)	Profits (billions of dollars)
Wal-Mart	315.6	11.2
Kroger	60.6	0.98
Home Depot	81.5	5.8
Costco	60.1	1.1
Target	52.6	2.4
Starbuck's	7.8	0.6
The Gap	16.0	1.1
Best Buy	30.8	1.1

Source: Fortune.com.

a. Construct a scatter diagram for the data.

b. Briefly characterize the correlation in words and estimate the correlation coefficient.

c. Discuss your observations. Does higher sales volume necessarily translate into greater earnings? Why or why not?

26. Calories and Infant Mortality. Consider the following table showing mean daily caloric intake (all residents) and infant mortality rate (per 1,000 births) for 10 countries.

Country	Mean daily calories	Infant mortality rate (per 1,000 births)
Afghanistan	1,523	154
Austria	3,495	6
Burundi	1,941	114
Colombia	2,678	24
Ethiopia	1,610	107
Germany	3,443	6
Liberia	1,640	153
New Zealand	3,362	7
Turkey	3,429	44
United States	3,671	7

a. Construct a scatter diagram for the data.

b. Briefly characterize the correlation in words and estimate the correlation coefficient.

c. Discuss any patterns you observe and any general conclusions that you can reach.

Properties of the Correlation Coefficient. For Exercises 27 and 28, determine whether the given property is true, and explain your answer.

27. Interchanging Variables. The correlation coefficient remains unchanged if we interchange the variables *x* and *y*.

28. Changing Units of Measurement. The correlation coefficient remains unchanged if we change the units used to measure *x*, *y*, or both.

Projects for the Internet and Beyond

For useful links, select "Links for Internet Projects" for Chapter 7 at www.aw.com/bbt.

29. Unemployment and Inflation. Use the Bureau of Labor Statistics Web page to find monthly unemployment rates and inflation rates over the past year. Make a scatter diagram for the data. Do you see any trends?

30. Success in the NFL. Find last season's NFL team statistics. Make a table showing the following for each team: number of wins, average yards gained on offense per game, and average yards allowed on defense per game. Make scatter diagrams to explore the correlations between offense and wins and between defense and wins. Discuss your findings. Do you think that there are other team statistics that would yield stronger correlations with the number of wins?

31. Statistical Abstract. Explore the "frequently requested tables" at the Web site for the *Statistical Abstract of the United States*. Choose data that are of interest to you and explore at least two correlations. Briefly discuss what you learn from the correlations.

32. Height and Arm Span. Select a sample of at least eight people and measure each person's height and arm span. (When you measure arm span, the person should stand with arms extended like the wings on an airplane.) Using the paired sample data, construct a scatter diagram and estimate or calculate the value of the correlation coefficient. What do you conclude?

33. Height and Pulse Rate. Select a sample of at least eight people and record each person's pulse rate by counting the number of heartbeats in 1 minute. Also record each person's height. Using the paired sample data, construct a scatter diagram and estimate or calculate the value of the correlation coefficient. What do you conclude?

IN THE NEWS

34. Correlations in the News. Find a recent news report that discusses some type of correlation. Describe the correlation. Does the article give any sense of the strength of the correlation? Does it suggest that the correlation reflects any underlying causality? Briefly discuss whether you believe the implications the article makes with respect to the correlation.

35. Your Own Positive Correlations. Give examples of two variables that you expect to be positively correlated. Explain why the variables are correlated and why the correlation is (or is not) important.

36. Your Own Negative Correlations. Give examples of two variables that you expect to be negatively correlated. Explain why the variables are correlated and why the correlation is (or is not) important.

7.2 Interpreting Correlations

Researchers sifting through statistical data are constantly looking for meaningful correlations, and the discovery of a new and surprising correlation often leads to a flood of news reports. You may recall hearing about some of these discovered correlations: oat bran consumption correlated with reduced risk of heart disease; cell phone use correlated with increased risk of auto accidents; or eating less correlated with increased longevity. Unfortunately, the task of *interpreting* such correlations is far more difficult than discovering them in the first place. Long after the news reports have faded, we may still be unsure of whether the correlations are significant and, if so, whether they tell us anything of practical importance. In this section, we discuss some of the common difficulties associated with interpreting correlations.

Statistics show that of those who contract the habit of eating, very few survive.

—Wallace Irwin

Beware of Outliers

Examine the scatter diagram in Figure 7.10. Your eye probably tells you that there is a positive correlation in which larger values of x tend to mean larger values of y. Indeed, if you calculate the correlation coefficient for these data, you'll find that it is a relatively high $r = 0.880$, suggesting a very strong correlation.

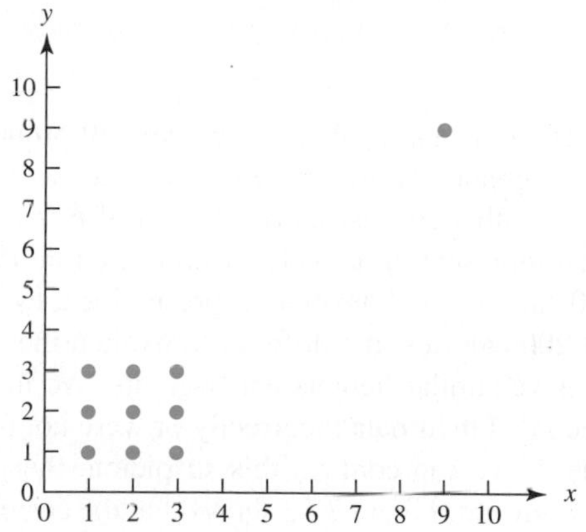

Figure 7.10 How does the outlier affect the correlation?

However, if you place your thumb over the data point in the upper right corner of Figure 7.10, the apparent correlation disappears. In fact, without this data point, the correlation coefficient is zero! In other words, removing this one data point changes the correlation coefficient from $r = 0.880$ to $r = 0$.

This example shows that correlations can be very sensitive to outliers. Recall that an *outlier* is a data value that is extreme compared to most other values in a data set (see Section 4.1). We must therefore examine outliers and their effects carefully before interpreting a correlation. On the one hand, if the outliers are mistakes in the data set, they can produce apparent correlations that are not real or mask the presence of real correlations. On the other hand, if the outliers represent real and correct data points, they may be telling us about relationships that would otherwise be difficult to see.

Note that while we should examine outliers carefully, we should *not* remove them unless we have strong reason to believe that they do not belong in the data set. Even in that case, good research principles demand that we report the outliers along with an explanation of why we thought it legitimate to remove them.

EXAMPLE 1 Masked Correlation

You've conducted a study to determine how the number of calories a person consumes in a day correlates with time spent in vigorous bicycling. Your sample consisted of ten women cyclists, all of approximately the same height and weight. Over a period of two weeks, you asked each woman to record the amount of time she spent cycling each day and what she ate on each of those days. You used the eating records to calculate the calories consumed each day. Figure 7.11 shows a scatter diagram with each woman's mean time spent cycling on the horizontal axis and mean caloric intake on the vertical axis. Do higher cycling times correspond to higher intake of calories?

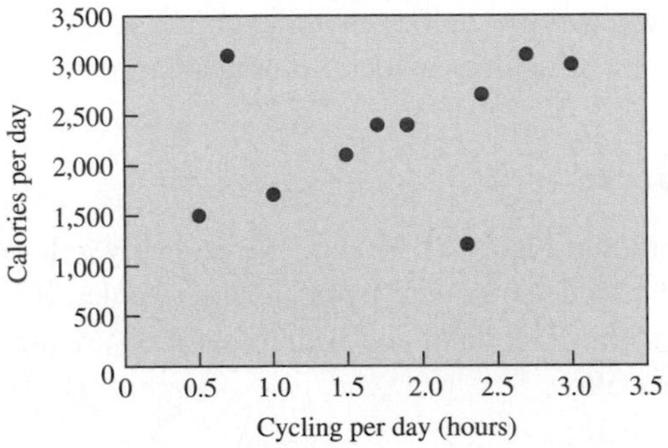

Figure 7.11 Data from the cycling study.

Solution If you look at the data as a whole, your eye will probably tell you that there is a positive correlation in which greater cycling time tends to go with higher caloric intake. But the correlation is very weak, with a correlation coefficient of $r = 0.374$. However, notice that two points are outliers: one representing a cyclist who cycled about a half-hour per day and consumed more than 3,000 calories, and the other representing a cyclist who cycled more than 2 hours per day on only 1,200 calories. It's difficult to explain the two outliers, given that all the women in the sample have similar heights and weights. We might therefore suspect that these two women either recorded their data incorrectly or were not following their usual habits during the two-week study. If we can confirm this suspicion, then we would have reason to delete the two data points as invalid. Figure 7.12 shows that the correlation is quite strong without those two outlier points, and suggests that the number of calories consumed rises by a little more than 500 calories for each hour of cycling. Of course, we should *not* remove the outliers without confirming our suspicion that they were invalid data points, and we should report our reasons for leaving them out.

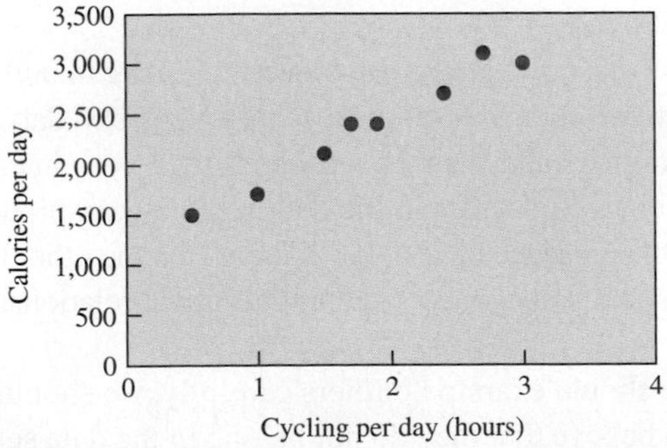

Figure 7.12 The data from Figure 7.11 without the two outliers.

Beware of Inappropriate Grouping

Correlations can also be misinterpreted when data are grouped inappropriately. In some cases, grouping data hides correlations. Consider a (hypothetical) study in which researchers seek a correlation between hours of TV watched per week and high school grade point average (GPA). They collect the 21 data pairs in Table 7.3.

The scatter diagram (Figure 7.13) shows virtually no correlation; the correlation coefficient for the data is about $r = -0.063$. The apparent conclusion is that TV viewing habits are unrelated to academic achievement. However, one astute researcher realizes that some of the students watched mostly educational programs, while others tended to watch comedies, dramas, and movies. She therefore divides the data set into two groups, one for the students who watched mostly educational television and one for the other students. Table 7.4 shows her results with the students divided into these two groups.

Table 7.3 Hours of TV and High School GPA (hypothetical data)

Hours per week of TV	GPA
2	3.2
4	3.0
4	3.1
5	2.5
5	2.9
5	3.0
6	2.5
7	2.7
7	2.8
8	2.7
9	2.5
9	2.9
10	3.4
12	3.6
12	2.5
14	3.5
14	2.3
15	3.7
16	2.0
20	3.6
20	1.9

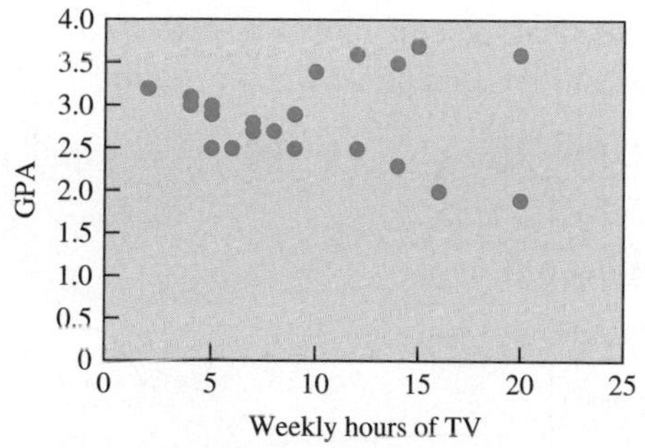

Figure 7.13 The full set of data concerning hours of TV and GPA shows virtually no correlation.

Table 7.4 Hours of TV and High School GPA—Grouped Data (hypothetical data)

Group 1: watched educational programs		Group 2: watched regular TV	
Hours per week of TV	**GPA**	**Hours per week of TV**	**GPA**
5	2.5	2	3.2
7	2.8	4	3.0
8	2.7	4	3.1
9	2.9	5	2.9
10	3.4	5	3.0
12	3.6	6	2.5
14	3.5	7	2.7
15	3.7	9	2.5
20	3.6	12	2.5
		14	2.3
		16	2.0
		20	1.9

Now we find two very strong correlations (Figure 7.14): a strong positive correlation for the students who watched educational programs ($r = 0.855$) and a strong negative correlation for the other students ($r = -0.951$). The moral of this story is that the original data set hid an important (hypothetical) correlation between TV and GPA: Watching educational TV correlated positively with GPA and watching non-educational TV correlated negatively with GPA. Only when the data were grouped appropriately could this discovery be made.

By the Way ...

Children aged 2-5 watch an average of 26 hours of television per week, while children aged 6-11 watch an average of 20 hours of television per week (Nielsen Media Research). Adult viewership averages more than 25 hours per week. If the average adult replaced television time with a job paying just $8 per hour, his or her annual income would rise by more than $10,000.

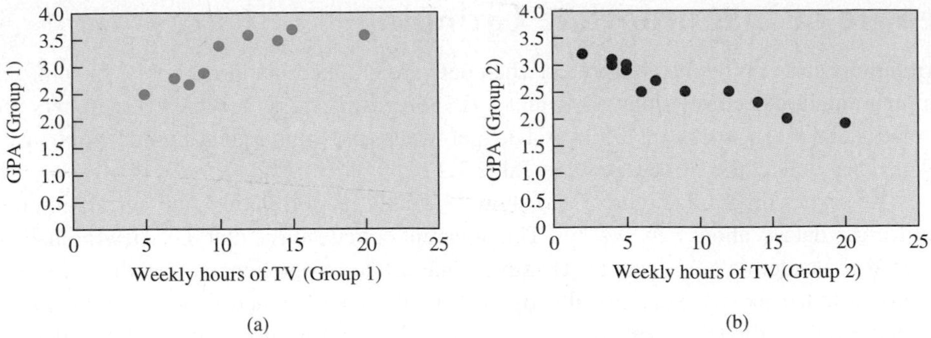

Figure 7.14 These scatter diagrams show the same data as Figure 7.13, separated into the two groups identified in Table 7.4.

In other cases, a data set may show a stronger correlation than actually exists among subgroups. Consider the (hypothetical) data in Table 7.5, collected by a consumer group studying the relationship between the weights and prices of cars. Figure 7.15 shows the scatter diagram.

Table 7.5 Car Weights and Prices (hypothetical data)

Weight (pounds)	Price (dollars)
1,500	9,500
1,600	8,000
1,700	8,200
1,750	9,500
1,800	9,200
1,800	8,700
3,000	29,000
3,500	25,000
3,700	27,000
4,000	31,000
3,600	25,000
3,200	30,000

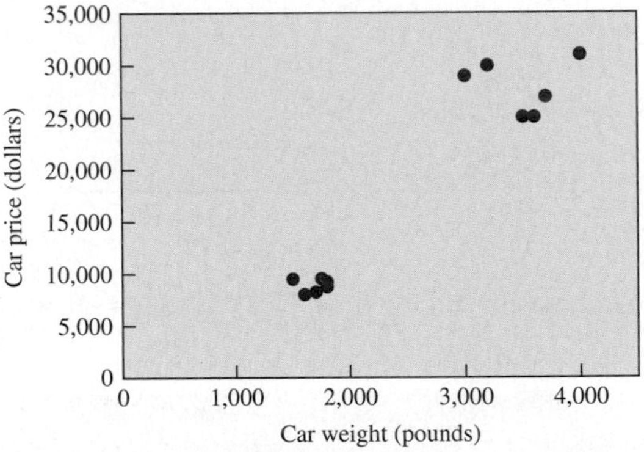

Figure 7.15 Scatter diagram for the car weight and price data in Table 7.5.

The data set as a whole shows a strong correlation; the correlation coefficient is $r = 0.949$. However, on closer examination, we see that the data fall into two rather distinct categories corresponding to light and heavy cars. If we analyze these subgroups separately, neither shows any correlation: The light cars alone (top six in Table 7.5) have a correlation coefficient $r = 0.019$ and the heavy cars alone (bottom six in Table 7.5) have a correlation coefficient $r = -0.022$. You can see the problem by looking at Figure 7.15. The apparent correlation of the full data set occurs because of the separation between the two clusters of points; there's no correlation within either cluster.

TIME OUT TO THINK

Suppose you were shopping for a compact car. If you looked at only the overall data and correlation coefficient from Figure 7.15, would it be reasonable to consider weight as an important factor in price? What if you looked at the data for light and heavy cars separately? Explain.

CASE STUDY Fishing for Correlations

Oxford physician Richard Peto submitted a paper to the British medical journal *Lancet* showing that heart-attack victims had a better chance of survival if they were given aspirin within a few hours after their heart attacks. The editors of *Lancet* asked Peto to break down the data into subsets, to see whether the benefits of the aspirin were different for different groups of patients. For example, was aspirin more effective for patients of a certain age or for patients with certain dietary habits?

Breaking the data into subsets can reveal important facts, such as whether men and women respond to the treatment differently. However, Peto felt that the editors were asking him to divide his sample into too many subgroups. He therefore objected to the request, arguing that it would result in purely coincidental correlations. Writing about this story in the *Washington Post*, journalist Rick Weiss said, "When the editors insisted, Peto capitulated, but among other things he divided his patients by zodiac birth signs and demanded that his findings be included in the published paper. Today, like a warning sign to the statistically uninitiated, the wacky numbers are there for all to see: Aspirin is useless for Gemini and Libra heart-attack victims but is a lifesaver for people born under any other sign."

The moral of this story is that a "fishing expedition" for correlations can often produce them. That doesn't make the correlations meaningful, even though they may appear significant by standard statistical measures.

Correlation Does *Not* Imply Causality

Perhaps the most important caution about interpreting correlations is one we've already mentioned: ***Correlation does not necessarily imply causality***. In general, correlations can appear for any of the following three reasons.

Possible Explanations for a Correlation

1. The correlation may be a *coincidence*.
2. Both correlation variables might be directly influenced by some *common underlying cause*.
3. One of the correlated variables may actually be a *cause* of the other. But note that, even in this case, it may be just one of several causes.

For example, the correlation between infant mortality and life expectancy in Figure 7.4 is a case of common underlying cause: Both variables respond to the underlying variable *quality of health care*. The correlation between smoking and lung cancer reflects the fact that smoking causes lung cancer (see the discussion in Section 7.4). Coincidental correlations are also quite common; Example 2 below discusses one such case.

Caution about causality is particularly important in light of the fact that many statistical studies are designed to look for causes. Because these studies generally begin with the search for correlations, it's tempting to think that the work is over as soon as a correlation is found. However, as we will discuss in Section 7.4, establishing causality can be very difficult.

By the Way ...

The Super Bowl indicator went into a slump with four straight years of incorrect predictions from 1998 through 2001. It has since recovered, with correct predictions in four of the next five years. This book went to press too early to check the prediction of an up market in 2007.

EXAMPLE 2 How to Get Rich in the Stock Market (Maybe)

Every financial advisor has a strategy for predicting the direction of the stock market. Most focus on fundamental economic data, such as interest rates and corporate profits. But an alternative strategy relies on a remarkable and well-known correlation between the Super Bowl winner in January and the direction of the stock market for the rest of the year: The stock market tends to rise when a team from the old, pre-1970 NFL wins the Super Bowl, and tends to fall when the winner is not from the old NFL. This correlation successfully matched 31 of the first 40 Super Bowls to the stock market; this success rate of 77.5% is far more than would be expected by chance. The winners of Super Bowl XLI in 2007, the Indianapolis Colts, were one of the old NFL teams before 1970. Based on this fact, should you have invested all your spare cash (and maybe even some that you borrow) in the stock market right after the 2007 Super Bowl?

Solution Based on the reported correlation, you might have been tempted to invest since the old-NFL winner suggests a rising stock market over the rest of the year. (The losing team, the Chicago Bears, also was from the old NFL, so the Super Bowl correlation would have predicted a rising market in 2007 no matter who won.) However, this investment would make sense only if you believed that the Super Bowl result actually *causes* the stock market to move in a particular direction. This is clearly preposterous, and the correlation is undoubtedly a coincidence. If you are going to invest, don't base your investment on this correlation.

CASE STUDY **Oat Bran and Heart Disease**

If you buy a product that contains oat bran, there's a good chance that the label will tout the healthful effects of eating oats. Indeed, several studies have found correlations in which people who eat more oat bran tend to have lower rates of heart disease. But does this mean that everyone should eat more oats?

Not necessarily. Just because oat bran consumption is correlated with reduced risk of heart disease does not mean that it *causes* reduced risk of heart disease. In fact, the question of causality is quite controversial in this case. Other studies suggest that people who eat a lot of oat bran tend to have generally healthful diets. Thus, the correlation between oat bran consumption and reduced risk of heart disease may be a case of a common underlying cause: Having a healthy diet leads people both to consume more oat bran and to have a lower risk of heart disease. In that case, for some people, adding oat bran to their diets might be a *bad* idea because it could cause them to gain weight, and weight gain is associated with *increased* risk of heart disease.

This example shows the importance of using caution when considering issues of correlation and causality. It may be a long time before medical researchers know for sure whether adding oat bran to your diet actually causes a reduced risk of heart disease.

Useful Interpretations of Correlation

In discussing uses of correlation that might lead to wrong interpretations, we have described the effects of outliers, inappropriate groupings, fishing for correlations, and incorrectly concluding that correlation implies causality. But there are many correct and useful interpretations of correlation. For example, in Section 7.1 we showed how correlation is used to determine the prices of diamonds. In other applications, correlation has been used to establish a relationship between population size and the weight of plastic discarded as garbage. Correlation has been used to establish a relationship between the durations of eruptions of Old Faithful geyser and the intervals between eruptions. In general, correlation plays a prominent and important role in a variety of fields, including meteorology, medical research, business, economics, market research, advertising, psychology, and computer science.

Section 7.2 Exercises

Statistical Literacy and Critical Thinking

1. **Correlation and Causality.** In interpreting correlation, it is extremely important to understand that correlation does not imply causality. What do we mean when we say that "correlation does not imply causality"?

2. **SIDS.** An article in the *New York Times* on infant deaths included a statement that, based on the study results, putting infants to sleep in the supine position decreased deaths due to SIDS (sudden infant death syndrome). What is wrong with that statement?

3. **Outliers.** What is an outlier? How might outliers affect conclusions about correlation?

4. **Scatter Diagram.** What is a scatter diagram and how is it useful?

Does It Make Sense? For Exercises 5–8, decide whether the statement makes sense (or is clearly true) or does not make sense (or is clearly false). Explain clearly; not all of these statements have definitive answers, so your explanation is more important than your chosen answer.

5. Weight and Fuel Consumption. Based on a study showing a significant correlation between the weight of a car (in pounds) and its fuel consumption rate (in miles per gallon), we can conclude that increasing a car's weight causes the fuel consumption rate to decrease.

6. Exercise and Health. Studies have shown that there is a correlation between exercise and health. But correlation does not imply causality. Thus, we should not exercise because doing so would not result in better health.

7. Drinking and Driving. Because studies show a correlation between drinking alcohol and car crashes, we can conclude that drinking alcohol causes car crashes.

8. Outlier. Given 20 pairs of data, suppose we add an additional pair that constitutes an outlier. The effect of the outlier on the correlation coefficient will be very small because the outlier represents only 1 of 21 pairs of data.

Concepts and Applications

Correlation and Causality. Exercises 9–16 make statements about a correlation. In each case, state the correlation clearly. (For example, there is a positive correlation between variable A and variable B.) Then state whether the correlation is most likely due to coincidence, a common underlying cause, or a direct cause. Explain your answer.

9. Guns and Crime Rate. In one state, the number of registered handguns steadily increased over the past several years, and the crime rate increased as well.

10. Running and Weight. It has been found that people who exercise regularly by running tend to weigh less than those who do not run, and those who run longer distances tend to weigh less than those who run shorter distances.

11. Toll and Distance. As a driver travels farther along the Massachusetts Turnpike, the cost of the toll increases.

12. Vehicles and Waiting Time. It has been found that as the number of registered vehicles increases, the time drivers spend sitting in traffic also increases.

13. Traffic Lights and Car Crashes. It has been found that as the number of traffic lights increases, the number of car crashes also increases.

14. Galaxies. Astronomers have discovered that, with the exception of a few nearby galaxies, all galaxies in the universe are moving away from us. Moreover, the farther the galaxy, the faster it is moving away. That is, the more distant a galaxy, the greater the speed at which it is moving away from us.

15. Gas and Driving. It has been found that as gas prices increase, the distance vehicles are driven tends to get shorter.

16. Melanoma and Latitude. Some studies have shown that, for certain ethnic groups, the incidence of melanoma (the most dangerous form of skin cancer) increases as latitude decreases.

17. Outlier Effects. Consider the scatter diagram in Figure 7.16.

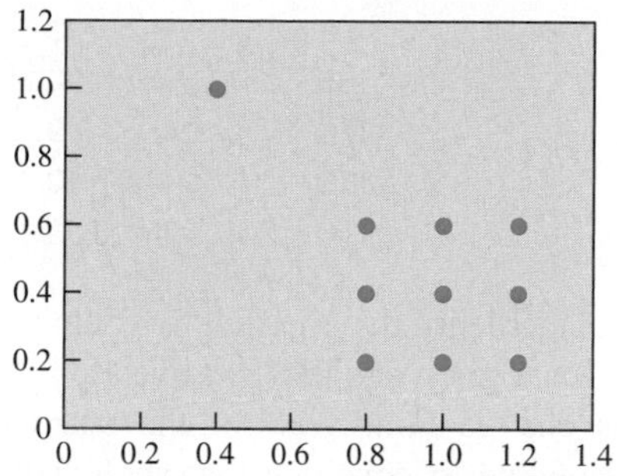

Figure 7.16

a. Which point is an outlier? Ignoring the outlier, estimate or compute the correlation coefficient for the remaining points.

b. Now include the outlier. How does the outlier affect the correlation coefficient? Estimate or compute the correlation coefficient for the complete data set.

18. Outlier Effects. Consider the scatter diagram in Figure 7.17.

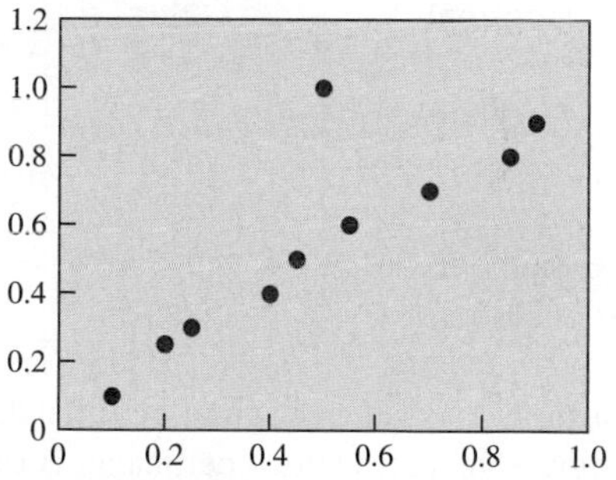

Figure 7.17

a. Which point is an outlier? Ignoring the outlier, estimate or compute the correlation coefficient for the remaining points.

b. Now include the outlier. How does the outlier affect the correlation coefficient? Estimate or compute the correlation coefficient for the complete data set.

19. Grouped Shoe Data. The following table gives measurements of weight and shoe size for 10 people (including both men and women).

a. Construct a scatter diagram for the data. Estimate or compute the correlation coefficient. Based on this correlation coefficient, would you conclude that shoe size and weight are correlated? Explain.

Weight (pounds)	Shoe size
105	6
112	4.5
115	6
123	5
135	6
155	10
165	11
170	9
180	10
190	12

b. You later learn that the first five data values in the table are for women and the next five are for men. How does this change your view of the correlation? Is it still reasonable to conclude that shoe size and weight are correlated?

20. Grouped Temperature Data. The following table shows the average January high temperature and the average July high temperature for 10 major cities around the world.

City	January high	July high
Berlin	35	74
Geneva	39	77
Kabul	36	92
Montreal	21	78
Prague	34	74
Auckland	73	56
Buenos Aires	85	57
Sydney	78	60
Santiago	85	59
Melbourne	78	56

a. Construct a scatter diagram for the data. Estimate or compute the correlation coefficient. Based on this correlation coefficient, would you conclude that January and July temperatures are correlated for these cities? Explain.

b. Notice that the first five cities in the table are in the Northern Hemisphere and the next five are in the Southern Hemisphere. How does this change your view of the correlation? Would you now conclude that January and July temperatures are correlated for these cities? Explain.

21. Birth and Death Rates. Figure 7.18 shows the birth and death rates for different countries, measured in births and deaths per 1,000 population.

a. Estimate the correlation coefficient and discuss whether there is a strong correlation between the variables.

b. Notice that there appear to be two groups of data points within the full data set. Make a reasonable guess as to the makeup of these groups. In which group might you find a relatively wealthy country like Sweden? In which group might you find a relatively poor country like Uganda?

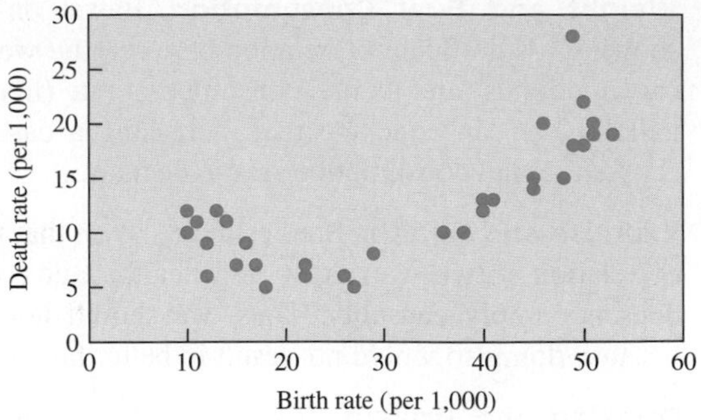

Figure 7.18 Birth and death rates for different countries. *Source:* United Nations.

c. Assuming that your guess about groups in part b is correct, do there appear to be correlations within the groups? Explain. How could you confirm your guess about the groups?

22. Reading and Test Scores. The following (hypothetical) data set gives the number of hours 10 sixth-graders read per week and their performance on a standardized verbal test (maximum of 100).

Reading time per week	Verbal test score
1	50
1	65
2	56
3	62
3	65
4	60
5	75
6	50
10	88
12	38

a. Construct a scatter diagram for these data. Estimate or compute the correlation coefficient. Based on this correlation coefficient, would you conclude that reading time and test scores are correlated? Explain.

b. Suppose you learn that five of the children read only comic books while the other five read regular books. Make a guess as to which data points fall in which group. How could you confirm your guess about the groups?

c. Assuming that your guess in part b is correct, how does it change your view of the correlation between reading time and test scores? Explain.

Projects for the Internet and Beyond

For useful links, select "Links for Internet Projects" for Chapter 7 at www.aw.com/bbt.

23. Football-Stock Update. Find data for recent years concerning the Super Bowl winner and the end-of-year change in the stock market (positive or negative). Do recent results still agree with the correlation described in Example 2? Explain.

24. Real Correlations.

a. Describe a real situation in which there is a positive correlation that is the result of coincidence.

b. Describe a real situation in which there is a positive correlation that is the result of a common underlying cause.

c. Describe a real situation in which there is a positive correlation that is the result of a direct cause.

d. Describe a real situation in which there is a negative correlation that is the result of coincidence.

e. Describe a real situation in which there is a negative correlation that is the result of a common underlying cause.

f. Describe a real situation in which there is a negative correlation that is the result of a direct cause.

IN THE NEWS

25. Misinterpreted Correlations. Find a recent news report in which you believe that a correlation may have been misinterpreted. Describe the correlation, the reported interpretation, and the problems you see in the interpretation.

26. Well-Interpreted Correlations. Find a recent news report in which you believe that a correlation has been presented with a reasonable interpretation. Describe the correlation and the reported interpretation, and explain why you think the interpretation is valid.

7.3 Best-Fit Lines and Prediction

Suppose you are lucky enough to win a 1.5-carat diamond in a contest. Based on the correlation between weight and price in Figure 7.1, it should be possible to predict the approximate value of the diamond. We need only study the graph carefully and decide where a point corresponding to 1.5 carats is most likely to fall. To do this, it is helpful to draw a **best-fit line** (also called a *regression line*) through the data, as shown in Figure 7.19. This line is a "best fit" in the sense that, according to a standard statistical measure (which we discuss shortly), the data points lie closer to this line than to any other straight line that we could draw through the data.

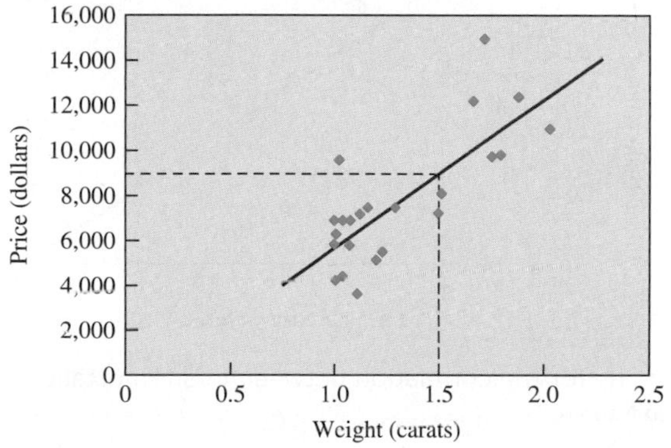

Figure 7.19 Best-fit line for the data from Figure 7.1.

Definition

The **best-fit line** (or *regression line*) on a scatter diagram is a line that lies closer to the data points than any other possible line (according to a standard statistical measure of closeness).

By the Way ...

The term *regression* comes from an 1877 study by Sir Francis Galton. He found that the heights of boys with short or tall fathers were closer to the mean than were the heights of their fathers. He therefore said that the heights of the children *regress* toward the mean, from which we get the term *regression*. The term is now used even for data that have nothing to do with a tendency to regress toward a mean.

Of all the possible straight lines that can be drawn on a diagram, how do you know which one is the best-fit line? In many cases, you can make a good estimate of the best-fit line simply by looking at the data and drawing the line that visually appears to pass closest to all the data points. This method involves drawing the best-fit line "by eye." As you might guess, there are methods for calculating the precise equation of a best-fit line (see the optional topic at the end of this section), and many computer programs and calculators can do these calculations automatically. For our purposes in this book, a fit by eye will generally be sufficient.

Predictions with Best-Fit Lines

It is a capital mistake to theorize before one has data.

—Arthur Conan Doyle

We can use the best-fit line in Figure 7.19 to predict the price of a 1.5-carat diamond. As indicated by the dashed lines in the figure, the best-fit line predicts that the diamond will cost about $9,000. Notice, however, that two actual data points in the figure correspond to 1.5-carat diamonds, and both of these diamonds cost less than $9,000. Thus, although the predicted price of $9,000 sounds reasonable, it is certainly not guaranteed. In fact, the degree of scatter among the data points in this case tells us that we should *not* trust the best-fit line to predict accurately the price for any individual diamond. Instead, the prediction is meaningful only in a statistical sense: It tells us that if we examined many 1.5-carat diamonds, their mean price would be about $9,000.

This is only the first of several important cautions about interpreting predictions with best-fit lines. A second caution is to beware of using best-fit lines to make predictions that go beyond the bounds of the available data. Figure 7.20 shows a best-fit line for the correlation between infant mortality and longevity from Figure 7.4. According to this line, a country with a life expectancy of more than about 80 years would have a *negative* infant mortality rate, which is impossible.

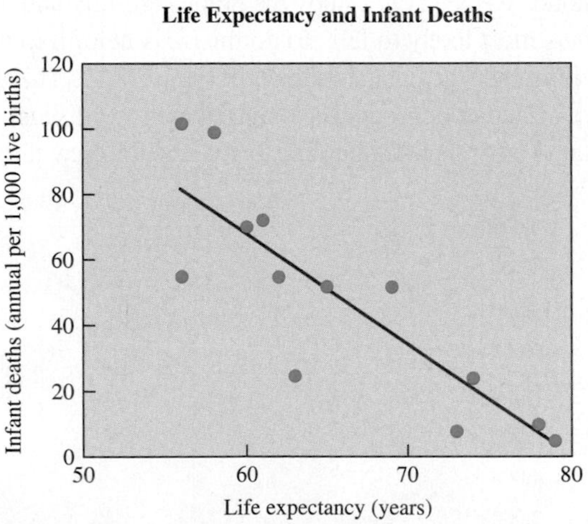

Figure 7.20 A best-fit line for the correlation between infant mortality and longevity from Figure 7.4. *Source:* United Nations.

It's tough to make predictions, especially about the future.

—Yogi Berra

A third caution is to avoid using best-fit lines from old data sets to make predictions about current or future results. For example, economists studying historical data found a strong negative correlation between unemployment and the rate of inflation. According to this correlation, inflation should have risen dramatically in recent years when the unemployment rate fell below 6%. But inflation remained low, showing that the correlation from old data did not continue to hold.

Fourth, a correlation discovered with a sample drawn from a particular population cannot generally be used to make predictions about other populations. For example, we can't expect that the correlation between aspirin consumption and heart attacks in an experiment involving only men will also apply to women.

Fifth, remember that we can draw a best-fit line through any data set, but that line is meaningless when the correlation is not significant or when the relationship is nonlinear. For example, there is no correlation between shoe size and IQ, so we could not use shoe size to predict IQ.

Cautions in Making Predictions from Best-Fit Lines

1. Don't expect a best-fit line to give a good prediction unless the correlation is strong and there are many data points. If the sample points lie very close to the best-fit line, the correlation is very strong and the prediction is more likely to be accurate. If the sample points lie away from the best-fit line by substantial amounts, the correlation is weak and predictions tend to be much less accurate.
2. Don't use a best-fit line to make predictions beyond the bounds of the data points to which the line was fit.
3. A best-fit line based on past data is not necessarily valid now and might not result in valid predictions of the future.
4. Don't make predictions about a population that is different from the population from which the sample data were drawn.
5. Remember that a best-fit line is meaningless when there is no significant correlation or when the relationship is nonlinear.

EXAMPLE 1 Valid Predictions?

State whether the prediction (or implied prediction) should be trusted in each of the following cases, and explain why or why not.

a. You've found a best-fit line for a correlation between the number of hours per day that people exercise and the number of calories they consume each day. You've used this correlation to predict that a person who exercises 18 hours per day would consume 15,000 calories per day.

b. There is a well-known but weak correlation between SAT scores and college grades. You use this correlation to predict the college grades of your best friend from her SAT scores.

c. Historical data have shown a strong negative correlation between national birth rates and affluence. That is, countries with greater affluence tend to have lower birth rates. These data predict a high birth rate in Russia.

d. A study in China has discovered correlations that are useful in designing museum exhibits that Chinese children enjoy. A curator suggests using this information to design a new museum exhibit for Atlanta-area school children.

e. Scientific studies have shown a very strong correlation between children's ingesting of lead and mental retardation. Based on this correlation, paints containing lead were banned.

f. Based on a large data set, you've made a scatter diagram for salsa consumption (per person) versus years of education. The diagram shows no significant correlation, but you've drawn a best-fit line anyway. The line predicts that someone who consumes a pint of salsa per week has at least 13 years of education.

By the Way ...

In the United States, lead was banned from house paint in 1978 and from food cans in 1991, and a 25-year phaseout of lead in gasoline was completed in 1995. Nevertheless, a 1997 report from the Centers for Disease Control still estimated that one million children under age 6 have enough lead in their blood to damage their health. Major sources of ongoing lead hazards include paint in older housing and soil near major roads, which has high lead content from past use of leaded gasoline.

Solution

a. No one exercises 18 hours per day on an ongoing basis, so this much exercise must be beyond the bounds of any data collected. Therefore, a prediction about someone who exercises 18 hours per day should not be trusted.

b. The fact that the correlation between SAT scores and college grades is weak means there is much scatter in the data. As a result, we should not expect great accuracy if we use this weak correlation to make a prediction about a single individual.

c. We cannot automatically assume that the historical data still apply today. In fact, Russia currently has a very low birth rate, despite also having a low level of affluence.

d. The suggestion to use information from the Chinese study for an Atlanta exhibit assumes that predictions made from correlations in China also apply to Atlanta. However, given the cultural differences between China and Atlanta, the curator's suggestion should not be considered without more information to back it up.

e. Given the strength of the correlation and the severity of the consequences, this prediction and the ban that followed seem quite reasonable. In fact, later studies established lead as an actual *cause* of mental retardation, making the rationale behind the ban even stronger.

f. Because there is no significant correlation, the best-fit line and any predictions made from it are meaningless.

EXAMPLE 2 Will Women Be Faster Than Men?

Figure 7.21 shows data and best-fit lines for both men's and women's world record times in the 1-mile race. Based on these data, predict when the women's world record will be faster than the men's world record. Comment on the prediction.

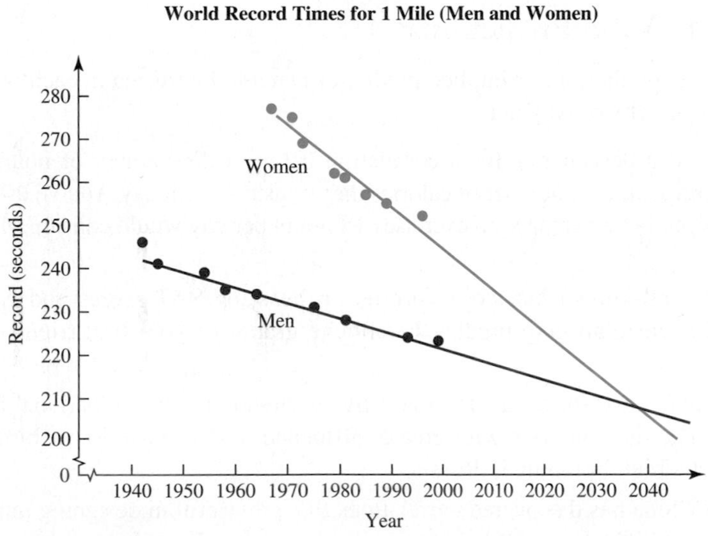

Figure 7.21 World record times in the mile (men and women).

Solution If we accept the best-fit lines as drawn, the women's world record will equal the men's world record by about 2040. However, this is *not* a valid prediction because it is based on extending the best-fit lines beyond the range of the actual data. It's certainly not out of the question that the women's record will be faster than the men's record in 2040. But by the same reasoning, both world records will eventually be zero, implying that someone will finish the race when the starting gun is fired!

The Correlation Coefficient and Best-Fit Lines

Earlier, we discussed the correlation coefficient as one way of measuring the strength of a correlation. We can also use the correlation coefficient to say something about the validity of predictions with best-fit lines.

For mathematical reasons (not discussed in this book), the *square* of the correlation coefficient, or r^2, is the proportion of the variation in a variable that is accounted for by the best-fit line (or, more technically, by the linear relationship that the best-fit line expresses). For example, the correlation coefficient for the diamond weight and price data (see Figure 7.19) turns out to be $r = 0.777$. If we square this value, we get $r^2 = 0.604$, which we can interpret as follows: About 0.6, or 60%, of the variation in the diamond prices is accounted for by the best-fit line relating weight and price. That leaves 40% of the variation in price that must be due to other factors, presumably such things as depth, table, color, and clarity—which is why predictions made with the best-fit line in Figure 7.19 are not very precise.

A best-fit line can give precise predictions only in the case of a perfect correlation ($r = 1$ or $r = -1$); we then find $r^2 = 1$, which means that 100% of the variation in a variable can be accounted for by the best-fit line. In this special case of $r^2 = 1$, predictions should be exactly correct, except for the fact that the sample data might not be a true representation of the population data.

Best-Fit Lines and r^2

The *square* of the correlation coefficient, or r^2, is the proportion of the variation in a variable that is accounted for by the best-fit line.

TECHNICAL NOTE

Statisticians often call r^2 the coefficient of determination.

EXAMPLE 3 Retail Hiring

You are the manager of a large department store. Over the years, you've found a reasonably strong correlation between your September sales and the number of employees you'll need to hire for peak efficiency during the holiday season. The correlation coefficient is 0.950. This year your September sales are fairly strong. Should you start advertising for help based on the best-fit line?

Solution In this case we find that $r^2 = 0.950^2 = 0.903$, which means that 90% of the variation in the number of peak employees can be accounted for by a linear relationship with September sales. That leaves only 10% of the variation in the number of peak employees unaccounted for. Because 90% is so high, we conclude that the best-fit line accounts for the data quite well, so it is a good idea to use it to predict the number of employees you'll need for this year's holiday season.

EXAMPLE 4 Voter Turnout and Unemployment

Political scientists are interested in knowing what factors affect voter turnout in elections. One such factor is the unemployment rate. Data collected in presidential election years since 1964 show a very weak negative correlation between voter turnout and the unemployment rate, with a correlation coefficient of about $r = -0.1$ (Figure 7.22). Based on this correlation, should we use the unemployment rate to predict voter turnout in the next presidential election?

Solution The square of the correlation coefficient is $r^2 = (-0.1)^2 = 0.01$, which means that only about 1% of the variation in the data is accounted for by the best-fit line. Nearly all of the variation in the data must therefore be explained by other factors. We conclude that unemployment is *not* a reliable predictor of voter turnout.

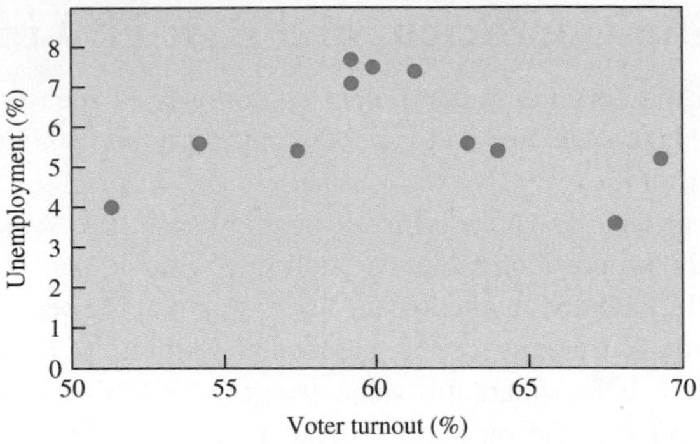

Figure 7.22 Data on voter turnout and unemployment, 1964–2004. *Source:* U.S. Bureau of Labor Statistics.

Multiple Regression

If you've ever purchased a diamond, you might have been surprised that we found such a weak correlation between color and price in Figure 7.2. Surely a diamond cannot be very valuable if it has poor color quality. Perhaps color helps to explain why the correlation between weight and price is not perfect. For example, maybe differences in color explain why two diamonds with the same weight can have different prices. To check this idea, it would be nice to look for a correlation between the price and some combination of *weight and color together.*

All who drink his remedy recover in a short time, except those whom it does not help, who all die. Therefore, it is obvious that it fails only in incurable cases.

—Galen, Roman "doctor"

TIME OUT TO THINK

Check this idea in Table 7.1. Notice, for example, that Diamonds 4 and 5 have nearly identical weights, but Diamond 4 costs only $4,299 while Diamond 5 costs $9,589. Can differences in their color explain the different prices? Study other examples in Table 7.1 in which two diamonds have similar weights but different prices. Overall, do you think that the correlation with price would be stronger if we used weight and color together instead of either one alone? Explain.

There is a method for investigating a correlation between one variable (such as price) and a *combination* of two or more other variables (such as weight and color). The technique is called **multiple regression**, and it essentially allows us to find a *best-fit equation* that relates three or more variables (instead of just two). Because it involves more than two variables, we cannot make simple diagrams to show best-fit equations for multiple regression. However, it is still possible to calculate a measure of how well the data fit a linear equation. The most common measure in multiple regression is the *coefficient of determination*, denoted R^2. It tells us how much of the scatter in the data is accounted for by the best-fit equation. If R^2 is close to 1, the best-fit equation should be very useful for making predictions within the range of the data values. If R^2 is close to zero, then predictions with the best-fit equation are essentially useless.

Definition

The use of **multiple regression** allows the calculation of a best-fit equation that represents the best fit between one variable (such as price) and a *combination* of two or more other variables (such as weight and color). The coefficient of determination, R^2, tells us the proportion of the scatter in the data accounted for by the best-fit equation.

In this book, we will not describe methods for finding best-fit equations by multiple regression. However, you can use the value of R^2 to interpret results from multiple regression. For example, the correlation between price and *weight and color together* results in a value of $R^2 = 0.79$.

This is somewhat higher than the $r^2 = 0.61$ that we found for the correlation between price and weight alone. Statisticians who study diamond pricing know that they can get stronger correlations by including additional variables in the multiple regression (such as depth, table, and clarity). Given the billions of dollars spent annually on diamonds, you can be sure that statisticians play prominent roles in helping diamond dealers realize the largest possible profits.

EXAMPLE 5 Alumni Contributions

You've been hired by your college's Alumni Association to help estimate how much the association can reasonably expect to gain in a new fund-raising drive. The director suggests that you use data concerning past contributions and alumni income level, which have a correlation coefficient of 0.6. Is this a good strategy? Can you suggest a better one?

Solution The correlation coefficient $r = 0.6$ means that $r^2 = 0.36$, so a linear relationship between donations and alumni income accounts for only 36% of the variation. Thus, using the best-fit line to predict the amounts of money that will be donated by individual alumni won't yield very accurate predictions. A better strategy would be to use a multiple regression equation that includes other factors that might influence donations, such as alumni majors, distance of current home from the college, years since graduation, and membership in a fraternity or sorority. It's possible that a well-chosen multiple regression equation will produce a much stronger correlation with donations than any correlation based on only two variables.

By the Way ...

One study of alumni donations found that, in developing a multiple regression equation, one should include these variables: income, age, marital status, whether the donor belonged to a fraternity or sorority, whether the donor is active in alumni affairs, the donor's distance from the college, and the nation's unemployment rate, used as a measure of the economy (Bruggink and Siddiqui, "An Econometric Model of Alumni Giving: A Case Study for a Liberal Arts College," *The American Economist*, Vol. 39, No. 2).

Finding Equations for Best-Fit Lines (Optional Section)

The mathematical technique for finding the equation of a best-fit line is based on the following basic ideas. If we draw *any* line on a scatter diagram, we can measure the *vertical* distance between each data point and that line. One measure of how well the line fits the data is the *sum of the squares* of these vertical distances. A large sum means that the vertical distances of data points from the line are fairly large and hence the line is not a very good fit. A small sum means the data points lie close to the line and the fit is good. Of all possible lines, the best-fit line is the line that minimizes the sum of the squares of the vertical distances. Because of this property, the best-fit line is sometimes called the *least squares line*.

You may recall that the equation of any straight line can be written in the general form

$$y = mx + b$$

where m is the *slope* of the line and b is the *y-intercept* of the line. The formulas for the slope and y-intercept of the best-fit line are as follows:

$$\text{slope} = m = r \cdot \frac{s_y}{s_x}$$

$$y\text{-intercept} = b = \bar{y} - (m \times \bar{x})$$

In the above expressions, r is the correlation coefficient, s_x denotes the standard deviation of the x values (or the values of the first variable), s_y denotes the standard deviation of the y values, $\bar{x}$ represents the mean of the values of the variable x, and $\bar{y}$ represents the mean of the values of the variable y. Because these formulas are tedious to use by hand, we usually use a calculator or computer to find the slope and y-intercept of best-fit lines.

When software or a calculator is used to find the slope and intercept of the best-fit line, results are commonly expressed in the format $y = b_0 + b_1 x$, where b_0 is the intercept and b_1 is the slope, so be careful to correctly identify those two values.

Section 7.3 Exercises

Statistical Literacy and Critical Thinking

1. **Best-Fit Line.** What is a best-fit line (also called a regression line)? How is a best-fit line useful?

2. **Multiple Regression.** What is multiple regression?

3. **r^2.** What does r^2 denote, and how can it be interpreted? That is, what does its value tell us about the variables?

4. **R^2.** What does R^2 denote, and what does it tell us about the variables? What does a value of R^2 close to 1 tell us about the data?

Does It Make Sense? For Exercises 5–8, decide whether the statement makes sense (or is clearly true) or does not make sense (or is clearly false). Explain clearly; not all of these statements have definitive answers, so your explanation is more important than your chosen answer.

5. **r^2 Value.** A value of $r^2 = 0.007$ is obtained from a sample of men, with each pair of data consisting of the height in inches and the height in centimeters for one man.

6. **r^2 Value.** A value of $r^2 = -0.040$ is obtained from a sample of men, with each pair of data consisting of the height in inches and the SAT score for one man.

7. **Height and Weight.** Using data from the National Health Survey, the equation of the best-fit line for women's heights and weights is obtained, and it shows that a woman 120 inches tall is predicted to weigh 430 pounds.

8. **Old Faithful.** Using paired sample data consisting of the duration time (in seconds) of eruptions of Old Faithful geyser and the time interval (in minutes) after the eruption, a value of $r^2 = 0.926$ is calculated, indicating that about 93% of the variation in the interval after eruption can be explained by the relationship between those two variables as described by the best-fit line.

Concepts and Applications

Best-Fit Lines on Scatter Diagrams. For Exercises 9–12, do the following.

a. Add a best-fit line to the given scatter diagram.

b. Estimate or compute r and r^2. Based on your value for r^2, determine how much of the variation in the variable can be accounted for by the best-fit line.

c. Briefly discuss whether you could make valid predictions from this best-fit line.

9. Use the scatter diagram for color and price in Figure 7.2.

10. Use the scatter diagram for life expectancy and infant mortality in Figure 7.4.

11. Use the scatter diagram for number of farms and size of farms in Figure 7.5.

12. Use both scatter diagrams for actual and predicted temperature in Figure 7.6.

Best-Fit Lines. Exercises 13–20 refer to the tables in the Section 7.1 Exercises. In each case, do the following.

a. Construct a scatter diagram and, based on visual inspection, draw the best-fit line by eye.

b. Briefly discuss the strength of the correlation. Estimate or compute r and r^2. Based on your value for r^2, tell how much of the variation in the variable can be accounted for by the best-fit line.

c. Identify any outliers on the diagram and discuss their effects on the strength of the correlation and on the best-fit line.

d. For this case, do you believe that the best-fit line gives reliable predictions outside the range of the data on the scatter diagram? Explain.

13. Use the data in Exercise 19 of Section 7.1.

14. Use the data in Exercise 20 of Section 7.1.

15. Use the data in Exercise 21 of Section 7.1.

16. Use the data in Exercise 22 of Section 7.1.

17. Use the data in Exercise 23 of Section 7.1. To locate the points, use the midpoint of each income category; use a value of \$25,000 for the category "less than \$30,000," and use a value of \$70,000 for the category "over \$60,000."

18. Use the data in Exercise 24 of Section 7.1.

19. Use the data in Exercise 25 of Section 7.1.

20. Use the data in Exercise 26 of Section 7.1.

Projects for the Internet and Beyond

For useful links, select "Links for Internet Projects" for Chapter 7 at www.aw.com/bbt.

21. **Lead Poisoning.** Research lead poisoning, its sources, and its effects. Discuss the correlations that have helped researchers understand lead poisoning. Discuss efforts to prevent it.

22. **Asbestos.** Research asbestos, its sources, and its effects. Discuss the correlations that have helped researchers

understand adverse health effects from asbestos exposure. Discuss efforts to prevent those adverse health effects.

23. Worldwide Population Indicators. The following table gives five population indicators for eleven selected countries. Study these data and try to identify possible correlations. Doing additional research if necessary, discuss the possible correlations you have found, speculate on the reasons for the correlations, and discuss whether they suggest a causal relationship. Birth and death rates are per 1,000 population; fertility rate is per woman.

Country	Birth rate	Death rate	Life expectancy	Percent urban	Fertility rate
Afghanistan	50	22	43	20	6.9
Argentina	21	8	72	88	2.6
Australia	15	7	78	85	1.9
Canada	14	7	78	77	1.6
Egypt	29	8	64	45	3.4
El Salvador	30	6	68	45	3.1
France	13	9	78	73	1.6
Israel	21	7	77	91	2.8
Japan	10	7	79	78	1.5
Laos	45	15	51	22	6.7
United States	16	9	76	76	2.0

Source: The New York Times Almanac.

IN THE NEWS

24. Predictions in the News. Find a recent news report in which a correlation is used to make a prediction. Evaluate the validity of the prediction, considering all of the cautions described in this section. Overall, do you think the prediction is valid? Why or why not?

25. Best-Fit Line in the News. Although scatter diagrams are rare in the news, they are not unheard of. Find a scatter diagram of any kind in a news article (recent or not). Draw a best-fit line by eye. Discuss what predictions, if any, can be made from your best-fit line.

26. Your Own Multiple Regression. Come up with an example from your own life or work in which a multiple regression analysis might reveal important trends. Without actually doing any analysis, describe in words what you would look for through the multiple regression and how the answers might be useful.

7.4 The Search for Causality

A correlation may suggest causality, but by itself a correlation *never* establishes causality. Much more evidence is required to establish that one factor *causes* another. Earlier, we found that a correlation between two variables may be the result of either (1) coincidence, (2) a common underlying cause, or (3) one variable actually having a direct influence on the other. The process of establishing causality is essentially a process of ruling out the first two explanations.

In principle, we can rule out the first two explanations by conducting experiments:

- We can rule out coincidence by repeating the experiment many times or using a large number of subjects in the experiment. Because coincidences occur randomly, they should not occur consistently in many subjects or experiments. Thus, while a coincidence may confuse the situation in a few trials of an experiment, it is unlikely to be confusing after many trials.
- We can rule out a common underlying cause by controlling and randomizing the experiment to eliminate the effects of confounding variables (see Section 1.3). In this way, *only* the variables of interest vary between the treatment and control groups. If the controls rule out confounding variables, any remaining effects must be caused by the variables being studied.

Unfortunately, these ideas are often difficult to put into practice. In the case of ruling out coincidence, it may be too time consuming or expensive to repeat an experiment a sufficient number of times. To rule out a common underlying cause, the experiment must control for *everything* except the variables of interest, and this is often impossible. Moreover, there are many cases in which experiments are impractical or unethical, so we can gather only observational data. Because observational studies cannot definitively establish causality, we must find other ways of trying to establish causality.

Establishing Causality

Suppose you have discovered a correlation and suspect causality. How can you test your suspicion? Let's return to the issue of smoking and lung cancer. The strong correlation between smoking and lung cancer did not by itself prove that smoking causes lung cancer. In principle, we could have looked for proof with a controlled experiment. But such an experiment would be unethical, since it would require forcing a group of randomly selected people to smoke cigarettes. So how was smoking established as a cause of lung cancer?

The answer involves several lines of evidence. First, researchers found correlations between smoking and lung cancer among many groups of people: women, men, and people of different races and cultures. Second, among groups of people that seemed otherwise identical, lung cancer was found to be rarer in nonsmokers. Third, people who smoked more and for longer periods of time were found to have higher rates of lung cancer. Fourth, when researchers accounted for other potential causes of lung cancer (such as exposure to radon gas or asbestos), they found that almost all the remaining lung cancer cases occurred among smokers (or people exposed to second-hand smoke).

These four lines of evidence made a strong case, but still did not rule out the possibility that some other factor, such as genetics, predisposes people both to smoking and to lung cancer. However, two additional lines of evidence made this possibility highly unlikely. One line of evidence came from animal experiments. In controlled experiments, animals were divided into randomly chosen treatment and control groups. The experiments still found a correlation between inhalation of cigarette smoke and lung cancer, which seems to rule out a genetic factor, at least in the animals. The final line of evidence came from biologists studying cell cultures (made from small samples of human lung tissue). The biologists discovered the basic process by which ingredients in cigarette smoke create cancer-causing mutations. This process does not appear to depend in any way on specific genetic factors, making it all but certain that lung cancer is caused by smoking and not by any preexisting genetic factor. The fact that second-hand smoke exposure is also associated with some cases of lung cancer further argues against a genetic factor (since second-hand smoke affects nonsmokers) but is consistent with the idea that ingredients in cigarette smoke create cancer-causing mutations.

The following box summarizes these ideas about establishing causality. Generally speaking, the case for causality is stronger when more of these guidelines are met.

By the Way ...

On the strength of the evidence that smoking causes cancer, Dr. David Sidransky of Johns Hopkins University says: "We have such strong molecular proof that we can take an individual cancer and potentially, based on the pattern of genetic change, determine whether cigarette smoking was the cause of that [particular] cancer."

Guidelines for Establishing Causality

If you suspect that a particular variable (the suspected cause) is causing some effect:

1. Look for situations in which the effect is correlated with the suspected cause even while other factors vary.
2. Among groups that differ only in the presence or absence of the suspected cause, check that the effect is similarly present or absent.
3. Look for evidence that larger amounts of the suspected cause produce larger amounts of the effect.
4. If the effect might be produced by other potential causes (besides your suspected cause), make sure that the effect still remains after accounting for these other potential causes.
5. If possible, test the suspected cause with an experiment. If the experiment cannot be performed with humans for ethical reasons, consider doing the experiment with animals, cell cultures, or computer models.
6. Try to determine the physical mechanism by which the suspected cause produces the effect.

By the Way ...

The first four guidelines to the right are called *Mill's methods* after John Stuart Mill (1806–1873). Mill was a leading scholar of his time and an early advocate of women's right to vote. In philosophy, the four methods are called, respectively, the methods of agreement, difference, concomitant variation, and residues.

TIME OUT TO THINK

There's a great deal of controversy concerning whether animal experiments are ethical. What is your opinion of animal experiments? Defend your opinion.

CASE STUDY Air Bags and Children

By the mid-1990s, passenger-side air bags had become commonplace in cars. Statistical studies showed that the air bags saved many lives in moderate- to high-speed collisions. But a disturbing pattern also appeared. In at least some cases, young children, especially infants and toddlers in child car seats, were killed by air bags in low-speed collisions.

At first, many safety advocates found it difficult to believe that air bags could be the cause of the deaths. But the observational evidence became stronger, meeting the first four guidelines for establishing causality. For example, the greater risk to infants in child car seats fit Guideline 3, because it indicated that being closer to the air bags increased the risk of death. (A child car seat sits on top of the built-in seat, thereby putting a child closer to the air bags than the child would be otherwise.)

To seal the case, safety experts undertook experiments using dummies. They found that children, because of their small size, often sit where they could be easily hurt by the explosive opening of an air bag. The experiments also showed that an air bag could impact a child car seat hard enough to cause death, thereby revealing the physical mechanism by which the deaths occurred.

With the physical mechanism understood, it was possible to develop strategies to prevent deaths. First, it was immediately obvious that the best prevention would be to keep children away from the air bags. This led to new guidelines telling parents that child car seats should *never* be used on the front seat, and that children under the age of 12 should sit in the back seat if possible. Second, the findings led to additional experiments suggesting that air bags could provide the same safety with less explosive openings. Air bags that open less forcefully are now required in all new cars. Third, the fact that the risk for children turned out to be related to their small size suggested that small adults would also be at risk. This led to recommendations that small adult drivers (such as short women) sit as far back as possible from the air bags and that they use only cars with the newer air bags that open less forcefully.

By the Way ...

Many states and countries have laws concerning where children sit in cars. In Belgium, for example, it is illegal for a child under the age of 12 to sit in the front seat if the back seat is open.

Hidden Causality

So far we have discussed how to establish causality after first discovering a correlation. Sometimes, however, correlations—or the lack of a correlation—can hide an underlying causality. As the next case study shows, such hidden causality often occurs because of confounding variables.

CASE STUDY Cardiac Bypass Surgery

Cardiac bypass surgery is performed on people who have severe blockage of arteries that supply the heart with blood (the coronary arteries). If blood flow stops in these arteries, a patient may suffer a heart attack and die. Bypass surgery essentially involves grafting new blood vessels onto the blocked arteries so that blood can flow around the blocked areas. By the mid-1980s, many doctors were convinced that the surgery was prolonging the lives of their patients.

However, a few early retrospective studies turned up a disconcerting result: Statistically, the surgery appeared to be making little difference. In other words, patients who had the surgery seemed to be faring no better on average than similar patients who did not have it. If this were true, it meant that the surgery was not worth the pain, risk, and expense involved.

Because these results flew in the face of what many doctors thought they had observed in their own patients, researchers began to dig more deeply. Soon, they found confounding variables that had not been accounted for in the early studies. For example, they found that patients getting the surgery tended to have more severe blockage of their arteries, apparently because doctors recommended the surgery more strongly to these patients. Because these patients were in worse shape to begin with, a comparison of longevity between them and other patients was not really valid.

More important, the research soon turned up substantial differences in the results among patients who had the surgery in different hospitals. In particular, a few hospitals were achieving remarkable success with bypass surgery and their patients fared far better than patients who did not have the surgery or had it at other hospitals. Clearly, the surgical techniques used by doctors at the successful hospitals were somehow different and superior. Doctors studied the differences to ensure that all doctors could be trained in the superior techniques.

In summary, the confounding variables of *amount of blockage* and *surgical technique* had prevented the early studies from finding a real correlation between cardiac bypass surgery and prolonged life. Today, cardiac bypass surgery is accepted as a *cause* of prolonged life in patients with blocked coronary arteries. It is now among the most common types of surgery, and it typically adds *decades* to the lives of the patients who undergo it.

By the Way ...

As you might guess, it is also difficult to define *reasonable doubt*. For criminal trials, the Supreme Court endorsed this guidance from Justice Ruth Bader Ginsburg: "Proof beyond a reasonable doubt is proof that leaves you firmly convinced of the defendant's guilt. There are very few things in this world that we know with absolute certainty, and in criminal cases the law does not require proof that overcomes every possible doubt. If, based on your consideration of the evidence, you are firmly convinced that the defendant is guilty of the crime charged, you must find him guilty. If on the other hand, you think there is a real possibility that he is not guilty, you must give him the benefit of the doubt and find him not guilty."

Confidence in Causality

The six guidelines offer us a way to examine the strength of a case for causality, but we often must make decisions before a case of causality is fully established. Consider, for example, the well-known case of global warming. It may never be possible to prove beyond all doubt that the burning of fossil fuels is causing global warming (see the Focus on Environment at the end of this chapter), so we must decide whether to act while we still face some uncertainty about causation. How much must we know before we decide to act?

In other areas of statistics, accepted techniques help us deal with this type of uncertainty by allowing us to calculate a numerical level of confidence or significance. But there are no accepted ways to assign such numbers to the uncertainty that comes with questions of causality. Fortunately, another area of study has dealt with practical problems of causality for hundreds of years: our legal system. You may be familiar with the following three broad ways of expressing a legal level of confidence.

Broad Levels of Confidence in Causality

Possible cause: We have discovered a correlation, but cannot yet determine whether the correlation implies causality. In the legal system, possible cause (such as thinking that a particular suspect possibly committed a particular crime) is often the reason for starting an investigation.

Probable cause: We have good reason to suspect that the correlation involves cause, perhaps because some of the guidelines for establishing causality are satisfied. In the legal system, probable cause is the general standard for getting a judge to grant a warrant for a search or wiretap.

Cause beyond reasonable doubt: We have found a physical model that is so successful in explaining how one thing causes another that it seems unreasonable to doubt the causality. In the legal system, cause beyond reasonable doubt is the usual standard for convictions and generally demands that the prosecution have shown how and why (essentially the physical model) the suspect committed the crime. Note that beyond *reasonable* doubt does *not* mean beyond *all* doubt.

While these broad levels remain fairly vague, they give us at least some common language for discussing confidence in causality. If you study law, you will learn much more about the subtleties of interpreting these terms. However, because statistics has little to say about them, we will not discuss them much further in this book.

Section 7.4 Exercises

Statistical Literacy and Critical Thinking

1. **Correlation.** Identify three different explanations for the presence of a correlation between two variables.

2. **Role of Experiments.** In theory, we can use experiments to rule out two of the three different explanations for the presence of a correlation between two variables. Which of the three explanations do we *not* want to rule out? Why would we not want to rule it out?

3. **Confounding Variable.** What is a confounding variable? How can a confounding variable create a situation in which an underlying causality is hidden?

4. **Correlation and Causality.** What is the difference between finding a correlation between two variables and establishing causality between two variables?

Does It Make Sense? For Exercises 5–8, decide whether the statement makes sense (or is clearly true) or does not make sense (or is clearly false). Explain clearly; not all of these statements have definitive answers, so your explanation is more important than your chosen answer.

5. **Value of *r*.** If a sample of paired data from two variables yields a "perfect" correlation coefficient of 1, then we can conclude that one of the variables has a direct causal link with the other variable.

6. **Value of *r*.** In a study of adverse reactions from different amounts of a drug treatment, the correlation coefficient is calculated from 20 pairs of data consisting of the amount of the drug and a measure of pain intensity on a standard scale. If $r = 0.013$, then the amount of the drug cannot be a single direct cause of the pain.

7. **Smoking and Cotinine.** A study showed that there is a correlation between exposure to second-hand smoke and the measured amount of cotinine in the body. We can establish that exposure to second-hand smoke is a cause of cotinine if we can rule out coincidence as a possible explanation of the correlation.

8. **Nicotine.** Assume that a study of nicotine patches established physical evidence that when nicotine is absorbed by the body, the body converts it to cotinine. It follows that in an experiment with five subjects, an analysis of the relationship between the amount of nicotine provided in patches and the amount of cotinine in the body will show the presence of a correlation.

Concepts and Applications

Physical Models. For Exercises 9–12, determine whether the stated causal connection is valid. If the causal connection appears to be valid, provide an explanation.

9. **Projectile Motion.** The distance that a golf ball travels is affected by the speed of the club when it strikes the ball.

10. **Magnet Treatment.** Heart disease can be cured by wearing a magnetic bracelet on your wrist.

11. **Drinking and Reaction Time.** Drinking greater amounts of alcohol decreases a person's reaction time.

12. **Altitude and Health.** When people climb to higher altitudes without supplemental oxygen, they tend to experience increased physiological problems, such as headaches or disorientation.

13. **Identifying Causes: Headaches.** You are trying to identify the cause of late-afternoon headaches that plague you several days each week. For each of the following tests and observations, explain which of the six guidelines for establishing causality you used and what you concluded. Then summarize your overall conclusion based on all the observations.

    a. The headaches occur only on days that you go to work.

    b. If you stop drinking Coke at lunch, the headaches persist.

    c. In the summer, the headaches occur less frequently if you open the windows of your office slightly. They occur even less often if you open the windows of your office fully.

14. **Smoking and Lung Cancer.** There is a strong correlation between tobacco smoking and incidence of lung cancer, and most physicians believe that tobacco smoking causes lung cancer. Yet, not everyone who smokes gets lung cancer. Briefly describe how smoking could cause cancer when not all smokers get cancer.

15. **Other Lung Cancer Causes.** Several things besides smoking have been shown to be probabilistic causal factors in lung cancer. For example, exposure to asbestos and exposure to radon gas, both of which are found in many homes, can cause lung cancer. Suppose that you meet a person who lives in a home that has a high radon level and insulation that contains asbestos. The person tells you, "I smoke, too, because I figure I'm doomed to lung cancer anyway." What would you say in response? Explain.

16. **Longevity of Orchestra Conductors.** A famous study in *Forum on Medicine* concluded that the mean lifetime of conductors of major orchestras was 73.4 years, about 5 years longer than that of all American males at the time. The author claimed that a life of music *causes* a longer life. Evaluate the claim of causality and propose other explanations for the longer life expectancy of conductors.

17. **Older Moms.** A study reported in *Nature* claims that women who give birth later in life tend to live longer. Of the 78 women who were at least 100 years old at the time of the study, 19% had given birth after their 40th birthday. Of the 54 women who were 73 years old at the time of the study, only 5.5% had given birth after their 40th birthday. A researcher stated that "if your reproductive system is aging slowly enough that you can have a child in your 40s, it probably bodes well for the fact that the rest of you is aging slowly too." Was this an observational study or an experiment? Does the study suggest that later child bearing *causes* longer lifetimes or that later child bearing reflects an underlying cause? Comment on how persuasive you find the conclusions of the report.

18. **High-Voltage Power Lines.** Suppose that people living near a high-voltage power line have a higher incidence of cancer than people living farther from the power line. Can you conclude that the high-voltage power line is the cause of the elevated cancer rate? If not, what other explanations might there be for it? What other types of research would you like to see before you concluded that high-voltage power lines cause cancer?

19. **Gun Control.** Those who favor gun control often point to a positive correlation between the availability of handguns and murder rates to support their position that gun control would save lives. Does this correlation, by itself, indicate that handgun availability causes a higher murder rate? Suggest some other factors that might support or weaken this conclusion.

20. **Vasectomies and Prostate Cancer.** An article entitled "Does Vasectomy Cause Prostate Cancer?" (*Chance*, Vol. 10, No. 1) reports on several large studies that found an increased risk of prostate cancer among men with vasectomies. In the absence of a direct cause, several researchers attribute the correlation to *detection bias*, in which men with vasectomies are more likely to visit the doctor and thereby are more likely to have any prostate cancer found by the doctor. Briefly explain how this detection bias could affect the claim that vasectomies cause prostate cancer.

Projects for the Internet and Beyond

For useful links, select "Links for Internet Projects" for Chapter 7 at www.aw.com/bbt.

21. **Air Bags and Children.** Starting from the Web site of the National Highway Traffic Safety Administration, research the latest studies on the safety of air bags, especially with regard to children. Write a short report summarizing your findings and offering recommendations for improving child safety in cars.

22. **Dietary Fiber and Coronary Heart Disease.** In the largest study of how dietary fiber prevents coronary heart disease (CHD) in women (*Journal of the American Medical Association*, Vol. 281, No. 21), researchers detected a reduced risk of CHD among women who have a high-fiber diet. Find the research paper, summarize its findings, and discuss whether a cause for the correlation is proposed.

23. **Coffee and Gallstones.** Writing in the *Journal of the American Medical Association* (Vol. 281, No. 22), researchers reported finding a negative correlation between incidence of gallstone disease and coffee consumption in men. Find the research paper, summarize its findings, and discuss whether a cause for the correlation is proposed.

24. Alcohol and Stroke. Researchers reported in the *Journal of the American Medical Association* (Vol. 281, No. 1) that moderate alcohol consumption is correlated with a decreased risk of stroke in people 40 years of age and older. (Heavy consumption of alcohol was correlated with deleterious effects.) Find the research paper, summarize its findings, and discuss whether a cause for the correlation is proposed.

25. Tobacco Lawsuits. Tobacco companies have been the subject of many lawsuits related to the dangers of smoking. Research one recent lawsuit. What were the plaintiffs trying to prove? What statistical evidence did they use? How well do you think they established causality? Did they win? Summarize your findings in one to two pages.

IN THE NEWS

26. Causation in the News. Find a recent news report in which a statistical study has led to a conclusion of causation. Describe the study and the claimed causation. Do you think the claim of causation is legitimate? Explain.

27. Legal Causation. Find a news report concerning an ongoing legal case, either civil or criminal, in which establishing causality is important to the outcome. Briefly describe the issue of causation in the case and how the ability to establish or refute causality will influence the outcome of the case.

Chapter Review Exercises

For Exercises 1–4, refer to the cigarette data in the table. All measurements are in milligrams per cigarette, and all cigarettes are 100 millimeters long, filtered, and not menthol or light types (based on data from the Federal Trade Commission).

Brand	Tar	Nicotine	Carbon monoxide
Camel	16	1.0	17
Kent	13	1.0	13
Lucky Strike	13	1.1	13
Malibu	15	1.2	15
Marlboro	16	1.2	15
Merit	9	0.7	11
Now	2	0.2	3
Old Gold	18	1.4	18
Pall Mall	15	1.2	15
Winston	16	1.1	18
Camel	16	1.0	17

1. The paired tar and nicotine data have a correlation coefficient of $r = 0.962$. What do you conclude about the strength of the correlation between tar and nicotine? What percentage of the variation in nicotine can be explained by the correlation between nicotine and tar?

2. The paired carbon monoxide and nicotine data have a correlation coefficient of $r = 0.909$. What do you conclude about the strength of the correlation between carbon monoxide and nicotine? What percentage of the variation in nicotine can be explained by the correlation between nicotine and carbon monoxide?

3. The paired tar and carbon monoxide data have a correlation coefficient of $r = 0.979$. What do you conclude about the strength of the correlation between tar and carbon monoxide? What percentage of the variation in tar can be explained by the correlation between carbon monoxide and tar?

4. Construct a scatter diagram that uses the given values of tar and nicotine. (Use the tar values for the horizontal scale.) What feature of the graph indicates that there is or is not a correlation between tar and nicotine?

5. Based on a study in Sweden, several newspapers reported that "living near power lines causes leukemia in children." What data are likely to be the basis for such a claim? What is fundamentally wrong with the claim that proximity to power lines causes leukemia?

6. For 10 pairs of sample data, the correlation coefficient is computed to be $r = -1$. What do you know about the scatter diagram?

7. In a study of randomly selected subjects, it is found that there is a strong correlation between household income and number of visits to dentists. Is it valid to conclude that higher incomes cause people to visit dentists more often? Is it valid to conclude that more visits to dentists cause people to have higher incomes? How might the correlation be explained?

8. You are considering the most expensive purchase that you are likely to make: the purchase of a home. Identify at least five different variables that are likely to affect the actual value of a home. Among the variables that you have identified, which single variable is likely to have the greatest influence on the value of the home? Identify a variable that is likely to have little or no effect on the value of a home.

9. A researcher collects paired sample data and computes the value of the linear correlation coefficient to be 0. Based on that value, he concludes that there is no relationship between the two variables. What is wrong with this conclusion?

10. Examine the scatter diagram in Figure 7.23 and estimate the value of the correlation coefficient.

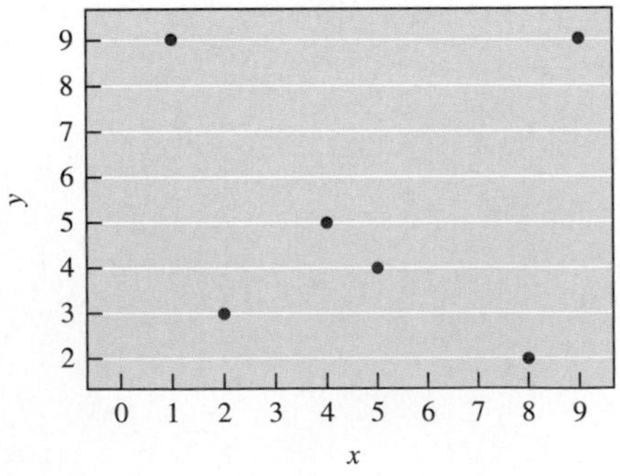

Figure 7.23

Chapter Quiz

1. Fill in the blanks: Every possible correlation coefficient must lie between the values of _____ and _____.

2. Which of the following are likely to have a correlation?

   a. SAT scores and weights of randomly selected subjects

   b. Reaction times and IQ scores of randomly selected subjects

   c. Height and arm span of randomly selected subjects

   d. Proportion of seats filled and amount of airline profit for randomly selected flights

   e. Value of cars owned and annual income of randomly selected car owners

3. For a collection of paired sample data, the correlation coefficient is found to be −0.988. Which of the following statements best describes the relationship between the two variables?

   a. There is no correlation.

   b. There is a weak correlation.

   c. There is a strong correlation.

   d. One of the variables is the direct cause of the other variable.

   e. Neither of the variables is the direct cause of the other variable.

4. Estimate the correlation coefficient for the data in Figure 7.24.

5. Refer again to the scatter diagram in Figure 7.24. Does there appear to be a significant correlation between the two variables?

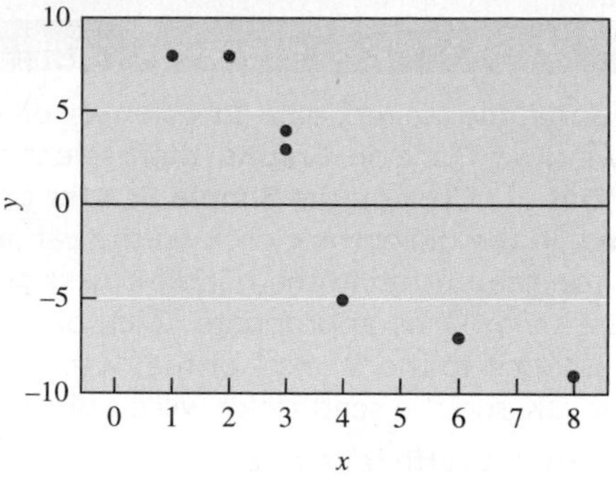

Figure 7.24

In Exercises 6–10, determine whether the given statement is true or false.

6. If there is a strong correlation between two variables, then one of the variables is the direct cause of the other variable.

7. A scatter diagram is one in which the points are scattered throughout, without any noticeable pattern.

8. If all of the points in a scatter diagram lie on a straight line starting at the lower left and rising to the upper right, then $r = 1$.

9. If paired income and tax data are collected from randomly selected subjects for the years 1995 through 2008, the presence of a correlation indicates that the relationship can be used to predict taxes based on income in the year 2015.

10. If $r^2 = 0.09$, then it is possible that the value of r is negative.

Using Technology

SPSS

Scatterplot: Enter the paired data in columns of the SPSS data editor window. Click on **Graphs**, then select the menu item **Scatter/Dot. . . .** Click on the **Simple Scatter** graph, then click on **Define**. In the dialog box, click on the variable at the left, and click on the button in the middle to assign it to the "Y axis" or the "X axis" as appropriate. Click on the second variable and assign it to the "Y axis" or the "X axis" as appropriate. Click on **OK** and the scatterplot will be displayed.

Linear correlation coefficient *r*: Enter the paired data in columns of the SPSS data editor window. Click on **Analyze**, select the menu item **Correlate**, then select **Bivariate**. In the dialog box, click on **Pearson**. Click on a variable at the left, and click on the middle button to move it to the "Variable" box. Also click on the other variable, and click on the middle button to move it to the "Variable" box. Click on **OK** to get the value of the linear correlation coefficient r, along with a statement about its significance.

Line of best fit: Enter the paired data in columns of the SPSS data editor window. Click on **Analyze**, select the menu item **Regression**, then select **Linear**. Click on a variable at the left, and click on the middle button to move it to the "Dependent" box or the "Independent(s)" box as appropriate. Click on **OK** to get several tables of information. See the "Coefficients" table for the y-intercept (identified as "Constant") and the slope of the line of best fit.

Excel

Linear correlation coefficient *r*: First enter the paired sample data in columns A and B. Click on the ***fx*** function key located on the main menu bar. Select the function category **Statistical** and the function name **CORREL,** then click on **OK.** In the dialog box, enter the cell range of values for x, such as A1:A8. Also enter the cell range of values for y, such as B1:B8. The value of the linear correlation coefficient r will be displayed.

Scatterplot: If you are using Excel 2003, click on the Chart Wizard on the main menu, then select the chart type identified as **XY(Scatter).** In the dialog box, enter the input range of the data, such as A1:B8. Click **Next** and proceed to use the dialog boxes to modify the graph as desired.

If you are using Excel 2007, use the mouse to highlight the paired data. Click on **Insert**, and in the Charts area select **Scatter**. Several graphs will be displayed; click on the one in the upper left corner, and the scatterplot will be displayed. Modify the graph as desired.

Line of best fit: Enter the paired data in columns A and B. If you are using Excel 2003, select **Tools** from the main menu, then select **Data Analysis** and **Regression,** then click on **OK.** If you are using Excel 2007, click on **Data**, then click on **Data Analysis**, select **Regression**, then click on **OK**. Enter the range for the y values, such as B1:B8. Enter the range for the x values, such as A1:A8. Click on the box labeled "Line Fit Plots," then click on **OK.** Among all of the information provided by Excel, the slope and intercept of the equation of the line of best fit can be found under the table heading "Coefficient." The displayed graph will include a scatterplot of the original sample points along with the points that would be predicted by the line of best fit.

STATDISK

First enter the paired data in columns of the STATDISK data window. Select **Analysis** from the main menu bar, then use the option **Correlation and Regression.** Enter a value for the significance level. Select the columns to be used. Click on the **Evaluate** button. The STATDISK display will include the value of the linear correlation coefficient along with the slope and intercept of the line of best fit. A scatterplot can also be obtained by clicking on the **Plot** button.

FOCUS ON EDUCATION

What Helps Children Learn to Read?

Everyone has an idea about how best to teach reading to children. Some advocate a phonetic approach, teaching students to "sound out" words. Some advocate a "whole language" approach, teaching students to recognize words from their context. Others advocate a combination of these approaches—or something else entirely. These differing ideas would be unimportant if they were merely opinions. But in a nation that spends roughly a *trillion dollars* per year on education, differing approaches to teaching reading involve major political confrontations among groups with different special interests. A change in politics can cause a sudden change in school policies. For example, in 1998, the California legislature passed laws making public school funding contingent upon the school's moving away from a whole language approach to reading.

The huge stakes involved in teaching reading demand statistics to measure the effectiveness of various approaches. Politically, at least, the most important educational statistics are those that come from the National Assessment of Educational Progress (NAEP), often known more simply as "the Nation's Report Card." The NAEP is an ongoing survey of student achievement conducted by a government agency, the National Center for Education Statistics, with authorization and funding from the U.S. Congress.

The NAEP uses stratified random sampling (see Chapter 1) to choose representative samples of 4th-, 8th-, and 12th-grade students of varying ethnicity, family income, type of school attended, and so on. Students chosen for the samples are given tests designed to measure their academic achievement in a particular subject area, such as reading, mathematics, or history. Samples are chosen on both state and national levels. Overall, a few thousand students are chosen for each test. Results from NAEP tests inevitably make the newspaper, with articles touting improvements or decrying drops in test scores. They also have political impact. For example, California's move away from whole language occurred after its rank on the NAEP tests was 45th among the 50 states.

But what really causes improvement in reading performance? Researchers begin by searching for correlations between reading performance and other factors. Sometimes the correlations are clear, but offer no direction for improving reading. For example, parental education is clearly correlated with reading achievement: Children with more highly educated parents tend to read more proficiently than those with uneducated parents. But this correlation doesn't offer much guidance for the schools, since children cannot replace their parents. Other times the correlations may suggest ways to improve reading. For example, students who report reading more pages daily in school and for homework tend to score higher than students who read fewer pages. This suggests that schools should assign more reading.

Of course, the high stakes involved in education make education statistics particularly prone to misinterpretation or misuse. Consider just a few of the problems that make the NAEP reading tests difficult to interpret:

- They are "standardized tests" that are the same for all students tested and tend to be mostly multiple choice. Some people believe that such tests are inevitably biased and cannot truly measure reading ability.

- Because the tests generally don't affect students' grades, some students may not take the tests seriously, in which case test results may not reflect actual reading ability. For example, some test administrators have reported students making designs with the multiple-choice "bubbles" on their tests, rather than trying to answer the questions to the best of their ability.
- State-by-state comparisons may not be valid if the makeup of the student population varies significantly among states. For example, California has a relatively large percentage of students for whom English is a second language. Some people believe that this explained the state's low test scores, rather than anything to do with teaching techniques.
- There is some evidence of cheating on the part of the *adults* involved in the NAEP tests by, for example, choosing samples that are not truly representative but instead skewed toward students who read better. This cheating may be motivated by the fact that individual schools, school districts, and states are ranked according to NAEP results. High scores can lead to rewards for teachers and administrators in the form of increased funding or higher salaries, while low scores may lead to various punitive actions.

You can probably think of a dozen other problems that make it difficult to interpret NAEP results. Thus, it should not be surprising that reading continues to be a huge political battleground. So what can you do, as an individual, to help a child to read? Fortunately, the NAEP studies also reveal a few correlations that are fairly uncontroversial and agree with common sense. For example, higher reading performance correlates with each of the following factors:

- more total reading, both for school and for pleasure
- more choice in reading—that is, allowing children to pick their own books to read
- more writing, particularly of extended pieces such as essays or long letters
- more discussion of reading material with friends and family
- less television watching

These correlations give at least some guidance on how to help a child learn to read and should be good starting points for discussions of how to increase literacy. Of course, politicians and special interest groups will probably find ways to make these results fit whatever preconceived agenda they might have. So, if you have strong opinions about teaching techniques, you can join the political battles that will probably continue for decades to come.

QUESTIONS FOR DISCUSSION

1. One clear result of the NAEP reading tests is that students in private schools tend to score significantly higher than students in public schools. Does this imply that private schools are "better" than public schools? Defend your opinion.

2. Do you think that standardized tests like those of the NAEP are valid ways to measure academic achievement? Why or why not?

3. Currently, the NAEP tests are given to only a few thousand of the millions of school children in the United States. Some people advocate giving similar tests to all students, on either a voluntary or a mandatory basis. Do you think such "standardized national testing" is a good idea? Why or why not?

4. One correlation that has not yet been studied carefully is the correlation between computer use and reading. Do you think that using a computer and the Internet helps or hurts children in terms of learning to read? Why?

5. Read the latest edition of the *NAEP Reading Report Card* (available online; link below). What are some of the latest results with regard to the teaching of reading in the United States?

SUGGESTED READING

Lemann, Nicholas. "The Reading Wars." *The Atlantic Monthly*, November 1997.

NAEP Reading Report Card. The latest edition is online at http://nces.ed.gov/nationsreportcard/reading.

Schemo, Diana. "In War over Teaching Reading, a U.S.-Local Clash," *New York Times*, March 9, 2007.

FOCUS ON ENVIRONMENT

What Causes Global Warming?

The Focus on Environment for Chapter 3 (page 140) presented visual data making the case that global warming is occurring. Here, we turn our attention to a deeper question: Is the observed global warming occurring for natural reasons, or are we causing it through our use of fossil fuels and other activities? This question has broad political and economic ramifications. If the warming is natural, then there may be no cause for alarm, because it might naturally reverse or there might be little we could do about it. But if the warming is being caused by us, then continuing to act as we have in the past will further exacerbate the warming, in which case it becomes imperative that we undertake political and economic changes to alleviate the problem before it threatens our civilization.

Today, there is a clear consensus among scientists who study Earth's climate in favor of the viewpoint that humans are the cause of global warming. In 2007, the Intergovernmental Panel on Climate Change (IPCC)—a panel of more than 600 climate scientists from around the world—concluded that there is at least a 90% certainty that the observed warming of the past 50 years has been caused by human emissions of gases such as carbon dioxide, and that continued emissions would warm the planet at least several more degrees during this century, causing substantial changes to climate patterns and raising sea level. How did the scientists reach this conclusion?

To understand the arguments, we must first look at the basic underlying science of global warming: the heat-trapping **greenhouse effect** caused by certain atmospheric gases. The physical mechanism of the greenhouse effect is simple and has been well understood for many decades. It works like this (Figure 7.25): Sunlight is a form of energy that warms a planet's surface. For the surface temperature to remain stable, this absorbed energy must be returned to space (otherwise, continued absorption of sunlight would rapidly raise the temperature by a huge amount), and planetary surfaces return this energy in the form of infrared light. (We cannot see infrared light with our eyes, but it can be easily detected with cameras.) The greenhouse effect occurs when gases such as water vapor (H_2O), carbon dioxide (CO_2) and methane (CH_4)—called **greenhouse gases**—absorb some of this infrared light on its way upward. These gases re-emit the infrared light in random directions, so their overall effect is to keep more energy concentrated near the planet's surface, making it warmer than it would be otherwise. The greater the abundance of greenhouse gases, the more the escape of the energy is slowed and the warmer the planet becomes.

Note that while there is still some uncertainty about the extent of human-induced global warming, there is no scientific doubt about the mechanism of the greenhouse effect. Indeed, the greenhouse effect occurs naturally on Earth, and that turns out to be a very good thing: Without the greenhouse effect, Earth's global average temperature would be close to 0°F, making our planet far too cold for liquid water or life. But because of the naturally occurring greenhouse effect, the actual global average temperature is close to 60°F. Studies of other planets further confirm our understanding of the greenhouse effect, and show that it can be even more important than the planet's distance from the Sun in determining a planet's surface temperature. Venus provides the most extreme example: Although Venus is closer to the Sun than Earth, its clouds are so reflective that less sunlight reaches Venus's surface than Earth's

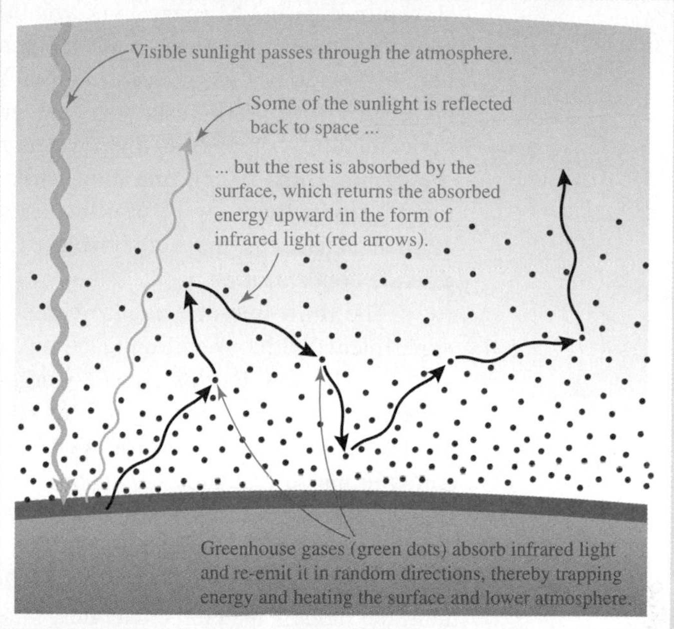
Visible sunlight passes through the atmosphere.
Some of the sunlight is reflected back to space ...
... but the rest is absorbed by the surface, which returns the absorbed energy upward in the form of infrared light (red arrows).
Greenhouse gases (green dots) absorb infrared light and re-emit it in random directions, thereby trapping energy and heating the surface and lower atmosphere.

Figure 7.25 This diagram shows the basic mechanism of the greenhouse effect. The greater the abundance of greenhouse gases, the more the escape of infrared light is slowed and the warmer the planet becomes.

surface. As a result, without the greenhouse effect, Venus would be colder than Earth. But because Venus has a thick atmosphere containing far more carbon dioxide than Earth's atmosphere (by a factor of about 170,000), Venus's actual surface temperature is a searing 870°F. Given that the naturally occurring greenhouse effect is a good thing for life on Earth, Venus offers proof that it's possible to have too much of a good thing.

It is this scientific certainty about the mechanism of the greenhouse effect that first led to concern about the possibility of human-induced global warming. Because human activity such as the burning of fossil fuels (along with other activities such as clearing or burning forests) releases carbon dioxide and other greenhouse gases into the atmosphere, we might expect an increase in the atmospheric greenhouse gas concentration, which in turn would warm our planet beyond its natural level of greenhouse warming. Indeed, there is no doubt that if we kept adding greenhouse gases to the atmosphere indefinitely, we would eventually cause enough global warming to threaten our very survival. The questions are whether (and how much) warming is already under way and whether we can separate out changes in greenhouse gas concentration due to human activity from changes that might be due to natural factors.

In an attempt to answer these questions, the United States and other nations have devoted billions of dollars over the past two decades to an unprecedented effort to understand Earth's climate. It is this research that has been used to make the case that human input of greenhouse gases is causing global warming. Three lines of evidence make the case particularly strong.

First, as we discussed in Chapter 3, data now show a clear warming trend over the past century, with the warming accelerating in the past three decades (see Figure 3.60, page 141). These data have effectively ended any debate over whether warming is under way, leaving us with the question of whether the warming is due to humans or due to natural factors.

The second line of evidence comes from careful measurements of past and present carbon dioxide concentrations in Earth's atmosphere. If you look back at these data, shown in Figure 3.61 (page 141), you can see that past changes in the carbon dioxide concentration correlate clearly with temperature changes, confirming our expectation that a higher greenhouse gas concentration goes with higher temperatures. (However, these data do not by themselves establish that greenhouse gases *cause* higher temperatures; that understanding comes from the physical mechanism.) Notice that while past data show that the carbon dioxide concentration varies naturally, they also show that the recent rise is much greater than any natural increase during the past several hundred thousand years. Human activity is the only viable explanation for the huge recent increase in carbon dioxide concentration.

The third line of evidence comes from experiments. We cannot perform controlled experiments with our entire planet, but we can run experiments with *computer models* that simulate the way Earth's climate works. Earth's climate is incredibly complex, and many uncertainties remain in attempts to model the climate on computers. However, today's models are the result of decades of work and refinement: Each time a model of the past failed to match real data, scientists sought to understand the missing (or incorrect) ingredients in the model and then tried again with improved models. Today's models are not perfect, but they match real climate data quite well, giving scientists confidence that the models have predictive value. Figure 7.26 compares model data and real data. Most important, the models match the data only when they include the effects of the greenhouse gases put into the atmosphere by humans. Models that consider only natural factors fail to match the observed data.

The conclusion is clear: Based on the models, the available data, and our understanding of the mechanism of the greenhouse effect, we can have great confidence that global warming is being caused primarily by the human release of carbon dioxide and other greenhouse gases.

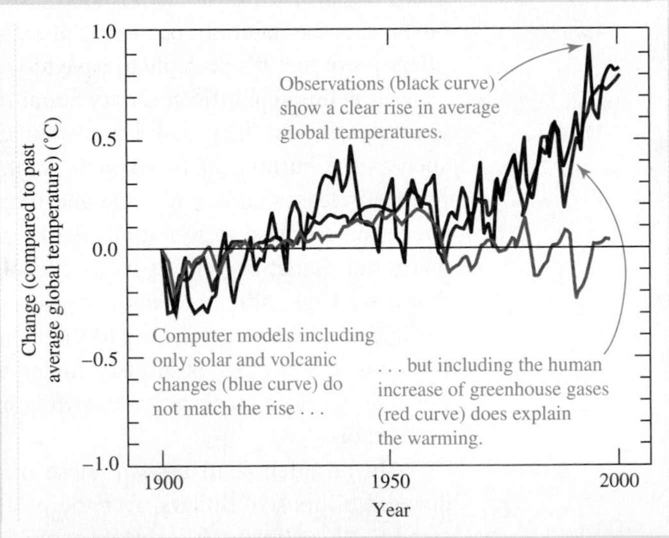

Figure 7.26 This graph compares observed temperature changes (black curve) with the predictions of climate models. The blue curve represents model predictions that include only natural factors, such as changes in the brightness of the Sun and effects of volcanoes. The red curve represents model predictions that include the human contribution due to increasing greenhouse gas concentration along with the natural factors. Notice that only the red curve matches the observations well, especially for recent decades, providing very strong evidence that global warming is a result of human activity. (The red and blue model curves are each averages of many scientists' independent models of global warming, which generally agree with each other to within 0.1°C–0.2°C.)

QUESTIONS FOR DISCUSSION

1. Look back at the six guidelines for establishing causality on page 316. Discuss whether or how each guideline is met by current data and understanding of global warming.
2. The IPCC concluded that global warming is "very likely" caused by us, but there are still some people who argue otherwise. Do a Web search to find some of the arguments made by those who disagree with the IPCC conclusion, and evaluate each argument carefully.
3. Look back at the legal levels of confidence in causality discussed in Section 7.4. Would you say that the case for human activity as the cause of global warming is now at the level of possible cause, probable cause, or cause beyond reasonable doubt? Defend your opinion.
4. Based on your view of global warming, describe what you think should be done about it.

SUGGESTED READING

Bennett, Jeffrey, and Shostak, Seth. *Life in the Universe*, 2nd ed. Addison Wesley, 2007 (see Section 10.5).

Flannery, Tim. *The Weather Makers*. Atlantic Monthly Press, 2005.

IPCC. *Climate Change 2007*. The full report is available online at http://www.ipcc.ch.

8

Some people hate the very name statistics, but I find them full of beauty and interest. Whenever they are not brutalized, but delicately handled by the higher methods, and are warily interpreted, their power of dealing with complicated phenomena is extraordinary.

—Sir Francis Galton (1822–1912)

From Samples to Populations

DID YOU EVER WONDER HOW THE OUTCOME OF A national election can be predicted hours before the polls close? Or how a large retailer can make critical marketing decisions based on a survey of only a few hundred people? These examples illustrate perhaps the most powerful aspect of statistics: the capability to use information gathered from a small sample to make predictions about a much larger population. This process, called *making inferences*, is the subject of the branch of statistics called *inferential statistics*.

LEARNING GOALS

8.1 Sampling Distributions

Understand the fundamental ideas of sampling distributions and how the distribution of sample means and the distribution of sample proportions are formed. Also learn the notation used to represent sample means and proportions.

8.2 Estimating Population Means

Learn to estimate population means and compute the associated margins of error and confidence intervals.

8.3 Estimating Population Proportions

Learn to estimate population proportions and compute the associated margins of error and confidence intervals.

8.1 Sampling Distributions

Consider the following statements taken from recent news articles or research reports.

- The mean daily protein consumption by Americans is 67 grams.
- Nationwide, the mean hospital stay after delivery of a baby decreased from 3.2 days in 1980 to the current mean of 2.0 days.
- Thirty percent of high school girls in this country believe they would be happier being married than not being married.
- About 5% of all American children live with a grandparent.

We hear or read statements like these every day. They make a claim about a large population of individuals, and yet it is not possible that everyone in the population could have been surveyed or measured. For example, the third statement was based on a random sample of high school girls, not based on a survey of every high school girl in the country. The responses of the girls in the sample were used to make a claim about the entire population of high school girls. How is it possible to *infer* from a random sample a very general conclusion about the population? These questions go to the heart of the branch of statistics called *inferential statistics*.

Public opinion in this country is everything.

—Abraham Lincoln

Notice that two different types of claims are made in the previous statements. The first two statements give estimates of a *mean* of a quantity—mean protein consumption of 67 grams and mean hospital stays of 3.2 days and 2.0 days. The last two statements say something about a *proportion* of the population—30% of high school girls and 5% of American children. These means and proportions pertain to the entire population, so they are *population parameters* (see Section 1.1). The topic of this chapter, and a major goal of inferential statistics, is learning how to estimate population parameters using data from samples.

Sample Means: The Basic Idea

Much of the work in this and the next chapter involves selecting a sample from a population, analyzing the sample, and then drawing conclusions about the population based on what was learned from the sample. In order to gain some experience with this important process, we will begin with an example.

Table 8.1 lists the weights of the five starting players (labeled A through E for convenience) on a professional basketball team. For the purposes of this example, we regard these five players as the *entire population* (with a mean of 242.4 pounds); in other words, it is the complete population of starters for the team. This is a very small population by statistical standards, but its small size will enable us to look carefully at the sampling process. Samples drawn from this population of five players can range in size from $n = 1$ (one player out of the five) to $n = 5$ (all five players).

With a sample size of $n = 1$, there are 5 different samples that could be selected: Each player is a sample. The mean of each sample of size $n = 1$ is simply the weight of the player in the sample. Figure 8.1 shows a histogram of the means of the 5 samples; it is called a **distribution of sample means**, because it shows the means of all 5 samples of size $n = 1$. The distribution of sample means created by this process is an example of a **sampling distribution**. This term simply refers to a distribution of a sample statistic, such as a mean, taken from *all* possible samples of a particular size. Notice that the mean of the 5 sample means is the mean of the entire population:

$$\frac{215 + 242 + 225 + 215 + 315}{5} = 242.4 \text{ pounds}$$

This demonstrates a general rule: *The mean of a distribution of sample means is the population mean.*

Table 8.1 Weights for the Population of Five Starters on a Basketball Team

Player	Weight (pounds)
A	215
B	242
C	225
D	215
E	315

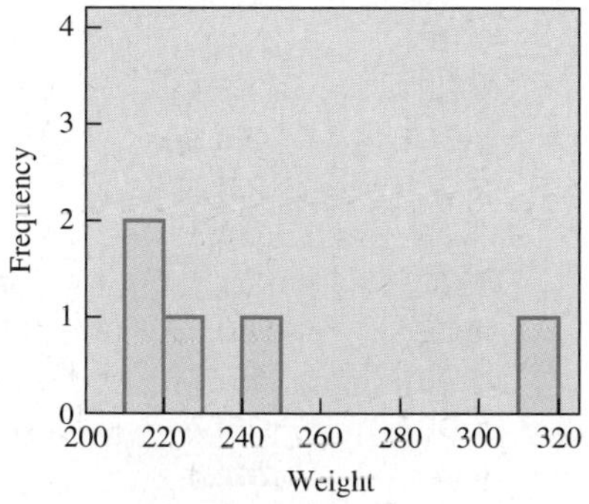

Figure 8.1 Sampling distribution for sample size $n = 1$.

Let's move on to samples of size $n = 2$, in which each sample consists of two different players. With five players, there are 10 different samples of size $n = 2$. Each sample has its own mean. For example, the sample of players A and B has a mean of $(215 + 242)/2 = 228.5$ pounds. Table 8.2 lists the 10 samples with their means, and Figure 8.2 shows the distribution of all 10 sample means. Again, notice that the mean of the distribution of sample means is equal to the population mean, 242.4 pounds.

Table 8.2 Sample Means for Basketball Example; Sample Size $n = 2$

Sample	Mean
AB	228.5
AC	220.0
AD	215.0
AE	265.0
BC	233.5
BD	228.5
BE	278.5
CD	220.0
CE	270.0
DE	265.0
Mean	**242.4**

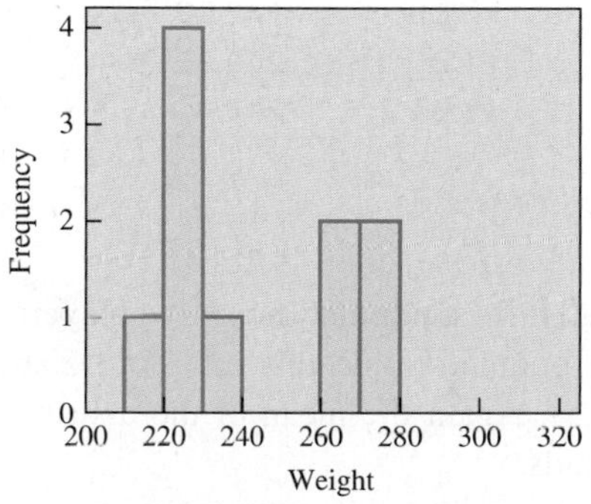

Figure 8.2 Sampling distribution for sample size $n = 2$.

> **TECHNICAL NOTE**
>
> Tables 8.2, 8.3, and 8.4 involve sampling *without replacement,* which has the advantage of avoiding the duplication that occurs whenever the same item is selected more than once. However, in statistics we are particularly interested in selection *with replacement* for these reasons: In large populations, it makes no significant difference whether we sample with replacement or without replacement, but sampling with replacement results in independent events that are unaffected by previous outcomes; independent events are easier to analyze and they result in simpler formulas.

Ten different samples of size $n = 3$ are possible in a population of five players. Table 8.3 shows these samples and their means, and Figure 8.3 shows the distribution of these sample means. Again, the mean of the distribution of sample means is equal to the population mean, 242.4 pounds.

Table 8.3 Sample Means for Basketball Example; Sample Size $n = 3$

Sample	Mean
ABC	227.3
ABD	224.0
ABE	257.3
ACD	218.3
ACE	251.7
ADE	248.3
BCD	227.3
BCE	260.7
BDE	257.3
CDE	251.7
Mean	**242.4**

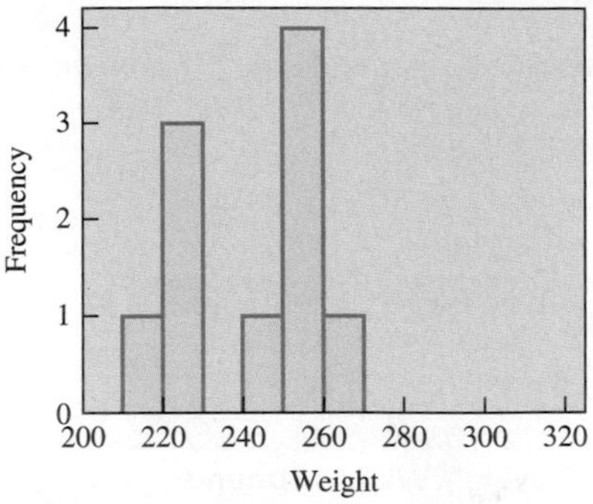

Figure 8.3 Sampling distribution for sample size $n = 3$.

With a sample size of $n = 4$, only 5 different samples are possible. Table 8.4 shows these samples and their means, and Figure 8.4 shows the distribution of these sample means.

Table 8.4 Sample Means for Basketball Example; Sample Size $n = 4$

Sample	Mean
ABCD	224.25
ABCE	249.25
ABDE	246.75
ACDE	242.50
BCDE	249.25
Mean	**242.4**

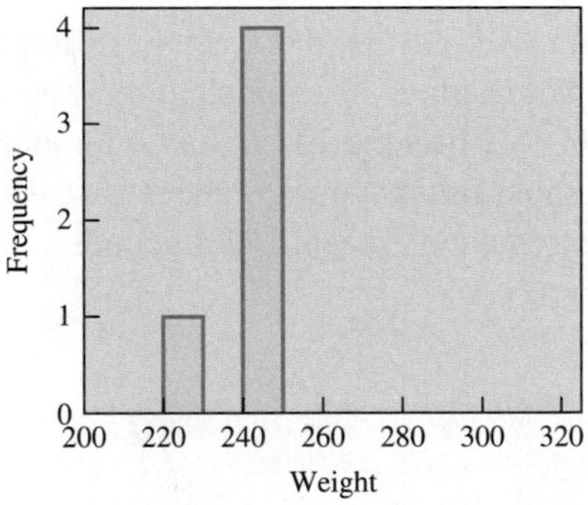

Figure 8.4 Sampling distribution for sample size $n = 4$.

Finally, for a population of five players, there is only 1 possible sample of size $n = 5$: the entire population. In this case, the distribution of sample means is just a single bar (Figure 8.5). Again the mean of the distribution of sample means is the population mean, 242.4 pounds.

To summarize, when we work with *all* possible samples of a population of a given size, the mean of the distribution of sample means is always the population mean. A closer look at the distributions in Figures 8.1 to 8.5 also reveals that as the sample size increases, the distribution narrows and clusters around the mean. In fact, if we looked at a large population and larger sample sizes, we would find that the distribution of sample means looks more and more like a normal distribution as the sample size increases (a consequence of the Central Limit Theorem, discussed in Section 5.3).

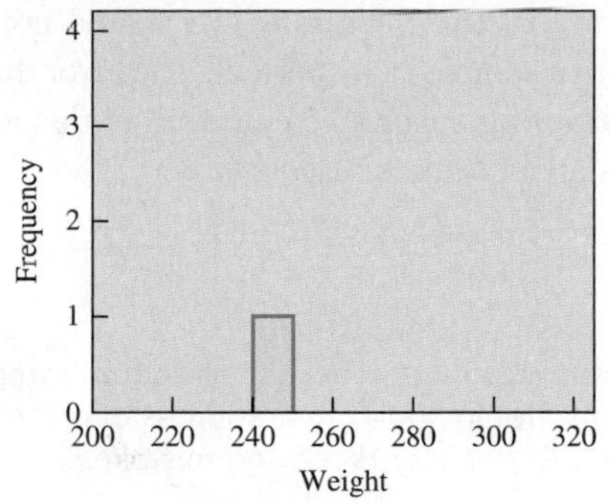

Figure 8.5 Sampling distribution for sample size $n = 5$.

Sample Means with Larger Populations

In real statistical studies, we rarely work with populations as small as five individuals. To explore the process of selecting samples and forming distributions of sample means in a more realistic setting, let's consider a somewhat larger population.

Imagine that you work for the computer services department of a small college. In order to develop networking strategies, you survey all 400 students at the college to determine how many hours per week they spend using a search engine on the Internet. The responses (hours per week) are shown below.

3.4	6.8	6.7	3.4	0.0	5.0	5.4	1.8	0.7	1.6	2.1	3.5	3.4	6.4	7.2	1.8	7.4	3.0	4.0	5.2
1.2	7.8	7.0	0.4	7.2	4.8	3.6	8.0	5.4	6.4	3.5	5.3	4.7	5.4	5.6	3.8	0.1	2.4	0.5	4.0
4.5	8.0	4.2	1.0	6.2	7.1	3.8	0.7	5.5	1.7	2.6	1.6	0.7	1.3	6.5	2.4	3.0	0.3	2.2	0.4
1.9	5.0	2.0	5.3	7.5	5.0	0.3	7.4	6.0	4.3	1.3	0.8	7.2	6.6	0.2	3.4	1.6	2.2	3.0	4.5
5.5	5.3	6.5	0.1	0.3	4.2	2.2	6.2	7.3	3.1	5.4	1.3	6.3	4.5	7.1	5.8	6.1	0.5	0.4	4.1
7.0	6.0	1.1	0.8	1.4	2.9	7.3	0.8	2.7	0.6	3.0	0.7	2.8	6.5	1.9	3.6	1.6	2.6	2.6	6.6
6.8	6.1	3.6	1.4	7.7	5.2	3.8	6.0	2.2	7.5	6.7	4.4	4.1	7.3	5.2	5.7	6.7	2.4	0.6	6.7
1.0	2.3	0.7	1.2	4.5	3.3	4.2	2.1	5.9	3.0	7.2	7.9	2.5	7.1	8.0	6.7	4.1	4.9	0.0	3.1
6.0	0.5	4.2	2.7	0.1	1.4	2.1	2.5	3.9	5.8	5.9	2.7	2.8	3.7	7.3	0.7	6.9	4.4	0.7	1.6
3.1	2.1	7.4	3.6	6.5	2.9	5.4	3.9	3.0	0.8	0.3	0.8	3.3	0.8	8.0	5.6	7.1	1.3	0.2	5.2
7.8	4.7	7.2	0.9	5.1	0.9	1.7	1.2	0.4	6.9	0.6	3.0	3.6	6.1	1.6	6.0	3.8	0.4	1.1	4.0
3.8	4.0	1.8	0.9	1.1	3.9	1.7	1.7	2.6	0.1	4.0	1.4	1.9	0.9	0.2	4.2	4.7	0.2	5.3	2.2
5.8	7.5	5.8	5.2	3.9	3.4	7.3	4.1	0.5	7.9	7.7	7.7	5.0	2.3	7.8	2.3	5.6	6.5	7.9	5.0
2.0	5.5	5.4	6.6	6.7	4.4	7.2	2.5	4.9	7.0	2.1	7.2	4.1	1.2	6.2	3.3	6.3	2.3	4.9	2.2
6.4	7.2	0.1	5.3	3.0	0.7	1.5	1.2	1.1	7.4	5.1	7.2	7.2	3.0	7.1	4.5	6.7	7.2	7.2	0.9
2.9	4.3	2.5	0.7	7.6	3.9	0.7	5.8	6.6	3.4	0.3	6.5	7.5	0.7	6.1	6.1	4.8	1.9	1.9	5.0
1.1	7.8	6.8	4.9	3.0	6.5	5.2	2.2	5.1	3.4	4.7	7.0	3.8	5.7	6.8	1.2	1.7	6.5	0.1	4.3
6.3	1.2	0.8	0.7	0.6	7.0	4.0	6.6	6.9	0.5	4.3	1.0	0.5	3.1	0.9	2.3	5.7	6.7	7.3	0.5
0.3	0.9	2.4	2.5	7.8	5.6	3.2	0.7	5.4	0.0	5.7	0.3	7.2	5.1	2.5	3.2	3.1	2.8	5.0	5.6
3.1	0.7	0.5	3.9	2.6	7.3	1.4	1.2	7.1	5.5	3.1	5.0	6.8	6.5	1.7	2.1	7.3	4.0	2.2	5.6

You could, of course, calculate the mean of all 400 responses. You would find that it is 3.88 hours per week. This mean is the true **population mean**, because it is the mean for the entire population of 400 students; we denote the population mean by the Greek letter μ (pronounced "mew"). Similarly, a calculation shows that the population standard deviation is $\sigma = 2.40$.

In typical statistical applications, populations are huge and it is impractical or expensive to survey every individual in the population; consequently, we rarely know the true population mean, μ. Therefore, it makes sense to consider using the mean of a *sample* to estimate the mean of the entire population. Although a sample is easier to work with, it cannot possibly represent the entire population exactly. Therefore, we should not expect an estimate of the population mean obtained from a sample to be perfect. The error that we introduce by working with a sample is called the **sampling error**. We can explore this idea by considering samples drawn from the 400 responses about Internet use.

Sampling Error

The **sampling error** is the error introduced because a random sample is used to estimate a population parameter. It does not include other sources of error, such as those due to biased sampling, bad survey questions, or recording mistakes.

TIME OUT TO THINK

Would you expect the sampling error to increase or decrease if the sample size were increased? Explain.

Suppose you select a random sample of $n = 32$ responses from the data set on page 337 and calculate their mean. For example, the random sample might be

1.1	7.8	6.8	4.9	3.0	6.5	5.2	2.2	5.1	3.4	4.7	7.0	3.8	5.7	6.5	2.7
2.6	1.4	7.1	5.5	3.1	5.0	6.8	6.5	1.7	2.1	1.2	0.3	0.9	2.4	2.5	7.8

The mean of this sample is $\bar{x} = 4.17$; we use the standard notation $\bar{x}$ to denote this mean. We say that $\bar{x}$ is a *sample statistic* because it comes from a sample of the entire population. Thus, $\bar{x}$ is called a **sample mean**.

Notation for Population and Sample Means

n = sample size

μ = population mean

$\bar{x}$ = sample mean

Our goal is to use a single sample mean to estimate the population mean, but first it's useful to explore other samples. Suppose you collect additional samples for the purpose of studying the behavior of sample means. Here is one such sample:

1.8	0.4	4.0	2.4	0.8	6.2	0.8	6.6	5.7	7.9	2.5	3.6	5.2	5.7	6.5	1.2
5.4	5.7	7.2	5.1	3.2	3.1	5.0	3.1	0.5	3.9	3.1	5.8	2.9	7.2	0.9	4.0

For this sample, the sample mean is $\bar{x} = 3.98$.

Now you have two sample means that don't agree with each other, and neither one agrees with the true population mean. Suppose you decide to select many more samples of 32 responses (an option that usually doesn't exist in practice). For each sample, you calculate the sample mean, $\bar{x}$. To get a good picture of all these sample means, you could make a histogram showing the number of samples that have various sample means. Figure 8.6 shows a histogram that results from 100 different samples, each with 32 students. Notice that this histogram is very close to a *normal distribution* and its mean is very close to the population mean, $\mu = 3.88$.

TIME OUT TO THINK

Suppose you choose only one sample of size $n = 32$. According to Figure 8.6, are you more likely to choose a sample with a mean less than 2.5 or a sample with a mean less than 3.5? Explain.

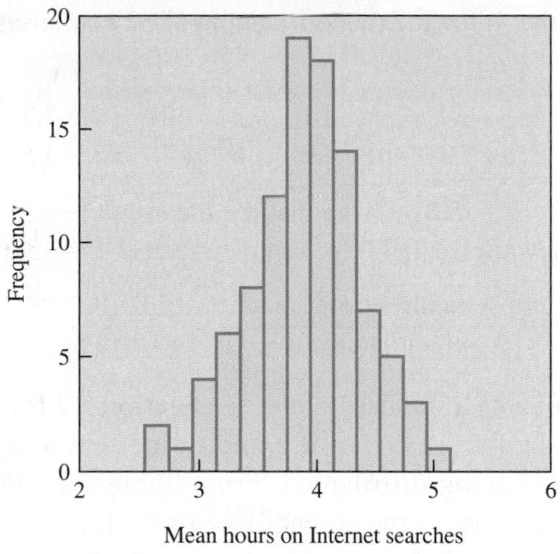

Figure 8.6 A distribution of 100 sample means, with a sample size of $n = 32$, appears close to a normal distribution with a mean of 3.88.

In our earlier example with the population of five basketball players, we were able to examine *all* possible samples of each size from $n = 1$ to $n = 5$. We cannot accomplish the same task for a population of 400, because the number of possible samples (combinations) of size $n = 32$ is nearly a trillion trillion trillion trillion (more precisely, it is about 2×10^{47}). However, even with the small number of samples in Figure 8.6, we can already see that the distribution of sample means is nearly normal. Once again, we are seeing an important property of the distribution of sample means: The distribution of sample means approaches a normal distribution for large sample sizes and the mean of the distribution of sample means equals the population mean, μ.

In practice, we often have only *one* sample available. In that case, its sample mean, $\bar{x}$, is our best estimate of the population mean, μ. Fortunately, as we will soon see, it is still possible to say something about how well $\bar{x}$ approximates μ.

The Distribution of Sample Means

The **distribution of sample means** is the distribution that results when we find the means of *all* possible samples of a given size. The larger the sample size, the more closely this distribution approximates a normal distribution. In all cases, the mean of the distribution of sample means equals the population mean. If only one sample is available, its sample mean, $\bar{x}$, is the best estimate for the population mean, μ.

TECHNICAL NOTE

A common guideline is to assume that the distribution of sample means is close to normal if the sample size is greater than 30.

We can extend this idea to see the important role that the normal distribution plays. Consider again the distribution of sample means in Figure 8.6. If we were to include *all* possible samples of size $n = 32$, this distribution would have these characteristics:

- The distribution of sample means is approximately a normal distribution.
- The mean of the distribution of sample means is 3.88 (the mean of the population).
- The standard deviation of the distribution of sample means depends on the population standard deviation and the sample size. The population standard deviation is $\sigma = 2.40$ and the sample size is $n = 32$, so the standard deviation of sample means is

$$\frac{\sigma}{\sqrt{n}} = \frac{2.40}{\sqrt{32}} = 0.42$$

Suppose we select the following random sample of 32 responses from the 400 responses given earlier:

5.8	7.5	5.8	5.2	3.9	3.4	7.3	4.1	0.5	7.9	7.7	7.7	5.0	2.3	7.8	2.3
5.0	6.8	6.5	1.7	2.1	7.3	4.0	2.2	5.6	4.7	5.3	3.5	6.5	3.4	6.6	5.0

The mean of this sample is $\bar{x} = 5.01$. Given that the mean of the distribution of sample means is 3.88 and the standard deviation is 0.42, the sample mean of $\bar{x} = 5.01$ has a *standard score* of

$$z = \frac{\text{sample mean} - \text{pop. mean}}{\text{standard deviation}} = \frac{5.01 - 3.88}{0.42} = 2.7$$

(Recall that we use z to denote a standard score; see Section 5.2 for review.) The sample we have selected has a standard score of $z = 2.7$, indicating that it is 2.7 standard deviations above the mean of the sampling distribution. From Table 5.1, this standard score corresponds to the 99.65th percentile, so the probability of selecting another sample with a mean *less* than 5.01 is about 0.9965. It follows that the probability of selecting another sample with a mean *greater* than 5.01 is about $1 - 0.9965 = 0.0035$. Apparently, the sample we selected is rather extreme within this distribution. This example shows that when we have a normal sampling distribution, we can determine whether a particular sample is somewhat ordinary or very rare.

TIME OUT TO THINK

Suppose a sample mean is in the 95th percentile. Explain why the probability of randomly selecting another sample with a mean greater than the first mean is 0.05.

By the Way ...

The total number of U.S. farms decreased from 2,440,000 in 1980 to 2,172,000 in 2000. The mean acreage per farm increased from 426 in 1980 to 434 in 2000.

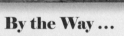

EXAMPLE 1 Sampling Farms

Texas has roughly 225,000 farms, more than any other state in the United States. The actual mean farm size is $\mu = 582$ acres and the standard deviation is $\sigma = 150$ acres. For random samples of $n = 100$ farms, find the mean and standard deviation of the distribution of sample means. What is the probability of selecting a random sample of 100 farms with a mean greater than 600 acres?

Solution Because the distribution of sample means is a normal distribution, its mean should be the same as the mean of the entire population, which is 582 acres. The standard deviation of the sampling distribution is $\sigma/\sqrt{n} = 150/\sqrt{100} = 15$. A sample mean of $\bar{x} = 600$ acres therefore has a standard score of

$$z = \frac{\text{sample mean} - \text{pop. mean}}{\text{standard deviation}} = \frac{600 - 582}{15} = 1.2$$

According to Table 5.1, this standard score is in the 88th percentile, so the probability of selecting a sample with a mean less than 600 acres is about 0.88. Thus, the probability of selecting a sample with a mean greater than 600 acres is about 0.12.

Sample Proportions

Much of what we have learned about distributions of sample means carries over to distributions of *sample proportions*. We can see the parallels by returning to the survey of 400 students described earlier. Suppose your goal is to determine the proportion (or percentage) of all 400 students who own a car. Each Y (for *yes*) or N (for *no*) below is one person's answer to the question "Do you own a car?"

Y N Y Y N Y N N Y N Y Y N N N Y Y Y Y N Y N Y N Y Y Y N Y Y N Y Y N N N Y Y Y N
Y N Y N Y N Y Y Y Y Y Y N Y Y N Y Y N N N Y Y Y N Y N Y N Y N Y Y Y Y Y N Y Y N
Y N N Y N Y Y N N N Y Y Y Y N Y N Y N Y N Y Y Y Y N Y N Y N Y Y Y N Y Y N Y Y N
N N Y Y Y N Y N Y N Y N Y Y Y Y Y N Y Y N Y N N Y N Y Y N N N N N Y Y Y N Y N Y
N Y N Y Y Y Y Y N Y Y N Y N N Y N Y Y N N N Y Y Y Y N Y N Y N Y Y Y N Y Y N Y Y
N Y Y N Y N N Y N Y Y N N N Y Y Y N N N Y Y Y N Y N Y N Y N Y Y Y Y Y Y N Y N Y
N Y Y Y N Y Y N Y Y N Y Y Y N Y Y N Y Y Y Y N N N Y Y Y N N N Y Y Y N Y N Y N Y
Y N Y N Y N N Y N Y N Y N Y Y Y Y Y Y N Y N Y N Y Y Y Y Y Y N Y N Y N Y Y N Y N
N Y N Y Y N N N Y Y Y N N N Y Y Y N Y N N Y Y Y N Y Y N Y Y Y Y Y N Y N Y N Y Y
Y N Y Y N Y N N Y N Y Y N N N Y Y Y N N N Y Y Y N Y N Y N Y N Y Y Y N Y Y N Y Y

If you counted carefully, you would find that 240 of the 400 responses are Y's, so the *exact* proportion of car owners in the 400-student population is

$$p = \frac{240}{400} = 0.6$$

This *population proportion, p* = 0.6, is another example of a *population parameter*. Of the 400 students in the population, it is the true proportion of car owners.

TIME OUT TO THINK

Give another survey question that would result in a population proportion rather than a population mean.

Once again, in typical statistical problems, it is impractical or prohibitively expensive to survey every individual in the population. Therefore, it's reasonable to consider the idea of using a random sample of, say, $n = 32$ people to estimate the population proportion. Suppose you randomly draw 32 responses from the list of Y's and N's to generate the following sample:

Y N Y Y N Y Y N Y N Y Y N Y Y Y N N Y Y Y N N Y Y N Y Y Y N Y Y

The proportion of Y responses in this list is

$$\hat{p} = \frac{21}{32} = 0.656$$

This proportion is another example of a *sample statistic*. In this case, it is a **sample proportion** because it is the proportion of car owners within a *sample;* we use the symbol $\hat{p}$ (read "*p*-hat") to distinguish this sample proportion from the population proportion, p.

Notation for Population and Sample Proportions

n = sample size

p = population proportion

$\hat{p}$ = sample proportion

Our goal is to determine how well a *single* sample proportion, $\hat{p}$, approximates the population proportion, p. For the moment, imagine that you have the luxury of selecting another sample of $n = 32$ responses; this sample produces the list

Y Y N N N Y Y N Y Y Y N N Y N Y Y N Y N Y N Y Y N Y Y N N Y N Y

The proportion of Y responses in this list is

$$\hat{p} = \frac{18}{32} = 0.563$$

There are three sorts of opinion: informed, uninformed and inconsequential. Uninformed opinions encompass most of American opinion.

—Paul Talmey, pollster

Suppose you decide to select *many* more samples of 32 responses (again, an option that usually doesn't occur in practice). For each sample, you calculate the sample proportion, $\hat{p}$. If you draw a histogram of the many sample proportions, it will show the number of samples that have particular values of $\hat{p}$ from 0 through 1. Figure 8.7 shows such a histogram, in this case resulting from 100 samples of size $n = 32$. As we found for sample means, this distribution of sample proportions is very close to a normal distribution. Furthermore, the mean of this distribution is very close to the population proportion of 0.6.

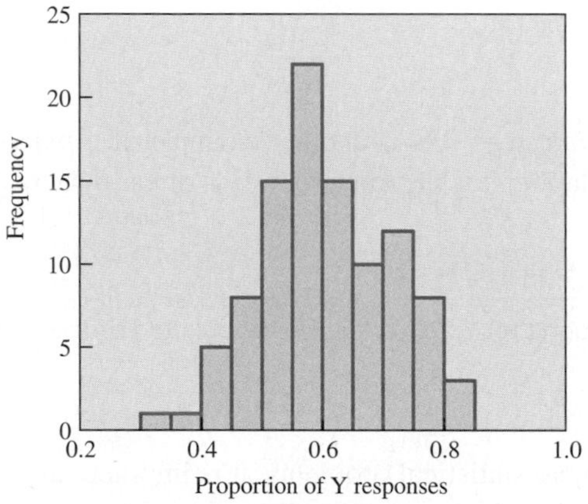

Figure 8.7 The distribution of 100 sample proportions, with a sample size of 32, appears to be close to a normal distribution.

Suppose it were possible to select *all* possible samples of size $n = 32$. The resulting distribution would be called a **distribution of sample proportions**. The mean of this distribution equals the population proportion exactly, and this distribution approaches a normal distribution as the sample size increases.

In practice, we often have only one sample to work with. In that case, the best estimate for the population proportion, p, is the sample proportion, $\hat{p}$.

The Distribution of Sample Proportions

The **distribution of sample proportions** is the distribution that results when we find the proportions ($\hat{p}$) in *all* possible samples of a given size. The larger the sample size, the more closely this distribution approximates a normal distribution. In all cases, the mean of the distribution of sample proportions equals the population proportion. If only one sample is available, its sample proportion, $\hat{p}$, is the best estimate for the population proportion, p.

EXAMPLE 2 Analyzing a Sample Proportion

Consider the distribution of sample proportions shown in Figure 8.7. Assume that its mean is $p = 0.6$ and its standard deviation is 0.1. Suppose you randomly select the following sample of 32 responses:

Y Y N Y Y Y Y N Y Y Y Y Y Y N Y Y N Y Y Y N Y Y N Y Y N Y N Y Y

Compute the sample proportion, $\hat{p}$, for this sample. How far does it lie from the mean of the distribution? What is the probability of selecting another sample with a proportion greater than the one you selected?

Solution The proportion of Y responses in this sample is

$$\hat{p} = \frac{24}{32} = 0.75$$

Using a mean of 0.6 and a standard deviation of 0.1, we find that the sample statistic, $\hat{p} = 0.75$, has a standard score of

$$z = \frac{\text{sample proportion} - \text{pop. proportion}}{\text{standard deviation}} = \frac{0.75 - 0.6}{0.1} = 1.5$$

The sample proportion is 1.5 standard deviations above the mean of the distribution. Using Table 5.1, we see that a standard score of 1.5 corresponds to the 93rd percentile. The probability of selecting another sample with a proportion less than the one we selected is about 0.93. Thus, the probability of selecting another sample with a proportion greater than the one we selected is about $1 - 0.93 = 0.07$. In other words, if we were to select 100 random samples of 32 responses, we should expect to see only 7 samples with a higher proportion than the one we selected.

TECHNICAL NOTE

The standard deviation of sample proportions is

$$\sqrt{\frac{p(1-p)}{n}}$$

If we don't know the value of the population proportion, p, we can estimate that standard deviation by using the sample proportion, $\hat{p}$. For example, if $\hat{p} = 0.75$ and $n = 32$, we get

$$\sqrt{\frac{0.75(1-0.75)}{32}} = 0.0765$$

or 0.1 rounded.

Section 8.1 Exercises

Statistical Literacy and Critical Thinking

1. **Sampling Distribution.** Pollsters often use randomly selected digits between 0 and 9 to generate parts of telephone numbers to be called. What is the distribution of such randomly selected digits? If we repeat the process of randomly generating 50 digits and finding the mean, what is the distribution of the resulting sample means?

2. **Best Estimates.** Given only one sample, what is the best estimate of the population mean? Given only one sample, what is the best estimate of the population proportion?

3. **Notation.** What does $\bar{x}$ denote, what does μ denote, and what is the difference between them?

4. **Notation.** What does $\hat{p}$ denote, what does p denote, and what is the difference between them?

Does It Make Sense? For Exercises 5–8, decide whether the statement makes sense (or is clearly true) or does not make sense (or is clearly false). Explain clearly. Not all of these statements have definitive answers, so your explanation is more important than your chosen answer.

5. **Large Sample.** Ted is a college student working on a project to estimate the proportion of adult Americans who have been involved in a car crash. He uses a convenience sample of students at his college, but he obtains a very large sample of size 3,000. Ted states that his estimate is good because the large sample size compensates for the fact that he is using a convenience sample.

6. **Recording Error.** A sampling error occurs when a pollster incorrectly records a survey response.

7. **Larger Sample Size.** When a random sample is used to estimate a population mean, the sample mean tends to become a better estimate of the population mean as the sample size increases.

8. **Survey.** In a Gallup poll of 1,012 randomly selected adults, 901 said that cloning of humans should not be allowed. If a larger sample were obtained, the sample proportion would be likely to be larger than 901/1012 (or 0.890).

Concepts and Applications

9. **Estimating Population Proportions.** In a survey of children 5 to 17 years old, 1,050 children were randomly selected from the nine states in the Northeast, and the proportion who spoke a language other than English in their home was 0.19. Based on that sample statistic, what is the best estimate of the proportion for all children in the Northeast? Is that sample proportion a good estimate for children in the United States? Why or why not?

10. **Estimating Population Means.** When 40 women were randomly selected and tested for their cholesterol levels, a mean of 240.9 milligrams was obtained. Based on this

sample statistic, what is the best estimate for the mean cholesterol level for all women? Would you be more confident of your estimate if the sample included measurements from 500 women? Explain.

11. Distribution of Sample Means. Assume that cans of Coke are filled so that the actual amounts have a mean of 12.00 ounces. A random sample of 36 cans has a mean amount of 12.19 ounces. The distribution of sample means of size 36 is normal with an assumed mean of 12.00 ounces and a standard deviation of 0.02 ounce.

a. How many standard deviations is the sample mean from the mean of the distribution of sample means?

b. In general, what is the probability that a random sample of size 36 has a mean of at least 12.19 ounces?

c. Does it appear that consumers are being cheated? Why or why not?

12. Distribution of Sample Means. Suppose you know that the distribution of sample means for household size for samples of 500 households is normal with a mean of 2.64 and a standard deviation of 0.06. Suppose you select a random sample of $n = 500$ households and determine that the mean number of people per household for this sample is 2.55.

a. How many standard deviations is the sample mean from the mean of the distribution of sample means?

b. What is the probability that a second sample selected would have a mean less than 2.55?

13. Sample and Population Proportions. Suppose that, in a suburb of 12,345 people, 6,523 people moved there within the last five years. You survey 500 people and find that 245 of the people in your sample moved to the suburb in the last five years.

a. What is the population proportion of people who moved to the suburb in the last five years?

b. What is the sample proportion of people who moved to the suburb in the last five years?

c. Does your sample appear to be representative of the population? Discuss.

14. Sample and Population Proportions. Suppose that, in a school with 1,348 students, 137 students are left-handed. You survey 100 students and find that 11 of the students in your sample are left-handed.

a. What is the population proportion of left-handed students?

b. What is the sample proportion of left-handed students?

c. Does your sample appear to be representative of the population? Discuss.

15. Estimating Population Proportions. You select a random sample of 150 people at a medical convention attended by 1,608 people. Within your sample, you find that 73 people have traveled from abroad. Based on this sample statistic, estimate how many people at the convention traveled from abroad. Would you be more confident of your estimate if you sampled 300 people? Explain.

16. Estimating Population Proportions. A random sample of 750 people is selected from the 74,512 people in attendance at a Miami Super Bowl game. Within the sample, 420 people are supporting the Chicago Bears. Based on this sample statistic, estimate how many people at the game support the Chicago Bears. Would you be more confident of your estimate if you sampled 2,000 people? Explain.

17. Distribution of Sample Proportions. Suppose you know that the distribution of sample proportions of nonresidents in samples of 200 students is normal with a mean of 0.34 and a standard deviation of 0.03. Suppose you select a random sample of 200 students and find that the proportion of nonresident students in the sample is 0.32.

a. How many standard deviations is the sample proportion from the mean of the distribution of sample proportions?

b. What is the probability that a second sample selected would have a proportion less than 0.32?

18. Distribution of Sample Proportions. Suppose you know that the distribution of sample proportions of women employees is normal with a mean of 0.42 and a standard deviation of 0.21. Suppose you select a random sample of employees and find that the proportion of women in the sample is 0.45.

a. How many standard deviations is the sample proportion from the mean of the distribution of sample proportions?

b. What is the probability that a second sample selected would have a proportion greater than 0.45?

19. Sampling Distribution. A quarterback threw 1 interception in his first game, 2 interceptions in his second game, 5 interceptions in his third game, and then he retired. Consider the values of 1, 2, and 5 to be a population. Assume that samples of size 2 are randomly selected *with replacement* from the population.

a. After listing the 9 different possible samples, find the mean of each sample.

b. What is the mean of the sample means from part a?

c. Is the mean of the sampling distribution from part b equal to the mean of the population of the three listed values? Are those means *always* equal?

20. Sampling Distribution. Here is the population of all five U.S. presidents who had professions in the military, along with their ages at inauguration: Eisenhower, 62; Grant, 46; Harrison, 68; Taylor, 64; and Washington, 57. Assume that samples of size 2 are randomly selected *with replacement* from the population of five ages.

a. After listing the 25 different possible samples, find the mean of each sample.

b. What is the mean of the sample means from part a?

c. Is the mean of the sampling distribution (from part b) equal to the mean of the population of the five listed values? Are those means *always* equal?

21. Forming Sampling Distributions. Five states and their areas (in thousands of square miles) are given in the following table. Consider these five states to be the entire population from which samples will be selected with replacement. Find the distribution of sample means for sample sizes $n = 1$, 2, 3, 4, and 5. Find the mean of the distribution of sample means in each case. Compare the means of the distributions of sample means to the population mean.

State	Area (thousands of square miles)
Alabama	52
Connecticut	5
Georgia	60
Maine	33
Oregon	97

22. Forming Sampling Distributions. At the end of a practice session, the five starters on a hockey team have scored the following numbers of points.

Player	Points
A	101
B	87
C	75
D	66
E	62

Find the distribution of sample means for sample sizes $n = 1, 2, 3, 4$, and 5, where the samples are selected with replacement. Find the mean of the distribution of sample means in each case. Compare the means of the distributions of sample means to the population mean.

Projects for the Internet and Beyond

For useful links, select "Links for Internet Projects" for Chapter 8 at www.aw.com/bbt.

23. Distributions of Sample Means. Consider the large data set of hours students spend using an Internet search engine, listed in this section on page 337. Discuss methods for selecting a random sample from this population. Let each person in the class select a random sample of $n = 10$ individuals from the population and find the mean of his or her sample. Find the mean of the individual sample means and compare it to the population mean. Repeat the process with samples of size $n = 20$. How does the mean of the sample means compare to the population mean?

24. Distributions of Sample Proportions. Consider the data set on page 341 showing 400 *yes* and *no* responses to a survey question ("Do you own a car?"). Discuss methods for selecting a random sample from this population. Let each person in the class select a random sample of $n = 10$ responses from the population and find the proportion of *yes* responses for his or her sample. Find the mean of the individual sample proportions and compare it to the population proportion of 0.6. Repeat the process with samples of size $n = 20$. How does the mean of the sample proportions compare to the population proportion?

IN THE NEWS

25. Sample Means in the News. Find a news or research report in which a sample mean is cited. Discuss how it is used to estimate a population mean.

26. Sample Proportions in the News. Find a news or research report in which a sample proportion is cited. Discuss how it is used to estimate a population proportion.

8.2 Estimating Population Means

In Section 8.1, we saw how distributions of sample means and sample proportions arise. We also saw that, if we have a single random sample from a population, its sample mean is our best estimate of the population mean. But how good is this "best estimate"? In this section, we investigate how to make quantitative statements about uncertainty when inferring a population mean from a sample mean. (We discuss population proportions in Section 8.3.)

By the Way ...

RDAs for protein are sometimes given in grams per kilogram of body weight. A typical figure for non-athletes and women who are not pregnant is 0.8 gram of protein per kilogram of body weight.

Estimating a Population Mean: The Basics

Plants and animals constantly produce new cells, not only for growth, but also to replace aging cells. The production of new cells requires proteins, the building blocks of living organisms. For this reason, proteins are an essential part of our diet. How much protein should you eat per day? Many nutritional organizations and government agencies provide recommended daily allowances (RDAs) not only for protein, but also for vitamins, minerals, carbohydrates, and fat. These RDAs differ according to the source, and they change over time as new research is done. Most recommendations for protein consumption are clustered around a value of 55–60 grams per day for men and 45–50 grams per day for women who are not pregnant or nursing.

Given these recommendations for how much protein we *should* eat, it is interesting to ask how much protein Americans actually *do* eat. Figure 8.8 shows partial results of a large survey of the nutritional habits of Americans (Third National Health and Nutrition Examination Survey, or NHANES III, conducted by the National Center for Health Statistics). The original study involved roughly 30,000 participants who were surveyed on many different aspects of their health and diet. The histogram shows the average daily intake of protein, in grams, for a sample of $n = 267$ men taken from the study.

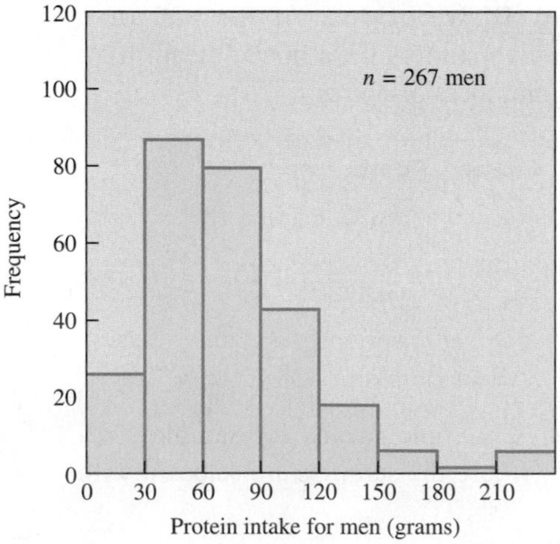

Figure 8.8 Histogram of daily protein intake for men taken from a sample of $n = 267$ men. *Source:* National Center for Health Statistics.

Our goal is to use the average protein intake for this sample of $n = 267$ men to make an inference about the average protein intake for the population of all American men. The sample data have a mean of $\bar{x} = 77.0$ grams and a standard deviation of $s = 58.6$ grams. As discussed in

Section 8.1, when we have only a single sample, the sample mean is the best estimate of the population mean, μ. However, we do not expect the sample mean to be equal to the population mean, because there is likely to be some sampling error. Therefore, in order to make an inference about the population mean, we need some way to describe how well we expect it to be represented by the sample mean. The most common method for doing this is by way of *confidence intervals*. We discussed the use of confidence intervals in Chapter 1. Now we are ready to look at how confidence intervals are computed.

The idea of a confidence interval comes directly from the work we did with sampling distributions. A precise calculation shows that if the distribution of sample means is normal with a mean of μ, then 95% of all sample means lie within 1.96 standard deviations of the population mean; for our purposes in this book, we will approximate this as 2 standard deviations. A **confidence interval** is a range of values likely to contain the true value of the population mean. In this book, we work only with 95% confidence levels, although other levels of confidence (such as 90% or 99%) are sometimes used. We find the 95% confidence interval from the sample mean by working with the **margin of error**, as defined in the following box.

95% Confidence Interval for a Population Mean

The **margin of error** for the 95% confidence interval is

$$\text{margin of error} = E \approx \frac{2s}{\sqrt{n}}$$

where s is the standard deviation of the sample. We find the **95% confidence interval** by adding and subtracting the margin of error from the sample mean. That is, the 95% confidence interval ranges

from ($\bar{x}$ − margin of error) to ($\bar{x}$ + margin of error)

We can write this confidence interval more formally as

$$\bar{x} - E < \mu < \bar{x} + E$$

or more briefly as

$$\bar{x} \pm E$$

TECHNICAL NOTE

The precise formula for the margin of error uses 1.96 rather than 2. Also, one consequence of the Central Limit Theorem is that the distribution of sample means has a standard deviation of $\sigma/\sqrt{n}$, but s may be used in place of σ as long as the sample is large enough (typically $n > 30$, but the size requirement depends on the nature of the actual population distribution). The formula for margin of error can be extended to other confidence levels. For example, for a 90% confidence level, replace the 2 in the formula by 1.645. For a 99% confidence level, replace the 2 in the formula by 2.575.

The sample mean, $\bar{x}$, lies at the center of the confidence interval, which extends a distance equal to the margin of error in either direction (Figure 8.9). We say that we are 95% confident that the confidence interval contains the true value of the population mean. This statement should be carefully interpreted as follows: If we were to repeat the process of obtaining samples and constructing such confidence intervals many times, 95% of the confidence intervals would contain the value of the population mean (μ), and 5% would not include it.

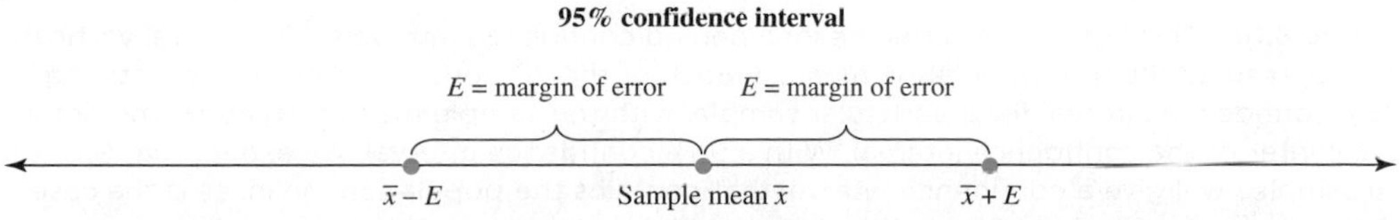

Figure 8.9 The 95% confidence interval extends a distance equal to the margin of error on either side of the sample mean.

EXAMPLE 1 Computing the Margin of Error

Compute the margin of error and find the 95% confidence interval for the protein intake sample of $n = 267$ men, which has a sample mean of $\bar{x} = 77.0$ grams and a sample standard deviation of $s = 58.6$ grams.

Solution The sample size is $n = 267$ and the standard deviation for the sample is $s = 58.6$, so the margin of error is

$$E \approx \frac{2s}{\sqrt{n}} = \frac{2 \times 58.6}{\sqrt{267}} = 7.2$$

TECHNICAL NOTE

If we use the precise formula (with 1.96 instead of 2), we find that the confidence interval limits are 70.0 and 84.0.

The sample mean is $\bar{x} = 77.0$ grams, so the 95% confidence interval extends approximately from $77.0 - 7.2 = 69.8$ grams to $77.0 + 7.2 = 84.2$ grams. We write this result more formally as

$$69.8 \text{ grams} < \mu < 84.2 \text{ grams}$$

or more simply as 77.0 ± 7.2 grams. We are 95% confident that the population mean for protein intake of all American men is between 69.8 and 84.2 grams. It is interesting to note that even the lower number in this confidence interval (69.8 grams) is greater than the recommended daily protein allowance for men of 55–60 grams, suggesting that actual protein consumption is significantly greater than recommended.

Interpreting the Confidence Interval

Figure 8.10 shows a visual interpretation of the confidence interval. Imagine that we have, say, 20 different samples. Each sample has a different sample mean, $\bar{x}$, with a confidence interval about the sample mean. We can never know for sure which intervals contain the population mean. However, on average, 95% of the samples, or 19 out of the 20 samples, will have a confidence interval that captures the true population mean. An occasional confidence interval (about 5% of those generated) does not capture the true population mean.

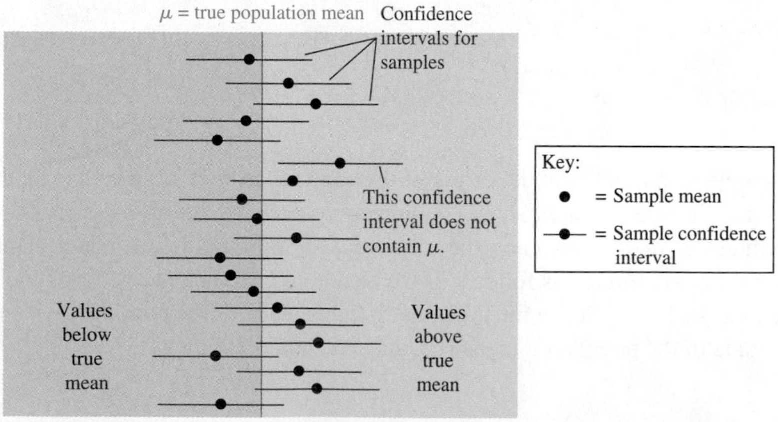

Figure 8.10 This figure illustrates the idea behind confidence intervals. The central vertical line represents the true population mean, μ. Each of the 20 horizontal lines represents the 95% confidence interval for a particular sample, with the sample mean marked by the dot in the center of the confidence interval. With a 95% confidence interval, we expect that 95% of all samples will give a confidence interval that contains the population mean, as is the case in this figure, for 19 of the 20 confidence intervals do indeed contain the population mean. We expect that the population mean will not be within the confidence interval in 5% of the cases; here, 1 of the 20 confidence intervals (the sixth from the top) does not contain the population mean.

Be careful—it is easy to interpret or state a confidence interval incorrectly. It is tempting to claim that a 95% confidence interval means there is a 0.95 probability that the population mean falls within the confidence interval. But this is not true: Even though we may not know the actual population mean, μ, it is a fixed number (not a variable), so either it *does* fall within the confidence interval or it *does not* fall within the confidence interval. We cannot talk about the probability that μ falls within the confidence interval. To recap, the correct interpretation of a 95% confidence interval is this: *If we repeat the process of obtaining samples and constructing confidence intervals, in the long run 95% of the confidence intervals will contain the true population mean.*

EXAMPLE 2 Constructing a Confidence Interval

A study finds that the average time spent by eighth-graders watching television is 6.7 hours per week, with a margin of error of 0.4 hour (for 95% confidence). Construct and interpret the 95% confidence interval.

Solution The best estimate of the population mean is the sample mean, $\bar{x} = 6.7$ hours. We find the confidence interval by adding and subtracting the margin of error from the sample mean, so the interval extends from $6.7 - 0.4 = 6.3$ hours to $6.7 + 0.4 = 7.1$ hours. We can therefore claim with 95% confidence that the average time spent watching television for the entire population of eighth-graders is between 6.3 and 7.1 hours, or

$$6.3 \text{ hours} < \mu < 7.1 \text{ hours}$$

If 100 random samples of the same size were taken, we would expect the confidence intervals of 95 of those samples to contain the population mean.

EXAMPLE 3 Protein Intake for Women

The NHANES III nutritional study also produced data on protein intake for women. Figure 8.11 shows a histogram for a random sample of $n = 264$ women (a small part of the entire study). The mean of these data is $\bar{x} = 59.6$ grams and the standard deviation is $s = 30.5$ grams. Estimate the population mean and give a 95% confidence interval. Comment on how these values compare to the recommended daily allowance (RDA) for women of 45–50 grams.

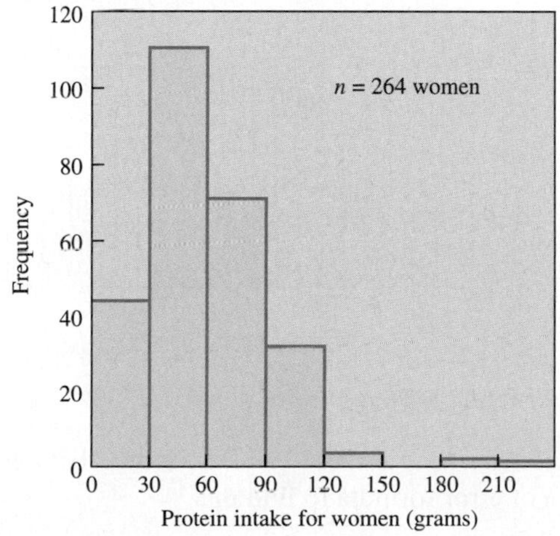

Figure 8.11 Histogram for daily protein intake for 264 women. *Source:* National Center for Health Statistics.

Solution The sample mean, $\bar{x} = 59.6$ grams, is our best estimate of the population mean. To find the 95% confidence interval, we must first find the margin of error using the sample size ($n = 264$) and sample standard deviation ($s = 30.5$ grams):

$$E \approx \frac{2s}{\sqrt{n}} = \frac{2 \times 30.5 \text{ grams}}{\sqrt{264}} = 3.8 \text{ grams}$$

The 95% confidence interval therefore ranges from $59.6 - 3.8 = 55.8$ grams to $59.6 + 3.8 = 63.4$ grams, or

$$55.8 \text{ grams} < \mu < 63.4 \text{ grams}$$

We can say with 95% confidence that the population mean—the mean protein intake for women—lies within the interval from 55.8 grams to 63.4 grams. We conclude that protein consumption by women is greater than the recommended daily allowance of 45–50 grams for women.

TIME OUT TO THINK

Recall that the standard deviation of the data on protein intake for the sample of men was $s = 58.6$ grams—almost double the standard deviation for the sample of women ($s = 30.5$ grams). How does this difference affect the margins of error? Speculate on why the standard deviations are different.

EXAMPLE 4 Garbage Production

A study conducted by the Garbage Project at the University of Arizona analyzed the contents of garbage discarded by $n = 62$ households; the households ranged in size from 2 to 11 members. The histogram in Figure 8.12 shows the total weekly garbage production (in pounds) for the households in the sample. (The complete study gives the breakdown of garbage by various categories.) The mean for the sample is $\bar{x} = 27.4$ pounds and the standard deviation is $s = 12.5$ pounds. Estimate the population mean for weekly garbage production with a 95% confidence interval. Comment on the conclusion of the study.

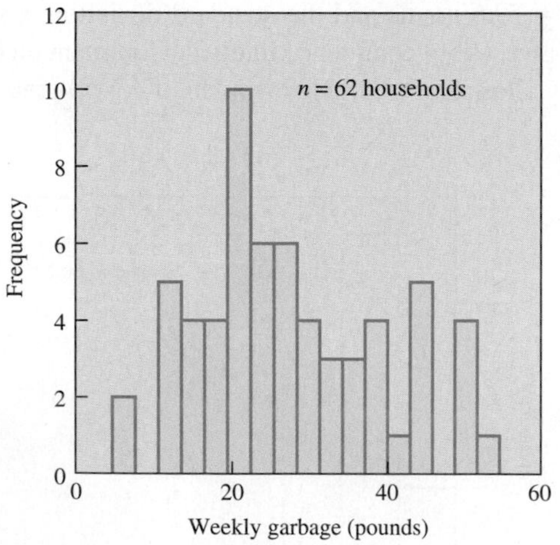

Figure 8.12 Histogram for total garbage production for $n = 62$ households.

Solution The sample mean, $\bar{x} = 27.4$ pounds, is our best estimate of the population mean. We use the margin of error formula to find that

$$E \approx \frac{2s}{\sqrt{n}} = \frac{2 \times 12.5 \text{ grams}}{\sqrt{62}} = 3.2 \text{ pounds}$$

By the Way ...

According to the Garbage Project data, the leading component of most household garbage is paper (comprising a third to a half of the total garbage), followed by food and glass. Also, there is a significant correlation between the weight of the discarded plastic and the size of a household, so discarded plastic could be used to estimate the population of a region.

The 95% confidence interval therefore ranges from 27.4 – 3.2 = 24.2 pounds to 27.4 + 3.2 = 30.6 pounds, or

$$24.2 \text{ pounds} < \mu < 30.6 \text{ pounds}$$

With 95% confidence, we can claim that the interval from 24.2 pounds to 30.6 pounds contains the mean amount of garbage discarded by American households in a week. Two observations might be made about this conclusion. First, the wide range of household sizes produces significant variation in the data. This variation is reflected in a large standard deviation that produces a large margin of error. The conclusion might be more meaningful if it were given in terms of individual household sizes. Second, the relatively small sample size (62 households) also contributes to the large margin of error. A more reliable conclusion could be obtained from a larger sample.

EXAMPLE 5 Mean Body Temperature

A study by University of Maryland researchers investigated the body temperatures of $n = 106$ subjects. The sample mean of the data set is $\bar{x} = 98.20°\text{F}$ and the standard deviation for the sample is $s = 0.62°\text{F}$. Estimate the population mean body temperature with a 95% confidence interval.

Solution The sample mean, $\bar{x} = 98.20°\text{F}$, is our best estimate of the population mean body temperature. The margin of error is

$$E \approx \frac{2s}{\sqrt{n}} = \frac{2 \times 0.62°\text{F}}{\sqrt{106}} = 0.12°\text{F}$$

The 95% confidence interval therefore ranges from 98.20°F – 0.12°F = 98.08°F to 98.20°F + 0.12°F = 98.32°F, or

$$98.08°\text{F} < \mu < 98.32°\text{F}$$

We interpret this result as follows: If we were to select many different samples of size $n = 106$ and compute confidence intervals for all of the samples, we expect that 95% of the confidence intervals would contain the true population mean. Notice that the commonly cited mean body temperature for humans (98.6°F) is *not* contained in the confidence interval. Based on this sample, it is likely that the accepted mean body temperature is wrong.

Choosing Sample Size

In planning statistical surveys and experiments, we often know in advance the margin of error we would like to achieve. For example, we might want to estimate the mean cost of a new car to within \$200. We can estimate the sample size, n, needed to ensure this margin of error by solving the margin of error formula $(E \approx 2s/\sqrt{n})$ for n. With a little bit of algebra, we find

$$n \approx \left(\frac{2s}{E}\right)^2$$

The exact sample size formula uses the population standard deviation, σ, in place of s. In practice, we rarely know the population standard deviation, because we study only samples. Therefore, to use the exact sample size formula, we usually estimate the population standard deviation based on previous studies, pilot studies, or educated guesses. The size of any actual sample must be a whole number, so we round the result of the sample size formula *up* to the nearest whole number. Any sample larger than this size will give us a margin of error as small as or smaller than the one we seek.

TECHNICAL NOTE

The given formula for sample size assumes we are working with a 95% level of confidence; a more accurate formula would use 1.96 instead of 2, as discussed earlier. For the methods of this section, if the population standard deviation is not known, we can use the sample standard deviation, *s*, as an estimate of that value. However, the results will be good only if the sample size is large. If a small sample is taken from a non-normal population, the use of *s* in place of σ may lead to very poor results.

Choosing the Correct Sample Size

In order to estimate the population mean with a specified margin of error of at most *E*, the size of the sample should be at least

$$n = \left(\frac{2\sigma}{E}\right)^2$$

where σ is the population standard deviation (often estimated by the sample standard deviation *s*).

EXAMPLE 6 Mean Housing Costs

You want to study housing costs in the country by sampling recent house sales in various (representative) regions. Your goal is to provide a 95% confidence interval estimate of the housing cost. Previous studies suggest that the population standard deviation is about \$7,200. What sample size (at a minimum) should be used to ensure that the sample mean is within

a. \$500 of the true population mean?

b. \$100 of the true population mean?

Solution

a. With E = \$500 and σ estimated as \$7,200, the minimum sample size that meets the requirements is

$$n = \left(\frac{2\sigma}{E}\right)^2 = \left(\frac{2 \times 7{,}200}{500}\right)^2 = 28.8^2 = 829.4$$

Because the sample size must be a whole number, we conclude that the sample should include *at least* 830 prices.

b. With E = \$100 and σ = \$7,200, the minimum sample size that meets the requirements is

$$n = \left(\frac{2\sigma}{E}\right)^2 = \left(\frac{2 \times 7{,}200}{100}\right)^2 = 144^2 = 20{,}736$$

Notice that to decrease the margin of error by a factor of 5 (from \$500 to \$100), we must increase the sample size by a factor of 25. That is why achieving greater accuracy generally comes with a high cost.

TIME OUT TO THINK

If you decide you want a smaller margin of error for a confidence interval, should you increase or decrease the sample size? Explain.

Section 8.2 Exercises

Statistical Literacy and Critical Thinking

1. **Confidence Interval.** Based on a random sample of measurements from women, we construct this 95% confidence interval estimate of the mean cholesterol level of all women:

$$183.3 \text{ milligrams} < \mu < 298.5 \text{ milligrams}$$

Interpret this confidence interval.

2. **Margin of Error.** Based on a random sample of blood pressure measurements from men, the sample mean is 73.2 mm Hg and the margin of error for a 95% confidence interval is 2.8 mm Hg. Identify the confidence interval.

3. **Confidence Intervals in the Media.** Here is a typical statement made by the media: "Based on a recent study, pennies weigh an average of 2.5 grams with a margin of error of 0.006 gram." What important and relevant piece of information is omitted from that statement? Is it okay to use the word "average"?

4. **Sample Size.** The National Health Examination involves measurements from about 25,000 people, and the results are used to estimate values of various population means. Is it valid to criticize this survey because the sample size is only about 0.01% of the population of all Americans? Explain.

Does It Make Sense? For Exercises 5–8, decide whether the statement makes sense (or is clearly true) or does not make sense (or is clearly false). Explain clearly. Not all of these statements have definitive answers, so your explanation is more important than your chosen answer.

5. **Estimating the Mean.** A survey of 200 randomly selected readers was obtained, and it was used to find the confidence interval of $47,200.

6. **Margin of Error.** The mean income of high school mathematics teachers was estimated to be $48,213 with a margin of error of 5%.

7. **Margin of Error.** When sample data are used to estimate the value of a population mean, the margin of error increases as the sample size increases.

8. **Best Estimate.** When sample data were used to estimate the value of the mean weight of all pennies, this 95% confidence interval was obtained:

$$2.495 \text{ grams} < \mu < 2.505 \text{ grams}$$

Based on that result, the best single-value estimate of the population mean is 2.500 grams.

Concepts and Applications

Finding Margins of Error and Confidence Intervals. For Exercises 9–12, assume that population means are to be estimated from the samples described. In each case, use the sample results to approximate the margin of error and 95% confidence interval.

9. Sample size = 100, sample mean = 75.0, sample standard deviation = 10.0

10. Sample size = 64, sample mean = 2.50, sample standard deviation = 1.08

11. Sample size = 1,068, sample mean = $46,205, sample standard deviation = $24,000

12. Sample size = 692, sample mean = 155 kilograms, sample standard deviation = 27 kilograms

Sample Sizes. For Exercises 13–16, assume that you want to construct a 95% confidence interval estimate of a population mean. Find an estimate of the sample size needed to obtain the specified margin of error for the 95% confidence interval. The sample standard deviation is given.

13. Margin of error = 1.0 centimeter, standard deviation = 3.5 centimeters

14. Margin of error = 2.0 centigrams, standard deviation = 16 centigrams

15. Margin of error = 0.01 kilometer, standard deviation = 0.11 kilometer

16. Margin of error = 24 millimeters, standard deviation = 416 millimeters

17. **Sample Size for TV Survey.** Nielsen Media Research wishes to estimate the mean number of hours that high school students spend watching TV on a weekday. A margin of error of 0.25 hour is desired. Past studies suggest that a population standard deviation of 1.7 hours is reasonable. Estimate the minimum sample size required to estimate the population mean with the stated accuracy.

18. **Sample Size for Housing Prices.** A government survey conducted to estimate the mean price of houses in a large metropolitan area is designed to have a margin of error of $10,000. Pilot studies suggest that the population standard deviation is $65,500. Estimate the minimum sample size needed to estimate the population mean with the stated accuracy.

19. **Sample Size for Mean IQ of Statistics Students.** The Wechsler IQ test is designed so that the mean is 100 and the standard deviation is 15 for the population of normal

adults. Find the sample size necessary to estimate the mean IQ score of Delaware residents. We want to be 95% confident that our sample mean is within 2 IQ points of the true mean. Assume then that $\sigma = 15$ and determine the required sample size.

20. Sample Size for Estimating Income. An economist wants to estimate mean annual income from the first year of work for college graduates who have had the profound wisdom to take a statistics course. How many such incomes must be found if she wants to be 95% confident that the sample mean is within $500 of the true population mean? Assume that a previous study has revealed that for such incomes, $\sigma = \$6{,}250$.

21. Weight of Quarters. You want to estimate the mean weight of quarters in circulation. A sample of 40 quarters has a mean weight of 5.639 grams and a standard deviation of 0.062 gram. Use a single value to estimate the mean weight of all quarters. Also, find the 95% confidence interval.

22. SAT Scores. A sample of 250 first-year students at a large university has a mean SAT score of 1722 with a standard deviation of 280. Use a single value to estimate the mean SAT mathematics score for the entire first-year class. Also, find the 95% confidence interval.

23. Time to Graduation. Data from the National Center for Education Statistics on 4,400 college graduates show that the mean time required to graduate with a bachelor's degree is 5.15 years with a standard deviation of 1.68 years. Use a single value to estimate the mean time required to graduate for all college graduates. Also, find the 95% confidence interval.

24. Garbage Production. Based on a sample of 62 households, the mean weight of discarded plastic is 1.91 pounds and the standard deviation is 1.07 pounds (data from the Garbage Project at the University of Arizona). Use a single value to estimate the mean weight of discarded plastic for all households. Also, find the 95% confidence interval.

25. Weight of Bears. The health of the bear population in Yellowstone National Park is monitored by periodic measurements taken from anesthetized bears. A sample of the weights of such bears is given below. Find a 95% confidence interval estimate of the mean of the population of all such bear weights.

80	344	416	348	166	220	262	360	204
144	332	34	140	180	105	166	204	26
120	436	125	132	90	40	220	46	154
116	182	150	65	356	316	94	86	150

26. Cotinine Levels of Smokers. When people smoke, the nicotine they absorb is converted to cotinine, which can be measured. A sample of cotinine levels of 40 smokers is listed below. Find a 95% confidence interval estimate of the mean cotinine level of all smokers.

1	0	131	173	265	210	44	277
32	3	35	112	477	289	227	103
222	149	313	491	130	234	164	198
17	253	87	121	266	290	123	167
250	245	48	86	284	1	208	173

27. Family Size. You select a random sample of $n = 31$ families in your neighborhood and find the following family sizes (number of people in the family):

2	3	6	5	4	2	3	3	1	2	3
2	3	4	5	3	1	3	3	4	7	3
2	3	2	2	3	4	1	5	2		

a. What is the mean family size for the sample?

b. What is the standard deviation for the sample?

c. What is the best estimate for the mean family size for the population of all American families?

d. What is the 95% confidence interval for the estimate?

e. Comment on the reliability of the estimate.

28. TV Sets. A random sample of $n = 31$ households is asked the number of TV sets in the household. The responses are as follows.

1	0	2	3	2	3	4	2	1	1	2
4	3	2	3	3	0	1	0	1	3	2
4	3	2	1	4	0	1	2	3		

a. What is the mean number of TVs for the sample?

b. What is the standard deviation for the sample?

c. What is the best estimate for the mean number of TVs for the population of all American households?

d. What is the 95% confidence interval for the estimate?

e. Comment on the reliability of the estimate.

Projects for the Internet and Beyond

For useful links, select "Links for Internet Projects" for Chapter 8 at www.aw.com/bbt.

29. Car Ages. Assume that you want to estimate the mean age of cars driven by students at your college. A previous study shows that the standard deviation of those ages is approximately 3.7 years. How many car ages must you randomly select in order to be 95% confident that your sample mean is within 1 year of the population mean? Using that sample size, collect your own sample data, consisting of the ages of cars driven by students at your college. Then use the methods of this section to construct a 95% confidence interval. Write a statement summarizing your results.

30. Polling Organizations. Three leading public polling organizations are the Gallup Organization, Harris Poll, and Yankelovich Partners. Visit their Web sites. Describe the history of each organization and the polling services it provides. Which organization has the best description on its Web site of its polling methods?

31. Network Polls. All of the major television networks conduct regular polls on a variety of issues. Visit the Web site of at least one major network and gather the results of a particular poll that involves estimation of a population mean. Be sure to include all information that is given about the sample size, margin of error, and confidence intervals.

32. International Corruption. Transparency International uses surveys to determine a Bribe Payer's Index and Corruption Perception Index that measure the degree of corruption in many countries worldwide. Visit the Transparency International Web site and review the extensive documentation describing the methods used by this organization. Discuss the results and their validity.

IN THE NEWS

33. Estimating Population Means. Find a news article or report in which a population mean is estimated from a sample. The article should include a margin of error and/or a confidence interval. Discuss the methods used in the study and how the conclusions were reached.

8.3 Estimating Population Proportions

In this section, we turn our attention to estimates of population *proportions*. Many well-known polls and surveys rely on the techniques that we will discuss. For example, the Nielsen ratings estimate the proportion of the population tuned in to certain radio and television shows, the monthly unemployment figures released by the Bureau of Labor Statistics are estimates of the proportion of Americans who are unemployed, and the opinion polls that dominate American politics estimate the proportion of the population that supports a particular candidate or measure.

The Basics of Estimating a Population Proportion

The Bureau of Labor Statistics estimates the unemployment rate from a monthly survey of 60,000 households (see Example 2 in Section 1.1). The unemployment rate for this sample is the *sample proportion* (the proportion of people in the sample who are unemployed), denoted $\hat{p}$. The sample proportion is the best estimate for the *population proportion* (the proportion of people in the population who are unemployed), denoted p.

Just as with population means, an estimate of the population proportion can be better understood if we say something about its accuracy. Again, we use margins of error and confidence intervals. The only change from estimating population means is in the definition of the margin of error, which is given in the following box. Figure 8.13 shows how we interpret the confidence interval.

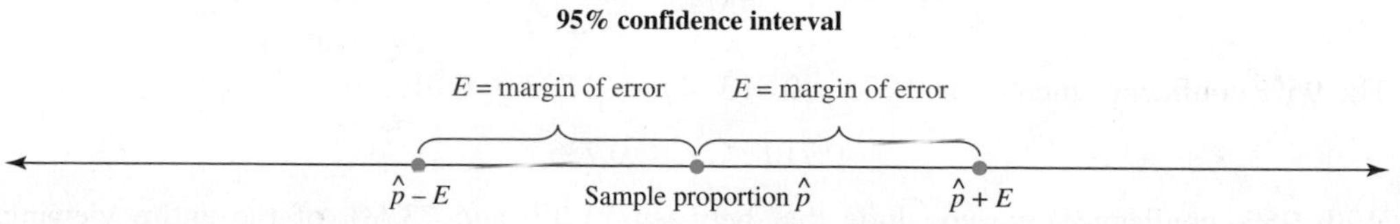

Figure 8.13 The confidence interval extends a distance equal to the margin of error on either side of the sample proportion, $\hat{p}$.

TECHNICAL NOTE

The precise formula for the margin of error uses 1.96 rather than 2. The formula can be extended to other confidence levels. For a 90% confidence level, replace the 2 in the formula by 1.645. For a 99% confidence level, replace the 2 in the formula by 2.575. The margin of error formula given here requires $np \geq 5$ and $n(1 - p) \geq 5$, conditions that are usually met easily in practice. (Techniques covered in more advanced texts work for small sample sizes.)

By the Way ...

Here's how the Bureau of Labor Statistics describes uncertainty in its unemployment survey: "A sample is not a total count and the survey may not produce the same results that would be obtained from interviewing the entire population. But the chances are 90 out of 100 that the monthly estimate of unemployment from the sample is within 230,000 of the figure obtainable from a total census. Since monthly unemployment totals have ranged between about 6 and 11 million in recent years, the possible error resulting from sampling is not large enough to distort the total unemployment picture."

95% Confidence Interval for a Population Proportion

For a population proportion, the **margin of error** for the 95% confidence interval is

$$E \approx 2\sqrt{\frac{\hat{p}(1-\hat{p})}{n}}$$

where $\hat{p}$ is the sample proportion. The **95% confidence interval** ranges

from $\hat{p}$ − margin of error to $\hat{p}$ + margin of error

We can write this confidence interval more formally as

$$\hat{p} - E < p < \hat{p} + E$$

EXAMPLE 1 Unemployment Rate

The Bureau of Labor Statistics finds 2,160 unemployed people in a sample of $n = 60{,}000$ people. Estimate the population unemployment rate and give a 95% confidence interval.

Solution The sample proportion is the unemployment rate for the sample:

$$\hat{p} = \frac{2{,}160}{60{,}000} = 0.036$$

This is the best estimate for the population unemployment rate. The margin of error is

$$E \approx 2\sqrt{\frac{\hat{p}(1-\hat{p})}{n}} = 2\sqrt{\frac{0.036(1-0.036)}{60{,}000}} = 0.0015$$

(The approximation is valid because of the large sample size.) The 95% confidence interval ranges from $0.0360 - 0.0015 = 0.0345$ to $0.0360 + 0.0015 = 0.0375$, or

$$0.0345 < p < 0.0375$$

We can have 95% confidence that the interval from 3.45% to 3.75% contains the true unemployment rate for the population. We interpret this result as follows: If we computed confidence intervals for many samples of size $n = 60{,}000$, we should expect 95% of the confidence intervals to contain the true population proportion.

EXAMPLE 2 TV Nielsen Ratings

The Nielsen ratings for television use a random sample of households. A Nielsen survey results in an estimate that a women's World Cup soccer game had 72.3% of the entire viewing audience. Assuming that the sample consists of $n = 5{,}000$ randomly selected households, find the margin of error and the 95% confidence interval for this estimate.

Solution The sample proportion, $\hat{p} = 72.3\% = 0.723$, is the best estimate of the population proportion. The margin of error is

$$E \approx 2\sqrt{\frac{\hat{p}(1-\hat{p})}{n}} = 2\sqrt{\frac{0.723(1-0.723)}{5{,}000}} = 0.013$$

The 95% confidence interval is $0.723 - 0.013 < p < 0.723 + 0.013$, or

$$0.710 < p < 0.736$$

With 95% confidence, we conclude that between 71.0% and 73.6% of the entire viewing audience watched the women's World Cup soccer game.

EXAMPLE 3 Gallup Polls

Since 1935, the Gallup Organization has been a leader in the measurement and analysis of people's attitudes, opinions, and behavior. Although Gallup is best known for the Gallup Poll, the company also provides marketing and management research for large corporations. In a recent survey, 1,016 randomly selected adults were asked:

> *As you may know, former major league player Pete Rose is ineligible for baseball's Hall of Fame due to charges that he had gambled on baseball games. Do you think he should or should not be eligible for admission to the Hall of Fame?*

Among those surveyed, 59% believed that Pete Rose should be eligible. A smaller sample of 628 adults identified themselves as "baseball fans." Among the baseball fans, 62% believed that he should be eligible for admission to the Hall of Fame. Find the margin of error and confidence interval for each sample. The survey of baseball fans cited a margin of error of "no more than 5 percentage points." Is this claim consistent with the sample size?

Solution A sample with $n = 1{,}016$ respondents and a sample proportion of $\hat{p} = 0.59$ has a margin of error of

$$E \approx 2\sqrt{\frac{\hat{p}(1-\hat{p})}{n}} = 2\sqrt{\frac{0.59(1-0.59)}{1{,}016}} = 0.031$$

or about 3 percentage points. Thus, the 95% confidence interval ranges from $59 - 3 = 56\%$ to $59 + 3 = 62\%$.

A sample with $n = 628$ respondents and a sample proportion of $\hat{p} = 0.62$ has a margin of error of

$$E \approx 2\sqrt{\frac{\hat{p}(1-\hat{p})}{n}} = 2\sqrt{\frac{0.62(1-0.62)}{628}} = 0.039$$

or about 4 percentage points. The 95% confidence interval extends from $62 - 4 = 58\%$ to $62 + 4 = 66\%$. Note that the cited margin of error of "no more than 5 percentage points" is, in fact, an overestimate.

Choosing Sample Size

Designers of surveys and polls often specify a certain level of accuracy for the results. For example, it might be desirable to estimate a population proportion with a 95% confidence interval and a margin of error of no more than 1.5 percentage points. In such situations, it's necessary to determine how large the sample must be to guarantee this accuracy. As long as we use a 95% level of confidence, we can work with this simplified, approximate formula for the margin of error:

$$E \approx \frac{1}{\sqrt{n}}$$

This formula gives a conservative (higher than necessary) estimate for the margin of error. Solving for n yields the sample size needed to achieve a margin of error E:

$$n \approx \frac{1}{E^2}$$

Any sample size equal to or larger than this value will suffice.

TECHNICAL NOTE

You can derive the $E \approx 1/\sqrt{n}$ formula from the more precise formula given earlier for the margin of error by replacing the product $\hat{p}(1-\hat{p})$ by its maximum possible value of 0.25. This approximation overestimates the actual margin of error and is most accurate when p is near 0.5.

Choosing the Correct Sample Size

In order to estimate a population proportion with a 95% degree of confidence and a specified margin of error of E, the size of the sample should be at least

$$n = \frac{1}{E^2}$$

EXAMPLE 4 Minimum Sample Size for Survey

You plan a survey to estimate the proportion of students on your campus who carry a cell phone regularly. How many students should be in the sample if you want (with 95% confidence) a margin of error of no more than 4 percentage points?

Solution Note that 4 percentage points means a margin of error of 0.04. From the given formula, the minimum sample size is

$$n = \frac{1}{E^2} = \frac{1}{0.04^2} = 625$$

You should survey at least 625 students.

EXAMPLE 5 Yankelovich Poll

Yankelovich Partners is an international public opinion and marketing research firm. The company does regular polls for *Time* magazine and CNN News. The results of its polls can be found in the monthly publication *Yankelovich Monitor*. A recent poll concluded that 61% of all households have a computer, with a margin of error of 3.5 percentage points. Approximately what sample size must have been used in this poll?

Solution A margin of error of 3.5 percentage points (or 0.035) could be achieved with a sample size of

$$n = \frac{1}{E^2} = \frac{1}{0.035^2} = 816.3$$

Because we must round up to the next larger whole number, we conclude that approximately 817 households were surveyed.

Section 8.3 Exercises

Statistical Literacy and Critical Thinking

1. **Confidence Interval.** Based on a poll conducted by the ICR Survey Research Group, the following 95% confidence interval for the population proportion p is obtained: $0.457 < p < 0.551$. Interpret that confidence interval.

2. **Margin of Error.** In a sample of 8,411 crash landings by general aviation aircraft, the proportion of crashes in which the pilots died is 0.052. When a 95% confidence interval is constructed for the population proportion of all such fatalities, the margin of error is found to be 0.005. Identify the confidence interval.

3. **Confidence Intervals in the Media.** Here is a typical statement made by the media: "Based on the poll, 38% of all likely voters will vote for Diaz. The margin of error for this poll is 3 percentage points." What important and relevant piece of information is omitted from that statement?

4. **Convenience Sample.** The president of the Student Government at Newport University manages to poll 3,250 students. Based on the results, she constructs the 95%

confidence interval for the proportion of all U.S. college students who do binge drinking. She claims that the large size of her sample compensates for the fact that all of her subjects attend the same college. Is that claim valid? Why or why not?

Does It Make Sense? For Exercises 5–8, decide whether the statement makes sense (or is clearly true) or does not make sense (or is clearly false). Explain clearly. Not all of these statements have definitive answers, so your explanation is more important than your chosen answer.

5. **Confidence Interval.** The *Kingston Chronicle* publishes an article stating that survey data were used to develop a confidence interval of 0.45.

6. **Confidence Interval in the Media.** The *Kingston Chronicle* publishes an article stating that, based on survey results, 82% of Orange County residents oppose an increase in the sales tax, with a margin of error of 4 percentage points. A reader says that this can be expressed as the confidence interval $0.78 < p < 0.86$.

7. **Sample Size.** The 95% confidence interval of $0.200 < p < 0.400$ is based on a sample size of 500. If the sample size is increased, the confidence interval will become smaller (or narrower).

8. **Sample Size.** A reporter for the *Kingston Chronicle* claims that any good confidence interval should be based on a sample that is at least 5% of the population size.

Concepts and Applications

Margins of Error and Confidence Intervals. For Exercises 9–12, assume that population proportions are to be estimated from the samples described. In each case, find the approximate margin of error and 95% confidence interval.

9. Sample size = 100, sample proportion = 0.25

10. Sample size = 400, sample proportion = 0.40

11. Sample size = 1,068, sample proportion = 0.228

12. Sample size = 1,492, sample proportion = 0.377

Sample Size. For Exercises 13–16, estimate the minimum sample size needed to achieve the given margin of error.

13. $E = 0.01$

14. $E = 0.03$

15. $E = 0.06$

16. $E = 0.025$

17. **Nielsen Ratings.** Nielsen Media Research uses samples of 5,000 households to rank TV shows. Nielsen reported that *60 Minutes* had 15% of the TV audience. What is the 95% confidence interval for this result?

18. **Nielsen Ratings.** Repeat Exercise 17 assuming that the sample size is doubled to 10,000. Given that the large cost and effort of conducting the Nielsen survey would be doubled, does this increase in sample size appear to be justified by the increased reliability?

19. **Hazing of Athletes.** A study done by researchers at Alfred University concluded that 80% of all student athletes in this country have been subjected to some form of hazing. The study is based on responses from 1,400 athletes. What are the margin of error and 95% confidence interval for the study?

20. **Student Opinions.** An annual survey of first-year college students, conducted by the Higher Education Research Institute at UCLA, asks approximately 276,000 students about their attitudes on a variety of subjects. According to a recent survey, 51% of first-year students believe that abortion should be legal (down from 65% in 1990) and 40% believe that casual sex is acceptable (down from 50% in 1975). What are the margins of error and 95% confidence intervals for these estimates?

21. **State Habits.** The Centers for Disease Control and Prevention conducts surveys to compare the habits and attitudes of citizens in various states. For each of the following results, give the 95% confidence interval.
    a. 17.2% of those surveyed in Colorado are sedentary (the lowest percentage in the country); sample size = 1,500.
    b. 13.2% of those surveyed in Utah smoke (the lowest percentage in the country); sample size = 2,500.
    c. 22.9% of those surveyed in Wisconsin are binge drinkers; sample size = 3,500.

22. **Important Issues.** In an ABC/*Washington Post* survey, 1,526 randomly selected Americans were asked to list the most important issues in recent elections. Education (79%), the economy (74%), and managing the budget (74%) were listed as very important issues. The margin of error cited for the survey was 3 percentage points. Is the margin of error consistent with the sample size?

23. **Drugs in Movies.** A study by Stanford University researchers for the Office of National Drug Control Policy and the Department of Health and Human Services concluded that 98% of the top rental films involve drugs, drinking, or smoking. Assume that this study is based on the top 400 rental films.
    a. Use the results of this sample to estimate the proportion of all films that involve drugs, drinking, or smoking.
    b. What is the 95% confidence interval?
    c. Do you believe that the top 400 films represent a random sample? Explain.

24. Teen Pressure. A study commissioned by the U.S. Department of Education concluded that 44% of teenagers cite grades as their greatest source of pressure. The study was based on responses from 1,015 teenagers. What is the 95% confidence interval?

25. Election Predictions. In a random sample of 1,600 people from a large city, it is found that 900 support the current mayor in the upcoming election. Based on this sample, would you claim that the mayor will win a majority of the votes? Explain.

26. Election Predictions. In a random sample of 2,500 people from a large city, it is found that 1,300 support the current mayor in the upcoming election. Based on this sample, would you claim that the mayor will win a majority of the votes? Explain. What conclusion would be reached from a sample of 250 people in which 130 supported the mayor?

27. Pre-Election Polls. Prior to a statewide election for the U.S. Senate, three polls are conducted. In the first poll, 780 of 1,500 voters favor candidate Martinez. In the second poll, 1,285 of 2,500 voters favor Martinez. In the third poll, 1,802 of 3,500 voters favor Martinez. Find the 95% confidence intervals for all three polls. Discuss Martinez's prospects for victory based on these polls.

28. Unemployment Survey. The Bureau of Labor Statistics estimates the unemployment rate in the United States monthly by surveying 60,000 individuals.

a. In one month, 3.4% of the 60,000 individuals surveyed are found to be unemployed. Find the margin of error for this estimate. Is the precision (nearest tenth of a percent) reasonable? Explain.

b. Suppose that the number of individuals surveyed were increased by a factor of four (to 240,000). By how much would the margin of error change?

c. Suppose that the number of individuals surveyed were decreased by a factor of one-fourth (to 15,000). By how much would the margin of error change?

29. Opinion Poll. A poll finds that 54% of the population approves of the job that the President is doing; the poll has a margin of error of 4% (assuming a 95% degree of confidence).

a. What is the 95% confidence interval for the true population percentage that approves of the President's performance?

b. What was the size of the sample for this poll?

30. Concealed Weapons. Two-thirds (or 66.6%) of 626 Colorado residents polled by Talmey-Drake Research & Strategy Inc. said they backed a bill pending in the legislature that would standardize laws on granting concealed-weapon permits to gun owners. The bill would force local law enforcement to grant such permits to anyone who can legally carry a gun. The margin of error in the poll was reported as 4 percentage points.

a. Is the reported margin of error consistent with the sample size for this estimate?

b. What sample size would be needed to give a margin of error of 2 percentage points?

Projects for the Internet and Beyond

For useful links, select "Links for Internet Projects" for Chapter 8 at www.aw.com/bbt.

31. Who's the Vice President? Assume that you want to estimate the proportion of students at your college who can correctly identify the Vice President of the United States. How many students must you randomly select in order to be 95% confident that your sample proportion is within 0.1 of the population proportion? Using that sample size, collect your own sample data by randomly selecting and surveying students at your college. Then use the methods of this section to construct a 95% confidence interval. Write a statement summarizing your results.

32. Nielsen Methods. Visit the Nielsen Media Research Web site and report on the actual methods used to estimate population proportions and confidence intervals in Nielsen ratings.

33. Network Polls. All of the major television networks conduct regular polls on a variety of issues. Visit the Web sites of the major networks and gather the results of a particular poll that involves estimation of a population proportion. Be sure to include all information that is given about the sample size, margin of error, and confidence intervals. Include any details about the actual polling procedure.

IN THE NEWS

34. Estimating Population Proportions. Find a news article or report in which a population proportion is estimated from a sample. The article should include a margin of error and/or a confidence interval. Discuss the methods used in the study and how the conclusions were reached.

Chapter Review Exercises

1. In a clinical study of the drug Ziac, 3.2% of 221 Ziac users experienced dizziness (based on data from Lederle Laboratories).

a. Use these sample results to construct a 95% confidence interval estimate of the population proportion of Ziac users who experience dizziness.

b. Write a statement that correctly interprets the confidence interval found in part a.

c. What is the margin of error?

d. The sample proportion of 0.032 was obtained from one specific sample of 221 subjects. Suppose many different samples of 221 subjects are obtained and the proportion of those who experience dizziness is found for each sample. What do we know about the shape of the distribution of the sample proportions?

e. If we could increase the sample size so that it became much larger than 221, what would be the effect on the confidence interval limits? Would they get closer together or farther apart, or would they not change?

f. Using a different set of sample data, we find a 95% confidence interval $0.400 < p < 0.500$. What is wrong with this interpretation: "There is a 95% chance that the population proportion will fall between 0.400 and 0.500"?

2. We want to estimate the mean IQ score on the Stanford-Binet test for the population of college students. We know that for people randomly selected from the general population, the standard deviation of IQ scores on the Stanford-Binet test is 16.

a. Using a standard deviation of 16, how many college students must we randomly select for IQ tests if we want to have 95% confidence that the sample mean is within 3 IQ points of the population mean?

b. How is our estimate of the mean IQ of college students affected if the sample size is larger than necessary? Smaller than necessary?

c. Is the actual standard deviation of IQ scores for college students likely to be equal to 16, more than 16, or less than 16? Explain. If we used the actual value instead of 16 in part a, how would the answer to part a be affected? Would it be the same, smaller, or larger?

3. A sample of 50 randomly selected subjects is obtained, and the white blood cell count of each subject is measured. The mean is 7.1 and the standard deviation is 2.5.

a. Use these sample results to construct a 95% confidence interval estimate of the population mean.

b. Write a statement that correctly interprets the confidence interval found in part a.

c. What is the margin of error?

d. If the standard deviation is 2.5, how many subjects must be included if we want 95% confidence that the sample mean is in error by at most 0.25?

4. a. You have been hired by Intel to determine the proportion of computer owners who plan to upgrade to a new operating system. Assuming that you want to be 95% confident that your sample proportion is within 0.02 of the true population proportion, how many people must you survey?

b. Suppose that, in conducting the survey described in part a, you find that half of the people called refuse to answer the survey questions because they believe that you are trying to sell them something. If you proceed by calling twice as many people so that your sample size is large enough, will your results be good? Explain.

Chapter Quiz

1. If many different random samples of size 100 are selected from the population of pulse rates of adult women, what is the shape of the distribution of the sample means?
2. If many different random samples of size 500 are selected from the population of college students, what is the shape of the distribution of the proportions of women?
3. What does the notation $\hat{p}$ represent?
4. A journal article provides a confidence interval in the format 0.60 ± 0.08. Express this confidence interval in the format $a < p < b$. (That is, rewrite $a < p < b$ using specific values in place of a and b.)
5. Assume that we want to estimate the mean grip strength of adult males in the United States. If a random sample of grip strengths is obtained, which of the following is the best estimate of the population mean?
   - **a.** Median of the sample
   - **b.** Mean of the sample
   - **c.** Standard deviation of the sample
   - **d.** Range of the sample
   - **e.** The sample proportion
6. Find the margin of error corresponding to this 95% confidence interval: $0.240 < p < 0.280$.
7. Find the margin of error corresponding to this 95% confidence interval: $240 < \mu < 280$.
8. When a 95% confidence interval is constructed for the population mean, a sample mean is found to be 4.60 and the margin of error is found to be 0.15. Identify the 95% confidence interval.
9. When a 95% confidence interval is constructed for the population proportion, a sample proportion is found to be 0.600 and the margin of error is found to be 0.200. Identify the 95% confidence interval.
10. Identify what is wrong with this 95% confidence interval for the population proportion: $65.0 < p < 70.0$.

Using Technology

In describing methods for finding a confidence interval estimate of a population mean, this chapter considers only cases in which the normal distribution applies. Chapter 10 introduces confidence interval estimates of a population mean that are found by using the *t* distribution. See the procedures at the end of Chapter 10 for those other methods.

SPSS

Confidence interval for a mean: SPSS does not have a procedure for obtaining a confidence interval estimate of a population mean with a known population standard deviation. See the technology instructions at the end of Chapter 10 for a procedure that applies to situations in which the population standard deviation is not known.

Confidence interval for a proportion: SPSS does not have a procedure for obtaining a confidence interval for a population proportion.

Excel

Confidence interval estimate of a population mean: Use the Data Desk XL add-in that is a supplement to Excel.

First enter the sample data in column A. If you are using Excel 2003, click on **DDXL.** If you are using Excel 2007, click on **Add-Ins,** then click on **DDXL.** Select **Confidence Intervals.** Under the function type options, select **1 Var z Interval.** Click on the pencil icon and enter the range of data, such as A1:A12 if you have 12 values listed in column A. Click on **OK.** In the dialog box, select the confidence level. Also enter the known value of the population standard deviation. Click on **Compute Interval** and the confidence interval will be displayed.

Confidence interval estimate of a population proportion: Use the Data Desk XL add-in that is a supplement to Excel.

First enter the number of successes in cell A1, then enter the total number of trials in cell B1. If you are using Excel 2003, click on **DDXL.** If you are using Excel 2007, click on **Add-Ins** and then click on **DDXL.** Select **Confidence Intervals.** Select **Summ 1 Var Prop Interval** (which is an abbreviated form of "confidence interval for a proportion using summary data for one variable"). Click on the pencil icon for "Num successes" and enter A1. Click on the pencil icon for "Num trials" and enter B1. Click on **OK.** In the dialog box, select the level of confidence, then click on **Compute Interval.**

STATDISK

Select **Analysis** and then **Confidence Intervals.** Then select either **Mean - One Sample** or **Proportion - One Sample.** In the dialog box that appears, first enter the significance level as a decimal number. Enter 0.95 for a 95% confidence level. Proceed to enter the other required items. Then click on **Evaluate** and the confidence interval will be displayed.

FOCUS ON HISTORY

Where Did Statistics Begin?

The origins of many disciplines are lost in antiquity, but the roots of statistics can be identified with some certainty. Systematic record keeping began in London in 1532 with weekly data collection on deaths. Later in the same decade, official data collection on baptisms, deaths, and marriages began in France. In 1608, the collection of similar vital statistics began in Sweden. Canada conducted the first official census in 1666.

Of course, statistics is more than the collection of data. If there is a founder of statistics, that person must be someone who worked with the data in clever and systematic ways and who used the data to reach conclusions that were not previously evident. Many experts believe that an Englishman named John Graunt deserves the title of the founder of statistics.

John Graunt was born in London in 1620. As the eldest child in a large family, he took up his father's business as a draper (a dealer in clothing and dry goods). He spent most of his life as a prominent London citizen, until he lost his house and possessions in the Fire of London in 1666. Eight years later, he died in poverty.

It's not clear how John Graunt became interested in the weekly records of baptisms and burials—known as bills of mortality—that had been kept in London since 1563. In the preface of his book *Natural and Political Observations on the Bills of Mortality*, he noted that others "made little other use of them" and wondered "what benefit the knowledge of the same would bring to the World." He must have worked on his statistical projects for many years before his book was first published in 1662.

Graunt worked primarily with the annual bills, which were year-end summaries of the weekly bills of mortality. Figure 8.14 shows the annual bill for 1665 (the year of the Great Plague). The top third of the bill shows the numbers of burials and baptisms (christenings) in each parish. Total burials and baptisms are noted in the middle of the bill, with deaths due to the plague recorded separately. The lower third of the bill shows deaths due to a variety of other causes, with totals given for males and females.

Graunt was aware of rough estimates of the population of London that were made periodically for taxation purposes, but he must have been skeptical of one estimate that put the population of London at 6 or 7 million in 1661. Using the annual bills, comparing burials and baptisms, and estimating the density of families in London (with an average family size of eight), he arrived at a population estimate of 460,000 by three different methods—quite a change from 6 or 7 million! He also found that the population of London was increasing while the populations of towns in the countryside were decreasing, showing an early trend toward urbanization. He raised awareness of the high rates of infant mortality. He also refuted a popular theory that plagues arrive with new kings.

Graunt's most significant contribution may have been his construction of the first life table. Although detailed data on age at death were not available, Graunt knew that of 100 new babies, "36 of them die before they be six years old, and that perhaps but one surviveth 76." With these two data points, he filled in the intervening years as shown in Table 8.5, using methods that he did not fully explain.

By the Way ...

Some historians claim that Graunt's book was actually written by his life-long friend and collaborator William Petty. The arguments are quite involved, but most statisticians believe Graunt wrote his own book. Either way, we know that Petty continued Graunt's work, publishing later editions of Graunt's book and creating the field of "political arithmetic," which we now call demography.

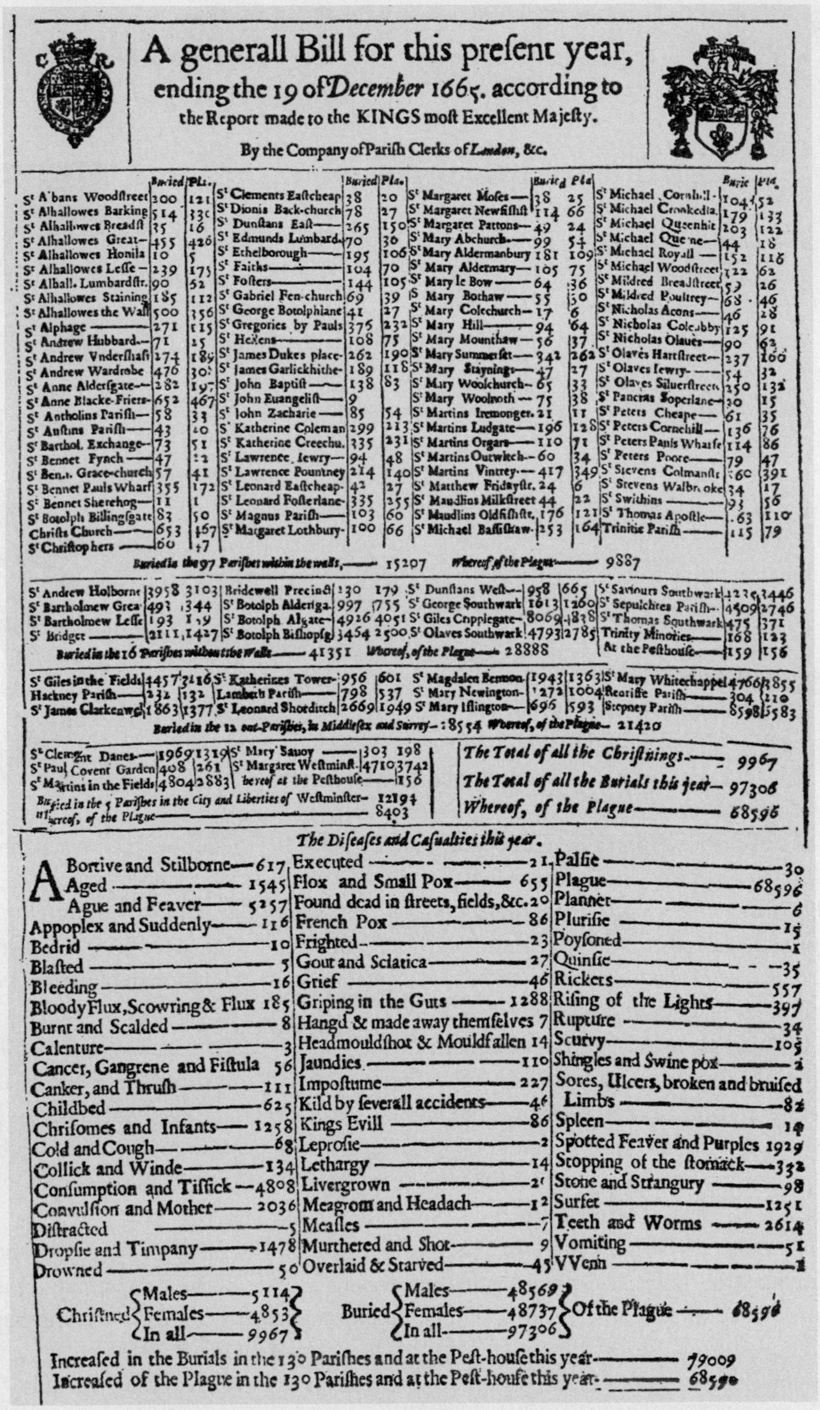

A generall Bill for this present year, ending the 19 of *December* 1665. according to the Report made to the KINGS most Excellent Majesty.

By the Company of Parish Clerks of *London*, &c.

	Buried	Pla.		Buried	Pla.		Buried	Pla.		Buried	Pla.
St Albans Woodstreet	200	121	St Clements Eastcheap	38	20	St Margaret Moses	38	25	St Michael Cornhill	104	52
St Alhallowes Barking	514	330	St Dionis Back-church	78	27	St Margaret Newfishst	114	66	St Michael Crookedla.	179	133
St Alhallowes Breadst	35	16	St Dunstans East	265	150	St Margaret Pattons	49	24	St Michael Queenhit	203	122
St Alhallowes Great	455	426	St Edmunds Lumbard.	70	36	St Mary Abchurch	99	54	St Michael Quene	44	18
St Alhallowes Honila	10	5	St Ethelborough	195	106	St Mary Aldermanbury	181	109	St Michael Royall	152	116
St Alhallowes Lesse	239	175	St Faiths	104	70	St Mary Aldermary	105	75	St Michael Woodstreet	122	62
St Alhall. Lumbardstr.	90	62	St Fosters	144	105	St Mary le Bow	64	36	St Mildred Breadstreet	59	26
St Alhallowes Staining	185	112	St Gabriel Fen-church	69	39	St Mary Bothaw	55	30	St Mildred Poultrey	68	46
St Alhallowes the Wall	500	356	St George Botolphlane	41	27	St Mary Colechurch	17	6	St Nicholas Acons	46	28
St Alphage	271	115	St Gregories by Pauls	376	232	St Mary Hill	94	64	St Nicholas Colcabby	125	91
St Andrew Hubbard	71	25	St Hellens	108	75	St Mary Mounthaw	56	37	St Nicholas Olaues	90	62
St Andrew Vndershaft	274	189	St James Dukes place	262	190	St Mary Summerset	342	262	St Olaves Hartstreet	237	160
St Andrew Wardrobe	476	308	St James Garlickhithe	189	118	St Mary Staynings	47	27	St Olaves Iewry	54	32
St Anne Aldersgate	282	197	St John Baptist	138	83	St Mary Woolchurch	65	33	St Olaves Siluerstreet	250	132
St Anne Blacke-Friers	652	467	St John Euangelist	9		St Mary Woolnoth	75	38	St Pancras Soperlane	30	15
St Antholins Parish	58	33	St Iohn Zacharie	85	54	St Martins Iremonger.	21	11	St Peters Cheape	61	35
St Austins Parish	43	20	St Katherine Coleman	299	213	St Martins Ludgate	196	128	St Peters Cornehill	136	76
St Barthol. Exchange	73	51	St Katherine Creechu.	335	231	St Martins Organs	110	71	St Peters Pauls Wharfe	114	86
St Bennet Fynch	47	22	St Lawrence Iewry	94	48	St Martins Outwitch	60	34	St Peters Poore	79	47
St Bennet Grace-church	57	41	St Lawrence Pountney	214	140	St Martins Vintrey	417	349	St Stevens Colmanstr.	560	391
St Bennet Pauls Wharf	355	172	St Leonard Eastcheap	42	27	St Matthew Fridaystr.	24	6	St Stevens Walbrooke	34	17
St Bennet Sherehog	11	1	St Leonard Fosterlane	335	255	St Maudlins Milkstreet	44	22	St Swithins	93	56
St Botolph Billingsgate	83	50	St Magnus Parish	103	60	St Maudlins Oldfishstr.	176	121	St Thomas Apostle	163	110
Christs Church	653	467	St Margaret Lothbury	100	66	St Michael Bassishaw	253	164	Trinitie Parish	115	79
St Christophers	60	47									

Buried in the 97 Parishes within the walls, — 15207 *Whereof, of the Plague* — 9887

St Andrew Holborne	3958	3103	Bridewell Precinct	230	179	St Dunstans West	958	665	St Saviours Southwark	4235	3446
St Bartholmew Great	493	344	St Botolph Aldersga.	997	755	St George Southwark	1613	1260	St Sepulchres Parish	4509	2746
St Bartholmew Lesse	193	139	St Botolph Algate	4926	4051	St Giles Cripplegate	8069	4838	St Thomas Southwark	475	371
St Bridget	2111	1427	St Botolph Bishopsg.	3454	2500	St Olaves Southwark	4793	2785	Trinity Minories	168	123
									At the Pesthouse	159	156

Buried in the 16 Parishes without the Walls — 41351 *Whereof, of the Plague* — 28888

St Giles in the Fields	4457	3216	St Katherines Tower	956	601	St Magdalen Bermon.	1943	1363	St Mary Whitechappel	4766	3855
Hackney Parish	232	132	Lambeth Parish	798	537	St Mary Newington	1272	1004	Rediriffe Parish	304	210
St James Clarkenwel	1863	1377	St Leonard Shorditch	2669	1949	St Mary Islington	696	593	Stepney Parish	8598	6583

Buried in the 12 out-Parishes, in Middlesex and Surrey — 28554 *Whereof, of the Plague* — 21420

St Clement Danes	1969	1319	St Mary Sauoy	303	198
St Paul Covent Garden	408	261	St Margaret Westminst.	4710	3742
St Martins in the Fields	4804	2883	whereof at the Pesthouse		156

Buried in the 5 Parishes in the City and Liberties of Westminster — 12194

Whereof, of the Plague — 8403

The Total of all the Christnings — 9967

The Total of all the Burials this year — 97306

Whereof, of the Plague — 68596

The Diseases and Casualties this year.

Abortive and Stilborne	617	Executed	21	Palsie	30
Aged	1545	Flox and Small Pox	655	Plague	68596
Ague and Feaver	5257	Found dead in streets, fields, &c.	20	Planner	6
Appoplex and Suddenly	116	French Pox	86	Plurisie	15
Bedrid	10	Frighted	23	Poysoned	1
Blasted	5	Gout and Sciatica	27	Quinsie	35
Bleeding	16	Grief	46	Rickets	557
Bloody Flux, Scowring & Flux	185	Griping in the Guts	1288	Rising of the Lights	397
Burnt and Scalded	8	Hang'd & made away themselves	7	Rupture	34
Calenture	3	Headmouldshot & Mouldfallen	14	Scurvy	105
Cancer, Gangrene and Fistula	56	Jaundies	110	Shingles and Swine pox	2
Canker, and Thrush	111	Impostume	227	Sores, Ulcers, broken and bruised Limbs	82
Childbed	625	Kild by severall accidents	46	Spleen	14
Chrisomes and Infants	1258	Kings Evill	86	Spotted Feaver and Purples	1929
Cold and Cough	68	Leprosie	2	Stopping of the stomack	332
Collick and Winde	134	Lethargy	14	Stone and Strangury	98
Consumption and Tissick	4808	Livergrown	20	Surfet	1251
Convulsion and Mother	2036	Meagrom and Headach	12	Teeth and Worms	2614
Distracted	5	Measles	7	Vomiting	51
Dropsie and Timpany	1478	Murthered and Shot	9	Wenn	1
Drowned	50	Overlaid & Starved	45		

Christned { Males — 5114, Females — 4853, In all — 9967 }
Buried { Males — 48569, Females — 48737, In all — 97306 } Of the Plague — 68596

Increased in the Burials in the 130 Parishes and at the Pest-house this year — 79009

Increased of the Plague in the 130 Parishes and at the Pest-house this year — 68590

Figure 8.14 Reproduction of 1665 annual bill of mortality. *Source:* From the Wellcome Historical Medical Library, reprinted in *Journal of the Royal Statistical Society,* Vol. 126, part 4, 1963, pp. 537–557.

With estimates of deaths for various ages, he was able to make the companion table of survivors shown in Table 8.6. While some modern statisticians have doubted the methods used to construct these tables, Graunt appears to have appreciated the value of these tables and anticipated actuarial tables used by life insurance companies of the future. It wasn't until 1693 that Edmund Halley, of comet fame, constructed life tables using age-based mortality rates.

Table 8.5 Graunt's Life Table: Number of Deaths by Age for 100 People

Age	Deaths
Within the first six years	36
The next ten years	24
The second decade	15
The third decade	9
The fourth	6
The next	4
The next	3
The next	2
The next	1

Table 8.6 Graunt's Table of Survivors at Various Ages

Age	Survivors
At sixteen years end	40
At twenty six	25
At thirty six	16
At forty six	10
At fifty six	6
At sixty six	3
At seventy six	1
At eighty	0

QUESTIONS FOR DISCUSSION

1. Is it surprising that accurate estimates of the population of London were not available in 1660? How would you have suggested making such estimates at that time?
2. Do you think that records of burials and baptisms would have given accurate counts of actual births and deaths? Why or why not?
3. Assuming that the estimates of deaths in Table 8.5 are accurate, are the estimates of survivors in Table 8.6 consistent? Note that the numbers in Table 8.6 do not total to 100; should they? Explain.
4. Based on Graunt's tables, estimate the average life expectancy in 1660. Explain your reasoning.

SUGGESTED READING

Johnson, N., and Kotz, S. (Eds.). *Leading Personalities in the Statistical Sciences*. John Wiley and Sons, 1997.

Sutherland, I. "John Graunt: A Tercentenary Tribute." *Journal of the Royal Statistical Society*, Vol. 126, Pt. 4, 1963, pp. 537–557.

FOCUS ON LITERATURE

How Many Words Did Shakespeare Know?

Imagine that you go to an orientation party for international students. During the course of the evening, you meet 12 Swedish people, 9 Chinese people, 6 French people, 4 Israelis, 3 Koreans, and 1 Iranian. You know there are people of other nationalities whom you did *not* meet at the party. Based only on the people you met, is it possible to estimate the *total* number of nationalities represented at the party—including those of the people you did not meet? Thanks to ideas of sampling, the answer is yes.

The party problem may be a bit frivolous. However, essentially the same question arose when Oxford University marine biologist Charles Paxton wondered how many "sea monsters" (creatures more than 2 meters in length) remain to be discovered. In this case, the nationalities at the party correspond to species of sea monsters. Using statistical methods, Paxton was able to estimate that, in addition to the roughly 220 sea monsters already known, another 47 wait to be discovered.

Similar methods have been used to analyze the works of Shakespeare. Statisticians Bradley Efron and Ronald Thisted wondered about the number of words Shakespeare actually knew, which must have been larger than the number he used in his writings. Here, the nationalities at the party correspond to different words in Shakespeare's plays and poems. The data collected at the party can be regarded as the *first sample*. For the Shakespeare question, the first sample consists of the complete known works of Shakespeare—specifically, the numbers of words that are used in these works once, twice, three times, and so forth. Table 8.7 shows a (small) part of the first sample. For example, the table says that in the works of Shakespeare, 14,376 words were used exactly once, 4,343 words were used exactly twice, and so forth. (The full table is much larger and continues far beyond 10 occurrences.)

Given the full table for the first sample, we can now ask a hypothetical question. Suppose a second, new and different sample of Shakespeare's works of the same size as the first sample were discovered. How many words could we expect to find in the second sample that were *not* used in the first sample? We would expect there to be fewer new words in the second sample, because in the first sample *every* first occurrence of a word is new, even a common word like "the"; in the second sample, those common words are no longer new. Efron and Thisted estimated that 11,430 words would appear in the second sample that did not appear in the first sample.

They repeated this argument with a third sample, fourth sample, fifth sample, and so on. With each new sample, the number of new words decreases, but the total number of words used (among all samples) increases. Efron and Thisted eventually found that the number of new words approached about 35,000. This means that in addition to the 31,534 words that Shakespeare knew and used, there were approximately 35,000 words that he knew but didn't use. Thus, they estimated that Shakespeare knew approximately 66,500 words.

The analysis of Efron and Thisted, which required advanced methods, was done in 1976. More than ten years later, the ideas were put to practical use when a new sonnet by an unknown author of Shakespeare's time period was discovered. As before, the complete volume of Shakespeare's works was considered the first sample. Now, the second sample was the new sonnet with 429 words. The same statistical method was used to predict that, if Shakespeare was the author, the new sonnet should have seven new words that did not appear in the complete works of Shakespeare. In fact, the new sonnet had nine new words that did not previously appear. Similarly, the method predicted that the new sonnet should have four words that were used exactly once in the complete works. In fact, there were seven words that were used exactly once in the complete works. And the number of words in the sonnet that were used exactly twice in the complete works was predicted to be three, when in fact there were five such words. The authors concluded that the agreement between the predictions and the actual word frequencies in the sonnet was good enough to attribute authorship to Shakespeare.

These statistical methods have been used to distinguish the works of Shakespeare from those of other Elizabethan writers such as Marlowe, Donne, and Jonson. They have been used to strengthen the belief that James Madison wrote certain of the Federalist Papers whose authorship was in doubt. They have even been used to establish the order of Plato's works.

This example illustrates the remarkable applicability of statistics to problems in a seemingly unrelated discipline. Perhaps more important, it demonstrates how two disciplines as distant from each other as literature and marine biology can be related by a common statistical thread.

Table 8.7 Numbers of Words in Complete Works of Shakespeare Used from One to Ten Times

Occurrences	Number of words
1	14,376
2	4,343
3	2,292
4	1,463
5	1,043
6	837
7	638
8	519
9	430
10	364

QUESTIONS FOR DISCUSSION

1. Comment on whether you believe that literature is enhanced by statistical analysis. For example, is your appreciation of Shakespeare improved by knowing how many words Shakespeare knew?
2. Do you believe that statistical analysis is useful for identifying authors of "lost works"?
3. Suggest another discipline, besides biology and literature, in which the ideas described in this Focus could be used to estimate an unknown quantity.

SUGGESTED READING

Efron, B., and Thisted, R. "Estimating the Number of Unknown Species: How Many Words Did Shakespeare Know?" *Biometrika*, Vol. 63, No. 3, 1976, pp. 435–437.

Morin, Richard. "Unconventional Wisdom: Statistics, the King's Deer, and Monsters of the Sea." *Washington Post*, March 7, 1999.

Snell, Laurie. Chance News 8.03.

9

It is the mark of an educated mind to be able to entertain a thought without accepting it.

—Aristotle

Hypothesis Testing

EVERYONE MAKES CLAIMS. ADVERTISERS MAKE CLAIMS about their products. Universities claim their programs are superb. Governments claim their programs are effective. Lawyers make claims about a suspect's guilt or innocence. Medical diagnoses are claims about the presence or absence of disease. Pharmaceutical companies make claims about the effectiveness of their drugs. But how do we know whether any of these claims are true? Statistics offers a way to test many claims, through a powerful set of techniques that go by the name of *hypothesis testing*.

LEARNING GOALS

9.1 Fundamentals of Hypothesis Testing
Understand the goal of hypothesis testing and the basic structure of a hypothesis test, including how to set up the null and alternative hypotheses, how to determine the possible outcomes of a hypothesis test, and how to decide between these possible outcomes.

9.2 Hypothesis Tests for Population Means
Understand and interpret one- and two-tailed hypothesis tests for claims made about population means, and learn to recognize and avoid common errors (type I and type II errors) in hypothesis tests.

9.3 Hypothesis Tests for Population Proportions
Understand and interpret hypothesis tests for claims made about population proportions.

9.1 Fundamentals of Hypothesis Testing

A company called ProCare Industries, Ltd. once claimed that its product, called Gender Choice, could increase a woman's chance of giving birth to a baby girl. The company claimed that the chance of a baby girl could be increased "up to 80%," but let's focus simply on the claim that the chance of having a baby girl is greater than the approximately 50%, or 0.5, expected under ordinary conditions. How could we test whether the Gender Choice claim is true?

One way would be to study a random sample of, say, 100 babies born to women who used the Gender Choice product. If the product does not work, we would expect about half of these babies to be girls. If it does work, we would expect significantly more than half of the babies to be girls. The key question, then, is what constitutes "significantly more." If there were 97 girls among the 100 births in the sample, we would all agree that this constituted significantly more than half and that the product probably worked. If there were only 52 girls among the 100 births, we'd probably agree that 52 was so close to half that we had no reason to think the product had any effect. But would we consider, say, 64 girls in the sample of 100 babies to be "significantly more" than half, and therefore think that the product might really work?

In statistics, we answer such questions through *hypothesis testing*. Before we discuss the specific terminology and procedures of hypothesis testing, let's look at the procedure we would use to draw a conclusion about Gender Choice based on a sample in which there are 64 girls among 100 babies born to women who used the product.

- We begin by assuming that Gender Choice does *not* work; that is, it does *not* increase the percentage of girls. If this is true, then we should expect about 50% girls among the *population* of all births to women using the product. That is, the proportion of girls is equal to 0.50.
- We now use our *sample* (with 64 girls among 100 births) to test the above assumption. We conduct this test by calculating the likelihood of drawing a random sample of 100 births in which 64% (or more) of the babies are girls (from a population in which the overall proportion of girls is only 50%).
- If we find that a random sample of births is fairly likely to have 64% (or more) girls, then we do not have evidence that Gender Choice works. However, if we find that a random sample of births is unlikely to have at least 64% girls, then we conclude that the result in the Gender Choice sample is probably due to something other than chance—meaning that the product may be effective. Note, however, that even in this case we will not have *proved* that the product is effective, as there could still be other explanations for the result (such as that we happened to choose an unusual sample or that there were confounding variables we did not account for).

By the Way ...

The Gender Choice product consisted of printed instructions and a thermometer. It was based on the principle that the gender of a baby can be determined through careful timing of conception. There was no evidence that the product actually worked, and it is no longer available. However, other products now on the market do indeed seem to have some effect, and they are therefore raising ethical questions about whether or when gender selection should be an option for parents.

TIME OUT TO THINK

Suppose Gender Choice is effective and you select a new random sample of 1,000 babies born to women who used the product. Based on the first sample (with 64 girls among 100 births), how many baby girls would you expect in the new sample? Next, suppose Gender Choice is *not* effective; how many baby girls would you expect in a random sample of 1,000 babies born to women who used the product?

Formulating the Hypotheses

The key remaining issues in our test of the effectiveness of Gender Choice are that we have not yet described how to calculate the likelihood of drawing a random sample like the one with 64% girls nor have we defined exactly what we mean when we ask whether such a sample is "fairly likely" or "unlikely" to be drawn at random. Most of this chapter is devoted to the formal procedures used to resolve these issues. The first step in resolving them is to define exactly what we are testing.

A **hypothesis** is a claim about a population parameter. For example, in the Gender Choice case, the population parameter is the proportion of girls born to all women who use the product. A **hypothesis test**, then, is a test of whether a particular claim is supported or unsupported by the available evidence.

Definitions

A **hypothesis** is a claim about a population parameter, such as a population proportion (p) or population mean (μ).

A **hypothesis test** is a standard procedure for testing a claim about a population parameter.

There are always at least two hypotheses in any hypothesis test. In the Gender Choice case, the two hypotheses are essentially (1) that Gender Choice works in raising the population proportion of girls from the 50% that we would normally expect and (2) that Gender Choice does *not* work in raising the population proportion of girls. As we've already discussed, the starting point for the hypothesis test is actually the second of those hypotheses—that it does *not* raise the proportion of baby girls. We call this starting point the **null hypothesis**, or H_0 (read "H-naught"). The other hypothesis (1) is called the **alternative hypothesis**, or H_a (read "H-a"). To summarize, for the Gender Choice example:

- The *null hypothesis* is the claim that Gender Choice does *not* work, in which case the population proportion of girls born to women who use the product should be 50%, or 0.50. Using p to represent the population proportion, we write this null hypothesis as

$$H_0 \text{ (null hypothesis): } p = 0.50$$

- The *alternative hypothesis* is the claim that Gender Choice *does* work, in which case the population proportion of girls born to women who use the product should be *greater than* 0.50. We write this alternative hypothesis as

$$H_a \text{ (alternative hypothesis): } p > 0.50$$

By the Way ...

The word *null* comes from the Latin *nullus*, meaning "nothing." A null hypothesis often states that there is no special effect or difference.

In this chapter, the null hypothesis will always include the condition of equality, just as the null hypothesis for the Gender Choice example is the equality $p = 0.50$. (We will see a few examples of other types of null hypotheses in Chapter 10.) However, the alternative hypothesis for the Gender Choice case ($p > 0.50$) is only one of three general forms of alternative hypotheses that we will encounter in this chapter:

$$\text{population parameter} < \text{claimed value}$$
$$\text{population parameter} > \text{claimed value}$$
$$\text{population parameter} \neq \text{claimed value}$$

As we will see in Section 9.2, these three different types of alternative hypotheses require slightly different calculations in the hypothesis test; for that reason, it is useful to give them names. The first form ("less than") leads to what is called a **left-tailed** hypothesis test, because it requires testing whether the population parameter lies to the *left* (lower values) of the claimed value. Similarly, the second form ("greater than") leads to a **right-tailed** hypothesis test, because it requires testing whether the population parameter lies to the *right* (higher values) of the claimed value. The third form ("not equal to") leads to a **two-tailed** hypothesis test, because it requires testing whether the population parameter lies significantly far to *either* side of the claimed value.

Null and Alternative Hypotheses

The **null hypothesis**, or H_0, is the starting assumption for a hypothesis test. For the types of hypothesis tests in this chapter, the null hypothesis always claims a specific value for a population parameter and therefore takes the form of an equality:

$$H_0 \text{ (null hypothesis): population parameter} = \text{claimed value}$$

The **alternative hypothesis**, or H_a, is a claim that the population parameter has a value that differs from the value claimed in the null hypothesis. It may take one of the following forms:

$$\text{(left-tailed) } H_a\text{: population parameter} < \text{claimed value}$$
$$\text{(right-tailed) } H_a\text{: population parameter} > \text{claimed value}$$
$$\text{(two-tailed) } H_a\text{: population parameter} \neq \text{claimed value}$$

EXAMPLE 1 Identifying Hypotheses

In each case, identify the population parameter about which a claim is made, state the null and alternative hypotheses for a hypothesis test, and indicate whether the hypothesis test will be left-tailed, right-tailed, or two-tailed.

a. The manufacturer of a new model of hybrid car advertises that the mean fuel consumption is equal to 62 miles per gallon on the highway. A consumer group claims that the mean is less than 62 miles per gallon.

b. The Ohio Department of Health claims that the average stay in Ohio hospitals after childbirth is greater than the national mean of 2.0 days.

c. A wildlife biologist working in the African savanna claims that the actual proportion of female zebras in the region is different from the accepted proportion of 50%.

Solution

a. The population parameter about which a claim is made is a population *mean* (μ)—the mean gas mileage of all new hybrid cars of this model. The null hypothesis must state that this population mean is *equal* to some specific value. We therefore recognize the null hypothesis as the advertised claim that the mean mileage for the cars is 62 miles per gallon. The alternative hypothesis is the consumer group's claim that the true gas mileage is *less* than advertised. To summarize:

$$H_0 \text{ (null hypothesis): } \mu = 62 \text{ miles per gallon}$$
$$H_a \text{ (alternative hypothesis): } \mu < 62 \text{ miles per gallon}$$

Because the alternative hypothesis has a "less than" form, the hypothesis test will be left-tailed.

b. The population is all women in Ohio who have recently given birth, and the population parameter is their mean (μ) stay after childbirth in Ohio hospitals. The null hypothesis must be an equality, so in this case it is the claim that the mean hospital stay equals the national average of 2.0 days. The alternative hypothesis is the Health Department's claim that the mean stay in Ohio is *greater than* this national average. To summarize:

$$H_0 \text{ (null hypothesis): } \mu = 2.0 \text{ days}$$
$$H_a \text{ (alternative hypothesis): } \mu > 2.0 \text{ days}$$

Because the alternative hypothesis has a "greater than" form, the hypothesis test will be right-tailed.

c. In this case, the claim is about a population *proportion* (p)— the proportion of female zebras among the zebra population of the region. The accepted population proportion is $p = 0.5$, which becomes the null hypothesis. The wildlife biologist claims that the actual population proportion is different from the value in the null hypothesis. Because "different" can be either greater or less than, the alternative hypothesis has a "not equal to" form:

$$H_0 \text{ (null hypothesis): } p = 0.5$$
$$H_a \text{ (alternative hypothesis): } p \neq 0.5$$

The "not equal to" form of this alternative hypothesis leads to a two-tailed test.

Possible Outcomes of a Hypothesis Test

A hypothesis test always begins with the assumption that the null hypothesis is true. We then test to see whether the data give us any reason to think otherwise. As a result, there are generally only two possible outcomes to a hypothesis test, summarized in the following box.

Two Possible Outcomes of a Hypothesis Test

There are two possible outcomes to a hypothesis test:

1. *Reject* the null hypothesis, H_0, in which case we have evidence in support of the alternative hypothesis.
2. *Not reject* the null hypothesis, H_0, in which case we do not have enough evidence to support the alternative hypothesis.

Notice that "accepting the null hypothesis" is *not* a possible outcome, because the null hypothesis is always the starting assumption. The hypothesis test may not give us reason to reject this starting assumption, but it cannot by itself give us reason to conclude that the starting assumption is true.

The fact that there are only two possible outcomes makes it extremely important that the null and alternative hypotheses be chosen in an unbiased way. In particular, both hypotheses must always be formulated *before* a sample is drawn from the population for testing. Otherwise, data from the sample might inappropriately bias the selection of the hypotheses to test.

By the Way ...

The first recorded example of hypothesis testing is attributed to the Scotsman John Arbuthnot (1667–1735). Using 82 years of data, he noticed that the annual number of male baptisms was consistently higher than the annual number of female baptisms. Knowing that there was no bias in baptisms based on gender, he argued that such a regular pattern could not be explained by chance and had to be due to "divine providence," meaning boys must represent slightly more than 50% of all births. In modern terminology, he rejected the null hypothesis that the data could be explained by chance alone.

TIME OUT TO THINK

The idea that a hypothesis test cannot lead us to accept the null hypothesis is an example of the old dictum that "absence of evidence is not evidence of absence." As an illustration of this idea, explain why it would be easy in principle to prove that some legendary animal (such as Bigfoot or the Loch Ness Monster) exists, but it is nearly impossible to prove that it does *not* exist.

EXAMPLE 2 Hypothesis Test Outcomes

For each of the three cases from Example 1, describe the possible outcomes of a hypothesis test and how we would interpret these outcomes.

Solution

a. Recall that the null hypothesis is the advertised claim that the mean mileage for the new cars is $\mu = 62$ miles per gallon. The alternative hypothesis is the consumer group's claim that the true gas mileage is *less* than advertised, or $\mu < 62$ miles per gallon. The possible outcomes are

- Reject the null hypothesis of $\mu = 62$ miles per gallon, in which case we have evidence in support of the consumer group's claim that the mileage is less than advertised.

- Do not reject the null hypothesis, in which case we lack evidence to support the consumer group's claim. Note, however, that this option does not imply that the advertised claim is true.

b. The null hypothesis is that the mean hospital stay is the national average of 2.0 days. The alternative hypothesis is the Health Department's claim that the mean stay in Ohio is *greater than* the national average. The possible outcomes are
- Reject the null hypothesis of $\mu = 2.0$ days, in which case we have evidence in support of the Health Department's claim that the mean stay in Ohio is greater than the national average.
- Do not reject the null hypothesis, in which case we lack evidence to support the Health Department's claim. Note, however, that this option does not imply that the Ohio average stay is actually equal to the national average stay of 2.0 days.

c. The null hypothesis is that the proportion of female zebras is the accepted population proportion of 50% ($p = 0.5$). The alternative hypothesis is the biologist's claim that the accepted value is wrong, meaning that the actual proportion of female zebras is not 50% (it may be either more or less than 50%). The possible outcomes are
- Reject the null hypothesis of $p = 0.5$, in which case we have evidence in support of the biologist's claim that the accepted value is wrong.
- Do not reject the null hypothesis, in which case we lack evidence to support the biologist's claim. Note, however, that this does not imply that the accepted value is correct.

Drawing a Conclusion from a Hypothesis Test

Let's return to the Gender Choice example in which we imagine drawing a random sample of 100 babies (born to women using the Gender Choice product) and find that 64 of these babies are girls. How do we decide whether this sample result should lead us to reject or not reject the null hypothesis? The answer comes down to deciding whether the sample result was likely or unlikely to have occurred by chance *if* the null hypothesis is true.

Remember that the null hypothesis for this case is that the true proportion of baby girls among the population of Gender Choice users is 50%, or $p = 0.50$. Using the notation introduced in Chapter 8, the sample we are studying has a sample size of $n = 100$ and a sample proportion of $\hat{p} = 0.64$. The precise question, then, is this: If the true population proportion is $p = 0.50$ (as the null hypothesis claims), what is the probability that by chance alone a sample of size $n = 100$ would have a sample proportion of at least $\hat{p} = 0.64$? If the probability is low, then it is highly unlikely that we would have found such a sample by chance; we therefore have reason to reject the null hypothesis. If the probability of observing the sample result is moderate or high, then it is fairly likely that we could have found such a result by chance, so we cannot reject the null hypothesis.

There are multiple ways to make the decision about rejecting or not rejecting the null hypothesis. Here, we'll look at two closely related options: making the decision based on the statistical significance of the result and making the decision based on the actual probability, or "*P*-value," of the test result.

Probable truth relies on statistical arguments to weigh which of several possibilities is more likely to be true. When something must be proven beyond a reasonable doubt, just what does that mean? What level of doubt is acceptable? One in twenty? One in a trillion?

—K. C. Cole

Statistical Significance

We introduced the idea of statistical significance in Section 6.1. Recall that if the probability of a particular result is 0.05 or less, we say that the result is statistically significant at the 0.05 level; if the probability is 0.01 or less, the result is statistically significant at the 0.01 level. The 0.01 level therefore means stronger significance than the 0.05 level. The following box summarizes how we can apply these ideas directly to hypothesis tests.

Hypothesis Test Decisions Based on Levels of Statistical Significance

We decide the outcome of a hypothesis test by comparing the actual sample result (mean or proportion) to the result expected *if* the null hypothesis is true. We must choose a significance level for the decision.

- If the chance of a sample result at least as extreme as the observed result is less than 1 in 100 (or 0.01), then the test is statistically significant at the 0.01 level and offers strong evidence for rejecting the null hypothesis.
- If the chance of a sample result at least as extreme as the observed result is less than 1 in 20 (or 0.05), then the test is statistically significant at the 0.05 level and offers moderate evidence for rejecting the null hypothesis.
- If the chance of a sample result at least as extreme as the observed result is greater than the chosen level of significance (0.05 or 0.01), then we do not reject the null hypothesis.

EXAMPLE 3 Statistical Significance in Hypothesis Testing

Consider the Gender Choice example in which the sample size is $n = 100$ and the sample proportion is $\hat{p} = 0.64$. Using techniques that we will discuss later in this chapter, it is possible to calculate the probability of randomly choosing such a sample (or a more extreme sample with $\hat{p} > 0.64$) under the assumption that the null hypothesis ($p = 0.50$) is true; the result is that the probability is 0.0026. Based on this result, should you reject or not reject the null hypothesis?

Solution The probability of 0.0026 means that if the null hypothesis is true (meaning that the true population proportion is 50%), the probability of drawing a random sample with a sample proportion of at least $\hat{p} = 0.64$ is just slightly under 3 in 1,000. Because this probability is less than 1 in 100, this result is statistically significant at the 0.01 level. It therefore gives us good reason to reject the null hypothesis, which means it provides support for the alternative hypothesis that Gender Choice raises the proportion of girls to more than 50%.

P-Values

In Example 3 above, we concluded that the sample result gave us reason to reject the null hypothesis, because the result was statistically significant at the 0.01 level. In fact, the result was even better than that: The computed probability of 0.0026 is about one-fourth of 0.01. The precise probability therefore gives us even more information than simply stating a level of statistical significance.

Published summaries or news reports of hypothesis tests often state only the level of significance, but determining that level always requires that we first compute a precise probability. This probability is called a ***P*-value** (short for *probability value*); notice the capital *P*, used to avoid confusion with the lowercase *p* that stands for a population proportion. In other words, for the case in Example 3 in which the probability of drawing a sample with a sample proportion of at least 64% is 0.0026, we say that the *P*-value for the hypothesis test is 0.0026. We will discuss the actual calculation of *P*-values in Sections 9.2 and 9.3; here, we focus only on their interpretation.

Hypothesis Test Decisions Based on *P*-Values

The ***P*-value** (probability value) for a hypothesis test of a claim about a population parameter is the probability of selecting a sample at least as extreme as the observed sample, assuming that the null hypothesis is true:

- A small *P*-value (such as less than or equal to 0.05) indicates that the sample result is unlikely, and therefore provides reason to reject the null hypothesis.
- A large *P*-value (such as greater than 0.05) indicates that the sample result could easily occur by chance, so we cannot reject the null hypothesis.

EXAMPLE 4 Fair Coin?

You suspect that a coin may have a bias toward landing tails more often than heads, and decide to test this suspicion by tossing the coin 100 times. The result is that you get 40 heads (and 60 tails). A calculation (not shown here) indicates that the probability of getting 40 or fewer heads in 100 tosses with a fair coin is 0.0228. Find the *P*-value and level of statistical significance for your result. Should you conclude that the coin is biased against heads?

Solution The null hypothesis is that the coin is fair, in which case the proportion of heads should be about 50% (H_0: $p = 0.50$). The alternative hypothesis is your suspicion that the coin is biased against heads, in which case the proportion of heads would be less than 50% (H_a: $p < 0.50$). Your 100 coin tosses represent a random sample of size $n = 100$, and the result of 40 heads is the sample proportion ($\hat{p} = 0.40$) for the hypothesis test. The *P*-value for the test is the probability of getting a sample proportion at least as extreme as the one you found ($\hat{p} \leq 0.40$), *assuming* that the coin is fair and the population proportion of heads is 0.5. The given probability for this occurrence is 0.0228, which is the *P*-value for the test. Because this *P*-value is smaller than 0.05, the result is statistically significant at the 0.05 level. Because it is not smaller than 0.01, the result is not significant at the 0.01 level. Statistical significance at the 0.05 level gives you moderate reason to reject the null hypothesis and conclude that the coin is biased against heads.

Putting It All Together

We have now covered all the basic ideas of hypothesis testing, except for the actual calculations that are required. We will discuss these calculations for hypothesis tests of population means in Section 9.2 and of population proportions in Section 9.3. The following box summarizes the steps that go into a hypothesis test.

The Hypothesis Test Process

Step 1. Formulate the null and alternative hypotheses, each of which must make a claim about a *population* parameter, such as a population mean (μ) or a population proportion (p); be sure this is done before drawing a sample or collecting data. Based on the form of the alternative hypothesis, decide whether you will need a left-, right-, or two-tailed hypothesis test.

Step 2. Draw a sample from the population and measure the sample statistics, including the sample size (n) and the relevant sample statistic, such as the sample mean ($\bar{x}$) or sample proportion ($\hat{p}$).

Step 3. Determine the likelihood of observing a sample statistic (mean or proportion) at least as extreme as the one you found *under the assumption that the null hypothesis is true*. The precise probability of such an observation is the *P*-value (probability value) for your sample result.

Step 4. Decide whether to reject or not reject the null hypothesis, based on your chosen level of significance (usually 0.05 or 0.01, but other significance levels are sometimes used).

Again, be sure to avoid confusion among the three different uses of the letter *p*:

- A lowercase *p* represents a *population proportion*—that is, the true proportion among a complete population.
- A lowercase $\hat{p}$ ("*p*-hat") represents a *sample proportion*—that is, the proportion found in a sample drawn from a population.
- An uppercase *P* stands for probability in a *P*-value.

EXAMPLE 5 Mean Rental Car Mileage

In the United States, the average car is driven about 12,000 miles each year. The owner of a large rental car company suspects that for his fleet, the mean distance is greater than 12,000 miles each year. He selects a random sample of $n = 225$ cars from his fleet and finds that the mean annual mileage for this sample is $\bar{x} = 12{,}375$ miles. A calculation shows that if you assume the fleet mean is the national mean of 12,000 miles, then the probability of selecting a 225-car sample with a mean annual mileage of at least 12,375 miles is 0.01. Based on these data, describe the process of conducting a hypothesis test and drawing a conclusion.

Solution We follow the four steps listed in the box above:

Step 1. The population parameter of interest is a *population mean* (μ)—the mean annual mileage for the population of all cars in the rental car fleet. The null hypothesis is that the population mean is the national average of 12,000 annual miles per car. The alternative hypothesis is the owner's claim that the population mean for his fleet is greater than this national average. That is,

$$H_0: \mu = 12{,}000 \text{ miles}$$
$$H_a: \mu > 12{,}000 \text{ miles}$$

Because the alternative hypothesis has a "greater than" form, the hypothesis test will be right-tailed.

Step 2. This step asks that we select the sample and measure the sample statistics; here, we are given the sample size of $n = 225$ and the sample mean of $\bar{x} = 12{,}375$ miles.

Step 3. This step asks us to determine the likelihood that, under the assumption that the fleet average is really 12,000 miles (the null hypothesis), we would by chance select a sample with a mean of at least 12,375 miles. Here, we don't need to do the calculation, because we are told that the probability is 0.01. (The calculation method is given in Section 9.2.) This probability is the *P*-value; it tells us that if the null hypothesis were true, there would be only a 0.01 probability of selecting a sample as extreme as the one observed.

Step 4. The *P*-value of 0.01 tells us that the result is significant at the 0.01 level. We therefore reject the null hypothesis and conclude that the test provides strong support for the alternative hypothesis, implying that the mean annual mileage for the rental car fleet is greater than the national average of 12,000 miles.

A Legal Analogy of Hypothesis Testing

A legal analogy might help clarify the idea of hypothesis testing. In American courts of law, the fundamental principle is that a defendant is presumed innocent until proven guilty. Because the starting assumption is innocence, this represents the null hypothesis:

H_0: The defendant is innocent.

H_a: The defendant is guilty.

The job of the prosecutor is to present evidence so compelling that the jury is persuaded to reject the null hypothesis and find the defendant guilty. If the prosecutor does not make a sufficiently compelling case, then the jury will not reject H_0 and the defendant will be found "not guilty." Note that finding a person innocent (accepting H_0) is not an option: A verdict of not guilty means the evidence is not sufficient to establish guilt, but it does not prove innocence.

TIME OUT TO THINK

Consider two situations. In one, you are a juror in a case in which the defendant could be fined a maximum of \$2,000. In the other, you are a juror in a case in which the defendant could receive the death penalty. Compare the significance levels you would use in the two situations. In each situation, what are the consequences of wrongly rejecting the null hypothesis?

Section 9.1 Exercises

Statistical Literacy and Critical Thinking

1. **Hypothesis Test.** What is a hypothesis test?

2. **Hypotheses.** What is a null hypothesis? What notation is used for a null hypothesis? What is an alternative hypothesis? What notation is used for an alternative hypothesis?

3. **Alternative Hypothesis.** A researcher wants to test a claim made about the mean body temperature of healthy adults; the claim refers to the value of 98.6°F. Identify the three different possible expressions that could be used for the alternative hypothesis.

4. ***P*-Value.** What is a *P*-value for a hypothesis test?

Does It Make Sense? For Exercises 5–12, decide whether the statement makes sense (or is clearly true) or does not make sense (or is clearly false). Explain clearly. Not all of these statements have definitive answers, so your explanation is more important than your chosen answer.

5. **Media Report.** The *Newport Chronicle* includes an article with the statement that "the study proved that the percentage of job applicants who fail drug tests is equal to 6%."

6. ***P*-Value.** A researcher is convinced that she can show that a new drug is effective in lowering LDL cholesterol. She claims that the *P*-value of 0.001 supports her claim of a lower mean level of LDL cholesterol.

7. **Null Hypothesis.** In testing a claim that the mean LDL cholesterol level is less than 130 mg/dL, the researcher states the null hypothesis as $\mu < 130$ mg/dL.

8. **Null Hypothesis.** After conducting a hypothesis test, a researcher forms an initial conclusion that there is sufficient sample evidence to support the null hypothesis.

9. **Drug Treatment.** In a test of the claim that, among patients treated with Ziac, the proportion who experience dizziness is less than 0.06, the alternative hypothesis is $p \geq 0.06$.

10. **Sample Mean.** A study is designed to determine the proportion of men who weigh more than 200 pounds, so the sample mean weight must be found.

11. ***P*-Value.** In a clinical test of a new drug, the project coordinator is happy to learn that the effectiveness of the drug is described with a *P*-value of 0.001.

12. **Pulse Rate.** In a test of the claim that those who exercise have a mean pulse rate of less than 74 beats per minute, the null hypothesis is $\mu = 74$ and the alternative hypothesis is $\mu < 74$.

Concepts and Applications

13. **What Is Significant?** Assume that you commute to college and fill your gasoline tank every Friday afternoon with 7 gallons of gasoline.
    a. One week, after normal driving and commuting, you find that your car requires 8 gallons. Would you consider this increase statistically significant or would you attribute it to random fluctuations?
    b. After another week of normal driving and commuting, you find that your car requires 12 gallons. Would you consider this increase statistically significant or would you attribute it to random fluctuations?
    c. How much of an increase would you need to see before you considered it statistically significant and looked for an explanation? Explain.

14. **What Is Significant?** Assume that your telephone long-distance charges are around \$7 each month.
    a. After one month of normal usage, your long-distance telephone charges amount to \$8.05. Would you consider this change statistically significant and look for an explanation, or is it the sort of variation that you would attribute to chance?
    b. After one month of normal usage, your long-distance telephone charges amount to \$40. Would you consider this change statistically significant and look for an explanation, or is it the sort of variation that you would attribute to chance?
    c. About how much of an increase would you need to see before you considered it statistically significant and looked for an explanation? Explain.

Formulating Hypotheses. In Exercises 15–22, formulate the null and alternative hypotheses for a hypothesis test. State in clear terms the two possible conclusions that address the given claim.

15. Cola Contents. A consumer complains that one-liter bottles of cola supplied by a supermarket actually contain less than one liter of cola.

16. SAT Scores. A high school principal claims that the mean SAT score of seniors at his school is less than the national average of 1518.

17. Gender Selection. The Chief Operations Officer of a medical facility claims that treatments can increase the probability that a baby will be a girl so that the proportion of girls is greater than 0.7.

18. Quality Control. The quality control manager at a treadmill manufacturing company claims that the proportion of defective treadmills is less than 0.02.

19. Vending Machines. A sales representative claims that her vending machines dispense coffee so that the mean amount supplied is equal to 10 ounces.

20. Aspirin. The Food and Drug Administration claims that a pharmaceutical company is producing aspirin tablets with a mean amount of aspirin that is less than 350 milligrams.

21. Holocaust. A high school teacher claims that the majority of her students do not know what the term *Holocaust* refers to.

22. Smoking. An educator claims that less than 20% of college graduates smoke.

***P*-Values and Births.** Assume that male births and female births are equally likely. The following table shows the probabilities of various numbers of male babies in a random sample of 100 births. Use this information for Exercises 23–28, and assume that we are testing for a bias against males.

Number of males among 100	Probability
35 or fewer	0.002
40 or fewer	0.028
45 or fewer	0.184
48 or fewer	0.382

23. A random sample of 100 births has 48 male babies. Is this result significant at the 0.05 level? What is the *P*-value for this result?

24. A random sample of 100 births has 45 male babies. Is this result significant at the 0.01 level? What is the *P*-value for this result?

25. A random sample of 100 births has 40 male babies. Is this result significant at the 0.01 level? What is the *P*-value for this result?

26. A random sample of 100 births has 40 male babies. Is this result significant at the 0.05 level? What is the *P*-value for this result?

27. A random sample of 100 births has 35 male babies. Is this result significant at the 0.01 level? What is the *P*-value for this result?

28. A random sample of 100 births has 32 male babies. Is this result significant at the 0.01 level? What is the *P*-value for this result?

Projects for the Internet and Beyond

For useful links, select "Links for Internet Projects" for Chapter 9 at www.aw.com/bbt.

29. Professional Journals. Many professional journals, such as *Journal of the American Medical Association*, contain articles that include information about formal tests of hypotheses. Find such an article and identify the null hypothesis and alternative hypothesis. In simple terms, state the objective of the hypothesis test and the conclusion that was reached.

30. Coin Activity. Select a particular quarter and test the claim that it favors heads when flipped. State the null and alternative hypotheses; then flip the quarter 100 times. Applying only common sense, what do you conclude about the claim that the quarter favors heads?

IN THE NEWS

31. Hypothesis Testing in the News. Find a news article or research report that describes (perhaps not explicitly) a hypothesis test for a population mean or proportion. Attach the article and summarize the method used.

9.2 Hypothesis Tests for Population Means

In Section 9.1, we sketched the basic outline of the hypothesis test process. Remember that in all cases, the ultimate goal of the test is to make a decision about whether to reject or not reject the null hypothesis, which is the starting assumption for the test. In this section, we describe the calculations used to make that decision for hypothesis tests with population means. (We'll cover population proportions in the next section.) We first investigate the procedure for one-tailed (left-tailed or right-tailed) tests and then discuss the small differences in procedure necessary for two-tailed tests.

One-Tailed Hypothesis Tests

Consider the following hypothetical situation. Columbia College advertises that the mean starting salary of its graduates is \$39,000. The Committee for Truth in Advertising, an independent organization, suspects that this claim is exaggerated and decides to conduct a hypothesis test to seek evidence to support its suspicion.

Notice that the parameter of interest is the mean starting salary of the population of all Columbia College graduates, so the hypothesis test will concern a *population mean* (μ). The null hypothesis is the College's claim that the mean starting salary is \$39,000. The alternative hypothesis is the Committee's claim that the College has exaggerated the mean starting salary, in which case the mean starting salary is *less than* \$39,000. In other words, the null and alternative hypotheses are

$$H_0: \mu = \$39{,}000$$
$$H_a: \mu < \$39{,}000$$

Because the alternative hypothesis has a "less than" form, we are dealing with a left-tailed hypothesis test. The general procedure for left- and right-tailed tests is the same, so we will consider both together as *one-tailed* tests.

Having formed the hypotheses, the Committee for Truth in Advertising selects a random sample of 100 recent graduates from the college. The mean salary of the graduates in the sample turns out to be \$37,000. In other words, the sample size is $n = 100$ and the sample mean is $\bar{x} = \$37{,}000$.

If you now look back at the four-step hypothesis test process on page 376, you'll see that the first two steps are already complete: The mean starting salary has been identified as the population parameter of interest, the null and alternative hypotheses have been stated, and the sample has been drawn and measured to determine its sample size and sample mean. We are therefore ready for Steps 3 and 4, in which we assume the null hypothesis is true and then determine whether the sample statistics give us reason to reject this assumption.

By the Way ...

As of 2006, data from the National Center for Education Statistics indicate that the median starting salary for *all* college graduates with a bachelor's degree is about \$35,000. The median is given rather than the mean, because salary distributions tend to be very skewed, with great variations by major. Graduates in engineering and business generally have the highest starting salaries.

The Sampling Distribution

Step 3 of the hypothesis test process is finding the likelihood of observing a sample mean as extreme as the one found, under the assumption that the null hypothesis is true. For the Columbia College example, the question becomes this: How likely are we to select a sample (of size $n = 100$) with a mean of \$37,000 or less when the mean for the whole population is \$39,000? To answer this question, we need a crucial observation based on our work with sampling distributions in Chapter 8: *The observed sample mean* ($\bar{x} = \$37{,}000$) *is just one point in a distribution of sample means*. Moreover, assuming the null hypothesis is true, this distribution of sample means is approximately normal with a mean of $\mu = \$39{,}000$.

To understand this observation, imagine that instead of drawing just one sample of size $n = 100$, the Committee for Truth in Advertising drew many samples of that size. Each sample would have a unique sample mean, $\bar{x}$, so we could make a graph of the distribution of all these sample means. As discussed in Section 8.1, for a reasonably large sample size such as $n = 100$, the resulting sampling distribution will be approximately normal with a mean that is equal to the population mean. Because the null hypothesis claims that the population mean is $\mu = \$39{,}000$, the sampling distribution will peak at this value if the null hypothesis is correct.

Figure 9.1 shows the idea graphically. The red curve represents the sampling distribution under the assumption that the null hypothesis is true. Remember that this curve is what we would expect to find if we plotted the distribution of means from *many* samples. When we have only a single sample mean (such as the sample mean $\bar{x} = \$37{,}000$ for the Columbia College example), it represents just one point along this curve. If this point lies close to the peak of the curve, it tells us that the sample mean is near the population mean expected with the null hypothesis. In that case, the probability of finding such a sample mean is not small and there is no reason to reject the null hypothesis. In contrast, if the sample mean lies far from the population mean claimed by the null hypothesis, then the probability of finding such a sample mean is small *if* the null hypothesis is true. We would therefore conclude that the true population mean probably is *not* what the null hypothesis claims, in which case we have reason to reject the null hypothesis.

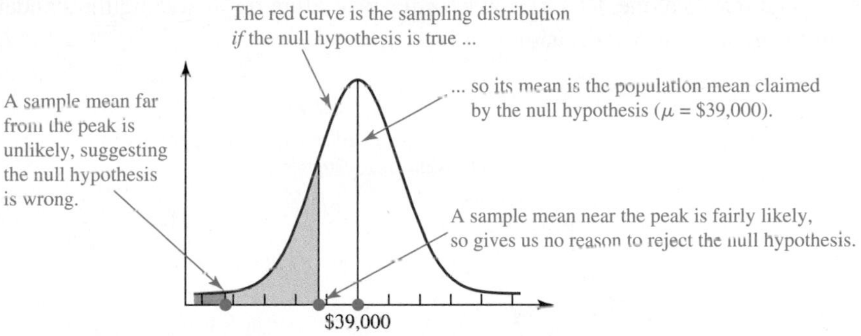

Figure 9.1 Graphical interpretation of the hypothesis test for the Columbia College example. If the null hypothesis is true, then the population mean is $\mu = \$39{,}000$. In that case, if we took many samples from the population, the distribution of sample means would be approximately normal with a mean of $39,000. Under this assumption, the distance of a single sample mean from the population mean allows us to decide whether to reject or not reject the null hypothesis.

Finding the Standard Score

As illustrated in Figure 9.1, the decision on whether to reject the null hypothesis depends on whether the sample mean is "near" to or "far" from the population mean claimed by the null hypothesis. Because the sampling distribution is a (nearly) normal distribution, the *standard score* of the sample mean represents a quantitative measure of the distance between the sample mean and the claimed population mean. Recall from Section 5.2 that the standard score (or z-score) of a data value in a normal distribution is the number of standard deviations that it lies above or below the mean of the distribution. From the Central Limit Theorem (Section 5.3), the standard deviation of a distribution of sample means is $\sigma/\sqrt{n}$, where σ is the population standard deviation and n is the sample size. Putting these ideas together, we find the following formula for the standard score of a sample mean $\bar{x}$:

$$z = \frac{\text{sample mean} - \text{population mean}}{\text{standard deviation of sampling distribution}} = \frac{\bar{x} - \mu}{\sigma/\sqrt{n}}$$

The only remaining problem is that we generally do not know the population standard deviation, σ. For now, let's assume that we can approximate the population standard deviation with the sample standard deviation, s, and that $s = \$6{,}150$ for the 100 salaries in the sample. (In Section 10.1 we'll discuss a better way to proceed when σ is not known.) In that case, we set $\sigma = \$6{,}150$ and the standard deviation for the distribution of sample means is

$$\frac{\sigma}{\sqrt{n}} = \frac{\$6{,}150}{\sqrt{100}} = \$615$$

Using this value in the previous equation tells us that the standard score for the sample mean of $\bar{x} = \$37{,}000$ in a sampling distribution with a population mean of $\mu = \$39{,}000$ is

$$z = \frac{\bar{x} - \mu}{\sigma/\sqrt{n}} = \frac{\$37{,}000 - \$39{,}000}{\$615} = -3.25$$

In other words, the sample mean of $\bar{x} = \$37{,}000$ lies 3.25 standard deviations below the mean of the sampling distribution.

Figure 9.2 shows that the standard score of −3.25 places the sample result far out on the left tail of the distribution, indicating that it would be a rare sample result *if* the population mean really were the $39,000 claimed by the null hypothesis. This suggests that the null hypothesis is incorrect and should be rejected. However, rather than rejecting the null hypothesis for this visual reason alone, let's look at a formal procedure for analyzing the standard score and making the hypothesis test decision.

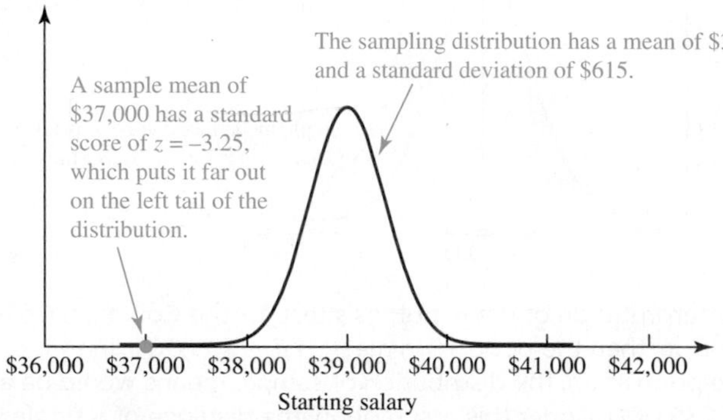

Figure 9.2 Assuming that the null hypothesis is true, the distribution of sample means for the Columbia College example has a mean of $39,000 and a standard deviation of $615. In that case, a sample mean of $\bar{x} = \$37{,}000$ has a standard score of −3.25.

TECHNICAL NOTE

The standard score for the sample mean is often called the *test statistic*.

Computing the Standard Score for the Sample Mean in a Hypothesis Test

When we draw a random sample for a hypothesis test, we can consider it to be one of many possible samples in the sampling distribution. Given the sample size (n), the sample mean ($\bar{x}$), the population standard deviation (σ), and the claimed population mean (μ), we make the following computations:

$$\text{standard deviation for the distribution of sample means} = \frac{\sigma}{\sqrt{n}}$$

$$\text{standard score for the sample mean, } z = \frac{\bar{x} - \mu}{\sigma/\sqrt{n}}$$

Note: In reality, it is rare that we know the population standard deviation σ; see Section 10.1 about how to deal with such cases.

Critical Values for Statistical Significance

Recall that the hypothesis test is significant at the 0.05 level if the probability of finding a result as extreme as the one actually observed is 0.05 or less (assuming the null hypothesis is true). For a left-tailed test, then, we are looking for a standard score that is at or below the 5th percentile of the sampling distribution. From Table 5.1 (page 211), the 5th percentile has a standard score between $z = -1.6$ and $z = -1.7$; the more precise standard score tables in Appendix A show that the 5th percentile has a standard score of $z = -1.645$. Therefore, a left-tailed hypothesis test is significant at the 0.05 level if the standard score of the sample mean is less than or equal to $z = -1.645$. This standard score represents the **critical value** for significance at the 0.05 level in a left-tailed hypothesis test.

A similar argument applies to right-tailed tests with alternative hypotheses of the form H_a: $\mu >$ claimed value. In such cases, significance at the 0.05 level requires that the sample mean lie at or above the 95th percentile—which requires a standard score greater than or equal to $z = 1.645$. Figure 9.3 illustrates these critical values for both left- and right-tailed tests. To find critical values for significance at the 0.01 level, we look for the standard scores of the 1st and 99th percentiles (rather than the 5th and 95th); Appendix A shows that these are −2.33 and 2.33, respectively.

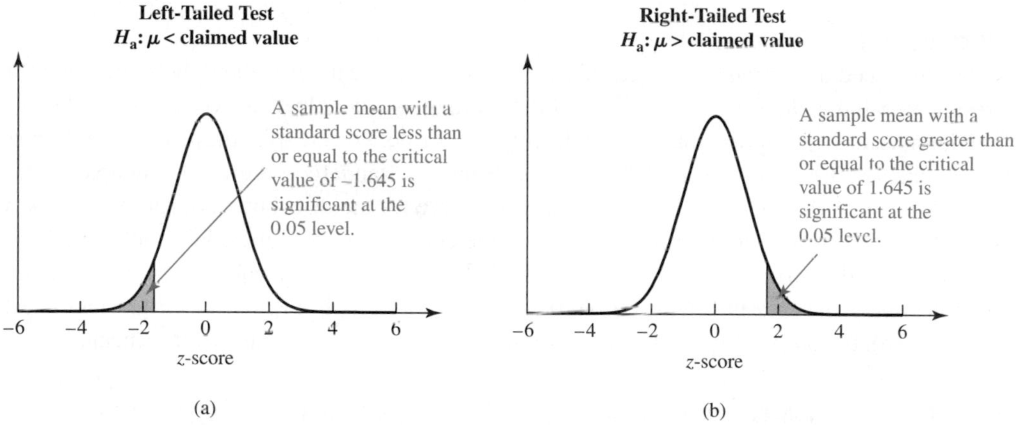

Figure 9.3 These graphs illustrate the meaning of critical values for one-tailed hypothesis tests when testing for significance at the 0.05 level. The locations of the critical values correspond to the 5th percentile for left-tailed tests and the 95th percentile for right-tailed tests.

Decisions Based on Statistical Significance for One-Tailed Hypothesis Tests

We decide whether to reject or not reject the null hypothesis by comparing the standard score (z) for a sample mean to **critical values** for significance at a given level. Table 9.1 summarizes the decisions for one-tailed hypothesis tests at the 0.05 and 0.01 levels of significance.

Table 9.1 Testing for Significance at the 0.05 and 0.01 Levels

Type of test	Form of H_a	For 0.05 level: Reject H_0 if standard score is	For 0.01 level: Reject H_0 if standard score is
Left-tailed test	H_a: $\mu <$ claimed value	$z \leq -1.645$	$z \leq -2.33$
Right-tailed test	H_a: $\mu >$ claimed value	$z \geq 1.645$	$z \geq 2.33$

TIME OUT TO THINK

Suppose that, in right-tailed tests, one study finds a sample mean with $z = 3$ and another study finds a sample mean with $z = 10$. Both are significant at the 0.01 level, but which result provides stronger evidence for rejecting the null hypothesis? Explain.

EXAMPLE 1 Columbia College Test Significance

Assuming that the null hypothesis is true and the mean starting salary for Columbia College graduates is \$39,000, is it statistically significant to find a sample in which the mean is only \$37,000? Based on your answer, should we reject or not reject the null hypothesis?

Solution Recall that the hypothesis test is left-tailed (because the alternative hypothesis has the "less than" form of $\mu < \$39{,}000$) and we already found that a sample mean of $\bar{x} = \$37{,}000$ has a standard score of -3.25. This result is significant at the 0.05 level because the standard score is less than the critical value of $z = -1.645$. In fact, it is also significant at the 0.01 level, because the standard score is less than -2.33. We therefore have strong reason to reject the null hypothesis and conclude that Columbia College officials did indeed exaggerate the mean starting salary of its graduates.

Finding the *P*-Value

As we discussed in Section 9.1, we can use *P*-values to be more precise about the significance of a result. Recall that the *P*-value is the probability of finding a sample mean as extreme as the one found, under the assumption that the null hypothesis is true. For the hypothesis tests we discuss in this chapter, we generally find the *P*-value from the standard score of the sample mean. Figure 9.4 shows the idea. The total area under the curve for the distribution of sample means is defined to be 1, so we can interpret areas under the curve as probabilities. For a left-tailed test (Figure 9.4a), the probability of finding a sample mean less than or equal to some particular value is simply the area under the curve to the *left* of the sample mean. For a right-tailed test (Figure 9.4b), the probability is the area under the curve to the *right* of the sample mean.

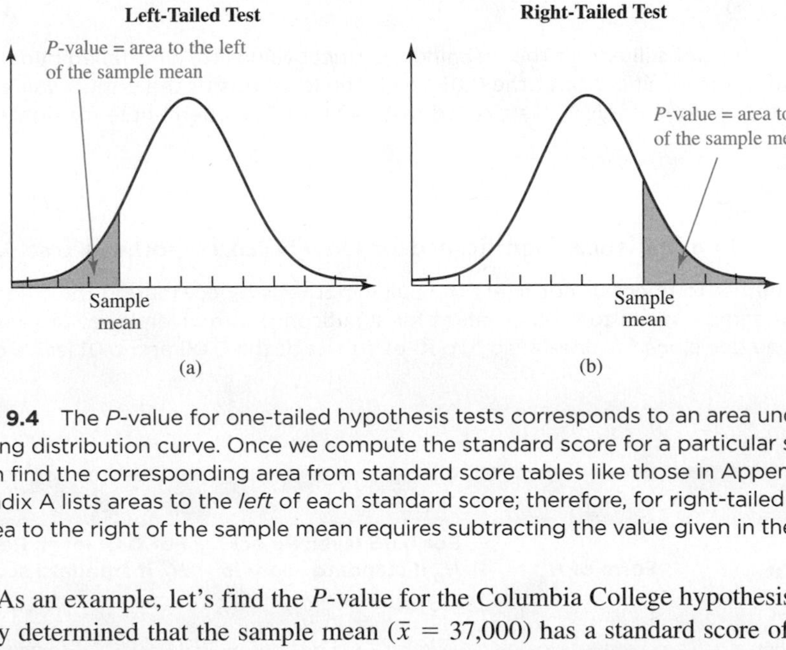

Figure 9.4 The *P*-value for one-tailed hypothesis tests corresponds to an area under the sampling distribution curve. Once we compute the standard score for a particular sample mean, we can find the corresponding area from standard score tables like those in Appendix A. (Note: Appendix A lists areas to the *left* of each standard score; therefore, for right-tailed tests, finding the area to the right of the sample mean requires subtracting the value given in the table from 1.)

The P-*value . . . is often used as a gauge of the degree of contempt in which the null hypothesis deserves to be held.*

—Robert Abelson, *Statistics as a Principled Argument*

As an example, let's find the *P*-value for the Columbia College hypothesis test. We have already determined that the sample mean ($\bar{x} = 37{,}000$) has a standard score of $z = -3.25$. If you look in Appendix A, you'll see that this standard score corresponds to a "cumulative area from the left" of 0.0006—so this is the P-value we are seeking. This small P-value provides strong evidence against the null hypothesis. Be sure to notice that this P-value is less than 0.01, which is why the result is significant at the 0.01 level; the fact that it is much less than 0.01 tells us that we have very strong evidence for rejecting the null hypothesis.

Summary of One-Tailed Tests for Population Means

We began this section with the goal of learning how to carry out Steps 3 and 4 of the four-step hypothesis test process on page 376. We have now succeeded in doing this for one-tailed hypothesis tests for population means. To summarize:

- Because we are dealing with population means, the null hypothesis has the form μ = claimed value. To decide whether to reject or not reject the null hypothesis, we must determine whether a sample as extreme as the one found in the hypothesis test is likely or unlikely to occur if the null hypothesis is true.
- We determine this likelihood from the standard score (z) of the sample mean, which we compute from the formula

$$z = \frac{\bar{x} - \mu}{\sigma/\sqrt{n}}$$

  where n is the sample size, $\bar{x}$ is the sample mean, μ is the population mean claimed by the null hypothesis, and σ is the population standard deviation. (In this section, we generally approximate σ with the sample standard deviation, s; a better approach is described in Section 10.1.)
- We can then assess the standard score in two ways:
  1. We can assess its level of statistical significance by comparing it to the critical values given in Table 9.1 (page 383).
  2. We can determine its P-value with standard score tables like those in Appendix A. For a left-tailed test, the P-value is the area under the normal curve to the left of the standard score; for a right-tailed test, it is the area under the normal curve to the right of the standard score.
- If the result is statistically significant at the chosen level (usually either the 0.05 or the 0.01 significance level), we reject the null hypothesis. If it is not statistically significant, we do not reject the null hypothesis.

EXAMPLE 2 Mean Rental Car Mileage (Revisited)

Recall the case of the rental car fleet owner (Example 5 of Section 9.1) who suspects that the mean annual mileage of his cars is greater than the national mean of 12,000 miles. He selects a random sample of $n = 225$ cars and calculates the sample mean to be $\bar{x} = 12{,}375$ miles and the sample standard deviation to be $s = 2{,}415$ miles. Determine the level of statistical significance and P-value for this hypothesis test, and interpret your findings.

Solution To determine the significance and P-value we must first calculate the standard score for the sample mean of $\bar{x} = 12{,}375$ miles. We are given the claimed population mean ($\mu = 12{,}000$ miles) and the sample size ($n = 225$); we do not know the population standard deviation, σ, but we will assume it is the same as the sample standard deviation and set $\sigma = 2{,}415$ miles. We find

$$z = \frac{\bar{x} - \mu}{\sigma/\sqrt{n}} = \frac{12{,}375 - 12{,}000}{2{,}415/\sqrt{225}} = 2.33$$

This standard score is greater than the critical value of $z = 1.645$ for significance at the 0.05 level and equal to the value of $z = 2.33$ for significance at the 0.01 level, giving us strong reason to reject the null hypothesis and conclude that the rental car fleet average really is greater than the national average. We can find the P-value from Appendix A, which shows that the area to the *left* of a standard score of $z = 2.33$ is 0.9901; because this is a right-tailed test, the probability is the area to the *right* (see Figure 9.4b), which is $1 - 0.9901 = 0.0099$. Therefore, the P-value is 0.0099, which is very close to 0.01.

Two-Tailed Tests

The same basic ideas apply to two-tailed hypothesis tests in which the alternative hypothesis has the "not equal to" form of H_a: $\mu \neq$ claimed value. However, the critical values and the calculations for P-values are slightly different for two-tailed tests.

As always, the hypothesis test is significant at the 0.05 level if the probability of finding a result as extreme as the one actually found is 0.05 or less. For one-tailed tests, a probability of 0.05 corresponds to standard scores at the 5th percentile for left-tailed tests and 95th percentile for right-tailed tests (see Figure 9.3). For two-tailed tests, however, a value "as extreme as the one actually found" can lie *either* on the left or on the right of the sampling distribution (Figure 9.5). A probability of 0.05, or 5%, therefore corresponds to standard scores either in the first 2.5% of the sampling distribution on the left or in the last 2.5% on the right. From Appendix A, the 2.5th percentile corresponds to a standard score of −1.96 and the 97.5th percentile corresponds to a standard score of 1.96. These standard scores become the critical values for two-tailed tests to be significant at the 0.05 level. Similarly, the two-tailed critical values for significance at the 0.01 level correspond to standard scores for the 0.5th and 99.5th percentiles; from Appendix A, these values are −2.575 and 2.575, respectively.

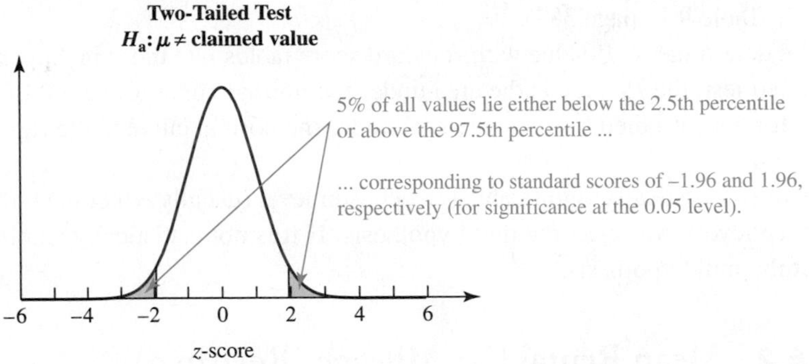

Figure 9.5 For two-tailed tests, the critical values for significance at the 0.05 level correspond to the 2.5th and 97.5th percentiles (as opposed to the 5th or 95th percentiles for one-tailed tests).

Similar considerations apply to P-values for two-tailed tests. Recall that for a one-tailed test, the P-value is the area under the curve to the left of the sample mean for a left-tailed test or to the right of the sample mean for a right-tailed test. Because a two-tailed test asks us to consider extremes on both sides of the claimed mean, the P-value for a two-tailed test must be twice what it would be if the test were one-tailed.

Two-Tailed Test (H_a: $\mu \neq$ claimed value)

Statistical significance: A two-tailed test is significant at the 0.05 level if the standard score of the sample mean is at or below a critical value of −1.96 *or* at or above a critical value of 1.96. For significance at the 0.01 level, the critical values are −2.575 and 2.575.

***P*-values**: To find the P-value for a sample mean in a two-tailed test, first use the standard score of the sample mean to find the P-value assuming the test is one-tailed; then double this value to find the P-value for the two-tailed test.

An example should help clarify these ideas. Consider a drug company that seeks to be sure that its "500-milligram" aspirin tablets really contain 500 milligrams of aspirin. If the tablets contain less than 500 milligrams, consumers are not getting the advertised dose. If the tablets contain more than 500 milligrams, consumers are getting too much of the drug. The null hypothesis says that the population mean of the aspirin content is 500 milligrams:

$$H_0: \mu = 500 \text{ milligrams}$$

The drug company is interested in the possibility that the mean weight is *either* less than *or* greater than 500 milligrams. Because the company is interested in possibilities on both sides of the claimed mean of 500 milligrams, we have a two-tailed test in which the alternative hypothesis is

$$H_a: \mu \neq 500 \text{ milligrams}$$

Suppose the company selects a random sample of $n = 100$ tablets and finds that they have a mean weight of $\bar{x} = 501.5$ milligrams; further suppose that the population standard deviation is $\sigma = 7.0$ milligrams. Then the standard score (z) for this sample mean is

$$z = \frac{\bar{x} - \mu}{\sigma/\sqrt{n}} = \frac{501.5 - 500}{7/\sqrt{100}} = 2.14$$

This standard score is above the critical value of 1.96 for a two-tailed test, so it is significant at the 0.05 level. (It is not significant at the 0.01 level, because it is not above the critical value of 2.575.) This gives us good reason to reject the null hypothesis and conclude that the mean weight of the "500-milligram" aspirin tablets is different from 500 milligrams.

We find the P-value by using Appendix A, which shows that the area to the *left* of a standard score of $z = 2.14$ is 0.9838. Therefore, if this were a one-tailed test, the P-value would be the area to the *right*, or $1 - 0.9838 = 0.0162$. But because this is a two-tailed test, we double this number to find that the P-value is $2 \times 0.0162 = 0.0324$. In other words, *if* the null hypothesis is true, the probability of drawing a sample as extreme as the one found is 0.0324.

EXAMPLE 3 What Is Human Mean Body Temperature?

Consider again the study in which University of Maryland researchers measured body temperatures in a sample of $n = 106$ healthy adults (see Example 5 in Section 8.2), finding a sample mean body temperature of $\bar{x} = 98.20°$F with a sample standard deviation of $s = 0.62°$F. We will assume that the population standard deviation is the same as the sample standard deviation. Determine whether this sample provides evidence for rejecting the common belief that mean human body temperature is $\mu = 98.6°$F.

Solution The null hypothesis is the claim that mean human body temperature is 98.6°F, or H_0: $\mu = 98.6°$F. We are asked to test the alternative hypothesis that mean body temperature is *not* 98.6°F, or H_a: $\mu \neq 98.6°$F. The "not equal to" form of the alternative hypothesis tells us that we need a two-tailed test. We are given the sample size $n = 106$, the sample mean $\bar{x} = 98.20°$F, and the assumed population standard deviation $\sigma = 0.62°$F. The standard score of the sample mean is

$$z = \frac{\bar{x} - \mu}{\sigma/\sqrt{n}} = \frac{98.20 - 98.60}{0.62/\sqrt{106}} = -6.64$$

This standard score is much less than the critical values of −1.96 for significance at the 0.05 level and −2.575 for significance at the 0.01 level (Figure 9.6). Because the result is significant at the 0.01 level, it provides strong evidence against the null hypothesis. We therefore reject the null hypothesis and conclude that human mean body temperature is *not* equal to 98.6°F. (In this case, the standard score of −6.64 is so extreme that we cannot find the precise P-value from Appendix A; a calculation shows that the P-value is about 3×10^{-11}.)

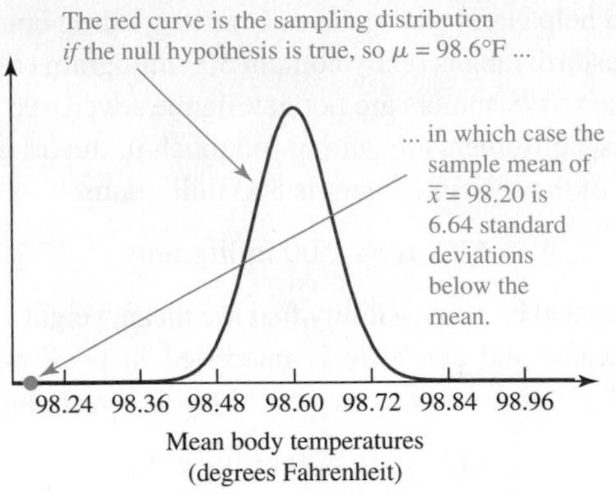
The red curve is the sampling distribution
if the null hypothesis is true, so μ = 98.6°F ...
... in which case the sample mean of x̄ = 98.20 is 6.64 standard deviations below the mean.
98.24 98.36 98.48 98.60 98.72 98.84 98.96
Mean body temperatures
(degrees Fahrenheit)

Figure 9.6 The null hypothesis states that human mean body temperature is 98.6°F. If this is true, then the distribution of sample means has a mean of 98.6°F. Based on the sample size and standard deviation, the sample mean of 98.20°F is more than 6 standard deviations below the claimed mean temperature—which makes it seem likely that the claim is wrong and human mean body temperature is *not* equal to 98.6°F.

Common Errors in Hypothesis Testing

We have now covered all the fundamentals of hypothesis testing, as well as specific methods used for hypothesis tests for population means. However, even if a hypothesis test is carried out correctly, two common types of error may affect the conclusions. To understand these errors, recall the legal analogy at the end of Section 9.1, in which the null hypothesis is H_0: The defendant is innocent. One type of error occurs if we conclude that the defendant is guilty when, in reality, he or she is innocent. In this case, the null hypothesis (innocence) has been wrongly rejected. The other type of error occurs if we find the defendant *not guilty* when he or she actually is guilty. In this case, we have wrongly failed to reject the null hypothesis.

For a statistical example, consider again the pharmaceutical company testing the claim that the mean amount of aspirin in its tablets is 500 milligrams (H_0: $\mu = 500$ milligrams). In drawing a conclusion from the test, the company could make the following two types of errors:

- The company might reject the null hypothesis and conclude that the mean amount is not 500 milligrams when it really is. This error can lead to wasting time and money trying to fix a process that isn't broken. An error of this type, in which H_0 is *wrongly rejected*, is called a **type I error**.
- The company might fail to reject the null hypothesis when, in fact, the mean amount of aspirin is not 500 milligrams. In this case, the company will distribute tablets that have too much or too little aspirin; consumers could suffer or sue. An error of this type, in which we *wrongly fail to reject* H_0, is called a **type II error**.

Table 9.2 summarizes the four possible cases.

Table 9.2 Decision Table for H_0 and H_a

		Reality	
		H_0 true	**H_a true**
Decision	**Reject H_0**	Type I error	Correct decision
	Do not reject H_0	Correct decision	Type II error

If you think about it, you'll realize that there is an important connection between the significance level of a hypothesis test and type I errors (wrongly rejecting H_0): ***The significance***

level is the probability of making a type I error. For example, if we conduct a hypothesis test using a 0.05 significance level, there is a 0.05 probability of making the mistake of rejecting the null hypothesis when it is actually true. If we conduct the test at a 0.01 significance level, there is a 0.01 probability of wrongly rejecting the null hypothesis. (The probability of a type II error can also be quantified but is beyond the scope of this book.)

EXAMPLE 4 Errors in the Body Temperature Test

Consider the null hypothesis from Example 3—that mean body temperature equals 98.6°F (H_0: $\mu = 98.6°F$).

a. What correct decisions are possible with this null hypothesis?

b. Explain the meaning of type I and type II errors in this case.

c. In Example 3, we rejected the null hypothesis. What is the probability that we made a mistake in doing so?

Solution

a. Any hypothesis test has two possible correct decisions. In this case, one correct decision occurs if the mean body temperature really is 98.6°F and we do *not* reject H_0. The other correct decision occurs if the mean body temperature is *not* 98.6°F and we reject H_0.

b. A type I error occurs if we reject H_0 when it is actually true. In this case, a type I error occurs if mean body temperature really is 98.6°F but we conclude that it is not. A type II error occurs if we do not reject H_0 when it is actually false. In this case, a type II error occurs if the mean body temperature is not 98.6°F but we fail to reach this conclusion.

c. We rejected the null hypothesis at a 0.01 significance level. Therefore, there is a 0.01 probability that we made a type I error of rejecting a null hypothesis that is actually true.

Bias in Choosing Hypotheses

As we have seen, a well-conducted hypothesis test follows a fairly strict process that tends to minimize the possibility of tainting the test with bias. Nevertheless, bias can be introduced in many ways, and in particular it can occur in the initial selection of the hypotheses. Consider a situation in which a factory is investigated for releasing pollutants into a nearby stream at a level that may (or may not) exceed the maximum level allowed by the government. The logical choice for the null hypothesis is that the actual mean level of pollutants equals the maximum allowed level:

$$H_0\text{: mean level of pollutants} = \text{maximum level allowed}$$

However, this leaves us with two reasonable choices for the alternative hypothesis, each of which introduces some bias into the ultimate conclusions.

If *factory officials* conducted the test, they would be inclined to claim that the mean level of pollutants is *less than* the maximum allowed level. That is, they would choose

$$H_a\text{: mean level of pollutants} < \text{maximum level allowed}$$

With this alternative hypothesis, the two possible outcomes are

- Reject H_0, in which case we conclude that the mean level of pollutants is less than the allowed maximum. This outcome would please the factory officials.
- Not reject H_0, in which case the test is inconclusive.

In other words, by choosing the left-tailed ("less than") hypothesis test, the factory officials ensure that at worst we end up with no evidence that they have exceeded the allowed maximum, while it's possible that the conclusion will be in their favor.

Now suppose an *environmental group* conducts the test. Because the group suspects that the factory is violating the government standards, its choice for the alternative hypothesis would be that the mean level of pollutants is *greater than* the maximum allowed level, or

$$H_a\text{: mean level of pollutants} > \text{maximum level allowed}$$

With this choice of alternative hypothesis, rejection of H_0 implies that the factory has violated the standards, while not rejecting H_0 is again inconclusive. In other words, choosing the right-tailed ("greater than") test creates a situation in which the factory may be found in violation but cannot be proved to be in compliance.

EXAMPLE 5 Mining Gold

The success of precious metal mines depends on the purity (or grade) of ore removed and the market price for the metal. Suppose the purity of gold ore must be at least 0.5 ounce of gold per ton of ore in order to keep a particular mine open. Samples of gold ore are used to estimate the purity of the ore for the entire mine. Discuss the impact of type I and type II errors on two of the possible alternative hypotheses:

$$H_a\text{: purity} < 0.5 \text{ ounce per ton}$$
$$H_a\text{: purity} > 0.5 \text{ ounce per ton}$$

Solution For the left-tailed case, the null and alternative hypotheses are

$$H_0\text{: purity} = 0.5 \text{ ounce per ton}$$
$$H_a\text{: purity} < 0.5 \text{ ounce per ton}$$

The two possible decisions in this case are

- Reject H_0, which means concluding that the purity of the ore is less than that needed to keep the mine open—so the mine is closed.
- Not reject H_0, which means we have insufficient evidence to conclude that the purity of the ore is less than that needed—so the mine stays open.

A type I error (wrongly rejecting a true H_0) means the mine is closed when, in fact, the purity of the gold ore is sufficient to operate the mine. For the mine's operators, this means the loss of potential profits from the mine; for the employees, it means unnecessary loss of jobs. A type II error (failing to reject a false H_0) means that the mine continues to operate, but is actually unprofitable.

Now suppose that the mine will be kept open only if the purity is greater than 0.5 ounce per ton. For this right-tailed case, the null and alternative hypotheses are

$$H_0\text{: purity} = 0.5 \text{ ounce per ton}$$
$$H_a\text{: purity} > 0.5 \text{ ounce per ton}$$

The two possible decisions are

- Reject H_0, which means concluding that the purity of the ore is *more* than that needed to keep the mine open—so the mine stays open.
- Not reject H_0, which means there is insufficient evidence to conclude that the purity of the ore is high enough to keep the mine open—so the mine is closed.

A type I error (wrongly rejecting a true H_0) means the mine is left open when it is actually unprofitable. A type II error (failing to reject a false H_0) means closing the mine when it is actually profitable—putting the employees out of work for no reason.

Section 9.2 Exercises

Statistical Literacy and Critical Thinking

1. **Notation.** When conducting hypothesis tests for a population mean, it is very important to clearly understand what n, $\bar{x}$, s, σ, and μ represent. Describe what each of those variables stands for.

2. **Pulse Rates.** A graduate student has been assigned to test the claim that the mean pulse rate of college instructors is greater than 70 beats per minute. She is able to obtain pulse rate measurements from the 225 instructors at her college, and she finds that there is sufficient evidence to support the claim. What is wrong with her project?

3. **Types of Errors.** What is a type I error? What is a type II error?

4. **Statistical Significance.** In a test of the Atkins diet, a P-value of 0.004 was obtained when the claim that there is a loss with the diet was tested. The mean weight loss after 12 months was 2.1 pounds. Is the mean weight loss of 2.1 pounds statistically significant? Does the mean weight loss of 2.1 pounds have practical significance? Why or why not?

Does It Make Sense? For Exercises 5–12, decide whether the statement makes sense (or is clearly true) or does not make sense (or is clearly false). Explain clearly. Not all of these statements have definitive answers, so your explanation is more important than your chosen answer.

5. **Hypothesis Test.** In a hypothesis test, the P-value is the same as the significance level.

6. ***P*-Value.** In testing a claim about a population mean, a test statistic of $z = 0.20$ corresponds to a larger P-value than a test statistic of $z = 2.00$.

7. ***P*-Value.** In a hypothesis test, a P-value of 0.999 indicates that you should support the alternative hypothesis.

8. **Alternative Hypothesis.** In testing the claim that the mean IQ score of prison inmates is less than 100, the alternative hypothesis is expressed as $\bar{x} < 100$.

9. **Significance.** The significance level in a hypothesis test is the probability of making a type I error.

10. **Type I Error.** When a consumer group is testing the claim that the mean amount of aspirin in tablets is 350 milligrams, it is extremely important not to reject a true null hypothesis wrongly. Thus, it is better to choose 0.01 than 0.05 for the significance level.

11. **Significance Level.** Because the significance level is the probability of making a type I error, it is wise to select a significance level of zero so that there is no probability of making that error.

12. **Mnemonic.** A handy mnemonic for interpreting the P-value in a hypothesis test is this: "If the P (value) is low, then the null must go."

Concepts and Applications

Alternative Hypothesis Supported? In Exercises 13–20, find the value of the standard score, z, and determine whether the alternative hypothesis is supported at a 0.05 significance level.

13. H_a: $\mu > 100$, $n = 100$, $\bar{x} = 102$, $\sigma = 15$
14. H_a: $\mu > 69.0$, $n = 400$, $\bar{x} = 75.2$, $\sigma = 12.0$
15. H_a: $\mu < 98.6$, $n = 64$, $\bar{x} = 98.2$, $\sigma = 0.6$
16. H_a: $\mu < 1{,}518$, $n = 81$, $\bar{x} = 1{,}516$, $\sigma = 14$
17. H_a: $\mu \neq 37.40$, $n = 400$, $\bar{x} = 37.30$, $\sigma = 0.80$
18. H_a: $\mu \neq 24.7$, $n = 100$, $\bar{x} = 24.0$, $\sigma = 1.9$
19. H_a: $\mu \neq 1{,}050$, $n = 49$, $\bar{x} = 1{,}058$, $\sigma = 20$
20. H_a: $\mu \neq 777$, $n = 900$, $\bar{x} = 779$, $\sigma = 42$

Finding *P*-Values. In Exercises 21–34, use Table 5.1 to find the P-value that corresponds to the standard z-score, and determine whether the alternative hypothesis is supported at the 0.05 significance level.

21. $z = -0.4$ for H_a: $\mu < 25$
22. $z = -2.4$ for H_a: $\mu < 727$
23. $z = 1.9$ for H_a: $\mu > 36.35$
24. $z = 1.5$ for H_a: $\mu > 0.227$
25. $z = -1.6$ for H_a: $\mu \neq 0.389$
26. $z = -2.0$ for H_a: $\mu \neq 172$
27. $z = 1.7$ for H_a: $\mu \neq 75$
28. $z = 1.9$ for H_a: $\mu \neq 25.7$
29. $z = 2.7$ for H_a: $\mu > 19.4$
30. $z = 3.5$ for H_a: $\mu > 75$
31. $z = -2.1$ for H_a: $\mu < 1{,}007$
32. $z = -1.1$ for H_a: $\mu < 149.6$
33. $z = 0.15$ for H_a: $\mu \neq 90.3$
34. $z = 3.5$ for H_a: $\mu \neq 1{,}022$

35. Interpreting *P*-Values. Assume that you are testing an alternative hypothesis of the form H_a: $\mu >$ claimed value. If the sample mean has a standard score of $z = -1.0$, what do you conclude? Why is it not necessary to actually conduct the formal hypothesis test?

36. Interpreting *P*-Values. Assume that you are testing an alternative hypothesis of the form H_a: $\mu <$ claimed value. If the sample mean has a standard score of $z = 0.5$, what do you conclude? Why is it not necessary to actually conduct the formal hypothesis test?

Hypothesis Tests for Means. For Exercises 37–50, use a 0.05 significance level and conduct a full hypothesis test using the four-step process described in the text. Be sure to state your conclusion.

37. Roper Poll. A Roper poll used a sample of 100 randomly selected car owners. Within the sample, the mean time of ownership for a single car was 7.01 years with a standard deviation of 3.74 years. Test the claim by the owner of a large dealership that the mean time of ownership for all cars is less than 7.5 years.

38. Hospital Time. According to a study by the Centers for Disease Control, the national mean hospital stay after childbirth is 2.0 days. Reviewing records at her own hospital, a hospital administrator calculates that the mean hospital stay for a sample of 81 women after childbirth is 2.2 days with a standard deviation of 1.2 days. Assuming that the patients represent a random sample of the population, test the claim that this hospital keeps new mothers longer than the national average. How does the result change if the standard deviation of the sample is 0.8 day?

39. Packaging. The makers of a leading brand of pasta need to be sure that the amount of pasta in every "16-ounce" package is neither too large (which leads to wasted product) nor too small (which leads to customer dissatisfaction). The packages in a sample of $n = 144$ packages of pasta have a mean weight of 15.8 ounces with a standard deviation of 1.6 ounces. Test the claim that the "16-ounce" packages actually have a mean weight of 16 ounces.

40. Motorcycle Manufacturing. The manufacturers of motorcycles must produce axles that meet specified dimensions. In particular, the diameters of the axles must be 8.50 centimeters. The axles in a sample of $n = 64$ axles have a mean diameter of 8.56 centimeters with a standard deviation of 0.24 centimeter. Test the claim that the axles actually have the specified mean diameter.

41. Drug Amounts. The cold medicine Dozenol lists 600 milligrams of acetominophen per fluid ounce as an active ingredient. The Food and Drug Administration tests 65 one-ounce samples of the medicine and finds that the mean amount of acetominophen for the sample is 589 milligrams with a standard deviation of 21 milligrams. Test the claim of the FDA that the medicine does not contain the required amount of acetominophen.

42. Income. The mean household income in Wasatch County, Utah, recently was \$41,045 with a standard deviation of \$1,605. Assume that the 13,267 residents of the county represent a random sample of all Utah residents. Based on this sample, test the claim that the mean household income in Utah differs from the national mean household income of \$35,100.

43. Fuel Consumption. According to the Energy Information Administration (Federal Highway Administration data), the average gas mileage of all automobiles is 21.4 miles per gallon. For a random sample of 40 sport utility vehicles (SUVs), the mean gas mileage is 19.8 miles per gallon with a standard deviation of 3.5 miles per gallon. Test the claim that the mean mileage of all SUVs is less than 21.4 miles per gallon.

44. Baseballs. A random sample of 40 new baseballs is obtained. Each ball is dropped onto a concrete surface, and the bounce heights have a mean of 92.67 inches and a standard deviation of 1.79 inches (based on data from *USA Today*). Test the claim that the new baseballs have a mean bounce height that is less than the mean bounce height of 92.84 inches found for older baseballs.

45. Compulsive Buyers. Researchers developed a questionnaire to identify compulsive buyers. A random sample of 32 subjects who identified themselves as compulsive buyers was obtained, and they had a mean questionnaire score of 0.83 with a standard deviation of 0.24 (based on data from "A Clinical Screener for Compulsive Buying," by Faber and Guinn, *Journal of Consumer Research*, Vol. 19). Test the claim that the population of self-identified compulsive buyers has a mean greater than the mean of 0.21 for the general population.

46. Coin Weight. According to the U.S. Department of the Treasury, the mean weight of a quarter is 5.670 grams. A random sample of 50 quarters has a mean weight of 5.622 grams with a standard deviation of 0.068 gram. Test the claim that the mean weight of quarters in circulation is 5.670 grams.

47. Birth Weight. The mean birth weight of male babies born to 121 mothers taking a vitamin supplement is 3.67 kilograms with a standard deviation of 0.66 kilogram (based on data from the New York State Department of Health). Test the claim that the mean birth weight of all babies born to mothers taking the vitamin supplement is equal to 3.39 kilograms, which is the mean for the population of all male babies.

48. Car Mileage. According to the Energy Information Administration (Federal Highway Administration data), the average annual mileage for automobiles in the United States is 11,725 miles. The owner of a rental car company selects a random sample of 225 cars from his entire fleet. For the sample, the mean annual mileage is 12,145 miles with a standard deviation of 3,000 miles. Test the claim that the mean mileage of the fleet is greater than the national average.

49. Convicted Embezzlers. When 70 convicted embezzlers were randomly selected, the mean length of prison terms was found to be 22.1 months and the standard deviation was 8.6 months (based on data from the U.S. Department of Justice). Jane Fleming is running for political office on a platform of tougher treatment of convicted criminals. Test her claim that prison terms for convicted embezzlers have a mean of less than 24 months.

50. Incomes. According to the Current Population Survey by the Census Bureau, men with bachelor's degrees earn an average of $46,700. Suppose a random sample of 400 women with bachelor's degrees has mean earnings of $28,700 (the national mean for women) with a standard deviation of $16,500. Test the claim that the mean for women's earnings is less than the mean for men's earnings.

Type I and Type II Errors. In Exercises 51–54, a null and alternative hypothesis are given. Without using the terms "null hypothesis" and "alternative hypothesis," identify the type I error and identify the type II error.

51. H_0: The patient is free of a particular disease.

H_a: The patient has the disease.

52. H_0: The defendant is not guilty.

H_a: The defendant is guilty.

53. H_0: The lottery is fair.

H_a: The lottery is biased.

54. H_0: The mean length of a bolt in the suspension system of new Audi cars is 3.456 centimeters.

H_a: The mean length of a bolt in the suspension system of new Audi cars is not equal to 3.456 centimeters.

Projects for the Internet and Beyond

For useful links, select "Links for Internet Projects" for Chapter 9 at www.aw.com/bbt.

55. Comparisons with National Averages. Choose several variables that are relatively easy to measure in a class or sample of students. The variables should involve a quantity that can be averaged (for example, height, weight, family size, blood pressure, heart rate, reaction time). Use the Internet or other references to determine national averages for these variables (by age categories, if appropriate). Collect data on the variables, using a random sample of at least 50 individuals. Carry out the relevant hypothesis test to determine whether the sample mean differs significantly from the population mean.

56. County Data. The *Statistical Abstract of the United States* and the Current Population Survey provide an inexhaustible supply of social, economic, and vital statistics at the county, state, and local levels. Use their Web sites to compare state data to national data in the following way.

a. Choose a variable of interest that involves a mean for a particular state (for example, the mean household size in Illinois).

b. Find the current national value for that variable (for example, the national mean household size).

c. Choose a particular county within the state and obtain the corresponding data for the county, as well as the sample size (for example, the mean household size in Cook County, Illinois).

d. Assuming that the county is a random sample of the state, test the claim that the state is above or below the national level in terms of that variable.

e. Discuss and interpret your results. The hypothesis test depends on the sample (the county) being a random sample of the population (the state). Be sure to discuss this factor in your conclusions.

57. Hypothesis Test Applet. Using a search engine such as Google, search for "hypothesis testing" and "applet." Find an applet and run it. Describe how the applet works and what it illustrates.

58. Power of a Test. Using a search engine such as Google, search for "power" of a hypothesis test. Describe what the power of a hypothesis test is.

9.3 Hypothesis Tests for Population Proportions

We now turn to the problem of hypothesis testing with *proportions* (rather than means). All the ideas from previous sections apply, except we need a different method for calculating the standard deviation of the sampling distribution. Let's use an example to illustrate the process.

Suppose a political candidate commissions a poll in advance of a close election. Using a random sample of $n = 400$ likely voters, the poll finds that 204 people support the candidate. Should the candidate be confident of winning? In Chapter 8, we discussed how to determine the margin of error and confidence interval for this type of poll. We now cast the question as a hypothesis test.

For a hypothesis test, we ask whether the poll results (which are the sample statistics) support the hypothesis that the candidate has more than 50% of the vote. As usual, we let p represent the proportion of people in the voting *population* who favor the candidate, and we let $\hat{p}$ denote the proportion of people in the *sample* who favor the candidate. Because 204 of the 400 people in the sample support the candidate, the sample proportion is

$$\hat{p} = \frac{204}{400} = 0.510$$

We can now formulate the null and alternative hypotheses. As usual, we set up our null hypothesis as an equality:

$$H_0\text{: } p = 0.5 \text{ (50\% of voters favor the candidate)}$$

The candidate wants to know if she has majority support, so the alternative hypothesis is right-tailed:

$$H_a\text{: } p > 0.5 \text{ (more than 50\% of voters favor the candidate)}$$

Calculations for Hypothesis Tests with Proportions

How do we determine whether there is enough evidence in the sample to reject the null hypothesis? Following the hypothesis test process on page 376, we see that we have completed the first two steps (formulating the hypotheses and collecting the sample data). Now, just as we did for sample means in Section 9.2, we must determine the likelihood that the sample result could have arisen by chance *if* the null hypothesis is true.

Proceeding as we did with sample means, we imagine selecting many samples of size $n = 400$. For each sample, we compute the proportion of people who favor the candidate. Again, because we have a reasonably large sample size, the distribution of sample proportions should be very close to a normal distribution. Under the starting assumption that the null hypothesis is true (that the proportion of people in the population who favor the candidate is 0.5), the peak of this distribution will be the population proportion claimed by the null hypothesis, $p = 0.5$. The standard deviation of the sampling distribution is given by the following formula (the derivation is beyond the scope of this book):

$$\text{standard deviation of distribution of sample proportions} = \sqrt{\frac{p(1-p)}{n}}$$

For this case, the standard deviation is

$$\sqrt{\frac{p(1-p)}{n}} = \sqrt{\frac{0.5(1-0.5)}{400}} = 0.025$$

TECHNICAL NOTE

This formula for the standard deviation of the distribution of sample proportions is accurate only when $np \geq 5$ and $n(1-p) \geq 5$. These conditions are met for the problems we consider in this book.

Figure 9.7 shows the distribution of sample proportions. We consider the sample proportion from the poll ($\hat{p} = 0.510$) as a single point in this sampling distribution. Much as in hypothesis tests with means (see Figure 9.1), we observe the following:

- If the sample result is close to the peak of the sampling distribution, then we have no reason to think the null hypothesis is wrong and we do not reject the null hypothesis.
- If the sample result is far from the peak of the sampling distribution, then the more likely explanation is that the sampling distribution does not really peak where the null hypothesis claims, in which case we reject the null hypothesis.

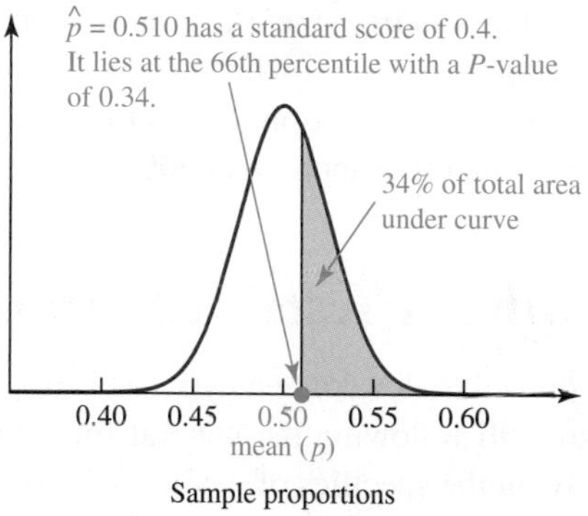
$\hat{p}$ = 0.510 has a standard score of 0.4.
It lies at the 66th percentile with a P-value of 0.34.
34% of total area under curve
0.40
0.45
0.50
0.55
0.60
mean (p)
Sample proportions

Figure 9.7 The distribution of sample proportions for an election poll with a mean of $p = 0.5$ and a standard deviation of 0.025. The sample proportion $\hat{p} = 0.510$ has a standard score of 0.4 and a *P*-value of 0.34.

Again as we did for sample means, we decide whether the sample proportion is "near" to or "far" from the peak of the sampling distribution (assuming the null hypothesis is true) by quantifying the distance with a standard score. The formula for the standard score (z) always has the same general form (see page 209); for the distribution of sample proportions, the formula becomes

$$z = \frac{\text{sample proportion} - \text{population proportion}}{\text{standard deviation of sampling distribution}} = \frac{\hat{p} - p}{\sqrt{p(1-p)/n}}$$

We can now compute and interpret the standard score for our current example. The sample size is the $n = 400$ people interviewed, the sample proportion is the 51% of the sample that supports the candidate ($\hat{p} = 0.510$), and the population proportion is the $p = 0.5$ claimed by the null hypothesis. Therefore, the standard score for the sample proportion is

$$z = \frac{\hat{p} - p}{\sqrt{p(1-p)/n}} = \frac{0.510 - 0.5}{0.025} = 0.4$$

In other words, the sample used in the poll has a proportion that is 0.4 standard deviation above the peak of the distribution of sample proportions. As you can see in Figure 9.7, this sample result is *not* very extreme, indicating that we should not reject the null hypothesis. That is, the candidate cannot be confident of majority support.

By the Way ...

In the 1948 presidential election, most pollsters and newspapers predicted a large victory for the Republican Dewey over the Democrat Truman. In fact, Truman won by a popular vote margin of 49.5% to 45.1%.

Standard Score for the Sample Proportion in a Hypothesis Test

Given the sample size (n), the sample proportion ($\hat{p}$), and the claimed population proportion (p), the standard score for the sample proportion is

$$z = \frac{\hat{p} - p}{\sqrt{p(1-p)/n}}$$

Significance Levels and *P*-Values

We can be more quantitative by using a significance level or *P*-value. Let's start with significance. Because the election poll example uses a right-tailed test, we find the critical values for significance in Table 9.1 (page 383). The standard score of $z = 0.4$ is *not* greater than the critical value of $z = 1.645$ for significance at the 0.05 level, confirming that the result should not cause us to reject the null hypothesis. To find the *P*-value, we use the tables in Appendix A. The tables show that the area to the *left* of a standard score of $z = 0.4$ is 0.6554. Because this is a right-tailed test, we subtract this value from 1 to find the area to the right (shown in Figure 9.7), which is the *P*-value for this hypothesis test. That is, the *P*-value is $1 - 0.6554 = 0.3446$, telling us that if the null hypothesis is true, there is a more than 0.34 chance of randomly selecting a sample as extreme as the one found in this poll. With such a high probability of drawing such a sample by chance, we have no reason to reject the null hypothesis, so the candidate cannot assume that she has the support of more than 50% of voters.

Summary of Hypothesis Tests with Proportions

We can now summarize the procedure for doing a hypothesis test with a population proportion. Be sure to notice that we are still following the general four-step hypothesis test process on page 376, and here focus only on the specifics of dealing with proportions.

- Because we are dealing with population proportions, the null hypothesis has the form p = claimed value. To decide whether to reject or not reject the null hypothesis, we must determine whether a sample as extreme as the one found in the hypothesis test is likely or unlikely to occur if the null hypothesis is true.
- We determine this likelihood from the standard score (z) of the sample proportion, which we compute from the formula

$$z = \frac{\hat{p} - p}{\sqrt{p(1 - p)/n}}$$

  where n is the sample size, $\hat{p}$ is the sample proportion, and p is the population proportion claimed by the null hypothesis.
- We then use the standard score in the same two ways it was used for hypothesis tests with means:
  1. We can assess its level of statistical significance by comparing the standard score to the critical values given in Table 9.1 (page 383) for one-tailed tests and in the box on page 386 for two-tailed tests.
  2. We can determine its *P*-value with standard score tables like those in Appendix A. For a left-tailed test, the *P*-value is the area under the normal curve to the left of the standard score; for a right-tailed test, it is the area under the normal curve to the right of the standard score; and for a two-tailed test, it is double the value we would find if we were calculating the *P*-value for a one-tailed test.
- If the result is statistically significant at the chosen level (usually either the 0.05 or the 0.01 significance level), we reject the null hypothesis. If it is not statistically significant, we do not reject the null hypothesis.

EXAMPLE 1 Local Unemployment Rate

Suppose the national unemployment rate is 3.5%. In a survey of $n = 450$ people in a rural Wisconsin county, 22 people are found to be unemployed. County officials apply for state aid based on the claim that the local unemployment rate is higher than the national average. Test this claim at a 0.05 significance level.

Solution Let's follow the original four-step process from page 376.

Step 1. The unemployment rate is a population proportion (as opposed to a population mean). The null hypothesis is the assumption that the local unemployment rate is equal to the national rate, or H_0: $p = 0.035$. The alternative hypothesis is the county's claim that the local unemployment rate is higher than the national average, or H_a: $p > 0.035$. Notice that this will be a right-tailed test.

Step 2. The sample statistics are the sample size, $n = 450$, and sample proportion, $\hat{p}$. From the given data, the sample proportion is

$$\hat{p} = \frac{22}{450} = 0.0489$$

Step 3. We now determine the likelihood that, *under the assumption that the null hypothesis is true*, chance alone would yield a sample proportion at least as extreme as the $\hat{p} = 0.0489$ found for this particular sample. To find this likelihood, we start by calculating the standard score for this sample proportion:

$$z = \frac{\hat{p} - p}{\sqrt{p(1 - p)/n}} = \frac{0.0489 - 0.035}{\sqrt{0.035(1 - 0.035)/450}} = 1.60$$

This standard score is close to the critical value of $z = 1.645$ for a right-tailed test, but it is not greater than this value, so the result is not significant at the 0.05 level. From Appendix A, the standard score of 1.60 has a P value of $1 - 0.9452 = 0.0548$, confirming that there is a greater than 0.05 probability that this sample would arise by chance if the null hypothesis is true. Figure 9.8 shows these ideas on the sampling distribution.

Step 4. Because the test does not quite meet the criterion for significance at the 0.05 level, we do not reject the null hypothesis. In other words, this sample does not provide sufficient evidence to support the claim that the county unemployment rate is above the national average.

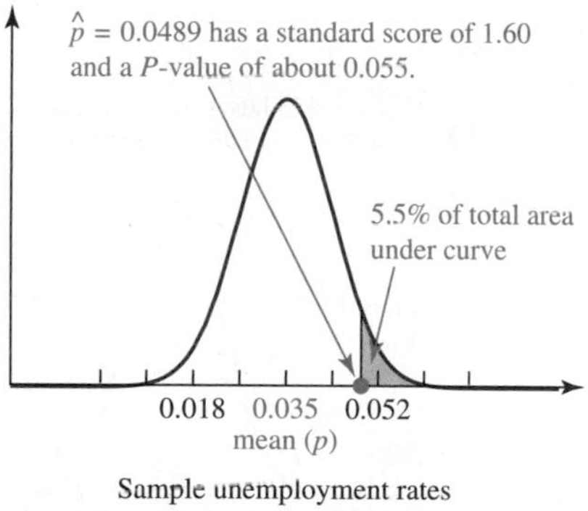

Figure 9.8 The distribution of sample proportions for Example 1. The sample proportion of $\hat{p} = 0.0489$ is not quite far enough to the right for significance at the 0.05 level.

EXAMPLE 2 Left-Handed Population

A random sample of $n = 750$ people is selected, of whom 92 are left-handed. Use these sample data to test the claim that 10% of the population is left-handed.

Solution We again follow the four-step process.

Step 1. The claim concerns the *proportion* of the population that is left-handed, so this is a test with a population proportion. The null hypothesis is the claim that 10% of the population is left-handed, or H_0: $p = 0.1$. To test this claim, we need to account for the possibility that the actual population proportion is either less than *or* greater than 10%. Therefore, the alternative hypothesis is H_a: $p \neq 0.1$, which calls for a two-tailed test.

Step 2. The sample statistics are the sample size, $n = 750$, and the proportion of left-handed people in the sample:

$$\hat{p} = \frac{92}{750} = 0.123$$

Step 3. The standard score for this sample proportion ($\hat{p} = 0.123$) is

$$z = \frac{\hat{p} - p}{\sqrt{p(1 - p)/n}} = \frac{0.123 - 0.1}{\sqrt{0.1(1 - 0.1)/750}} = 2.09$$

From the box on page 386, the critical values for significance at the 0.05 level in a two-tailed test are standard scores less than −1.96 or greater than 1.96. The standard score of 2.09 for this test is greater than 1.96, so we conclude that the test is significant at the 0.05 level. From Appendix A, the area to the right of a standard score of 2.09 is $1 - 0.9817 = 0.0183$. This would be the *P*-value if we were conducting a one-tailed test. Because we have a two-tailed test, we double it to find that the *P*-value is $2 \times 0.0183 = 0.0366$. Figure 9.9 shows the meaning of this *P*-value on the sampling distribution.

Step 4. Because the test is significant at the 0.05 level, we reject the null hypothesis and conclude that the proportion of the population that is left-handed is *not* equal to 10%. Remembering that the significance level is the probability of a type I error, we recognize that there is a 0.05 probability that we made a type I error of rejecting a hypothesis that is actually true.

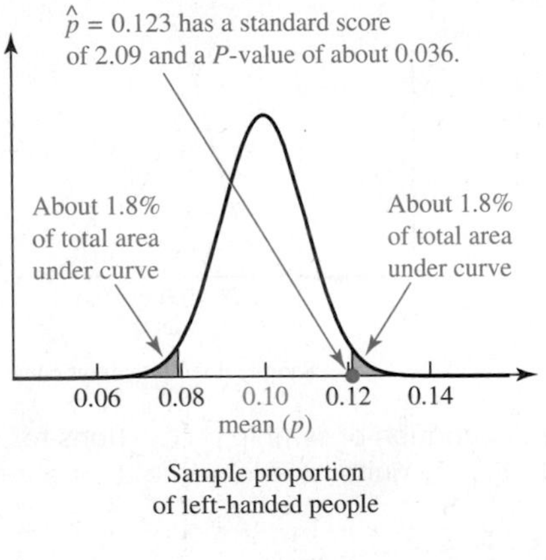

Figure 9.9 This graph shows the position of the sample proportion ($\hat{p} = 0.123$) on the distribution of sample proportions for Example 2. Because this is a two-tailed test, the *P*-value corresponds to the area under the curve more than 2.09 standard deviations from the peak in *either* direction.

Section 9.3 Exercises

Statistical Literacy and Critical Thinking

1. **Notation.** What do p, $\hat{p}$, and P-value represent?

2. **Distribution.** In conducting a hypothesis test as described in this section, what particular distribution is used? Why?

3. **Sampling.** America OnLine conducts a survey in which Internet users are asked to respond to a question. Among the 96,772 responses, there are 76,885 *yes* responses. Is it valid to use these sample results for testing the claim that the majority of the general population answers *yes* to this question? Why or why not?

4. ***P*-Value.** A P-value of 0.00001 is obtained when sample data are used to test the claim that the majority of car crashes occur within 5 miles of home. What does this P-value tell us?

Does It Make Sense? For Exercises 5–8, decide whether the statement makes sense (or is clearly true) or does not make sense (or is clearly false). Explain clearly. Not all of these statements have definitive answers, so your explanation is more important than your chosen answer.

5. **Null Hypothesis.** In a test of the claim that a majority of Americans favor registration of all handguns, the null hypothesis is $p = 0.5$.

6. **Alternative Hypothesis.** In a test of the claim that a majority of Americans favor registration of all handguns, the alternative hypothesis is $p > 0.5$.

7. **Alternative Hypothesis.** In a two-tailed hypothesis test of a claim about a proportion, the P-value is the area to the right of the standard score, z.

8. **Alternative Hypothesis.** It is claimed that $p > 0.5$ can never be supported if the sample proportion $\hat{p}$ is less than 0.5.

Concepts and Applications

Hypothesis Tests. For Exercises 9–18, use a 0.05 significance level to conduct a hypothesis test using the four-step procedure described in the text. Be sure to state your conclusion.

9. **Voter Poll.** In a pre-election poll, a candidate for district attorney receives 205 of 400 votes. Assuming that the people polled represent a random sample of the voting population, test the claim that a majority of voters support the candidate.

10. **Women in College.** According to the U.S. Census Bureau, 58% of college students 25 years of age or older are women. Suppose that at Clarion College, in a random sample of 2,100 students who are 25 years of age or older, 1,234 are women. Test the claim that the percentage of older women at the college is above the national average.

11. **Grade Pressure.** A study commissioned by the U.S. Department of Education, based on responses from 1,015 randomly selected teenagers, concluded that 44% of teenagers cite grades as their greatest source of pressure. Test the claim that fewer than half of all teenagers in the population feel that grades are their greatest source of pressure.

12. **Married Adults.** In a recent year, 125.8 million adults, or 58.6% of the adult American population, were married. In a New England town, a simple random sample of 1,445 adults includes 56.0% who are married. Test the claim that this sample comes from a population with a married percentage of less than 58.6%.

13. **Drug Use.** A Department of Health and Human Services study of illegal drug use among 12- to 17-year-olds reported a decrease in use (from 11.4% in 1997) to 9.9% now. Suppose a survey in a large high school reveals that, in a random sample of 1,050 students, 98 report using illegal drugs. Test the principal's claim that illegal drug use in her school is below the current national average.

14. **Poverty.** According to recent estimates, 12.1% of the 4,342 people in Custer County, Idaho, live in poverty. Assume that the people in this county represent a random sample of all people in Idaho. Based on this sample, test the claim that the poverty rate in Idaho is less than the national rate of 13.3%.

15. **Abortion Survey.** An annual survey of first-year college students, conducted by the Higher Education Research Institute at UCLA, asks approximately 276,000 students about their attitudes on a variety of subjects. According to a recent survey, 51% of first-year students believe that abortion should be legal (down from 65% in 1990). Test the claim that over half of all first-year students believe that abortion should be legal.

16. **Smoking.** The smoking rate for the entire U.S. population of adults (over 21 years of age) is 32% (U.S. National Institute on Drug Abuse). Suppose that for a sample of 75 fine arts students over 21 years old, the smoking rate is 35%. Use a 0.05 significance level to test the claim that the smoking rate for all fine arts students is higher than the national average.

17. **Natural Gas Use.** According to the Energy Information Administration, 53.0% of households nationwide used

natural gas for heating in 1997. A recent survey of 3,600 randomly selected households showed that 54.0% used natural gas. Use a 0.05 significance level to test the claim that the 53.0% national rate has changed.

18. Clinical Test. In clinical tests of the drug Lipitor, 863 patients were treated with the drug and 19 of them experienced flu symptoms (based on data from Parke-Davis). Test the claim that the percentage of treated patients with flu symptoms is greater than the 1.9% rate for patients not given treatments.

Projects for the Internet and Beyond

For useful links, select "Links for Internet Projects" for Chapter 9 at www.aw.com/bbt.

19. Left-Handedness. Given the claim that 10% of Americans are left-handed, randomly select at least 50 students at your college and determine whether they are left-handed. Test the claim with a formal hypothesis test.

20. Smoking. Use the Internet or library references to determine the proportion of Americans who smoke. Test the claim that the proportion of students at your college who smoke is different from the proportion of all Americans. Collect sample data from at least 50 randomly selected students.

21. Women College Students. Use the Internet or library references to find the proportion of college students in the United States who are women. Test the claim that the proportion of women students at your college is different from the proportion of all U.S. college students. Collect sample data from at least 100 randomly selected students.

22. County Data. The *Statistical Abstract of the United States* and the Current Population Survey provide an extensive supply of social, economic, and vital statistics at the county, state, and local levels. Use their Web sites to compare state data to national data in the following way.

a. Choose a variable of interest that involves a proportion for a particular state (for example, the percentage of people living in poverty in Arizona).

b. Find the current national value for that variable (for example, the national poverty rate).

c. Choose a particular county within the state and obtain the corresponding data for the county (for example, the poverty rate in Pima County, Arizona).

d. Assuming that the county is a random sample of the state, test the claim that the state is above or below the national level in terms of that variable.

e. Discuss and interpret your results. The hypothesis test depends on the sample (the county) being a random sample of the population (the state). Be sure to discuss this factor in your conclusions.

IN THE NEWS

23. Hypothesis Testing in the News. Find a news article or research report that describes (perhaps not explicitly) a hypothesis test for a population proportion. Attach the article and summarize the method used.

Chapter Review Exercises

1. Randomly selected cans of Coke are measured for the amount of cola, in ounces. The sample values listed below have a mean of 12.19 ounces and a standard deviation of 0.11 ounce. Assume that we want to use a 0.05 significance level to test the claim that cans of Coke have a mean amount of cola greater than 12 ounces. Assume that the population has a standard deviation given by $\sigma = 0.115$ ounce.

12.3	12.1	12.2	12.3	12.2	12.3	12.0	12.1	12.2
12.1	12.3	12.3	11.8	12.3	12.1	12.1	12.0	12.2
12.2	12.2	12.2	12.2	12.2	12.4	12.2	12.2	12.3
12.2	12.2	12.3	12.2	12.2	12.1	12.4	12.2	12.2

a. What is the null hypothesis?

b. What is the alternative hypothesis?

c. What is the value of the standard score for the sample mean of 12.19 ounces?

d. What is the critical value?

e. What is the *P*-value?

f. What do you conclude? (Be sure to address the original claim that the mean is greater than 12 ounces.)

g. Describe a type I error for this test.

h. Describe a type II error for this test.

i. Find the *P*-value if the test is modified to test the claim that the mean is *different from* 12 ounces (instead of being greater than 12 ounces).

2. In a study of smokers who tried to quit smoking with nicotine patch therapy, 39 were smoking one year after the treatment, and 32 were not smoking one year after the treatment (based on data from "High Dose Nicotine Patch Therapy," by Dale et al., *Journal of the American Medical Association*, Vol. 274, No. 17). We want to use a 0.05 significance level to test the claim that among smokers who try to quit with nicotine patch therapy, the majority are smoking a year after the treatment.

a. What is the null hypothesis?

b. What is the alternative hypothesis?

c. What is the value of the standard score for the sample proportion?

d. What is the critical value?

e. What is the *P*-value?

f. What do you conclude? (Be sure to address the original claim that among smokers who try to quit with nicotine patch therapy, the majority are smoking a year the treatment.)

g. Describe a type I error for this test.

h. Describe a type II error for this test.

i. What is the *P*-value if the claim is modified to state that the proportion is *equal* to 0.5 ?

3. We want to test the claim that the Clarke method of gender selection is effective in increasing the likelihood that a newborn baby will be a girl. In a random sample of 80 couples who use the Clarke method, it is found that there are 35 girls among the 80 newborn babies. What should we conclude about the claim? Why is it *not* necessary to go through all the steps of a formal hypothesis test?

4. Mike Shanley wants to test the hypothesis that adult men have a mean weight greater than 150 pounds. He surveys the adult male members of his family and obtains 40 sample values, which lead him to support the alternative hypothesis that the population mean is greater than 150 pounds. What is the fundamental flaw in his procedure?

Chapter Quiz

1. What is the alternative hypothesis that results from the claim that the proportion of men who exercise is greater than 0.2?
2. What is the null hypothesis that is used for testing the claim that the proportion of men who exercise is greater than 0.2?
3. What is the alternative hypothesis that results from the claim that the proportion of convicted felons who serve time in prison is equal to 0.6?
4. Is a test of the claim that $p > 0.75$ left-tailed, right-tailed, or two-tailed?

In Exercises 5–10, assume that we want to use a 0.05 significance level to test the claim that the mean IQ score of professional comedians is greater than 110.

5. What is the null hypothesis?
6. What is the alternative hypothesis?
7. If the test results in a *P*-value of 0.2500, what do you conclude about the given claim?
8. What are the two possible conclusions that can be reached about the null hypothesis?
9. What are the two possible conclusions that can be reached about the claim being tested?
10. If you incorrectly conclude that professional comedians have a mean IQ score greater than 110 when their actual mean IQ score is 100, have you made a type I error or a type II error?

Using Technology

In describing methods for testing hypotheses about a population mean, this chapter considers only cases in which the normal distribution applies. Chapter 10 introduces methods that use a *t* distribution. See the procedures at the end of Chapter 10 for those other methods.

SPSS

Hypothesis test for mean: SPSS does not have a procedure for conducting a hypothesis test for a claim about a population mean when the population standard deviation is known. See the technology section at the end of Chapter 10 for a procedure for testing a claim about a population mean when the population standard deviation is not known.

Hypothesis test for proportion: SPSS does not have a procedure for conducting a hypothesis test for a claim about a population proportion.

Excel

Testing a claim about a population mean: Use the Data Desk XL add-in that is a supplement to Excel.

First enter the sample data in column A. If you are using Excel 2003, click on **DDXL.** If you are using Excel 2007, click on **Add-Ins,** then click on **DDXL.** Select **Hypothesis Tests.** Under the function type options, select **1 Var z Test.** Click on the pencil icon and enter the range of data values, such as A1:A45 if you have 45 values listed in column A. Click on **OK.** Follow the four steps listed in the dialog box. After clicking on **Compute** in Step 4, you will get the *P*-value, test statistic, and conclusion.

Testing a claim about a population proportion: Use the Data Desk XL add-in that is a supplement to Excel.

First enter the number of successes in cell A1, and enter the total number of trials in cell B1. If you are using Excel 2003, click on **DDXL.** If you are using Excel 2007, click on **Add-Ins,** then click on **DDXL.** Select **Hypothesis Tests.** Under the function type options, select **Summ 1 Var Prop Test** (for testing a claimed proportion using summary data for one variable). Click on the pencil icon for "Num successes" and enter A1. Click on the pencil icon for "Num trials" and enter B1. Click on **OK.** Follow the four steps listed in the dialog box. After clicking on **Compute** in Step 4, you will get the *P*-value, test statistic, and conclusion.

STATDISK

Select **Analysis,** then select **Hypothesis Testing.** For the methods discussed in this chapter, select either **Mean - One Sample** or **Proportion - One Sample.** A dialog box will appear. Click on the box in the upper left corner and select the item that matches the claim being tested. Proceed to enter the other items in the dialog box, then click on **Evaluate.** The results will include the test statistic and *P*-value.

FOCUS ON HEALTH & EDUCATION

Will Your Education Help You Live Longer?

In this chapter, we focused on hypothesis testing in its simplest form, in which we test a single claim about a population and determine whether it is supported by evidence collected from a single sample. Not surprisingly, many of the most interesting statistical problems require much more complex analysis. Consider, for example, the question of what factors contribute to longevity (longer life).

We often hear about factors that can shorten our lives. For example, smoking, excessive alcohol consumption, and obesity are all factors that tend to make people die young. Knowing such facts can help you decide what to avoid in your lifestyle. But are there things you can do to increase your life span?

Although this question is much more difficult to answer than anything we encountered in this chapter, it can be addressed through the same basic principles. Suppose, for example, that a researcher suspects that eating oats can make you live longer. To test this suspicion, the researcher conducts a hypothesis test. He or she starts with the null hypothesis that oat consumption has no effect on life span, then examines data to look for evidence that would support rejecting this null hypothesis and concluding that oat consumption really does increase life span. The difficult parts of this research are collecting the data—for example, finding people who consume oats to be compared to people who don't—and then finding a way to separate the effects of oat consumption from those of the huge number of other variables that may also affect life span.

Over the past few decades, researchers have identified many factors that appear to contribute to longevity. For example, greater wealth is correlated with longer life. Race plays a role in life span. Diet has numerous health effects. Exercise is generally a positive factor for longer life. Somewhat surprisingly, however, one factor appears to be more important than all the others: years of education. The longer you stay in school, the longer you'll live, at least on average. Figures 9.10 and 9.11 show some of the data that support this conclusion.

Why might more education lead to longer life? Researchers have suggested many possible reasons. For example, people with more education tend to be less likely to engage in life-shortening behaviors such as smoking or excessive drinking, while also undertaking exercise and other activities that can increase life span. One particularly interesting hypothesis is that getting an education always involves some measure of short-term sacrifice for longer-term gain (such as paying for your education now in hopes of getting a higher-paying job later), and this willingness to accept delayed gratification helps people make all kinds of decisions that contribute to longer life. Of course, no one yet knows for sure—so you can be sure there is much more hypothesis testing to be done.

Nevertheless, based on the current data, we can draw at least one very interesting conclusion directly relevant to your reading of this book: You may be taking a statistics class because it is required, but it may also help you live longer, especially if it allows you to stay in school.

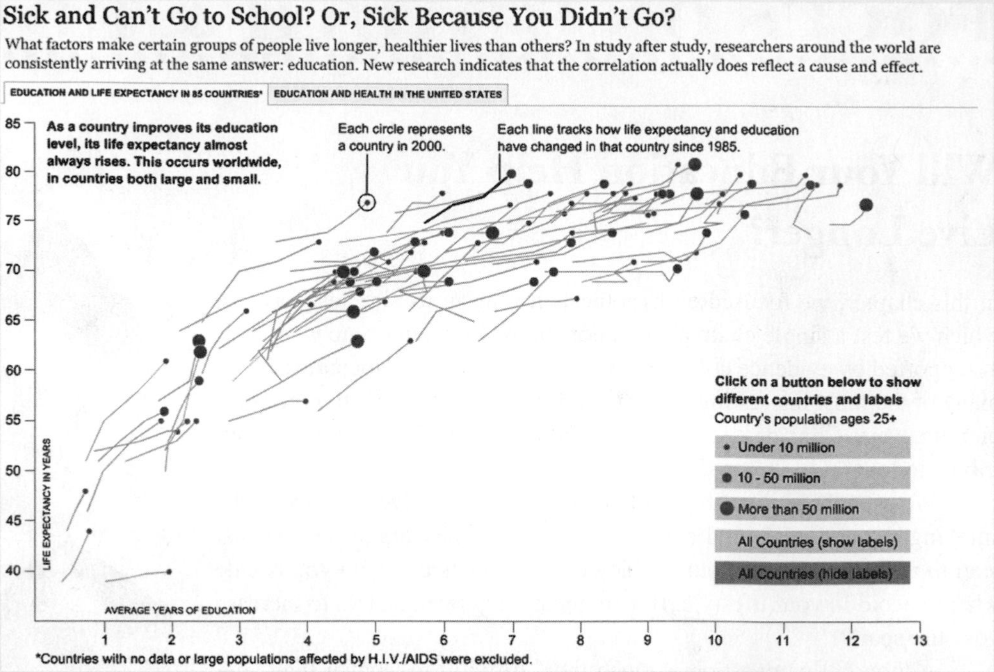

Figure 9.10 This graph plots life expectancy against years of education for many different countries. The line for each country starts at the left with its average life expectancy and years of education in 1985 and ends at the right (at a dot) with those averages for 2000. The fact that nearly all the lines slant upward shows that as years of education have increased, so has life expectancy. *Source: New York Times.*

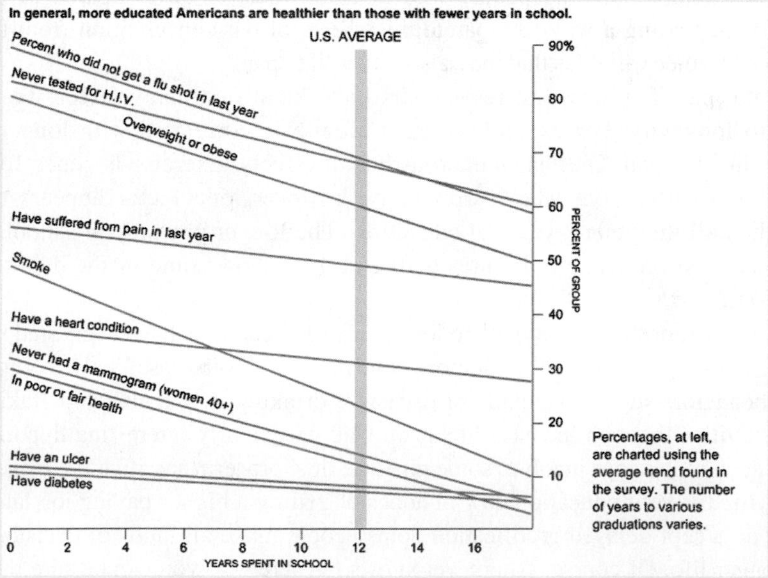

Figure 9.11 This graph shows that the percentages of people with a variety of health risk factors all tend to go down with the number of years spent in school. For example, the top line shows that some 90% of people with no formal education did not get a flu shot in the past year, while the same was true of only about 60% of people with 16 years of education. The fact that so many risk factors go downward with education suggests that education may be the common underlying cause that leads to healthier lifestyles. *Source: New York Times.*

QUESTIONS FOR DISCUSSION

1. Figures 9.10 and 9.11 both present a wealth of data. Interpret each graph carefully. Be sure you understand the meaning of each line (or curve) shown. How does each graph support the hypothesis that more education leads to longer life? Explain.

2. Why do *you* think education contributes to longer life? For example, do you think the mere fact of staying in school makes people live longer, or is it the extra learning associated with more years of schooling that makes the difference? How would you test your hypothesis?

3. Smoking is well known to be one of the biggest risk factors in disease and early death. As a result, the government and health care groups have often undertaken expensive advertising campaigns in hopes of convincing people to stop smoking (or never start). But Figure 9.11 shows that smoking drops dramatically with years of education. Therefore, if the goal is to reduce smoking, one might argue that it would make more sense to spend money on programs to help people stay in school than to spend money on anti-smoking campaigns. Do you think this is a good idea? Defend your opinion.

4. Based on the finding that education contributes to longevity—along with the fact that longevity correlates with lower health care costs and higher productivity—some people have proposed that the government should increase taxes in order to provide more people with further education. Take a position for or against this proposal, and outline the arguments that support your position.

SUGGESTED READING

Kolata, G. "A Surprising Secret to a Long Life: Stay in School." *New York Times*, January 3, 2007.

Lleras-Muney, Adriana. "The Relationship Between Education and Adult Mortality in the United States." *Review of Economic Studies*, Vol. 72, No. 1, 2005.

FOCUS ON AGRICULTURE

Are Genetically Modified Foods Safe?

British newspapers call them "Frankenfoods" (a word play on *Frankenstein*). Europeans, by and large, don't eat them at all. But today, many foods sold in the United States already fall in this category, including an estimated 36% of American corn and 55% of American soybeans. We are speaking of "genetically modified foods," or GM foods. GM foods are a fairly recent agricultural invention, dating only from about the mid-1990s. Nevertheless, they are at the forefront of one of the biggest debates in agricultural history.

A genetically modified organism is an organism into which scientists have inserted a gene that does not naturally exist in the organism or its close relatives. Genetically modified organisms are being tested and used for many purposes. For example, scientists have developed bacteria containing genes that produce drugs like insulin.

In agriculture, genetic modification allows scientists to create in crops traits that would be difficult or impossible to achieve by traditional breeding techniques. One of the first widely used genetically modified crops, often called "Bt corn," illustrates the idea behind genetic modification. Corn is usually susceptible to destruction by a variety of insect pests. As a result, farmers usually spray crops with pesticides. Unfortunately, many pesticides are toxic to animals besides the insect pests and hence can cause environmental damage (and may not be ideal for humans to consume).

By the Way ...

More than 50 new GM crops have been approved for sale in the United States, including corn and soybeans that produce their own pesticide and tomatoes engineered for longer shelf life. GM crops under development include potatoes that resist bruising, better-tasting soybeans, and grains containing vitamins and other nutrients that could improve nutrition in impoverished countries.

Here's where Bt, a bacterium known as *bacillus thuringiensis*, comes in. Bt lives naturally in soil. As early as 1911, scientists discovered that Bt produces a toxin that kills certain types of insects. Different strains of Bt kill different insects, but are generally harmless to other animals and humans. By the 1960s, these traits had led to the use of Bt as an "environmentally friendly" pesticide that could be sprayed on crops (in the form of killed bacteria). Unfortunately, Bt pesticide proved to be expensive and often ineffective, mainly because it works only if insects eat it (more effective pesticides kill on contact). And, because it breaks down rapidly in sunlight and washes off plants in rain, it kills pests only if its application is timed just right.

Genetic modification solves the problems inherent in Bt pesticide. The pesticide action of Bt bacteria arises from particular proteins that the bacteria produce. Scientists identified the genes responsible for these proteins, then transferred these genes into plants such as corn. Once the corn contains the necessary genes, it produces the same pest-killing proteins as the Bt bacteria. The application of a sprayed pesticide is no longer necessary because the corn itself is now toxic to the pests. Moreover, because the corn continually produces the pest-killing proteins, there are no more concerns about the pesticide breaking down or washing away.

The advantages of the Bt corn over traditional strains of corn are clear, but are there any disadvantages? This is where the great debate over GM foods begins. On one side, many scientists argue that GM foods are completely safe and that their benefits will help improve nutrition for people throughout the world. On the other side stand the people, including some scientists, who label GM foods as "Frankenfoods" and argue that they are one of the most dangerous technologies ever invented.

Broadly speaking, the issue of GM food safety can be broken down into three major questions:

1. Do GM foods have any toxic effects in humans?
2. Do the new proteins contained in GM foods cause allergic reactions in some people?
3. Can the GM crops cause any unforeseen environmental damage, such as transferring their genes into weeds (thereby making "superweeds") or killing animals besides the insect pests?

These questions can be addressed through hypothesis testing. In each case, we begin with a null hypothesis that states that there is no safety difference between traditional foods and GM foods. For example, the null hypothesis for Question 1 says that GM foods are no more toxic than traditional foods (which often contain low levels of toxic chemicals). The alternative hypothesis states that there is a difference in toxicity. Scientists then develop experiments to test the hypotheses. If the evidence provided by the experiments reveals a significant difference in toxicity between the food groups (beyond what would be expected by chance alone), there is reason to reject the null hypothesis.

Unfortunately, experiments to date have been unable to resolve the controversy. For example, some genetically modified crops *do* contain products toxic to humans—and therefore never receive approval for sale. To proponents of GM foods, this fact provides an argument in their favor, because it appears to show that current regulatory procedures (for example, requiring approval of GM foods by the U.S. Food and Drug Administration) adequately ensure that only safe foods enter the marketplace. Similarly, while some of the proteins in GM foods undoubtedly can cause allergic reactions, many scientists think they understand these reactions well enough to ensure that only safe foods are approved.

Opponents of GM foods use the very same experimental results to support their case. They argue that even if scientists have identified toxicity that is obvious in experiments, they may still be missing long-term effects that might not show up in the population for many years. Similarly, they claim that we cannot be sure we understand *all* allergic reactions and therefore might inadvertently approve GM foods that could have severe allergic consequences in at least some people.

The environmental issues are even more difficult to study. For example, one study has shown that Bt corn is toxic to the Monarch butterfly, a species that is not a pest and that no one wants to kill. But the study was conducted in a laboratory and may not accurately represent what occurs in the real environment. Similarly, the possibility of "gene jumping," in which the Bt genes might spread from corn to other plants, is not well understood and therefore is very difficult to study.

The debate over GM foods is likely to continue, and even to intensify, for many years to come. After all, when it comes to food, everyone has an interest.

By the Way ...

Opponents of GM foods often point to the pesticide DDT as an example of how hard it may be to discover long-term toxic or environmental effects. DDT was used for some 60 years before its detrimental effects were discovered.

QUESTIONS FOR DISCUSSION

1. Propose an experiment that could be used to test the safety of a GM food product, such as corn. Describe your experiment in detail, and discuss any practical difficulties that might be involved in carrying it out or interpreting its results.

2. Ignoring any safety issues of GM foods, make a list of as many ways as you can think of in which GM foods might be beneficial to humanity. Then, ignoring any benefits of GM foods, make a list of as many ways as you can think of in which GM foods might be dangerous. Overall, do you think that further use of GM foods should be encouraged or discouraged? Defend your opinion.

3. Some people advocate giving the choice about GM foods to consumers by requiring labeling on all products that contain genetically modified ingredients. Do you think this is a good idea? Why or why not?

4. Investigate recent developments in the debate over GM foods. Does the new information alter any of the major arguments in the debate? Explain.

SUGGESTED READING

The debate over genetically modified foods generates frequent headlines. To find the latest studies, try a Web search on "GM foods."

Belsie, Laurent. "Superior Crops or 'Frankenfood'?" *Christian Science Monitor*, March 1, 2000.

Barrionuevo, Alexei. "Can Gene-Altered Rice Rescue the Farm Belt?" *New York Times*, August 16, 2005.

Rosenthal, Elisabeth. "Biotech Food Tears Rifts in Europe." *New York Times*, June 6, 2006.

Rosenthal, Elisabeth. "Questions on Biotech Crops with No Clear Answers." *New York Times*, June 6, 2006.

Yoon, Carol K. "Pollen from Genetically Altered Corn Threatens Monarch Butterfly, Study Finds." *New York Times*, May 20, 1999.

10

The web of this world is woven of necessity and chance. Woe to him who has accustomed himself to find something capricious in what is necessary, and who would ascribe something like reason to chance.

—Johann Goethe

t Tests, Two-Way Tables, and ANOVA

LEARNING GOALS

10.1 *t* Distribution for Inferences about a Mean

Understand when it is appropriate to use the Student *t* distribution rather than the normal distribution for constructing confidence intervals or conducting hypothesis tests for population means, and know how to make proper use of the *t* distribution.

10.2 Hypothesis Testing with Two-Way Tables

Interpret and carry out hypothesis tests for independence of variables with data organized in two-way tables.

10.3 Analysis of Variance (One-Way ANOVA)

Interpret and carry out hypothesis tests using the method of one-way analysis of variance.

WE HAVE EXPLORED MANY CORE IDEAS AND APPLICATIONS of statistics in Chapters 1–9, but you will run across many more applications if you continue your study of statistics. In this final chapter, we explore three particularly common applications of statistics that you may encounter in future course work; they will also help you further understand the role of statistics in your daily life. All three of these applications build upon the important technique of hypothesis testing introduced in Chapter 9. We begin with the *t* distribution, which applies to confidence intervals as well as hypothesis tests, and then investigate hypothesis tests with two variables and with the method known as analysis of variance, or ANOVA.

10.1 *t* Distribution for Inferences about a Mean

Section 8.2 discussed confidence interval estimates of a population mean. After making the assumption that the distribution of sample means is a normal distribution, we estimated the margin of error, E, to be

$$E \approx \frac{2s}{\sqrt{n}}$$

It was stated in a Technical Note that the precise formula for the margin of error uses 1.96 rather than 2. The value of 1.96 is the standard score z with an area of 0.025 to its right (see Appendix A).

Section 9.2 discussed hypothesis tests for claims about a population mean, again based on the assumption that the sampling distribution is normal. We used the following formula for the standard score of the sample mean:

$$z = \frac{\bar{x} - \mu}{\sigma/\sqrt{n}}$$

Section 9.2 provides these criteria for rejecting the null hypothesis (at the 0.05 level of significance): $z \leq -1.645$ for a left-tailed test, $z \geq 1.645$ for a right-tailed test, and $z \leq -1.96$ or $z \geq 1.96$ for a two-tailed test. The values of -1.645, 1.645, -1.96, and 1.96 are all derived from the standard normal distribution.

In real applications, professional statisticians do not always assume a normal distribution for confidence intervals and hypothesis tests. (In fact, a review of articles in professional journals shows that the normal distribution is rarely used.) A major reason for this is that, as you can see in the standard score formula above, using the normal distribution requires that we know the population standard deviation σ. Because we generally do not know σ, we must estimate it with the sample standard deviation s. Statisticians therefore prefer an approach that does not require knowing σ. Such is the case with the **Student *t* distribution**, or ***t* distribution** for short, which can be used when we do not know the population standard deviation and either the sample size is greater than 30 or the population has a normal distribution.

By the Way ...

The Student *t* distribution was developed by William Gosset (1876–1937), a Guinness Brewery employee who needed a distribution that could be used with relatively small samples. The Irish brewery where he worked did not allow publication of research results, so Gosset published under the pseudonym *Student*.

Inferences about a Population Mean: Choosing between *t* and Normal Distributions

***t* distribution:**	Population standard deviation is not known and the population is normally distributed.
or	Population standard deviation is not known and the sample size is greater than 30.
Normal distribution:	Population standard deviation is known and the population is normally distributed.
or	Population standard deviation is known and the sample size is greater than 30.

We won't go into detail about the nature of the *t* distribution in this book, but the basic idea is easy to understand: The *t* distribution is very similar in shape and symmetry to the normal distribution, but it accounts for the greater variability that is expected with small samples. Figure 10.1 contrasts the *t* distribution with the standard normal distribution for sample sizes of

$n = 3$ and $n = 12$. Notice that for the larger sample size, the t distribution is closer to the normal distribution. In fact, the larger the sample, the more closely the t distribution matches the normal distribution.

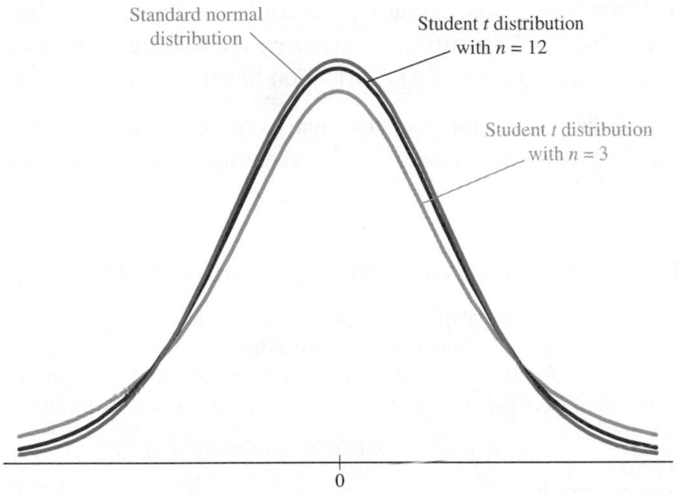

Figure 10.1 This figure compares the standard normal distribution to the t distribution for two different sample sizes. Notice that as the sample size gets larger, the t distribution more closely approximates the normal distribution.

The real value of the t distribution, then, is that it allows us to extend ideas of confidence intervals or hypothesis tests to many cases in which we cannot use the normal distribution because we do not know the population standard deviation. Keep in mind, however, that it still does not work for all cases. For example, if we have a small sample of size 30 or less or the sample data suggest that the population has a distribution which is radically different from a normal distribution, then neither the t distribution nor the normal distribution applies. Such cases require other methods not discussed in this book.

Confidence Intervals Using the *t* Distribution

Much as we did in Sections 8.2 and 9.2 under the assumption of a normal distribution, we can use the t distribution to construct a confidence interval for a population mean or to conduct a hypothesis test for a population mean. However, instead of determining significance based on standard z-scores such as -1.645, 1.645, -1.96, and 1.96, we use values of t, which are shown in tables similar to Table 10.1. (Statistical software can also generate t values.)

Let's start with confidence intervals. To specify a confidence interval, we must first calculate the margin of error, E. With a t distribution, the formula is

$$E = t \cdot \frac{s}{\sqrt{n}}$$

where n is the sample size, s is the sample standard deviation, and t is a value that we look up in Table 10.1. The only "trick" is in finding the correct value of t from the table, which we do as follows:

- First, determine the number of **degrees of freedom** for the sample data, defined to be the sample size minus 1:

$$\text{degrees of freedom for } t \text{ distribution} = n - 1$$

Table 10.1 Critical Values of t

Degrees of freedom ($n - 1$)	Area in one tail 0.025	0.05
	Area in two tails 0.05	**0.10**
1	12.706	6.314
2	4.303	2.920
3	3.182	2.353
4	2.776	2.132
5	2.571	2.015
6	2.447	1.943
7	2.365	1.895
8	2.306	1.860
9	2.262	1.833
10	2.228	1.812
11	2.201	1.796
12	2.179	1.782
13	2.160	1.771
14	2.145	1.761
15	2.131	1.753
16	2.120	1.746
17	2.110	1.740
18	2.101	1.734
19	2.093	1.729
20	2.086	1.725
21	2.080	1.721
22	2.074	1.717
23	2.069	1.714
24	2.064	1.711
25	2.060	1.708
26	2.056	1.706
27	2.052	1.703
28	2.048	1.701
29	2.045	1.699
30	2.042	1.697
31	2.040	1.696
32	2.037	1.694
34	2.032	1.691
36	2.028	1.688
38	2.024	1.686
40	2.021	1.684
50	2.009	1.676
100	1.984	1.660
Large	1.960	1.645

- The table shows degrees of freedom in column 1. Find the row corresponding to the number of degrees of freedom in your sample data, and then look across the row to find the appropriate t value. For confidence intervals with population means, the t values correspond to 95% confidence in column 2 and 90% confidence in column 3. (We use the table values for the "area in two tails" because the margin of error can be either below the mean or above it; for example, 95% confidence means we are looking for a total area of 0.05 both to the far left and to the far right of a t distribution like those shown in Figure 10.1.)

Once you find the t value for your data and confidence level, you can determine the confidence interval just as we did in Section 8.2, except using the new formula for the margin of error, E.

Confidence Interval for a Population Mean (μ) with the t Distribution

If conditions require use of the t distribution (σ not known and $n > 30$ *or* population normally distributed), the confidence interval for the true value of the population mean (μ) extends from the sample mean minus the margin of error ($\bar{x} - E$) to the sample mean plus the margin of error ($\bar{x} + E$). That is, the confidence interval for the population mean is

$$\bar{x} - E < \mu < \bar{x} + E \text{ (or, equivalently, } \bar{x} \pm E)$$

where the margin of error is

$$E = t \cdot \frac{s}{\sqrt{n}}$$

and we find t from Table 10.1.

EXAMPLE 1 Confidence Interval for Diastolic Blood Pressure

Here are five measures of diastolic blood pressure from randomly selected adult men: 78, 54, 81, 68, 66. These five values result in these sample statistics: $n = 5$, $\bar{x} = 69.4$, $s = 10.7$. Using this sample, construct the 95% confidence interval estimate of the mean diastolic blood pressure level for the population of all adult men.

Solution Because the population standard deviation is not known and because it is reasonable to assume that blood pressure levels of adult men are normally distributed, we use the t distribution instead of the normal distribution. With a sample of size $n = 5$, the number of degrees of freedom is

$$\text{degrees of freedom for } t \text{ distribution} = n - 1 = 5 - 1 = 4$$

For 95% confidence, we use column 2 in Table 10.1 to find that $t = 2.776$. We now use this value along with the given sample size ($n = 5$) and sample standard deviation ($s = 10.7$) to calculate the margin of error, E:

$$E = t \cdot \frac{s}{\sqrt{n}} = 2.776 \cdot \frac{10.7}{\sqrt{5}} = 13.3$$

Finally, we use the margin of error and the sample mean to find the 95% confidence interval:

$$\bar{x} - E < \mu < \bar{x} + E$$
$$69.4 - 13.3 < \mu < 69.4 + 13.3$$
$$56.1 < \mu < 82.7$$

Based on the five sample measurements, we have 95% confidence that the limits of 56.1 and 82.7 contain the mean diastolic blood pressure level for the population of all adult men.

Hypothesis Tests Using the t Distribution

When the t distribution is used for a hypothesis test of a claim about a population mean (H_0: μ = claimed value), the t value plays the role that the standard score z played when we studied these hypothesis tests in Section 9.2. Recall that we determined statistical significance by comparing the standard score of the sample mean to critical values or by finding its P-value. With the t distribution, instead of calculating the standard score z, we use the following formula to calculate t:

$$t = \frac{\bar{x} - \mu}{s / \sqrt{n}}$$

where n is the sample size, $\bar{x}$ is the sample mean, s is the sample standard deviation, and μ is the population mean claimed by the null hypothesis.

Once we have calculated t, we decide whether to reject or not reject the null hypothesis by comparing our value of t to the critical values of t found in Table 10.1. The critical values depend on the type of test as follows.

Right-tailed test: Reject the null hypothesis if the computed test statistic t is greater than or equal to the value of t found in the column of Table 10.1 labeled "Area in one tail." Notice that for the one-tailed test, column 2 gives critical values for significance at the 0.025 level and column 3 gives critical values for significance at the 0.05 level.

Left-tailed test: Reject the null hypothesis if the computed test statistic t is less than or equal to the negative of the value of t found in the column of Table 10.1 labeled "Area in one tail." Again, because this is a one-tailed test, column 2 gives critical values for significance at the 0.025 level and column 3 gives critical values for significance at the 0.05 level.

Two-tailed test: Reject the null hypothesis if the absolute value of the computed test statistic t is greater than or equal to the value of t found in the column of Table 10.1 labeled "Area in two tails." For this case, column 2 gives critical values for significance at the 0.05 level and column 3 gives critical values for significance at the 0.10 level.

The computed test statistic t can also be used to find a P-value; however, that is usually done with the aid of statistical software rather than with tables.

EXAMPLE 2 Right-Tailed Hypothesis Test for a Mean

Listed below are ten randomly selected IQ scores of statistics students:

111 115 118 100 106 108 110 105 113 109

Using methods from Chapter 4, you can confirm that these data have the following sample statistics: $n = 10$, $\bar{x} = 109.5$, $s = 5.2$. Using a 0.05 significance level, test the claim that statistics students have a mean IQ score greater than 100, which is the mean IQ score of the general population.

Solution Based on the claim that the mean IQ of statistics students is greater than 100, we use the null hypothesis H_0: $\mu = 100$ and the alternative hypothesis H_a: $\mu > 100$. Because the standard deviation of all IQ scores for the population of all statistics students is not known and because it is reasonable to assume that IQ scores of statistics students are normally distributed, we use the t distribution instead of the normal distribution. The value of the t test statistic is computed as follows:

$$t = \frac{\bar{x} - \mu}{s / \sqrt{n}} = \frac{109.5 - 100}{5.2 / \sqrt{10}} = 5.777$$

We now need to compare this value to the appropriate critical value from Table 10.1:

- We find the correct row by recognizing that this data set has $n - 1 = 10 - 1 = 9$ degrees of freedom.
- Because it is a one-tailed test and we are asked to test for significance at the 0.05 level, we use the values from column 3.
- Looking in the row for 9 degrees of freedom and column 3, we find that the critical value for significance at the 0.05 level is $t = 1.833$.

Because the sample test statistic $t = 5.777$ is greater than the critical value $t = 1.833$, we reject the null hypothesis. We conclude that there is sufficient evidence to support the claim that the mean IQ score is greater than 100.

We can be more precise by using software to compute the *P*-value for this hypothesis test, which turns out to be 0.000135. Notice that this *P*-value is much less than 0.05, so we can be quite confident in the decision to reject the null hypothesis and support the claim that the mean IQ score is greater than 100.

EXAMPLE 3 Two-Tailed Hypothesis Test for a Mean

Using the sample data in Example 2 and the same significance level of 0.05, test the claim that the mean IQ score of statistics students is *equal to* 100.

Solution Based on the claim that the mean IQ score of statistics students is equal to 100, we use the null hypothesis H_0: $\mu = 100$ and the alternative hypothesis H_a: $\mu \neq 100$. The "not equal to" form of the alternative hypothesis means that this is a two-tailed test.

The value of the sample test statistic is still the same as in Example 2 ($t = 5.777$), but the critical values are different for the two-tailed test. Because we are using the same data set, the number of degrees of freedom is still 9, telling us which row to look at in Table 10.1. But for a two-tailed test, we find the critical value for significance at the 0.05 level in column 2 rather than in column 3. The critical value in row 9 and column 2 is $t = 2.262$.

Notice that the absolute value of the test statistic $t = 5.777$ is greater than the critical value $t = 2.262$, so again we reject the null hypothesis. We conclude that there is sufficient evidence to reject the claim that the mean IQ score of statistics students is equal to 100. (With software, we find that the test statistic has a *P*-value of 0.000269.)

EXAMPLE 4 Left-Tailed Hypothesis Test for a Mean

Because of the expense involved, car crash tests often use small samples. In one study, five BMW cars are crashed under standard conditions, and the repair costs (in dollars) are used to test the claim that the mean repair cost for all BMW cars is less than \$3,000. The five sample repair costs have a mean of \$2,835 with a standard deviation of \$883. Use a 0.05 significance level to test the claim that the population of BMW repair costs has a mean less than \$3,000.

Solution Based on the claim that the mean repair cost is less than \$3,000, the null hypothesis is H_0: $\mu = \$3{,}000$ and the alternative hypothesis is H_a: $\mu < \$3{,}000$. The population standard deviation is not known. We will assume that repair costs are normally distributed. (The hypothesis test for *t* is not very sensitive to small departures from normal distributions, so this is a reasonable assumption.) The value of the *t* test statistic is

$$t = \frac{\bar{x} - \mu}{s / \sqrt{n}} = \frac{2835 - 3000}{883 / \sqrt{5}} = -0.418$$

For this example, the sample size is $n = 5$, so the number of degrees of freedom is $n - 1 = 5 - 1 = 4$. Because this is a one-tailed test, we find the critical values for significance at the 0.05 level in column 3. If you look at row 4 and column 3, you will find the critical value $t = 2.132$. For a *left*-tailed test, we are looking for values less than or equal to the negative of this value, or $t = -2.132$.

Because the sample test statistic $t = -0.418$ is *not* less than or equal to the critical value $t = -2.132$, we do not reject the null hypothesis. There is not sufficient evidence to support the claim that the mean repair cost is less than \$3,000. With software, you would find that the *P*-value for the test statistic is 0.3488. In other words, if the null hypothesis is true, the probability of selecting a sample at least as extreme as the one found here is about 35%. This is a fairly high probability that gives us no reason to think the null hypothesis is wrong.

Section 10.1 Exercises

Statistical Literacy and Critical Thinking

1. ***t* Distribution.** Assume that you want to construct a confidence interval or test a hypothesis about a population mean. Describe the conditions indicating that the *t* distribution should be used.

2. ***t* Distribution.** In constructing a confidence interval estimate of a population mean or testing a hypothesis about a population mean, why is the *t* distribution used so much more often than the normal distribution?

3. **Terminology.** What is the difference between a Student *t* distribution and a *t* distribution?

4. **Sampling Method.** Assume that you want to estimate the mean amount of cash carried by people in the United States. If you survey 12 of your best friends, can you use the *t* distribution to develop a confidence interval? Will the result be good for estimating the population mean?

Does It Make Sense? For Exercises 5–8, decide whether the statement makes sense (or is clearly true) or does not make sense (or is clearly false). Explain clearly. Not all of these statements have definitive answers, so your explanation is more important than your chosen answer.

5. ***t* Test.** In testing a hypothesis about a population mean with a sample that has fewer than 30 values, the *t* distribution is always used.

6. **Sampling Method.** In testing a hypothesis about a population mean with a sample that appears to be from a normally distributed population with an unknown population standard deviation, the *t* distribution can be used regardless of the method that was used to collect the sample data.

7. ***t* Test.** When using the *t* distribution to test the claim that students have a mean IQ score greater than 105, if the sample mean is 103, the hypothesis test is not necessary and we can conclude that there is not sufficient data to support the claim.

8. ***t* vs. Normal Distribution.** Because the *t* test does not require a known value of the population standard deviation and because the value of the population standard deviation is rarely known, the *t* test is used in real situations much more often than the hypothesis test using the normal distribution.

Concepts and Applications

Confidence Intervals. In Exercises 9–16, use the *t* distribution to construct the confidence interval estimate of the population mean.

9. **IQ Scores.** A simple random sample of IQ scores is selected from a normally distributed population. The sample statistics are $n = 25$, $\bar{x} = 102.5$, $s = 12.8$. Construct the 95% confidence interval estimate of the population mean.

10. **Heights of NBA Players.** A simple random sample of heights of basketball players in the NBA is obtained, and the population has a distribution that is approximately normal. The sample statistics are $n = 16$, $\bar{x} = 77.9$ inches, $s = 3.50$ inches. Construct the 95% confidence interval estimate of the population mean.

11. **Elbow to Fingertip Length of Men.** A simple random sample of men is obtained, and the elbow to fingertip length of each man is measured. The population of those lengths has a distribution that is normal. The sample statistics are $n = 35$, $\bar{x} = 14.5$ inches, $s = 0.7$ inch. Construct the 95% confidence interval estimate of the population mean.

12. SAT Scores. A simple random sample of SAT scores is obtained, and the population has a distribution that is approximately normal. The sample statistics are $n = 41$, $\bar{x} = 1503$, $s = 352$. Construct the 95% confidence interval estimate of the population mean.

13. Crash Hospital Costs. A study was conducted to estimate hospital costs for accident victims who wore seat belts. Twenty randomly selected cases have a distribution that appears to be approximately bell-shaped with a mean of \$9,004 and a standard deviation of \$5,629 (based on data from the U.S. Department of Transportation).

a. Construct the 95% confidence interval for the mean of all such costs.

b. If you are a manager for an insurance company that provides lower rates for drivers who wear seat belts and you want a conservative estimate for a worst-case scenario, what amount should you use as the possible hospital cost for an accident victim who wears a seat belt?

14. Forecast and Actual Temperatures. One of the authors compiled a list of actual high temperatures and the corresponding list of three-day-forecast high temperatures. The difference for each day was then found by subtracting the three-day-forecast high temperature from the actual high temperature; the result was a list of 31 values with a mean of −0.419° and a standard deviation of 3.704°.

a. Construct a 95% confidence interval estimate of the mean difference between all actual high temperatures and three-day-forecast high temperatures.

b. Does the confidence interval include 0°? If the confidence interval does include 0°, can we conclude that the three-day-forecast temperatures are inaccurate?

15. Estimating Car Pollution. Each car in a sample of seven cars was tested for nitrogen-oxide emissions (in grams per mile), and the following results were obtained: 0.06, 0.11, 0.16, 0.15, 0.14, 0.08, 0.15 (based on data from the Environmental Protection Agency).

a. Assuming that this sample is representative of cars in use, construct a 95% confidence interval estimate of the mean amount of nitrogen-oxide emissions for all cars.

b. The Environmental Protection Agency requires that nitrogen-oxide emissions be less than 0.165 gram/mile. Someone claims that nitrogen-oxide emissions have a mean equal to 0.165 gram/mile. Does the confidence interval suggest that this claim is not valid? Why or why not?

16. Monitoring Lead in Air. Listed below are measured amounts of lead (in micrograms per cubic meter, or $\mu g/m^3$) in the air.

5.40 1.10 0.42 0.73 0.48 1.10

The Environmental Protection Agency has established an air quality standard for lead of 1.5 $\mu g/m^3$. The measurements were recorded at Building 5 of the World Trade Center site on different days immediately following the destruction caused by the terrorist attacks of September 11, 2001. After the collapse of the two World Trade Center buildings, there was considerable concern about the quality of the air. Use the given values to construct a 95% confidence interval estimate of the mean amount of lead in the air. Is there anything about this data set suggesting that the confidence interval might not be very good? Explain.

Hypothesis Tests. In Exercises 17–24, test the given claim.

17. Sugar in Cereal. A simple random sample of 16 different cereals is obtained, and the sugar content (in grams of sugar per gram of cereal) is measured for each cereal selected. Those amounts have a mean of 0.295 gram and a standard deviation of 0.168 gram. Use a 0.05 significance level to test the claim of a cereal lobbyist that the mean for all cereals is less than 0.3 gram.

18. Testing Wristwatch Accuracy. Students randomly selected 30 people and measured the accuracy of their wristwatches, with positive errors representing watches that were ahead of the correct time and negative errors representing watches that were behind the correct time. The 30 values have a mean of 117.3 seconds and a standard deviation of 185.0 seconds. Use a 0.05 significance level to test the claim that the population of all watches has a mean equal to 0 seconds. What can be concluded about the accuracy of people's wristwatches?

19. Baseballs. In previous tests, baseballs were dropped 24 feet onto a concrete surface, and they bounced an average of 92.84 inches. In a test of a sample of 20 new balls, the bounce heights had a mean of 92.67 inches and a standard deviation of 1.79 inches (based on data from Brookhaven National Laboratory and *USA Today*). Use a 0.05 significance level to determine whether there is sufficient evidence to support the claim that the new balls have bounce heights with a mean different from 92.84 inches. Does it appear that the new baseballs are different?

20. Reliability of Aircraft Radios. The mean time between failures for a Telektronic Company radio used in light aircraft is 420 hours. After 15 new radios were modified in an attempt to improve reliability, tests were conducted to measure the times between failures. The 15 radios had a mean time between failures of 442 hours with a standard deviation of 44.0 hours. Use a 0.05 significance level to test the claim that modified radios have a mean time between failures that is greater than 420 hours. Does it appear that the modifications improved reliability?

21. Effect of Vitamin Supplement on Birth Weight. When birth weights were recorded for a simple random sample of 16 male babies born to mothers taking a special vitamin supplement, the sample had a mean of 3.675 kilograms and a standard deviation of 0.657 kilogram (based on data from the New York State Department of Health). Use a 0.05 significance level to test the claim that the mean birth weight for all male babies of mothers given vitamins is different from 3.39 kilograms, which is the mean for the population of all males. Based on these results, does the vitamin supplement appear to have an effect on birth weight?

22. Pulse Rates. One of the authors claimed that his pulse rate was lower than the mean pulse rate of statistics students. The author's pulse rate was measured and found to be 60 beats per minute, and the 20 students in his class measured their pulse rates. The 20 students had a mean pulse rate of 74.5 beats per minute, and their standard deviation was 10.0 beats per minute. Is there sufficient evidence to support the claim that the mean pulse rate of statistics students is greater than 60 beats per minute? Use a 0.05 significance level.

23. Olympic Winners. Listed below are the winning times (in seconds) of men in the 100-meter dash for consecutive summer Olympic games, listed in order by row.

12.0	11.0	11.0	11.2	10.8	10.8	10.8	10.6
10.8	10.3	10.3	10.3	10.4	10.5	10.2	10.0
9.95	10.14	10.06	10.25	9.99	9.92	9.96	

Assuming that these results are sample data randomly selected from the population of all past and future Olympic games, use a 0.05 significance level to test the claim that the mean time is less than 10.5 seconds. What do you observe about the precision of the numbers? What extremely important characteristic of the data set is not considered in this hypothesis test?

24. Nicotine in Cigarettes. The Carolina Tobacco Company advertised that its best-selling nonfiltered cigarettes contain 40 milligrams of nicotine or less, but *Consumer Advocate* magazine ran tests of 10 randomly selected cigarettes and found the amounts (in milligrams) shown below.

47.3	39.3	40.3	38.3	46.3
43.3	42.3	49.3	40.3	46.3

Use a significance level of 0.05 to test the magazine's claim that the mean nicotine content is greater than 40 milligrams.

Projects for the Internet and Beyond

For useful links, select "Links for Internet Projects" for Chapter 10 at www.aw.com/bbt.

25. Noted Personalities. The *World Almanac and Book of Facts* includes a section called "Noted Personalities," with subsections for architects, artists, business leaders, cartoonists, and several other categories. Select a sample from one group and find the mean and standard deviation of the life spans. Test the claim that the group has a mean life span that is different from 77 years, which is the current mean life span for the general population.

26. Gosset. The t distribution was originally developed by William Sealey Gosset. Use the Internet to search for "William Sealey Gosset" or "William S. Gosset." Write a paragraph describing important information about Gosset and his accomplishments.

10.2 Hypothesis Testing with Two-Way Tables

All the hypothesis tests we have considered so far have had null hypotheses in which a population mean (μ) or proportion (p) is claimed to be *equal* to some value. But there are many situations in which the null hypothesis takes a different form. In this section, we examine hypothesis tests designed to look for a relationship between two variables. The basic process is the same as always: Identify the null and alternative hypotheses, test the null hypothesis with sample data, and then decide whether the evidence from the sample supports rejecting or not rejecting the null hypothesis. If we reject the null hypothesis, it means we accept the alternative hypothesis.

Identifying the Hypotheses with Two Variables

Suppose that administrators at a college are concerned that there may be bias in the way degrees are awarded to men and women in different departments. They therefore collect data on the number of degrees awarded to men and women in different departments. These data concern two variables: *major* and *gender*. The variable *major* can take on many values, such as biology, business, mathematics, and music. The variable *gender* can take on only two values: male or female. To test whether there is bias in the awarding of degrees, the administrators ask the following question:

Do the data suggest a relationship between the two variables?

If there is a relationship, then men and women are choosing majors at different rates, suggesting that a person's gender somehow influences his or her choice of major (either by choice or because of bias within different departments). If there is no relationship, it means there is no evidence that gender influences a person's choice of major.

This idea suggests the following choices for the two hypotheses. The null hypothesis, H_0, states that the two variables are *independent* (there is *no relationship* between them); in our current example, it states that there is no relationship between gender and major. The alternative hypothesis, H_a, states the opposite: There *is* a relationship between the variables, which in this case implies that gender influences a person's choice of major.

Null and Alternative Hypotheses with Two Variables

The **null hypothesis**, H_0, states that the variables are independent (there is *no relationship* between them).

The **alternative hypothesis**, H_a, states that there *is* a relationship between the two variables.

Displaying the Data in Two-Way Tables

With the hypotheses identified, the next step in the hypothesis test is to examine the data set to see if it supports rejecting or not rejecting the null hypothesis. Collecting the data for our current example means finding the numbers of men and women awarded degrees in various majors. Once the data have been collected, we need to find an efficient way to display them. Because we are dealing with two variables, we can display the data efficiently with a **two-way table** (also called a **contingency table**), so named because it displays two variables.

Table 10.2 shows what the two-way table might look like for data on the variables *major* and *gender*. Each cell shows a frequency (or count) for one combination of the two variables. For example, the cell in row *Women* and column *Biology* shows that 32 bachelor's degrees were awarded to women in biology. Similarly, the cell in row *Men* and column *Business* shows that 87 bachelor's degrees were awarded to men in business.

Table 10.2 Two-Way Table for the Variables *Major* and *Gender*

variable 1 *major* →

	Biology	Business	Mathematics	Psychology	...
Women	32	110	18	75	...
Men	21	87	15	70	...

↑ variable 2 *gender*

Note: One variable is displayed along the columns and the other along the rows. Here, there are only two rows because gender can be only either male or female. There are many columns for the majors, with just the first few shown here.

I cannot do it without counters.

—William Shakespeare, *The Winter's Tale*

Two-Way Tables

A **two-way table** shows the relationship between two variables by listing one variable in the rows and the other variable in the columns. The entries in the table's cells are called *frequencies* (or *counts*).

If we were looking for *any* relationship between major and gender, we would need a complete set of data for all majors, which means Table 10.2 would have dozens of columns. In addition, as we'll see shortly, carrying out the calculations for the hypothesis test requires that we find totals for all the rows and columns. Here, to simplify the calculations, let's focus on just two majors, biology and business. That is, instead of asking if there is a relationship between major and gender across all majors, we will look only at this simpler question: Does a person's gender influence whether he or she chooses to major in biology or business? Table 10.3 shows the biology and business data extracted from Table 10.2, along with row and column totals.

Table 10.3 Two-Way Table for Biology and Business Degrees

	Biology	Business	Total
Women	32	110	**142**
Men	21	87	**108**
Total	**53**	**197**	**250**

By the Way ...

Across all American colleges and universities, men outnumber women in declared majors of engineering, computer science, and architecture. Women outnumber men in psychology, fine art, accounting, biology, and elementary education.

TIME OUT TO THINK

Use Table 10.3 to answer the following questions: (a) How many business degrees were awarded to men? (b) How many business degrees were awarded in total? (c) Compare the total number of degrees awarded to men and women to the total number of degrees awarded in business and biology. Are these totals the same or different? Why?

EXAMPLE 1 A Two-Way Table for a Survey

Table 10.4 shows the results of a pre-election survey on gun control. Use the table to answer the following questions.

Table 10.4 Two-Way Table for Gun Control Survey (with totals)

	Favor stricter laws	Oppose stricter laws	Undecided	Total
Democrat	456	123	43	**622**
Republican	332	446	21	**799**
Total	**788**	**569**	**64**	**1,421**

a. Identify the two variables displayed in the table.

b. What percentage of Democrats favored stricter laws?

c. What percentage of all voters favored stricter laws?

d. What percentage of those who opposed stricter laws are Republicans?

Solution Note that the total of the row totals and the total of the column totals are equal.

a. The rows show the variable *survey response*, which can be either "favor stricter laws," "oppose stricter laws," or "undecided." The columns show the variable *party affiliation*, which in this table can be either Democrat or Republican.

b. Of the 622 Democrats polled, 456 favored stricter laws. The percentage of Democrats favoring stricter laws is 456/622 = 0.733, or 73.3%.

c. Of the 1,421 people polled, 788 favored stricter laws. The percentage of all respondents favoring stricter laws is 788/1,421 = 0.555, or 55.5%.

d. Of the 569 people polled who opposed stricter laws, 446 are Republicans. Since 446/569 = 0.783, 78.3% of those opposed to stricter laws are Republicans.

Carrying Out the Hypothesis Test

With the hypotheses identified and the data organized in the two-way table, we are now ready to carry out the hypothesis test. Again, the basic idea of the hypothesis test is the same as always—to decide whether the data provide enough evidence to reject the null hypothesis. For the case of a test with a two-way table, the specific steps are as follows:

- As always, we start by assuming that the null hypothesis is true, meaning there is no relationship between the two variables. In that case, we would expect the frequencies (the numbers in the individual cells) in the two-way table to be those that would occur by pure chance. Our first step, then, is to find a way to calculate the frequencies we would expect by chance.
- We next compare the frequencies expected by chance to the observed frequencies from the sample, which are the frequencies displayed in the table. We do this by calculating something called the *chi-square statistic* (pronounced "ky-square") for the sample data, which here plays a role similar to the role of the standard score z in the hypothesis tests we carried out in Chapter 9 or the role of the t test statistic in Section 10.1.
- Recall that for the hypothesis tests in Chapter 9, we made the decision about whether to reject or not reject the null hypothesis by comparing the computed value of the standard score for the sample data to critical values given in tables; similarly, in Section 10.1 we compared computed values of the t test statistic to values found in a table. Here, we do the same thing, except rather than using critical values for the standard score or t, we use critical values for the chi-square statistic.

As an example of the process, let's work through these steps with the data in Table 10.3.

Finding the Frequencies Expected by Chance

Our first step is to find the frequencies we would expect in Table 10.3 *if there were no relationship* between the variables, which is equivalent to the frequency expected by chance alone. Let's start by finding the frequency we would expect by chance for male business majors. To do this, we first calculate the fraction of *all* students in the sample who received business degrees:

$$\frac{\text{total business degree}}{\text{total degrees}} = \frac{197}{250}$$

As discussed in Chapter 6, we can interpret this result as a *relative frequency probability*. That is, if we select a student *at random* from the sample, the probability that he or she earned a business degree is 197/250. Using the notation for probability, we write

$$P(\text{business}) = \frac{197}{250}$$

Similarly, if we select a student at random from the sample, the probability that this student is a man is

$$P(\text{man}) = \frac{\text{total men}}{\text{total men and women}} = \frac{108}{250}$$

We now have all the information needed to find the frequency we would expect by chance for male business majors. Recall from Section 6.5 that if two events A and B are independent (the outcome of one does not affect the probability of the other), then

$$P(A \text{ and } B) = P(A) \times P(B)$$

We can apply this rule to determine the probability that a student is *both* a man and a business major (assuming the null hypothesis that gender is independent of major):

$$P(\text{man and business}) = P(\text{man}) \times P(\text{business}) = \frac{108}{250} \times \frac{197}{250} \approx 0.3404$$

This probability is equivalent to the fraction of the total students whom we expect to be male business majors *if there is no relationship* between gender and major. We therefore multiply this probability by the total number of students in the sample (250) to find the number (or frequency) of male business majors that we expect by chance:

$$\frac{108}{250} \times \frac{197}{250} \times 250 \approx 85.104$$

We call this value the **expected frequency** for the number of male business majors. (Notice the similarity between the idea of expected frequency and that of *expected value* discussed in Section 6.3.)

Definition

The **expected frequencies** in a two-way table are the frequencies we would expect by chance *if there were no relationship* between the row and column variables.

EXAMPLE 2 Expected Frequencies for Table 10.3

Find the frequencies expected by chance for the three remaining cells in Table 10.3. Then construct a table showing both observed frequencies and frequencies expected by chance.

Solution We follow the procedure used above to find the expected number of male business majors. We already have P(man) and P(business). We'll also need P(woman) and P(biology):

$$P(\text{woman}) = \frac{\text{total women}}{\text{total men and women}} = \frac{142}{250} = 0.5680$$

$$P(\text{biology}) = \frac{\text{total biology degrees}}{\text{total degrees}} = \frac{53}{250} = 0.2120$$

We combine the individual probabilities to find the probability for each of the three remaining cells:

Millions saw the apple fall, but Newton was the one who asked why.

—Bernard Baruch

$$P(\text{woman and business}) = P(\text{woman}) \times P(\text{business}) = \frac{142}{250} \times \frac{197}{250} \approx 0.4476$$

$$P(\text{man and biology}) = P(\text{man}) \times P(\text{biology}) = \frac{108}{250} \times \frac{53}{250} \approx 0.0916$$

$$P(\text{woman and biology}) = P(\text{woman}) \times P(\text{biology}) = \frac{142}{250} \times \frac{53}{250} \approx 0.1204$$

Notice that, as we should expect, the total of the probabilities for all four cells is $0.3404 + 0.4476 + 0.09158 + 0.1204 = 1.0000$.

We now find the expected frequencies by multiplying the cell probabilities by the total number of students (250):

$$\text{Expected frequency of women business majors} = 250 \times \frac{142}{250} \times \frac{197}{250} \approx 111.896$$

$$\text{Expected frequency of men biology majors} = 250 \times \frac{108}{250} \times \frac{53}{250} \approx 22.896$$

$$\text{Expected frequency of women biology majors} = 250 \times \frac{142}{250} \times \frac{53}{250} \approx 30.104$$

Table 10.5 repeats the data from Table 10.3, but this time it also shows the expected frequency for each cell (in parentheses). To check that we did our work correctly, we confirm that the total of all four expected frequencies equals the total of 250 students in the sample:

$$85.104 + 111.896 + 22.896 + 30.104 = 250.000$$

Notice also that the values in the "Total" row and "Total" column are the same for both the observed frequencies and the frequencies expected by chance. This should always be the case, providing another good check on your work.

Table 10.5 Observed Frequencies and Expected Frequencies (in parentheses) for Table 10.3

	Biology	Business	Total
Women	32 (30.104)	110 (111.896)	**142 (142.000)**
Men	21 (22.896)	87 (85.104)	**108 (108.000)**
Total	**53 (53.000)**	**197 (197.000)**	**250 (250.000)**

Computing the Chi-Square Statistic

Notice that the expected frequencies in Table 10.5 appear to agree fairly well with the observed frequencies. For example, the expected frequency of about 85.1 for male business majors is quite close to the observed frequency of 87. We might therefore already guess that the data do not give us any reason to reject the null hypothesis of no relationship between the variables *gender* and *major*. However, we can be more specific by finding a way to quantify the difference between the observed and expected frequencies.

Let's denote the observed frequencies by O and the expected frequencies by E. With this notation, $O - E$ ("O minus E") tells us the difference between the observed frequency and the expected frequency for each cell. We are looking for a measure of the *total* difference for the whole table. We cannot get such a measure by simply adding the individual differences, $O - E$, because they always sum to zero. Instead, we consider the *square* of the difference in each cell, $(O - E)^2$. We then make each value of $(O - E)^2$ a relative difference by dividing it by the corresponding expected frequency; this gives us the quantity $(O - E)^2/E$ for each cell. Summing the individual values of $(O - E)^2/E$ gives us the **chi-square statistic**, denoted χ^2 (χ is the Greek letter chi).

Finding the Chi-Square Statistic

Step 1. For each cell in the two-way table, identify O as the observed frequency and E as the expected frequency if the null hypothesis is true (no relationship between the variables).

Step 2. Compute the value $(O - E)^2/E$ for each cell.

Step 3. Sum the values from step 2 to get the chi-square statistic:

$$\chi^2 = \text{sum of all values } \frac{(O - E)^2}{E}$$

The larger the value of χ^2, the greater the average difference between the observed and expected frequencies in the cells.

To do this calculation in an organized way, it's best to make a table such as Table 10.6, with a row for each of the cells in the original two-way table. As shown in the lower right cell, the result for the gender/major data is $\chi^2 = 0.350$.

Table 10.6 Calculation of χ^2 Statistic for Data in Table 10.5

Outcome	O	E	$O - E$	$(O - E)^2$	$(O - E)^2/E$
Women/business	110	111.896	−1.896	3.595	0.032
Women/biology	32	30.104	1.896	3.595	0.119
Men/business	87	85.104	1.896	3.595	0.042
Men/biology	21	22.896	−1.896	3.595	0.157
Totals	**250**	**250.000**	**0.000**	**14.380**	$\chi^2 = 0.350$

TIME OUT TO THINK

Why must the numbers in the $O - E$ column always sum to zero?

Making the Decision

The value of χ^2 gives us a way of testing the null hypothesis of no relationship between the variables. If χ^2 is small, then the average difference between the observed and expected frequencies is small and we should *not* reject the null hypothesis. If χ^2 is large, then the average difference between the observed and expected frequencies is large and we have reason to reject the null hypothesis of independence. To quantify what we mean by "small" or "large," we compare the χ^2 value found for the sample data to critical values:

- If the calculated value of χ^2 is *less than* the critical value, the differences between the observed and expected values are small and there is *not* enough evidence to reject the null hypothesis.
- If the calculated value of χ^2 is *greater than or equal to* the critical value, then there is enough evidence in the sample to reject the null hypothesis (at the given level of significance).

Table 10.7 gives the critical values of χ^2 for two significance levels, 0.05 and 0.01. Notice that the critical values differ for different table sizes, so you must make sure you read the critical values for a data set from the appropriate table size row. For the gender/major data we have been studying in Tables 10.3 and 10.5, there are two rows and two columns (do not count the "total" rows or columns), which means a table size of 2×2. Looking in the first row of Table 10.7, we see that the critical value of χ^2 for significance at the 0.05 level is 3.841. The chi-square value that we found for the gender/major data is $\chi^2 = 0.350$; because this is less than the critical value of 3.841, we cannot reject the null hypothesis. Of course, failing to reject the null hypothesis does not *prove* that major and gender are independent. It simply means that we do not have enough evidence to justify rejecting the null hypothesis of independence.

Table 10.7 Critical Values of χ^2: Reject H_0 Only If $\chi^2 >$ Critical Value

Table size (rows × columns)	Significance level	
	0.05	**0.01**
2×2	3.841	6.635
2×3 or 3×2	5.991	9.210
3×3	9.488	13.277
2×4 or 4×2	7.815	11.345
2×5 or 5×2	9.488	13.277

TECHNICAL NOTE

The χ^2 test statistic is technically a discrete variable, whereas the actual χ^2 distribution is continuous. That discrepancy does not cause any substantial problems as long as the expected frequency for every cell is at least 5. We will assume that this condition is met for all the examples in this book.

EXAMPLE 3 Vitamin C Test

A (hypothetical) study seeks to determine whether vitamin C has an effect in preventing colds. Among a sample of 220 people, 105 randomly selected people took a vitamin C pill daily for a period of 10 weeks and the remaining 115 people took a placebo daily for 10 weeks. At the end of 10 weeks, the number of people who got colds was recorded. Table 10.8 summarizes the results. Determine whether there is a relationship between taking vitamin C and getting colds.

Table 10.8 Two-Way Table for Observed Number in Each Category

	Cold	No cold	Total
Vitamin C	45	60	**105**
Placebo	75	40	**115**
Total	**120**	**100**	**220**

Solution We begin by stating the null and alternative hypotheses.

H_0 (null hypothesis): There is no relationship between taking vitamin C and getting colds; that is, vitamin C has no more effect on colds than the placebo.

H_a (alternative hypothesis): There is a relationship between taking vitamin C and getting colds; that is, the numbers of colds in the two groups are not what we would expect if vitamin C and the placebo were equally effective (or equally ineffective).

As always, we assume that the null hypothesis is true and calculate the expected frequency for each cell in the table. Noting that the sample size is 220 and proceeding as in Example 2, we find the following expected frequencies:

$$\textit{Vitamin C and cold:} \quad 220 \times \underbrace{\frac{105}{220}}_{P(\text{vit. C})} \times \underbrace{\frac{120}{220}}_{P(\text{cold})} = 57.273$$

$$\textit{Vitamin C and no cold:} \quad 220 \times \underbrace{\frac{105}{220}}_{P(\text{vit. C})} \times \underbrace{\frac{100}{220}}_{P(\text{no cold})} = 47.727$$

$$\textit{Placebo and cold:} \quad 220 \times \underbrace{\frac{115}{220}}_{P(\text{placebo})} \times \underbrace{\frac{120}{220}}_{P(\text{cold})} = 62.727$$

$$\textit{Placebo and no cold:} \quad 220 \times \underbrace{\frac{115}{220}}_{P(\text{placebo})} \times \underbrace{\frac{100}{220}}_{P(\text{no cold})} = 52.273$$

Table 10.9 shows the two-way table with the expected frequencies in parentheses.

Table 10.9 Observed and Expected Frequencies for Vitamin C Study

	Cold	No cold	Total
Vitamin C	45 (57.273)	60 (47.727)	**105 (105.000)**
Placebo	75 (62.727)	40 (52.273)	**115 (115.000)**
Total	**120 (120.000)**	**100 (100.000)**	**220 (220.000)**

We now compute the chi-square statistic for the sample data. Table 10.10 shows how we organize the work; you should confirm all the calculations shown.

Table 10.10 Table for Computing χ^2 Statistic for Vitamin C Study

Outcome	O	E	$O - E$	$(O - E)^2$	$(O - E)^2/E$
Vitamin C/cold	45	57.273	−12.273	150.627	2.630
Vitamin C/no cold	60	47.727	12.273	150.627	3.156
Placebo/cold	75	62.727	12.273	150.627	2.401
Placebo/no cold	40	52.273	−12.273	150.627	2.882
Totals	**220**	**220.000**	**0.000**	**602.508**	χ^2 **= 11.069**

To make the decision about whether to reject the null hypothesis, we compare the value of chi-square for the sample data, $\chi^2 = 11.069$, to the critical values from Table 10.7. We look in the row for a table size of 2 × 2, because the original data in Table 10.8 have two rows and two columns (not counting the "total" values). We see that the critical value of χ^2 for significance at the 0.01 level is 6.635. Because our sample value of $\chi^2 = 11.069$ is greater than this critical value, we reject the null hypothesis and conclude that there is a relationship between vitamin C and colds. That is, based on the data from this sample, there is reason to believe that vitamin C *does* have more effect on colds than a placebo.

By the Way ...

Dozens of careful studies have been conducted on the question of vitamin C and colds. Some have found high levels of confidence in the effects of vitamin C, but others have not. Because of these often conflicting results, the issue of whether vitamin C helps to prevent colds remains controversial.

EXAMPLE 4 To Plead or Not to Plead

The two way table in Table 10.11 shows how a plea of guilty or not guilty affected the sentence in 1,028 randomly selected burglary cases in the San Francisco area. Test the claim that the sentence (prison or no prison) is independent of the plea.

Table 10.11 Observed Frequencies for Plea and Sentence

	Prison	No prison	Total
Guilty plea	392	564	**956**
Not-guilty plea	58	14	**72**
Total	**450**	**578**	**1,028**

Source: Law and Society Review, Vol. 16, No. 1.

Solution The null and alternative hypotheses for the problem are

H_0 (null hypothesis): The sentence in burglary cases is independent of the plea.

H_a (alternative hypothesis): The sentence in burglary cases depends on the plea.

We find the following *expected* number of people in each category, assuming that the row variables are independent of the column variables:

Guilty and prison: $$1{,}028 \times \frac{956}{1{,}028} \times \frac{450}{1{,}028} = 418.482$$

Guilty and no prison: $$1{,}028 \times \frac{956}{1{,}028} \times \frac{578}{1{,}028} = 537.518$$

Not guilty and prison: $$1{,}028 \times \frac{72}{1{,}028} \times \frac{450}{1{,}028} = 31.518$$

Not guilty and no prison: $$1{,}028 \times \frac{72}{1{,}028} \times \frac{578}{1{,}028} = 40.482$$

Table 10.12 summarizes the observed frequencies and expected frequencies.

Table 10.12 Observed Frequencies and Frequencies Expected by Chance (in parentheses) for Plea and Sentence

	Prison	No prison	Total
Guilty	392 (418.482)	564 (537.518)	**956**
Not guilty	58 (31.518)	14 (40.482)	**72**
Total	**450**	**578**	**1,028**

As usual, we calculate χ^2 by organizing our work as shown in Table 10.13, where O denotes observed frequency and E denotes expected frequency. As shown in the lower right cell, the chi-square statistic for these data is $\chi^2 = 42.556$. This value is much greater than the critical values (for a 2×2 table) for significance at both the 0.05 level ($\chi^2 = 3.841$) and the 0.01 level ($\chi^2 = 6.635$). The data therefore support rejecting the null hypothesis and accepting the alternative hypothesis. Based on these data, there is reason to believe that the sentence given in a burglary case is associated with the plea. Specifically, of the people who pled guilty, fewer actually went to prison than expected (by chance) and more avoided prison than expected. Of the people who pled not guilty, more actually went to prison than expected and fewer avoided prison than expected. Remember that the test does not prove a causal relationship between the plea and the sentence.

Table 10.13 Calculation of χ^2

Outcome	O	E	$O - E$	$(O - E)^2$	$(O - E)^2/E$
Guilty/prison	392	418.482	−26.482	701.296	1.676
Guilty/no prison	564	537.518	26.482	701.296	1.305
Not guilty/prison	58	31.518	26.482	701.296	22.251
Not guilty/no prison	14	40.482	−26.482	701.296	17.324
Totals	**1,028**	**1,028.000**	**0.000**	**2805.184**	**$\chi^2 = 42.556$**

TIME OUT TO THINK

If you were the lawyer for a burglary suspect, how might the results of the previous example affect your strategy in defending your client? Explain.

Section 10.2 Exercises

Statistical Literacy and Critical Thinking

1. **Two-Way Tables.** What is a two-way table and what are the values entered in a two-way table?

2. **Relationship Between Variables.** Assume that we have analyzed the frequencies in a two-way table with gender (male/female) corresponding to rows and dominant hand (left/right) corresponding to columns. Also assume that we reject independence between gender and dominant hand. Can we conclude that the gender of a person has an effect on whether the person is left-handed or right-handed? Why or why not?

3. **Expected Frequency.** Assume that we know frequencies in a two-way table with gender (male/female) as the row variable and speeding tickets (yes/no) as the other variable. Given that the observed frequencies must all be whole numbers, must the expected frequencies also be whole numbers? Explain.

4. **Notation.** In the context of two-way tables, describe the following notations: O, E, and χ^2.

Does It Make Sense? For Exercises 5–8, decide whether the statement makes sense (or is clearly true) or does not make sense (or is clearly false). Explain clearly. Not all of these statements have definitive answers, so your explanation is more important than your chosen answer.

5. **Survey.** In a quick survey, subjects are asked if they can identify the year in which the United States declared its independence. The results from that question are summarized in a two-way table.

6. **χ^2 Test Statistic.** A two-way table is used to calculate the test statistic, and the value $\chi^2 = -2.500$ is obtained.

7. **χ^2 Test Statistic.** In a two-way table, all of the observed frequencies are very close to the expected frequencies, so the χ^2 test statistic is very small, and we fail to reject the null hypothesis of independence between the row and column variables.

8. **Null Hypothesis.** Assume that a two-way table is configured so that *gender* (male/female) represents the row variable and *response* (yes/no) to a survey question represents the column variable. The null hypothesis is the statement that gender and response are independent.

Concepts and Applications

Survey Results. In Exercises 9–12, assume that a yes/no survey question is presented to a simple random sample of male and female subjects and the results are summarized in a two-way table with the format of the table below. Use the given value of the χ^2 test statistic and the given significance level to test for independence between gender and response.

	Yes	No
Female		
Male		

9. Test statistic: $\chi^2 = 0.051$; significance level: 0.01

10. Test statistic: $\chi^2 = 10.785$; significance level: 0.01

11. Test statistic: $\chi^2 = 2.924$; significance level: 0.05

12. Test statistic: $\chi^2 = 15.238$; significance level: 0.05

Complete Hypothesis Test. In Exercises 13–20, carry out the following steps.

a. State the null and alternative hypotheses.

b. Assuming independence between the two variables, find the expected frequency for each cell of the table.

c. Find the value of the χ^2 test statistic.

d. Use the given significance level to find the χ^2 critical value.

e. Using the given significance level, complete the test of the claim that the two variables are independent. State the conclusion that addresses the original claim.

13. **College Demographics.** The following table (consistent with national data) shows the distribution of part-time and full-time college students for a sample of 148 men and women. Use a 0.05 significance level to test the claim of independence between gender and student status (full-time or part-time).

	Full-time	Part-time
Women	47	37
Men	38	26

14. **Voter Turnout.** The following table shows the number of citizens in a sample who voted in the last presidential election, according to gender (consistent with national population data). Use a 0.05 significance level to test the claim that gender is independent of voter turnout.

	Voted	Did not vote
Women	140	120
Men	130	110

15. **E-Mail and Privacy.** Workers and senior-level bosses were asked if it was seriously unethical to monitor employee e-mail; the results are summarized in the table below (based on data from a Gallup poll). Use a 0.05 significance level to test the claim that the response is independent of whether the subject is a worker or a senior-level boss.

	Yes	No
Workers	192	244
Bosses	40	81

16. **Effectiveness of Bicycle Helmets.** A study was conducted of 531 persons injured in bicycle crashes; randomly selected sample results are summarized in the table below (based on data from "A Case-Control Study of the Effectiveness of Bicycle Safety Helmets in Preventing Facial Injury," by Thompson, Thompson, Rivara, and Wolf, *American Journal of Public Health,* Vol. 80, No. 12). Using a 0.01 significance level, test the claim that wearing a helmet is independent of whether facial injuries are received.

	Helmet worn	No helmet
Facial injuries received	30	182
No facial injuries	83	236

17. **Arthritis Treatment.** Of the 98 participants in a drug trial who were given a new experimental treatment for arthritis, 56 showed improvement. Of the 92 participants given a placebo, 49 showed improvement. Construct a two-way table for these data, and then use a 0.05 significance

level to test the claim that improvement is independent of whether the participant was given the drug or a placebo.

18. Drinking and Pregnancy. A simple random sample of 1,252 pregnant women under the age of 25 includes 13 who were drinking alcohol during their pregnancy. A simple random sample of 2,029 pregnant women of age 25 and over includes 37 who were drinking alcohol during their pregnancy. (The data are based on results from the U.S. National Center for Health Statistics.) Use a 0.05 significance level to test the claim that the age category (under 25 and 25 or over) is independent of whether the pregnant woman was drinking during pregnancy.

19. Crime and Strangers. The table below lists survey results obtained from a random sample of different crime victims (based on data from the U.S. Department of Justice). Use a 0.01 significance level to test the claim that the type of crime is independent of whether the criminal was a stranger.

	Homicide	Robbery	Assault
Criminal was a stranger	12	379	727
Criminal was acquaintance or relative	39	106	642

20. Smoking in China. The table below summarizes results from a survey of males aged 15 or older living in the Minhang District of China (based on data from "Cigarette Smoking in China" by Gong, Koplan, Feng, et al., *Journal of the American Medical Association,* Vol. 274, No. 15). The males are categorized by their current educational status and whether they smoke. Using a 0.05 significance level, test the claim that smoking is independent of education level.

	Primary school	Middle school	College
Smoker	606	1234	100
Never smoked	205	505	137

Projects for the Internet and Beyond

For useful links, select "Links for Internet Projects" for Chapter 10 at www.aw.com/bbt.

21. Constructing Two-Way Tables. Choose two variables that appear to have a relationship that is worth investigating. One variable should have at least two categories of individuals—for example, two or more age categories, racial categories, or geographical locations. The other variable should have at least two categories for some social, economic, or health factor—for example, two or more income categories, drinking categories, or educational attainment categories. Find the required population or sample data needed to fill in a two-way table for the two variables. Discuss whether there appears to be a relationship between the variables. A good place to start is the Web site for the *Statistical Abstract of the United States* of the U.S. Census Bureau.

22. Analyzing Two-Way Tables. Choose two variables that appear to have a relationship that is worth investigating. One variable should have at least two categories of individuals—for example, two or more age categories, racial categories, or geographical locations. The other variable should have at least two categories for some social, economic, or health factor—for example, two or more income categories, drinking categories, or educational attainment categories. Find the frequency data needed to fill in a two-way table for the two variables. Carry out a hypothesis test to determine whether there is a relationship between the variables. A good data source is the Web site for the *Statistical Abstract of the United States* of the U.S. Census Bureau.

IN THE NEWS

23. Two-Way Tables in the News. It's unusual (but not impossible) to see a two-way table in a news article. But often a news story provides information that could be expressed in a two-way table. Find an article that discusses a relationship between two variables that could be expressed in a two-way table. Create the table.

24. Hypothesis Testing in the News. News reports often describe results of statistical studies in which the conclusions came from a hypothesis test involving two-way tables. However, the reports rarely give the actual table or describe the details of the hypothesis test. Find a recent news report in which you think the conclusions *probably* were based on a hypothesis test with a two-way table. Assuming you are correct, describe in words how the hypothesis test probably worked. That is, describe the null and alternative hypotheses and the procedure by which the researchers probably carried out their test.

25. Your Own Hypothesis Test. Think of an example of something you'd like to know that could be tested with a hypothesis test on a two-way table. Without actually collecting data or doing any calculations, describe how you would go about conducting your study. That is, describe how you would collect the data, explain how you would organize them into a two-way table, state the null and alternative hypotheses that would apply, and describe how you would conduct the hypothesis test and reach a conclusion.

10.3 Analysis of Variance (One-Way ANOVA)

So far we have examined hypothesis testing with three different types of claims for the null hypothesis: claims that a population mean equals some value (H_0: μ = claimed value), which we examined with a normal distribution in Section 9.2 and with the t distribution in Section 10.1; claims that a population proportion equals some value (H_0: p = claimed value), which we examined in Section 9.3; and claims that two variables are independent of each other (H_0: no relationship), which we discussed in Section 10.2. Statisticians have developed techniques for considering many other types of null hypotheses, making it possible to apply statistics to an incredible range of applications. To give you a taste of what is possible with statistics—and perhaps to encourage you to study statistics further—we briefly consider one more type of hypothesis testing in this final section of the book.

Hypothesis Testing for Variance

A simple random sample of 12 pages was obtained from each of three different books: Tom Clancy's *The Bear and the Dragon*, J. K. Rowling's *Harry Potter and the Sorcerer's Stone*, and Leo Tolstoy's *War and Peace*. The Flesch Reading Ease score was obtained for each of those pages, and the results are listed in Table 10.14. The Flesch Reading Ease scoring system results in *higher* scores for text that is *easier* to read. Low scores are associated with works that are difficult to read. Our goal in this section is to use these sample data from just 12 pages of each book to make inferences about the readability of the *population* of all pages in each book.

We can informally explore the sample data by investigating center, variation, distribution, and outliers. Table 10.15 shows the important sample statistics for our case. If you compare the original data in Table 10.14 and the sample means in Table 10.15, you'll notice that although a few scores are farther from the mean than most others (such as the lowest Clancy score of 43.9 and the lowest Rowling score of 70.9), no values seem so extreme that we would consider them outliers. Moreover, detailed study of the data suggests that the samples come from populations with distributions that are close to normal.

Table 10.14 Flesch Reading Ease Scores

Clancy	Rowling	Tolstoy
58.2	85.3	69.4
73.4	84.3	64.2
73.1	79.5	71.4
64.4	82.5	71.6
72.7	80.2	68.5
89.2	84.6	51.9
43.9	79.2	72.2
76.3	70.9	74.4
76.4	78.6	52.8
78.9	86.2	58.4
69.4	74.0	65.4
72.9	83.7	73.6

Table 10.15 Statistics for Readability Scores

	Flesch Reading Ease Score		
	Clancy	Rowling	Tolstoy
Sample size *n*	12	12	12
Sample mean $\bar{x}$	70.73	80.75	66.15
Sample standard deviation *s*	11.33	4.68	7.86

Even before studying the data, we might expect Rowling's book to be the easiest to read of the three books because it is the only one written for children. Similarly, we might expect Tolstoy's book to be the most difficult because it is a translation of a Russian classic. Now look at the mean readability scores in Table 10.15. Recalling that a higher Flesch score indicates an easier reading level, the data appear to support our expectations: Rowling has the highest readability score and Tolstoy has the lowest. Still, the three sample means are not wildly different, ranging only from 66.15 for Tolstoy to 80.75 for Rowling, and the sample size is a relatively small n = 12 for each case. We therefore arrive at our key statistical question for this section: Do these sample data provide sufficient evidence for us to conclude that the books by Clancy, Rowling, and Tolstoy really do have different mean Flesch scores?

To answer this question, we follow the same general principles laid out for hypothesis testing in Section 9.1. To begin with, we identify the null hypothesis. Because we want to know whether the three books really do have different mean Flesch scores, we start with the assumption that they do *not* have different means. In other words, our null hypothesis is that the mean Flesch scores for all three books are equal. The alternative hypothesis, then, is that the three population means are different. The hypothesis test must tell us whether to reject or not reject the null hypothesis. Rejecting the null hypothesis would allow us to conclude that the books really do have different mean Flesch scores, as we expect. Not rejecting the null hypothesis would tell us that the data do not provide sufficient evidence for concluding that the mean Flesch scores are different.

Remember that for this example each population mean (μ) represents the mean Flesch score we would obtain if we measured the score for *all* the pages in each book. Using more formal notation, we can therefore write the null hypothesis as

$$H_0: \mu_{\text{Clancy}} = \mu_{\text{Rowling}} = \mu_{\text{Tolstoy}}$$

As you can see, we need a hypothesis test that will allow us to determine whether three different populations have the same mean. The method we use is called **analysis of variance**, commonly abbreviated **ANOVA**. The name comes from the formal statistic known as the *variance* of a set of sample values; as we noted briefly in Section 4.3, variance is defined as the square of the sample standard deviation, or s^2. (For example, if a sample of heights has a standard deviation $s = 3.0$ cm, its variance is $s^2 = 9.0$ cm^2.)

TECHNICAL NOTE

The method of analysis of variance can be used with two means, but it is equivalent to a *t* test that pools the two sample variances. There is another *t* test that can be used with two independent samples, and it generally performs better. Neither of these *t* tests is included in this book.

Definition

Analysis of variance (ANOVA) is a method of testing the equality of three or more population means by analyzing sample variances.

More specifically, the method used to analyze data like those from Table 10.14 is called *one-way* analysis of variance (one-way ANOVA), because the sample data are separated into groups according to just *one* characteristic (or factor). In this example, the characteristic is the author (Clancy, Rowling, or Tolstoy). There is also a method referred to as *two-way analysis of variance* that allows comparisons among populations separated into categories by two characteristics (or factors). For example, we might separate heights of people using the following two characteristics: (1) gender (male or female) and (2) right- or left-handedness. We do not consider two-way analysis of variance in this book.

Conducting the Test

Analysis of variance is based on this fundamental concept: We *assume* that the populations all have the same variance, and we then compare the variance *between* the samples to the variance *within* the samples. More specifically, the test statistic (usually called *F*) for one-way analysis of variance is the ratio of those two variances:

$$\text{test statistic } F \text{ (for one-way ANOVA)} = \frac{\text{variance between samples}}{\text{variance within samples}}$$

The actual calculation of this test statistic is tedious, so these days it is almost always done with statistical software (see Using Technology, page 437). However, we can interpret the statistic as follows, using our example of the readability of the three books:

- The variance *between* samples is a measure of how much the three sample means (from Table 10.15) differ from one another.
- The variance *within* samples is a measure of how much the Flesch Reading Ease scores for the 12 pages in each individual sample (from Table 10.14) differ from one another.

- *If* the three population means were really all equal—as the null hypothesis claims—then we would expect the sample mean from any one individual sample to fall well within the range of variation for any other individual sample. The test statistic (F = variance between samples/variance within samples) tells us whether that is the case:

  A large test statistic tells us that the sample means differ *more* than the data within the individual samples, which would be *unlikely* if the populations means really were equal (as the null hypothesis claims). That is, a large test statistic provides evidence for rejecting the null hypothesis that the population means are equal.

  A small test statistic tells us that the sample means differ *less* than the data within the individual samples, suggesting that the difference among the sample means could easily have arisen by chance. Therefore, a small test statistic does not provide evidence for rejecting the null hypothesis that the population means are equal.

Notice that the test statistic F for analysis of variance plays a role similar to that of the standard score z or the t test statistic in hypothesis tests we considered earlier. Therefore, just as we did in those earlier cases, we quantify the interpretation of the test statistic by finding its P-value, which tells us the probability of getting sample results at least as extreme as those obtained, assuming that the null hypothesis is true (the population means are all equal). A small P-value shows that it is unlikely that we would get the sample results by chance with equal population means. A large P-value shows that we could easily get the sample results by chance with equal population means.

Like the test statistic itself, the P-value calculation is generally done with the aid of software. Once we have found the P-value, we use it to make the final decision just as we have in other cases: If the P-value is small (such as 0.05 or less), reject the null hypothesis of equal means; if the P-value is large (such as greater than 0.05), do not reject the null hypothesis of equal means. The following box summarizes the requirements for one-way analysis of variance and the software-aided procedure outlined in this section.

One-Way ANOVA for Testing H_0: $\mu_1 = \mu_2 = \mu_3 = \ldots$

Step 1. Enter sample data into a statistical software package, and use the software to determine the test statistic (F = variance between samples/variance within samples) and the P-value of the test statistic.

Step 2. Make a decision to reject or not reject the null hypothesis based on the P-value of the test statistic:

- If the P-value is less than or equal to the significance level, reject the null hypothesis of equal means and conclude that at least one of the means is different from the others.
- If the P-value is greater than the significance level, do not reject the null hypothesis of equal means.

This method is valid as long as the following requirements are met: The populations have distributions that are approximately normal with the same variance, and the samples from each population are simple random samples that are independent of each other.

EXAMPLE 1 Readability of Clancy, Rowling, Tolstoy

Given the readability scores listed in Table 10.14 and a significance level of 0.05, test the null hypothesis that the three samples come from populations with means that are all the same.

Solution We begin by checking the requirements for using one-way analysis of variance. As noted earlier, close examination of the data suggests that each sample comes from a distribution that is approximately normal. The sample standard deviations are not dramatically different, so it is reasonable to assume that the three populations have the same variance. The samples are simple random samples and they are all independent. The requirements are therefore satisfied.

We now test the null hypothesis that the population means are all equal (H_0: $\mu_1 = \mu_2 = \mu_3$). The Using Technology section on page 437 describes how to compute the test statistic and *P*-value with various software packages. The table below shows the resulting display from Excel; other software packages will give similar displays.

Source of Variation	***SS***	***df***	***MS***	***F***	***P-value***	***F crit***
Between Groups	1338.002222	2	669.0011111	9.469487401	0.000562133	3.284924333
Within Groups	2331.386667	33	70.64808081			
Total	3669.388889	35				

Notice that the display includes columns for *F* and for the *P*-value. These are the two items of interest to us here, which we interpret as follows:

- *F* is the test statistic for the one-way analysis of variance (F = variance between samples/variance within samples). Notice that it is much greater than 1, indicating that the sample means differ more than we would expect if all the population means were equal.
- The *P*-value tells us the probability of having obtained such an extreme result by chance if the null hypothesis is true. Notice that the *P*-value is extremely small—much less than the value of 0.05 necessary to reject the null hypothesis at the 0.05 level of significance (and also much less than the 0.01 necessary to reject at the 0.01 level of significance).

We conclude that there is sufficient evidence to reject the null hypothesis, which means the sample data support the claim that the three population means are not all the same. Based on randomly selected pages from Clancy's *The Bear and the Dragon*, Rowling's *Harry Potter and the Sorcerer's Stone*, and Tolstoy's *War and Peace*, we conclude that those books have readability levels that are not all the same. Note that we have *not* concluded that the three books have the readability order that we expect—Rowling as easiest and Tolstoy as hardest—because the hypothesis test shows only that the readabilities are unequal. Nevertheless, our expectation seems reasonable since the sample means in Table 10.15 go in the expected order.

Section 10.3 Exercises

Statistical Literacy and Critical Thinking

1. **ANOVA.** What method does ANOVA represent, and what is the purpose of the method?

2. **Variance.** What is the variance of a sample?

3. **Comparing Majors.** A student at the College of Newport administers a test of abstract reasoning to randomly selected English, mathematics, and science majors at her college. She then uses analysis of variance to conclude that the mean scores are not all equal. Can she conclude that, in the United States, English, mathematics, and science majors have mean abstract reasoning scores that are not all the same? Why or why not?

4. **One-Way ANOVA.** Why is the method of this section referred to as *one-way* analysis of variance?

Does It Make Sense? For Exercises 5–8, decide whether the statement makes sense (or is clearly true) or does not make sense (or is clearly false). Explain clearly. Not all of these statements have definitive answers, so your explanation is more important than your chosen answer.

5. ***P*-Value.** In a test for equality of the mean cholesterol levels of people from the United States, Norway, and China, the sample data result in a *P*-value of 0.99, so we reject equality of the three population means.

6. ***P*-Value.** In a test for equality of the mean ages of audience members at an animated movie, a comedy, and a thriller, the *P*-value is 0.009, so we reject equality of the three population means.

7. **ANOVA.** For a study of IQ scores of children, families with three children are randomly selected. One sample includes IQ scores of the first-born child, the second sample includes IQ scores of the second-born child, and the third sample includes IQ scores of the third-born child. The method of analysis of variance can be used to test for equality of the three population means.

8. **ANOVA.** A manufacturer of auto shock absorbers is testing three different machines, and analysis of variance is used to compare the amounts of variation resulting from the three different machines.

Concepts and Applications

9. **Readability of Authors.** The example in this section used the Flesch Reading Ease scores for randomly selected pages from books by Tom Clancy, J. K. Rowling, and Leo Tolstoy. When the Flesch–Kincaid Grade Level scores are used instead, the analysis of variance results from STATDISK are as shown in Figure 10.2. Assume that we want to use a 0.05 significance level in testing the null hypothesis that the three authors have Flesch–Kincaid Grade Level scores with the same mean.

   a. What is the null hypothesis?

   b. What is the alternative hypothesis?

   c. Identify the *P*-value.

   d. Based on the preceding results, what do you conclude about equality of the population means?

Source:	DF:	SS:	MS:	Test Stat, F:	Critical F:	P-Value:
Treatment:	2	68.187222	34.093611	8.978506	3.284914	0.00077
Error:	33	125.309167	3.797247			
Total:	35	193.496389	5.528468			

Reject the Null Hypothesis
Reject equality of means

Figure 10.2

10. **Fabric Flammability Tests in Different Laboratories.** The Vertical Semirestrained Test was used to conduct flammability tests on children's sleepwear. Pieces of fabric were burned under controlled conditions. After the burning stopped, the length of the charred portion was measured and recorded. The same fabric samples were tested at five different laboratories. The analysis of variance results from Excel are shown below.

   a. What is the null hypothesis?

   b. What is the alternative hypothesis?

   c. Identify the *P*-value.

   d. Is there sufficient evidence to support the claim that the means for the different laboratories are not all the same? Assume that a 0.05 significance level is used.

Source of Variation	*SS*	*df*	*MS*	*F*	*P-value*	*F crit*
Between Groups	2.087194264	4	0.521798566	2.949333035	0.030665893	2.588834036
Within Groups	7.607597403	43	0.17692087			
Total	9.694791667	47				

11. **Marathon Times.** A random sample of males who finished the New York marathon is partitioned into three categories with ages of 21–29, 30–39, and 40 or over. The times (in seconds) are obtained for the selected males. The analysis of variance results obtained from Excel are shown below.

    a. What is the null hypothesis?

    b. What is the alternative hypothesis?

    c. Identify the *P*-value.

    d. Is there sufficient evidence to support the claim that men in the different age categories have different mean times?

Source of Variation	*SS*	*df*	*MS*	*F*	*P-value*	*F crit*
Between Groups	3532063.284	2	1766031.642	0.188679406	0.828324293	3.080387501
Within Groups	1010875649	108	9359959.71			
Total	1014407712	110				

12. **Systolic Blood Pressure in Different Age Groups.** A random sample of 40 women is partitioned into three categories with ages of below 20, 20 through 40, and over 40. The analysis of variance results obtained from SPSS are shown below.

    a. What is the null hypothesis?

    b. What is the alternative hypothesis?

    c. Identify the *P*-value.

    d. Is there sufficient evidence to support the claim that women in the different age categories have different mean blood pressure levels?

	Sum of Squares	***df***	**Mean Square**	***F***	**Sig.**
Between Groups	937.930	2	468.965	1.655	.205
Within Groups	10484.470	37	283.364		
Total	11422.400	39			

In Exercises 13–16, use software to conduct the analysis of variance test.

13. **Head Injury in a Car Crash.** In car crash experiments conducted by the National Transportation Safety Administration, new cars were purchased and crashed into a fixed barrier at 35 miles per hour. The subcompact cars were the Ford Escort, Honda Civic, Hyundai Accent, Nissan Sentra, and Saturn SL4. The compact cars were the Chevrolet Cavalier, Dodge Neon, Mazda 626 DX, Pontiac Sunfire, and Subaru Legacy. The midsize cars were the Chevrolet Camaro, Dodge Intrepid, Ford Mustang, Honda Accord, and Volvo S70. The full-size cars were the Audi A8, Cadillac Deville, Ford Crown Victoria, Oldsmobile Aurora, and Pontiac Bonneville. Head injury data (in hic) for the dummies in the driver's seat are listed below. Use a 0.05 significance level to test the null hypothesis that the different weight categories have the same mean. Do the sample data suggest that larger cars are safer?

Subcompact:	681	428	917	898	420
Compact:	643	655	442	514	525
Midsize:	469	727	525	454	259
Full-size:	384	656	602	687	360

14. Chest Deceleration in a Car Crash. The chest deceleration data (in *g*'s) from the tests described in Exercise 13 are given below. Use a 0.05 significance level to test the null hypothesis that the different weight categories have the same mean. Do the data suggest that larger cars are safer?

Subcompact:	55	47	59	49	42
Compact:	57	57	46	54	51
Midsize:	45	53	49	51	46
Full-size:	44	45	39	58	44

15. Archeology: Skull Breadths from Different Epochs. The values in the table below are measured maximum breadths (in millimeters) of male Egyptian skulls from different epochs (based on data from *Ancient Races of the Thebaid*, by Thomson and Randall-Maciver). Changes in head shape over time suggest that interbreeding occurred with immigrant populations. Use a 0.05 significance level to test the claim that the different epochs all have the same mean.

4000 B.C.	1850 B.C.	150 A.D.
131	129	128
138	134	138
125	136	136
129	137	139
132	137	141
135	129	142
132	136	137
134	138	145
138	134	137

16. Solar Energy in Different Weather. A student of one of the authors lives in a home with a solar electric system. At the same time each day, she collected voltage readings from a meter connected to the system; the results are listed in the table below. Use a 0.05 significance level to test the claim that the mean voltage reading is the same for the three different types of days. We might expect that a solar system would provide more electrical energy on sunny days than on cloudy or rainy days. Is there sufficient evidence to support a claim of different population means?

Sunny days	Cloudy days	Rainy days
13.5	12.7	12.1
13.0	12.5	12.2
13.2	12.6	12.3
13.9	12.7	11.9
13.8	13.0	11.6
14.0	13.0	12.2

Projects for the Internet and Beyond

For useful links, select "Links for Internet Projects" for Chapter 10 at www.aw.com/bbt.

17. Noted Personalities. The *World Almanac and Book of Facts* includes a section called "Noted Personalities," with subsections for architects, artists, business leaders, cartoonists, and several other categories. Design and conduct a study that begins with selection of samples from select groups, followed by a comparison of mean life spans of people from the different categories. Do any particular groups appear to have life spans that are different from those of the other groups?

18. ANOVA. Find a journal article that refers to use of analysis of variance. Identify the test being used and describe the conclusion. Did the test result in rejection of equal means? What was the *P*-value? What was the role of the method of analysis of variance?

IN THE NEWS

19. Sports. Find a recent news article discussing salaries of players on different professional sports teams, such as baseball teams. Find the salaries and use analysis of variance to test the null hypothesis of equal means. Summarize your findings and write a brief report that includes your conclusions.

Chapter Review Exercises

1. A study sponsored by AT&T and the Automobile Association of America included the sample data in the following table.

	Had accident in last year	Had no accident in last year
Cell phone user	23	282
Not a cell phone user	46	407

a. Compare the percentage of cell phone users who had an accident to the percentage of those who did not use a cell phone and had an accident. On the basis of these results, do cell phones appear to be dangerous?

b. Identify the null and alternative hypotheses for a test of the claim that having an accident is independent of cell phone use.

c. Find the expected value for each cell of the table by assuming that having an accident is independent of cell phone use.

d. Find the value of the χ^2 statistic for a hypothesis test of the claim that having an accident is independent of cell phone use.

e. Based on the result from part d and the size of the table, refer to Table 10.7 (on page 423) and determine what is known about the *P*-value.

f. Based on the preceding results, what can you conclude from the hypothesis test about whether the two variables (*cell phone use* and *having an accident*) are independent?

g. Is the conclusion from part f consistent with what is now known about cell phone use and driving?

2. The axial load of a can is the maximum weight supported by its side, and it must be greater than 165 pounds because that is the maximum force applied when the top lid is pressed into place. Listed below are axial loads (in pounds) for a random sample of 12-ounce aluminum cans.

270 273 258 204 254 228 282

Use a 0.05 significance level to test the claim that the sample is from a population with a mean that is greater than 165 pounds.

3. Using the sample data from Exercise 2, construct a 95% confidence interval estimate of the mean axial load of all cans.

4. Listed in the table below are body temperatures (°F) of randomly selected subjects from three different age groups. The STATDISK display in Figure 10.3 results from these sample values. Assume that we want to use a 0.05 significance level to test the claim that the three age groups have the same mean body temperature.

18–20	21–29	30 and older
98.0	99.6	98.6
98.4	98.2	98.6
97.7	99.0	97.0
98.5	98.2	97.5
97.1	97.9	97.3

a. What is the null hypothesis?

b. What is the alternative hypothesis?

c. Identify the *P*-value.

d. Is there sufficient evidence to reject the claim that the three age groups have the same mean body temperature?

Source:	DF:	SS:	MS:	Test Stat, F:	Critical F:	P-Value:
Treatment:	2	1.729333	0.864667	1.87971	3.88529	0.194915
Error:	12	5.52	0.46			
Total:	14	7.249333	0.51781			

Figure 10.3

Chapter Quiz

1. A simple random sample of 15 values is obtained from a normally distributed population with an unknown standard deviation. Which of the following distributions is most appropriate for a hypothesis test involving a claim about a population mean?

   a. normal distribution **c.** chi-square distribution

   b. *t* distribution **d.** uniform distribution

2. A simple random sample of 45 values is obtained from a normally distributed population with an unknown standard deviation. Which of the following distributions is most appropriate for a hypothesis test involving a claim about a population mean?

   a. normal distribution **c.** chi-square distribution

   b. *t* distribution **d.** uniform distribution

3. A simple random sample of 45 values is obtained from a normally distributed population with a known standard deviation. Which of the following distributions is most appropriate for a hypothesis test involving a claim about a population mean?

   a. normal distribution **c.** chi-square distribution

   b. *t* distribution **d.** uniform distribution

4. What is the null hypothesis for a claim that the mean score on an SAT test is greater than 1500?

5. What is the alternative hypothesis for a claim that the mean score on an SAT test is greater than 1500?

6. Determine whether the following statement is true or false: A *t* test is used for testing the claim that, in a two-way table, the row variable and column variable are somehow related.

7. Assume that you want to test the claim that students majoring in science, literature, and business all have the same mean IQ score. What method would you use to test that claim?

8. If the hypothesis test of the claim described in Exercise 7 results in a *P*-value of 0.3500, what do you conclude about the null hypothesis?

9. A two-way table, constructed from survey results, consists of two rows representing sex (male/female) and two columns representing the response to a question (yes/no). What is the null hypothesis for a test to determine whether there is some relationship between sex and response?

10. If the hypothesis test described in Exercise 9 results in a *P*-value of 0.001, what do you conclude about the null hypothesis?

Using Technology

Confidence Intervals Using the *t* Distribution

SPSS: Enter the sample data in the SPSS data editor window. Click on **Analyze**, then select **Descriptive Statistics**, then select the menu item **Explore**. Click on the variable at the left of the dialog box and click on the middle button to paste it into the "Dependent List" box. In the same dialog box, click on the **Statistics** button and be sure that the "Descriptives" box is checked and the confidence level is the value desired. Click on **Continue**. Click on **OK**. The "Lower Bound" and the "Upper Bound" of the confidence interval will be displayed in the box labeled "Descriptives."

Excel: Use the Data Desk XL add-in that is a supplement to Excel.

First enter the sample data in column A. If you are using Excel 2003, click on **DDXL**. If you are using Excel 2007, click on **Add-Ins**, then on **DDXL**. Select **Confidence Intervals**. Under the Function Type options, select **1 Var t Interval** if σ is not known. (If σ is known, select **1 Var z Interval.**) Click on the pencil icon and enter the range of data, such as A1:A12 if you have 12 values listed in column A. Click on **OK**. In the dialog box, select the level of confidence. (If you are using **1 Var z Interval,** also enter the value of σ.) Click on **Compute Interval** and the confidence interval will be displayed.

STATDISK: Select **Analysis,** then **Confidence Intervals,** then **Mean - One Sample.** In the dialog box that appears, first enter the significance level as a decimal number. Enter 0.95 for a 95% confidence level. Proceed to enter the other required items and then click on **Evaluate;** the confidence interval will be displayed.

Hypothesis Tests with the *t* Distribution

SPSS: Enter the sample data in the SPSS data editor window. Click on **Analyze**, then select the menu item **Compare Means**, and then select **One-Sample T Test**. Click on the variable at the left of the dialog box and click on the middle button to paste it into the "Test Variable(s)" box. Enter the assumed value of the population mean in the "Test Value" box. (This is the same value used in the null hypothesis.) Click on **OK**. The display will include the key components in the box labeled "One-Sample Test." That box will include the value of the *t* test statistic in the column with the heading "*t*". That same box will include a *P*-value in the column labeled "Sig. (2-tailed)." Note that this is the correct *P*-value for a two-tailed test, but it must be adjusted for a one-tailed test. For one-tailed tests, you can usually divide the displayed value by 2, but this does not work if the alternative hypothesis is inconsistent with the sample mean, as it is in either of these two cases: (1) a left-tailed test with a sample mean greater than the mean assumed from H_0 or (2) a right-tailed test with a sample mean less than the mean assumed from H_0.

Excel: Excel does not have a built-in function for a *t* test, so use the Data Desk XL add-in that is a supplement to Excel.

First enter the sample data in column A. If you are using Excel 2003, click on **DDXL.** If you are using Excel 2007, click on **Add-Ins,** then click on **DDXL.** Select **Hypothesis Tests.** Under the function type options, select **1 Var t Test.** Click on the pencil icon and enter the range of data values, such as A1:A12 if you have 12 values listed in column A. Click on **OK.** Follow the four steps listed in the dialog box. After clicking on **Compute** in Step 4, you will get the *P*-value, test statistic, and conclusion.

STATDISK: Select **Analysis,** then select **Hypothesis Testing.** For the methods discussed in Section 10.1, select **Mean - One Sample.** A dialog box will appear. Click on the box in the upper left corner and select the item that matches the claim being tested. Proceed to enter the other items in the dialog box, and then click on **Evaluate.** The results will include the test statistic and *P*-value.

Hypothesis Testing with Two-Way Tables

SPSS: In the SPSS data window, enter all of the frequency counts in the first column, enter the corresponding column names in the second column, and enter the corresponding row names in the third column. The frequency counts must be weighted as follows: Click on **Data,** then select the menu item **Weight Cases.** Click on the button labeled "Weight cases by" and paste the variable representing frequencies into the box labeled "Frequency Variable." Now click on **Analyze,** select **Descriptive Statistics,** and select **Crosstabs.** Click on the variable representing the row names and paste it into the box labeled "Row(s)." Click on the variable representing column names and paste it into the box labeled "Column(s)." Now click on the **Statistics** button and proceed to check the box labeled "Chi-Square." Then click on **Continue.** Click on the **Cells** button and check the boxes labeled "Observed" and "Expected." Click on the **Continue** button. Finally, click on **OK.** One of the displayed boxes will show the observed and expected frequencies. The box labeled "Chi-Square Tests" shows the χ^2 test statistic in the row labeled "Pearson Chi-Square," and the *P*-value is in that same row.

Excel: Use the DDXL add-in.

First, enter the *names* of the column categories in column A. Second, enter the *names* of the row categories in column B. Third, enter the corresponding observed frequency counts in column C. If you are using Excel 2003, click on **DDXL.** If you are using Excel 2007, click on **Add-Ins,** then click on **DDXL.** Now select **Tables,** then **Indep. Test for Summ Data.** Click on the pencil icon for **Variable One Names** and enter the range of cells containing the names of the *column* categories, such as A1:A6. Click on the pencil icon for **Variable Two Names** and enter the range of cells containing the names of the *row* categories, such as B1:B6. Click on the pencil icon for **Counts** and enter the range of cells containing the observed frequency counts, such as C1:C6. Click on **OK** to get the test results, which include the χ^2 test statistic and the *P*-value. (The expected frequencies will also be displayed.)

STATDISK: First enter the observed frequencies in columns of the data window. Select **Analysis** from the main menu bar, then select **Contingency Tables,** and proceed to identify the columns containing the frequencies. Click on **Evaluate.** The STATDISK results include the test statistic, critical value, *P*-value, and conclusion.

Analysis of Variance

SPSS: Enter all of the data values in the first column of the SPSS data window. In the second column, enter the *numbers* corresponding to the categories. Enter 1 for the values from the first category, enter 2 for the values from the second category, and so on. Click on **Analyze,** select **Compare Means,** then select **One-Way ANOVA.** In the dialog box that appears, click on the variable for the first column of data values and click on the button in the middle to move it into the box labeled "Dependent List." Also click on the variable containing the category numbers and click on the middle button to move it into the box labeled "Factor." Click on **OK** to obtain the results, which will include the *P*-value.

Excel: First enter the data in columns A, B, C, ... If you are using Excel 2003, click on **Tools** from the main menu bar, then select **Data Analysis.** If you are using Excel 2007, click on **Data,** then click on **Data Analysis.** Select **Anova: Single Factor.** In the dialog box, enter the range containing the sample data. (For example, enter A1:C12 if the first value is in row 1 of column A and the last entry is in row 12 of column C.) The results will include the *P*-value for the analysis of variance test.

STATDISK: Enter the data in columns of the data window. Select **Analysis** from the main menu bar, then select **One-Way Analysis of Variance**, and proceed to select the columns of sample data. Click on **Evaluate** when you are done. The results will include the *P*-value for the analysis of variance test.

Can You Tell a Fraud When You See One?

Suppose your professor gives you a homework assignment in which you are supposed to toss a coin 200 times and record the results in order. The two data sets below represent the results turned in by two students. Now suppose you learn that one of the students really did the assignment, while the other faked the data. Can you tell which one is fake?

Data Set 1 (H = heads; T = tails)

H T H T H H T T T T T T H T H T T T T H
H H T T T T H T T H T T H H H T T H H T
H T H T H H H H T T H T H T H H H H T H
T T H H T T H H H T T T T T T H H T H T
T H T H T T H H T T H T H T H H T T H T
T T H T H T H H T T T H H T T H H H H T
H T H H T H T T T T H T T T H T H T H H
T H T H H H H T H T H H H T T T H T T H
T H T T T H T H H T H H H H T T T H H T
T H T H H T T H T H H H T H H T H T T H

Data Set 1 (H = heads; T = tails)

T H H T T H H T T H T H H T T H H T H T
H T T H T T H H T T H T T T H H T H T H
H H T T H T H H T H T T H T T H H T T H
H T H H T T H T H T H H T H T H T H H T
H T H T T H T T H H T T H H T H T H H T
T H T H H T T H T H T T H T T H T H H T
H T H T T H T H H T H T H T H T H H T H
T T H T H T H H T T H T T H T H H T H H
H T H H T T H T H T H T H T H H T H T T
T H H T H T H H T H H T T H H T H T H T

To make the job a little easier, the following table summarizes a few characteristics of the two data sets that might help you decide which one is fake.

Characteristics of Data Set 1	Characteristics of Data Set 2
Total of 97 H, 103 T Two cases of 6 T in a row Five cases of 4 H in a row Three cases of 4 T in a row	Total of 101 H, 99 T No case of more than 3 H or 3 T in a row

If you are like most people, you will probably guess that Data Set 1 is the fake one. After all, its total numbers of heads and tails are farther from the 100 of each that many people expect, and it has two cases in which there were 6 tails in a row, plus several more cases in which there were 4 heads or tails in a row.

But consider this: The probability of getting 6 heads in a row is $(1/2)^6$, or 1 in 64. The chance of 6 consecutive tails is also 1 in 64. Thus, with 200 tosses, the chance of getting at least one case of 6 heads or tails in a row is quite good, so the strings of consecutive heads and tails in Data Set 1 really are not surprising. In contrast, Data Set 2 has no string as long as 4 heads or tails in a row, even though the probability of such a string is only $(1/2)^4$, or 1 in 16; we conclude that Data Set 2 is almost certainly a fake.

This simple example reveals an important application of statistics in criminology: It is often possible to catch people who have faked data of any kind. For example, statistics helps bank regulators and financial auditors to catch fraudulent financial statements, the Internal Revenue Service to identify people with fraudulent tax returns, and scientists to catch other scientists who have falsified data.

One of the most powerful tools for detecting fraud was identified by physicist Frank Benford. In the 1930s, Benford noticed that tables of logarithms (which scientists and engineers used regularly in the days before calculators) tended to be more worn on the early pages, where the numbers started with the digit 1, than on later pages. Following up on this observation, he soon discovered that many sets of numbers from everyday life, such as stock market values, baseball statistics, and the areas of lakes, include more numbers starting with the digit 1 than with the digit 2, and more starting with 2 than with 3, and so on. He eventually published a formula describing how often numbers begin with different digits, and this formula is now called *Benford's law.* Figure 10.4 shows what his law predicts for the first digits of numbers, along with actual results from several real sets of numbers. Notice how well Benford's law describes the results. (Interestingly, Benford's law was first discovered more than 50 years earlier, and published by astronomer and mathematician Simon Newcomb in 1881. However, Newcomb's article had been forgotten by the time Benford did his work.)

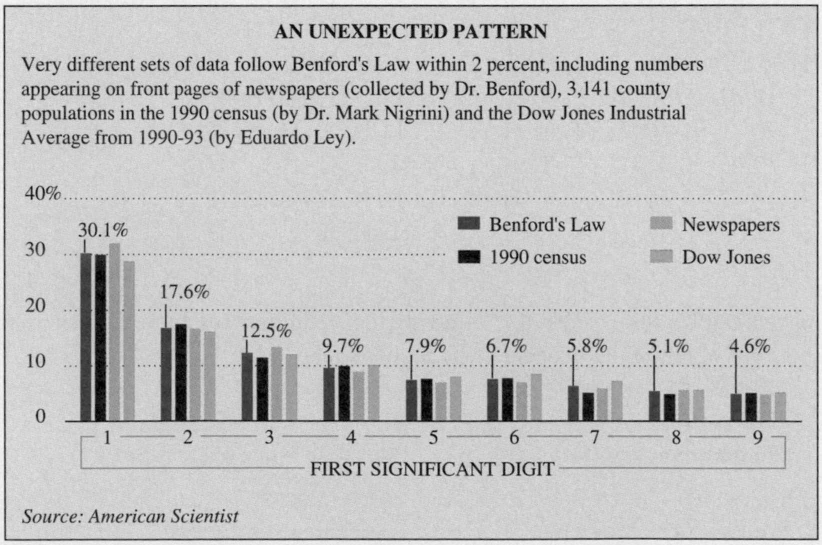

Figure 10.4 *Source:* Malcolm W. Browne, "Following Benford's Law, or Looking Out for No. 1," *New York Times,* August 4, 1998, p. B10.

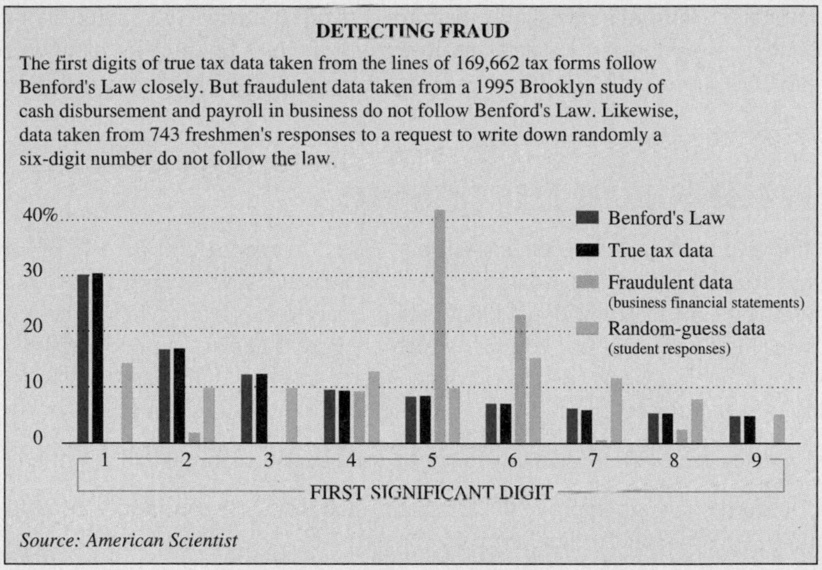

Figure 10.5 *Source:* Adapted from Malcom W. Browne, "Following Benford's Law, or Looking Out for No. 1," *New York Times,* August 4, 1998, p. B10.

Benford's law is surprising because most people guess that every digit (1 through 9) would be equally likely as a starting digit. Indeed, this is the case for sets of random numbers such as lottery numbers, which are no more likely to start with 1 than with any other digit. (So Benford's law should not be used for picking lottery numbers.) However, because Benford's law does apply to many *real data* sets, it can be used to detect fraud. As shown in Figure 10.5, the data from real tax forms (dark red bars) follow Benford's law (dark blue bars) closely. In contrast, the data from a set of financial statements (light blue bars) examined in a 1995 study do not follow Benford's law. Based on this fact, the District Attorney suspected fraud, which he eventually was able to prove. The "random guess data" (green bars) came from students of Professor Theodore P. Hill at Georgia Institute of Technology (the source of much of the information for this Focus). The guesses do not follow Benford's law at all, which is why people who try to fake data can often be caught.

Benford's law mystified scientists and mathematicians for decades. Today it seems to be fairly well understood, though still difficult to explain. Here is one explanation of why Benford's law applies to the Dow Jones Industrial Average, thanks to Dr. Mark J. Nigrini of Southern Methodist University (as reported in the *New York Times*): Imagine that the Dow is at 1,000, so the first digit is a 1, and rises at a rate of about 20% per year. The doubling time at this rate of increase is a little less than four years, so the Dow would remain in the 1,000s, still with a first digit of 1, for almost four years, until it hit 2,000. It would then have a first digit of 2 until it hit 3,000. However, moving from 2,000 to 3,000 requires only a 50% increase, which takes only a little over two years. Thus, the first digit of 2 would occur for only a little more than half the number of days that a first digit of 1 occurred. Subsequent changes of first digit take even shorter times. By the time the Dow hits 9,000, it takes only an 11% increase and just seven months to reach the 10,000 mark, so a first digit of 9 occurs for only seven months. At the 10,000 mark, however, the Dow is back to a first digit of 1 again, and this would not change until the Dow doubled again to 20,000, which means another almost four years at the rate of increase of 20% per year. Thus, if you graphed the number of days the Dow had each starting digit 1 through 9, you'd find that the number 1 would be the starting digit for longer periods of time than the number 2, and so on down the line.

In summary, Benford's law shows that numbers don't always arise with the frequencies that most people would guess. As a result, it not only helps explain a lot of mysteries about numbers (such as those in the Dow), but also has become a valuable tool for the detection of criminal fraud.

QUESTIONS FOR DISCUSSION

1. Try this experiment with friends: Ask one friend to record 200 actual coin tosses and another friend to try to fake data for 200 coin tosses. Have them give you their results anonymously, so that you don't know which sheet came from which friend. Using the ideas in this Focus, try to determine which set is real and which one is fake. After you make your guess, check with your friends to see whether you were right. Discuss your ability to detect the fakes.
2. Briefly discuss how Benford's law might be used to detect fraudulent tax returns.
3. Can Benford's law alone prove that data are fraudulent? Or can it only point to data that should be investigated further? Explain.
4. Find a set of data similar to the sets shown in Figure 10.4 and make a bar chart showing the frequencies of the first digits 1 through 9. Do the data follow Benford's law? Explain.

SUGGESTED READING

Browne, Malcolm W. "Following Benford's Law, or Looking Out for No. 1." *New York Times*, August 4, 1998.

Hill, Theodore P. "The First Digit Phenomenon." *American Scientist*, Vol. 86, No. 358, July–August 1998.

Nigrini, Mark J. *Digital Analysis Using Benford's Law*. Global Audit Publications, Vancouver, 2000.

FOCUS ON EDUCATION

What Can a Fourth-Grader Do with Statistics?

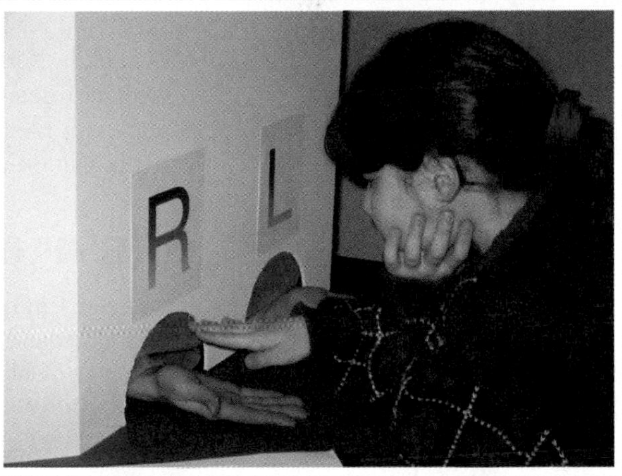

Nine-year-old Emily Rosa was in fourth grade, trying to decide what to do for her school science fair project. She was thinking of doing a project on the colors of M&M candies, when she noticed her mother, a nurse, watching a videotape about a practice called "therapeutic touch" or "TT." TT is a leading alternative medical treatment, practiced in many places throughout the world and even taught at some schools of nursing. But no statistically valid test had ever clearly demonstrated whether it actually works. Emily told her mother that she had an idea for testing TT and wanted to make it her science fair project.

Despite the name, TT therapists do *not* actually touch their patients. Instead, they move their hands a few inches above a patient's body. Therapeutic touch supporters claim that these hand movements allow trained therapists to feel and manipulate what they call a "human energy field." By doing these manipulations properly, the therapists can supposedly cure many different ailments and diseases. Emily Rosa's science fair project sought to find out whether trained TT therapists could really feel a human energy field.

To do her project, Emily recruited 21 TT therapists to participate in a simple experiment. Each therapist sat across a table from Emily, laying his or her arms out flat, palms up. Emily then put up a cardboard partition with cutouts for the therapist's arms. This prevented Emily and the therapist from seeing each other's face, but allowed Emily to see the therapist's hands.

Emily then placed one of her hands a few inches above *one* of the therapist's two hands. If the therapist could truly feel Emily's "human energy field," then the therapist should have been able to tell whether his or her right or left hand was closest to Emily's hand. Thus, each trial of the experiment ended with Emily's recording whether the therapist was right or wrong in identifying the hand.

Emily took several precautions to make sure her experiment would be statistically valid. For example, to ensure that her choices between the two hands were random, Emily used the outcome of a coin toss to determine whether she placed her hand over the therapist's left or right hand in each case. And to make sure she had enough data to evaluate statistical significance, 14 of the 21 therapists got 10 tries each, while 7 got 20 tries each.

The results were a miserable failure for the TT therapists. Because there were only two possible answers in each trial—left hand or right hand—by pure chance the therapists ought to have been able to guess the correct hand about 50% of the time. But the overall results showed that they got the correct answer only 44% of the time. Moreover, none of the therapists performed better than expected by chance in a statistically significant way. Emily also checked to see whether therapists with more experience did better than those with less experience. They did not. Emily's conclusion: If there is such a thing as a "human energy field" (which she doubts), the TT therapists can't feel it. And even if the "human energy field" exists, it's difficult to imagine how TT therapists could use it for healing if they can't even detect its presence.

One of the most interesting aspects of this study was that Emily was able to do it at all. Other skeptics of TT had hoped to conduct similar studies in the past, but TT therapists had refused to participate. One famous skeptic, magician James Randi, had even offered a $1.1 million prize to any TT therapist who could pass a test similar to Emily's. Only one person accepted Randi's challenge, and she succeeded in only 11 of 20 trials, about the same as would be expected by chance. So why was Emily able to succeed where more experienced researchers had failed? Apparently, the therapists agreed to participate in Emily's experiment because they did not feel threatened by a fourth-grader.

The novelty of Emily's science fair project drew media attention, and it was not long before word reached retired Pennsylvania psychiatrist Stephen Barrett. Dr. Barrett specialized in debunking "quack" therapies, and he convinced Emily and her mother to report her results in a medical research paper. The paper was published in the *Journal of the American Medical Association* (April 1, 1998) when Emily was 11, making her the youngest-ever author of a paper in that prestigious journal.

QUESTIONS FOR DISCUSSION

1. After Emily's results were published, many TT supporters claimed that her experiment was invalid because she and her mother were biased against TT. Based on the way her experiment was designed, do you think that her personal bias could have affected her results? Why or why not?
2. Another objection to Emily's experiment was that it was only single-blind rather than double-blind. That is, the therapist could not see what Emily was doing, but Emily could see what the therapist was doing. Do you think this objection is valid in this case? Can you think of a way that Emily's experiment might be repeated but be made double-blind?
3. Emily's experiment was not a direct test of whether TT treatment works, because it did not check to see whether patients actually improved when treated by TT. Suggest a statistically valid way to test whether TT is more effective than a placebo.
4. Based on the results of Emily's study, skeptics now say that TT is so clearly invalid that it should no longer be used or funded. Do you agree? Why or why not?

SUGGESTED READING

Ball, T. S., and Alexander, D. D. "Catching Up with Eighteenth Century Science in the Evaluation of Therapeutic Touch." *Skeptical Inquirer*, July/August 1998.

Kolata, Gina. "4th Grader Challenges Alternative Therapy." *New York Times*, April 2, 1998.

Rosa, E. "TT and Me." *Skeptic Magazine*, September 1998.

Rosa, L., Rosa, E., Sarner, L., and Barrett, S. "A Close Look at Therapeutic Touch." *Journal of the American Medical Association*, Vol. 279, No. 1005, April 1, 1998.

Epilogue: A Perspective on Statistics

A single introductory statistics course cannot transform you into an expert statistician. After studying statistics in this book, you may feel that you have not yet mastered the material to the extent necessary to use statistics confidently in real applications. Nevertheless, by now you should understand enough about statistics to interpret critically the reports of statistical research that you see in the news and to converse with experts in statistics when you need more information. And, if you go on to take further course work in statistics, you should be well prepared to understand important topics that are beyond the scope of this introductory book.

Most importantly, while this book is not designed to make you an expert statistician, it is designed to make you a better-educated person with improved job marketability. You should know and understand the basic concepts of probability and chance. You should know that in attempting to gain insight into a data set, it's important to investigate measures of center (such as mean and median), measures of variation (such as range and standard deviation), the nature of the distribution (via a frequency table or graph), and the presence of outliers. You should know and understand the importance of estimating population parameters (such as a population mean or proportion), as well as testing hypotheses about population parameters. You should understand that a correlation between two variables does not necessarily imply that there is also some cause-and-effect relationship. You should know the importance of good sampling. You should recognize that many surveys and polls obtain very good results, even though the sample sizes might seem to be relatively small. Although many people refuse to believe it, a nationwide survey of only 1,700 voters can provide good results if the sampling is carefully planned and executed.

There once was a time when a person was considered educated if he or she could read, but we are in a new millennium that is much more demanding. Today, an educated person must be able to read, write, understand the significance of the Renaissance, operate a computer, and apply statistical reasoning. The study of statistics helps us see truths that are sometimes distorted by a failure to approach a problem carefully or concealed by data that are disorganized. Understanding statistics is now essential for both employers and employees—for all citizens. H. G. Wells once said, "Statistical thinking will one day be as necessary for efficient citizenship as the ability to read and write." That day is now.

Appendix A: z-Score Tables

This table is a more detailed version of Table 5.1; note that the areas under the curve shown here correspond to *percentiles*. To read this table, find the first two digits of the *z*-score in the left column, then read across the rows for third digit. Negative *z*-scores are on the left page and positive *z*-scores are on the right page.

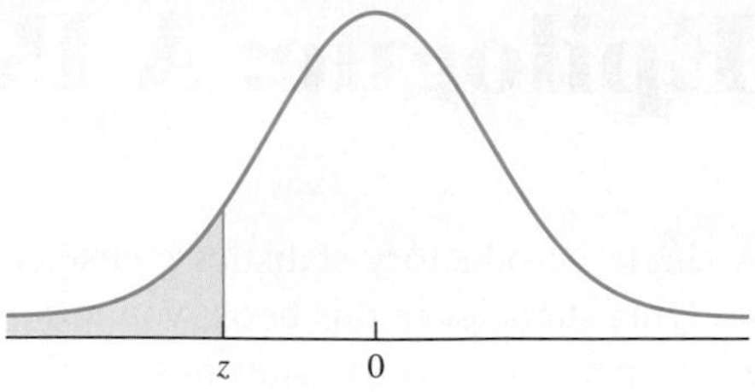

Negative *z*-scores

Table A-1 Standard Normal (*z*) Distribution: Cumulative Area from the LEFT

z	.00	.01	.02	.03	.04	.05	.06	.07	.08	.09
−3.50 and lower	.0001									
−3.4	.0003	.0003	.0003	.0003	.0003	.0003	.0003	.0003	.0003	.0002
−3.3	.0005	.0005	.0005	.0004	.0004	.0004	.0004	.0004	.0004	.0003
−3.2	.0007	.0007	.0006	.0006	.0006	.0006	.0006	.0005	.0005	.0005
−3.1	.0010	.0009	.0009	.0009	.0008	.0008	.0008	.0008	.0007	.0007
−3.0	.0013	.0013	.0013	.0012	.0012	.0011	.0011	.0011	.0010	.0010
−2.9	.0019	.0018	.0018	.0017	.0016	.0016	.0015	.0015	.0014	.0014
−2.8	.0026	.0025	.0024	.0023	.0023	.0022	.0021	.0021	.0020	.0019
−2.7	.0035	.0034	.0033	.0032	.0031	.0030	.0029	.0028	.0027	.0026
−2.6	.0047	.0045	.0044	.0043	.0041	.0040	.0039	.0038	.0037	.0036
−2.5	.0062	.0060	.0059	.0057	.0055	.0054	.0052	.0051 *	.0049	.0048
−2.4	.0082	.0080	.0078	.0075	.0073	.0071	.0069	.0068	.0066	.0064
−2.3	.0107	.0104	.0102	.0099	.0096	.0094	.0091	.0089	.0087	.0084
−2.2	.0139	.0136	.0132	.0129	.0125	.0122	.0119	.0116	.0113	.0110
−2.1	.0179	.0174	.0170	.0166	.0162	.0158	.0154	.0150	.0146	.0143
−2.0	.0228	.0222	.0217	.0212	.0207	.0202	.0197	.0192	.0188	.0183
−1.9	.0287	.0281	.0274	.0268	.0262	.0256	.0250	.0244	.0239	.0233
−1.8	.0359	.0351	.0344	.0336	.0329	.0322	.0314	.0307	.0301	.0294
−1.7	.0446	.0436	.0427	.0418	.0409	.0401	.0392	.0384	.0375	.0367
−1.6	.0548	.0537	.0526	.0516	.0505 *	.0495	.0485	.0475	.0465	.0455
−1.5	.0668	.0655	.0643	.0630	.0618	.0606	.0594	.0582	.0571	.0559
−1.4	.0808	.0793	.0778	.0764	.0749	.0735	.0721	.0708	.0694	.0681
−1.3	.0968	.0951	.0934	.0918	.0901	.0885	.0869	.0853	.0838	.0823
−1.2	.1151	.1131	.1112	.1093	.1075	.1056	.1038	.1020	.1003	.0985
−1.1	.1357	.1335	.1314	.1292	.1271	.1251	.1230	.1210	.1190	.1170
−1.0	.1587	.1562	.1539	.1515	.1492	.1469	.1446	.1423	.1401	.1379
−0.9	.1841	.1814	.1788	.1762	.1736	.1711	.1685	.1660	.1635	.1611
−0.8	.2119	.2090	.2061	.2033	.2005	.1977	.1949	.1922	.1894	.1867
−0.7	.2420	.2389	.2358	.2327	.2296	.2266	.2236	.2206	.2177	.2148
−0.6	.2743	.2709	.2676	.2643	.2611	.2578	.2546	.2514	.2483	.2451
−0.5	.3085	.3050	.3015	.2981	.2946	.2912	.2877	.2843	.2810	.2776
−0.4	.3446	.3409	.3372	.3336	.3300	.3264	.3228	.3192	.3156	.3121
−0.3	.3821	.3783	.3745	.3707	.3669	.3632	.3594	.3557	.3520	.3483
−0.2	.4207	.4168	.4129	.4090	.4052	.4013	.3974	.3936	.3897	.3859
−0.1	.4602	.4562	.4522	.4483	.4443	.4404	.4364	.4325	.4286	.4247
−0.0	.5000	.4960	.4920	.4880	.4840	.4801	.4761	.4721	.4681	.4641

Note: For values of *z* below −3.49, use 0.0001 for the area.
*Use these common values that result from interpolation:

z-score	Area
−1.645	0.0500
−2.575	0.0050

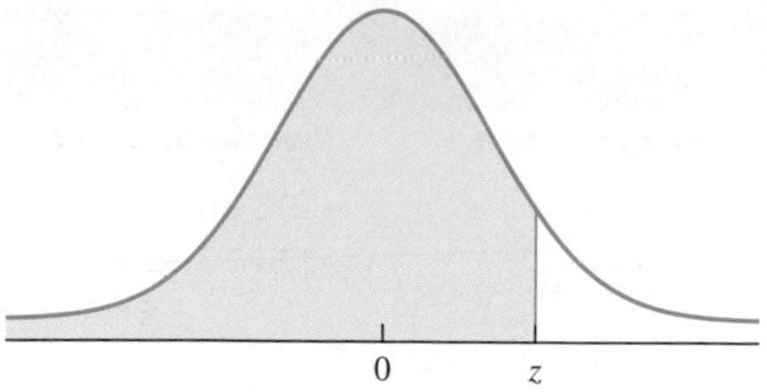

Positive z-scores

Table A-1 (*continued*) Cumulative Area from the LEFT

z	.00	.01	.02	.03	.04	.05	.06	.07	.08	.09
0.0	.5000	.5040	.5080	.5120	.5160	.5199	.5239	.5279	.5319	.5359
0.1	.5398	.5438	.5478	.5517	.5557	.5596	.5636	.5675	.5714	.5753
0.2	.5793	.5832	.5871	.5910	.5948	.5987	.6026	.6064	.6103	.6141
0.3	.6179	.6217	.6255	.6293	.6331	.6368	.6406	.6443	.6480	.6517
0.4	.6554	.6591	.6628	.6664	.6700	.6736	.6772	.6808	.6844	.6879
0.5	.6915	.6950	.6985	.7019	.7054	.7088	.7123	.7157	.7190	.7224
0.6	.7257	.7291	.7324	.7357	.7389	.7422	.7454	.7486	.7517	.7549
0.7	.7580	.7611	.7642	.7673	.7704	.7734	.7764	.7794	.7823	.7852
0.8	.7881	.7910	.7939	.7967	.7995	.8023	.8051	.8078	.8106	.8133
0.9	.8159	.8186	.8212	.8238	.8264	.8289	.8315	.8340	.8365	.8389
1.0	.8413	.8438	.8461	.8485	.8508	.8531	.8554	.8577	.8599	.8621
1.1	.8643	.8665	.8686	.8708	.8729	.8749	.8770	.8790	.8810	.8830
1.2	.8849	.8869	.8888	.8907	.8925	.8944	.8962	.8980	.8997	.9015
1.3	.9032	.9049	.9066	.9082	.9099	.9115	.9131	.9147	.9162	.9177
1.4	.9192	.9207	.9222	.9236	.9251	.9265	.9279	.9292	.9306	.9319
1.5	.9332	.9345	.9357	.9370	.9382	.9394	.9406	.9418	.9429	.9441
1.6	.9452	.9463	.9474	.9484	.9495	.9505	.9515	.9525	.9535	.9545
1.7	.9554	.9564	.9573	.9582	.9591	.9599	.9608	.9616	.9625	.9633
1.8	.9641	.9649	.9656	.9664	.9671	.9678	.9686	.9693	.9699	.9706
1.9	.9713	.9719	.9726	.9732	.9738	.9744	.9750	.9756	.9761	.9767
2.0	.9772	.9778	.9783	.9788	.9793	.9798	.9803	.9808	.9812	.9817
2.1	.9821	.9826	.9830	.9834	.9838	.9842	.9846	.9850	.9854	.9857
2.2	.9861	.9864	.9868	.9871	.9875	.9878	.9881	.9884	.9887	.9890
2.3	.9893	.9896	.9898	.9901	.9904	.9906	.9909	.9911	.9913	.9916
2.4	.9918	.9920	.9922	.9925	.9927	.9929	.9931	.9932	.9934	.9936
2.5	.9938	.9940	.9941	.9943	.9945	.9946	.9948	.9949	.9951	.9952
2.6	.9953	.9955	.9956	.9957	.9959	.9960	.9961	.9962	.9963	.9964
2.7	.9965	.9966	.9967	.9968	.9969	.9970	.9971	.9972	.9973	.9974
2.8	.9974	.9975	.9976	.9977	.9977	.9978	.9979	.9979	.9980	.9981
2.9	.9981	.9982	.9982	.9983	.9984	.9984	.9985	.9985	.9986	.9986
3.0	.9987	.9987	.9987	.9988	.9988	.9989	.9989	.9989	.9990	.9990
3.1	.9990	.9991	.9991	.9991	.9992	.9992	.9992	.9992	.9993	.9993
3.2	.9993	.9993	.9994	.9994	.9994	.9994	.9994	.9995	.9995	.9995
3.3	.9995	.9995	.9995	.9996	.9996	.9996	.9996	.9996	.9996	.9997
3.4	.9997	.9997	.9997	.9997	.9997	.9997	.9997	.9997	.9997	.9998
3.50 and up	.9999									

Note: For values of z above 3.49, use 0.9999 for the area.
*Use these common values that result from interpolation:

z-score	Area
1.645	0.9500
2.575	0.9950

Table A-2 Select Critical Values of z

	Left-tailed test	Right-tailed test	Two-tailed test
0.05 significance level	−1.645	1.645	−1.96 and 1.96
0.01 significance level	−2.33	2.33	−2.576 and 2.576

Appendix B: Table of Random Numbers

For many statistical applications it is useful to generate a set of randomly chosen numbers. You can generate such numbers with most calculators and computers, but sometimes it is easier to use a table such as the one given below. This table was generated by randomly selecting one of the digits 0, 1, 2, 3, 4, 5, 6, 7, 8, or 9 for each position in the table; that is, each of these digits is equally likely to appear in any position. Thus, you can generate a sequence of random digits simply by starting at any point in the table and taking the digits in the order in which they appear. A larger set of random numbers is available on the text Web site (www.aw.com/bbt).

Example 1: Generate a random list of yes/no responses.
Solution: Start at an arbitrary point in the table. If the digit is 0, 1, 2, 3, or 4, call it a *yes* response. If the digit is 5, 6, 7, 8, or 9, call it a *no* response. Continue through the table from your starting point, using each digit shown to determine either a *yes* or a *no* response for your list.

Example 2: Generate a random list of letter grades A, B, C, D, or F.
Solution: Let 0 or 1 be an A, 2 or 3 be a B, 4 or 5 be a C, 6 or 7 be a D, and 8 or 9 be an F. Start at an arbitrary point in the table and use each digit shown to determine a grade for your list.

9 9 3 2 7	5 6 0 8 1	6 0 2 3 2	8 3 3 1 2	4 7 6 3 4
9 7 1 8 1	6 6 7 6 6	5 4 4 7 7	6 8 1 7 1	0 8 4 9 9
8 1 7 5 0	7 8 5 2 0	9 4 3 9 0	7 6 1 9 1	0 8 7 3 4
0 5 1 0 2	8 7 4 0 3	9 2 6 2 5	8 4 2 2 5	1 9 8 3 4
2 7 8 0 8	1 8 5 6 6	4 4 5 5 4	9 3 5 2 8	6 5 5 4 3
4 8 8 3 3	8 4 6 9 1	8 2 5 7 6	9 7 1 2 3	6 5 1 8 2
5 4 5 8 7	3 8 4 5 7	4 5 2 2 2	7 7 0 2 3	0 2 4 8 6
4 1 9 4 3	0 7 1 9 0	7 3 1 4 0	8 3 2 8 0	5 0 1 0 1
2 8 7 4 6	5 7 7 6 0	0 8 9 5 5	4 0 7 3 9	1 6 3 3 2
5 8 6 8 6	9 6 7 7 5	1 3 5 2 9	7 6 6 3 5	9 4 6 0 5
8 0 9 4 8	5 0 5 6 9	1 0 6 9 5	9 0 7 8 9	9 4 8 3 7
6 1 0 4 1	7 4 0 0 3	5 6 4 2 1	5 1 1 9 0	0 2 5 0 7
3 4 0 9 1	9 3 9 5 1	0 7 4 8 1	1 9 7 0 7	1 4 5 2 6
9 9 6 5 0	7 8 6 1 0	4 9 8 7 7	4 4 7 4 0	7 8 6 4 9
1 5 0 7 8	1 6 3 6 9	5 4 9 5 4	2 4 6 0 4	3 2 6 8 4
5 7 2 8 5	8 3 1 6 4	4 2 2 3 7	0 6 6 3 2	3 5 0 4 6
1 2 3 5 7	1 9 7 7 6	5 5 3 1 9	6 0 5 1 6	3 8 0 3 4
3 4 9 3 5	0 3 7 4 6	2 1 8 5 1	4 1 7 0 2	1 4 9 4 9
2 4 2 6 6	9 9 7 2 1	7 7 3 2 0	6 7 0 7 3	9 3 3 7 5
9 9 1 5 2	8 3 9 9 4	8 9 6 0 6	7 4 6 3 1	3 6 9 8 3
2 6 3 3 9	3 4 1 9 3	0 6 9 1 2	4 1 9 9 8	7 3 2 6 9
1 3 7 8 0	1 0 6 1 6	5 4 9 3 0	7 0 7 2 3	1 7 8 9 2
8 8 4 8 4	0 2 8 5 5	7 3 7 1 2	1 8 3 5 2	0 1 5 3 2
4 2 9 2 8	8 8 9 1 2	4 6 7 1 7	5 1 5 1 9	3 2 2 8 0
1 3 7 2 1	0 6 4 7 6	1 0 8 4 8	9 7 6 3 5	5 1 2 2 9

Suggested Readings

Against the Gods: The Remarkable Story of Risk, P. Bernstein, Wiley, 1996.

All the Math That's Fit to Print: Articles from the Manchester Guardian, K. Devlin, Mathematical Association of America, 1994.

The Arithmetic of Life, G. Shaffner, Ballantine Books, 1999.

The Arithmetic of Life and Death, G. Shaffner, Ballantine Books, 2001.

The Bell Curve Debate, R. Jacoby and N. Glauberman (eds.), Times Books, 1995.

Beyond Numeracy: Ruminations of a Number Man, J. A. Paulos, Vintage, 1992.

Beyond the Limits: Confronting Global Collapse, Envisioning a Sustainable Future, D. H. Meadows, D. L. Meadows, and J. Randers, Chelsea Green Publishing Company, 1992.

Billions and Billions, C. Sagan, Random House, 1997.

The Broken Dice and Other Mathematical Tales of Chance, I. Ekeland, University of Chicago Press, 1993.

Can You Win?, M. Orkin, Freeman, 1991.

The Cartoon Guide to Statistics, L. Gonick and W. Smith, HarperCollins, 1993.

The Complete How to Figure It, D. Huff, Norton, 1996.

Damned Lies and Statistics, J. Best, University of California Press, 2001.

Ecological Numeracy: Quantitative Analysis of Environmental Issues, R. Herendeen, Wiley, 1998.

Elementary Statistics, 10th ed., M. Triola, Addison-Wesley, 2006.

Emblems of Mind: The Inner Life of Music and Mathematics, E. Rothstein, Random House, 1995.

Envisioning Information, E. Tufte, Connecticut Graphics Press, Cheshire, 1983.

Flaws and Fallacies in Statistical Thinking, S. Campbell, Prentice-Hall, 1974.

Games, Gods, and Gambling, F. N. David, Dover, 1998.

Go Figure! The Numbers You Need for Everyday Life, N. Hopkins and J. Mayne, Visible Ink Press, 1992.

The Golden Mean, C. F. Linn, Doubleday, 1974.

The Honest Truth About Lying with Statistics, C. Holmes, Charles C Thomas, 1990.

How Many People Can the Earth Support, J. E. Cohen, Norton, 1995.

How to Lie with Statistics, D. Huff, Norton, 1993.

How to Tell the Liars from the Statisticians, R. Hooke, Dekker, 1983.

How to Use (and Misuse) Statistics, G. Kimble, Prentice-Hall, 1978.

Innumeracy, J. A. Paulos, Hill and Wang, 1988.

The Jungles of Randomness, I. Peterson, Wiley, 1998.

Lady Luck: The Theory of Probability, W. Weaver, Anchor Books, 1963.

The Lady Tasting Tea: How Statistics Revolutionized Science in the 20th Century, D. Salsburg, Freeman, 2001.

Life by the Numbers, K. Devlin, Wiley, 1998.

The Mathematical Tourist, I. Peterson, Freeman, 1988.

A Mathematician Reads the Newspaper, J. A. Paulos, Basic Books, 1995.

Mathematics, the Science of Patterns: The Search for Order in Life, Mind, and the Universe, K. Devlin, Scientific American Library, 1994.

The Mismeasure of Man, S. J. Gould, Norton, 1981.

The Mismeasure of Woman, C. Tavris, Touchstone Books, 1993.

Misused Statistics: Straight Talk for Twisted Numbers, A. Jaffe and H. Spirer, Dekker, 1987.

Nature's Numbers, I. Stewart, Basic Books, 1995.

Number: The Language of Science, T. Dantzig, Macmillan, 1930.

Once Upon a Number, J. A. Paulos, Basic Books, 1998.

Overcoming Math Anxiety, S. Tobias, Houghton Mifflin, 1978. Revised edition, Norton, 1993.

Pi in the Sky: Counting, Thinking, and Being, J. D. Barrow, Little, Brown, 1992.

The Population Explosion, P. R. Ehrlich and A. H. Ehrlich, Simon and Schuster, 1990.

Probabilities in Everyday Life, J. McGervey, Ivy Books, 1989.

The Psychology of Judgment and Decision Making, S. Plous, McGraw-Hill, 1993.

Randomness, D. Bennett, Harvard University Press, 1998.

The Statistical Exorcist: Dispelling Statistics Anxiety, M. Hollander and F. Proschan, Dekker, 1984.

Statistics, 3rd ed., D. Freedman, R. Pisani, R. Purves, and A. Adhikari, Norton, 1997.

Statistics: Concepts and Controversies, 5th ed., D. Moore, Freeman, 2001.

Statistics with a Sense of Humor, F. Pyrczak, Fred Pyrczak Publisher, 1989.

Tainted Truth: The Manipulation of Fact in America, C. Crossen, Simon and Schuster, 1994.

The Tipping Point: How Little Things Can Make a Big Difference, M. Gladwell, Little, Brown, 2000.

200% of Nothing, A. K. Dewdney, Wiley, 1993.

The Universe and the Teacup: The Mathematics of Truth and Beauty, K. C. Cole, Harcourt Brace, 1998.

The Visual Display of Quantitative Information, E. Tufte, Connecticut Graphics Press, Cheshire, 1983.

Vital Signs, compiled by the Worldwatch Institute, Norton, 1998 (updated annually).

What Are the Odds? Chance in Everyday Life, L. Krantz, Harper Perennial, 1992.

Credits

FIGURE AND TEXT CREDITS

Chapter 1 Figure 1.1: *New York Times,* November 1, 1999. Copyright © 1999 The New York Times Co. Reprinted with permission.

Chapter 2 Figure 2.4: *New York Times,* January 3, 2007. Copyright © 2007 The New York Times Co. Reprinted with permission.

Chapter 3 Figure 3.15: Portions from *The Wall Street Journal Almanac* by Editors of *The Wall Street Journal,* 1999, p. 540. © 1999 Dow Jones & Co. Figure 3.16: From *The Wall Street Journal Almanac* by Editors of *The Wall Street Journal,* 1999, p. 694. © 1999 Dow Jones & Co. Reprinted with permission from Ballantine Books, an imprint of Random House, Inc. Figure 3.18: From *The Wall Street Journal Almanac* by Editors of *The Wall Street Journal,* 1999, p. 238. © 1999 Dow Jones & Co. Reprinted with permission from Ballantine Books, an imprint of Random House, Inc. Figure 3.19: *New York Times,* September 30, 1995. Copyright © 1995 The New York Times Co. Reprinted with permission. Figure 3.22: From Bennett/Briggs, *Using and Understanding Mathematics,* 4th ed., p. 365. Copyright © 2008 Pearson Education, Inc. Reprinted by permission of Pearson Education, Inc. Figure 3.24: From Bennett/Briggs, *Using and Understanding Mathematics,* 4th ed., p. 367. Copyright © 2008 Pearson Education, Inc. Reprinted by permission of Pearson Education, Inc. Figure 3.25: From Bennett/Briggs, *Using and Understanding Mathematics,* 4th ed., p. 377. Copyright © 2008 Pearson Education, Inc. Reprinted by permission of Pearson Education, Inc. Figure 3.27: *New York Times,* October 3, 1995. Copyright © 1995 The New York Times Co. Reprinted with permission. Figure 3.28: Portions from *New York Times,* August 20, 2000. Copyright © 2000 The New York Times Co. Reprinted with permission. Figure 3.33: From Bennett/Briggs, *Using and Understanding Mathematics,* 4th ed., p. 376. Copyright © 2008 Pearson Education, Inc. Reprinted by permission of Pearson Education, Inc. Figure 3.34: From Bennett/Briggs, *Using and Understanding Mathematics,* 4th ed., p. 375. Copyright © 2008 Pearson Education, Inc. Reprinted by permission of Pearson Education, Inc. Figure 3.35: From Bennett/Briggs, *Using and Understanding Mathematics,* 2nd ed., p. 330. Copyright © 2002 Pearson Education, Inc. Reprinted by permission of Pearson Education, Inc. Figure 3.36: *New York Times,* April 2, 2000. Copyright © 2000 The New York Times Co. Reprinted with permission. Figure 3.40: Portions from *The Wall Street Journal Almanac* by Editors of *The Wall Street Journal,* 1999, p. 323. © 1999 Dow Jones & Co. Figure 3.41: From Bennett/Briggs, *Using and Understanding Mathematics,* 4th ed., p. 371. Copyright © 2008 Pearson Education, Inc. Reprinted by permission of Pearson Education, Inc. Figure 3.42: From Bennett/Briggs, *Using and Understanding Mathematics,* 4th ed., p. 371. Copyright © 2008 Pearson Education, Inc. Reprinted by permission of Pearson Education, Inc. Figure 3.43: From Bennett et al., *The Cosmic Perspective,* 4th ed., p. 378. © 2007 Pearson Education, Inc. Reprinted with permission. Figure 3.34: From Bennett/Briggs, *Using and Understanding Mathematics,* 4th ed., p. 372. Copyright © 2008 Pearson Education, Inc. Reprinted by permission of Pearson Education, Inc. Figure 3.50: From *The Wall Street Journal Almanac* by Editors of *The Wall Street Journal,* 1999, p. 577. © 1999 Dow Jones & Co. Figure 3.52: From Mortensen, Pedersen, Westergaard, Wohlfahrt, Ewald, Mors, Andersen, Melbye, "Effects of Family History and Place and Season of Birth on the Risk of Schizophrenia," *The New England Journal of Medicine,* 2/25/99, p. 606. Copyright © 1999 Massachusetts Medical Society. All rights reserved. Figure 3.54: From *The Wall Street Journal Almanac* by Editors of *The Wall Street Journal,* 1999, p. 545. © 1999 Dow Jones & Co. Figure 3.59: Adapted from Edward R. Tufte, *The Visual Display of Quantitative Information,* Cheshire, CT: Graphics Press, 1983. Reprinted with permission. Figure 3.58: Portions from *The Wall Street Journal Almanac* by Editors of *The Wall Street Journal,* 1999, p. 662. © 1999 Dow Jones & Co. Figure 3.60: From Bennett et al., *Essential Cosmic Perspective,* 4th ed., p. 216. © 2009 Pearson Education, Inc. Reprinted with permission. Figure 3.61: From Bennett et al., *Essential Cosmic Perspective,* 4th ed., p. 217. © 2009 Pearson Education, Inc. Reprinted with permission.

Chapter 4 Figure 4.1: From Mario F. Triola, Elementary Statistics, 7th ed., p. 61. © 1998 Addison Wesley Longman Inc. Reprinted by permission of Pearson Education. Figure 4.9: From Bennett/Briggs, *Using and Understanding Mathematics,* 4th ed., p. 411. Copyright © 2008 Pearson Education, Inc. Reprinted by permission of Pearson Education, Inc. Figure 4.10: From Bennett/Briggs, *Using and Understanding Mathematics,* 4th ed., p. 411. Copyright © 2008 Pearson Education, Inc. Reprinted by permission of Pearson Education, Inc. Figure 4.11: From Bennett/Briggs, *Using and Understanding Mathematics,* 4th ed., p. 411. Copyright © 2008 Pearson Education, Inc. Reprinted by permission of Pearson Education, Inc. Figure 4.17: From Bennett/Briggs, *Using and Understanding Mathematics,* 4th ed., p. 258. Copyright © 2008 Pearson Education, Inc. Reprinted by permission of Pearson Education, Inc.

Chapter 5 Figure 5.26: From Mario F. Triola, *Elementary Statistics,* 7th ed., p. 256. © 1998 Addison Wesley Longman Inc. Reprinted by permission of Pearson Education. Figure 5.28: *New York Times,* February 24, 1998. Copyright © 1998 The New York Times Co. Reprinted with permission.

Chapter 6 Epigraph: Ogden Nash Poems & Stories by Linell Nash Smith and Isabel Nash Eberstadt. Figure 6.5: From *The Virginia Pilot,* 9/14/99. Figure 6.13: Portions from *New York Times,* December 11, 1999. Copyright © 1999 The New York Times Co. Reprinted with permission. Figure 6.19: From *Statistical Science* by Kathryn Roeder, May 15, 1994. Copyright 1994 by the Institute of Mathematical Statistics. All rights reserved. Reprinted with permission via Copyright Clearance Center.

Chapter 7 Figure 7.21: From Bennett/Briggs, *Using and Understanding Mathematics,* p. 501. Copyright © 1999 Addison Wesley Longman Inc. Reprinted by permission of Pearson Education, Inc. Figure 7.25: From Bennett et al., *The Cosmic Perspective,* 4th ed., p. 293. © 2007 Pearson Education, Inc. Reprinted with permission. Figure 7.26: From Bennett et al., *The Cosmic Perspective,* 4th ed., p. 321. © 2007 Pearson Education, Inc. Reprinted with permission.

Chapter 9 Figure 9.10: *New York Times,* January 3, 2007. Copyright © 2007 The New York Times Co. Reprinted with permission. Figure 9.11: *New York Times,* January 3, 2007. Copyright © 2007 The New York Times Co. Reprinted with permission.

Chapter 10 Figure 10.4: *New York Times,* August 4, 1998, p. B10. Copyright © 1998 The New York Times Co. Reprinted with permission. Figure 10.5: *New York Times,* August 4, 1998, p. B10. Copyright © 1998 The New York Times Co. Reprinted with permission.

PHOTO CREDITS

Chapter 1 p. 1, © Time & Life Pictures/Getty Images; p. 3, © Getty Images Sport; p. 6, PhotoDisc; p. 8, © AFP/Corbis; p. 12, © Taxi/Getty Images; p. 14, courtesy of Jeff Bennett; p. 16, PhotoDisc; p. 25, Corbis Royalty Free; p. 28, © Digital Vision; p. 30, PhotoDisc; p. 35, PhotoDisc; p. 38, NASA; p. 41, © Michael J. Okoniewski, 143 Peck Ave, Syracuse NY; 13206; p. 48, PhotoDisc; p. 50, © Digital Vision

Chapter 2 p. 53, PhotoDisc; p. 60, Corbis Royalty Free; p. 63, NASA; p. 64, PhotoDisc; p. 70, © Peter Turnley/Corbis; p. 71, © Digital Vision; p. 75, © Gene Blevins/Corbis; p. 78, © Bettmann/Corbis; p. 84, Corbis Royalty Free; p. 86, Image Source/Getty Royalty Free

Chapter 3 p. 89, © AFP/Getty Images; p. 92, Corbis Royalty Free; p. 95, © Stone/Getty Images; p. 103, © Ralf-Finn Hestoft/Corbis; p. 106, © Everett Collection; p. 125, © AP Wideworld Photo; p. 126, PhotoDisc Red; p. 128, © Charles & Josette Lenars/Corbis; p. 137, © Archivo Iconografico, S.A./Corbis; p. 140, PhotoDisc

Chapter 4 p. 145, PhotoDisc; p. 149, © PCN Photography; p. 161, © Laurent Rebours/AP Wideworld Photos; p. 168, photo by Beth Anderson; p. 173, © Toru Hanai/Reuters/Corbis; p. 181, © Richard T. Nowitz/Corbis; p. 188, © John-Marshall Mantel/Corbis; p. 191, © James L. Amos/Corbis

Chapter 5 p. 195, (left) © Dimitri Lundt, TempSport/Corbis; (right) © Getty Images Sport; p. 198, photo by Beth Anderson; p. 206, PhotoDisc; p. 210, U.S. Army; p. 226, photo by Beth Anderson; p. 229, PhotoDisc

Chapter 6 p. 233, PhotoDisc; p. 240, (top) PhotoDisc; (bottom) PhotoDisc; p. 241, PhotoDisc; p. 245, Gary Holscher/Stone/Getty Images; p. 252, © Getty Images News; p. 260, Stockbyte/Getty Royalty Free; p. 268, © Greg Fiume/NewSport/Corbis; p. 270, © Bettmann/Corbis; p. 279, Photographer's Choice Getty Royalty Free; p. 282, Corbis Royalty Free

Chapter 7 p. 285, PhotoDisc; p. 287, PhotoDisc; p. 300, PhotoDisc; p. 303, © Tomasso DeRosa/Corbis; p. 307, PhotoDisc Red; p. 313, StockDisc CD SD174; p. 317, PhotoDisc; p. 318, © AP Wideworld Photos; p. 325, Blend Images/Getty Royalty Free; p. 328, © Digital Vision

Chapter 8 p. 333, © Digital Vision; p. 340, © James A Sugar/Corbis; p. 346, © Digital Vision; p. 363, PhotoDisc

Chapter 9 p. 369, PhotoDisc; p. 370, PhotoDisc; p. 377, © Digital Vision; p. 390, © Layne Kennedy/Corbis; p. 403, © Digital Vision; p. 406, PhotoDisc CD Vol 12

Chapter 10 p. 409, © Photonica/Getty Images; p. 439, photo by Beth Anderson; p. 443, photo by Beth Anderson

Glossary

absolute change The actual increase or decrease from a reference value to a new value:

absolute change = new value − reference value

absolute difference The actual difference between the compared value and the reference value:

absolute difference = compared value − reference value

absolute error The actual amount by which a measured value differs from the true value:

absolute error = measured value − true value

accident rate The number of accidents due to some particular cause, expressed as a fraction of all people at risk for the same cause. For example, an accident rate of "5 per 1,000 people" means that an average of 5 in 1,000 people suffer an accident from this particular cause.

accuracy How closely a measurement approximates a true value. An accurate measurement is very close to the true value.

alternative hypothesis (H_a) A statement that can be accepted only if the null hypothesis is rejected.

analysis of variance (ANOVA) A method of testing the equality of three or more population means by analyzing sample variances.

***and* probability** The probability that event A *and* event B will both occur. How it is calculated depends on whether the events are independent or dependent. Also called *joint probability*.

ANOVA See *analysis of variance*.

***a priori* method** See *theoretical method*.

bar graph A diagram consisting of bars representing the frequencies (or relative frequencies) for particular categories. The bar lengths are proportional to the frequencies.

best-fit line The line on a scatter diagram that lies closer to the data points than all other possible lines (according to a standard statistical measure of closeness). Also called *regression line*.

bias In a statistical study, any problem in the design or conduct of the study that tends to favor certain results. See also *participation bias*; *selection bias*.

bimodal distribution A distribution with two peaks, or modes.

binning Grouping data into categories (bins), each of which covers a range of possible data values.

blinding The practice of keeping experimental subjects and/or experimenters in the dark about who is in the treatment group and who is in the control group. See also *double-blind experiment*; *single-blind experiment*.

boxplot A graphical display of a five-number summary. A number line is used for reference, the values from the lower to the upper quartiles are enclosed in a box, a line is drawn through the box for the median, and two "whiskers" are extended to the low and high data values. Also called *box-and-whisker plot*.

case-control study An observational study that resembles an experiment because the sample naturally divides into two (or more) groups. The participants who engage in the behavior under study form the cases, like the treatment group in an experiment. The participants who do not engage in the behavior are the *controls*, like the control group in an experiment.

causality A relationship present when one variable is a cause of another.

census The collection of data from every member of a population.

Central Limit Theorem Theorem stating that, for random samples (all of the same size) of a variable with any distribution (not necessarily a normal distribution), the distribution of the means of the samples will, as the sample size increases, tend to be approximately a normal distribution.

chi-square statistic (χ^2) A number used to determine the statistical significance of a hypothesis test in a contingency table (or two-way table). If it is less than a critical value (which depends on the table size and the desired significance level), the differences between the observed frequencies and the expected frequencies are not significant.

cluster sampling Dividing the population into groups, or clusters; selecting some of these clusters at random; and then obtaining the sample by choosing all the members within each cluster.

coefficient of determination (R^2) A number that describes how well data fit a best-fit equation found through multiple regression.

compared value A number that is compared to a reference value in computing a relative difference.

complement For an event A, all outcomes in which A does *not* occur, expressed as $\bar{A}$. Its probability is $P(\bar{A}) = 1 - P(A)$.

conditional probability The probability of one event given the occurrence of another event, written $P(B \text{ given } A)$or $P(B|A)$.

confidence interval A range of values associated with a confidence level, such as 95%, that is likely to contain the true value of a population parameter.

confounding Confusion in the interpretation of statistical results that occurs when the effects of different factors are mixed such that the effects of the individual factors being studied cannot be determined.

confounding factors Any factors or variables in a statistical study that can lead to confounding. Also called *confounding variables*.

Consumer Price Index (CPI) An index number designed to measure the rate of inflation. It is computed and reported monthly, based on a sample of more than 60,000 goods, services, and housing costs.

contingency table See *two-way table.*

continuous data Quantitative data that can take on any value in a given interval.

contour map A map that uses curves (contours) to connect geographical regions with the same data values.

control group The group of subjects in an experiment who do not receive the treatment being tested.

convenience sampling Selecting a sample that is readily available.

correlation A statistical relationship between two variables. See also *negative correlation; no correlation; positive correlation.*

correlation coefficient (r) A measure of the strength of the relationship between two variables. Its value is always between -1 and 1 (that is, $-1 \leq r \leq 1$).

cumulative frequency For any data category, the number of data values in that category and all preceding categories.

death rate The number of deaths due to some particular cause, expressed as a fraction of all people at risk for the same cause. For example, a death rate of "5 per 1,000 people" means that an average of 5 in 1,000 people die from this particular cause.

degrees of freedom (for a t distribution) The sample size minus one ($n - 1$).

dependent events Two events for which the outcome of one affects the probability of the other.

deviation How far a particular data value lies from the mean of a data set, used to compute standard deviation.

discrete data Quantitative data that can take on only particular values and not other values in between (for example, the whole numbers 0, 1, 2, 3, 4, 5).

distribution The way the values of a variable are spread over all possible values. It can be displayed with a table or with a graph.

distribution of sample means The distribution that results when the means ($\bar{x}$) of all possible samples of a given size are found.

distribution of sample proportions The distribution that results when the proportions ($\hat{p}$) in all possible samples of a given size are found.

dotplot A diagram similar to a bar graph except that each individual data value is represented with a dot.

double-blind experiment An experiment in which neither the participants nor the experimenters know who belongs to the treatment group and who belongs to the control group.

either/or probability The probability that *either* event A *or* event B will occur. How it is calculated depends on whether the events are overlapping or non-overlapping.

empirical method See *relative frequency method.*

event In probability, a collection of one or more outcomes that share a property of interest. See also *outcome.*

expected frequency In a two-way table, the frequency one would expect in a given cell of the table if the row and column variables were independent of each other.

expected value The mean value of the outcomes for some random variable.

experiment A study in which researchers apply a treatment and then observe its effects on the subjects.

experimenter effect An effect that occurs when a researcher or experimenter somehow influences subjects through such factors as facial expression, tone of voice, or attitude.

five-number summary A description of the variation of a data distribution in terms of the minimum value, lower quartile, median, upper quartile, and maximum value.

frequency For a data category, the number of times data values fall within that category.

frequency table A table that lists all the categories of data in one column and the frequency for each category in another column.

gambler's fallacy The mistaken belief that a streak of bad luck makes a person "due" for a streak of good luck.

geographical data Data that can be assigned to different geographical locations.

histogram A bar graph showing a distribution for quantitative data (at the interval or ratio level of measurement). The bars have a natural order, and the bar widths have specific meaning.

hypothesis In statistics, a claim about a population parameter, such as a population proportion, p, or population mean, μ. See also *alternative hypothesis*; *null hypothesis.*

hypothesis test A standard procedure for testing a claim about the value of a population parameter.

independent events Two events for which the outcome of one does not affect the probability of the other.

index number A number that provides a simple way to compare measurements made at different times or in different places. The value at one particular time (or place) must be chosen to be the reference value (or base value). The index number for any other time (or place) is

$$\text{index number} = \frac{\text{value}}{\text{reference value}} \times 100$$

inflation The increase over time in prices and wages. Its overall rate is measured by the CPI.

interval level of measurement A level of measurement for quantitative data in which differences, or intervals, are meaningful but ratios are not. Data at this level have an arbitrary starting point.

joint probability See *and probability.*

law of large numbers An important result in probability that applies to a process for which the probability of an event A is $P(A)$ and the results of repeated trials are independent. It states: If the process is repeated through many trials, the larger the number of trials, the closer the proportion should be to $P(A)$. Also called *law of averages.*

left-skewed distribution A distribution in which the values are more spread out on the left side.

left-tailed test A hypothesis test that involves testing whether a population parameter lies to the left (lower values) of a claimed value.

level of measurement See *nominal level of measurement*; *ordinal level of measurement*; *interval level of measurement*; *ratio level of measurement.*

life expectancy The number of years a person of a given age today can be expected to live, on average. It is based on current health and medical statistics and does not take into account future changes in medical science or public health.

line chart A graph showing the distribution of quantitative data as a series of dots connected by lines. The horizontal position of each dot corresponds to the center of the bin it represents and the vertical position corresponds to the frequency value for the bin.

lower quartile See *quartile, lower*.

margin of error The maximum likely difference between an observed sample statistic and the true value of a population parameter. Its size depends on the desired level of confidence.

mean The sum of all values divided by the total number of values. It's what is most commonly called the average value.

median The middle value in a sorted data set (or halfway between the two middle values if the number of values is even).

median class For binned data, the bin into which the median data value falls.

meta-analysis A study in which researchers analyze many individual studies (on a particular topic) as a combined group, with the aim of finding trends that were not evident in the individual studies.

middle quartile See *quartile, middle*.

mode The most common value (or group of values) in a distribution.

multiple bar graph A simple extension of a regular bar graph, in which two or more sets of bars allow comparison of two or more data sets.

multiple line chart A simple extension of a regular line chart, in which two or more lines allow comparison of two or more data sets.

multiple regression A technique that allows the calculation of a best-fit equation that represents the best fit between one variable (such as price) and a *combination* of two or more other variables (such as weight and color).

negative correlation A correlation in which the two variables tend to change in opposite directions, with one increasing while the other decreases.

no correlation Absence of any apparent relationship between two variables.

nominal level of measurement A level of measurement for qualitative data that consist of names, labels, or categories only and cannot be ranked or ordered.

nonlinear relationship A relationship between two variables that cannot be expressed with a linear (straight-line) equation.

non-overlapping events Two events for which the occurrence of one precludes the occurrence of the other.

normal distribution A special type of symmetric, bell-shaped distribution with a single peak that corresponds to the mean, median, and mode of the distribution. Its variation can be characterized by the standard deviation. See also *68-95-99.7 rule*.

null hypothesis (H_0) A specific claim (such as a specific value for a population parameter) against which an alternative hypothesis is tested.

observational study A study in which researchers observe or measure characteristics of the sample members, but do not attempt to influence or modify these characteristics.

***of* versus *more than (less than)* rule** A rule for comparisons. It states: If the compared value is *P% more than* the reference value, then it is $(100 + P)\%$ *of* the reference value. If the compared value is *P% less than* the reference value, then it is $(100 - P)\%$ *of* the reference value.

one-tailed test See *left-tailed test*; *right-tailed test*.

ordinal level of measurement A level of measurement for qualitative data that can be arranged in some order. It generally does not make sense to do computations with the data.

outcome In probability, the most basic possible result of an observation or experiment. See also *event*.

outlier A value in a data set that is much higher or much lower than almost all other values.

overlapping events Two events that could possibly both occur.

Pareto chart A bar graph of data at the nominal level of measurement, with the bars arranged in frequency order.

participants People (as opposed to objects) who are the subjects of a study.

participation bias Bias that occurs any time participation in a study is voluntary.

peer review A process by which several experts in a field evaluate a research report before it is published.

percentiles Values that divide a data distribution into 100 segments, each representing about 1% of the data values.

pictograph A graph embellished with artwork.

pie chart A circle divided so that each wedge represents the relative frequency of a particular category. The wedge size is proportional to the relative frequency, and the entire pie represents the total relative frequency of 100%.

placebo Something that lacks the active ingredients of a treatment but is identical in appearance to the treatment. Thus, participants in a study cannot distinguish the placebo from the real treatment.

placebo effect An effect in which patients improve simply because they believe they are receiving a useful treatment, when in fact they may be receiving only a placebo.

population The complete set of people or things being studied.

population mean The true mean of a population, denoted by the Greek letter μ (pronounced "mew").

population parameters Specific characteristics of the population that a statistical study is designed to estimate.

population proportion The true proportion of some characteristic in a population, denoted by p.

positive correlation A type of correlation in which two variables tend to increase (or decrease) together.

practical significance In a statistical study, significance in the sense that the result is associated with some meaningful course of action.

precision The amount of detail in a measurement.

probability For an event, the likelihood that the event will occur. The probability of an event, written as *P*(event), is always between

0 and 1 inclusive. A probability of 0 means the event is impossible and a probability of 1 means the event is certain. See also *relative frequency method; subjective method; theoretical method.*

probability distribution The complete distribution of the probabilities of all possible events associated with a particular variable. It may be shown as a table or as a graph.

***P*-value** In a hypothesis test, the probability of selecting a sample at least as extreme as the observed sample, assuming that the null hypothesis is true.

qualitative data Data consisting of values that describe qualities or nonnumerical categories.

quantitative data Data consisting of values representing counts or measurements. Quantitative data may be either discrete or continuous.

quartile, lower The median of the data values in the lower half of a data set. Also called *first quartile.*

quartile, middle The overall median of a data set. Also called *second quartile.*

quartile, upper The median of the data values in the upper half of a data set. Also called *third quartile.*

quartiles Values that divide a data distribution into four equal parts.

random errors Errors that occur because of random and inherently unpredictable events in the measurement process.

randomization The process of ensuring that the subjects of an experiment are assigned to the treatment or control group at random and in such a way that each subject has an equal chance of being assigned to either group.

range For a distribution, the difference between the lowest and highest data values.

range rule of thumb A guideline stipulating that, for a data set with no outliers, the standard deviation is approximately equal to range/4.

rare event rule Rule stating that it is appropriate to conclude that a given assumption (such as the null hypothesis) is probably not correct if the probability of a particular event at least as extreme as the observed event is very small.

ratio level of measurement A level of measurement for quantitative data in which both intervals and ratios are meaningful. Data at this level have a true zero point.

raw data The actual measurements or observations collected from a sample.

reference value The number that is used as the basis for a comparison.

regression line See *best-fit line.*

relative change The size of an absolute change in comparison to the reference value, expressed as a percentage:

$$\text{relative change} = \frac{\text{new value} - \text{reference value}}{\text{reference value}} \times 100\%$$

relative difference The size of an absolute difference in comparison to the reference value, expressed as a percentage:

$$\text{relative difference} = \frac{\text{compared value} - \text{reference value}}{\text{reference value}} \times 100\%$$

relative error The relative amount by which a measured value differs from the true value, expressed as a percentage:

$$\text{relative error} = \frac{\text{measured value} - \text{true value}}{\text{true value}} \times 100\%$$

relative frequency For any data category, the fraction or percentage of the total frequency that falls in that category:

$$\text{relative frequency} = \frac{\text{frequency in category}}{\text{total frequency}}$$

relative frequency method A method of estimating a probability based on observations or experiments by using the observed or measured relative frequency of the event of interest. Also called *empirical method.*

representative sample A sample in which the relevant characteristics of the members are generally the same as the characteristics of the population.

right-skewed distribution A distribution in which the values are more spread out on the right side.

right-tailed test A hypothesis test that involves testing whether a population parameter lies to the right (higher values) of a claimed value.

rounding rule For statistical calculations, the practice of stating answers with one more decimal place of precision than is found in the raw data. For example, the mean of 2, 3, and 5 is 3.3333 . . . , which would be rounded to 3.3.

sample A subset of the population from which data are actually obtained.

sample mean The mean of a sample, denoted $\bar{x}$ ("*x*-bar").

sample proportion The proportion of some characteristic in a sample, denoted $\hat{p}$ ("*p*-hat").

sample statistics Characteristics of the sample that are found by consolidating or summarizing the raw data.

sampling The process of choosing a sample from a population.

sampling distribution The distribution of a sample statistic, such as a mean or proportion, taken from all possible samples of a particular size.

sampling error Error introduced when a random sample is used to estimate a population parameter; the difference between a sample result and a population parameter.

sampling methods See *cluster sampling*; *convenience sampling*; *simple random sampling*; *stratified sampling*; *systematic sampling*.

scatter diagram A graph, often used to investigate correlations, in which each point corresponds to the values of two variables. Also called *scatterplot.*

selection bias Bias that occurs whenever researchers select their sample in a biased way. Also called *selection effect.*

self-selected survey A survey in which people decide for themselves whether to be included. Also called *voluntary response survey.*

simple random sampling A sample of items chosen in such a way that every possible sample of the same size has an equal chance of being selected.

Simpson's paradox A statistical paradox that arises when the results for a whole group seem inconsistent with those for its subgroups; it can occur whenever the subgroups are unequal in size.

single-blind experiment An experiment in which the participants do not know whether they are members of the treatment group or the control group but the experimenters do know—or, conversely, the participants do know but the experimenters do not.

single-peaked distribution A distribution with a single mode. Also called *unimodal distribution.*

68-95-99.7 rule Guideline stating that, for a normal distribution, about 68% (actually, 68.3%) of the data values fall within 1 standard deviation of the mean, about 95% (actually, 95.4%) of the data values fall within 2 standard deviations of the mean, and about 99.7% of the data values fall within 3 standard deviations of the mean.

skewed See *left-skewed distribution*; *right-skewed distribution.*

stack plot A type of bar graph or line chart in which two or more different data sets are stacked vertically.

standard deviation A single number commonly used to describe the variation in a data distribution, calculated as

$$\text{standard deviation} = \sqrt{\frac{\text{sum of all (deviations from the mean)}^2}{\text{total number of data values} - 1}}$$

standard score For a particular data value, the number of standard deviations (usually denoted by z) between it and the mean of the distribution:

$$z = \text{standard score} = \frac{\text{data value} - \text{mean}}{\text{standard deviation}}$$

Also called *z-score*.

statistical significance A measure of the likelihood that a result is meaningful.

statistically significant result A result in a statistical study that is unlikely to have occurred by chance. The most commonly quoted levels of statistical significance are the 0.05 level (the probability of the result's having occurred by chance is 5% or less, or less than 1 in 20) and the 0.01 level (the probability of the result's having occurred by chance is 1% or less, or less than 1 in 100).

statistics (plural) The data that describe or summarize something.

statistics (singular) The science of collecting, organizing, and interpreting data.

stem-and-leaf plot A graph that looks much like a histogram turned sideways, with lists of the individual data values in place of bars. Also called *stemplot*.

stratified sampling A sampling method that addresses differences among subgroups, or strata, within a population. First the strata are identified, and then a random sample is drawn within each stratum. The total sample consists of all the samples from the individual strata.

subjective method A method of estimating a probability based on experience or intuition.

subjects In a statistical study, the people or objects chosen for the sample. See also *participants*.

symmetric distribution A distribution in which the left half is a mirror image of the right half.

systematic errors Errors that occur when there is a problem in the measurement system that affects all measurements in the same way.

systematic sampling Using a simple system to choose the sample, such as selecting every 10th or every 50th member of the population.

***t* distribution** A distribution that is very similar in shape and symmetry to the normal distribution but that accounts for the greater variability expected with small samples. It approaches the normal distribution for large sample sizes.

theoretical method A method of estimating a probability based on a theory, or set of assumptions, about the process in question. Assuming that all outcomes are equally likely, the theoretical probability of a particular event is found by dividing the number of ways the event can occur by the total number of possible outcomes. Also called *a priori method.*

time-series diagram A histogram or line chart in which the horizontal axis represents time.

treatment Something given or applied to the members of the treatment group in an experiment.

treatment group The group of subjects in an experiment that receive the treatment being tested.

two-tailed test A hypothesis test that involves testing whether a population parameter lies to either side of a claimed value.

two-way table A table showing the relationship between two variables by listing the values of one variable in its rows and the values of the other variable in its columns. Also called *contingency table*.

type I error In a hypothesis test, the mistake of rejecting the null hypothesis, H_0, when it is true.

type II error In a hypothesis test, the mistake of failing to reject the null hypothesis, H_0, when it is false.

uniform distribution A distribution in which all data values have the same frequency.

unimodal distribution See *single-peaked distribution*.

unusual values In a data distribution, values that are not likely to occur by chance, such as those values that are more than 2 standard deviations away from the mean.

upper quartile See *quartile, upper*.

variable Any item or quantity that can vary or take on different values.

variables of interest In a statistical study, the items or quantities that the study seeks to measure.

variation How widely data are spread out about the center of a distribution. See also *five-number summary*; *range*; *standard deviation*.

vital statistics Data concerning births and deaths of people.

voluntary response survey See *self-selected survey*.

weighted mean A mean that accounts for differences in the relative importance of data values. Each data value is assigned a weight, and then

$$\text{weighted mean} = \frac{\text{sum of (each data value} \times \text{its weight)}}{\text{sum of all weights}}$$

***z*-score** See *standard score*.

Answers

SECTION 1.1

1. A population is the complete set of people or things being studied, while a sample is a subset of the population. The difference is that the sample is only a part of the complete population.

3. A sample statistic is a characteristic of a *sample* found by consolidating or summarizing raw data. A population parameter is a characteristic of a *population.* Because it is usually impractical to directly measure population parameters for large populations, we usually infer likely values of the population parameters from the measured sample statistics.

5. Does not make sense. **7.** Does not make sense.

9. Does not make sense.

11. *Sample*: The 1,002 adults selected. *Population*: All adults in the United States. *Sample statistic*: 48%. The value of the population parameter is not known, but it is the percentage of all adults in the United States who would say, if asked, that they were in favor.

13. *Sample*: The distances of the selected stars. *Population*: The distances for all stars in the galaxy. *Sample statistic*: The mean of the distances from the selected stars. The value of the population parameter is not known, but it is the mean distance that would be obtained if all stars within the galaxy were used.

15. 45% to 51% **17.** 22.6 to 29.4

19. Although there is no guarantee, the results suggest that the Republican is likely to win a solid majority because he or she will most likely get between 55% and 61% of the vote.

21. Based on the survey, the actual percentage of voters is expected to be between 67% and 73%, which does not include the 61% value based on actual voter results. If the survey was conducted well, it is unlikely that its results would be so different from the actual results, implying either that respondents intentionally lied to appear favorable to the pollsters or that their memories were inaccurate.

23. **a.** *Goal*: Determine how satisfied executives are with their career choices. *Population*: The complete set of all executives. *Population parameter*: The percentage of all executives who would say, if asked, that if they could start their careers again, they would would choose a different field. **b.** *Sample*: The 1,733 executives selected for the survey. *Raw data*: Individual responses to the question. *Sample statistic*: 51%. **c.** 48% to 54%

25. **a.** *Goal*: Determine the percentage of adults who are unemployed. *Population*: The complete set of all adults. *Population parameter*: The percentage of all adults who are unemployed. **b.** *Sample*: The sample of adults selected for the survey. *Raw data*: Individual responses to the question. *Sample statistic*: 4.6%. **c.** 4.4% to 4.8%

27. *Step 1.* Goal: Identify the percentage of all licensed drivers who used a cell phone at least once while they were driving during the last week. *Step 2.* Choose a sample of licensed drivers. *Step 3.* Survey a sample of licensed drivers to determine how many of them used a cell phone at least once while they were driving during the last week. For this sample, find the percentage of licensed drivers who used a cell phone at least once while they were driving during the last week. *Step 4.* Use the techniques of the science of statistics to make an inference about the percentage of licensed drivers who used a cell phone at least once while they were driving during the last week. *Step 5.* Based on the likely value of the population parameter, form a conclusion about the percentage of licensed drivers who used a cell phone at least once during the past week.

29. *Step 1.* Goal: Identify the mean weight of all commercial airline passengers. *Step 2.* Choose a sample of airline passengers. *Step 3.* Weigh each selected airline passenger, and then find the mean of these weights. *Step 4.* Use the techniques of the science of statistics to make an inference about the mean weight of all airline passengers. *Step 5.* Based on the likely value of the population mean, form a conclusion about the mean weight of all airline passengers.

SECTION 1.2

1. A census is the collection of data from every member of the population; a sample is a collection of data from only part of the population.

3. With cluster sampling, we select all members of randomly selected subgroups (or clusters); with stratified sampling, we select samples from each of the different subgroups (or strata).

5. Does not make sense. **7.** Makes sense.

9. A census is practical. The population consists of the small number of players on the LA Lakers team, and it is easy to obtain their heights.

11. A census is not practical. The number of statistics instructors is large, and it would be extremely difficult to get them all to take an IQ test.

13. The sample is the service times of the four selected Senators. The population is the service times of all 100 Senators. This is an example of random sampling (or simple random sampling). However, because the sample is so small, it is not likely to be representative of the population.

15. The sample is the 1,012 randomly selected adults. The population is the complete set of all adult Americans. Because the sample is fairly large and it was obtained by a reputable firm, the sample is likely to be representative of the population.

17. Sample c is the most representative because the list is likely to represent the most people and there is no reason to think that the people with the first 1,000 numbers would differ in any particular way from other people. Sample a is biased because it involves only owners of expensive vehicles. Sample b is biased because it involves people from only one geographic region. Sample d is biased because it involves a self-selected sample consisting of people who are more likely to have strong feelings about the issue of credit card debt.

19. Because the film critic indirectly works for Disney, she may be more inclined to submit a favorable review.

21. The university scientists receive payment from Monsanto, so they might be inclined to please the company in the hope of getting projects in the future. Thus, there may be an inclination to provide favorable results.

23. The sample is a simple random sample that is likely to be representative because there is no bias in the selection process.

25. The sample is a cluster sample that is likely to be representative, although the exact method of selecting the polling stations could affect whether the sample is biased.

27. The sample is a convenience sample. It is likely to be biased because the sample consists of family members likely to have similar physical characteristics and those characteristics are not likely to be representative of the general population.

29. The sample is a stratified sample. It is likely to be biased because people from those age groups are not evenly distributed throughout the population. However, the results could be adjusted to reflect the age distribution of the population.

31. The sample is a systematic sample. It is likely to be representative because there is nothing about the alphabetical ordering that is likely to result in a biased sample.

33. The sample is a stratified sample. It is likely to be biased because the population does not have an equal number of people in each of the three categories. However, the results could be weighted to reflect the actual distribution of the population.

35. The sample is a convenience sample. It is likely to be biased because it is a self-selected sample and consists of those with strong feelings about the issue.

37. The sample is a simple random sample, so it is likely to be representative.

39. Simple random sampling of the student body, with a large enough sample size to get meaningful results.

41. Cluster sampling of death records should give good data. Clusters could be based on gender or geographical location or both.

SECTION 1.3

1. A placebo is physically similar to a treatment, but it lacks active ingredients so it should not by itself produce any effects. A placebo is important so that the results from subjects given a real treatment can be compared to the results from subjects given a placebo.

3. Confounding is the mixing of effects from different factors so that specific effects cannot be attributed to specific factors. For example, if males are given the treatment and females are given placebos, it cannot be determined whether effects are due to the treatment or the sex of the participant.

5. It makes sense to use a double-blind experiment, but the experimental design is tricky because the subjects can clearly see the clothing that they are wearing and the evaluators can also see those colors. Blinding might be used by not telling the subjects about the experiment, so their knowledge of their clothes colors would not affect their results. Because it is difficult to use blinding for the evaluators, it is important to obtain results based on objective measures not subject to the judgments of the evaluators.

7. The experimenter effect occurs when a researcher or experimenter somehow influences subjects through such factors as facial expression, tone of voice, or attitude. The experimenter effect can be avoided by using blinding so that those who evaluate the results do not know which subjects were given an actual treatment and which subjects were given a placebo.

9. Observational study because the subjects were tested but were not given any treatment. The variable of interest represents the result of either *correct* or *incorrect* for each trial.

11. Experiment because the treatment group consists of the subjects given the magnetic bracelets and the control group consists of the subjects given the bracelets that have no magnetism. It is probably not possible to use blinding with this study, since the passengers could easily test their bracelets to see if they were magnetic (by holding them near metal). The variable of interest is whether the passenger experienced motion sickness.

13. Observational, retrospective study examining how a characteristic determined before birth (identical or fraternal twins) affected mental skills later. The variable of interest is the difference between the measured level of mental skills in the two people who are twins.

15. Meta-analysis. The variable of interest is whether the subject developed prostate cancer.

17. Experiment. The treatment group consists of the genetically modified corn; the control group consists of corn not genetically modified. The variable of interest is the measured amount of insecticide that is released through roots.

19. Experiment. The treatment group consists of those treated with magnets; the control group consists of those given the nonmagnetic devices. The variable of interest is the measured level of back pain.

21. Confounding is likely to occur. If there are differences in effects from the two groups, there is no way to know if those differences are attributable to the treatment (fertilizer or irrigation) or the type of region (moist or dry). This confounding can be avoided by using blocks of fertilized trees in both the moist region and the dry region and using blocks of irrigated trees in both the moist region and the dry region.

23. Confounding is likely. If there are differences in the amounts of gasoline consumed, there is no way to know whether those differences are due to the octane rating of the gasoline or the type of vehicle. Confounding can be avoided by using 87 octane gasoline in half of the vans and half of the sport utility vehicles and using 91 octane gasoline in the other vehicles.

25. Confounding is possible from a placebo effect and/or experimenter effect. It can be avoided with a double-blind experiment.

27. Confounding is possible because of a placebo effect and/or an experimenter effect, with the tennis balls playing the role of placebos. It would be better to use the heavy weights and the tennis balls with the same subjects at different times, with the order mixed.

29. The control group consists of those who do not listen to Beethoven, and the treatment group consists of those who do listen to Beethoven. Blinding could be used by coding subjects so that those who measure intelligence are not influenced by their knowledge about the participants.

31. The control group consists of cars using gasoline without the ethanol additive, and the treatment group consists of cars using gasoline with the ethanol additive. Blinding is not necessary for the cars, and it is probably unnecessary for the researchers because the mileage will likely be measured with objective tools. Also, the same cars could be used with and without the additive so that extraneous differences were eliminated.

SECTION 1.4

1. Peer review is a process in which experts in a field evaluate a research report before the report is published. It is useful for lending credibility to research because it implies that other experts agree that the research was carried out properly.

3. When participants select themselves for a survey, those with strong opinions about the topic being surveyed are more likely to participate, and this group is typically not representative of the general population.

5. Does not make sense. **7.** Does not make sense.

9. Because the researchers are from the public relations department, the researchers may be biased, so Guideline 2 is most relevant.

11. Because "good" is not well defined and because it is probably difficult to measure "good ethics," Guideline 4 is most relevant.

13. The sample is self-selected, so participation bias is a serious issue. Guideline 3 is most relevant.

15. The wording of the question is biased and tends to elicit negative responses, so Guideline 6 is most relevant.

17. Because much of the funding was provided by Mars and the Chocolate Manufacturers Association, the researchers may have been more inclined to provide favorable results. The bias could have been avoided if the researchers had not been paid by the chocolate manufacturers. If that was the only way the research could be done, the researchers should have instituted procedures to ensure that all results, including negative ones, were published.

19. Because the respondents are self-selected, they are likely to represent those who have strong feelings about the issue. A better sampling method, such as the simple random sampling used by most polling companies, is needed.

21. The word *wrong* in the first question could be misleading. Some people might believe that abortion is wrong but still favor choice. The second question could also be confusing, as some people might think that *advice of her doctor* means that the woman's life is in danger, which could alter their opinion about abortion in this situation. Groups opposed to abortion would be likely to cite the results of the first question, while groups favoring choice would be more likely to cite the results of the second question.

23. The first question requires a study of Internet dates. The second question involves a study of married people to determine whether their first date was an Internet date. The second group is much more limited than the first.

25. The first question involves a study of college students in general. The second question involves a study of those who binge drink. The first question might be addressed by surveying college students. The second question would be addressed by surveying binge drinkers, a group that would be much more difficult to survey.

27. The headline refers to drugs, whereas the story refers to drug use, drinking, or smoking. Because *drugs* is generally considered to consist of drugs other than cigarettes or alcohol, the headline is very misleading.

29. No information is given about the meaning of *confidence*. The sample size and margin of error are not provided.

31. No information is given to justify the statement that "more" companies try to bet on weather forecasting. If only the four cited companies are new, the increase is relatively insignificant.

CHAPTER 1 REVIEW EXERCISES

1. a. 35% to 41% **b.** All adults in the United States **c.** Observational study because the subjects were not treated or modified in any way. The variable of interest is *gun in home*, which for this study can take on two values: *yes* or *no*. **d.** The value is a sample statistic because it is based on the sample of 1,012 adults, not the population of all adults. **e.** No, because it would be a self-selected sample with a likely participation bias. **f.** There are many different procedures for selecting a simple random sample of adults in the United States, but any procedure should be designed to ensure that all samples of 1,012 adults have the same chance of being selected. One simple procedure is to compile a numbered list of all adults in the United States, use a computer to randomly generate 1,012 different numbers, and then select the subjects corresponding to the selected numbers. **g.** Select a sample of households in each state. **h.** Select all of the households in several randomly selected election precincts. **i.** Select every 10th household, by address, on each street in a city. **j.** Select the households of your classmates.

2. a. A random sample is a sample chosen in such a way that every sample of the same size has the same chance of being selected. **b.** No, because not every sample of 2,007 people has the same chance of being selected. For example, it is impossible to select the sample with 2,007 people in the same primary sampling unit. **c.** Repeat the process of randomly selecting a primary sampling unit and then randomly selecting one of its members. If anyone is selected more than once, ignore the subsequent selections.

3. a. No, because there is no information about the occurrence of headaches among people who do not use Zocor. **b.** Because the headache rate is lower among Zocor users, it appears that headaches are not an adverse reaction to Zocor use. **c.** With blinding, the trial participants do not know whether they are getting Zocor or a placebo and those who evaluate the results also do not know. **d.** Experiment because subjects are given a treatment. **e.** An experimenter effect occurs if the experimenter somehow influences subjects through such factors as facial expression, tone of voice, or attitude. It can be avoided through the use of blinding.

4. a. The second question, because the word *welfare* has negative connotations. **b.** The first question, because it is more likely to elicit negative responses. **c.** This is largely a subjective judgment. Some professional pollsters are opposed to all questions that are deliberately biased, but others believe that such questions can be used. An important consideration is that survey questions can modify how people think, and such modification should not occur without their awareness or agreement.

CHAPTER 1 QUIZ

1. a **2.** c **3.** a **4.** b **5.** c **6.** a **7.** c **8.** c **9.** b **10.** b **11.** b **12.** b **13.** b **14.** c **15.** b

SECTION 2.1

1. Qualitative data consist of values that can be placed into different nonnumerical categories, whereas quantitative data consist of values representing counts or measurements.

3. Yes. Data consist of either qualities or quantities (numbers), so all data are either qualitative or quantitative.

5. Blood groups are qualitative because they don't measure or count anything.

7. Heights are quantitative because they consist of measurements.

9. The lengths of movies are quantitative because they consist of measurements.

11. The television shows are qualitative because they don't measure or count anything.

13. The number is quantitative because it consists of a count.

15. The salaries are quantitative because they consist of counts of money.

17. The numbers are discrete because only the counting numbers are used, and no values between counting numbers are possible.

19. The numbers are discrete because they are counts. Only the counting numbers are used, and no values between counting numbers are possible.

21. The times are continuous data because they can have any value within some range of values.

23. The numerical test scores are discrete data because they can be counting numbers only.

25. The speeds are continuous data because they can have any value within some range of values.

27. The numbers are discrete data because they can be counting numbers only.

29. Ratio **31.** Nominal **33.** Ordinal **35.** Ordinal

37. Nominal **39.** Ratio

41. The ratio level does not apply. The ratio is not meaningful because the stars don't measure or count anything. Differences between star values are not meaningful.

43. The ratio level applies. The 450-mi/h speed is three times the 150-mi/h speed.

45. The ratio level applies. The ratio of "twice" is meaningful.

47. The ratio level applies. The $150,000 salary amount is twice the salary of $75,000, so the ratio of "twice" is meaningful.

49. The data are quantitative and are at the ratio level of measurement. The data are continuous. The times have a natural zero starting point and can be any values within a particular range.

51. The data are qualitative and are at the nominal level of measurement. The numbers are different ways to express the names, and they don't measure or count anything.

53. The data are quantitative and are at the interval level of measurement. The data are discrete because they consist of whole numbers only. The years are measured from an arbitrary reference (the year 0), not a natural zero starting point. Differences between the years are meaningful values, but ratios are not meaningful.

55. The data are qualitative and are at the ordinal level of measurement. The ratings consist of an ordering, but they do not represent counts or measurements.

SECTION 2.2

1. Because it is the result of a recording problem in the measurement system, this is a systematic error. It is not a random error because it is not the result of random and inherently unpredictable events in the measurement process.

3. Because the recorded height has so many decimal places, it is very precise. Because the recorded height is not very close to the true height, it is not very accurate.

5. Does not make sense. **7.** Makes sense.

9. Mistakes tend to result from random errors, but dishonesty tends to result in systematic errors that benefit the taxpayer.

11. Because about half of the screws are longer than 25 mm and half are shorter than 25 mm, this appears to be a random error.

13. Random errors occur when reported incomes are recorded incorrectly or when survey respondents don't know their exact incomes. Systematic errors occur when people report higher incomes so that they appear to be more successful.

15. Random errors occur with inconsistent scales or mistakes in reading the scales or recording the data. Systematic errors occur with a scale that consistently reads too high or too low.

17. Random errors occur with an inconsistent radar gun or with honest mistakes by the officer recording speeds. Systematic errors occur with a radar gun that is incorrectly calibrated so that it consistently reads too high or too low.

19. Random errors occur with mistakes in calculations. Systematic errors occur with underreporting of packs of cigarettes that are illegally obtained without tax stamps.

21. *Absolute error*: $1,750. *Relative error*: 141% (using the error of $1,750 and the correct bill amount of $1,245).

23. *Absolute error*: −2 mi/gal. *Relative error*: −8.3%.

25. **a.** These errors are random. **b.** The average is the better choice for minimizing random errors. **c.** Systematic errors might arise from, for example, a problem with the measuring device or a problem in defining the "length" of a room. **d.** Averaging measurements will not reduce systematic errors.

27. The Department of Transportation scale is more precise because its weight of 3,298.2 lb is more detailed than the other weight of 3,250 lb. The manufacturer's scale is more accurate because its weight of 3,250 lb is closer to the true weight. (The weight of 3,250 lb is in error by 23 lb, but the weight of 3,298.2 lb is in error by 25.2 lb.)

29. The digital scale at the gym is more precise and more accurate than the health clinic scale (assuming that your actual weight is what you thought).

31. The given number is very precise, but it is not likely to be accurate. We cannot measure the population this precisely even today, and the uncertainties were greater in 1860.

33. The given number is very precise, but it is not likely to be accurate. There are many people in China who are not counted. The census of any nation is likely to be in error by a considerable amount, because of the difficulties inherent in conducting a national census. The population of China likely changed during the course of the year.

35. It is easy to measure accurately the height of a structure with a reasonable degree of precision, such as the nearest 1/10 foot, but the given number has far too much precision (it implies knowing the height to a size smaller than that of an atom!), so the claim is not believable.

37. The number of college students is constantly changing with new enrollments and dropouts, so the given number must be an estimate. The given number suggests that it is precise only to the nearest million, which seems quite possible for a good estimate. So it is believable, though we would need more information to know if it really is accurate.

SECTION 2.3

1. For 10th graders, the new rate of smoking is 18.3%, and the percentage increase is 45%. Therefore, the old rate times 1.45 equals the new rate of 18.3%, so the old rate for 10th graders is 12.6%. For 8th graders, the new rate of smoking is 10.4%, and the percentage increase is 44%. Therefore, the old rate times 1.44 equals the new rate of 10.4%, so the old rate for 8th graders is 7.2%.

3. The statement incorrectly implies that the error can be up to 1.2% of 5%. Because 1.2% of 5% is 0.06%, the statement incorrectly implies that the error can be up to 0.06% (from 4.94% to 5.06%); the actual error can be up to 1.2 percentage points away from 5% (from 3.8% to 6.2%).

5. Does not make sense. **7.** Makes sense.

9. **a.** 2/5; 0.4; 40% **b.** 150/100, or 3/2; 1.50; 150% **c.** 1/4; 0.25; 25% **d.** 30/100, or 3/10; 0.30; 30%

11. **a.** 93% **b.** 44% **c.** 41% **d.** 81%

13. −35% (There is a 35% decrease from 1990.)

15. 115% (There is a 115% increase from 1980.)

17. 61%. The *Wall Street Journal* has 61% more circulation than the *New York Times*.

19. −19%. O'Hare handled 19% fewer passengers than Hartsfield.

21. 66 **23.** 834

25. 140%. The truck weighs 100% of the car's weight plus another 40%.

27. 80%. Montana's population is 100% of New Hampshire's population minus 20% of New Hampshire's population.

29. Yes. Three percentage points corresponds to a range from 86% to 92%, which was intended. A margin of error of 3% would correspond to 3% of 89%, which is $0.03 \times 0.89 = 0.0267$, but this is not what was meant.

31. −15.5 percentage points; −22.7%

33. 22 percentage points; 56.4%

SECTION 2.4

1. An index number is a ratio without any units. The number appears to be the actual cost of gasoline in 2007, not an index number.

3. Yes. The Consumer Price Index is based on the prices of goods, services, and housing, so increases in those prices will result in an increase in the Consumer Price Index.

5. 573.2 **7.** $1.12

9. 93.3, 100, 181.7, 383.3, 386.2, 496.8, 740.4

11. $21.69

13. The tuition in 2004 is 466.4% of the 1980 tuition, but the CPI in 2004 is 229.2% of the 1980 CPI, so the cost of tuition rose much more than the cost of typical goods, services, and housing.

15. The home prices in 2004 are 206.5% of the amount in 1990, but the CPI in 2004 is 144.5% of the 1990 CPI, so the cost of homes rose at a lower rate than the cost of typical goods, services, and housing.

17. $582,000; $180,000 **19.** $1,591,667; $1,491,667

CHAPTER 2 REVIEW EXERCISES

1. a. 706 **b.** Discrete; a person cannot fractionally survive. **c.** 23.89% **d.** 64 **e.** Ratio **f.** Nominal

2. a. 860 **b.** 37% **c.** Ordinal, because there is an ordering. **d.** Because the poll uses respondents who themselves chose to participate, the sample is a self-selected sample and is not likely to accurately reflect the opinion of the population.

3. The health care spending in 2004 is 2,150% of the amount in 1973 CPI, but the CPI in 2004 is 325.5% of the 1973 CPI, so health care spending grew at a much greater rate than the general rate for goods, services, and housing.

4. a. $2.78 **b.** $5.77 **c.** The minimum wage increases did not keep up with inflation, so the actual 2006 minimum wage of $5.15 has the same purchasing power as $4.14 in 1996. Workers earning the minimum wage of $4.75 in 1996 had more relative income than workers earning $5.15 in 2006.

CHAPTER 2 QUIZ

1. Nominal **2.** Continuous **3.** Ratio

4. −10 cm **5.** −18.1% **6.** 52% **7.** 52

8. 178 cm, 179.18 cm **9.** 123.7 **10.** $52,972

SECTION 3.1

1. A frequency table has two columns, one for categories and one for frequencies. Categories are the different values that a variable may have (for example, different eye colors or flavors of ice cream). Frequencies are the number of data points (counts) in each category.

3. 2, 11, 25, 37, 40 **5.** Does not make sense.

7. Does not make sense.

9.

Grade	Frequency	Relative frequency	Cumulative frequency
A	4	16.7%	4
B	7	29.2%	11
C	8	33.3%	19
D	3	12.5%	22
E	2	8.3%	24
Total	**24**	**1 = 100%**	**24**

11.

Weight (pounds)	Frequency	Relative frequency	Cumulative frequency
0.7900–0.7949	1	1/36	1
0.7950–0.7999	0	0	1
0.8000–0.8049	1	1/36	2
0.8050–0.8099	3	3/36	5
0.8100–0.8149	4	4/36	9
0.8150–0.8199	17	17/36	26
0.8200–0.8249	6	6/36	32
0.8250–0.8299	4	4/36	36
Total	**36**	**100%**	**36**

13.

Age	No. of actors
20–29	0
30–39	12
40–49	13
50–59	5
60–69	3
70–79	1

15.

Category	Frequency	Relative frequency
A	13	24%
B	9	18%
C	12	24%
D	11	22%
F	6	12%
Total	**50**	**100%**

17. a. 200 **b.** 142 **c.** 16% **d.** 0.135, 0.155, 0.210, 0.200, 0.140, 0.160 **e.** 27, 58, 100, 140, 168, 200

19. a.

Rating	Frequency	Relative frequency
0–2	20	38.5%
3–5	14	26.9%
6–8	15	28.8%
9–11	2	3.8%
12–14	1	1.9%
Total	**52**	**100%**

b.

Rating	Frequency	Relative frequency
0–2	33	63.5%
3–5	19	36.5%
6–8	0	0%
9–11	0	0%
12–14	0	0%
Total	**52**	**100%**

c. The Dvorak keyboard appears more efficient because it has more lower ratings and fewer high ones.

SECTION 3.2

1. The distribution of data is the way data values are spread over all possible values. The histogram provides a graph that has a shape, and it is much easier to understand the shape of the distribution from the graph than from a list of data values.

3. The data must be ordered in some time sequence, and the time series graph helps to reveal trends or patterns over time.

5. Does not make sense. **7.** Makes sense.

9. A histogram would work well to show the frequencies of the different categories of scores.

11. A Pareto chart or pie chart would work well, but the Pareto chart would do a better job of showing the most common causes of death.

13. a.

Reading category	Relative frequency
Popular fiction	50.4%
Cooking/Crafts	10.2%
General nonfiction	8.9%
Religious	8.6%
Psychology/Recovery	6.3%
Technical/Science/Education	5.6%
Art/Literature/Poetry	3.7%
Reference	2.6%
All other categories	2.5%
Travel/Regional	1.3%

b.

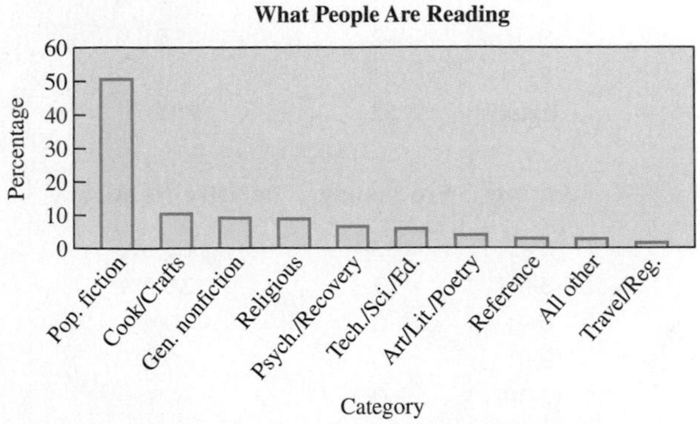

c. The Pareto chart makes it easier to see which categories are the most popular.

15.

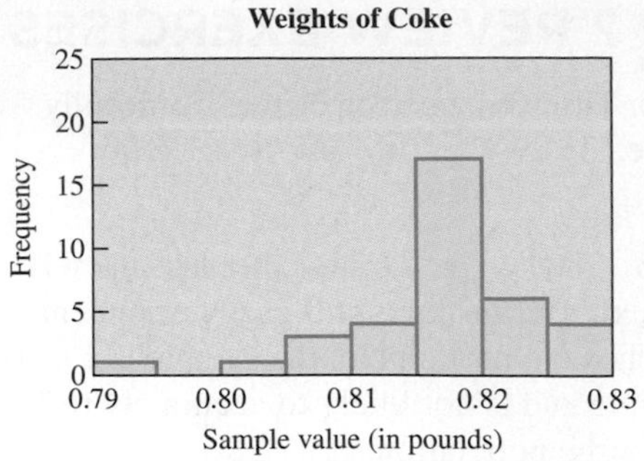

17.

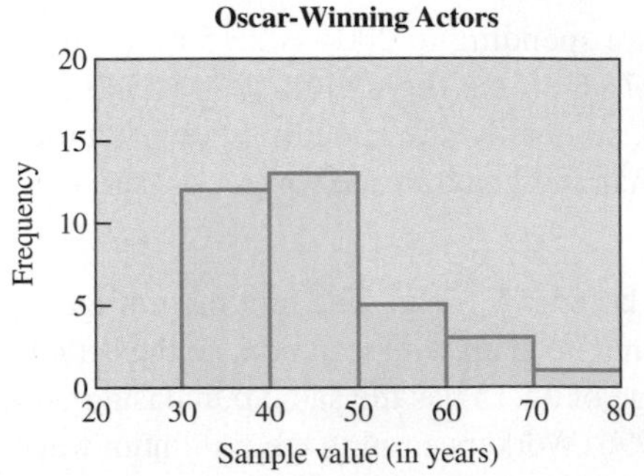

19.

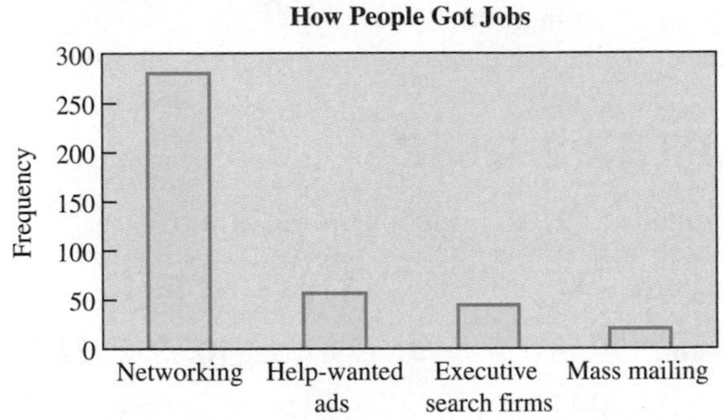

21.

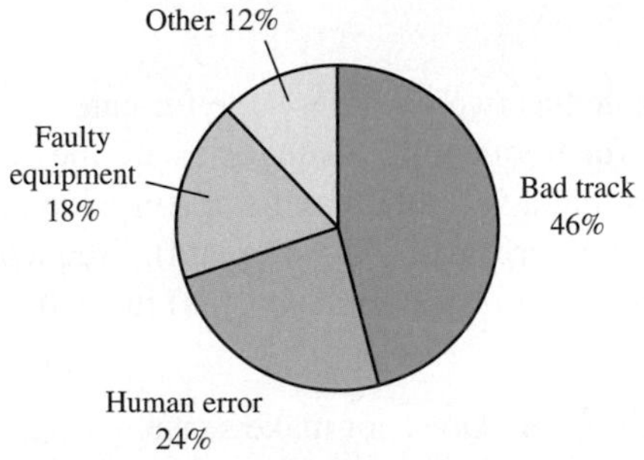

23.

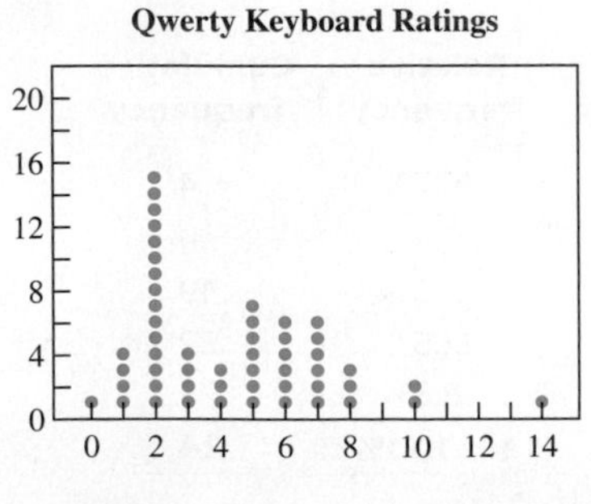

25.

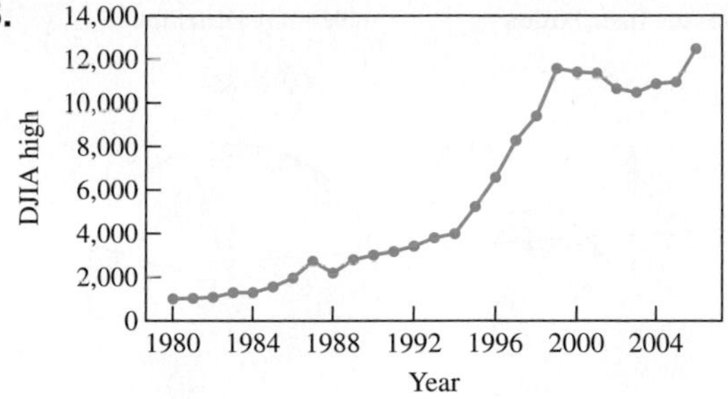

Over the years, there is a trend of increasing stock prices, so the stock market appears to be a good investment.

27.

Stem	Leaves
6	7
7	25
8	5899
9	09
10	0

The lengths of the rows are similar to the heights of bars in a histogram; longer rows of data correspond to higher frequencies.

SECTION 3.3

1. The use of the oil barrels is not appropriate because the data are not three-dimensional. It would be better to illustrate the data with a simple bar graph.

3. Geographical data are raw data corresponding to different geographic locations. Two examples of displays of geographical data are color-coded maps and contour maps.

5. Does not make sense. **7.** Does not make sense.

9. a. Females consistently outnumber males. The numbers of both genders are increasing gradually over time.

b.

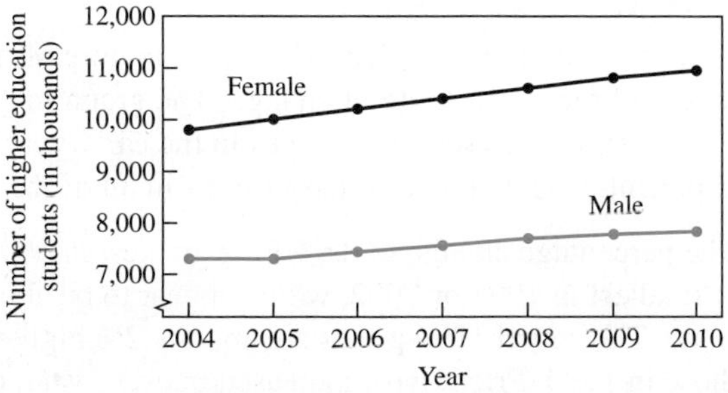

11. a. Males consistently have much higher median incomes than females, and both males and females have steadily increasing incomes over time. Comparing the heights of the bars from left to right, we see that the ratios of male income to female income appear to be decreasing. The leftmost pair shows that male income is more than twice female income, but the rightmost pair of bars seems to show that male income is not as much as twice female income. **b.** The line graph makes it easier to examine the trend over time. Also, the line graph is less cluttered, so the information is easier to understand and interpret.

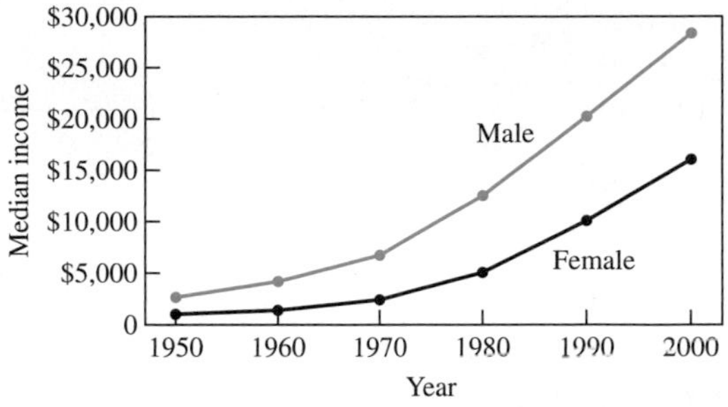

13. This stack plot shows *percentages* of the total budget. Thus, its total height is always 100%, and we cannot directly determine the actual amount of money spent by the government. However, it is easy to see long-term trends in how the government spends its money. For example, the proportion of the federal budget going to national defense fell substantially, from about 50% of the federal budget in 1960 to about 13% in 2005. During the same period, the proportion of the budget going to payments to individuals (e.g., Social Security, Medicare, and welfare) rose dramatically, from about 30% to about 70%. The proportion spent on net interest more than doubled.

15. There appears to be a general trend of higher melanoma mortality rates in southern states and western states. This might be the result of people in these states spending more time outdoors, exposed to sunlight. As a researcher, you might be particularly interested in regions that deviate from general trends. For example, a county in eastern Washington state stands out with a very high melanoma mortality rate. You might first want to verify that the data point is accurate, and not an error of some type. If it is accurate, you might want to find out why this one county has a higher melanoma mortality rate than surrounding counties.

17.

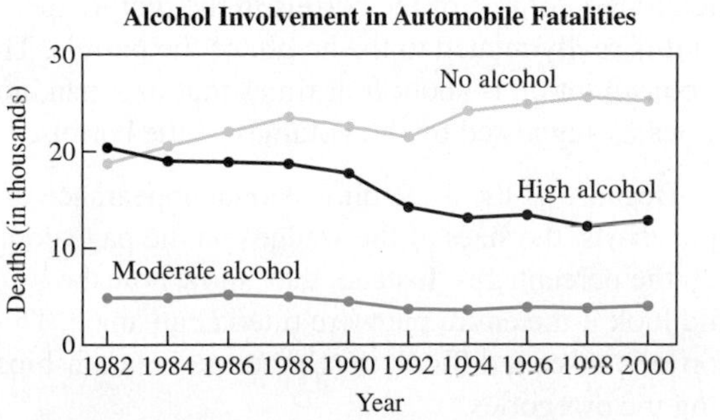

19.

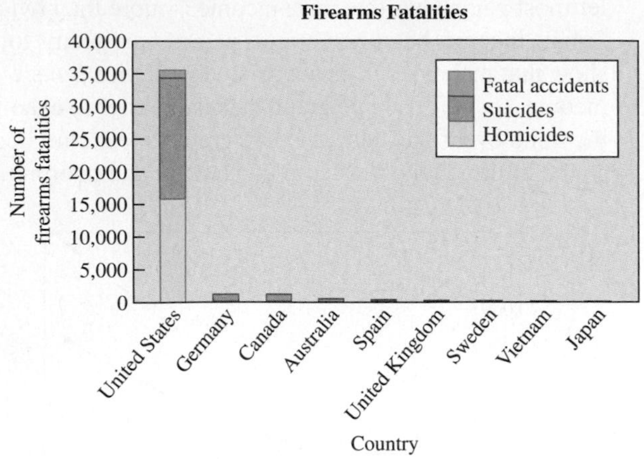

SECTION 3.4

1. Cutting off the bottom of the graph distorts the reader's perception, causing the reader to view only part of the graph instead of the complete graph, which gives a better perspective on all of the data. Changes tend to be exaggerated.

3. Instead of using intervals that change by the same amount, an exponential scale uses intervals that change by powers of some number (often 10). For example, if powers of 10 are used, equal intervals on the scale represent 1, 10, 100, and so on. Exponential scales are helpful for displaying data that vary over a huge range.

5. The graph is misleading because the horizontal scale does not start at zero. The braking distance for the Acura is 42% greater than that for the Volvo.

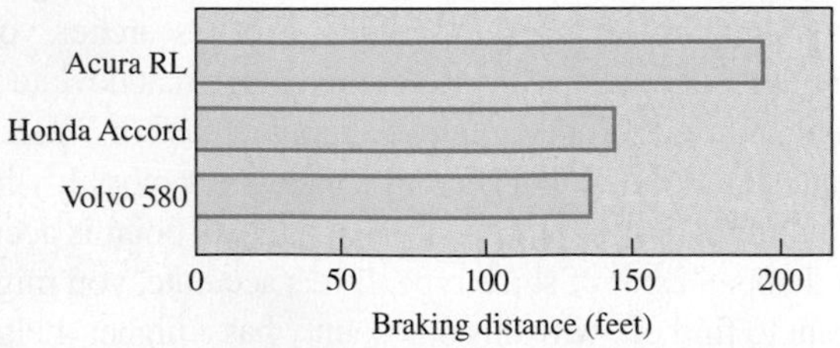

7. The amount of oil used by each country appears to be related to the volume of the barrels in the pictograph, when it is really related to the height of the barrels. The U.S. consumption is about four times that of Japan, not 64 times as suggested by the volumes of the barrels.

9. **a, b.** Because of the three-dimensional appearance of the pie charts, the sizes of the wedges on the page do not match the percentages. Instead, they show how the wedges would look if the entire pie were tilted at an angle. This distortion makes it difficult to see the true relationships among the categories.

c.

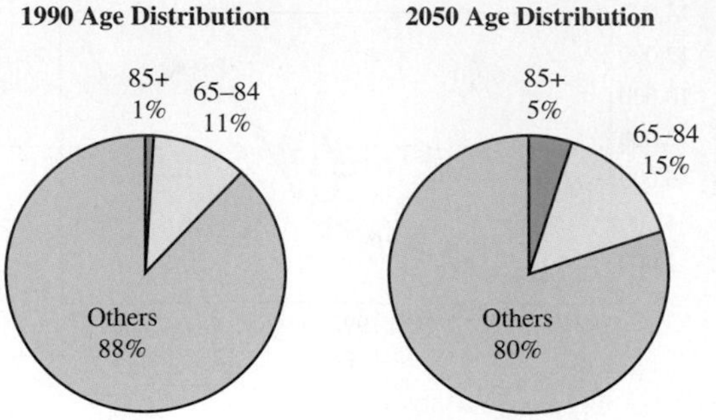

d. In 2050, there will be relatively more older people and fewer younger people in the U.S. population than in 1990. Note, however, that because of population growth, all age groups are expected to be larger in number in 2050 than in 1990.

11. **a.**

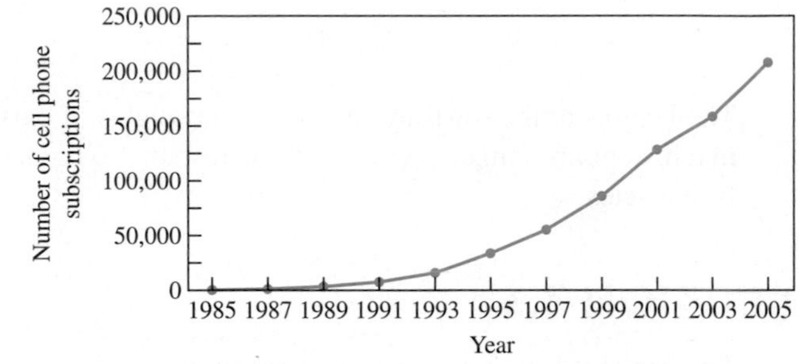

b.

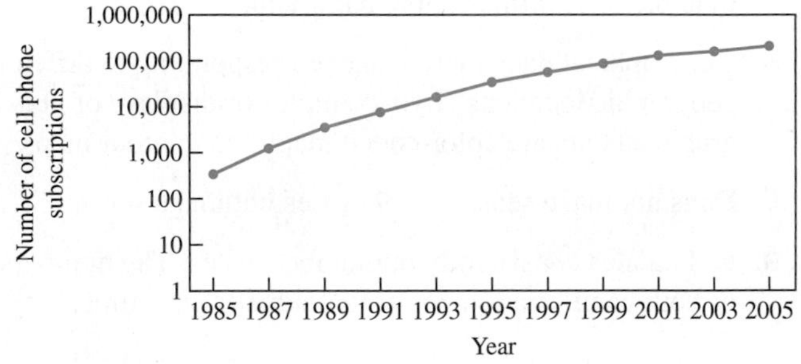

c. The graph in part a gives us a better picture of the true nature of the overall rate of change. The graph in part b makes it easier to see the changes in the early years; the exponential scale makes it easier to fit in all of the data.

13. The percentage change in the CPI is greatest in 1990 and is smallest in 1998 or 2002, which appear to be about the same. The actual 1991 prices are about 4.2% higher than those in 1990. Prices have increased in every year, but the increases near the end of the time period are lower than those near the beginning.

15. The actual minimum wage in unadjusted dollars has either remained constant or risen steadily since 1955, but the purchasing power has been decreasing since about

1980. The purchasing power in 2006 is less than it was in 1955.

CHAPTER 3 REVIEW EXERCISES

1. a. The frequencies are 5, 12, 12, 5, 0, 2. **b.** The frequencies are 5, 5, 16, 8, 2. **c.** The weights of regular Pepsi are consistently larger than those of diet Pepsi. The weights of regular Pepsi are larger because of the sugar that is not included in the diet Pepsi.

2. a. The relative frequencies are 0.139, 0.333, 0.333, 0.139, 0, 0.056. **b.** The cumulative frequencies are 5, 17, 29, 34, 34, 36.

3. a.

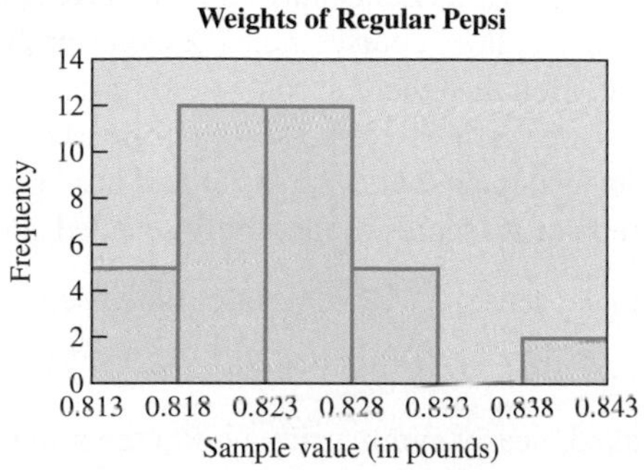

b.

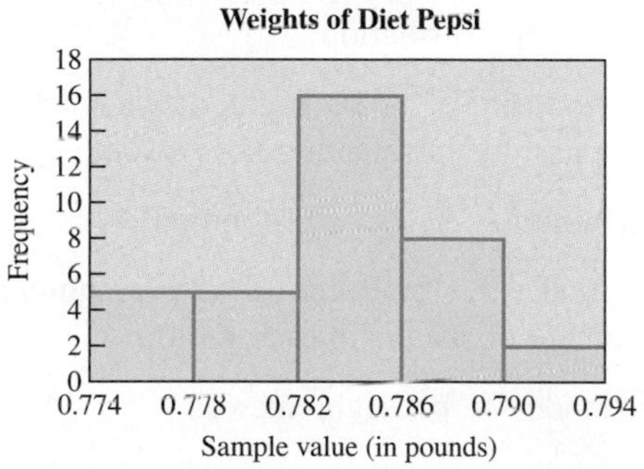

c. The shapes of the histograms are not dramatically different, indicating that the distributions of the weights are similar. However, the range of values is very different, indicating that the weights of regular Pepsi are considerably higher than those of diet Pepsi.

4.

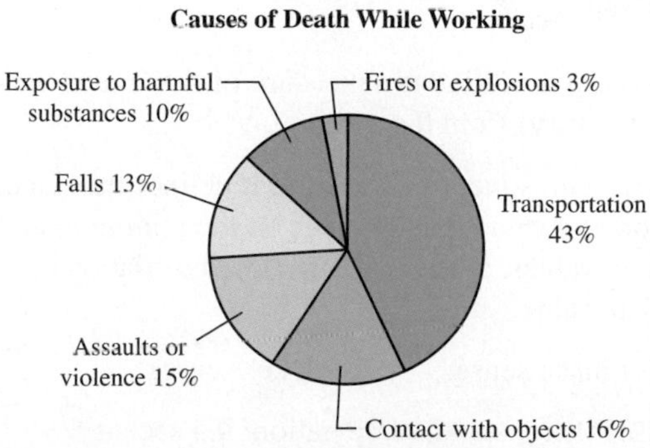

5. By drawing attention to the most serious causes of death, the Pareto chart is more effective. The Pareto chart also does a better job of showing the relative importance of the different causes of death.

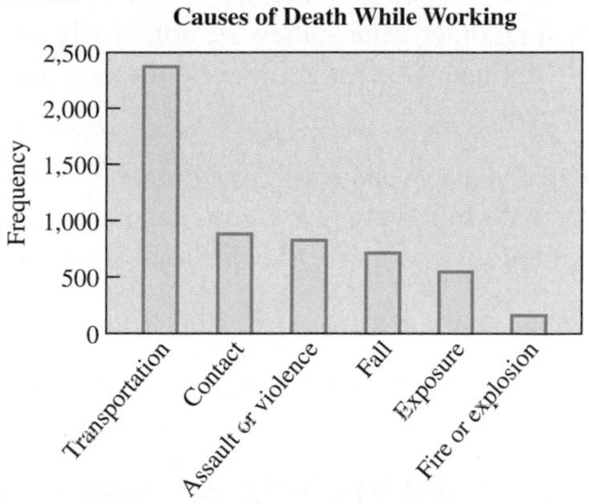

6. The graph has a vertical scale that does not begin with zero, so the difference between the two frequencies is exaggerated. The graph makes it appear that adoptions more than doubled in 2005, but this is not the case.

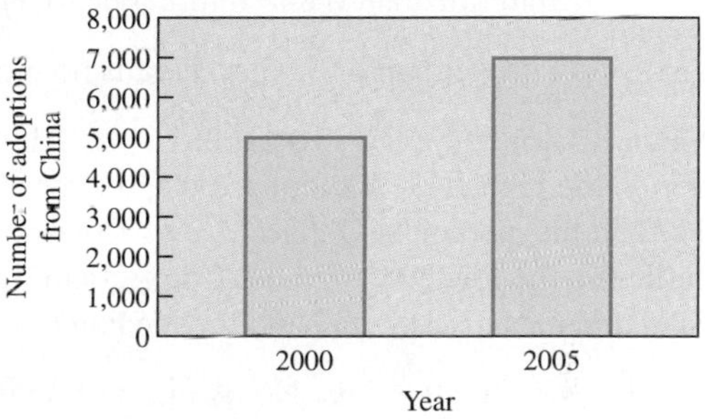

CHAPTER 3 QUIZ

1. Bar graph **2.** Multiple bar graph **3.** Histogram

4. There are 7 values between 50 and 59.

5. Among all the values, the proportion of values between 50 and 59 is 0.2. That is, 20% of the values are between 50 and 59.

6. 5, 10, 12, 12, 14

7. Using a vertical scale that does not start at zero causes the difference between the two frequencies to be exaggerated.

8. The largest data value is 9. The largest frequency is 3 (for the data value 6).

9. 25,000 in 1982; 16,652 in 2000

10. Unlike amounts such as prices or incomes, the annual amounts of gasoline are unaffected by inflation, so the actual unadjusted amounts should be used.

SECTION 4.1

1. An outlier in a data set is a value that is much higher or much lower than almost all other values. Because outliers are defined as being "much higher" or "much lower" than almost all other values, they are not clearly and objectively defined. Determination of outliers requires some judgment.

3. Not necessarily. The result would be distorted by small states with relatively few commuters. The mean should be weighted to account for the different state populations.

5. Does not make sense. 7. Makes sense.

9. Income data often contain outliers on the high side, so the median is best.

11. This distribution is probably fairly symmetric, so either the mean or the median should work.

13. Mean: 58.3 sec; median: 55.5 sec; mode: 49 sec.

15. Mean: 0.188; median: 0.165; mode: 016.

17. Mean: 0.807 mm; median: 0.840 mm; mode: 0.84 mm.

19. Mean: 0.9194 g; median: 0.9200 g; there is no mode.

21. **a.** Mean: 157,586; median: 104,100. **b.** Alaska is an outlier on the high end. Without Alaska, the mean is 81,317 and the median is 78,650. **c.** Connecticut is an outlier on the low end. Without Connecticut (but with Alaska), the mean is 182,933 and the median is 109,050.

23. **a.** 73.75 **b.** b. 80 **c.** No. A mean of 80 for five quizzes requires a total of 400 points, which you cannot reach even with 100 on the fifth exam.

25. With a 90, your new mean is 81.4. The best you could do is score 100, which would make your new mean 82.9. The worst you could do is score 0, which would make your new mean 68.6.

27. Mean score of your students: 74.7; median score: 70. So it depends on the meaning of "average." If it is the mean, then your students are above average in their scores. If it is the median, then your students are below average.

29. 0.39 pound.

31. Each student is taking three classes with enrollments of 20 each and one class with an enrollment of 100, so the mean class size for each student is 160/4 = 40. There are three classes with 100 students each and 45 classes with 20 students each, for a total enrollment of 1,200 students in 48 classes. Thus, the mean enrollment per class is 1,200/48 = 25.

33. .417; it is the average number of hits per at-bat.

35. No, unless the inspector tested the same number of eggs at both farms.

37. The outcome is no, by 600 votes to 400 votes.

39. Mean: 1.9; median: 2; mode: 1. The mode of 1 correctly indicates that the smooth yellow peas occur more than any other phenotype, but the mean and median don't make sense with these data at the nominal level of measurement.

SECTION 4.2

1. The shape of the distribution is such that the left half is a mirror image of the right half.

3. Because the statistics students have satisfied prerequisites and are taking a college course, their IQ scores probably have less variation than the IQ scores of randomly selected adults. The lower variation of IQ scores of the statistics students results in a graph that is narrower and has less spread than the graph of IQ scores of randomly selected adults.

5. Does not make sense. 7. Does not make sense.

9. Two modes, left-skewed, wide variation

11. Single-peaked, nearly symmetric, moderate variation

13. **a.** Right-skewed **b.** 150 (half of 300) **c.** No; it depends on the distribution.

15. **a.** One mode of $0 **b.** Right-skewed, because there will be many students making little or nothing.

17. **a.** One mode **b.** Nearly symmetric

19. **a.** Two modes **b.** Right-skewed (assuming a roughly equal number of skaters in each group)

21. **a.** One mode **b.** Right-skewed

23. **a.** One mode **b.** Right-skewed

25. **a.** One mode **b.** Symmetric

27. **a.** One mode **b.** Right-skewed

29. **a.** One mode **b.** Right-skewed

SECTION 4.3

1. The standard deviation is a measure of how much values deviate (or vary) from the mean.

3. The statement is incorrect because it defines the standard deviation as a value that depends on the minimum and maximum values, whereas the standard deviation uses every data value.

5. Does not make sense. 7. Makes sense.

9. Range: 26.0 sec; standard deviation: 9.5 sec

11. Range: 0.170; standard deviation: 0.057

13. Range: 0.280 mm; standard deviation: 0.094 mm

15. Range: 0.1160 g; standard deviation: 0.0336 g

17. *Cat on a Hot Tin Roof:* range = 10.0; standard deviation = 2.6. *The Cat in the Hat:* range = 3.0; standard deviation = 0.9. There is much less variation among the word lengths in *The Cat in the Hat.*

19. One day: range = 11.0 degrees; standard deviation = 2.6 degrees. Five day: range = 15.0 degrees; standard deviation = 4.5 degrees. There appears to be greater variation in errors from the five-day forecast.

21. **a.** 5th percentile **b.** 69th percentile **c.** 48th percentile.

23. Answers for data set 1 only: **a.** Histogram has frequency of 7 for data value of 9; no other values. **b.** Low value = 9; lower quartile = 9; median = 9; upper quartile = 9; high value = 9 **c.** 0

25. **a.** Faculty: mean = 2, median = 2, range = 4. Student: mean = 6.18, median = 6, range = 9 **b.** Faculty: low = 0, lower quartile = 1, median = 2, upper quartile = 3, high = 4. Student: low = 1, lower quartile = 4, median = 6, upper quartile = 9, high = 10 **c.** Faculty: standard deviation = 1.18. Student: standard deviation = 3.03 **d.** Faculty: range/4 = 1.0. Student: range/4 = 2.25

27. **a.** First 7: mean = 58.3, median = 57, range = 4. Last 7: mean = 57.6, median = 56, range = 23 **b.** First 7: low = 57, lower quartile = 57, median = 57, upper quartile = 61, high = 61. Last 7: low = 46, lower quartile = 52, median = 56, upper quartile = 64, high = 69 **c.** First 7: standard deviation = 1.9. Last 7: standard deviation = 7.7 **d.** First 7: range/4 = 1. Last 7: range/4 = 5.75

29. Order from the first shop (standard deviation of 3 minutes).

31. For long-term investment, the higher mean growth rate will pay out more in the long term. For short-term investment, the lower standard deviation is less risky.

SECTION 4.4

1. A false positive occurs when the test indicates drug use for someone who does not actually use drugs. A false negative occurs when the test indicates that drugs are not used by someone who actually uses drugs.

3. The fraction of liars is usually very small, which means that the proportion of truthful people is large. If even a small percentage of the truthful people are given inaccurate results, this could amount to a relatively large number of people.

5. Makes sense. **7.** Does not make sense.

9. Josh, Josh, Jude **11.** **a.** New Jersey, Nebraska

13. **a.** Whites: 0.18%; nonwhites: 0.54%; total 0.19% **b.** Whites: 0.16%; nonwhites: 0.34%; total 0.23% **c.** The rate for both whites and nonwhites was higher in New York than in Richmond, yet the overall rate was higher in Richmond than in New York. The percentage of nonwhites was significantly lower in New York than in Richmond.

15. **a.** Spelman has a better record for home games (34.5% vs. 32.1%) and away games (75.0% vs. 73.3%), individually. **b.** Morehouse has a better overall average. **c.** Morehouse has a better team, as teams are generally rated on overall records.

17. **b.** 216 people were accused of lying. Of these, 18 were actually lying and 198 were telling the truth. **c.** 1,784 people were telling the truth, according to the polygraph. Of these, 1,782 were actually telling the truth and two were lying; 99.9% of those found to be telling the truth were actually telling the truth.

19. A higher percentage of women than men were hired in both the white-collar and the blue-collar positions, suggesting a hiring preference for women. Overall, 20% of the 200 females who applied for white-collar positions (40) were hired, and 85% of the 100 females who applied for blue-collar positions (85) were hired. Thus, 40 + 85 = 125 of the 200 + 100 = 300 females who applied were hired, a percentage of 41.7. Overall, 15% of the 200 males who applied for white-collar positions (30) were hired, and 75% of the 400 males who applied for blue-collar positions (300) were hired. Thus, 30 + 300 = 330 of the 200 + 400 = 600 males who applied were hired, a percentage of 55.0.

21. **a.** In the general population, 57 + 3 = 60 of the 20,000 in the sample are infected. This is an incidence rate of 60/20,000 = 0.3%. In the at-risk population, 475 + 25 = 500 of the 5,000 in the sample are infected. This is an incidence rate of 500/5,000 = 10.0%. **b.** In the at-risk category, 475 out of the 500 infected with HIV test positive, or 95%. Of those who test positive, 475 out of 475 + 225 = 700 have HIV, a percentage of 475/700 = 67.9%. These two figures are different because they measure different things. The 700 who test positive include 225 who were false positives. While the test correctly identifies 95% of those who have HIV, it also incorrectly identifies some who do not have HIV. Thus, only 67.9% of those who test positive actually have HIV. **c.** In the at-risk population, a patient who tests positive for the disease has about a 68% chance of actually having the disease. This is nearly 7 times the incidence rate (10%) of the disease in the at-risk category. Thus, the test is very valuable in identifying those with HIV. **d.** In the general population, patients with HIV test positive 57 times out of 60, or 95% of the time. Of those who test positive, 57 out of 57 + 997 = 1,054 actually have HIV, a percentage of 57/1,054 = 5.4%. These two figures are different because they measure different things. The 1,054 who test positive include 997 who were false positives. While the test correctly identifies 95% of those who have HIV, it also incorrectly identifies some who do not have HIV. Thus, only 5.4% of those who test positive actually have HIV. **e.** In the general population, a patient who tests positive for the disease has about a 5.4% chance of actually having the disease. This is 18 times the incidence rate (0.3%) of the disease in the general population. Thus, the test is very valuable in identifying those with HIV.

CHAPTER 4 REVIEW EXERCISES

1. a. Mean for red: 0.8635 g; mean for green: 0.8635 g. Median for red: 0.8590 g; median for green: 0.8650 g. **b.** Range for red: 0.2150 g; range for green: 0.2370 g. Standard deviation for red: 0.0576 g; standard deviation for green: 0.0570 g. **c.** Five-number summary for red: 0.7510 g, 0.8250 g, 0.8590 g, 0.8975 g, 0.9660 g. Five-number summary for green: 0.7780 g, 0.8140 g, 0.8650 g, 0.8810 g, 1.0150 g.

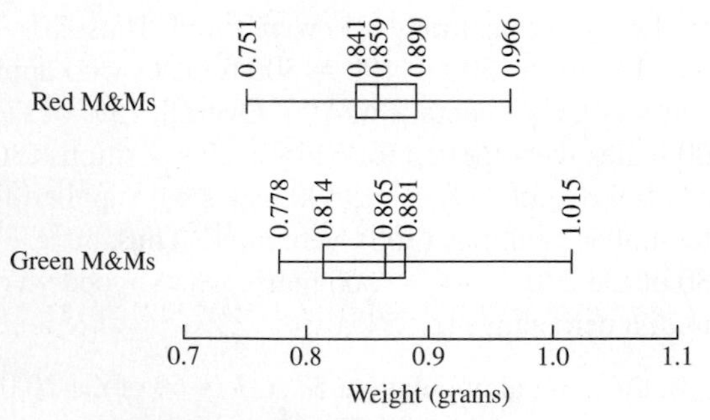

d. Red: Standard deviation is estimated to be 0.2150/4 = 0.05375 g; actual standard deviation is 0.0576 g. Green: Standard deviation is estimated to be 0.2370/4 = 0.05925 g; actual standard deviation is 0.0570 g. Estimating the standard deviation by using the range rule of thumb works reasonably well in both cases. **e.** A comparison of the results from the two data sets suggests that they are not very different. It appears that weights of red M&Ms and green M&Ms are about the same.

2. a. 31st percentile **b.** 0.865 g

3. a. 0 **b.** Although both batteries have the same mean lifetime, the batteries with the smaller standard deviation are better, because their lifetimes will be closer to the mean and fewer of them will strand drivers by failing sooner than expected. **c.** The outlier pulls the mean either up or down, depending on whether it is above or below the mean, respectively. **d.** The outlier has no effect on the median. **e.** The outlier increases the range. **f.** The outlier increases the standard deviation.

CHAPTER 4 QUIZ

1. Mean **2.** Standard deviation

3. No. It is an estimate that rarely yields the exact value.

4. Any one of the statements could be correct.

5. Only the second, third, and fourth statements could be correct.

6. Low: 30; high: 70 **7.** 2 **8.** 0 **9.** 7.0

10. Minimum value, first quartile, median (or second quartile), third quartile, maximum value

SECTION 5.1

1. The word *normal* has a special meaning in statistics. It refers to a specific category of distributions, all of them bell-shaped.

3. No. The ten different possible digits are all equally likely, so the graph of the distribution will be flat, not bell-shaped.

5. Does not make sense. **7.** Does not make sense.

9. Distribution b is not normal. Distribution c has the larger standard deviation.

11. Normal. It is common for a manufactured product, such as CDs, to have a distribution that is normal. The weights typically vary above and below the mean weight by about the same amounts, so the distribution has one peak and is symmetric.

13. Not normal. The outcomes of 1, 2, 3, 4, 5, and 6 are all equally likely, so the distribution is uniform, not normal. A graph of the distribution will tend to be very flat, not bell-shaped.

15. Normal. Such physical measurements generally tend to be normally distributed. A small number of males will have extremely high grip strength measurements, and a very small number will have low measurements. The distribution will tend to peak around the value of the mean.

17. Not normal. The waiting times will tend to be uniformly distributed.

19. Not normal. Movie lengths have a minimum length (zero) and no maximum length.

21. Nearly normal. The deviations from the mean are evenly distributed around the mean.

23. a. 1 **b.** 0.20 **c.** 0.80 **d.** 0.35 **e.** 0.45

25. a. 115 **b.** 15% **c.** 45% **d.** 15%

SECTION 5.2

1. 0

3. No. The rule applies to normal distributions, but the outcomes from a die roll have a uniform distribution, not a normal distribution.

5. Does not make sense. **7.** Makes sense.

9. a. 50% **b.** 0.84 **c.** 97.5% **d.** 16% **e.** 0.025 **f.** 16% **g.** 2.5% **h.** 0.84 **i.** 68% **j.** 81.5%

11. a. 68% **b.** 95% **c.** 99.7% **d.** 47.5%

13. 50% **15.** 15.87% **17.** 2.28% **19.** 1.39%

21. 68.26% **23.** 80.64% **25.** 50.00%

27. 84.13% **29.** 99.38% **31.** 0.13% **33.** 68.26%

35. 89.25%

37. In all cases, 5% of coins are rejected. Cents: 2.44 g to 2.56 g; nickels: 4.88 g to 5.12 g; dimes: 2.208 g to 2.328 g; quarters: 5.530 g to 5.810 g; half dollars: 11.060 g to 11.620 g

39. a. 6.68% **b.** 48.01% **c.** 36.54%

41. a. 0.47% **b.** Approximately 4% **c.** 30.055 to 30.745 in. **d.** The best estimate would be the mean of the readings, or 30.4 in.

43. Approximately 96.33%

SECTION 5.3

1. No, the sample is a convenience sample and is subject to a bias that would not be expected with a random sample. The student cannot assume that her convenience sample has the same characteristics as a sample that was randomly selected.

3. No. The Central Limit Theorem indicates that the distribution of the sample means will be approximately normal for large sample sizes, but samples of size 2 may not be sufficient, unless the original population is itself normally distributed.

5. a. The mean is 100 and the standard deviation is 2. **b.** The mean is 100 and the standard deviation is 1.6. **c.** With larger sample sizes (as in part b), the means tend to be closer together, so they have less variation, which results in a smaller standard deviation.

7. a. The mean is 6.5 and the standard deviation is 0.384. **b.** The mean is 6.5 and the standard deviation is 0.345. **c.** With larger sample sizes (as in part b), the means tend to be closer together, so they have less variation, which results in a smaller standard deviation.

9. 40%, 4% **11.** 29.6%, 99.96%

13. a. 0.35% **b.** No, it appears that the mean is greater than 12.00 oz, but consumers are not being cheated because the cans are being overfilled, not underfilled.

15. a. 57.93% chance **b.** 97.72% chance **c.** Although the mean head breadth of 100 men is very likely to be less than 6.2 in., there could be many individual men who could not use the helmets because they had head breadths greater than 6.2 in. Based on the result from part a, these helmets would not fit about 42% of men.

17. a. 52.91% **b.** Approximately a 73% chance **c.** Part a, because the seats will be occupied by individual women, not groups of women.

19. a. 16 (5.74% of 280) **b.** The standard z-scores are beyond the range of those included in Table 5.1, but Table 5.1 shows that it is very unlikely (less than a 0.04% chance) that the mean falls between the limits of 5.550 g and 5.790 g. **c.** Part a, because the individual rejected quarters could result in lost sales and lower profits.

21. a. The standard deviation of the distribution of sample means is $\sigma/\sqrt{n} = \sigma/10$. **b.** The standard deviation of the distribution of sample means is approximately $\sigma/\sqrt{1{,}000} \cong \sigma/32$, which is about one-third of the standard deviation of part a. **c.** The standard deviation of the distribution of sample means decreases as n increases.

CHAPTER 5 REVIEW EXERCISES

1. a. Because the 38 outcomes are all equally likely, the distribution is uniform, not normal. **b.** Weights of a homogeneous population, such as Golden Retriever dogs, typically have a normal distribution. **c.** Approximately normal. Human physical and psychological measurements tend to be normally distributed. (However, the distribution could be slightly skewed, because there can't be any negative reaction times, but there may be a few people with very high reaction times.)

2. a. 95% **b.** 99.7% **c.** Yes, because such a high score occurs only about 0.3% of the time.

3. a. 90 percentile **b.** 1.29 **c.** No, the data value lies less than 2 standard deviations from the mean. **d.** 0.0065 **e.** The temperature is unusual; it lies more than 2 standard deviations above the mean. **f.** 99.22 **g.** 97.18 **h.** Fewer than 0.01% **i.** The sample mean is 6.6 standard deviations below the mean; the chance of selecting such a sample is extremely small. The assumed mean (98.60) may be incorrect.

CHAPTER 5 QUIZ

1. b and e are correct. **2.** 95% **3.** 50 **4.** 0.5 **5.** 2 **6.** −1 **7.** 50% **8.** 2.28% **9.** 2.28%

10. Part c

SECTION 6.1

1. The claim is not valid, because it is easy to get 11 girls among 20 births by chance.

3. No. Statistical significance at the 0.05 level means that there is less than a 0.05 probability that the result occurred by chance, but a probability less than 0.05 does not necessarily mean that there is less than a 0.01 chance.

5. Does not make sense. **7.** Makes sense.

9. Not statistically significant. The results are close to the 250 tails expected, so the results could easily occur by chance.

11. Statistically significant. With six possible outcomes, we expect that about 10 of the 60 outcomes will be 3, so not getting any 3s is very unlikely to occur by chance.

13. Statistically significant. It is very unlikely that when 20 adults are randomly selected, they are all women.

15. Not statistically significant. The result could easily occur by chance.

17. While the sample size is small, a 21% improvement in mileage is significant.

19. With 325 babies, the number of girls would usually be around 162, so the result of 295 girls is a substantial departure from the results expected by chance. The results appear to be statistically significant.

21. **a.** If 100 samples were selected, the mean temperature would be 98.20 or less in 5 or fewer of the samples. **b.** Selecting a sample with a mean this small is extremely unlikely and would not be expected by chance.

23. This result is not significant at the 0.05 level because the probability of its occurring by chance when there is no real improvement is greater than 0.05.

SECTION 6.2

1. $P(A)$ represents the probability that event A occurs. $P(\text{not } A)$, or $P(\bar{A})$, is the probability that event A does not occur.

3. The reasoning is wrong because it assumes that the two possible outcomes of rain and no rain are equally likely, but they are not equally likely.

5. Makes sense. **7.** Makes sense.

9. Does not make sense.

11. **a.** Outcome **b.** Event, since it can occur in three different ways. **c.** Outcome **d.** Event, since it can occur in three different ways. **e.** Event, since it can occur in six different ways. **f.** Event, since it can occur in six different ways.

13. 1/3, assuming that the die is fair and the outcomes are equally likely.

15. 2/38, or 1/19, assuming that the outcomes are equally likely.

17. 1/365, assuming that births on the 365 days are equally likely.

19. 1/2, or 0.5 **21.** 4/6, or 2/3 **23.** 36/38, or 18/19 **25.** 0.45 **27.** 0.720 **29.** 0.22, 0.33, 0.44, 0.56

31. **a.** 1/8 = 0.125 (GGG) **b.** 3/8 = 0.375 (BBG, BGB, GBB) **c.** 1/8 = 0.125 (GBB) **d.** 7/8 = 0.875 (GGG, BGG, GBG, GGB, BBG, BGB, GBB) **e.** 4/8 = 0.5 (BBG, BGB, GBB, BBB)

33. 0.40 **35.** P(success) = 0.86

37. The probability that a person you meet at random is over 65 will be 34.7 million/281 million = 0.123 in 2000 and will be 78.9/394 = 0.200 in 2050. Thus, your chances will be greater in 2050.

39. **a.** Outcomes for rolling four fair coins:

Coin 1	Coin 2	Coin 3	Coin 4	Outcome	Probability
H	H	H	H	HHHH	1/16
H	H	H	T	HHHT	1/16
H	H	T	H	HHTH	1/16
H	H	T	T	HHTT	1/16
H	T	H	H	HTHH	1/16
H	T	H	T	HTHT	1/16
H	T	T	H	HTTH	1/16
H	T	T	T	HTTT	1/16
T	H	H	H	THHH	1/16
T	H	H	T	THHT	1/16
T	H	T	H	THTH	1/16
T	H	T	T	THTT	1/16
T	T	H	H	TTHH	1/16
T	T	H	T	TTHT	1/16
T	T	T	H	TTTH	1/16
T	T	T	T	TTTT	1/16

b. Probability distribution for the number of heads in rolling four fair coins:

Result	Probability
4 heads (0 tails)	1/16
3 heads (1 tail)	4/16
2 heads (2 tails)	6/16
1 head (3 tails)	4/16
0 heads (4 tails)	1/16
Total	**1**

c. 6/16 = 0.375 **d.** 15/16 = 0.9375 **e.** 2 heads (probability is 0.375)

SECTION 6.3

1. If some process is repeated over many trials, the proportion of trials in which some event occurs will be close to the probability of the event. As the number of trials increases, the proportion of trials in which the event occurs gets closer to the probability of the event.

3. His betting strategy is unwise because the casino continues to have an advantage. His past performance does not affect future events. His reasoning is not correct. Although his proportion of winning bets is likely to increase, it could increase while he continues to lose more money. This flawed reasoning has caused many gamblers to lose large amounts of money.

5. Makes sense. **7.** Makes sense.

9. No. You should not expect to get exactly 5,000 heads, since the probability of that particular outcome is extremely small. The proportion of heads should approach 0.5 as the number of tosses increases.

11. Expected value to you for one game is ($5 × 0.25) + (−$1 × 0.75) = $1.25 − 0.75 = $0.50. Because you must either win $5 or lose $1 on the first game, you cannot win the expected value of $0.50. If you play 100 games, you should expect to win 100 × $0.50 = $50.

13. 0.94, 0.74

15. 15 minutes **17.** −$0.78, −$285 **19.** −7.07¢, −1.4¢

21. a. Decision 1—Option A: expected value = $1,000,000; Option B: expected value = $1,140,000. **Decision 2**—Option A: expected value = $110,000; Option B: expected value = $250,000 **b.** Responses are not consistent with expected values in Decision 1, but they are in Decision 2. **c.** It appears that people choose the certain outcome ($1,000,000) in Decision 1.

23. a. 0.5, 0.5 **b.** Loss of $10 **c.** Loss of $16 **d.** Loss of $20 **e.** 45%, 46%, 48%. The percentage of even numbers approaches 50%, but the difference between the numbers of even and odd numbers increases.
f. 60 even numbers

SECTION 6.4

1. Because they are expressed as the numbers of births for each 1,000 people in the population, the birth rates can be compared directly. The actual numbers of births should be considered in the context of the population sizes of the countries, so a comparison would require the numbers of births along with population sizes, and even then the comparison would not be easy because of the numbers involved.

3. Life expectancy is the number of years a person with a given age can expect to live on average. A 30-year-old person will have a shorter life expectancy than a 20-year-old person. The 30-year-old person is not expected to live as many additional years as a 20-year-old person.

5. Does not make sense. **7.** Makes sense.

9. 1995: 0.0208; 2000: 0.0102; 2004: 0.0013. The year 2004 was the safest because it had the lowest number of fatalities per 1,000 departures.

11. 1995: 3.1; 2000: 1.4; 2004: 0.2. The year 2004 was the safest because it had the lowest number of fatalities per 10 million passengers.

13. 58.8 years **15.** 8.3 deaths per 10,000 people

17. a. Utah: 137; Maine: 38 **b.** Utah: 20.8; Maine: 10.7

19. a. 4,230,000 **b.** 2,520,000 **c.** 1,710,000
d. 0.0057; 0.57%

SECTION 6.5

1. The occurrence of one of the events does not affect the probability of the other event.

3. Sampling with replacement. The second outcome is independent of the first.

5. Does not make sense. **7.** Does not make sense.

9. 1/8, or 0.125

11. 1/260; the password is too short to be effective.

13. a. 0.0039 **b.** 0.00098 **c.** 0.125 **d.** 0.0625
e. There are 60 equally likely songs available for each selection. No matter which song is played first, the probability that the next one is the same is 1/60 = 0.0167.

15. P(guilty plea or sent to prison) = 1,014/1,028 = 0.986

17. 0.865 **19.** 0.381 **21.** 0.410 **23.** 0.920

25. 0.0195

27. a. 0.733 **b.** 1 **c.** 0.643 **d.** 0.22

29. a. 0.5625 **b.** 0.375 **c.** 0.0625

d.

Event	Probability
AA	0.5625
Aa	0.1875
aA	0.1875
aa	0.0625

e. 0.9375

31. a. 0.518 **b.** 0.491

CHAPTER 6 REVIEW EXERCISES

1. 85/99, or 0.859 **2.** 83/99, or 0.838

3. 88/99, or 0.889 **4.** 96/99, or 0.970

5. 0.702 **6.** 0.736

7. a. 0.73 **b.** 0.073 **c.** 1.35 **d.** 0.0014; you should doubt the stated yield.

8. a. 0.10 **b.** No, the probability is greater than 0.05.

9. a. 1/38 = 0.026 **b.** Yes, the probability is less than 0.05.

10. a. 0.14 **b.** No, the probability is greater than 0.05.

CHAPTER 6 QUIZ

1. 1/6 **2.** 1/216, or 0.00463 **3.** 0.55 **4.** 7.5

5. Yes **6.** 0.502 **7.** 0.141 **8.** 0.613

9. 0.251 **10.** 0.123

SECTION 7.1

1. A correlation exists between two variables when higher values of one consistently tend to be associated with higher values of another or when higher values of one consistently tend to be associated with lower values of another. The meaning of the term *correlation* in statistics is much more specific than its meaning in general usage.

3. The points are quite close to a linear (or straight-line) pattern, and they have greater height the farther to the right they are located.

5. Does not make sense. **7.** Does not make sense.

9. Positive correlation because taller women tend to weigh more.

11. Negative correlation because cars that weigh more tend to use more fuel, so fuel consumption in miles per gallon will be lower.

13. Positive correlation because running the marathon faster (in less time) will result in a lower finish order.

15. The variables are not correlated.

17. There is a strong positive correlation, with the correlation coefficient approximately 0.8. Much of this correlation is due to the fact that a large fraction of the grain produced is used to feed livestock.

19. a.

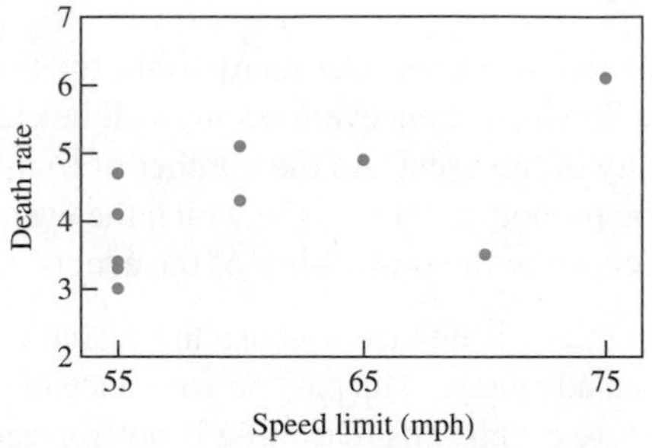

b. There is a moderate positive correlation ($r = 0.59$ exactly). **c.** With the exception of Britain, higher speed limits are generally associated with higher death rates. Death rates are also influenced by other factors besides speed limits.

21. a.

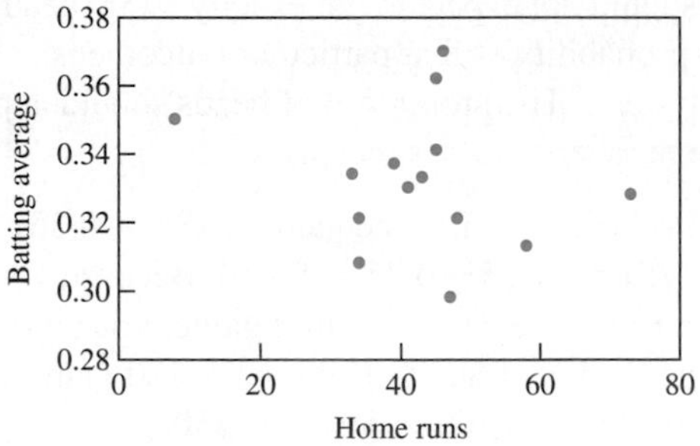

b. There is a moderate negative correlation ($r = -0.29$ exactly). **c.** No

23. a.

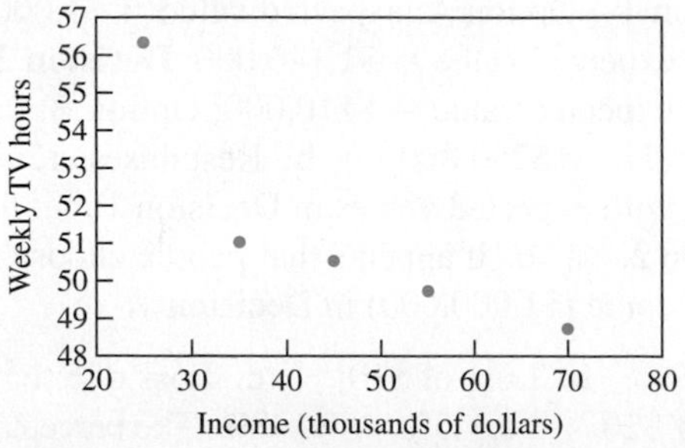

b. There is a strong negative correlation between income and the number of TV hours per week ($r = -0.86$ exactly). **c.** Families with more income have more opportunities to do other things. No.

25. a.

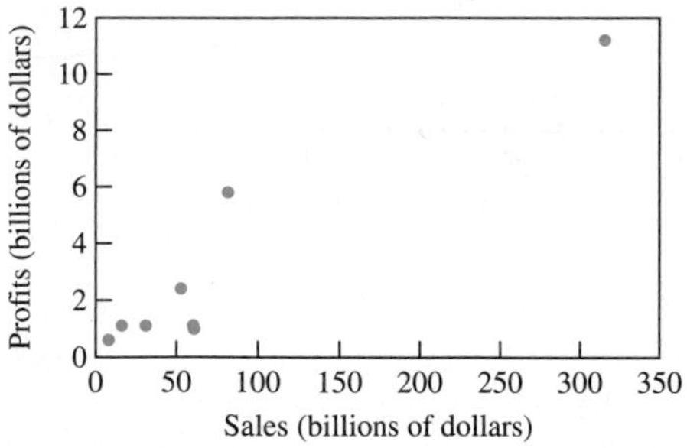

b. There is a strong correlation between sales and earnings ($r = 0.92$ exactly). The strong correlation in this case is highly affected by the outlier. **c.** Higher sales do not necessarily translate into higher earnings. Some companies have larger expenses, driving earnings down.

27. The variables x and y appear symmetrically in the formula (interchanging x and y does not change the formula).

SECTION 7.2

1. If there is a significant correlation between two variables, there is a statistical association, but this association does not imply that one of the variables is somehow the cause or direct effect of the other variable.

3. Outliers are values that are very far away from almost all of the other values in a data set. Outliers might make it appear that there is a significant correlation when there is not, or they might mask a real correlation.

5. Does not make sense. **7.** Does not make sense.

9. There is a positive correlation that is probably due to a common underlying cause. Many crimes are committed with handguns that are not registered.

11. There is a positive correlation that is due to a direct cause. Tolls along the Massachusetts Turnpike are based on the distance traveled.

13. There is a positive correlation that is probably due to a common cause, such as the general increase in the number of cars and traffic.

15. There is a negative correlation that is probably due to a direct cause. As gas prices increase by large amounts, people can't afford to drive as much, so they cut costs by driving less.

17. a. The outlier is the upper left-hand point (0.4, 1.0). Without the outlier, the correlation coefficient is 0.0. **b.** With the outlier, the correlation coefficient is -0.58.

19. a. The actual correlation coefficient is $r = 0.92$, which is significant at the 0.01 level, so there is a very strong correlation between weight and shoe size.

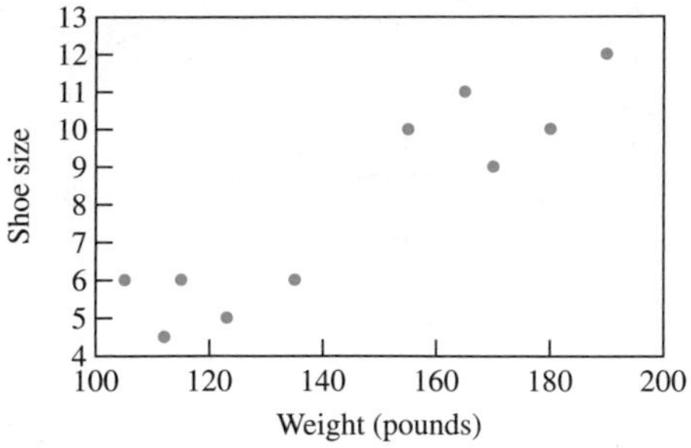

b. Within each of the two groups, the correlation is less strong than the overall correlation.

21. a. The actual correlation coefficient is $r = 0.77$, which is significant at the 0.01 level, indicating a strong correlation. **b.** The 16 points to the right correspond to relatively poor countries, such as Uganda. The remaining points correspond to relatively affluent countries, such as Sweden. **c.** There appears to be a negative correlation between the variables for the poorer countries and a positive correlation for the wealthier countries.

SECTION 7.3

1. A best-fit line (or regression line) is a line on a scatter diagram that lies closer to the data points than any other possible line (according to the standard statistical measure of closeness). A best-fit line is useful for predicting the value of a variable given some value of the other variable.

3. r^2 is the square of the correlation coefficient, and it is the proportion of variation in a variable that can be explained by the best-fit line.

5. Does not make sense. **7.** Does not make sense.

9. a.

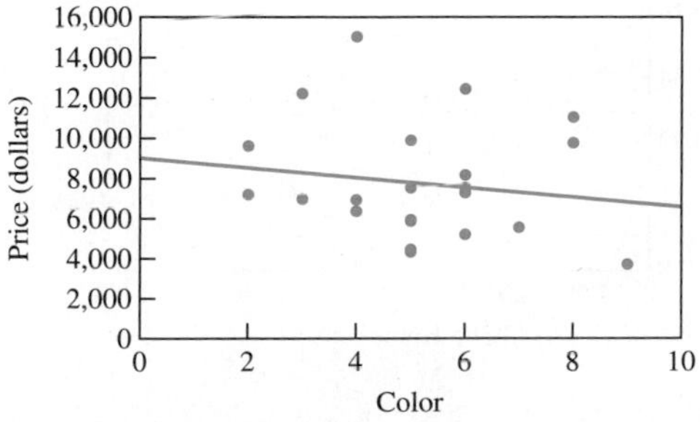

b. Actual $r = -0.16$; $r^2 = 0.026$. About 3% of the variation in price can be explained by the best-fit line. **c.** The best-fit line should not be used to make predictions.

11. a.

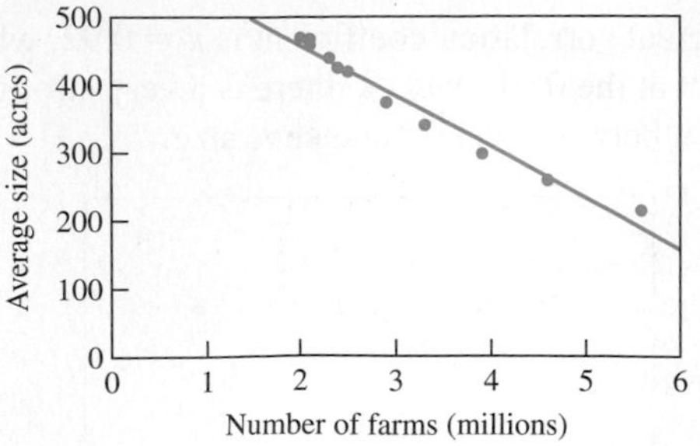

b. Actual $r = -0.99$; $r^2 = 0.97$. About 97% of the variation in farm size can be explained by the best-fit line. **c.** The best-fit line could be used to make predictions within the range of the number of farms included in the data. Because there does appear to be a slight curvature to the points, predictions should not be made outside that range.

13. a.

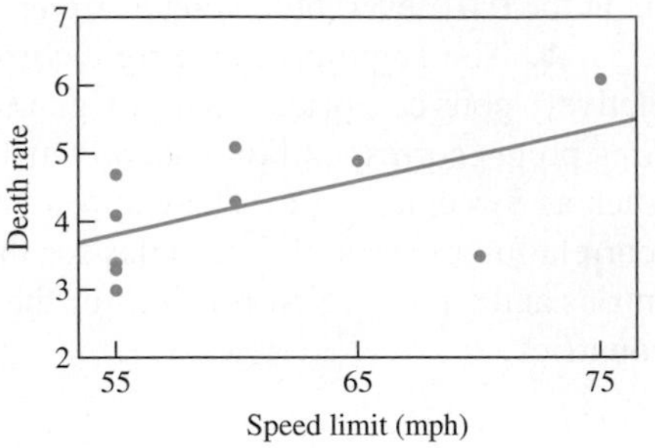

b. Actual $r = 0.59$; $r^2 = 0.35$; 35% of the variation can be accounted for by the best-fit line. **c.** (75, 6.1) and (70, 3.5) are both possible outliers, the former because it is away from most of the data points and the latter because the death rate is lower than might be expected considering the rest of the data. Because one point is above the best-fit line and one is below, the net effect of the two points is probably to cancel each other out. **d.** The value of r is too small for predictions based on the best-fit line to be considered reliable.

15. a.

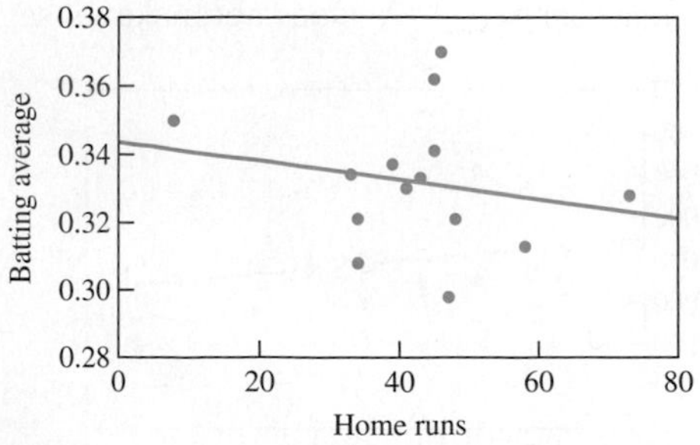

b. Actual $r = -0.29$; $r^2 = 0.08$. Only 8% of the variation can be accounted for by the best-fit line. **c.** (.366, 49) is an outlier, since it is far from the other data points. **d.** Predictions based on the best-fit line are not reliable.

17. a.

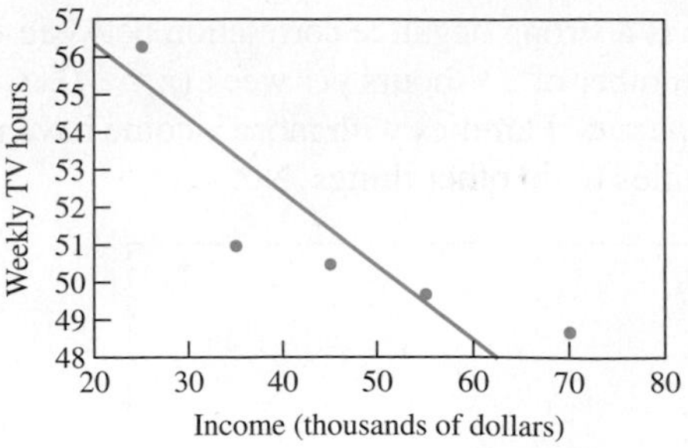

b. Actual $r = -0.86$; $r^2 = 0.74$; 74% of the variation can be accounted for by the best-fit line. **c.** (25,000, 56.3) is an outlier. The best-fit line would have a less steep downward slope if that point were removed.
d. Predictions based on the best-fit line could be reliable, but the presence of the outlier and its effect on the best-fit line make the reliability questionable.

19. a. Actual $r = 0.92$; $r^2 = 0.84$; 84% of the variation can be accounted for by the best-fit line.

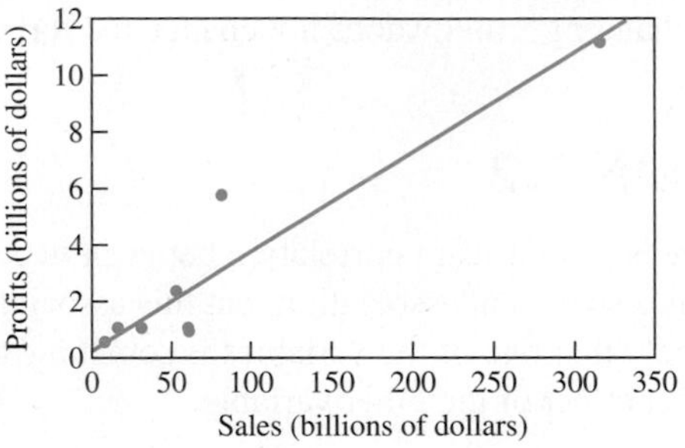

b. (81.5, 5.8) is an outlier. **c.** Predictions based on the best-fit line could be reliable, but the strong effect of the outlier in determining the best-fit line makes the reliability questionable. More data are needed.

SECTION 7.4

1. (1) The correlation is the result of a coincidence; (2) the correlation is due to a common cause; (3) the correlation is due to a direct influence of one of the variables on the other.

3. A confounding variable is a variable that is not included in the analysis but that affects the variables included in the analysis. Failure to include or account for a confounding variable might cause a researcher to miss an underlying causality.

5. Does not make sense. **7.** Does not make sense.

9. The causal connection is valid. As the speed of the club head increases, more force is applied to the golf ball, so it travels farther.

11. The causal connection is valid. Alcohol is a depressant to the central nervous system, and it has several effects that include decreased reaction time. This is one important reason why drinking and driving is so dangerous.

13. Guideline 1; guidelines 2 and 5; guidelines 3 and 5. The headaches are associated with work days in some way. The headaches are not associated with Coke or the caffeine in Coke. The headaches are possibly the result of bad ventilation in the building.

15. Smoking can only increase the risk already present.

17. This was an observational study. Later child bearing reflects an underlying cause. While it's possible that the conclusions are correct, there are other possible explanations for the findings. For example, it's possible that the younger women lived during a time when having babies after age 40 was less likely (by choice). It is still possible for them to live to be 100.

19. Availability is not itself a cause. Social, economic, or personal conditions cause individuals to use the available weapons.

CHAPTER 7 REVIEW EXERCISES

1. Correlation is strong and positive; 92.5%.

2. Correlation is strong and positive; 82.6%.

3. Correlation is strong and positive; 95.8%.

4. Because the pattern of the points is close to a straight-line pattern, the scatter diagram indicates that there is a correlation between the variables.

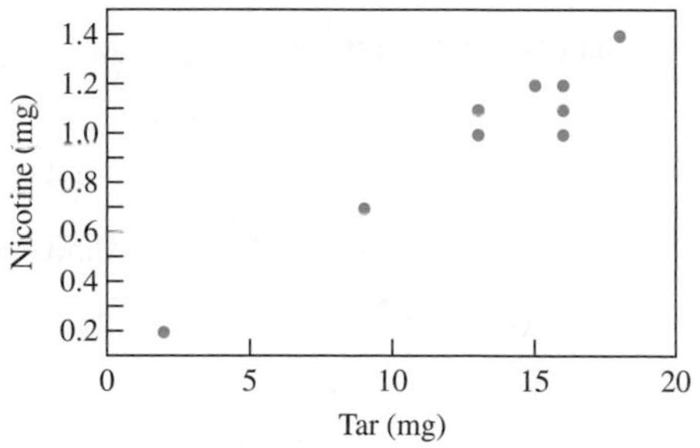

5. Data would consist of the death rate in a neighborhood and the distance between the neighborhood and power lines for many neighborhoods. It may be possible to establish a correlation between power lines and leukemia deaths, but it would be very difficult to establish a causal relationship.

6. The points on the scatter diagram lie on a straight line with negative slope (falling to the right).

7. Correlation alone never implies causation, and in this case, certainly more trips to the dentist do not cause higher incomes. Households with more disposable income can afford more trips to the dentist or can afford dental insurance which covers the costs of the trips.

8. Variables affecting the value of a home might include its location, size, age, condition, and lot size. Location is often cited as the most important factor, with considerations including nearness to schools, shopping, and churches. The age of the previous owner would be unrelated to the value.

9. The data values that were collected were uncorrelated. It's still possible that the variables represented by the data values are related in some nonlinear way—i.e., the scatter diagram forms a curve instead of a straight line.

10. The correlation coefficient is -0.056, but any positive or negative value near 0 is reasonable.

CHAPTER 7 QUIZ

1. $-1, 1$ **2.** c, d, and e **3.** c **4.** The actual value of r is -0.934, but any value between -0.5 and -0.99 is a reasonable answer.

5. Yes **6.** False **7.** False **8.** True

9. False **10.** True

SECTION 8.1

1. Uniform; approximately normal.

3. $\bar{x}$ denotes the mean of a sample, whereas μ denotes the mean of a population.

5. Does not make sense. **7.** Makes sense.

9. The best estimate of the population proportion for the Northeast is the sample proportion: 0.19. The given sample proportion is not likely to be a good estimate of the proportion for the population of children in the United States. There are regional factors that could have a strong influence on that proportion in different parts of the country.

11. **a.** 9.5 standard deviations above the mean ($z = 9.5$)
b. The probability is very small, such as 0.01%.
c. No. It appears that the cans are being filled with an amount that is greater than 12.00 ounces.

13. **a.** 0.528 **b.** 0.49 **c.** The sample proportion is slightly too low.

15. 783 people. The larger sample is likely to provide a more reliable estimate of the population proportion.

17. **a.** -0.67 standard deviation below the mean ($z = -0.67$)
b. 0.253

19. **a.** 1.0, 1.5, 3.0, 1.5, 2.0, 3.5, 3.0, 3.5, 5.0 **b.** 2.7
c. Yes; yes

21. **Samples of size $n = 1$**

Sample	Mean
A	52
C	5
G	60
M	33
O	97
Mean	**49.4**

Samples of size $n = 2$

Sample	Mean
AC	28.5
AG	56.0
AM	42.5
AO	74.5
CG	32.5
CM	19.0
CO	51.0
GM	46.5
GO	78.5
MO	65.0
Mean	**49.4**

Samples of size $n = 3$

Sample	Mean
ACG	39.00
ACM	30.00
ACO	51.33
AGM	48.33
AGO	69.67
AMO	60.67
CGM	32.67
CGO	54.00
CMO	45.00
GMO	63.33
Mean	**49.4**

Samples of size $n = 4$

Sample	Mean
ACGM	37.50
ACGO	53.50
ACMO	46.75
AGMO	60.50
CGMO	48.75
Mean	**49.4**

Samples of size $n = 5$

Sample	Mean
ACGMO	49.4
Mean	**49.4**

SECTION 8.2

1. We have 95% confidence that the limits of 183.3 g and 298.5 g actually do contain the true population mean cholesterol level of all women. We expect that 95% of such samples will result in confidence interval limits that contain the population mean.

3. The media often omit reference to the confidence level, which is typically 95%. The word "mean" should be used instead of the word "average."

5. Does not make sense. **7.** Does not make sense.

9. Margin of error: 2.0. $73.0 < \mu < 77.0$

11. Margin of error: $1,469. $\$44,736 < \mu < \$47,674$

13. 49 **15.** 484 **17.** 185 **19.** 225

21. 5.639 g; $5.619 \text{ g} < \mu < 5.659 \text{ g}$

23. 5.15 yr; $5.10 \text{ yr} < \mu < 5.20 \text{ yr}$

25. $146 \text{ lb} < \mu < 221 \text{ lb}$

27. **a.** 3.1 **b.** 1.4 **c.** 3.1 **d.** $2.6 < \mu < 3.6$
e. The sample is small and was concentrated in one neighborhood, so it is unlikely to be representative of all American families.

SECTION 8.3

1. We have 95% confidence that the limits of 0.457 and 0.551 actually contain the true population proportion. We expect that 95% of such samples will result in confidence interval limits that contain the population proportion.

3. The media often omit reference to the confidence level, which is typically 95%.

5. Does not make sense. **7.** Does not make sense.

9. Margin of error: 0.0866. $0.163 < p < 0.337$

11. Margin of error: 0.0257. $0.202 < p < 0.254$

13. 10,000 **15.** 278 **17.** $0.140 < p < 0.160$

19. $E = 0.021$; $0.779 < p < 0.821$

21. **a.** $0.153 < p < 0.191$ **b.** $0.118 < p < 0.146$
c. $0.215 < p < 0.243$

23. **a.** 0.98 **b.** $0.966 < p < 0.994$ **c.** For the sample to be a random sample, all films must have an equal chance of being chosen.

25. The 95% confidence interval is $0.538 < p < 0.587$. A prediction of victory is reasonable.

27. Poll 1: $0.494 < p < 0.546$; Poll 2: $0.494 < p < 0.534$; Poll 3: $0.498 < p < 0.532$. Because all of these confidence intervals include values less than 0.5, Martinez cannot be confident of winning a majority.

29. a. $0.50 < p < 0.58$ **b.** 621

CHAPTER 8 REVIEW EXERCISES

1. a. $0.008 < p < 0.056$ **b.** There is 95% confidence that the limits of 0.008 and 0.056 contain the true value of the population proportion. If such samples of size 221 were randomly selected many times, the resulting confidence intervals would contain the true population proportion in 95% of those samples. **c.** 0.0237 **d.** The distribution will be approximately normal, or bell-shaped. **e.** They would get closer together. **f.** The population proportion is a fixed value, not a random variable, and either it is contained within the confidence interval limits or it is not, so there is no probability involved with it.

2. a. 114 **b.** If the sample size is larger than necessary, the confidence interval will be better in the sense that it will be narrower than necessary. If the sample size is smaller than necessary, the confidence interval will be worse in the sense that it will be wider than it should be. **c.** Because college students are a more homogeneous group than the population as a whole, the standard deviation for college students is likely to be smaller than 16. If we use the actual value that is smaller than 16, the sample size will be smaller than 114.

3. a. $6.4 < \mu < 7.8$ **b.** There is 95% confidence that the limits of 6.4 and 7.8 contain the true value of the population mean. If such samples of size 50 were randomly selected many times, the resulting confidence intervals would contain the true population mean in 95% of those samples. **c.** 0.707 **d.** 400

4. a. 2,500 **b.** No. Your sample size will be large enough, but it has a high potential for being biased. The people who refuse to answer could well constitute a segment of the population with a different perspective, and that perspective will be incorrectly excluded from the sample.

CHAPTER 8 QUIZ

1. Approximately normal, or bell-shaped

2. Approximately normal, or bell-shaped

3. The sample proportion **4.** $0.52 < p < 0.68$ **5.** b

6. 0.02 **7.** 20 **8.** $4.45 < \mu < 4.75$

9. $0.400 < p < 0.800$

10. The confidence interval limits should be proportions between 0 and 1, but they are not.

SECTION 9.1

1. A hypothesis test is a standard procedure for testing a claim about the value of a population parameter.

3. $\mu \neq 98.6$, $\mu < 98.6$, $\mu > 98.6$

5. Does not make sense. **7.** Does not make sense.

9. Does not make sense. **11.** Does not make sense.

13. a. The increase is relatively small, so it does not appear to be statistically significant. **b.** The increase is quite large, so it appears to be statistically significant. **c.** Answers will vary, but an increase of about three or more gallons would seem to be statistically significant.

15. H_0: $\mu = 1$ L. H_a: $\mu < 1$ L. There is not sufficient sample evidence to support the claim that the mean amount of cola is less than 1 L. There is sufficient sample evidence to support the claim that the mean amount of cola is less than 1 L.

17. H_0: $p = 0.7$. H_a: $p > 0.7$. There is not sufficient sample evidence to support the claim that the proportion of girls is greater than 0.7. There is sufficient sample evidence to support the claim that the proportion of girls is greater than 0.7.

19. H_0: $\mu = 10$ oz. H_a: $\mu \neq 10$ oz. There is not sufficient sample evidence to warrant rejection of the claim that the mean amount of coffee is equal to 10 oz. There is sufficient sample evidence to warrant rejection of the claim that the mean amount of coffee is equal to 10 oz.

21. H_0: $p = 0.5$. H_a: $p > 0.5$. There is not sufficient sample evidence to support the claim that the proportion of students is greater than 0.5. There is sufficient sample evidence to support the claim that the proportion of students is greater than 0.5.

23. No. The P-value of 0.382 is greater than the significance level of 0.05, and this high value suggests that there is a high likelihood of getting 48 or fewer males by chance.

25. No. The P-value of 0.028 is greater than the significance level of 0.01, and this high value suggests that there is a high likelihood of getting 40 or fewer males by chance.

27. Yes. The P-value of 0.002 is less than the significance level of 0.01, and this low value suggests that there is a small likelihood of getting 35 or fewer males by chance.

SECTION 9.2

1. n represents the sample size, $\bar{x}$ represents the mean of the sample values, s represents the standard deviation of the sample values, σ represents the standard deviation of the population values, and μ represents the mean of the population.

3. A type I error is the mistake of rejecting the null hypothesis when that null hypothesis is actually true. A type II error is the mistake of failing to reject the null hypothesis when that null hypothesis is actually false.

5. Does not make sense. **7.** Does not make sense.

9. Makes sense. **11.** Does not make sense.

13. $z = 1.33$. The alternative hypothesis is not supported.

15. $z = -5.33$. The alternative hypothesis is supported.

17. $z = -2.50$. The alternative hypothesis is supported.

19. $z = 2.80$. The alternative hypothesis is supported.

21. 0.3446. H_a is not supported.

23. 0.0287. H_a is supported.

25. 0.1096. H_a is not supported.

27. 0.0892. H_a is not supported.

29. 0.0035. H_a is supported.

31. 0.0179. H_a is supported.

33. 0.8808. H_a is not supported.

35. The test is right-tailed, but the standard score is to the left of the center of the distribution, so the P-value must be greater than 0.5. The alternative hypothesis is not supported. The formal hypothesis test is not necessary because there is no way that we could support a claim that the mean is greater than some value when the sample mean is less than that value.

37. H_0: $\mu = 7.5$ years, H_a: $\mu < 7.5$ years; $n = 100$, $\bar{x} = 7.01$ years, $s = 3.74$ years, $\sigma = 0.374$ year, $z = -1.3$; not significant at the 0.05 level, P-value $= 0.0968$; there is not sufficient evidence to support the claim that the mean time of ownership is less than 7.5 years.

39. H_0: $\mu = 16$ oz, H_a: $\mu \neq 16$ oz; $n = 144$, $\bar{x} = 15.8$ oz, $s = 1.6$ oz, $\sigma \approx s$, $z = -1.5$; not significant at the 0.05 level, P-value $= 0.0668$; there is not sufficient evidence to support the claim that the mean weight of the packages is different from the advertised 16 oz.

41. H_0: $\mu = 600$ mg, H_a: $\mu \neq 600$ mg; $n = 65$, $\bar{x} = 589$ mg, $s = 21$ mg, $\sigma \approx s$, $z = -4.2$; significant at the 0.05 level, P-value $= 0.0004$; there is sufficient evidence to support the claim that the mean amount of acetominophen is different from the listed 600 mg.

43. H_0: $\mu = 21.4$ mpg, H_a: $\mu < 21.4$ mpg; $n = 40$, $\bar{x} = 19.8$ mpg, $s = 3.5$ mpg, $\sigma \approx s$, $z = -2.9$; significant at the 0.05 level, P-value $= 0.0019$; there is sufficient evidence to support the claim that the mean gas mileage for SUVs is less than 21.4 mpg.

45. H_0: $\mu = 0.21$, H_a: $\mu > 0.21$; $n = 32$, $\bar{x} = 0.83$, $s = 0.24$, $\sigma \approx s$, $z = 14.6$; significant at the 0.05 level, P-value < 0.0002; there is sufficient evidence to support the claim that the mean is greater than 0.21.

47. H_0: $\mu = 3.39$ kg, H_a: $\mu \neq 3.39$ kg; $n = 121$, $\bar{x} = 3.67$ kg, $s = 0.66$ kg, $\sigma \approx s$, $z = 4.7$; significant at the 0.05 level, P-value < 0.0004; there is sufficient evidence to support the claim that the mean birth weight of male babies born to mothers on a vitamin supplement is different from the national average for all male babies.

49. H_0: $\mu = 24$ mo, H_a: $\mu < 24$ mo; $n = 70$, $\bar{x} = 22.1$ mo, $s = 8.6$ mo, $\sigma \approx s$, $z = -1.8$; significant at the 0.05 level, P-value $= 0.0359$; there is sufficient evidence to support the claim that the mean is less than 24 mo.

51. Type I error: Reject the claim that the patient is free of a disease when the patient is actually free of the disease. Type II error: Fail to reject the claim that the patient is free of a disease when the patient is not free of the disease.

53. Type I error: Reject the claim that the lottery is fair when the lottery is fair. Type II error: Fail to reject the claim that the lottery is fair when the lottery is biased.

SECTION 9.3

1. The symbol p represents the proportion in a population, $\hat{p}$ represents the proportion in a sample, and a P-value is the probability of getting a sample proportion that is at least as extreme as the sample proportion being considered.

3. No. The sample is a voluntary response sample, not a simple random sample. It is likely that those who responded are not representative of the general population.

5. Makes sense. **7.** Does not make sense.

9. H_0: $p = 0.5$, H_a: $p > 0.5$; $n = 400$, $\hat{p} = 0.51$, standard deviation of distribution of sample proportions is 0.025, $z = 0.5$; not significant at the 0.05 level, P-value $=$ 0.3446; there is not sufficient evidence to support the claim that a majority of the voters support the candidate.

11. H_0: $p = 0.50$, H_a: $p < 0.50$; $n = 1015$, $\hat{p} = 0.44$, standard deviation of distribution of sample proportions is 0.0157, $z = -3.8$; significant at the 0.05 level, P-value < 0.0002; there is sufficient evidence to support the claim that fewer than half of all teenagers in the population feel that grades are the greatest source of pressure.

13. H_0: $p = 0.099$, H_a: $p < 0.099$; $n = 1050$, $\hat{p} = 0.0933$, standard deviation of distribution of sample proportions is 0.0092, $z = -0.6$; not significant at the 0.05 level, P-value $= 0.2743$; there is not sufficient evidence to support the claim that illegal drug use is below the national average.

15. H_0: $p = 0.5$, H_a: $p > 0.5$; $n = 276{,}000$, $\hat{p} = 0.51$, standard deviation of distribution of sample proportions is 0.000952, $z = 10.5$; significant at the 0.05 level, P-value < 0.0002; there is sufficient evidence to support the claim that over half of all first-year students believe that abortion should be legal.

17. H_0: $p = 0.53$, H_a: $p \neq 0.53$; $n = 3{,}600$, $\hat{p} = 0.54$, standard deviation of distribution of sample proportions is 0.0083, $z = 1.2$; not significant at the 0.05 level, P-value $= 0.2302$; there is not sufficient evidence to support the claim that the percentage of households using natural gas has changed.

CHAPTER 9 REVIEW EXERCISES

1. a. H_0: $\mu = 12$ oz **b.** H_a: $\mu > 12$ oz **c.** $z = 10.4$ **d.** $z = 1.645$ **e.** Less than 0.0002 **f.** Reject the null hypothesis and claim that cans of Coke have a mean amount of cola greater than 12 oz. **g.** A type I error would result if we concluded that the population mean was greater than 12 oz when, in fact, it was not. **h.** A type II error would result if we did not conclude that the population mean was greater than 12 oz when, in fact, it was greater than 12 oz. **i.** The P-value would be the probability that z is greater than 10.4 or less than -10.4. Since the probability that z is greater than 10.4 is less than 0.0002, the P-value is less than $2 \times 0.0002 = 0.0004$.

2. a. H_0: $p = 0.5$ **b.** H_a: $p > 0.5$ **c.** $z = 0.8$ **d.** $z = 1.645$ **e.** 0.2119 **f.** There is not sufficient evidence to support the claim that the majority are smoking a year after the nicotine patch treatment. **g.** A type I error would be the mistake of supporting the claim that the majority of the subjects are smoking a year later when the proportion of smokers a year later is 0.5 (or less). **h.** A type II error would be the mistake of not supporting the claim that the majority are smoking a year later when a majority of the subjects actually are smoking a year later. **i.** The P-value doubles to become 0.4238.

3. The claim of an increase in the likelihood of a girl is not supported. With a sample proportion *less than* 0.5, there is no way we could ever support the claim that $p > 0.5$.

4. The sample is not random. It is possible that the family is not representative of the population, so the results are biased.

CHAPTER 9 QUIZ

1. $p > 0.2$ **2.** $p = 0.2$ **3.** $p \neq 0.6$

4. Right-tailed **5.** $\mu = 110$ **6.** $\mu > 110$

7. Do not support the claim that the mean is greater than 110.

8. Reject the null hypothesis. Do not reject the null hypothesis.

9. Support the claim that the mean is greater than 110. Do not support the claim that the mean is greater than 110.

10. Type I error

SECTION 10.1

1. The population standard deviation is not known and the population is normally distributed, or the population standard deviation is not known and the sample size is greater than 30.

3. There is no difference. The term "t distribution" is commonly used instead of Student t distribution, but the two terms are equivalent.

5. Does not make sense. **7.** Makes sense.

9. $97.2 < \mu < 107.8$ **11.** 14.3 in. $< \mu <$ 14.7 in.

13. a. $\$6{,}370 < \mu < \$11{,}638$ **b.** \$11,638

15. a. $0.085 < \mu < 0.157$ **b.** The claim does not appear to be valid. Because 0.165 grams per mile is not included in the confidence interval, that value does not appear to be the mean.

17. H_0: $\mu = 0.3$, H_a: $\mu < 0.3$. The test statistic $t = -0.119$ is not less than the value of $t = -1.753$ found in Table 10.1, so do not reject the null hypothesis. (P-value $= 0.4534$.) There is not sufficient evidence to support the claim that the mean amount of sugar in all cereals is less than 0.3 g.

19. H_0: $\mu = 92.84$, H_a: $\mu \neq 92.84$. The test statistic $t = -0.425$ is not less than the value of $t = -2.093$ found in Table 10.1, so do not reject the null hypothesis. (P-value $= 0.6758$.) There is not sufficient evidence to support the claim that the new baseballs have a mean bounce height that is different from 92.84 in. The new baseballs do not appear to be different.

21. H_0: $\mu = 3.39$, H_a: $\mu \neq 3.39$. The test statistic $t = 1.735$ is not greater than the value of $t = 2.131$ found in

Table 10.1, so do not reject the null hypothesis. (P-value = 0.1032.) There is not sufficient evidence to support the claim that the mean birth weight for all male babies of mothers given vitamins is different from 3.39 kg. Based on the given results, the vitamin supplement does not appear to have an effect on birth weight.

23. H_0: $\mu = 10.5$, H_a: $\mu < 10.5$. The test statistic $t = -0.095$ is not less than or equal to the value of $t = -1.717$ found in Table 10.1, so do not reject the null hypothesis. (P-value = 0.4627.) There is not sufficient evidence to support the claim that the mean time is less than 10.5 s. The times have been measured with greater precision in the later years. The hypothesis test does not take into account the pattern of decreasing times.

SECTION 10.2

1. A two-way table includes entries that are frequency counts corresponding to two different variables. One variable is used for the row categories and the other variable is used for the column categories.

3. No. It is common to find that expected values are not whole numbers. They are frequencies expected on average, not frequencies that would be observed in specific cases.

5. Does not make sense. **7.** Makes sense.

9. Critical value: $\chi^2 = 6.635$. Do not reject the null hypothesis of independence. Gender and response do not appear to be related.

11. Critical value: $\chi^2 = 3.841$. Do not reject the null hypothesis of independence. Gender and response do not appear to be related.

13. a. H_0: Gender and student status (full-time or part-time) are independent. H_a: Gender and student status (full-time or part-time) are somehow related. **b.** 48.24, 35.76, 36.76, 27.24 **c.** 0.174 **d.** 3.841 **e.** Do not reject the null hypothesis of independence. There does not appear to be a relationship between gender and student status (full-time or part-time).

15. a. H_0: Response is independent of whether the subject is a worker or a senior-level boss. H_a: There is some relationship between response and whether the subject is a worker or a senior-level boss. **b.** 181.60, 254.40, 50.40, 70.60 **c.** 4.698 **d.** 3.841 **e.** Reject the null hypothesis of independence. There does appear to be a relationship between response and whether the subject is a worker or a senior-level boss.

17. a. H_0: Improvement is independent of whether the participant was given the drug or a placebo. H_a: Improvement and treatment (drug or placebo) are somehow related. **b.** 54.16, 43.84, 50.84, 41.16 **c.** 0.289 **d.** 3.841 **e.** Do not reject the null hypothesis of independence. There does not appear to be a relationship between improvement and treatment (drug or placebo).

19. a. H_0: The type of crime is independent of whether the criminal is a stranger. H_a: The type of crime is somehow related to whether the criminal is a stranger. **b.** 29.93, 284.64, 803.43, 21.07, 200.36, 565.57 **c.** 119.330 **d.** 9.210 **e.** Reject the null hypothesis of independence. The type of crime appears to be somehow related to whether the criminal is a stranger.

SECTION 10.3

1. ANOVA represents analysis of variance. It is a method for testing the equality of three or more population means.

3. No. Students at the College of Newport are not necessarily representative of students at all colleges in the United States.

5. Does not make sense. **7.** Does not make sense.

9. a. H_0: Pages from the three books have the same mean Flesch–Kincaid Grade Level scores. **b.** H_a: Pages from the three books have mean Flesch–Kincaid Grade Level scores that are not all equal. **c.** 0.00077 **d.** Reject the null hypothesis of equal means. The pages from the three books do not have the same mean Flesch–Kincaid Grade Level scores.

11. a. The three age groups have the same mean times. **b.** The three age groups have times with means that are not all the same. **c.** 0.828324293 **d.** No. The large P-value suggests that we should not reject the null hypothesis of equal means. The three population means do not appear to be different.

13. With a P-value of 0.4216, do not reject the null hypothesis of equal means. There is not sufficient evidence to reject the claim that the different weight categories have the same mean. Based on these sample data, we cannot conclude that larger cars are safer.

15. With a P-value of 0.0305, reject the null hypothesis of equal means. There is sufficient evidence to reject the claim that the different epochs have the same mean.

CHAPTER 10 REVIEW EXERCISES

1. a. 7.5% of those who used a cell phone had an accident, compared to 10.2% of those who did not use a cell phone. Based on these results, cell phone use does not appear to be dangerous. **b.** H_0: Having an accident in the last year is independent of whether the person uses a cell phone. H_a: Having an accident in the last year and using a cell phone are dependent. **c.** 27.76, 277.24, 41.24, 411.76 **d.** 1.505 **e.** Because the test statistic of 1.505 is less than 3.841, the *P*-value is greater than 0.05. **f.** There is not sufficient evidence to warrant rejection of the claim that having an accident in the last year is independent of whether the person uses a cell phone. Based on these results, cell phones do not appear to be a factor in having accidents. **g.** More data suggest that cell phone usage and driving are somehow related, with cell phones causing a distraction that becomes dangerous. Some states have banned cell phone use by drivers.

2. H_0: $\mu = 165$, H_a: $\mu > 165$. The test statistic $t = 8.398$ is greater than the value of $t = 1.943$ found in Table 10.1, so reject the null hypothesis. (P-value $= 0.00008$.) There is sufficient evidence to support the claim that the mean axial load is greater than 165 pounds.

3. $227.2 \text{ lb} < \mu < 278.3 \text{ lb}$

4. **a.** The three different age groups have the same mean body temperature. **b.** The three different age groups have mean body temperatures that are not all the same. **c.** 0.194915 **d.** No. The *P*-value is greater than 0.05, which suggests that we should not reject the null hypothesis of equal means.

CHAPTER 10 QUIZ

1. b **2.** b **3.** a **4.** H_0: $\mu = 1500$

5. H_a: $\mu > 1500$ **6.** False **7.** Analysis of variance

8. Do not reject the null hypothesis that the three populations have the same mean.

9. H_0: Sex and response are independent.

10. Reject the null hypothesis.

Index